Studies in Logic

Logic and Cognitive Systems

Volume 10

Automated Reasoning in Higher-Order Logic

Set Comprehension and Extensionality in Church's Type Theory

Volume 1
Proof Theoretical Coherence
Kosta Dosen and Zoran Petric

Volume 2
Model Based Reasoning in Science and Engineering
Lorenzo Magnani, editor

Volume 3
Foundations of the Formal Sciences IV: The History of the Concept of the Formal Sciences
Benedikt Löwe, Volker Peckhaus and Thoralf Räsch, editors

Volume 4
Algebra, Logic, Set Theory. Festschrift für Ulrich Felgner zum 65. Geburtstag
Benedikt Löwe, editor

Volume 5
Incompleteness in the Land of Sets
Melvin Fitting

Volume 6
How to Sell a Contradiction: The Logic and Metaphysics of Inconsistency
Francesco Berto

Volume 7
Fallacies — Selected Papers 1972-1982
John Woods and Douglas Walton, with a Foreword by Dale Jacquette

Volume 8
A New Approach to Quantum Logic
Kurt Engesser, Dov M. Gabbay and Daniel Lehmann

Volume 9
Handbook of Paraconsistency
Jean-Yves Béziau, Walter Carnielli and Dov Gabbay, editors

Volume 10
Automated Reasoning in Higher-Order Logic. Set Comprehension and Extensionality in Church's Type Theory
Chad E. Brown

Studies in Logic Series Editor
Dov Gabbay dov.gabbay@kcl.ac.uk

Logic and Cognitive Systems editors
D. Gabbay, J. Siekmann, J. van Benthem, J. Woods

Automated Reasoning in Higher-Order Logic

Set Comprehension and Extensionality in Church's Type Theory

Chad E. Brown

ISBN 978-1-904987-57-4
College Publications
Scientific Director: Dov Gabbay
Managing Director: Jane Spurr
Department of Computer Science
King's College London
Strand, London WC2R 2LS, UK

Original cover design by orchid creative www.orchidcreative.co.uk

Contents

List of Figures

List of Tables

Foreword

Church's type theory is an elegant and expressive formulation of higher-order logic in which much of mathematics, and disciplines which use mathematics, can be formalized in a very natural way. The material in this book contributes significantly to our understanding of how to conduct automated searches for proofs of theorems of this logical language, and how to develop and apply semantical ideas in order to answer questions related to issues of search.

One of the fundamental problems in theorem proving is to choose the terms with which to instantiate quantifiers when one is trying to prove a particular theorem. In higher-order logic, a particularly important problem is to choose the terms with which to instantiate quantifiers on set variables. These terms will denote sets (or relations), and will have a form such as $[\lambda \mathbf{v}\, \mathbf{A}]$ (in the notation of Church's Type Theory), or $\{\mathbf{v} |\, \mathbf{A}\}$ in traditional mathematical notation. The set comprehension schema in Church's type theory asserts that $\exists \mathbf{u}_{o\beta} \forall \mathbf{v}_\beta [\mathbf{u}_{o\beta} \mathbf{v}_\beta \equiv \mathbf{A}_o]$, where $\mathbf{A}_o$ is a wff in which $\mathbf{v}_\beta$ may occur free, but $\mathbf{u}_{o\beta}$ does not occur free. Choosing appropriate instances of this schema corresponds to choosing appropriate wffs of the form $[\lambda \mathbf{v}_\beta\, \mathbf{A}_o]$ with which to instantiate quantifiers on set variables in a proof. This is a very important and difficult problem, since in general the wff $\mathbf{A}_o$ may contain quantifiers and propositional connectives and be arbitrarily complex. The formulas which are thus incorporated into a proof often express the key ideas in the proof.

Another problem which arises in proving theorems of higher-order logic is that of efficiently incorporating into search procedures appropriate consequences of the principles of functional extensionality

$$\forall f_{\alpha\beta}\forall g_{\alpha\beta}[\forall x_\beta[f_{\alpha\beta}x_\beta = g_{\alpha\beta}x_\beta] \supset [f_{\alpha\beta} = g_{\alpha\beta}]]$$

(at all types) and Boolean extensionality

$$\forall p_\circ\forall q_\circ[[p_\circ \equiv q_\circ] \supset [p_\circ = q_\circ]].$$

These problems are not independent of each other. The material in this book is the result of a broad and multifaceted attack on these problems. It greatly extends the study of the syntax and semantics of various forms of Church's type theory, providing new completeness and cut-elimination proofs, and new methods of constructing models for Church's type theory; these are essential background for anyone wishing to do further research on the model theory of classical type theory.

Different people will find different parts of it most interesting. Without trying to summarize the work, let us note a few of its highlights.

One of the ways to generate trial substitution terms for set variables is to introduce formulas such as

$$[\lambda \mathrm{w}^1_{o\beta}\lambda \mathrm{w}^2_{\beta}\forall \mathrm{w}^6_{o\beta}[\mathrm{r}^3_{o(o\beta)\beta(o\beta)}\mathrm{w}^1\mathrm{w}^2\mathrm{w}^6 \vee \mathrm{r}^4_{o(o\beta)\beta(o\beta)}\mathrm{w}^1\mathrm{w}^2\mathrm{w}^6]],$$

which have a certain structure involving quantifiers and propositional connectives (here the main operator is a universal quantifier whose scope is a disjunctive formula), and let the search procedure determine what wffs to substitute for the free variables ($\mathrm{r}^3_{o(o\beta)\beta(o\beta)}$ and $\mathrm{r}^4_{o(o\beta)\beta(o\beta)}$ in this example) by applying higher-order unification to formulas involving these variables as dictated by other aspects of the search procedure. If one can restrict the choices of quantifiers and connectives which one needs to introduce in such formulas in order to prove a particular theorem, one can significantly reduce the size of the space one must search to find a proof of that theorem. This problem motivates much of the work in Part I of the book. In Chapter 6 (see Corollary 6.7.9 and Corollary 6.7.25) one finds explicit examples of theorems which cannot be proven without using instantiation terms involving certain connectives and quantifiers, even when principles of extensionality are available. The results essentially confirm that the usual proofs of these theorems provide a good indication of what connectives and quantifiers are needed in instantiation terms in their proofs, and in that sense these results are not surprising. What is important is the metatheoretical machinery which makes it possible to give really rigorous proofs of these results. Here we have wonderful examples of how intricate semantic constructions can be used (and are needed!) to obtain significant syntactic results.

Part II of the book focuses on structures and procedures directly related to the search for proofs in type theory which may involve extensionality. The concept of an extensional expansion proof which is introduced here is a very significant extension of the concept of an expansion proof. In retrospect, it seems very natural, and it seems destined to play a fundamental role in the development of automated theorem proving in type theory.

The ideas about proof search which are discussed in this book have been implemented, tested, and shown to be fruitful, as indicated by the examples in Chapter 10. In particular, the variants of the Knaster-Tarski fixed point theorem discussed in Section 10.4 are well beyond the reach of other currently available systems for proving theorems of type theory automatically.

The ideas in this book open up new vistas for research in both model theory and automated theorem proving for Church's type theory.

Peter B. Andrews
Professor of Mathematics
Carnegie Mellon University

Preface

The purpose of this book is to provide a firm foundation for automated reasoning in fragments of Church's simple theory of types (a form of higher-order logic). Higher-order logic warrants this attention since many mathematical theorems can be naturally expressed and proven in this theory. The primary reason for the expressive power of higher-order logic is the ability to quantify over sets and functions. On the other hand, many proof techniques used to study first-order logic fail in the higher-order case. For this reason it is important to carefully study proof techniques that do apply in the higher-order case and recognize techniques which do not.

The book is written primarily for computer scientists interested in studying higher-order automated reasoning. We focus on showing how to construct a calculus appropriate for automated search and proving such a calculus is complete. In the process the distinctions between the higher-order case and first-order case become clear. We also describe search procedures implemented in the higher-order theorem prover TPS and experimental results. Although the book is written with a practical purpose in mind, the model constructions could also be of interest to mathematicians curious about the expressive power of simple type theory and to logicians working on the model theory of type theory.

The content of this book is largely derived from my Ph.D. thesis written at Carnegie Mellon University during the years 2001-2004. I wish to thank my Ph.D. advisor Peter B. Andrews for his support during my time as a graduate student at Carnegie Mellon University. He gave me the freedom to explore many ideas as well as expert guidance as I searched for a research topic. His patience and encouragement over the years have been invaluable. I also thank Peter B.

Andrews for his financial support during my years at Carnegie Mellon. I am also grateful to several other professors at Carnegie Mellon including Frank Pfenning, Rick Statman, Dana Scott, Steve Awodey, Jeremy Avigad and John Reynolds. Their guidance and discussions were very helpful.

During my Ph.D. studies, Christoph Benzmüller arranged for me to spend some time working with him at Universität des Saarlandes in Saarbrücken, Germany. The discussions we had during those periods (as well as remote discussions) were very helpful. Over the years, I have also had many productive discussions with Michael Kohlhase, Hongwei Xi, Matthew Bishop, Serge Autexier, James Lipton, Thomas Forster, Brigitte Pientka, Andrej Bauer, Lars Birkedal, John Krueger and Ksenija Simic.

Since my Ph.D. studies I have been fortunate enough to work under Jörg Siekmann at Universität des Saarlandes. I wish to thank Professor Siekmann for his support of me during this period and for his help publishing this book.

I would also like to thank my parents Carl and Sharia for support over the years. Aimee Smith proofread early versions of this book, for which I owe her a special thanks. Finally, I am especially grateful to Elena Tregubova for her partnership: *Danke*.

This work was supported by the National Science Foundation through grants CCR-9624683, CCR-9732312, and CCR-0097179, by the "Deutsche Forschungsgemeinschaft" (DFG) under Grant HOTEL, by Intel Corporation through an equipment grant, and by SFB 378 Project DIALOG.

Part I

Theory

CHAPTER 1

Introduction

There is a certain tension in automated theorem proving. On the one hand, the logical system should be powerful enough to express (and prove) interesting theorems. On the other hand, the logical system should be simple enough to provide reasonable (even if undecidable) search spaces. For this reason, much of the work in automated theorem proving has concentrated on the first-order case. In first-order logic, the *subformula property* provides strong restrictions on the search space for proofs. Also, one can formalize (much of) mathematics in first-order logic by using some form of axiomatic set theory.

However, this is a bit misleading. For example, a theorem **A** of Gödel-Bernays set theory in first-order logic is represented by the statement that the (finite) axioms of Gödel-Bernays set theory imply the statement **A**. The axioms of Gödel-Bernays set theory (as well as the other common first-order axiomatizations of set theory) include comprehension axioms. The subformulas of comprehension axioms include any combination of set constructors. In essence, one is forced to search in an enormous space of terms which represent sets and classes. Even worse, different terms may represent the *same* set or class. Nevertheless, some successes in automated reasoning have been reported for first-order Gödel-Bernays set theory (cf. [24; 58; 16]).

An alternative system capable of formalizing (much of) mathematics is Church's type theory. Church's type theory is a form of higher-order logic. In contrast to first-order axiomatizations of set theory, one need not assume comprehension axioms in Church's type theory. Instead, the levels of propositions and terms are intertwined in a way that directly allows quantification over sets defined by formulas of the language. "Definability" in this context combines λ-definability with logical constants. One pays a price for this expressibility: the subformula property for higher-order logic is far less restrictive than the subformula property for first-order logic. (See [10] for a comparison of first order logic with type theory on two examples.)

Automated reasoning in first-order logic has been extensively studied. Also, the semantics of first-order logic (model theory) and Zermelo-Fraenkel set theory have blossomed into very rich topics in mathematical logic. In contrast, Church's type theory has yet to receive similar attention.

As soon as one decides to study higher-order logic, one quickly realizes that many proof techniques from the first-order case no longer apply. In the first-order case, for example, for any formula $\mathbf{A}$ and term t, the formula $[t/x]\mathbf{A}$ where t is substituted for x can be considered less complex than $[\forall x\, \mathbf{A}]$ (in the sense that $[t/x]\mathbf{A}$ has one less quantifier than $[\forall x\, \mathbf{A}]$). In the higher-order case, one may have a formula such as $[\forall p\, p]$ (where p is a propositional variable). In general $[t/p]p$ will not be less complex than $[\forall p\, p]$. In particular, t may be $[\forall p\, p]$ in which case $[\forall p\, p]$ and $[t/p]p$ are actually the same. Gentzen's proof of cut-elimination in the first-order case does not generalize to a proof of cut-elimination in the higher-order case for this very reason. This is a well-known fact among experts in the field but is rarely explicitly pointed out in the literature.

1.1. Goal-Directed Higher-Order Calculi

Since our motivation is automated reasoning, we need a logical calculus that is reasonably goal directed. Consider the cut rule in a sequent calculus:

$$\frac{\Gamma, \mathbf{C} \quad \Gamma, \neg\mathbf{C}}{\Gamma}\, Cut$$

If we are trying to prove a goal sequent Γ and need to apply the cut rule, then we must determine the cut formula $\mathbf{C}$. For this reason the cut rule is quite undirected (with respect to backwards search). Consequently, we would like to base automated search on a cut-free calculus and thereby avoid making such arbitrary choices.

Once we define a cut-free calculus, we can consider whether the cut rule is *admissible*. That is, given any derivations $\mathcal{D}_1$ and $\mathcal{D}_2$ of the sequents $\Gamma, \mathbf{C}$ and $\Gamma, \neg\mathbf{C}$, respectively, is there a derivation $\mathcal{D}$ of the sequent Γ? Gentzen's proof not only shows admissibility of cut in the first-order case, but also provides an algorithm for computing the derivation $\mathcal{D}$ using the given derivations $\mathcal{D}_1$ and $\mathcal{D}_2$. The higher-order case is not so simple. It is not merely the case that Gentzen's proof does not apply in the higher-order case, but in fact there can be *no* proof of higher-order cut-elimination which can be formalized in higher-order logic (assuming consistency of analysis, see [5]).

On the other hand, semantic proofs of higher-order cut-elimination are possible. First one shows the calculus is complete (with respect to an appropriate class of models). Suppose there are derivations $\mathcal{D}_1$ and $\mathcal{D}_2$ of the sequents $\Gamma, \mathbf{C}$ and $\Gamma, \neg\mathbf{C}$, respectively. Since in any model the interpretation of $\mathbf{C}$ must be either true or false, the sequent Γ must be valid in any model. Since the calculus is complete, there must be a derivation $\mathcal{D}$ of Γ. Note that this does not provide any method for calculating $\mathcal{D}$ from $\mathcal{D}_1$ and $\mathcal{D}_2$. In this manner, one can show a cut-elimination result by showing a cut-free sequent calculus is complete.

The first semantic proofs of cut-elimination for higher-order logic (without extensionality) were discovered independently by Takahashi (cf. [61]) and Prawitz (cf. [56]). In both cases, the version of higher-order logic did not consider the complete set of function types provided by Church's type theory (cf. Remark 2.1.44). Andrews (cf. [2]) used a similar method to prove cut-elimination for a nonextensional fragment of Church's type theory (elementary type theory) which includes all function types. The higher-order cut-elimination results in this book are obtained using similar semantic methods (cf. Corollaries 5.6.8 and 5.7.20). For the nonextensional case, we generalize the construction of [2] to an arbitrary signature of logical constants. For the extensional case, we combine the construction of [2] with partial equivalence relations.

1.2. Higher-Order Automated Theorem Proving

TPS is an automated theorem prover for Church's simple type theory. Actually, TPS has been designed to search in elementary type theory, which is Church's simple type theory without axioms of extensionality, infinity, descriptions, or choice. Within this fragment, TPS has had success as a theorem prover by combining mating search (a method which makes sense in the first-order case) with Huet's higher-order preunification algorithm for the simply typed λ-calculus. In essence, the search procedure combines first-order search with solving equations up to β or $\beta\eta$-conversion. In this sense, the search techniques traditionally employed by TPS combine first-order methods with comprehension of λ-definable sets and functions. Although Huet's preunification algorithm provides a way to handle λ-definable instantiations, it does not provide a complete method for finding instantiations definable using logical constants.

To ensure completeness, a higher-order theorem prover must also somehow ensure that instantiations for sets definable using logical constants can be discovered. The higher-order theorem provers TPS[12] and LEO[20] ensure completeness using primitive (and general) substitutions to enumerate terms with a certain logical structure. A *primitive substitution* (primsub) for a set variable v is a term introducing a single logical constant or quantifier, or projecting an argument to the head. A simple theorem requiring a primsub is the theorem expressing the existence of the union v of two sets X and Y. An appropriate primsub for the set variable v would introduce a disjunction (corresponding to the disjunction in the definition of union). A *general substitution* (gensub) for a set variable v is a term introducing several logical constants, quantifiers, and projections. A gensub can be thought of as a composition of several primsubs. Primitive and general substitutions in TPS are discussed in [12] on page 331. It should not be surprising that such enumeration techniques place practical limitations on the search space.

1.3. Set Comprehension and Logical Constants

One primary topic of this book is the role of sets definable using logical constants. First, we develop a collection of logical systems which have a restricted collection of logical constants available for instantiations. Next, we develop (Henkin-style) semantics for these logical systems and prove soundness and completeness results. This allows us to determine semantically whether all proofs of a theorem require a set instantiation involving a certain collection of logical constants. We will give a number of independence results of this sort.

We can also use semantic methods to determine that certain sets of logical constants are sufficient to prove any statement which is provable if all logical constants are available. Thus we can restrict which logical constants must be considered.

1.4. Extensionality

Another topic of the book is the role of extensionality. By including extensionality one can obtain more restrictions on set comprehension. Consequently, one can limit the logical constants on which a theorem prover must perform primsubs. Furthermore, in extensional type theory some mathematical properties have a more natural representation than in elementary type theory.

Benzmüller [18] has defined a higher-order resolution calculus including extensionality. Here we adapt these methods for sequent calculi and expansion proofs so our search methods will be complete with respect to extensional type theory.

Benzmüller [18] also included techniques for integrating equational reasoning via paramodulation into the search space. We include equational reasoning in our extensional calculi, but use a form of E-unification instead of paramodulation. This provides a complete method for handling primitive equality, allowing one to avoid the use of Leibniz equality. Since Leibniz equality involves a set variable, we can completely avoid the problem of instantiating set variables associated with Leibniz equality.

1.5. Compact Representations of Proofs

While sequent calculi are valuable theoretical tools, they are not the most practical representations of proofs for search. Miller defined expansion proofs in order to obtain compact representations of proofs in elementary type theory appropriate for automated search (cf. [43] and [44]). The latter part of the book concerns extensional expansion proofs. Extensional expansion proofs provide a compact representation of proofs in extensional type theory (relative to a signature). Consequently, extensional expansion proofs form a basis for automated search for proofs in extensional type theory.

After giving precise mathematical descriptions of extensional expansion proofs, we establish lifting results. Using these results we describe a search procedure and prove this procedure is complete. In particular, the lifting results imply we can search while delaying consideration of *flex-flex* equations (cf. Definition 8.3.1). Benzmüller conjectured a similar kind of completeness in [18] for his resolution calculus $\mathcal{ER}$ (cf. Remark 8.3.4). At this time, Benzmüller's conjecture remains open (cf. Remark 8.3.4).

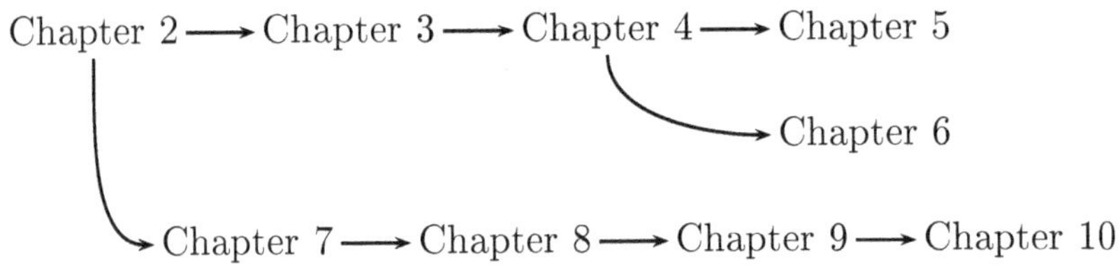

FIGURE 1.1. Dependence of Chapters

1.6. Outline

The book is divided into three parts. The first part (Chapters 3-6) analyzes the theoretical nature of elementary type theory and extensional type theory with respect to general signatures of logical constants. The second part (Chapters 7-10) concerns automated proof search with respect to extensional type theory with restricted signatures. The third part is the appendix which contains many proofs omitted in the main body of the text. Certain portions of the book can be read (almost) independently of others. The diagram in Figure 1.1 indicates when chapters strongly depend on the material in previous chapters.

CHAPTER 2

Preliminaries

We begin by presenting the syntax of simply-typed λ-terms and propositions. Relative to different signatures of logical constants and different amounts of extensionality, we introduce a proof theory for propositions in the form of a sequent calculus.

2.1. Syntax of Higher Order Logic

We consider (classical) higher-order logic formulated as Church's simple type theory [26] based on the simply typed λ-calculus. We first define the set of (simple) types $\mathcal{T}$.

DEFINITION 2.1.1 (Types). The set of types $\mathcal{T}$ is the least set such that

- $o \in \mathcal{T}$,
- $\iota \in \mathcal{T}$, and
- $(\alpha\beta) \in \mathcal{T}$ whenever $\alpha, \beta \in \mathcal{T}$.

We assume types are generated freely. That is, $o \neq \iota$, $o \neq (\alpha\beta)$ for any $\alpha, \beta \in \mathcal{T}$, $\iota \neq (\alpha\beta)$ for any $\alpha, \beta \in \mathcal{T}$, and if $(\alpha\beta) = (\gamma\delta)$, then $\alpha = \gamma$ and $\beta = \delta$.

REMARK 2.1.2 (Signatures, Variables, Parameters). To define terms, we start with a signature $\mathcal{S}$ of typed (logical) constants, a collection of variables $\mathcal{V}$, and a collection of parameters $\mathcal{P}$. For each type α, let $\mathcal{S}_\alpha \subseteq \mathcal{S}$ [$\mathcal{V}_\alpha \subseteq \mathcal{V}$, $\mathcal{P}_\alpha \subseteq \mathcal{P}$] be the set of constants [variables, parameters] of type α. Let $\aleph_s$ be a fixed infinite cardinal. We assume the set of parameters $\mathcal{P}_\alpha$ has cardinality $\aleph_s$ for each type α. Similarly, we assume $\mathcal{V}_\alpha$ is countably infinite for each type α.

To formulate higher-order logic, Church (cf. [26]) assumed the signature included logical constants $\neg_{oo}$, $\vee_{ooo}$ and $\Pi^\alpha_{o(o\alpha)}$ for each type α. From these, one can define the other logical operators as in [26]. Equality at type α can be defined by Leibniz equality.

The alternative pursued in [9] is to have primitive equality $=^{\alpha}_{o\alpha\alpha}$ in the signature for each type α. The logical connectives and quantifiers can be defined from primitive equality (assuming full extensionality).

In the general case, we will not assume any logical constants to be in the signature $\mathcal{S}$. Without (enough) logical constants we should not consider the type theory to be "higher-order logic" since one cannot define the same sets using the restricted language as one can define using higher-order logic. Each collection of logical constants yields a fragment of higher-order logic. To avoid confusion with other formulations of higher-order logic, we will call these *fragments of elementary type theory* (*with* η) or *fragments of extensional type theory* depending on what forms extensionality are assumed. These fragments allow us to study what logical constants must be available for instantiations to prove particular theorems. In other words, these fragments allow us to measure how much (set) comprehension is necessary to prove particular theorems (in the presence or absence of extensionality).

DEFINITION 2.1.3 (Wffs). The set of *well-formed formulas* (or *terms*) of type α over a signature $\mathcal{S}$ is denoted by $\mathit{wff}_\alpha(\mathcal{S})$ (or wff_α when the signature $\mathcal{S}$ is clear in context). These sets are defined inductively as follows.

- $x_\alpha \in \mathit{wff}_\alpha(\mathcal{S})$ for each variable $x_\alpha \in \mathcal{V}_\alpha$.
- $W_\alpha \in \mathit{wff}_\alpha(\mathcal{S})$ for each parameter $W_\alpha \in \mathcal{P}_\alpha$.
- $c_\alpha \in \mathit{wff}_\alpha(\mathcal{S})$ for each constant $c_\alpha \in \mathcal{S}_\alpha$.
- $[\mathbf{F}_{\alpha\beta}\,\mathbf{B}_\beta] \in \mathit{wff}_\alpha(\mathcal{S})$ for each $\mathbf{F} \in \mathit{wff}_{\alpha\beta}(\mathcal{S})$ and $\mathbf{B} \in \mathit{wff}_\beta(\mathcal{S})$.
- $[\lambda x_\beta \mathbf{A}_\alpha] \in \mathit{wff}_{\alpha\beta}(\mathcal{S})$ for each variable $x_\beta \in \mathcal{V}_\beta$ and $\mathbf{A} \in \mathit{wff}_\alpha(\mathcal{S})$.

We define $\mathbf{Free}(\mathbf{A}_\alpha) \subset \mathcal{V}$ to be the set of free variables in $\mathbf{A}$ in the usual way. We also define $\mathbf{Consts}(\mathbf{A}_\alpha) \subset \mathcal{S}$ [$\mathbf{Params}(\mathbf{A}_\alpha) \subset \mathcal{P}$] to be the set of constants [parameters] that occur in $\mathbf{A}$. We call a term $\mathbf{A}_\alpha$ *closed* if there are no free variables in $\mathbf{A}$. Let $\mathit{cwff}_\alpha(\mathcal{S})$ (or cwff_α) be the set of all closed terms of type α. For any set of formulas $\Phi \subseteq \mathit{wff}_\alpha(\mathcal{S})$, we define $\mathbf{Free}(\Phi) := \bigcup_{\mathbf{A}\in\Phi} \mathbf{Free}(\mathbf{A})$ and $\mathbf{Params}(\Phi) := \bigcup_{\mathbf{A}\in\Phi} \mathbf{Params}(\mathbf{A})$.

We next make precise which logical constants we will consider.

DEFINITION 2.1.4 (Logical Constants). We define the set of *logical constants* to be

$$\{\top_o, \bot_o, \neg_{oo}, \wedge_{ooo}, \vee_{ooo}, \supset_{ooo}, \equiv_{ooo}\}$$
$$\cup \{\Pi^\alpha_{o(o\alpha)} \mid \alpha \in \mathcal{T}\} \cup \{\Sigma^\alpha_{o(o\alpha)} \mid \alpha \in \mathcal{T}\} \cup \{=^\alpha_{o\alpha\alpha} \mid \alpha \in \mathcal{T}\}.$$

We now define a few signatures which will be commonly used.

DEFINITION 2.1.5.

- Let $\mathcal{S}_{Ch}$ be $\{\neg, \vee\} \cup \{\Pi^\alpha \mid \alpha \in \mathcal{T}\}$.
- Let $\mathcal{S}_{Bool}$ be $\{\top, \bot, \neg, \vee, \wedge, \supset, \equiv\}$.
- Let $\mathcal{S}^{elem}_{all}$ be $\mathcal{S}_{Ch} \cup \mathcal{S}_{Bool} \cup \{\Sigma^\alpha \mid \alpha \in \mathcal{T}\}$.
- Let $\mathcal{S}^{=}$ be $\{=^\alpha \mid \alpha \in \mathcal{T}\}$.
- Let $\mathcal{S}^{=}_{Ch}$ be $\mathcal{S}_{Ch} \cup \mathcal{S}^{=}$.
- Let $\mathcal{S}_{all}$ be the set of all logical constants.

Whenever we refer to a signature $\mathcal{S}$, we mean a signature of logical constants. That is, we always assume $\mathcal{S} \subseteq \mathcal{S}_{all}$.

DEFINITION 2.1.6 (Equality-Free Signatures). A signature $\mathcal{S}$ is called *equality-free* if $=^\alpha \notin \mathcal{S}$ for every α.

For example, $\mathcal{S}_{Ch}$, $\mathcal{S}_{Bool}$ and $\mathcal{S}^{elem}_{all}$ are equality-free, but $\mathcal{S}^{=}_{Ch}$ and $\mathcal{S}_{all}$ are not. A signature $\mathcal{S}$ is equality-free iff $\mathcal{S} \subseteq \mathcal{S}^{elem}_{all}$.

We adopt the following conventions for the presentation of well-formed formulas:

- We use Church's dot convention. A dot stands for a left bracket whose mate is as far to the right as possible without changing the pairing of existing brackets. For example, $[\mathbf{F}\,\lambda x_\alpha \centerdot \mathbf{G}\,\mathbf{X}]\,\mathbf{Y}$ stands for $[\mathbf{F}\,[\lambda x_\alpha[\mathbf{G}\,\mathbf{X}]]]\,\mathbf{Y}$.
- We may indicate a well-formed formula $\mathbf{A}$ has type α by writing $\mathbf{A}_\alpha$. However, we may omit type symbols when they are clear in context.
- When Π^α or Σ^α is in the signature, we may write $\forall x_\alpha\,\mathbf{M}_o$ or $\exists x_\alpha\,\mathbf{M}_o$ for $\Pi^\alpha\,[\lambda x_\alpha\,\mathbf{M}]$ or $\Sigma^\alpha\,[\lambda x_\alpha\,\mathbf{M}]$, respectively.

- Quantifiers and negation have the smallest possible scope.
- When any of the logical constants $\wedge$, $\vee$, $\supset$, $\equiv$ or $=^{\alpha}$ are in the signature, we may use infix notation. That is, $[\wedge_{ooo} \mathbf{M}_o \mathbf{N}_o]$ may be written $[\mathbf{M}_o \wedge \mathbf{N}_o]$. When $\wedge$, $\vee$, $\supset$ and $\equiv$ are written in infix, brackets may be omitted. Omitted brackets should be restored in order $\wedge$, $\vee$, $\supset$ and $\equiv$, assuming association to the left.
- We usually use capital letters $W_\alpha \in \mathcal{P}_\alpha$ for parameters and small letters $x_\alpha \in \mathcal{V}_\alpha$ for variables.
- We use the vector notation $[\lambda \overline{x^n}\, \mathbf{A}]$ to abbreviate n abstractions $[\lambda x^1 \cdots \lambda x^n\, \mathbf{A}]$. Similarly, $[h\, \overline{\mathbf{A}^n}]$ abbreviates $[h\, \mathbf{A}^1 \cdots \mathbf{A}^n]$.

REMARK 2.1.7 (α-conversion). To define notions such as substitution, reduction and convertibility, one needs to be careful about alphabetic change of bound variables (α-conversion). We consider alphabetic change of bound variables (α-conversion) built into the language. That is, we consider α-convertible terms to be identical. Since relations will be defined on representatives of these equivalence classes, we will only consider n-ary relations R which are *α-respecting*. That is, we assume $R(\mathbf{A}^1_\alpha, \ldots, \mathbf{A}^n_\alpha)$ holds iff $R(\mathbf{B}^1_\alpha, \ldots, \mathbf{B}^n_\alpha)$ holds whenever $\mathbf{A}^i$ is α-convertible to $\mathbf{B}^i$ for each i such that $1 \leq i \leq n$. We will also only consider relations R such that there is a finite set of *vulnerable variables* $R^{\mathcal{V}}$. Since we assume $R^{\mathcal{V}}$ is finite, for any terms $\mathbf{C}^1_\alpha, \ldots, \mathbf{C}^n_\alpha$, there are α-equivalent $\mathbf{A}^1, \ldots, \mathbf{A}^n$ such that no variable in $R^{\mathcal{V}}$ occurs bound in $\mathbf{A}^i$ for some i ($1 \leq i \leq n$). Hence it is enough to define such a relation R on representatives such that no variable in $R^{\mathcal{V}}$ occurs bound in the representatives and extend R so that $R(\mathbf{C}^1_\alpha, \ldots, \mathbf{C}^n)$ holds iff $R(\mathbf{A}^1_\alpha, \ldots, \mathbf{A}^n)$ holds for some $\mathbf{A}^1, \ldots, \mathbf{A}^n$ where $\mathbf{A}^i$ is α-convertible to $\mathbf{C}^i$ and no variable in $R^{\mathcal{V}}$ occurs bound in $\mathbf{A}^i$. We will often do this without explicit mention.

Functions are a special case of relations. Technically, we consider functions on terms to be relations that are functional up to α-conversion.

We use the notation $[\mathbf{B}_\beta/x_\beta]\mathbf{A}_\alpha$ to denote the result of substituting $\mathbf{B}$ for the free occurrences of the variable x in $\mathbf{A}$. Note that $[\mathbf{B}/x]$ is a typed functional relation from $\mathit{wff}(\mathcal{S})$ to $\mathit{wff}(\mathcal{S})$. In the inductive definition of $[\mathbf{B}/x]\mathbf{A}$, we will ensure that neither x nor any free variable of $\mathbf{B}$ occur bound in $\mathbf{A}$. As in Remark 2.1.7, we let the set of vulnerable variables for $[\mathbf{B}/x]$ be the set $\mathbf{Free}(\mathbf{B}) \cup \{x\}$, define $[\mathbf{B}/x]$ on terms $\mathbf{A}$ such that no variable in $\mathbf{Free}(\mathbf{B}) \cup \{x\}$ occurs bound in $\mathbf{A}$, and extend $[\mathbf{B}/x]$ to all terms up to α-conversion. The following clauses inductively define $[\mathbf{B}/x]\mathbf{A}$ for terms $\mathbf{A}$ such that no variable in $\mathbf{Free}(\mathbf{B}) \cup \{x\}$ occurs bound in $\mathbf{A}$:

- $[\mathbf{B}/x]x := \mathbf{B}$.
- $[\mathbf{B}/x]w := \mathbf{B}$ for all $w \in \mathcal{P} \cup \mathcal{S} \cup (\mathcal{V} \setminus \{x\})$.
- $[\mathbf{B}/x][\mathbf{F}\,\mathbf{D}] := [[\mathbf{B}/x]\mathbf{F}\,[\mathbf{B}/x]\mathbf{D}]$.
- $[\mathbf{B}/x][\lambda y\,\mathbf{D}] := [\lambda y\,[\mathbf{B}/x]\mathbf{D}]$ (where y is assumed not to be in the set $\mathbf{Free}(\mathbf{B}) \cup \{x\}$ of vulnerable variables).

For any term $\mathbf{C}$, we define $[\mathbf{B}/x]\mathbf{C}$ to be $[\mathbf{B}/x]\mathbf{A}$ for some $\mathbf{A}$ which is the same as $\mathbf{C}$ up to the names of bound variables and such that no variable in $\mathbf{Free}(\mathbf{B}) \cup \{x\}$ occurs bound in $\mathbf{A}$.

We also use the notation $[\mathbf{B}_\beta/c_\beta]\mathbf{A}_\alpha$ and $[\mathbf{B}_\beta/W_\beta]\mathbf{A}_\alpha$ to denote the result of substituting $\mathbf{B}$ for the constant c or parameter W. For these functional relations, the set of vulnerable variables is $\mathbf{Free}(\mathbf{B})$. The clauses defining each of these functions are analogous to those defining $[\mathbf{B}/x]$.

We also consider substitutions θ which substitute simultaneously for the variables, parameters and constants in the (finite) domain

$$\mathbf{Dom}(\theta) \subset \mathcal{V} \cup \mathcal{P} \cup \mathcal{S}$$

of θ. The clauses defining simultaneous substitution are also analogous to those defining $[\mathbf{B}/x]$ where the set of vulnerable variables here is given by

$$\{y \in \mathbf{Free}(\theta(w)) | w \in \mathbf{Dom}(\theta)\} \cup (\mathbf{Dom}(\theta) \cap \mathcal{V}).$$

Given a substitution θ, we use $\theta, [\mathbf{B}/x]$ to denote the substitution with $\mathbf{Dom}(\theta, [\mathbf{B}/x]) := \mathbf{Dom}(\theta) \cup \{x\}$ such that

$$(\theta, [\mathbf{B}/x])(y) := \begin{cases} \mathbf{B} & \text{if } y = x \\ \theta(y) & \text{otherwise.} \end{cases}$$

Note that while we do allow substitution of terms for constants and parameters, we allow binding only of variables.

DEFINITION 2.1.8 (Ground Substitutions). A substitution θ is a *ground substitution* if $\theta(w_\alpha) \in \mathit{cwff}_\alpha(\mathcal{S})$ for every $w_\alpha \in \mathbf{Dom}(\theta)$.

Sometimes we consider simultaneous substitutions which map parameters to parameters. We call these *parameter substitutions*. A parameter substitution may send two distinct parameters of the same type to the same parameter.

DEFINITION 2.1.9 (Parameter Substitutions). A substitution θ is a *parameter substitution* if $\mathbf{Dom}(\theta) \subseteq \mathcal{P}$ and $\theta(A) \in \mathcal{P}$ for every $A \in \mathbf{Dom}(\theta)$. A substitution θ *respects parameters* if $\theta(A) \in \mathcal{P}$ for every $A \in (\mathbf{Dom}(\theta) \cap \mathcal{P})$.

Later we will consider certain restricted collections of terms. We will often want these collections to be closed under parameter substitutions.

DEFINITION 2.1.10. Let $\mathcal{S}$ be a signature and $\mathcal{W} \subseteq wff(\mathcal{S})$ be a set of typed terms. We say $\mathcal{W}$ is *closed under parameter substitutions* if $\theta(\mathbf{A}_\alpha) \in \mathcal{W}$ whenever θ is a parameter substitution and $\mathbf{A}_\alpha \in \mathcal{W}$.

We now turn to the task of extending relations to congruence relations.

DEFINITION 2.1.11 (Congruence Closure). The *congruence closure* of a typed relation $\mathcal{R}_\alpha \subseteq wff_\alpha(\mathcal{S}) \times wff_\alpha(\mathcal{S})$ is the least relation $\mathcal{R}'$ such that

- $\mathcal{R}'$ is α-respecting,
- $\mathcal{R}_\alpha \subseteq \mathcal{R}'_\alpha$ for each type α,
- $\mathcal{R}'_\alpha([\mathbf{F}_{\alpha\beta}\,\mathbf{A}_\beta], [\mathbf{G}_{\alpha\beta}\,\mathbf{A}_\beta])$ whenever $\mathcal{R}'_{\alpha\beta}(\mathbf{F}, \mathbf{G})$,
- $\mathcal{R}'_\alpha([\mathbf{F}_{\alpha\beta}\,\mathbf{A}_\beta], [\mathbf{F}_{\alpha\beta}\,\mathbf{B}_\beta])$ whenever $\mathcal{R}'_\beta(\mathbf{A}, \mathbf{B})$, and
- $\mathcal{R}'_{\alpha\beta}([\lambda x_\beta\,\mathbf{A}_\alpha], [\lambda x_\beta\,\mathbf{B}_\alpha])$ whenever $\mathcal{R}'_\alpha(\mathbf{A}, \mathbf{B})$.[1]

We say $\mathcal{R}$ is a *congruence relation* if the congruence closure of $\mathcal{R}$ is $\mathcal{R}$.

DEFINITION 2.1.12. A term of the form $[[\lambda x_\beta\,\mathbf{C}_\alpha]\,\mathbf{D}_\beta]$ is called a β-redex. The β-reduct of this redex is $[\mathbf{D}/x]\mathbf{C}$. We define $\xrightarrow{\beta}_1$ be the congruence closure of the relation consisting of pairs $(\mathbf{A}, \mathbf{B})$ where $\mathbf{A}$ is a β-redex with reduct $\mathbf{B}$. That is, $\mathbf{A} \xrightarrow{\beta}_1 \mathbf{B}$ holds if $\mathbf{B}$ is the result of β-reducing a single β-redex in $\mathbf{A}$. We define $\xrightarrow{\beta}$ to be the reflexive, transitive closure of $\xrightarrow{\beta}_1$. That is, $\mathbf{A} \xrightarrow{\beta} \mathbf{B}$ holds when $\mathbf{A}$ reduces to $\mathbf{B}$ by zero or more β-reductions on subterms.

[1] Here we use α-conversion to assume the bound variables are both x_β and assume x is not vulnerable.

The symmetric transitive closure of the β-reduction relation is β-conversion. We will write $\mathbf{A} \stackrel{\beta}{=} \mathbf{B}$ when $\mathbf{A}$ β-converts to $\mathbf{B}$. We say $\mathbf{A}$ is *β-normal* if no subterm of $\mathbf{A}$ is a β-redex.

A term of the form $\lambda x_\alpha \mathbin{\scriptstyle\blacksquare} \mathbf{F}_{\beta\alpha}\, x$ where $x \notin \mathbf{Free}(\mathbf{F})$ is called an η-redex. The η-reduct of this redex is $\mathbf{F}$. We define $\stackrel{\eta}{\longrightarrow}_1$ to be the congruence closure of the relation consisting of pairs $(\mathbf{A}, \mathbf{B})$ where $\mathbf{A}$ is an η-redex with reduct $\mathbf{B}$. We define $\stackrel{\eta}{\longrightarrow}$ to be the reflexive, transitive closure of $\stackrel{\eta}{\longrightarrow}_1$. That is, $\mathbf{A} \stackrel{\eta}{\longrightarrow} \mathbf{B}$ holds when $\mathbf{A}$ reduces to $\mathbf{B}$ by zero or more η-reductions on subterms. We will write $\mathbf{A} \stackrel{\eta}{=} \mathbf{B}$ when $\mathbf{A}$ η-converts to $\mathbf{B}$.

Similarly, we let $\stackrel{\beta\eta}{\longrightarrow}_1$ be the congruence closure of the relation consisting of pairs $(\mathbf{A}, \mathbf{B})$ where $\mathbf{A}$ is a β-redex with reduct $\mathbf{B}$ or $\mathbf{A}$ is an η-redex with reduct $\mathbf{B}$. Also, we let $\stackrel{\beta\eta}{\longrightarrow}$ be the reflexive, transitive closure of $\stackrel{\beta\eta}{\longrightarrow}_1$. That is, $\mathbf{A} \stackrel{\beta\eta}{\longrightarrow} \mathbf{B}$ holds when $\mathbf{A}$ reduces to $\mathbf{B}$ by zero or more β-reductions and η-reductions on subterms. The symmetric transitive closure of the $\beta\eta$-reduction relation is $\beta\eta$-conversion. We will write $\mathbf{A} \stackrel{\beta\eta}{=} \mathbf{B}$ when $\mathbf{A}$ $\beta\eta$-converts to $\mathbf{B}$. We say $\mathbf{A}$ is *$\beta\eta$-normal* if no subterm of $\mathbf{A}$ is a β-redex or an η-redex.

The following result is well known:

LEMMA 2.1.13. *The reduction relations $\stackrel{\beta}{\longrightarrow}$ and $\stackrel{\beta\eta}{\longrightarrow}$ above have the following property: for every wff $\mathbf{A}$ there is a unique (up to renaming of bound variables) β-normal [$\beta\eta$-normal] $\mathbf{B}$ such that $\mathbf{A} \stackrel{\beta}{\longrightarrow} \mathbf{B}$ [$\mathbf{A} \stackrel{\beta\eta}{\longrightarrow} \mathbf{B}$].*

PROOF. See (for example) [14; 33]. □

NOTATION *We use $\mathbf{A}^{\downarrow\beta}$ [$\mathbf{A}^{\downarrow}$] to denote the β-normal [$\beta\eta$-normal] form of $\mathbf{A}$. In some situations we do not wish to be specific about which normal form we mean. To allow for these situations, we use the notation $\mathbf{A}^{\downarrow *}$ where ${}^{\downarrow *}$ means either ${}^{\downarrow\beta}$ or ${}^{\downarrow}$.*

An easy induction verifies that every term $\mathbf{A} \in \mathit{wff}_\alpha(\mathcal{S})$ must be of the form $\lambda \overline{x^n}\, [\mathbf{H}\, \overline{\mathbf{A}^m}]$ where $\mathbf{H}$ is not an application and if $m = 0$ then $\mathbf{H}$ is not a λ-abstraction. This $\mathbf{H}$ is called the *head* of $\mathbf{A}$.

DEFINITION 2.1.14 (Flexible and Rigid Terms). Suppose $\mathbf{A}$ is a term of the form $\lambda \overline{x^n}\, [h\, \overline{\mathbf{A}^m}]$ where $h \in \mathcal{V} \cup \mathcal{P} \cup \mathcal{S}$. The term $\mathbf{A}$ is called *rigid* if $h \in \mathcal{P} \cup \mathcal{S} \cup \{x^1, \ldots, x^n\}$. The term $\mathbf{A}$ is called *flexible* if $h \in \mathcal{V} \setminus \{x^1, \ldots, x^n\}$.

Note that every β-normal form has the form $\lambda\overline{x^n}\,[h\,\overline{\mathbf{A}^m}]$ for some variable, parameter or logical constant $h \in \mathcal{V} \cup \mathcal{P} \cup \mathcal{S}$.

DEFINITION 2.1.15 (Atoms). A formula $\mathbf{A} \in \mathit{wff}_o(\mathcal{S})$ is an *atom* if the head of $\mathbf{A}^{\downarrow\beta}$ is not a logical constant in $\mathcal{S}$.

We will also consider long $\beta\eta$-forms (cf. [36; 57]). These are terms in β-normal form which are fully η-expanded.

DEFINITION 2.1.16 (Long $\beta\eta$-Forms). We define the set of *long $\beta\eta$-forms* inductively. Let $n \geq 0$, $\alpha^1, \ldots, \alpha^n$ be types, and $\beta \in \{o, \iota\}$ be a base type. A term $\mathbf{A}$ of type $(\beta\alpha^n \cdots \alpha^1)$ is a *long $\beta\eta$-form* if it has the form

$$\lambda x^1_{\alpha^1} \cdots \lambda x^n_{\alpha^n}\,[h_{\beta\gamma^m\cdots\gamma^1}\,\mathbf{A}^1_{\gamma^1} \cdots \mathbf{A}^m_{\gamma^m}]$$

(or $\lambda\overline{x^n}\,[h\,\overline{\mathbf{A}^m}]$ in vector notation) for some $h_{\beta\gamma^m\cdots\gamma^1} \in \mathcal{V} \cup \mathcal{P} \cup \mathcal{S}$, $m \geq 0$ and long $\beta\eta$-forms $\mathbf{A}^1_{\gamma^1}, \ldots, \mathbf{A}^m_{\gamma^m}$. (The base case of this inductive definition is when $m = 0$.)

Note that if a term $\lambda\overline{x^n}\,[h\,\overline{\mathbf{A}^m}]$ is a long $\beta\eta$-form, then $[h\,\overline{\mathbf{A}^m}]$ is of base type. A variable $y_{o\iota}$ is not in long $\beta\eta$-form. The corresponding long $\beta\eta$-form of y would be $[\lambda x_\iota\,[y\,x]]$.

LEMMA 2.1.17. *For each type α and variable x_α, there is a unique long $\beta\eta$-form $\mathbf{B}^x$ such that $\mathbf{B}^x \xrightarrow{\eta} x$.*

PROOF. We prove this result by induction on α. The type α can be written $(\beta\gamma^m \cdots \gamma^1)$ for some base type β, $m \geq 0$ and types γ^j for each $j \in \{1, \ldots, m\}$. Choose distinct variables $z^j_{\gamma^j}$. Note that $[\lambda\overline{z^m}[x\,\overline{z^m}]] \xrightarrow{\eta} x$. By the inductive hypothesis, there is a unique long $\beta\eta$-form $\mathbf{B}^j$ such that $\mathbf{B}^j \xrightarrow{\eta} z^j$ for each j $(1 \leq j \leq m)$. Hence $[\lambda\overline{z^m}[x\,\overline{\mathbf{B}^m}]] \xrightarrow{\eta} x$ where $[\lambda\overline{z^m}[x\,\overline{\mathbf{B}^m}]]$ is a long $\beta\eta$-form.

To determine uniqueness of $[\lambda\overline{z^m}[x\,\overline{\mathbf{B}^m}]]$, suppose $\mathbf{C}$ is a long $\beta\eta$-form with $\mathbf{C} \xrightarrow{\eta} x$. Since $\mathbf{C}$ is a long $\beta\eta$-form, it must have the form $[\lambda\overline{z^m}[h\,\overline{\mathbf{C}^n}]]$ (up to α-conversion) where each $\mathbf{C}^i$ is a long $\beta\eta$-form. Since η-reductions do not change the head of a term, the head of $\mathbf{C}^\downarrow$ is h. Since $\mathbf{C}^\downarrow = x$, we know $h = x$ and $n = m$. Hence $\mathbf{C}$ is $[\lambda\overline{z^m}[x\,\overline{\mathbf{C}^m}]]$. For $\mathbf{C}$ to η-normalize to x, we must have $\mathbf{C}^j \xrightarrow{\eta} z^j$ for each j $(1 \leq j \leq m)$. By uniqueness of $\mathbf{B}^j$, we must have $\mathbf{C}^j = \mathbf{B}^j$. We conclude $\mathbf{C} = [\lambda\overline{z^m}[x\,\overline{\mathbf{B}^m}]]$. $\square$

LEMMA 2.1.18. *For every β-normal term* $\mathbf{A}$, *there is a unique long $\beta\eta$-form* $\mathbf{B}$ *such that* $\mathbf{B} \xrightarrow{\eta} \mathbf{A}$.

PROOF. We prove this by induction on the size of $\mathbf{A}$. Suppose $\mathbf{A}$ is a β-normal term of type $(\beta\alpha^n \cdots \alpha^1)$ where β is a base type. The β-normal term $\mathbf{A}$ has the form

$$\lambda x^1_{\alpha^1} \cdots \lambda x^k_{\alpha^k} [h\, \mathbf{A}^1 \cdots \mathbf{A}^m]$$

for some $k \leq n$, $h_{\delta\gamma^m\cdots\gamma^1} \in \mathcal{V} \cup \mathcal{P} \cup \mathcal{S}$, $m \geq 0$, $\delta, \gamma^1, \ldots, \gamma^m \in \mathcal{T}$ and β-normal $\mathbf{A}^j_{\gamma^j}$ for each j $(1 \leq j \leq m)$. By the inductive hypothesis, there is a long $\beta\eta$-form $\mathbf{B}^j$ such that $\mathbf{B}^j \xrightarrow{\eta} \mathbf{A}^j$ for each j $(1 \leq j \leq m)$. Hence $\lambda\overline{x^k}\, [h\, \overline{\mathbf{B}^m}] \xrightarrow{\eta} \mathbf{A}$. Since $\mathbf{A}$ has type

$$(\beta\alpha^n \cdots \alpha^{k+1}\alpha^k \cdots \alpha^1),$$

the type δ of $[h\, \overline{\mathbf{B}^m}]$ is $(\beta\alpha^n \cdots \alpha^{k+1})$. Choose distinct variables $x^{k+1}_{\alpha^{k+1}}, \ldots, x^n_{\alpha^n}$ which do not occur in $\lambda\overline{x^k}\, [h\, \overline{\mathbf{B}^m}]$. Consequently, we have

$$\lambda\overline{x^n}\, [h\, \overline{\mathbf{B}^m}\, x^{k+1} \cdots x^n] \xrightarrow{\eta} \lambda\overline{x^k}\, [h\, \overline{\mathbf{B}^m}] \xrightarrow{\eta} \mathbf{A}.$$

By Lemma 2.1.17, for each $i \in \{k+1, \ldots, n\}$, there is a long $\beta\eta$-form $\mathbf{C}^i$ with $\mathbf{C}^i \xrightarrow{\eta} x^i$. We compute

$$\lambda\overline{x^n}[h\, \overline{\mathbf{B}^m}\, \mathbf{C}^{k+1} \cdots \mathbf{C}^n] \xrightarrow{\eta} \lambda\overline{x^n}\, [h\, \overline{\mathbf{B}^m}\, x^{k+1} \cdots x^n] \xrightarrow{\eta} \mathbf{A}$$

where $\lambda\overline{x^n}[h\, \overline{\mathbf{B}^m}\, \mathbf{C}^{k+1} \cdots \mathbf{C}^n]$ is a long $\beta\eta$-form.

For uniqueness, suppose $\mathbf{D}$ is a long $\beta\eta$-form such that $\mathbf{D} \xrightarrow{\eta} \mathbf{A}$. Since $\mathbf{D}$ is a long $\beta\eta$-form, it must (up to α-conversion) be $\lambda\overline{x^n}\, [k\, \overline{\mathbf{D}^p}]$ for long $\beta\eta$-forms $\mathbf{D}^j$ with $1 \leq j \leq p$. Since $\mathbf{D} \xrightarrow{\eta} \mathbf{A}$, we must have $k = h$, $p = n$, $\mathbf{D}^j \xrightarrow{\eta} \mathbf{A}^j$ for each $j \in \{1, \ldots, m\}$, and $\mathbf{D}^j \xrightarrow{\eta} x^j$ for each $j \in \{k+1, \ldots, n\}$. By the inductive hypothesis, $\mathbf{D}^j = \mathbf{B}^j$ for each $j \in \{1, \ldots, m\}$. By Lemma 2.1.17, we know $\mathbf{D}^j = \mathbf{C}^j$ for each $j \in \{k+1, \ldots, n\}$. Hence $\mathbf{D}$ is $\lambda\overline{x^n}[h\, \overline{\mathbf{B}^m}\, \mathbf{C}^{k+1} \cdots \mathbf{C}^n]$, as desired. □

LEMMA 2.1.19 (Long $\beta\eta$-Forms). *For every term* $\mathbf{A}$, *there is a unique long $\beta\eta$-form* $\mathbf{B}$ *such that* $\mathbf{A} \stackrel{\beta\eta}{=} \mathbf{B}$.

PROOF. Applying Lemma 2.1.18 to $\mathbf{A}{\downarrow}$, there is a unique long $\beta\eta$-form $\mathbf{B}$ such that $\mathbf{B} \xrightarrow{\eta} \mathbf{A}{\downarrow}$. Hence $\mathbf{A} \stackrel{\beta\eta}{=} \mathbf{B}$. For uniqueness, suppose $\mathbf{C}$ is a long $\beta\eta$-form such that $\mathbf{A} \stackrel{\beta\eta}{=} \mathbf{C}$. Thus $\mathbf{C} \xrightarrow{\beta\eta} \mathbf{A}{\downarrow}$. Since $\mathbf{A}$ and $\mathbf{C}$ are both β-normal, $\mathbf{C} \xrightarrow{\eta} \mathbf{A}{\downarrow}$. By uniqueness of $\mathbf{B}$, we have $\mathbf{B} = \mathbf{C}$ (up to α-conversion). □

NOTATION *We use* $\mathbf{A}^{\updownarrow}$ *to denote the (unique) long* $\beta\eta$*-form with* $\mathbf{A}^{\updownarrow} \stackrel{\beta\eta}{=} \mathbf{A}$.

2.1.1. Propositions. If we assume the signature includes enough logical constants, we can express any proposition (or sentence) internally as a term in wff_o (or cwff_o). For example, we can express the surjective version of Cantor's theorem using the term

$$\neg \exists g_{o\iota\iota} \forall f_{o\iota} \exists j_{\iota} \centerdot g\, j =^{o\iota} f$$

in $\mathit{wff}_o(\mathcal{S})$ as long as

$$\{\neg, \Sigma^{o\iota\iota}, \Pi^{o\iota}, \Sigma^{\iota}, =^{o\iota}\} \subseteq \mathcal{S}$$

TPS has long been able to prove this form of Cantor's theorem automatically. A natural deduction proof for this theorem is shown in Figure 2.1.[2] Note that instantiation in line (3) of the proof involves a negation. We would like to determine that the formula is not provable unless one makes use of such an instantiation. However, without negation in the signature, we cannot even *express* the theorem. For this reason, it is useful to distinguish between terms in $\mathit{wff}_o(\mathcal{S})$ over a signature $\mathcal{S}$ and *external propositions* constructed from these terms (cf. Definition 2.1.20). Cantor's theorem can be expressed as an external proposition even when the signature is empty. In such a case, the instantiation in line (3) is not a term in $\mathit{wff}_{o\iota}(\emptyset)$, and hence is not available for use in a proof. In fact, we will determine there is no proof of the surjective Cantor theorem if the signature is empty (cf. Corollary 6.7.9).

We define the set of external propositions over a signature inductively in Definition 2.1.20. To distinguish the logical constants in Definition 2.1.4 (e.g., $\vee$) from the operators defining the set of propositions we use a modified notation (e.g., $\dot{\vee}$).

DEFINITION 2.1.20. The set $\mathit{prop}(\mathcal{S})$ of *external propositions* over a signature $\mathcal{S}$ is defined inductively.

- If $\mathbf{A} \in \mathit{wff}_o(\mathcal{S})$, then $\mathbf{A} \in \mathit{prop}(\mathcal{S})$.
- If α is a type and $\mathbf{A}, \mathbf{B} \in \mathit{wff}_\alpha(\mathcal{S})$, then $[\mathbf{A} \doteq^{\alpha} \mathbf{B}] \in \mathit{prop}(\mathcal{S})$.
- $\dot{\top} \in \mathit{prop}(\mathcal{S})$.
- If $\mathbf{M} \in \mathit{prop}(\mathcal{S})$, then $[\dot{\neg}\mathbf{M}] \in \mathit{prop}(\mathcal{S})$.

[2]This natural deduction proof is a simplified version of a proof found by TPS automatically.

(1)	1	$\vdash$	$\exists g_{o\iota\iota} \forall f_{o\iota} \exists j_\iota . g\, j = f$	Hyp
(2)	1,2	$\vdash$	$\forall f_{o\iota} \exists j_\iota . g_{o\iota\iota}\, j = f$	Choose: $g_{o\iota\iota}$ 1
(3)	1,2	$\vdash$	$\exists j_\iota . g_{o\iota\iota}\, j = \lambda x_\iota . \neg g\, x\, x$	UI: $[\lambda x_\iota . \neg g_{o\iota\iota}\, x\, x]$ 2
(4)	1,2,4	$\vdash$	$g_{o\iota\iota}\, j_\iota = \lambda x_\iota . \neg g\, x\, x$	Choose: j_ι 3
(5)	1,2,4	$\vdash$	$\forall x_\iota . g_{o\iota\iota}\, j_\iota\, x = [\lambda x . \neg g\, x\, x]\, x$	Ext=: 4
(6)	1,2,4	$\vdash$	$g_{o\iota\iota}\, j_\iota\, j = [\lambda x_\iota . \neg g\, x\, x]\, j$	UI: j_ι 5
(7)	1,2,4	$\vdash$	$g_{o\iota\iota}\, j_\iota\, j = \neg g\, j\, j$	Lambda: 6
(8)	1,2,4	$\vdash$	$g_{o\iota\iota}\, j_\iota\, j \equiv \neg g\, j\, j$	Ext=: 7
(9)	1,2,4	$\vdash$	$\bot$	RuleP: 8
(10)	1,2	$\vdash$	$\bot$	RuleC: 3 9
(11)	1	$\vdash$	$\bot$	RuleC: 1 10
(12)		$\vdash$	$\neg \exists g_{o\iota\iota} \forall f_{o\iota} \exists j_\iota . g\, j = f$	NegIntro: 11

FIGURE 2.1. Proof of the Surjective Cantor Theorem

- If $\mathbf{M}, \mathbf{N} \in$ *prop*$(\mathcal{S})$, then $[\mathbf{M} \mathbin{\underset{\cdot}{\vee}} \mathbf{N}] \in$ *prop*$(\mathcal{S})$.
- If $\mathbf{M} \in$ *prop*$(\mathcal{S})$, then $[\underset{\cdot}{\forall} x_\alpha \mathbf{M}] \in$ *prop*$(\mathcal{S})$.

We will usually simply refer to *external propositions* as *propositions* or $\mathcal{S}$*-propositions*, leaving the adjective "external" implicit. We sometimes refer to terms in $wff_\alpha(\mathcal{S})$ as *internal terms* to emphasize the distinction between terms and external propositions. Suppose $\mathcal{S}$ is a signature with $\neg \in \mathcal{S}$ and $\mathbf{A} \in wff_o(\mathcal{S})$. Then $\neg\mathbf{A} \in wff_o(\mathcal{S})$ is both an internal term of type o and an external proposition. Also, $\mathbf{A}$ is an internal term of type o and an external proposition. Hence $\underset{\cdot}{\neg}\mathbf{A}$ is an external proposition, but is *not* a term of type o.

One should also note that we have not introduced (and will not introduce) a notion of λ-abstraction at the level of propositions. That is, if $\mathbf{M} \in$ *prop*$(\mathcal{S})$ but is not in $wff_o(\mathcal{S})$, then $[\lambda x_\alpha \mathbf{M}]$ will never be considered. The only uses of λ-abstractions occur in the construction of (internal) terms.

DEFINITION 2.1.21. We extend the notions of free and bound variables to propositions in the obvious way. Furthermore, we consider propositions syntactically identical if they are the same up to alphabetic changes in the names of bound variables. An $\mathcal{S}$*-sentence* (or, *sentence*) is a closed $\mathcal{S}$-proposition. Let *sent*$(\mathcal{S})$ denote the set of sentences over $\mathcal{S}$. Furthermore, we introduce the following shorthand notations for propositions.

- $\dot{\bot}$ means $\dot{\neg}\dot{\top}$.
- If $\mathbf{M}, \mathbf{N} \in prop(\mathcal{S})$, then $[\mathbf{M} \dot{\wedge} \mathbf{N}]$ means $\dot{\neg}[\dot{\neg}\mathbf{M} \dot{\vee} \dot{\neg}\mathbf{N}]$.
- If $\mathbf{M}, \mathbf{N} \in prop(\mathcal{S})$, then $[\mathbf{M} \dot{\supset} \mathbf{N}]$ means $[\dot{\neg}\mathbf{M} \dot{\vee} \mathbf{N}]$.
- If $\mathbf{M}, \mathbf{N} \in prop(\mathcal{S})$, then $[\mathbf{M} \dot{\equiv} \mathbf{N}]$ stands for the proposition $[\mathbf{M} \dot{\supset} \mathbf{N}] \dot{\wedge} [\mathbf{N} \dot{\supset} \mathbf{M}]$.
- If $\mathbf{M} \in prop(\mathcal{S})$, then $[\dot{\exists} x_\alpha \mathbf{M}]$ means $\dot{\neg}[\dot{\forall} x_\alpha \dot{\neg}\mathbf{M}]$.
- If α is a type and $\mathbf{A}, \mathbf{B} \in \mathit{wff}_\alpha(\mathcal{S})$, then $[\mathbf{A} \dot{\doteq}^\alpha \mathbf{B}]$ means Leibniz equality (at the level of propositions), i.e., $\dot{\forall} q_{o\alpha} \blacksquare q\,\mathbf{A} \dot{\supset} q\,\mathbf{B}$.

We say a proposition $\mathbf{M}$ is *equality-free* if it is constructed without using any proposition $\mathbf{A} \dot{=}^\alpha \mathbf{B}$ for $\mathbf{A}, \mathbf{B} \in \mathit{wff}_\alpha(\mathcal{S})$ and no logical constant $=^\alpha$ occurs in $\mathbf{M}$. We say a set Φ of propositions is *equality-free* if every $\mathbf{M} \in \Phi$ is equality-free.

Suppose $\mathcal{S}$ is equality-free. We know $=^\alpha$ does not occur in $\mathbf{M}$ for any $\mathbf{M} \in prop(\mathcal{S})$ since $=^\alpha \notin \mathcal{S}$. Consequently, if the signature is equality-free, then a proposition is equality-free iff it is constructed without using any proposition $\mathbf{A} \dot{=}^\alpha \mathbf{B}$ where $\mathbf{A}, \mathbf{B} \in \mathit{wff}_\alpha(\mathcal{S})$. Note also that any proposition of the form $[\mathbf{A} \dot{\doteq}^\alpha \mathbf{B}]$ is equality-free. One can always use Leibniz equality instead of $\dot{=}^\alpha$ in order to express a proposition in an equality-free way. However, the proposition $[\mathbf{A} \dot{\doteq}^\alpha \mathbf{B}]$ need not be equivalent to the proposition $[\mathbf{A} \dot{=}^\alpha \mathbf{B}]$ (cf. Remark 6.6.3). In general, we will restrict ourselves to equality-free propositions (i.e., only allowing Leibniz equality) in the non-extensional case, and allow unrestricted propositions in the fully extensional case. The main reason for making this distinction is pragmatic. Equality is vital for handling extensionality appropriately. On the other hand, trying to incorporate equality into the non-extensional cases requires a radical departure from the way equality has historically been handled in TPS. TPS historically has always considered equality as an abbreviation, either for Leibniz equality or some form of "extensional" equality (without building in full extensionality).

We can extend any typed relation or typed function on the set $\mathit{wff}(\mathcal{S})$ of formulas to the set of propositions by an obvious recursion. For example, we can extend the notion of $\beta\eta$-normalization to propositions. We define $[\mathbf{A} \dot{=} \mathbf{B}]^\downarrow$ to be $[\mathbf{A}^\downarrow \dot{=} \mathbf{B}^\downarrow]$. We define $[\dot{\neg}\mathbf{N}]^\downarrow$ to be $[\dot{\neg}\mathbf{N}^{\downarrow}]$. We define $\mathbf{M}^{\downarrow}$ similarly when $\mathbf{M}$ is constructed using $\dot{\top}$, $\dot{\vee}$ or $\dot{\forall}$.

In Definition 2.1.22, we give a general technique for extending n-ary typed relations (and typed functions) on *wff*$(\mathcal{S})$ to n-ary relations on *prop*$(\mathcal{S})$. For the example of $\beta\eta$-reduction, we can define the binary (functional) relation $R(\mathbf{A}_\alpha, \mathbf{B}_\alpha)$ to hold when $\mathbf{B}$ is $\mathbf{A}^{\downarrow}$. We use Definition 2.1.22 to extend R to a binary relation R^* on *prop*$(\mathcal{S})$. We then define $\mathbf{M}^{\downarrow}$ for $\mathbf{M} \in$ *prop*$(\mathcal{S})$ to be the unique $\mathbf{N} \in$ *prop*$(\mathcal{S})$ (up to α-conversion) such that $R^*(\mathbf{M}, \mathbf{N})$ holds. The intuition is that $\mathbf{M}$ and $\mathbf{N}$ must have similar constructions and the terms used to construct $\mathbf{M}$ must $\beta\eta$-normalize to the corresponding terms used to construct $\mathbf{N}$. We use the same technique described in Remark 2.1.7 to handle α-conversion implicitly.

DEFINITION 2.1.22. For any $n \geq 1$ and typed n-ary relation R on *wff*$(\mathcal{S})$, we define the *syntactic extension R^* of R to propositions* to be the least (α-respecting) n-ary relation on *prop*$(\mathcal{S})$ such that the following hold:

- If $\mathbf{M}^1, \ldots, \mathbf{M}^n \in$ *wff*$_o(\mathcal{S})$ and $R(\mathbf{M}^1, \ldots, \mathbf{M}^n)$ holds, then $R^*(\mathbf{M}^1, \ldots, \mathbf{M}^n)$ holds.
- If $\mathbf{A}^1, \ldots, \mathbf{A}^n, \mathbf{B}^1, \ldots, \mathbf{B}^n \in$ *wff*$_\alpha(\mathcal{S})$ and $R(\mathbf{A}^1, \ldots, \mathbf{A}^n)$ and $R(\mathbf{B}^1, \ldots, \mathbf{B}^n)$ hold, then $R^*([\mathbf{A}^1 \mathrel{\dot{=}^\alpha} \mathbf{B}^1], \ldots, [\mathbf{A}^n \mathrel{\dot{=}^\alpha} \mathbf{B}^n])$ holds.
- $R^*(\dot{\top}, \ldots, \dot{\top})$ holds.
- If $R^*(\mathbf{M}^1, \ldots, \mathbf{M}^n)$ holds, then $R^*([\dot{\neg}\mathbf{M}^1], \ldots, [\dot{\neg}\mathbf{M}^n])$ holds.
- If $R^*(\mathbf{M}^1, \ldots, \mathbf{M}^n)$ holds and $R^*(\mathbf{N}^1, \ldots, \mathbf{N}^n)$ holds, then $R^*([\mathbf{M}^1 \mathrel{\dot{\vee}} \mathbf{N}^1], \ldots, [\mathbf{M}^n \mathrel{\dot{\vee}} \mathbf{N}^n])$ holds.
- If $R^*(\mathbf{M}^1, \ldots, \mathbf{M}^n)$ and x_α is not vulnerable for R (cf. Remark 2.1.7), then $R^*([\dot{\forall} x_\alpha \, \mathbf{M}^1], \ldots, [\dot{\forall} x_\alpha \, \mathbf{M}^n])$ holds.

It is easy to check that if R is a functional relation on *wff*$(\mathcal{S})$ (up to α-conversion), then R^* is a functional relation on *prop*$(\mathcal{S})$ (up to α-conversion).

DEFINITION 2.1.23. We extend the notation for substitution, simultaneous substitution by θ, β-normal forms, $\beta\eta$-normal forms and long $\beta\eta$-forms by taking the syntactic extension of each corresponding binary functional relation to propositions. We use the same notations

$[\mathbf{C}/x]\mathbf{M}$, $\theta(\mathbf{M})$, $\mathbf{M}^{\downarrow\beta}$, $\mathbf{M}^{\downarrow}$ and $\mathbf{M}^{\updownarrow}$ for the syntactic extensions to propositions.

DEFINITION 2.1.24. For any set $\mathcal{W} \subseteq prop(\mathcal{S}) \cup wff(\mathcal{S})$ of terms and propositions, we define

$$\mathcal{W}^{\downarrow *} := \{\mathbf{M} \in \mathcal{W} \mid \mathbf{M}^{\downarrow *} = \mathbf{M}\}$$

and

$$\mathcal{W}^{\updownarrow} := \{\mathbf{M} \in \mathcal{W} \mid \mathbf{M}^{\updownarrow} = \mathbf{M}\}.$$

For example, $wff_\alpha(\mathcal{S})^{\downarrow}$ is the set of $\beta\eta$-normal forms of type α and $prop(\mathcal{S})^{\downarrow}$ is the set of $\beta\eta$-normal propositions.

LEMMA 2.1.25. *For any $\mathbf{A}_\alpha \in wff_\alpha(\mathcal{S})$, $\mathbf{M} \in prop(\mathcal{S})$ and variable x_α, $([\mathbf{A}/x](\mathbf{M}^{\downarrow *}))^{\downarrow *} = ([\mathbf{A}/x]\mathbf{M})^{\downarrow *}$.*

PROOF. This follows by induction on $\mathbf{M}$. The base cases follow from uniqueness of normal forms. The remaining cases follow from the inductive hypothesis and the definitions of substitution and normalization at the level of external propositions. □

2.1.2. Measures. For some inductive proofs on propositions, we use an *external measure* $|\mathbf{M}|_{\mathbf{e}}$ and an *internal measure* $|\mathbf{M}|_{\mathbf{i}}$ on propositions $\mathbf{M}$. The *external measure* of the proposition depends on the number of times $\doteq^\alpha$, $\dot{\neg}$, $\dot{\vee}$ and $\dot{\forall}$ are used in the construction of $\mathbf{M}$. The *internal measure* of the proposition depends on the occurrences of constants, parameters and variables in the long $\beta\eta$-normal form of $\mathbf{M}$. We start by defining a natural number measure $|\alpha|_{\mathbf{t}}$ on types $\alpha \in \mathcal{T}$.

DEFINITION 2.1.26. We inductively define $|\alpha|_{\mathbf{t}} \in \mathbb{N}$ for each type $\alpha \in \mathcal{T}$.

- $|o|_{\mathbf{t}} = 0$
- $|\iota|_{\mathbf{t}} = 0$
- $|\alpha\beta|_{\mathbf{t}} = |\alpha|_{\mathbf{t}} + |\beta|_{\mathbf{t}} + 1$

We next define $|\mathbf{M}|_{\mathbf{e}}$. Note that the definition on equations depends on the type of the equation.

DEFINITION 2.1.27. We define the *external measure* $|\mathbf{M}|_{\mathbf{e}} \in \mathbb{N}$ of $\mathbf{M}$ by recursion on the proposition $\mathbf{M}$.

- If $\mathbf{M} \in wff_o(\mathcal{S})$, then $|\mathbf{M}|_{\mathbf{e}} := 0$.

- $|[\mathbf{A}_\alpha \doteq^\alpha \mathbf{B}_\alpha]|_\mathbf{e} := 2|\alpha|_\mathbf{t} + 5$.
- $|\dot{\top}|_\mathbf{e} := 0$.
- $|[\dot{\neg}\mathbf{M}]|_\mathbf{e} := |\mathbf{M}|_\mathbf{e} + 1$.
- $|[\mathbf{M} \dot{\vee} \mathbf{N}]|_\mathbf{e} := |\mathbf{M}|_\mathbf{e} + |\mathbf{N}|_\mathbf{e} + 1$.
- $|[\dot{\forall} x_\alpha \mathbf{M}]|_\mathbf{e} := |\mathbf{M}|_\mathbf{e} + 1$.

LEMMA 2.1.28. *For any* $\mathbf{A}, \mathbf{B} \in \mathit{wff}_o(\mathcal{S})$, $\mathbf{F}_{\alpha\beta}, \mathbf{G}_{\alpha\beta} \in \mathit{wff}_{\alpha\beta}(\mathcal{S})$, *variable* y_β *and* $\mathcal{S}$*-propositions* $\mathbf{M}$ *and* $\mathbf{N}$, *we have the following:*

1. $|\dot{\neg}\mathbf{M}|_\mathbf{e} < |\dot{\neg}[\mathbf{M} \vee \mathbf{N}]|_\mathbf{e}$.
2. $|\dot{\neg}\mathbf{N}|_\mathbf{e} < |\dot{\neg}[\mathbf{M} \vee \mathbf{N}]|_\mathbf{e}$.
3. $|[\dot{\forall} y_\beta \centerdot \mathbf{F}\, y \doteq^\alpha \mathbf{G}\, y]|_\mathbf{e} < |[\mathbf{F} \doteq^{\alpha\beta} \mathbf{G}]|_\mathbf{e}$.
4. $|[\dot{\neg}\mathbf{A} \dot{\vee} \mathbf{B}]|_\mathbf{e} + |[\dot{\neg}\mathbf{B} \dot{\vee} \mathbf{A}]|_\mathbf{e} < |[\mathbf{A} \doteq^o \mathbf{B}]|_\mathbf{e}$.
5. $|[\mathbf{A} \dot{\vee} \mathbf{B}]|_\mathbf{e} + |[\dot{\neg}\mathbf{A} \dot{\vee} \dot{\neg}\mathbf{B}]|_\mathbf{e} < |[\mathbf{A} \doteq^o \mathbf{B}]|_\mathbf{e}$.

PROOF. Each inequality is easily verified using Definition 2.1.27. □

LEMMA 2.1.29. *Let* $n \geq 1$, R *be a typed* n*-ary relation* R *on* $\mathit{wff}(\mathcal{S})$ *and let* R^* *be the syntactic extension of* R *to propositions. If* $R^*(\mathbf{M}^1, \ldots, \mathbf{M}^n)$, *then*

$$|\mathbf{M}^1|_\mathbf{e} = \cdots = |\mathbf{M}^n|_\mathbf{e}.$$

for all $\mathbf{M}^1, \ldots, \mathbf{M}^n \in \mathit{prop}(\mathcal{S})$.

PROOF. We define M to be the n-ary relation on $\mathit{prop}(\mathcal{S})$ so that $M(\mathbf{M}^1, \ldots, \mathbf{M}^n)$ holds if $|\mathbf{M}^1|_\mathbf{e} = \cdots = |\mathbf{M}^n|_\mathbf{e}$. We know R^* is the least (α-respecting) relation satisfying the conditions in Definition 2.1.22. We verify R^* is a subrelation of M by verifying M satisfies these conditions.

If $\mathbf{M}^1, \ldots, \mathbf{M}^n \in \mathit{wff}_o(\mathcal{S})$ and $R(\mathbf{M}^1, \ldots, \mathbf{M}^n)$ holds, then

$$|\mathbf{M}^1|_\mathbf{e} = \cdots = |\mathbf{M}^n|_\mathbf{e}$$

since $|\mathbf{M}^i|_\mathbf{e} = 0$ for every i $(1 \leq i \leq n)$.

Suppose $\mathbf{A}^1, \ldots, \mathbf{A}^n, \mathbf{B}^1, \ldots, \mathbf{B}^n \in \mathit{wff}_\alpha(\mathcal{S})$ and $R(\mathbf{A}^1, \ldots, \mathbf{A}^n)$.

$$|[\mathbf{A}^i \doteq^\alpha \mathbf{B}^i]|_\mathbf{e} = 2|\alpha|_\mathbf{t} + 5$$

for each i $(1 \leq i \leq n)$ and so

$$|[\mathbf{A}^1 \doteq^\alpha \mathbf{B}^1]|_\mathbf{e} = \cdots = |[\mathbf{A}^n \doteq^\alpha \mathbf{B}^n]|_\mathbf{e}.$$

We trivially know $M(\dot{\top}, \ldots, \dot{\top})$ holds.

If $M(\mathbf{M}^1, \ldots, \mathbf{M}^n)$ holds, then we have

$$|\dot{\neg}\mathbf{M}^1|_\mathbf{e} = \cdots = |\dot{\neg}\mathbf{M}^n|_\mathbf{e}$$

since $|\dot{\neg}\mathbf{M}^i|_\mathbf{e} = |\mathbf{M}^i|_\mathbf{e} + 1 = |\mathbf{M}^1|_\mathbf{e} + 1$ for each $i \in \{1, \ldots, n\}$.

If $M(\mathbf{M}^1, \ldots, \mathbf{M}^n)$ and $M(\mathbf{N}^1, \ldots, \mathbf{N}^n)$ hold, then

$$|[\mathbf{M}^1 \dot{\vee} \mathbf{N}^1]|_\mathbf{e} = \cdots = |[\mathbf{M}^n \dot{\vee} \mathbf{N}^n]|_\mathbf{e}$$

holds since $|[\mathbf{M}^i \dot{\vee} \mathbf{N}^i]|_\mathbf{e} = |\mathbf{M}^i|_\mathbf{e} + |\mathbf{N}^i|_\mathbf{e} + 1 = |\mathbf{M}^1|_\mathbf{e} + |\mathbf{N}^1|_\mathbf{e} + 1$ for each i $(1 \leq i \leq n)$.

Finally, if $M(\mathbf{M}^1, \ldots, \mathbf{M}^n)$ (and x_α is not vulnerable for R), then

$$|\dot{\forall} x_\alpha \mathbf{M}^1|_\mathbf{e} = \cdots = |\dot{\forall} x_\alpha \mathbf{M}^n|_\mathbf{e}$$

since $|\dot{\forall} x_\alpha \mathbf{M}^i|_\mathbf{e} = |\mathbf{M}^i|_\mathbf{e} + 1 = |\mathbf{M}^1|_\mathbf{e} + 1$ for each i $(1 \leq i \leq n)$.

Also, note that M respects α-conversion since the particular bound variables play no role in the definition of $|\mathbf{M}|_\mathbf{e}$. Hence R^* is a subrelation of M. That is, we have

$$|\mathbf{M}^1|_\mathbf{e} = \cdots = |\mathbf{M}^n|_\mathbf{e}$$

whenever $R^*(\mathbf{M}^1, \ldots, \mathbf{M}^n)$. □

LEMMA 2.1.30. *For any* $\mathbf{A}_\alpha \in \mathit{wff}_\alpha(\mathcal{S})$, $\mathbf{F}_{\alpha\beta}, \mathbf{G}_{\alpha\beta} \in \mathit{wff}_{\alpha\beta}(\mathcal{S})$, *variables* x_α *and* y_β *and* $\mathcal{S}$*-propositions* $\mathbf{M}$ *and* $\mathbf{N}$*, we have the following:*

1. $|\mathbf{M}^\downarrow|_\mathbf{e} = |\mathbf{M}|_\mathbf{e}$.
2. $|[\mathbf{A}/x]\mathbf{M}|_\mathbf{e} = |\mathbf{M}|_\mathbf{e}$.
3. $|[\mathbf{A}/x]\mathbf{M}|_\mathbf{e} < |[\dot{\forall} x_\alpha \mathbf{M}]|_\mathbf{e}$.
4. $|\dot{\neg}[[\mathbf{A}/x]\mathbf{M}]|_\mathbf{e} < |\dot{\neg}[\dot{\forall} x_\alpha \mathbf{M}]|_\mathbf{e}$.

PROOF. The first two equations follow from Lemma 2.1.29 since $\mathbf{M}^\downarrow$ and $[\mathbf{A}/x]\mathbf{M}$ are defined to be syntactic extensions of functional relations on $\mathit{wff}(\mathcal{S})$ (cf. Definition 2.1.23). Using the second equation we have

$$|[\mathbf{A}/x]\mathbf{M}|_\mathbf{e} = |\mathbf{M}|_\mathbf{e} < |[\dot{\forall} x_\alpha \mathbf{M}]|_\mathbf{e}$$

and

$$|\dot{\neg}[[\mathbf{A}/x]\mathbf{M}]|_{\mathbf{e}} = |[\mathbf{A}/x]\mathbf{M}|_{\mathbf{e}} + 1 < |[\dot{\forall} x_\alpha \mathbf{M}]|_{\mathbf{e}} + 1 = |\dot{\neg}[\dot{\forall} x_\alpha \mathbf{M}]|_{\mathbf{e}}$$

□

We now define the internal measure $|\mathbf{M}|_{\mathbf{i}}$ on propositions $\mathbf{M}$ by defining $|\mathbf{A}|_l$ for long $\beta\eta$-forms $\mathbf{A} \in \mathit{wff}_\alpha(\mathcal{S})^{\updownarrow}$, and then defining $|\mathbf{M}|_{\mathbf{i}}$ inductively for propositions $\mathbf{M}$.

Definition 2.1.31. We define the *length* $|\mathbf{A}|_l$ of long $\beta\eta$-forms by induction on $\mathbf{A}$. Since $\mathbf{A}$ is in long $\beta\eta$-form, it must have the form $\lambda\overline{x^n}\,[h\,\overline{\mathbf{A}^m}]$ for some $h \in \mathcal{V} \cup \mathcal{P} \cup \mathcal{S}$, $m \geq 0$ and long $\beta\eta$-forms $\mathbf{A}^1, \ldots, \mathbf{A}^m$. By the inductive hypothesis, we may assume $|\mathbf{A}^i|_l$ is defined for each i $(1 \leq i \leq m)$ and define

$$|\mathbf{A}|_l := 1 + 2|\mathbf{A}^1|_l + \cdots + 2|\mathbf{A}^m|_l.$$

(The base case of this definition is when $m = 0$ and $|\lambda\overline{x^n}\,h_\beta|_l = 1$ where $\beta \in \{o, \iota\}$.)

We define the *internal measure* $|\mathbf{M}|_{\mathbf{i}}$ of propositions $\mathbf{M}$ by induction as follows:

- If $\mathbf{M} \in \mathit{wff}_o(\mathcal{S})$, then let $|\mathbf{M}|_{\mathbf{i}} := |\mathbf{M}^{\updownarrow}|_l$.
- If $\mathbf{M}$ is $[\mathbf{A} \mathrel{\dot{=}^\alpha} \mathbf{B}]$ for some type α and terms $\mathbf{A}, \mathbf{B} \in \mathit{wff}_\alpha(\mathcal{S})$, then let $|\mathbf{M}|_{\mathbf{i}} := 2|\mathbf{A}^{\updownarrow}|_l + 2|\mathbf{B}^{\updownarrow}|_l$.
- If $\mathbf{M}$ is $\dot{\top}$, then let $|\mathbf{M}|_{\mathbf{i}} := 0$.
- If $\mathbf{M}$ is $[\dot{\neg}\mathbf{N}]$ for some proposition $\mathbf{N}$, then let $|\mathbf{M}|_{\mathbf{i}} := |\mathbf{N}|_{\mathbf{i}}$.
- If $\mathbf{M}$ is $[\mathbf{N} \mathbin{\dot{\vee}} \mathbf{P}]$ for propositions $\mathbf{N}$ and $\mathbf{P}$, then define $|\mathbf{M}|_{\mathbf{i}} := |\mathbf{N}|_{\mathbf{i}} + |\mathbf{P}|_{\mathbf{i}}$.
- If $\mathbf{M}$ is $[\dot{\forall} x_\alpha \mathbf{N}]$ for some variable x_α and proposition $\mathbf{N}$, then let $|\mathbf{M}|_{\mathbf{i}} := |\mathbf{N}|_{\mathbf{i}}$.

Lemma 2.1.32. *For base types $\beta \in \{o, \iota\}$, $h_{\beta\alpha^n\cdots\alpha^1} \in \mathcal{V} \cup \mathcal{P} \cup \mathcal{S}$, variables $z^1_{\gamma^1}, \ldots, z^m_{\gamma^m}$ and terms $\mathbf{A}^1_{\alpha^1}, \ldots, \mathbf{A}^n_{\alpha^n}$, we have*

$$\sum_{i=1}^{n} |[\lambda\overline{z^m}\,\mathbf{A}^i]^{\updownarrow}|_l < |[\lambda\overline{z^m}\,[h\,\overline{\mathbf{A}^n}]]^{\updownarrow}|_l.$$

PROOF. Since $[h\,\overline{\mathbf{A}^n}]$ is of base type, we have

$$[\lambda\overline{z^m}\,[h\,\overline{\mathbf{A}^n}]]^{\updownarrow} = [\lambda\overline{z^m}\,[h\,(\mathbf{A}^1)^{\updownarrow}\cdots(\mathbf{A}^n)^{\updownarrow}]].$$

For each i $(1 \leq i \leq n)$,

$$[\lambda\overline{z^m}\,\mathbf{A}^i]^{\updownarrow} = [\lambda\overline{z^m}\,(\mathbf{A}^i)^{\updownarrow}].$$

We compute

$$\begin{aligned}\sum_{i=1}^{n} |[\lambda\overline{z^m}\,\mathbf{A}^i]^{\updownarrow}|_l &= \sum_{i=1}^{n} |(\mathbf{A}^i)^{\updownarrow}|_l \\ &< 1 + 2|(\mathbf{A}^1)^{\updownarrow}|_l + \cdots + 2|(\mathbf{A}^n)^{\updownarrow}|_l \\ &= |[\lambda\overline{z^m}\,[h\,\overline{\mathbf{A}^n}]]^{\updownarrow}|_l.\end{aligned}$$

□

LEMMA 2.1.33. *For base types $\beta \in \{o, \iota\}$, $h_{\beta\alpha^n\cdots\alpha^1} \in \mathcal{V} \cup \mathcal{P} \cup \mathcal{S}$ and terms $\mathbf{A}^1_{\alpha^1}, \ldots, \mathbf{A}^n_{\alpha^n}, \mathbf{B}^1_{\alpha^1}, \ldots, \mathbf{B}^n_{\alpha^n}$, we have*

$$\sum_{i=1}^{n} |[\mathbf{A}^i \doteq^{\alpha^i} \mathbf{B}^i]|_{\mathbf{i}} < |[[h\,\overline{\mathbf{A}^n}] \doteq^{\beta} [h\,\overline{\mathbf{B}^n}]]|_{\mathbf{i}}.$$

PROOF. Using Lemma 2.1.32, we compute

$$\begin{aligned}\textstyle\sum_{i=1}^{n} |[\mathbf{A}^i \doteq^{\alpha^i} \mathbf{B}^i]|_{\mathbf{i}} &= \textstyle\sum_{i=1}^{n} \left(2|(\mathbf{A}^i)^{\updownarrow}|_l + 2|(\mathbf{B}^i)^{\updownarrow}|_l\right) \\ &= 2\left(\textstyle\sum_{i=1}^{n} |(\mathbf{A}^i)^{\updownarrow}|_l\right) + 2\left(\textstyle\sum_{i=1}^{n} |(\mathbf{B}^i)^{\updownarrow}|_l\right) \\ &< 2|[h\,\overline{\mathbf{A}^n}]^{\updownarrow}|_l + 2|[h\,\overline{\mathbf{B}^n}]^{\updownarrow}|_l \\ &= |[[h\,\overline{\mathbf{A}^n}] \doteq^{\beta} [h\,\overline{\mathbf{B}^n}]]|_{\mathbf{i}}.\end{aligned}$$

□

LEMMA 2.1.34. *Let $\mathbf{A}_\alpha$ be a term, $x_\beta \in \mathcal{V}$ and $b_\beta \in \mathcal{V} \cup \mathcal{P} \cup \mathcal{S}$. If $\mathbf{A}_\alpha$ is a long $\beta\eta$-form, then $[b/x]\mathbf{A}$ is a long $\beta\eta$-form and*

$$|[b/x]\mathbf{A}|_l = |\mathbf{A}|_l.$$

PROOF. Since $\mathbf{A}$ is in long $\beta\eta$-form, it must have the form $\lambda\overline{x^n}\,[h\,\overline{\mathbf{A}^m}]$ for some $h \in \mathcal{V} \cup \mathcal{P} \cup \mathcal{S}$, $n \geq 0$, $m \geq 0$ and long $\beta\eta$-forms $\mathbf{A}^1, \ldots, \mathbf{A}^m$. By the inductive hypothesis, $[b/x]\mathbf{A}^i$ is a long $\beta\eta$-form and $|[b/x]\mathbf{A}^i|_l = |\mathbf{A}^i|_l$ for each i $(1 \leq i \leq m)$. Let $\mathbf{B}^i$ be $[b/x]\mathbf{A}^i$ for each i $(1 \leq i \leq m)$. If $h \neq x$, then $[b/x]\mathbf{A}$ is the long $\beta\eta$-form $\lambda\overline{x^n}\,[h\,\overline{\mathbf{B}^m}]$ and

$$\begin{aligned}|[b/x]\mathbf{A}|_l &= 1 + 2|\mathbf{B}^1|_l + \cdots + 2|\mathbf{B}^m|_l \\ &= 1 + 2|\mathbf{A}^1|_l + \cdots + 2|\mathbf{A}^m|_l = |\mathbf{A}|_l.\end{aligned}$$

If $h = x$, then $[b/x]\mathbf{A}$ is the long $\beta\eta$-form $\lambda\overline{x^n}\,[b\,\overline{\mathbf{B}^m}]$ (which is a long $\beta\eta$-form since $b \in \mathcal{V} \cup \mathcal{P} \cup \mathcal{S}$) and

$$\begin{aligned}|[b/x]\mathbf{A}|_l &= 1 + 2|\mathbf{B}^1|_l + \cdots + 2|\mathbf{B}^m|_l \\ &= 1 + 2|\mathbf{A}^1|_l + \cdots + 2|\mathbf{A}^m|_l = |\mathbf{A}|_l.\end{aligned}$$

□

LEMMA 2.1.35. *Let* $\mathbf{M} \in prop(\mathcal{S})$, $x_\beta \in \mathcal{V}$ *and* $b_\beta \in \mathcal{V} \cup \mathcal{P} \cup \mathcal{S}$. *Then* $|[b/x]\mathbf{M}|_{\mathbf{i}} = |\mathbf{M}|_{\mathbf{i}}$.

PROOF. An easy induction on $\mathbf{M}$ using Lemma 2.1.34 for the base cases verifies the result. □

LEMMA 2.1.36. *Let* $\mathbf{F}_{\alpha\beta} \in wff_{\alpha\beta}(\mathcal{S})$ *be a long* $\beta\eta$*-form. For any* $b_\beta \in \mathcal{V} \cup \mathcal{P} \cup \mathcal{S}$, $|[\mathbf{F}\,b]^{\downarrow}|_l = |\mathbf{F}|_l$.

PROOF. The type α must be of the form $(\gamma\alpha^n \cdots \alpha^1)$ for some $n \geq 0$, types $\alpha^1, \ldots, \alpha^n$ and base type $\gamma \in \{o, \iota\}$. As a long $\beta\eta$-form of type $(\gamma\alpha^n \cdots \alpha^1\beta)$, $\mathbf{F}$ must have the form $\lambda x_\beta\, \lambda\overline{x^n}\,[h\,\overline{\mathbf{A}^m}]$ for some $h \in \mathcal{V} \cup \mathcal{P} \cup \mathcal{S}$, $m \geq 0$ and long $\beta\eta$-forms $\mathbf{A}^1, \ldots, \mathbf{A}^m$. Let $\mathbf{A}$ be the long $\beta\eta$-form $\lambda\overline{x^n}\,[h\,\overline{\mathbf{A}^m}]$. Note that $[\mathbf{F}\,b] \stackrel{\beta}{=} [b/x]\mathbf{A}$. Using Lemma 2.1.34, we know $[b/x]\mathbf{A}$ is a long $\beta\eta$-form and hence $[b/x]\mathbf{A}$ is the unique long $\beta\eta$-normal form of $[\mathbf{F}\,b]$. Using Lemma 2.1.34, we compute

$$|[\mathbf{F}\,b]^{\downarrow}|_l = |[b/x]\mathbf{A}|_l = |\mathbf{A}|_l = 1 + 2|\mathbf{A}^1|_l + \cdots + 2|\mathbf{A}^m|_l = |\mathbf{F}|_l$$

□

Since the internal measure of a proposition is defined in terms of long $\beta\eta$-forms, two $\beta\eta$-equivalent propositions have the same internal measure.

LEMMA 2.1.37. *Let* $\mathbf{M}, \mathbf{N} \in prop(\mathcal{S})$ *be propositions. If* $\mathbf{M} \stackrel{\beta\eta}{=} \mathbf{N}$, *then* $|\mathbf{M}|_{\mathrm{i}} = |\mathbf{N}|_{\mathrm{i}}$

PROOF. The proof proceeds by induction on the proposition $\mathbf{M}$. The base case for terms of type o follows by computing

$$|\mathbf{M}|_{\mathrm{i}} = |\mathbf{M}^{\updownarrow}|_l = |\mathbf{N}^{\updownarrow}|_l = |\mathbf{N}|_{\mathrm{i}}$$

when $\mathbf{M}, \mathbf{N} \in \mathit{wff}_o$ since $\mathbf{M}^{\updownarrow} = \mathbf{N}^{\updownarrow}$. The base case for $\beta\eta$-equal equations $[\mathbf{A} \mathrel{\dot{=}^{\alpha}} \mathbf{B}]$ and $[\mathbf{C} \mathrel{\dot{=}^{\alpha}} \mathbf{D}]$ follows by computing

$$|[\mathbf{A} \mathrel{\dot{=}^{\alpha}} \mathbf{B}]|_{\mathrm{i}} = 2|\mathbf{A}^{\updownarrow}|_l + 2|\mathbf{B}^{\updownarrow}|_l = 2|\mathbf{C}^{\updownarrow}|_l + 2|\mathbf{D}^{\updownarrow}|_l = |[\mathbf{C} \mathrel{\dot{=}^{\alpha}} \mathbf{D}]|_{\mathrm{i}}$$

since $\mathbf{A}^{\updownarrow} = \mathbf{C}^{\updownarrow}$ and $\mathbf{B}^{\updownarrow} = \mathbf{D}^{\updownarrow}$. The remaining cases follow directly from the inductive hypothesis. □

Whenever a logical constant c is in the signature $\mathcal{S}$ and an internal term $\mathbf{A} \in \mathit{wff}_o(\mathcal{S})$ has c at the head, there is a corresponding external proposition where we have lifted c to the level of external propositions. We make this precise by defining an operation taking internal terms $\mathbf{A} \in \mathit{wff}_o(\mathcal{S})$ to external propositions $\mathbf{A}^{\sharp} \in prop(\mathcal{S})$ (cf. Definition 2.1.38). For example, we will define $[\neg\mathbf{A}]^{\sharp}$ to be $\dot{\neg}\mathbf{A}$. This operation will only act on the logical constant at the head of $\mathbf{A}$. The operation is *not* recursive. For instance, $[\neg\neg\mathbf{A}]^{\sharp}$ will be defined to be $\dot{\neg}[\neg\mathbf{A}]$ instead of $\dot{\neg}\dot{\neg}\mathbf{A}$. The intuition is that we only lift an internal logical constant to the level of external propositions when the constant is at the head of the term.

DEFINITION 2.1.38. For each $\mathbf{A} \in \mathit{wff}_o(\mathcal{S})$, we define the proposition $\mathbf{A}^{\sharp}$ by cases.

variables: For any variable $x_{o\alpha^n\cdots\alpha^1}$, $[x\,\overline{\mathbf{A}^n}]^{\sharp} := [x\,\overline{\mathbf{A}^n}]$.

parameters: For any parameter $W_{o\alpha^n\cdots\alpha^1}$, $[W\,\overline{\mathbf{A}^n}]^{\sharp} := [W\,\overline{\mathbf{A}^n}]$.

λ: For any term, $[\lambda x_\beta\, \mathbf{C}_{o\alpha^n\cdots\alpha^1}]$,

$$[[\lambda x_\beta\, \mathbf{C}_{o\alpha^n\cdots\alpha^1}]\,\mathbf{B}\,\overline{\mathbf{A}^n}]^{\sharp} := [[\lambda x_\beta\, \mathbf{C}_{o\alpha^n\cdots\alpha^1}]\,\mathbf{B}\,\overline{\mathbf{A}^n}].$$

$\top$: $\top^{\sharp} := \dot{\top}$.

$\bot$: $\bot^{\sharp} := \dot{\bot}$.

$\neg$: $[\neg \mathbf{B}_o]^\sharp := \dot{\neg} \mathbf{B}$.

$\vee$: $[\mathbf{B}_o \vee \mathbf{C}_o]^\sharp := [\mathbf{B} \dot{\vee} \mathbf{C}]$.

$\wedge$: $[\mathbf{B} \wedge \mathbf{C}]^\sharp := [\mathbf{B} \dot{\wedge} \mathbf{C}]$.

$\supset$: $[\mathbf{B} \supset \mathbf{C}]^\sharp := [\mathbf{B} \dot{\supset} \mathbf{C}]$.

$\equiv$: $[\mathbf{B} \equiv \mathbf{C}]^\sharp := [\mathbf{B} \dot{\equiv} \mathbf{C}]$.

Π^α: $[\Pi^\alpha \mathbf{F}_{o\alpha}]^\sharp := [\dot{\forall} x_\alpha \centerdot \mathbf{F}\, x]$ (where $x_\alpha \notin \mathbf{Free}(\mathbf{F})$).

Σ^α: $[\Sigma^\alpha \mathbf{F}]^\sharp := [\dot{\exists} x_\alpha \centerdot \mathbf{F}\, x]$ (where $x_\alpha \notin \mathbf{Free}(\mathbf{F})$).

$=^\alpha$: $[\mathbf{B}_\alpha =^\alpha \mathbf{C}_\alpha]^\sharp := [\mathbf{B} \dot{=}^\alpha \mathbf{C}]$.

A few comments about $\mathbf{A}^\sharp$ are in order. Note that $\mathbf{A}^\sharp$ is *only* defined when $\mathbf{A}$ is an internal term in $\mathit{wff}_o(\mathcal{S})$. Hence there is no scoping ambiguity for an external proposition of the form $\dot{\neg}\mathbf{A}^\sharp \dot{\vee} \mathbf{B}^\sharp$ since this only makes sense if $\mathbf{A}, \mathbf{B} \in \mathit{wff}_o(\mathcal{S})$ and the $^\sharp$ operation is applied to $\mathbf{A}$ and $\mathbf{B}$. For clarity, we may write $\dot{\neg}(\mathbf{A}^\sharp) \dot{\vee} (\mathbf{B}^\sharp)$, though this is not necessary. Also, in the first three cases of Definition 2.1.38 where the head of $\mathbf{A} \in \mathit{wff}_o(\mathcal{S})$ is a variable, parameter, or an abstraction, $\mathbf{A}^\sharp \in \mathit{wff}_o(\mathcal{S}) \subseteq \mathit{prop}(\mathcal{S})$. In every other case, $\mathbf{A}^\sharp \in \mathit{prop}(\mathcal{S})$ but $\mathbf{A}^\sharp \notin \mathit{wff}_o(\mathcal{S})$.

Substitutions which do not change the logical constant at the head of a term will commute with the operation taking terms to propositions.

LEMMA 2.1.39. *Suppose $\mathcal{S}$ is a signature of logical constants, $c \in \mathcal{S}$, θ is a substitution and $[c\, \overline{\mathbf{A}^m}] \in \mathit{wff}_o(\mathcal{S})$. If $\theta(c) = c$, then*

$$\theta([c\, \overline{\mathbf{A}^m}]^\sharp) = (\theta([c\, \overline{\mathbf{A}^m}]))^\sharp.$$

PROOF. The proof is by a simple case analysis on $c \in \mathcal{S}$. □

We do not in general have $((\mathbf{A}^{\downarrow *})^{\sharp})^{\downarrow *} = (\mathbf{A}^{\sharp})^{\downarrow *}$ for $\mathbf{A} \in \mathit{wff}_o(\mathcal{S})$ because of β-reduction. For example,

$$(([[\lambda x_o x_o]\, \top]^{\downarrow *})^{\sharp})^{\downarrow *} = (\top^{\sharp})^{\downarrow *} = \underset{\cdot}{\top}$$

$$\neq \top = [[\lambda x_o x_o]\, \top]^{\downarrow *} = ([[\lambda x_o x_o]\, \top]^{\sharp})^{\downarrow *}.$$

On the other hand, when the head of $\mathbf{A}$ is a logical constant, we can prove $((\mathbf{A}^{\downarrow *})^{\sharp})^{\downarrow *} = (\mathbf{A}^{\sharp})^{\downarrow *}$.

LEMMA 2.1.40. *Suppose $\mathcal{S}$ is a signature of logical constants, $c \in \mathcal{S}$ and $[c\,\overline{\mathbf{A}^m}] \in \mathit{wff}_o(\mathcal{S})$. Then*

$$(([c\,\overline{\mathbf{A}^m}]^{\downarrow *})^{\sharp})^{\downarrow *} = ([c\,\overline{\mathbf{A}^m}]^{\sharp})^{\downarrow *}.$$

PROOF. This also follows by a simple case analysis on $c \in \mathcal{S}$. □

We can also prove that the same free variables and parameters occur in both $\mathbf{A}$ and $(\mathbf{A}^{\sharp})^{\downarrow}$ when $\mathbf{A}$ is $\beta\eta$-normal.

LEMMA 2.1.41. *Suppose $\mathcal{S}$ is a signature of logical constants. Then* $\mathbf{Free}(\mathbf{A}) = \mathbf{Free}((\mathbf{A}^{\sharp})^{\downarrow})$ *and* $\mathbf{Params}(\mathbf{A}) = \mathbf{Params}((\mathbf{A}^{\sharp})^{\downarrow})$ *for any $\beta\eta$-normal $\mathbf{A} \in \mathit{wff}_o(\mathcal{S})$.*

PROOF. The proof is by a case analysis on the head of $\mathbf{A}$. □

Furthermore, if $\mathbf{A} \in \mathit{wff}_o$ has a logical constant at the head, then the internal measure of $\mathbf{A}^{\sharp}$ is less than the internal measure of $\mathbf{A}$.

LEMMA 2.1.42. *Suppose $\mathcal{S}$ is a signature of logical constants, $c_{o\alpha^n \cdots \alpha^1} \in \mathcal{S}$ and $\mathbf{A}^i \in \mathit{wff}_{\alpha^i}(\mathcal{S})$ for each i $(1 \leq i \leq n)$. Then*

$$|[c\,\overline{\mathbf{A}^n}]^{\sharp}|_{\mathbf{i}} < |[c\,\overline{\mathbf{A}^n}]|_{\mathbf{i}}.$$

PROOF. First, note that $[c\,\overline{\mathbf{A}^n}]^{\updownarrow}$ is $[c\,\mathbf{A}^{1\updownarrow} \cdots \mathbf{A}^{n\updownarrow}]$ and so

$$|[c\,\overline{\mathbf{A}^n}]|_{\mathbf{i}} = |[c\,\overline{\mathbf{A}^n}]^{\updownarrow}|_l = 1 + 2|\mathbf{A}^{1\updownarrow}|_l + \cdots + 2|\mathbf{A}^{n\updownarrow}|_l.$$

Unless $c \notin \{\Pi^\alpha, \Sigma^\alpha, =^\alpha \mid \alpha \in \mathcal{T}\}$, we may easily compute

$$|[c\,\overline{\mathbf{A}^n}]^{\sharp}|_{\mathbf{i}} = |\mathbf{A}^{1\updownarrow}|_l + \cdots + |\mathbf{A}^{n\updownarrow}|_l < |[c\,\overline{\mathbf{A}^n}]|_{\mathbf{i}}$$

as desired. If c is $=^\alpha$ for some type α, then $n = 2$ and

$$|[c\,\mathbf{A}^1\,\mathbf{A}^2]^{\sharp}|_{\mathbf{i}} = |[\mathbf{A}^1 \underset{\cdot}{=}^\alpha \mathbf{A}^2]|_{\mathbf{i}} = 2|(\mathbf{A}^1)^{\updownarrow}|_l + 2|(\mathbf{A}^2)^{\updownarrow}|_l$$

$$< 1 + 2|(\mathbf{A}^1)^{\updownarrow}|_l + 2|(\mathbf{A}^2)^{\updownarrow}|_l = |[c\,\mathbf{A}^1\,\mathbf{A}^2]|_{\mathbf{i}}.$$

Next assume c is Π^α or Σ^α for some type α. Then $n = 1$, $\mathbf{A}^1 \in \mathit{wff}_{o\alpha}(\mathcal{S})$ and $[c\,\mathbf{A}^1]^\sharp$ is either $[\dot{\forall} x_\alpha \centerdot \mathbf{A}^1\, x]$ or $[\dot{\exists} x_\alpha \centerdot \mathbf{A}^1\, x]$ for some variable x_α not in $\mathbf{Free}(\mathbf{A}^1)$. By Lemma 2.1.36, $|[\mathbf{A}^1\, x]^\updownarrow|_l = |\mathbf{A}^{1\updownarrow}|_l$. We compute

$$|[c\,\mathbf{A}^1]^\sharp|_\mathbf{i} = |[\mathbf{A}^1\, x]|_\mathbf{i} = |[\mathbf{A}^1\, x]^\updownarrow|_l = |(\mathbf{A}^1)^\updownarrow|_l < 1 + 2|(\mathbf{A}^1)^\updownarrow|_l = |[c\,\mathbf{A}^1]|_\mathbf{i}$$

as desired. □

2.1.3. Special Sets of Types. It is worthwhile to distinguish certain simple types in order to prove special results about them. First, we distinguish the class of *propositional types* (cf. [32; 1]). These types make no reference to the type ι of individuals.

DEFINITION 2.1.43 (Propositional Types). We define the set $\mathcal{T}_o$ of *propositional types* by induction.

- The type o is a propositional type.
- If α and β are propositional types, then $(\alpha\beta)$ is a propositional type.

We will see that in the extensional case, the domains of propositional type must be finite (cf. Lemma 6.5.1). As a consequence, we can often avoid using quantifiers and equalities at such types in proofs (cf. Corollary 6.5.3).

REMARK 2.1.44 (Relational Formulations of Simple Type Theory). Often in the literature, higher-order logic is formulated only using the "relation" types. For example, Schütte's formulation in [59] is of this form. (Schütte also includes also a type of truth values.) We could map Schütte's types to simple types inductively by $0 \mapsto \iota$, $1 \mapsto o$ and $(\tau^1, \cdots, \tau^n) \mapsto (o\alpha^n \cdots \alpha^1)$ where $\tau^i \mapsto \alpha^i$ for each i $(1 \leq i \leq n)$. Takahashi in [61] and Prawitz in [56] prove cut-elimination for Schütte's formulation of type theory. Andrews also briefly discusses such a formulation called $\mathcal{F}^\omega$ in [9]. In the presence of extensionality, equality and descriptions at all types, one can argue that such a formulation is equivalent to higher-order logic formulated using simple types. Since we are not discussing the description operator here, such a formulation is not equivalent to the logics discussed here.

A "relational" formulation of type theory can be further restricted to only use the "power" types

$$\{\iota, o\iota, o(o\iota), \ldots\}.$$

The justification for this restriction is that (under certain conditions) one can use Kuratowski pairing to reduce n-ary relations to sets (of higher type).

DEFINITION 2.1.45 (Power Types). Let α be a type. For each $n \geq 0$, we define the n-th power type $\mathcal{P}^n(\alpha)$ of α recursively. Let $\mathcal{P}^0(\alpha) := \alpha$ and $\mathcal{P}^{n+1}(\alpha) := (o\mathcal{P}^n(\alpha))$. In particular, we define the set of *power types* (over individuals) to be

$$\mathcal{T}_{\mathcal{P}} := \{\mathcal{P}^n(\iota) \mid n \geq 0\}.$$

Finally, we give a general definition of a set of types generated by a given set of types.

DEFINITION 2.1.46. Let $B \subseteq \mathcal{T}$ be a set of types. We define $\mathcal{T}_{gen}(B)$ to be the least set of types such that $B \subseteq \mathcal{T}_{gen}(B)$ and

$$(\alpha\beta) \in \mathcal{T}_{gen}(B) \text{ whenever } \alpha, \beta \in \mathcal{T}_{gen}(B).$$

We will study, in particular, the set of type $\mathcal{T}_{gen}(\{(o\iota), \iota\})$. Note that no propositional type is in the set $\mathcal{T}_{gen}(\{(o\iota), \iota\})$. There are also types such as (ιo) which are not included in $\mathcal{T}_{gen}(\{(o\iota), \iota\})$.

2.1.4. Church's Type Theory. The type theory introduced by Church in [26] used the signature $\mathcal{S}_{Ch}$ (i.e., the logical constants $\neg$, $\vee$ and Π^α for each type α). The other logical operators were defined in terms of these constants. For example, equality $\dot{=}^\alpha$ is defined as Leibniz equality $\lambda x_\alpha \lambda y_\beta \forall q_{o\alpha} . q\, x \supset q\, y$ where $q\, x \supset q\, y$ abbreviates $\neg q\, x \vee q\, y$. Church also included description operators $\iota_{\alpha(o\alpha)}$ in his signature. We ignore description operators here.

Church introduced a Hilbert-style proof system for formulas in wff_o over this signature. Axioms 1-4 of Church's system are propositional axioms. Axioms 5^α and 6^α are quantificational axioms. Axioms $10^{\alpha\beta}$ are axioms of functional extensionality. We will ignore Church's additional axioms of infinity (Axioms 7 and 8), axioms of descriptions (Axioms 9^α) and axioms of choice (Axioms 11^α).

Church comments that one may introduce an axiom of Boolean extensionality $[p \equiv q] \supset \centerdot p \dot{=}^o q$, but does not do so. In [31], Henkin does add this as part of his Axiom 10 to obtain the system for which he proves completeness with respect to Henkin Models. Furthermore, Henkin did not include Church's axioms of infinity or descriptions (but did include choice).

In this book, we will not study axioms of infinity, descriptions or choice. While we will study extensionality, we will not concentrate on the case in which functional extensionality is assumed but Boolean extensionality is not. We take *extensional type theory* to mean the system used by Henkin in [31] without the axioms of choice (Axioms 11^α). Also, we will generally work with sequent style proof systems (and expansion proofs) instead of Hilbert-style proof systems.

Elementary type theory is Church's simple type theory without axioms of extensionality, descriptions, choice, or infinity. Elementary type theory also uses the signature $\mathcal{S}_{Ch}$. Andrews introduced a Hilbert-calculus $\mathscr{T}$ for elementary type theory in [2]. This calculus consists of Axioms 1-6 of Church's (and Henkin's) calculus. Also, in [2] Andrews introduces an abstract consistency method for establishing completeness relative to $\mathscr{T}$. Using this abstract consistency method, Miller [43] was able to demonstrate that every proof in elementary type theory can be represented as an *expansion proof* as described in [43; 44; 12; 8]. An expansion proof consists of an expansion tree and a complete mating. Expansion proofs are the representation TPS uses when searching for a proof.

2.2. Sequents

In [2], Andrews establishes completeness (relative to $\mathscr{T}$) of a cut-free sequent calculus for elementary type theory with the signature $\mathcal{S}_{Ch}$. Here, we define sequent calculi for a general signature $\mathcal{S}$ of logical constants. We also define sequent calculi corresponding to elementary type theory (with and without η-conversion) as well as extensional type theory.

We define a sequent to be a multiset of sentences.

DEFINITION 2.2.1 (Sequents). Let $\mathcal{S}$ be a signature. A *finite multiset* Γ *of propositions over* $\mathcal{S}$ is a function from $prop(\mathcal{S})$ to the natural numbers $\mathbb{N}$ such that $\Gamma(\mathbf{M}) = 0$ for all but finitely many $\mathbf{M} \in prop(\mathcal{S})$. For any finite multiset Γ of propositions, we define the set $|\Gamma| := \{\mathbf{M} \mid \Gamma(\mathbf{M}) > 0\}$. By $\Gamma, \mathbf{M}$ we mean the finite multiset where $(\Gamma, \mathbf{M})(\mathbf{M}) := \Gamma(\mathbf{M}) + 1$ and $(\Gamma, \mathbf{M})(\mathbf{N}) := \Gamma(\mathbf{N})$ for $\mathbf{N} \neq \mathbf{M}$. We write $\mathbf{M} \in \Gamma$ if $\mathbf{M} \in |\Gamma|$ and $\mathbf{M} \notin \Gamma$ otherwise. For any finite set

Φ of propositions, we define Φ^{ms} to be the finite multiset

$$\Phi^{ms}(\mathbf{M}) := \begin{cases} 1 & \text{if } \mathbf{M} \in \Phi \\ 0 & \text{otherwise} \end{cases}$$

A *one-sided* $\mathcal{S}$*-sequent* (or, *sequent*) is a finite multiset of sentences over $\mathcal{S}$ (i.e., a finite multiset Γ of propositions where $\mathbf{M} \in \mathit{sent}(\mathcal{S})$ for every $\mathbf{M} \in \Gamma$). A one-sided sequent Γ is *equality-free* if every sentence in Γ is equality-free. A *two-sided* $\mathcal{S}$*-sequent* is of the form $\Gamma \rightarrow \Delta$ where Γ and Δ are finite multisets of sentences over $\mathcal{S}$. A two-sided sequent $\Gamma \rightarrow \Delta$ is *equality-free sequent* if every sentence in $|\Gamma| \cup |\Delta|$ is equality-free.

Since we are working in classical logic, we can use one-sided sequents. We mean one-sided sequents when we refer to sequents. However, in the presentation of set constraints (cf. Chapter 9) we will prefer two-sided sequents.

NOTATION *Let* Γ *be a finite multiset* $\mathbf{M}^1, \ldots, \mathbf{M}^n$ *of propositions over* $\mathcal{S}$*. For any substitution* θ*, we use* $\theta(\Gamma)$ *to denote the finite multiset* $\theta(\mathbf{M}^1), \ldots, \theta(\mathbf{M}^n)$ *of propositions over* $\mathcal{S}$*. Similarly, we use* $\Gamma^{\downarrow}$ *to denote the finite multiset* $\mathbf{M}^{1\downarrow}, \ldots, \mathbf{M}^{n\downarrow}$ *of propositions over* $\mathcal{S}$ *and* $\Gamma^{\updownarrow}$ *to denote the finite multiset* $\mathbf{M}^{1\updownarrow}, \ldots, \mathbf{M}^{n\updownarrow}$ *of propositions over* $\mathcal{S}$*. Note that if* Γ *is a sequent, then* $\Gamma^{\downarrow}$ *and* $\Gamma^{\updownarrow}$ *are sequents.*

DEFINITION 2.2.2 (Sequent Rules). The sequent rules we will consider are in Figure 2.2, Figure 2.3, Figure 2.4, Figure 2.5 and Figure 2.6. Each sequent rule is of the form

$$\frac{\Gamma, \Delta_1 \quad \cdots \quad \Gamma, \Delta_n}{\Gamma, \Delta_{n+1}}\, \mathit{rule}$$

For each instance of a rule, we call each sentence in Γ a *side formula* and each sentence in Δ_{n+1} a *principal formula.* Note that the rules $\mathcal{G}(\mathit{Init})$, $\mathcal{G}(\mathit{Init}^{\dot{=}})$, $\mathcal{G}(\mathit{EUnif}_1)$ and $\mathcal{G}(\mathit{EUnif}_2)$ have two principal formulas. Every other rule has a single principal formula.

DEFINITION 2.2.3 (Elementary Sequent Calculi). Let $\mathcal{S}$ be an equality-free signature. We define the one-sided equality-free sequent calculus $\mathcal{G}_\beta^{\mathcal{S}}$ to be given by the rule $\mathcal{G}(\mathit{Init})$ in Figure 2.2, the rules in Figure 2.3 and the rule $\mathcal{G}(\beta)$ in Figure 2.4. We call this calculus the *elementary sequent calculus for* $\mathcal{S}$. We call the set of sentences which can be derived in $\mathcal{G}_\beta^{\mathcal{S}}$ the $\mathcal{S}$ *fragment of elementary type theory.*

$$\frac{\mathbf{A} \in cwff_o(\mathcal{S}) \text{ atom}}{\Gamma, \mathbf{A}, \dot{\neg}\mathbf{A}} \mathcal{G}(Init)$$

FIGURE 2.2. Initial Sequent Rule

$$\frac{\Gamma, \mathbf{M}, \mathbf{M}}{\Gamma, \mathbf{M}} \mathcal{G}(Contr) \qquad \frac{}{\Gamma, \dot{\top}} \mathcal{G}(\dot{\top}) \qquad \frac{\Gamma, \mathbf{M}}{\Gamma, \dot{\neg}\dot{\neg}\mathbf{M}} \mathcal{G}(\dot{\neg}\dot{\neg})$$

$$\frac{\Gamma, \mathbf{M}, \mathbf{N}}{\Gamma, [\mathbf{M} \dot{\vee} \mathbf{N}]} \mathcal{G}(\dot{\vee}) \qquad \frac{\Gamma, \dot{\neg}\mathbf{M} \quad \Gamma, \dot{\neg}\mathbf{N}}{\Gamma, \dot{\neg}[\mathbf{M} \dot{\vee} \mathbf{N}]} \mathcal{G}(\dot{\neg}\dot{\vee})$$

$$\frac{\Gamma, [[W/x]\mathbf{M}] \quad W \in \mathcal{P}_\alpha \setminus \mathbf{Params}(|\Gamma, [\dot{\forall} x_\alpha \mathbf{M}]|)}{\Gamma, [\dot{\forall} x_\alpha \mathbf{M}]} \mathcal{G}(\dot{\forall}^W)$$

$$\frac{\Gamma, \dot{\neg}[[\mathbf{C}/x]\mathbf{M}] \quad \mathbf{C} \in cwff_\alpha(\mathcal{S})}{\Gamma, \dot{\neg}[\dot{\forall} x_\alpha \mathbf{M}]} \mathcal{G}(\dot{\neg}\dot{\forall}, \mathcal{S})$$

$$\frac{\Gamma, \mathbf{A}^\sharp \quad \mathbf{A} \in cwff_o(\mathcal{S})}{\Gamma, \mathbf{A}} \mathcal{G}(\sharp) \qquad \frac{\Gamma, \dot{\neg}(\mathbf{A}^\sharp) \quad \mathbf{A} \in cwff_o(\mathcal{S})}{\Gamma, \dot{\neg}\mathbf{A}} \mathcal{G}(\dot{\neg}\sharp)$$

FIGURE 2.3. Basic Sequent Rules

$$\frac{\Gamma, \mathbf{M}^{\downarrow\beta}}{\Gamma, \mathbf{M}} \mathcal{G}(\beta) \qquad \frac{\Gamma, \mathbf{M}^{\downarrow}}{\Gamma, \mathbf{M}} \mathcal{G}(\beta\eta)$$

FIGURE 2.4. λ Sequent Rules

We define the one-sided equality-free sequent calculus $\mathcal{G}^{\mathcal{S}}_{\beta\eta}$ to be given by the rule $\mathcal{G}(Init)$ in Figure 2.2, the rules in Figure 2.3 and the rule $\mathcal{G}(\beta\eta)$ in Figure 2.4. We call this calculus the *elementary* *sequent calculus with η for $\mathcal{S}$.* We call the set of sentences which can be derived in $\mathcal{G}^{\mathcal{S}}_{\beta\eta}$ the *$\mathcal{S}$ fragment of elementary type theory with η.*

We study the sequent calculi $\mathcal{G}^{\mathcal{S}}_{\beta}$ and $\mathcal{G}^{\mathcal{S}}_{\beta\eta}$ because they correspond to TPS in historical and practical ways. In particular, $\mathcal{G}^{\mathcal{S}_{Ch}}_{\beta}$ essentially corresponds to Andrews' $\mathscr{T}$ calculus [2] for elementary type theory. This is why we refer to the sequent calculi as *elementary*. Furthermore, proofs in $\mathcal{G}^{\mathcal{S}}_{\beta}$ and $\mathcal{G}^{\mathcal{S}}_{\beta\eta}$ can be represented as expansion proofs. (Actually, one must generalize expansion proofs slightly to allow for external propositions. We will not pursue this generalization here.) Finally, given certain settings of TPS flags, one can have TPS search for such expansion proofs.

We only study equality-free sequents in the non-extensional calculi $\mathcal{G}^{\mathcal{S}}_{\beta}$ and $\mathcal{G}^{\mathcal{S}}_{\beta\eta}$ since we can consider equality simply as an abbreviation for Leibniz equality. Under some flag settings, TPS can expand equality using extensionality instead of Leibniz equality. In such cases, we may consider equality as an abbreviation for extensional abbreviation. There are other alternative ways to view extensional expansions of equality. We could consider extensional expansions of equality to be (unsound) extensions of the calculi $\mathcal{G}^{\mathcal{S}}_{\beta}$ and $\mathcal{G}^{\mathcal{S}}_{\beta\eta}$. Also, we could consider extensional expansions of equality as an incomplete method of search for extensional higher-order logic.

The role of the $\mathcal{G}(\sharp)$ and $\mathcal{G}(\neg\sharp)$ rules is to lift logical constants at the head of an internal formula to the external level of propositions. Application of the $\mathcal{G}(\sharp)$ and $\mathcal{G}(\neg\sharp)$ rules allow the rules for the appropriate (external) connective or quantifier to apply. For demonstrations of these rules, see the derivations in Figures 2.12 and 6.1.

There are two forms of extensionality: functional and Boolean. As in [19], we use $\mathfrak{f}$ to indicate functional extensionality and $\mathfrak{b}$ to indicate Boolean extensionality. The corresponding semantic properties will be defined later (cf. Definition 3.3.5).

DEFINITION 2.2.4 (Extensional Sequent Calculi). Let $\mathcal{S}$ be a signature. We define the (one-sided) sequent calculus $\mathcal{G}^{\mathcal{S}}_{\beta\mathfrak{f}\mathfrak{b}}$ to be given by the rules in Figure 2.3, the rule $\mathcal{G}(\beta\eta)$ in Figure 2.4, the equational rules in Figure 2.5, and the extensional rules in Figure 2.6. We call this calculus the *extensional sequent calculus for $\mathcal{S}$*. We call the set of sentences which can be derived in $\mathcal{G}^{\mathcal{S}}_{\beta\mathfrak{f}\mathfrak{b}}$ the *$\mathcal{S}$ fragment of extensional type theory.*

$\boxed{n \geq 0:}$

$$\frac{\Gamma, [\mathbf{A}^1 \doteq \mathbf{B}^1] \quad \cdots \quad \Gamma, [\mathbf{A}^n \doteq \mathbf{B}^n] \quad H_{o\alpha^n \cdots \alpha^1} \in \mathcal{P}}{\Gamma, [H\,\overline{\mathbf{A}^n}], \neg[H\,\overline{\mathbf{B}^n}]}\ \mathcal{G}(Init^{\doteq})$$

$$\frac{\Gamma, [\mathbf{A}^1 \doteq \mathbf{B}^1] \quad \cdots \quad \Gamma, [\mathbf{A}^n \doteq \mathbf{B}^n] \quad H_{\iota\alpha^n \cdots \alpha^1} \in \mathcal{P}}{\Gamma, [[H\,\overline{\mathbf{A}^n}] \doteq^{\iota} [H\,\overline{\mathbf{B}^n}]]}\ \mathcal{G}(Dec)$$

$$\frac{\Gamma, [\mathbf{A} \doteq^{\iota} \mathbf{C}] \quad \Gamma, [\mathbf{B} \doteq^{\iota} \mathbf{D}]}{\Gamma, \neg[\mathbf{A} \doteq^{\iota} \mathbf{B}], [\mathbf{C} \doteq^{\iota} \mathbf{D}]}\ \mathcal{G}(EUnif_1)$$

$$\frac{\Gamma, [\mathbf{A} \doteq^{\iota} \mathbf{D}] \quad \Gamma, [\mathbf{B} \doteq^{\iota} \mathbf{C}]}{\Gamma, \neg[\mathbf{A} \doteq^{\iota} \mathbf{B}], [\mathbf{C} \doteq^{\iota} \mathbf{D}]}\ \mathcal{G}(EUnif_2)$$

$$\frac{\Gamma, \mathbf{A}, \mathbf{B} \quad \Gamma, \neg\mathbf{A}, \neg\mathbf{B}}{\Gamma, \neg[\mathbf{A} \doteq^{o} \mathbf{B}]}\ \mathcal{G}(\neg \doteq^{o})$$

$$\frac{\Gamma, \neg[[\mathbf{G}\,\mathbf{B}] \doteq^{\alpha} [\mathbf{H}\,\mathbf{B}]] \quad \mathbf{B} \in cwff_{\beta}(\mathcal{S})}{\Gamma, \neg[\mathbf{G} \doteq^{\alpha\beta} \mathbf{H}]}\ \mathcal{G}(\neg \doteq^{\rightarrow}, \mathcal{S})$$

FIGURE 2.5. Sequent Rules for Equality

$$\frac{\Gamma, [[\mathbf{G}\,W] \doteq^{\alpha} [\mathbf{H}\,W]] \quad W \in \mathcal{P}_{\beta} \setminus \mathbf{Params}(|\Gamma, [\mathbf{G} \doteq \mathbf{H}]|)}{\Gamma, [\mathbf{G} \doteq^{\alpha\beta} \mathbf{H}]}\ \mathcal{G}(\mathfrak{f}^{W})$$

$$\frac{\Gamma, \neg\mathbf{A}, \mathbf{B} \quad \Gamma, \neg\mathbf{B}, \mathbf{A}}{\Gamma, [\mathbf{A} \doteq^{o} \mathbf{B}]}\ \mathcal{G}(\mathfrak{b})$$

FIGURE 2.6. Extensional Sequent Rules

It is not surprising that the rule $\mathcal{G}(\beta)$ could be used in place of $\mathcal{G}(\beta\eta)$ in $\mathcal{G}^{\mathcal{S}}_{\beta\mathfrak{fb}}$ since the rules for functional extensionality imply that η-conversion is sound. However, in practice we will assume η-conversion is built into the language. The rule $\mathcal{G}(\beta\eta)$ more closely models this practice. Also, note that we do not include the initial rule $\mathcal{G}(Init)$ in $\mathcal{G}^{\mathcal{S}}_{\beta\mathfrak{fb}}$ since this has been replaced by the $\mathcal{G}(Init^{\overline{=}})$ rule. The only rules which are "initial" (in the sense of having no premises) in $\mathcal{G}^{\mathcal{S}}_{\beta\mathfrak{fb}}$ are $\mathcal{G}(Init^{\overline{=}})$ and $\mathcal{G}(Dec)$ in the special cases where $n = 0$ (i.e., the parameter H is of base type).

There are multiple reasons for considering the extensional calculus $\mathcal{G}^{\mathcal{S}}_{\beta\mathfrak{fb}}$. First, the extensional version of higher-order logic reflects mathematical reasoning in which sets and functions are considered extensionally. For example, few mathematicians would consider a theory in which the set $A \cup B$ is not the same as $B \cup A$. The problem was that before Benzmüller [18] there was no practical way to automate extensional reasoning. Benzmüller (cf. [18]) presented a resolution calculus with rules for handling extensionality in a goal-directed manner. Some of the sequent calculus rules in Figures 2.5 and 2.6 correspond closely to Benzmüller's resolution rules.

Unfortunately, the completeness proof in [18] has a technical flaw. The abstract consistency class corresponding to the resolution calculus is claimed to be saturated. However, verifying saturation is just as hard as proving cut elimination. We prove a model existence theorem in Chapter 5 (cf. Theorem 5.7.17) without using saturation. This theorem can be used to repair the proof in [18] though we will not explicitly do this. We instead use the model existence theorem to show completeness of $\mathcal{G}^{\mathcal{S}}_{\beta\mathfrak{fb}}$ (cf. Theorem 5.7.19). Cut elimination will follow from this completeness result (cf. Corollary 5.7.20).

Finally, the most important reason we consider extensional higher-order logic is to study what restrictions on set instantiations one can make without sacrificing completeness.

We use the indices β, $\beta\eta$ and $\beta\mathfrak{fb}$ to indicate which forms of extensionality are included. Note that we have only defined the sequent calculi $\mathcal{G}^{\mathcal{S}}_{\beta}$ and $\mathcal{G}^{\mathcal{S}}_{\beta\eta}$ for equality-free signatures $\mathcal{S}$.

DEFINITION 2.2.5. Let $\mathcal{S}_{\beta}$ and $\mathcal{S}_{\beta\eta}$ be the signature $\mathcal{S}^{elem}_{all}$ and let $\mathcal{S}_{\beta\mathfrak{fb}}$ be the signature $\mathcal{S}_{all}$. That is, for each $* \in \{\beta, \beta\eta, \beta\mathfrak{fb}\}$ we let $\mathcal{S}_*$

denote the largest signature of logical constants which is equality-free if $* \in \{\beta, \beta\eta\}$.

DEFINITION 2.2.6. Let $\mathcal{S}$ be a signature. For each $* \in \{\beta, \beta\eta\}$ let $prop_*(\mathcal{S})$ be the set of equality-free propositions over $\mathcal{S}$. Let $prop_{\beta\mathfrak{fb}}(\mathcal{S})$ be the set $prop(\mathcal{S})$ of all propositions over $\mathcal{S}$. For each $* \in \{\beta, \beta\eta, \beta\mathfrak{fb}\}$ let $sent_*(\mathcal{S})$ be the set of all closed propositions in $prop_*(\mathcal{S})$.

Note that for any $* \in \{\beta, \beta\eta, \beta\mathfrak{fb}\}$ and signature $\mathcal{S} \subseteq \mathcal{S}_*$ we have defined a sequent calculus $\mathcal{G}_*^{\mathcal{S}}$ for sequents Γ with $|\Gamma| \subseteq sent_*(\mathcal{S})$.

Sequent calculi are more appropriate for theoretical work than for practical automated theorem proving. Historically, TPS searches for expansion proofs. We will clarify how the notion of expansion proofs can be extended to represent extensional proofs (see Chapter 7). Extensional expansion proofs will provide compact representations of proofs that are better suited for automated theorem proving than derivations in $\mathcal{G}_{\beta\mathfrak{fb}}^{\mathcal{S}}$.

THEOREM 2.2.7 (Consistency). *There is no $\mathcal{G}_*^{\mathcal{S}}$-derivation of the empty sequent.*

PROOF. The conclusion of every rule is a nonempty sequent. □

In order to explore examples, we define two sentences. One can be proven using extensionality and the other can be proven using equality reasoning.

DEFINITION 2.2.8. Let A_o, B_o, P_{oo}, C_ι, D_ι and $Q_{o\iota}$ be parameters. We define the following sentences:

THM617: $[\neg A_o \dot\vee \neg B_o \dot\vee \neg[P_{oo}\, A] \dot\vee [P\, B]]$

THM618: $[\neg[C_\iota \doteq^\iota D_\iota] \dot\vee \neg[Q_{o\iota}\, C] \dot\vee [Q\, D]]$

Note that these are sentences in any signature $\mathcal{S}$ and **THM617** is equality-free.

EXAMPLE 2.2.9. Let $\mathcal{S}$ be a signature. The sentences **THM617** and **THM618** are derivable in $\mathcal{G}_{\beta\mathfrak{fb}}^{\mathcal{S}}$. Derivations are shown in Figures 2.7 and 2.8.

LEMMA 2.2.10 (Weakening). *If there is a $\mathcal{G}_*^{\mathcal{S}}$-derivation of Γ, then there is a $\mathcal{G}_*^{\mathcal{S}}$-derivation of Γ, Δ.*

$$\dfrac{\dfrac{\dfrac{\dfrac{\dfrac{\dfrac{}{\dot\neg A,\ \dot\neg B,\ \dot\neg B,\ A}\,\mathcal{G}(Init^{\dot=}) \qquad \dfrac{}{\dot\neg A,\ \dot\neg B,\ \dot\neg A,\ B}\,\mathcal{G}(Init^{\dot=})}{\dot\neg A,\ \dot\neg B,\ [B \doteq A]}\,\mathcal{G}(\mathfrak{b})}{\dot\neg A,\ \dot\neg B,\ \dot\neg[P\,A],\ [P\,B]}\,\mathcal{G}(Init^{\dot=})}{[\dot\neg A \mathbin{\dot\vee} \dot\neg B],\ \dot\neg[P\,A],\ [P\,B]}\,\mathcal{G}(\dot\vee)}{[\dot\neg A \mathbin{\dot\vee} \dot\neg B \mathbin{\dot\vee} \dot\neg[P\,A]],\ [P\,B]}\,\mathcal{G}(\dot\vee)}{[\dot\neg A_o \mathbin{\dot\vee} \dot\neg B_o \mathbin{\dot\vee} \dot\neg[P_{oo}\,A] \mathbin{\dot\vee} [P\,B]]}\,\mathcal{G}(\dot\vee)$$

FIGURE 2.7. Derivation of **THM617**

$$\dfrac{\dfrac{\dfrac{\dfrac{\dfrac{}{[C \doteq^{\iota} C]}\,\mathcal{G}(Dec) \qquad \dfrac{}{[D \doteq^{\iota} D]}\,\mathcal{G}(Dec)}{\dot\neg[C, \doteq D],\ [D \doteq C]}\,\mathcal{G}(EUnif_2)}{\dot\neg[C \doteq D],\ \dot\neg[Q\,C],\ [Q\,D]}\,\mathcal{G}(Init^{\dot=})}{[\dot\neg[C \doteq D] \mathbin{\dot\vee} \dot\neg[Q\,C]],\ [Q\,D]}\,\mathcal{G}(\dot\vee)}{[\dot\neg[C_\iota \doteq^{\iota} D_\iota] \mathbin{\dot\vee} \dot\neg[Q_{o\iota}\,C] \mathbin{\dot\vee} [Q\,D]]}\,\mathcal{G}(\dot\vee)$$

FIGURE 2.8. Derivation of **THM618**

PROOF. By a simple induction on derivations we can show that for every parameter substitution θ if there is a $\mathcal{G}_*^{\mathcal{S}}$-derivation of $\theta(\Gamma)$, then there is a $\mathcal{G}_*^{\mathcal{S}}$-derivation of $\theta(\Gamma), \Delta$. (The parameter substitution is required to handle the rules $\mathcal{G}(\dot\forall^W)$ and $\mathcal{G}(\mathfrak{f}^W)$. The assumption that θ maps parameters to parameters is necessary to handle the rules $\mathcal{G}(Dec)$ and $\mathcal{G}(Init^{\dot=})$.) The lemma then follows using the identity substitution for θ. □

LEMMA 2.2.11. *For any sequent* Γ*, there is a* $\mathcal{G}_*^{\mathcal{S}}$*-derivation of* Γ *iff there is a* $\mathcal{G}_*^{\mathcal{S}}$*-derivation of* $|\Gamma|^{ms}$.

PROOF. We prove this by induction on

$$\#(\Gamma) := \sum_{\mathbf{M}\in\Gamma} (\Gamma(\mathbf{M}) - 1).$$

If $\#(\Gamma)$ is 0, then this is trivial since Γ is $|\Gamma|^{ms}$. If $\#(\Gamma)$ is $n+1$, then there is some $\mathbf{M} \in \Gamma$ such that $\Gamma(\mathbf{M}) \geq 2$. We can write Γ as $\Gamma_1, \mathbf{M}, \mathbf{M}$. By contraction ($\mathcal{G}(Contr)$) and weakening (Lemma 2.2.10), $\Gamma_1, \mathbf{M}, \mathbf{M}$ is derivable iff $\Gamma_1, \mathbf{M}$ is derivable. By the inductive hypothesis, $\Gamma_1, \mathbf{M}$ is derivable iff $|\Gamma_1, \mathbf{M}|^{ms}$ is derivable. Note that $|\Gamma_1, \mathbf{M}|$ equals $|\Gamma|$. Thus Γ is derivable iff $|\Gamma|^{ms}$ is derivable. $\square$

LEMMA 2.2.12 (Inversion of $\mathcal{G}(\neg\neg)$). *If there is a $\mathcal{G}_*^{\mathcal{S}}$-derivation of the sequent $\Gamma, \neg\neg\mathbf{M}$, then there is a $\mathcal{G}_*^{\mathcal{S}}$-derivation of $\Gamma, \mathbf{M}$.*

PROOF. We prove by induction on derivations that if $\mathcal{D}$ is a derivation of

$$\Gamma, \underbrace{\neg\neg\mathbf{M}, \ldots, \neg\neg\mathbf{M}}_{n}$$

then

$$\Gamma, \underbrace{\mathbf{M}, \ldots, \mathbf{M}}_{n}$$

is derivable. If the last rule of $\mathcal{D}$ is $\mathcal{G}(Init)$, then since $\neg\neg\mathbf{M}$ is neither an atom nor the negation of an atom, the rule application must be of the form

$$\frac{\mathbf{A} \in cwff_o(\mathcal{S}) \text{ atom}}{\Gamma_1, \mathbf{A}, \neg\mathbf{A}, \neg\neg\mathbf{M}, \ldots, \neg\neg\mathbf{M}} \mathcal{G}(Init)$$

The desired derivation is

$$\frac{\mathbf{A} \in cwff_o(\mathcal{S}) \text{ atom}}{\Gamma_1, \mathbf{A}, \neg\mathbf{A}, \mathbf{M}, \ldots, \mathbf{M}} \mathcal{G}(Init)$$

For any other rule, if the principal formula is a member of Γ, then we can simply apply the inductive hypothesis and then apply the same rule. (We use the fact that $\mathbf{Params}(\neg\neg\mathbf{M}) = \mathbf{Params}(\mathbf{M})$ for the rules $\mathcal{G}(\dot{\forall}^W)$ and $\mathcal{G}(\mathfrak{f}^W)$.)

If the principal formula is a copy of $\neg\neg\mathbf{M}$, then the rule must be $\mathcal{G}(Contr)$ or $\mathcal{G}(\neg\neg)$. If the last rule application is

$$\frac{\begin{matrix}\mathcal{D}_1 \\ \Gamma, \overbrace{\neg\neg\mathbf{M}, \neg\neg\mathbf{M}, \ldots, \neg\neg\mathbf{M}}^{n+1}\end{matrix}}{\Gamma, \underbrace{\neg\neg\mathbf{M}, \ldots, \neg\neg\mathbf{M}}_{n}} \mathcal{G}(Contr)$$

then we apply the inductive hypothesis to $\mathcal{D}_1$ to obtain $\mathcal{E}_1$ and apply contraction as

$$\dfrac{\begin{array}{c}\mathcal{E}_1\\ \Gamma, \overbrace{\mathbf{M}, \mathbf{M}, \ldots, \mathbf{M}}^{n+1}\end{array}}{\Gamma, \underbrace{\mathbf{M}, \ldots, \mathbf{M}}_{n}}\ \mathcal{G}(Contr)$$

If the last rule application is

$$\dfrac{\begin{array}{c}\mathcal{D}_1\\ \Gamma, \mathbf{M}, \overbrace{\dot{\neg}\dot{\neg}\mathbf{M}, \ldots, \dot{\neg}\dot{\neg}\mathbf{M}}^{n}\end{array}}{\Gamma, \dot{\neg}\dot{\neg}\mathbf{M}, \underbrace{\dot{\neg}\dot{\neg}\mathbf{M}, \ldots, \dot{\neg}\dot{\neg}\mathbf{M}}_{n}}\ \mathcal{G}(\dot{\neg}\dot{\neg})$$

then we apply the inductive hypothesis to $\mathcal{D}_1$ giving a derivation of

$$\Gamma, \mathbf{M}, \underbrace{\mathbf{M}, \ldots, \mathbf{M}}_{n}$$

□

2.2.1. Derived Rules. Since we are using $\dot{\bot}$, $\dot{\wedge}$, $\dot{\supset}$, $\dot{\equiv}$ and $\dot{\exists}$ as abbreviations for propositions, it is useful to verify the rules of inference shown in Figure 2.9 appropriate for these abbreviations. We prove below that each of these rules is a derived rule of $\mathcal{G}_*^{\mathcal{S}}$. That is, any application of the rule can be replaced by a particular finite sequence of rule applications from $\mathcal{G}_*^{\mathcal{S}}$.

THEOREM 2.2.13. *Each application of a rule in Figure 2.9 can be replaced by a derivation using $\mathcal{G}_*^{\mathcal{S}}$ rules (i.e., the rules are* derived rules *of $\mathcal{G}_*^{\mathcal{S}}$).*

PROOF. We will implicitly use the shorthand notation given in Definition 2.1.21. For example, $\dot{\bot}$ means $\dot{\neg}\dot{\top}$. Furthermore, once we have verified one rule is a derived rule, we may use the rule to verify another rule is derived.

$\mathcal{G}(\dot{\neg}\dot{\bot})$:

$$\dfrac{\dfrac{}{\Gamma, \dot{\top}}\ \mathcal{G}(\dot{\top})}{\Gamma, \dot{\neg}\dot{\bot}}\ \mathcal{G}(\dot{\neg}\dot{\neg})$$

$$\frac{}{\Gamma, \dot{\neg}\dot{\bot}}\ \mathcal{G}(\dot{\neg}\dot{\bot})$$

$$\frac{\Gamma, \mathbf{M} \quad \Gamma, \mathbf{N}}{\Gamma, [\mathbf{M} \dot{\wedge} \mathbf{N}]}\ \mathcal{G}(\dot{\wedge}) \qquad \frac{\Gamma, \dot{\neg}\mathbf{M}, \dot{\neg}\mathbf{N}}{\Gamma, \dot{\neg}[\mathbf{M} \dot{\wedge} \mathbf{N}]}\ \mathcal{G}(\dot{\neg}\dot{\wedge})$$

$$\frac{\Gamma, \dot{\neg}\mathbf{M}, \mathbf{N}}{\Gamma, [\mathbf{M} \dot{\supset} \mathbf{N}]}\ \mathcal{G}(\dot{\supset}) \qquad \frac{\Gamma, \mathbf{M} \quad \Gamma, \dot{\neg}\mathbf{N}}{\Gamma, \dot{\neg}[\mathbf{M} \dot{\supset} \mathbf{N}]}\ \mathcal{G}(\dot{\neg} \dot{\supset})$$

$$\frac{\Gamma, \dot{\neg}\mathbf{M}, \mathbf{N} \quad \Gamma, \dot{\neg}\mathbf{N}, \mathbf{M}}{\Gamma, [\mathbf{M} \dot{\equiv} \mathbf{N}]}\ \mathcal{G}(\dot{\equiv})$$

$$\frac{\Gamma, [[\mathbf{C}/x]\mathbf{M}] \quad \mathbf{C} \in cwff_\alpha(\mathcal{S})}{\Gamma,\ [\dot{\exists} x_\alpha \mathbf{M}]}\ \mathcal{G}(\dot{\exists}, \mathcal{S})$$

$$\frac{\Gamma, \dot{\neg}[[W/x]\mathbf{M}] \quad W \in \mathcal{P}_\alpha \setminus \mathbf{Params}(|\Gamma, [\dot{\exists} x_\alpha\, \mathbf{M}]|)}{\Gamma, \dot{\neg}[\dot{\exists} x_\alpha \mathbf{M}]}\ \mathcal{G}(\dot{\neg}\dot{\exists}^W)$$

FIGURE 2.9. Basic Derived Sequent Rules

$\mathcal{G}(\dot{\wedge})$:

$$\frac{\dfrac{\Gamma, \mathbf{M}}{\Gamma, \dot{\neg}\dot{\neg}\mathbf{M}}\ \mathcal{G}(\dot{\neg}\dot{\neg}) \qquad \dfrac{\Gamma, \mathbf{N}}{\Gamma, \dot{\neg}\dot{\neg}\mathbf{N}}\ \mathcal{G}(\dot{\neg}\dot{\neg})}{\Gamma, [\mathbf{M} \dot{\wedge} \mathbf{N}]}\ \mathcal{G}(\dot{\neg}\dot{\vee})$$

$\mathcal{G}(\dot{\neg}\dot{\wedge})$:

$$\frac{\dfrac{\Gamma, \dot{\neg}\mathbf{M}, \dot{\neg}\mathbf{N}}{\Gamma, [\dot{\neg}\mathbf{M} \dot{\vee} \dot{\neg}\mathbf{N}]}\ \mathcal{G}(\dot{\vee})}{\Gamma, \dot{\neg}[\mathbf{M} \dot{\wedge} \mathbf{N}]}\ \mathcal{G}(\dot{\neg}\dot{\neg})$$

$\mathcal{G}(\dot{\supset})$:

$$\frac{\Gamma, \dot{\neg}\mathbf{M}, \mathbf{N}}{\Gamma, [\mathbf{M} \dot{\supset} \mathbf{N}]}\ \mathcal{G}(\dot{\vee})$$

$$\frac{}{\Gamma, \mathbf{M}, \underset{\cdot}{\neg}\mathbf{M}}\,\mathcal{G}(GInit) \qquad\qquad \frac{\Gamma, \mathbf{M} \quad \Gamma, \underset{\cdot}{\neg}\mathbf{M}}{\Gamma}\,\mathcal{G}(Cut)$$

$$\frac{\Gamma, \underset{\cdot}{\neg}\mathbf{M}, \underset{\cdot}{\neg}\mathbf{N} \quad \Gamma, \mathbf{M}, \mathbf{N}}{\Gamma, \underset{\cdot}{\neg}[\mathbf{M} \underset{\cdot}{\equiv} \mathbf{N}]}\,\mathcal{G}(\underset{\cdot}{\neg} \underset{\cdot}{\equiv})$$

FIGURE 2.10. Basic Admissible Sequent Rules

$\mathcal{G}(\underset{\cdot}{\neg} \underset{\cdot}{\supset})$:

$$\frac{\dfrac{\Gamma, \mathbf{M}}{\Gamma, \underset{\cdot}{\neg}\underset{\cdot}{\neg}\mathbf{M}}\,\mathcal{G}(\underset{\cdot}{\neg}\underset{\cdot}{\neg}) \qquad \Gamma, \underset{\cdot}{\neg}\mathbf{N}}{\Gamma, \underset{\cdot}{\neg}[\mathbf{M} \underset{\cdot}{\supset} \mathbf{N}]}\,\mathcal{G}(\underset{\cdot}{\neg}\underset{\cdot}{\vee})$$

$\mathcal{G}(\underset{\cdot}{\equiv})$:

$$\frac{\dfrac{\Gamma, \underset{\cdot}{\neg}\mathbf{M}, \mathbf{N}}{\Gamma, [\mathbf{M} \underset{\cdot}{\supset} \mathbf{N}]}\,\mathcal{G}(\underset{\cdot}{\supset}) \qquad \dfrac{\Gamma, \underset{\cdot}{\neg}\mathbf{N}, \mathbf{M}}{\Gamma, [\mathbf{N} \underset{\cdot}{\supset} \mathbf{M}]}\,\mathcal{G}(\underset{\cdot}{\supset})}{\Gamma, [\mathbf{M} \underset{\cdot}{\equiv} \mathbf{N}]}\,\mathcal{G}(\underset{\cdot}{\wedge})$$

$\mathcal{G}(\underset{\cdot}{\exists}, \mathcal{S})$:

$$\frac{\dfrac{\Gamma, [[\mathbf{C}/x]\mathbf{M}]}{\Gamma, \underset{\cdot}{\neg}\underset{\cdot}{\neg}[[\mathbf{C}/x]\mathbf{M}]}\,\mathcal{G}(\underset{\cdot}{\neg}\underset{\cdot}{\neg}) \qquad \mathbf{C} \in cwff_\alpha(\mathcal{S})}{\Gamma,\ [\underset{\cdot}{\exists} x_\alpha \mathbf{M}]}\,\mathcal{G}(\underset{\cdot}{\neg}\underset{\cdot}{\forall}, \mathcal{S})$$

$\mathcal{G}(\underset{\cdot}{\neg}\underset{\cdot}{\exists}^W)$:

$$\cfrac{\Gamma, \underset{\cdot}{\neg}[[W/x]\mathbf{M}] \qquad W \in \mathcal{P}_\alpha \setminus \mathbf{Params}(|\Gamma, [\underset{\cdot}{\exists} x_\alpha\, \mathbf{M}]|)}{\dfrac{\Gamma, \underset{\cdot}{\forall} x_\alpha \underset{\cdot}{\neg}\mathbf{M}}{\Gamma, \underset{\cdot}{\neg}[\underset{\cdot}{\exists} x_\alpha \mathbf{M}]}\,\mathcal{G}(\underset{\cdot}{\neg})}\,\mathcal{G}(\underset{\cdot}{\forall}^W)$$

□

2.2.2. Admissible Rules. In Figure 2.10, we present *admissible rules* for each $\mathcal{G}_*^{\mathcal{S}}$. That is, if there is a proof using these rules (in addition to the rules in $\mathcal{G}_*^{\mathcal{S}}$), then there is a proof which does not use these rules.

$$\frac{}{\Gamma, [\mathbf{A} \doteq^{\alpha} \mathbf{A}]} \mathcal{G}(Refl)$$

FIGURE 2.11. Extensional Admissible Sequent Rule

The reflexivity rule $\mathcal{G}(Refl)$ shown in Figure 2.11 is admissible for $\mathcal{G}^{\mathcal{S}}_{\beta\mathfrak{fb}}$. (It does not even make sense for $\mathcal{G}^{\mathcal{S}}_{\beta}$ and $\mathcal{G}^{\mathcal{S}}_{\beta\eta}$ since it involves an equation.)

It is well-known that proving that $\mathcal{G}(Cut)$ is admissible requires strong proof methods. An elementary proof would contradict Gödel's Second Incompleteness Theorem (cf. [5]). We will prove admissibility of cut as a consequence of completeness (cf. Corollaries 5.6.8 and 5.7.20).

On the other hand, for $\mathcal{G}^{\mathcal{S}}_{\beta}$ and $\mathcal{G}^{\mathcal{S}}_{\beta\eta}$, one can prove admissibility of the general (nonatomic) initial rule $\mathcal{G}(GInit)$ (for equality-free propositions $\mathbf{M}$) by induction on the sentence $\mathbf{M}$ after proving it for all $\mathbf{M} \in cwff_o(\mathcal{S})$ by induction on the number of occurrences of logical constants in $\mathbf{M}$. We instead establish admissibility of $\mathcal{G}(GInit)$ at the abstract consistency level (cf. Lemma 5.3.1, Theorem 5.3.4 and Corollary 5.4.8).

For the extensional calculus $\mathcal{G}^{\mathcal{S}}_{\beta\mathfrak{fb}}$ the situation is far more complex since we have not assumed the initial rule $\mathcal{G}(Init)$. Admissibility of $\mathcal{G}(GInit)$ and $\mathcal{G}(Refl)$ can be proven by a mutual induction using both the internal and external measures of the principal formulas (cf. Definitions 2.1.27 and 2.1.31). In Figure 2.12, we show a derivation of an initial sequent in $\mathcal{G}^{\mathcal{S}}_{\beta\mathfrak{fb}}$ that demonstrates the need for such a mutual induction. We will also verify admissibility of $\mathcal{G}(GInit)$ and $\mathcal{G}(Refl)$ at the abstraction consistency level using a logical relations-style argument (cf. Lemmas 5.3.2 and 5.3.3, Theorems 2.3.2 and 5.3.4 and Corollaries 5.4.8 and 5.4.9).

In any of the sequent calculi, admissibility of $\mathcal{G}(\neg \doteq)$ can be proven from admissibility of $\mathcal{G}(GInit)$.

LEMMA 2.2.14. *If $\mathcal{G}(GInit)$ is admissible in $\mathcal{G}^{\mathcal{S}}_{*}$, then $\mathcal{G}(\neg \doteq)$ is also admissible.*

$$
\dfrac{
\dfrac{\dfrac{\dfrac{\dfrac{\dfrac{\dfrac{}{[W \dot{=}^{\iota} W]}\,\mathcal{G}(Dec)}{[F\,W], \dot{\neg}[F\,W]}\,\mathcal{G}(Init^{\dot{=}})}{[F\,W_{\iota}], \dot{\neg}[\dot{\forall} x_{\iota} \centerdot F\,x]}\,\mathcal{G}(\dot{\neg}\dot{\forall}, \mathcal{S})}{[\dot{\forall} x_{\iota} \centerdot F\,x], \dot{\neg}[\dot{\forall} x_{\iota} \centerdot F\,x]}\,\mathcal{G}(\dot{\forall}^{W})}{[\dot{\forall} x_{\iota} \centerdot F\,x], \dot{\neg}[\Pi^{\iota}\,F]}\,\mathcal{G}(\dot{\neg}\sharp)}{[\Pi^{\iota}\,F], \dot{\neg}[\Pi^{\iota}\,F]}\,\mathcal{G}(\sharp)
\qquad
\dfrac{\dfrac{\dfrac{\dfrac{\dfrac{\dfrac{}{[U \dot{=}^{\iota} U]}\,\mathcal{G}(Dec)}{[F\,U], \dot{\neg}[F\,U]}\,\mathcal{G}(Init^{\dot{=}})}{[F\,U_{\iota}], \dot{\neg}[\dot{\forall} x_{\iota} \centerdot F\,x]}\,\mathcal{G}(\dot{\neg}\dot{\forall}, \mathcal{S})}{[\dot{\forall} x_{\iota} \centerdot F\,x], \dot{\neg}[\dot{\forall} x_{\iota} \centerdot F\,x]}\,\mathcal{G}(\dot{\forall}^{U})}{\dot{\neg}[\Pi^{\iota}\,F], [\dot{\forall} x_{\iota} \centerdot F\,x]}\,\mathcal{G}(\dot{\neg}\sharp)}{\dot{\neg}[\Pi^{\iota}\,F], [\Pi^{\iota}\,F]}\,\mathcal{G}(\sharp)
}{
\dfrac{\dfrac{[[\Pi^{\iota}\,F_{o\iota}] \dot{=}^{o} [\Pi^{\iota}\,F]]}{[\Pi^{\iota} \dot{=}^{o(o\iota)} \Pi^{\iota}]}\,\mathcal{G}(\mathfrak{f}^{F})}{[H_{o(o(o\iota))}\,\Pi^{\iota}], \dot{\neg}[H\,\Pi^{\iota}]}\,\mathcal{G}(Init^{\dot{=}})
}\,\mathcal{G}(\mathfrak{b})
$$

FIGURE 2.12. Extensional Derivation of an "Initial" Sequent

PROOF. This follows from the derivation

$$
\dfrac{\dfrac{\begin{matrix}\mathcal{D}\\ \Gamma, \mathbf{M}, \dot{\neg}[\mathbf{N} \dot{\supset} \mathbf{M}]\end{matrix} \qquad \begin{matrix}\mathcal{E}\\ \Gamma, \dot{\neg}\mathbf{N}, \dot{\neg}[\mathbf{N} \dot{\supset} \mathbf{M}]\end{matrix}}{\Gamma, \dot{\neg}[\mathbf{M} \dot{\supset} \mathbf{N}], \dot{\neg}[\mathbf{N} \dot{\supset} \mathbf{M}]}\,\mathcal{G}(\dot{\neg} \dot{\supset})}{\Gamma, \dot{\neg}[\mathbf{M} \dot{\equiv} \mathbf{N}]}\,\mathcal{G}(\dot{\neg}\dot{\wedge})
$$

where $\mathcal{D}$ is

$$
\dfrac{\Gamma, \mathbf{M}, \mathbf{N} \qquad \dfrac{}{\Gamma, \mathbf{M}, \dot{\neg}\mathbf{M}}\,\mathcal{G}(GInit)}{\Gamma, \mathbf{M}, \dot{\neg}[\mathbf{N} \dot{\supset} \mathbf{M}]}\,\mathcal{G}(\dot{\neg} \dot{\supset})
$$

and $\mathcal{E}$ is

$$
\dfrac{\dfrac{}{\Gamma, \dot{\neg}\mathbf{N}, \mathbf{N}}\,\mathcal{G}(GInit) \qquad \Gamma, \dot{\neg}\mathbf{N}, \dot{\neg}\mathbf{M}}{\Gamma, \dot{\neg}\mathbf{N}, \dot{\neg}[\mathbf{N} \dot{\supset} \mathbf{M}]}\,\mathcal{G}(\dot{\neg} \dot{\supset})
$$

□

REMARK 2.2.15 (Lifting). One may (reasonably) wonder why we have complicated the situation by not simply assuming the rules $\mathcal{G}(Init)$ and $\mathcal{G}(Refl)$ to be rules in $\mathcal{G}^{\mathcal{S}}_{\beta\mathfrak{f}\mathfrak{b}}$. The reason has to do with

$$\frac{\Gamma, \mathbf{M}}{\Gamma, \mathbf{M}^{\downarrow}}\,\mathcal{G}(\beta\eta^*) \qquad \frac{\Gamma, [\mathbf{A} \doteq^{\alpha} \mathbf{B}]}{\Gamma, [\mathbf{B} \doteq^{\alpha} \mathbf{A}]}\,\mathcal{G}(Sym)$$

$$\frac{\Gamma, [\forall y_\beta \centerdot \mathbf{F}_{\alpha\beta}\, y \doteq^{\alpha} \mathbf{G}\, y]^{\downarrow} \quad \mathbf{F}, \mathbf{G} \in cwff_{\alpha\beta}(\mathcal{S})^{\downarrow}}{\Gamma, [\mathbf{F} \doteq^{\alpha\beta} \mathbf{G}]}\,\mathcal{G}(\mathfrak{f}\forall)$$

$$\frac{\Gamma, \neg[\forall y_\beta \centerdot \mathbf{F}_{\alpha\beta}\, y \doteq^{\alpha} \mathbf{G}\, y]^{\downarrow} \quad \mathbf{F}, \mathbf{G} \in cwff_{\alpha\beta}(\mathcal{S})^{\downarrow}}{\Gamma, \neg[\mathbf{F} \doteq^{\alpha\beta} \mathbf{G}]}\,\mathcal{G}(\neg \doteq^{\rightarrow} \forall)$$

FIGURE 2.13. Extensional Constructively Admissible Rules

lifting of (ground) sequent proofs to proofs with *expansion variables* (cf. Definition 7.2.15). For either of the rules $\mathcal{G}(Init)$ and $\mathcal{G}(Refl)$ to apply, we must check that two formulas are syntactically equal. While this is trivial in the ground case, it involves higher-order unification in the case with variables. As a result, we might need to "unify" terms like $[H\,\Pi^{\alpha}]$ and $[H\,x_{o(o\alpha)}]$. Defining the sequent calculus without using these rules, we avoid needing to unify against logical constants. We also avoid needing to apply any rule to a *flex-flex* equation corresponding to an application of the $\mathcal{G}(Refl)$ rule (cf. Remark 8.3.4).

2.2.3. Constructively Admissible Rules. Some rules can be proven admissible by giving an algorithm which takes derivations of the premises to construct a derivation of the conclusion. As discussed above, one can give mutually recursive algorithms for the admissible rules $\mathcal{G}(GInit)$ and $\mathcal{G}(Refl)$ in the extensional sequent calculus $\mathcal{G}^{\mathcal{S}}_{\beta\mathfrak{fb}}$, though we do not do so here.

There are a few admissible rules for the calculus $\mathcal{G}^{\mathcal{S}}_{\beta\mathfrak{fb}}$ we will use later to establish soundness of extensional expansion proofs (cf. Theorem 7.9.1). In order to know the soundness proof is constructive, we now prove the rules in Figure 2.13 are admissible using constructive proofs.

LEMMA 2.2.16. *If* Γ *is derivable in* $\mathcal{G}^{\mathcal{S}}_{\beta\mathfrak{fb}}$, *then* $\Gamma^{\downarrow}$ *is derivable in* $\mathcal{G}^{\mathcal{S}}_{\beta\mathfrak{fb}}$.

PROOF. The proof is by a simple induction on derivations. □

LEMMA 2.2.17. *If* $\Gamma^{\downarrow}, \Delta$ *is* $\mathcal{G}^{\mathcal{S}}_{\beta\mathfrak{fb}}$*-derivable, then so is* Γ, Δ.

PROOF. Let Γ be the multiset $\mathbf{M}^1, \ldots, \mathbf{M}^n$. We prove the lemma by induction on n. If $n = 0$, then $\Gamma^{\downarrow}, \Delta$ is the same as Γ, Δ. Suppose $n \geq 1$ and let Γ' be $\mathbf{M}^2, \ldots, \mathbf{M}^n$. By the inductive hypothesis, there is a $\mathcal{G}^{\mathcal{S}}_{\beta\mathfrak{fb}}$-derivation $\mathcal{D}$ of $\mathbf{M}^{1\downarrow}, \Gamma', \Delta$. Thus

$$\frac{\begin{array}{c}\mathcal{D}\\ \mathbf{M}^{1\downarrow}, \Gamma', \Delta\end{array}}{\mathbf{M}^1, \Gamma', \Delta}\,\mathcal{G}(\beta\eta)$$

is a $\mathcal{G}^{\mathcal{S}}_{\beta\mathfrak{fb}}$-derivation of Γ, Δ. □

LEMMA 2.2.18. *The rule* $\mathcal{G}(\beta\eta^*)$ *is admissible for* $\mathcal{G}^{\mathcal{S}}_{\beta\mathfrak{fb}}$.

PROOF. Suppose $\mathcal{D}$ is a derivation of $\Gamma, \mathbf{M}$. By Lemma 2.2.16, there is a derivation $\mathcal{D}_1$ of $\Gamma^{\downarrow}, \mathbf{M}^{\downarrow}$. By Lemma 2.2.17, there is a derivation $\mathcal{D}_2$ of $\Gamma, \mathbf{M}^{\downarrow}$, as desired. □

LEMMA 2.2.19. *The rule* $\mathcal{G}(Sym)$ *is admissible for* $\mathcal{G}^{\mathcal{S}}_{\beta\mathfrak{fb}}$.

PROOF. By a simple induction on derivations one can show that if $\mathcal{D}$ is a derivation

$$\Gamma, \underbrace{[\mathbf{A}^1 \doteq^{\alpha^1} \mathbf{B}^1], \ldots, [\mathbf{A}^n \doteq^{\alpha^n} \mathbf{B}^n]}_{n}$$

then

$$\Gamma, \underbrace{[\mathbf{B}^1 \doteq^{\alpha^1} \mathbf{A}^1], \ldots, [\mathbf{B}^n \doteq^{\alpha^n} \mathbf{A}^n]}_{n}$$

is derivable. If the last rule of $\mathcal{D}$ is $\mathcal{G}(EUnif_1)$ (respectively, $\mathcal{G}(EUnif_2)$), then we apply the inductive hypothesis followed by the rule $\mathcal{G}(EUnif_2)$ (respectively, $\mathcal{G}(EUnif_1)$). □

LEMMA 2.2.20. *The rule* $\mathcal{G}(\mathfrak{f}\dot\forall)$ *is admissible for* $\mathcal{G}^{\mathcal{S}}_{\beta\mathfrak{fb}}$.

PROOF. Let $\mathbf{F}, \mathbf{G} \in cwff_{\alpha\beta}(\mathcal{S})^{\downarrow}$ (i.e., $\beta\eta$-normal forms) be given and let $\mathbf{M}$ be $[\dot\forall y_\beta \,\blacksquare\, \mathbf{F}\, y \doteq^{\alpha} \mathbf{G}\, y]^{\downarrow}$. We prove by induction on derivations that if $\mathcal{D}$ is a derivation

$$\Gamma, \underbrace{\mathbf{M}, \ldots, \mathbf{M}}_{n}$$

then there is a derivation $\mathcal{E}$ of

$$\Gamma, \underbrace{[\mathbf{F} \underset{\cdot}{=}^{\alpha\beta} \mathbf{G}], \ldots, [\mathbf{F} \underset{\cdot}{=}^{\alpha\beta} \mathbf{G}]}_{n}$$

If none of the n occurrences of $\mathbf{M}$ are principal in the last rule of $\mathcal{D}$, then we can simply apply the inductive hypothesis to the derivations of the premises and reapply the same rule. Otherwise, assume some of the n occurrences are principal. The only rules of $\mathcal{G}^{\mathcal{S}}_{\beta\mathfrak{fb}}$ for which $\mathbf{M}$ can be principal are $\mathcal{G}(\beta\eta)$, $\mathcal{G}(Contr)$ and $\mathcal{G}(\underset{\cdot}{\forall}^{W})$. Let Γ' be

$$\Gamma, \underbrace{\mathbf{M}, \ldots, \mathbf{M}}_{n-1}$$

and Γ'' be

$$\Gamma, \underbrace{[\mathbf{F} \underset{\cdot}{=}^{\alpha\beta} \mathbf{G}], \ldots, [\mathbf{F} \underset{\cdot}{=}^{\alpha\beta} \mathbf{G}]}_{n-1}$$

If the last rule is $\mathcal{G}(\beta\eta)$, then since $\mathbf{M}$ is already assumed to be $\beta\eta$-normal this is a trivial application of the $\mathcal{G}(\beta\eta)$. We can simply apply the inductive hypothesis to the derivation of the premise (with the same n occurrences of $\mathbf{M}$) to obtain $\mathcal{E}$.

Suppose $\mathcal{D}$ is of the form

$$\dfrac{\begin{matrix}\mathcal{D}_1 \\ \Gamma', \mathbf{M}, \mathbf{M}\end{matrix}}{\Gamma', \mathbf{M}}\ \mathcal{G}(Contr)$$

Applying the inductive hypothesis to $\mathcal{D}_1$ with the $n+1$ occurrences of $\mathbf{M}$, we have a derivation $\mathcal{E}_1$ of $\Gamma'', [\mathbf{F} \underset{\cdot}{=}^{\alpha\beta} \mathbf{G}], [\mathbf{F} \underset{\cdot}{=}^{\alpha\beta} \mathbf{G}]$. We complete the derivation $\mathcal{E}$ as follows:

$$\dfrac{\begin{matrix}\mathcal{E}_1 \\ \Gamma'', [\mathbf{F} \underset{\cdot}{=}^{\alpha\beta} \mathbf{G}], [\mathbf{F} \underset{\cdot}{=}^{\alpha\beta} \mathbf{G}]\end{matrix}}{\Gamma'', [\mathbf{F} \underset{\cdot}{=}^{\alpha\beta} \mathbf{G}]}\ \mathcal{G}(Contr)$$

The most important case is when $\mathcal{D}$ is of the form

$$\dfrac{\begin{matrix}\mathcal{D}_1 \\ \Gamma', [W/y]([\mathbf{F}\, y \underset{\cdot}{=} \mathbf{G}\, y]^{\downarrow})\end{matrix} \qquad W \notin \mathbf{Params}(|\Gamma', [\mathbf{F} \underset{\cdot}{=} \mathbf{G}]|)}{\Gamma', [\underset{\cdot}{\forall} y_{\beta} \centerdot [\mathbf{F}\, y]^{\downarrow} \underset{\cdot}{=}^{\alpha} [\mathbf{G}\, y]^{\downarrow}]}\ \mathcal{G}(\underset{\cdot}{\forall}^{W})$$

Since W is a parameter, $[W/y]([\mathbf{F}\, y \doteq \mathbf{G}\, y]^{\downarrow})$ is $[\mathbf{F}\, W \doteq \mathbf{G}\, W]^{\downarrow}$. Applying the inductive hypothesis to $\mathcal{D}_1$ with the remaining $n-1$ occurrences of $\mathbf{M}$, there is a derivation $\mathcal{E}_1$ of $\Gamma'', [\mathbf{F}\, W \doteq \mathbf{G}\, W]^{\downarrow}$. Note that

$$\begin{aligned}\mathbf{Params}([\mathbf{F} \doteq \mathbf{G}]) &= \mathbf{Params}([\dot{\forall} y_\beta \centerdot [\mathbf{F}\, y]^{\downarrow} \doteq [\mathbf{G}\, y]^{\downarrow}]) \\ &= \mathbf{Params}(\mathbf{M})\end{aligned}$$

since $\mathbf{F}$ and $\mathbf{G}$ are $\beta\eta$-normal. Consequently, the same parameters occur in Γ' as in Γ''. Hence we can apply the $\mathcal{G}(\mathfrak{f}^W)$ rule to complete the derivation $\mathcal{E}$ by

$$\dfrac{\dfrac{\begin{array}{c}\mathcal{E}_1 \\ \Gamma'', [\mathbf{F}\, W \doteq \mathbf{G}\, W]^{\downarrow}\end{array}}{\Gamma'', [\mathbf{F}\, W \doteq \mathbf{G}\, W]}\,\mathcal{G}(\beta\eta)}{\Gamma'', [\mathbf{F} \doteq \mathbf{G}]}\,\mathcal{G}(\mathfrak{f}^W)$$

□

LEMMA 2.2.21. *The rule* $\mathcal{G}(\dot{\neg} \doteq^{\rightarrow} \dot{\forall})$ *in Figure 2.13 is admissible for* $\mathcal{G}^{\mathcal{S}}_{\beta\mathfrak{f}\mathfrak{b}}$.

PROOF. This is analogous to the proof of Lemma 2.2.20 except we make use of the admissible rule $\mathcal{G}(\beta\eta^*)$ (cf. Lemma 2.2.18). □

REMARK 2.2.22. The proofs of Lemmas 2.2.16, 2.2.17, 2.2.19, and 2.2.20 are constructive. Hence the proofs of Lemmas 2.2.18 and 2.2.21 are also constructive. That is, the proofs of admissibility of $\mathcal{G}(\beta\eta^*)$, $\mathcal{G}(Sym)$, $\mathcal{G}(\mathfrak{f}\dot{\forall})$ and $\mathcal{G}(\dot{\neg} \doteq^{\rightarrow} \dot{\forall})$ provide algorithms for constructing a derivation of the conclusion given derivations of the premises.

2.2.4. Restricted Derivations. We now define $\mathcal{W}$-restricted derivations of sequents by only allowing instantiations with closed terms from some set $\mathcal{W}$. When we define other representations of proofs we can relate $\mathcal{W}$-restricted derivations to these other representations with appropriate restrictions. For example, in Chapter 7 we define extensional expansion proofs (cf. Definition 7.4.18) and $\mathcal{W}$-restricted extensional expansion proofs (cf. Definition 7.5.1). In Theorem 7.10.12 we demonstrate that every $\mathcal{W}$-restricted derivation induces a $\mathcal{W}$-restricted extensional expansion proof (assuming $\mathcal{W}$ is closed under parameter substitutions).

$$\frac{\Gamma, \dot{\neg}[[\mathbf{C}/x]\mathbf{M}] \quad \mathbf{C} \in \mathcal{W}_\alpha}{\Gamma, \dot{\neg}[\dot{\forall} x_\alpha \mathbf{M}]} \mathcal{G}(\dot{\neg}\dot{\forall}_{re}, \mathcal{W})$$

$$\frac{\Gamma, \dot{\neg}[[\mathbf{G}\,\mathbf{B}] \dot{=}^{\alpha} [\mathbf{H}\,\mathbf{B}]] \quad \mathbf{B} \in \mathcal{W}_\beta}{\Gamma, \dot{\neg}[\mathbf{G} \dot{=}^{\alpha\beta} \mathbf{H}]} \mathcal{G}(\dot{\neg} \dot{=}^{\rightarrow}_{re}, \mathcal{W})$$

FIGURE 2.14. $\mathcal{W}$-Restricted Instantiation Rules

DEFINITION 2.2.23 (Restricted Derivations). Let $\mathcal{S}$ be a signature and $\mathcal{W} \subseteq cwff(\mathcal{S})$ be a set of closed formulas. A *$\mathcal{W}$-restricted $\mathcal{S}$-derivation* of an $\mathcal{S}$-sequent Γ is a $\mathcal{G}^{\mathcal{S}}_{\beta\mathfrak{fb}}$-derivation $\mathcal{D}$ of Γ where every application of a $\mathcal{G}(\dot{\neg}\dot{\forall}, \mathcal{S})$ rule is of the form $\mathcal{G}(\dot{\neg}\dot{\forall}_{re}, \mathcal{W})$ and every application of a $\mathcal{G}(\dot{\neg} \dot{=}^{\rightarrow}, \mathcal{S})$ rule is of the form $\mathcal{G}(\dot{\neg} \dot{=}^{\rightarrow}_{re}, \mathcal{W})$ (cf. Figure 2.14).

Note that every derivation is a $cwff(\mathcal{S})$-restricted derivation.

2.3. Syntactic Logical Relations

Logical relations (cf. [55; 60]) are a powerful tool for studying properties of typed λ-calculi. We can use logical relations to prove certain properties of terms using two simple steps:

1. Define a relation on closed terms of base types, then extend this to a relation on function types.
2. Check that constants and parameters are related to themselves in order to conclude all closed terms are related to themselves.

In this section, we consider relations on formulas. In Section 4.4 we will construct models using logical relations on semantic structures (cf. Definition 4.4.2).

DEFINITION 2.3.1. Let $\mathcal{S}$ be a signature and n be a natural number. An *n-ary logical relation on $cwff(\mathcal{S})$* is a typed family of of n-ary relations $\mathcal{R}_\alpha \subseteq (cwff_\alpha(\mathcal{S}))^n$ such that for each function type $(\alpha\beta)$ and $\mathbf{F}^1, \ldots, \mathbf{F}^n \in cwff_{\alpha\beta}(\mathcal{S})$, $\mathcal{R}_{\alpha\beta}(\mathbf{F}^1_{\alpha\beta}, \cdots, \mathbf{F}^n_{\alpha\beta})$ iff we have

$$\mathcal{R}_\alpha([\mathbf{F}^1\,\mathbf{B}^1], \ldots, [\mathbf{F}^n\,\mathbf{B}^n])$$

whenever $\mathcal{R}_\beta(\mathbf{B}^1, \ldots, \mathbf{B}^n)$ for all $\mathbf{B}^1, \ldots, \mathbf{B}^n \in cwff_\beta(\mathcal{S})$. A logical relation $\mathcal{R}$ is *admissible* if $\mathcal{R}$ is closed under head expansions. That is, for all $\mathbf{B}^1, \ldots, \mathbf{B}^n$ in $cwff_\beta(\mathcal{S})$ and $[\lambda x_\beta\, \mathbf{A}^1], \ldots, [\lambda x_\beta\, \mathbf{A}^n]$ in $cwff_{\alpha\beta}(\mathcal{S})$, if $\mathcal{R}_\alpha([\mathbf{B}^1/x]\mathbf{A}^1, \ldots, [\mathbf{B}^n/x]\mathbf{A}^n)$, then

$$\mathcal{R}_\beta([[\lambda x_\beta\, \mathbf{A}^1]\, \mathbf{B}^1], \ldots, [[\lambda x_\beta\, \mathbf{A}^n]\, \mathbf{B}^n]).$$

We now prove the Fundamental Theorem of Logical Relations.

THEOREM 2.3.2 (Fundamental Theorem of Logical Relations). *Let $\mathcal{S}$ be a signature and $\mathcal{R}$ be an n-ary admissible logical relation on $cwff(\mathcal{S})$. Suppose for each constant or parameter $b_\beta \in \mathcal{S} \cup \mathcal{P}$, we have $\mathcal{R}_\beta(b, \ldots, b)$. Then $\mathcal{R}_\beta(\mathbf{A}, \ldots, \mathbf{A})$ for every closed term $\mathbf{A} \in cwff_\alpha(\mathcal{S})$.*

PROOF. We prove the stronger statement that

$$\mathcal{R}_\alpha(\theta^1(\mathbf{A}), \ldots, \theta^n(\mathbf{A}))$$

holds for any term $\mathbf{A}_\alpha \in wff_\alpha(\mathcal{S})$ and ground substitutions $\theta^1, \ldots, \theta^n$ defined on the free variables of $\mathbf{A}$ such that $\mathcal{R}_\beta(\theta^1(x_\beta), \ldots, \theta^n(x_\beta))$ for every $x_\beta \in \mathbf{Free}(\mathbf{A})$. We prove this by induction on the term $\mathbf{A}$.

If $\mathbf{A}$ is a variable, this follows from the assumption on $\theta^1, \ldots, \theta^n$. If $\mathbf{A} \in \mathcal{S} \cup \mathcal{P}$, then this follows by the assumption of the theorem.

For the application case, suppose $\mathbf{A}$ is of the form $[\mathbf{G}\,\mathbf{B}]$. By the inductive hypothesis, we have $\mathcal{R}_{\alpha\beta}(\theta^1(\mathbf{G}), \ldots, \theta^n(\mathbf{G}))$ and $\mathcal{R}_\beta(\theta^1(\mathbf{B}), \ldots, \theta^n(\mathbf{B}))$. Since $\mathcal{R}$ is a logical relation, we have $\mathcal{R}_\alpha([\theta^1(\mathbf{G})\, \theta^1(\mathbf{B})], \ldots, [\theta^n(\mathbf{G})\, \theta^n(\mathbf{B})])$. That is,

$$\mathcal{R}_\alpha(\theta^1([\mathbf{G}\,\mathbf{B}]), \ldots, \theta^n([\mathbf{G}\,\mathbf{B}])).$$

Finally, suppose $\mathbf{A}$ is of the form $\lambda x_\beta \mathbf{D}_\gamma$. This case is where we need the stronger inductive hypothesis. Let $\mathbf{B}^1, \ldots, \mathbf{B}^n \in cwff_\beta(\mathcal{S})$ be such that $\mathcal{R}_\beta(\mathbf{B}^1, \ldots, \mathbf{B}^n)$. We verify $\mathcal{R}_\gamma([\theta^1(\mathbf{A})\, \mathbf{B}^1], \ldots, [\theta^n(\mathbf{A})\, \mathbf{B}^n])$. Let i such that $1 \le i \le n$ be given. Note that

$$[\theta^i(\mathbf{A})\, \mathbf{B}^i] \;=\; [\theta^i([\lambda x_\beta\, \mathbf{D}])\, \mathbf{B}^i]$$

is a β-redex with reduct $(\theta^i, [\mathbf{B}^i/x])(\mathbf{D})$. Let $\psi^i := \theta^i, [\mathbf{B}^i/x]$. By the inductive hypothesis, we have $\mathcal{R}_\gamma(\psi^1(\mathbf{D}), \ldots, \psi^n(\mathbf{D}))$. Since $\mathcal{R}$ is admissible,

$$\mathcal{R}_\gamma([\theta^1(\mathbf{A})\, \mathbf{B}^1], \ldots, [\theta^n(\mathbf{A})\, \mathbf{B}^n]).$$

Generalizing over $\mathbf{B}^1, \ldots, \mathbf{B}^n$, we know $\mathcal{R}_{\gamma\beta}(\theta^1(\mathbf{A}), \ldots, \theta^n(\mathbf{A}))$. □

CHAPTER 3

Semantics

In [19], a semantics for higher-order logic with different forms of extensionality is presented. The presentation in [19] assumes the signature contains the logical constants $\neg$, $\vee$, Π^α for each type α, and optionally $=^\alpha$ for each type α. In this chapter, we generalize the semantics to arbitrary signatures of logical constants. We will begin by defining applicative structures and evaluations to give semantics for the simply typed λ-calculus. To obtain a model for higher-order logic, we use a valuation to determine whether propositions are true or false.

Many of the basic definitions and results in this chapter are very similar to the article [19] which can be consulted for additional motivation, explanation, and examples.

3.1. Applicative Structures

DEFINITION 3.1.1. A *typed family of sets* $\mathcal{D}$ is a collection of sets $\mathcal{D}_\alpha$ indexed by types $\alpha \in \mathcal{T}$. A *typed family of nonempty sets* $\mathcal{D}$ is a typed family of sets such that $\mathcal{D}_\alpha$ is nonempty for each $\alpha \in \mathcal{T}$. Given two typed families of sets $\mathcal{D}$ and $\mathcal{D}'$, a *typed function* $f : \mathcal{D} \longrightarrow \mathcal{D}'$ is a collection of functions $f_\alpha : \mathcal{D}_\alpha \longrightarrow \mathcal{D}'_\alpha$ indexed by types $\alpha \in \mathcal{T}$. We use $\mathfrak{F}_\mathcal{T}(\mathcal{D}, \mathcal{D}')$ to denote the set of typed functions from a typed family of sets $\mathcal{D}$ to a typed family of sets $\mathcal{D}'$.

DEFINITION 3.1.2 ((Typed) Applicative Structure). A (*typed*) *applicative structure* is a pair $(\mathcal{D}, @)$ where $\mathcal{D}$ is a typed family of nonempty sets and $@^{\alpha\beta} : \mathcal{D}_{\alpha\beta} \times \mathcal{D}_\beta \longrightarrow \mathcal{D}_\alpha$ for each function type $(\alpha\beta)$.

Each (nonempty) set $\mathcal{D}_\alpha$ is called the *domain of type* α and the family of functions $@$ is called the *application operator*. We write simply $\mathsf{f}@\mathsf{b}$ for $\mathsf{f}@^{\alpha\beta}\mathsf{b}$ when $\mathsf{f} \in \mathcal{D}_{\alpha\beta}$ and $\mathsf{b} \in \mathcal{D}_\beta$ are clear in context.

REMARK 3.1.3. Often an applicative structure is defined to also include an interpretation of the constants in a given signature (for example, in [46]). We prefer this signature-independent definition (as in [35]) for our purposes.

NOTATION *Let* $(\mathcal{D}, @)$ *be an applicative structure. We use the convention that* $@$ *associates to the left, i.e., we write* $\mathsf{f}@\mathsf{a}^1@\cdots@\mathsf{a}^n$ *for* $(\cdots(\mathsf{f}@\mathsf{a}^1)@\cdots@\mathsf{a}^n)$ *where* $\mathsf{f} \in \mathcal{D}_{\beta\alpha^n\cdots\alpha^1}$, $\mathsf{a}^1 \in \mathcal{D}_{\alpha^1}, \ldots, \mathsf{a}^n \in \mathcal{D}_{\alpha^n}$ *for some types* $\beta, \alpha^1, \ldots, \alpha^n \in \mathcal{T}$.

DEFINITION 3.1.4 (Frame). A typed family $\mathcal{D}$ of nonempty sets is called a *frame* if $\mathcal{D}_{\alpha\beta} \subseteq \mathcal{D}_\alpha^{\mathcal{D}_\beta}$ for all types α and β. The frame is *standard* if $\mathcal{D}_{\alpha\beta} = \mathcal{D}_\alpha^{\mathcal{D}_\beta}$ for all types α and β. Likewise, we call an applicative structure $(\mathcal{D}, @)$ a *frame structure* if $\mathcal{D}$ is a frame and $@^{\alpha\beta}$ is application for functions for all types α and β.

Frames are the applicative structures which correspond to Henkin models.

DEFINITION 3.1.5 (Functional/Full/Standard Applicative Structures). Let $\mathcal{A} := (\mathcal{D}, @)$ be an applicative structure. We say $\mathcal{A}$ is *functional* if for all types α and β and objects $\mathsf{f}, \mathsf{g} \in \mathcal{D}_{\alpha\beta}$, $\mathsf{f} = \mathsf{g}$ whenever $\mathsf{f}@\mathsf{a} = \mathsf{g}@\mathsf{a}$ for every $\mathsf{a} \in \mathcal{D}_\beta$. (This is called "extensional" in [46]. We use the term "functional" to distinguish it from other forms of extensionality.) We say $\mathcal{A}$ is *full* if for all types α and β and every function $f\colon \mathcal{D}_\beta \longrightarrow \mathcal{D}_\alpha$ there is an object $\mathsf{f} \in \mathcal{D}_{\alpha\beta}$ such that $\mathsf{f}@\mathsf{b} = f(\mathsf{b})$ for every $\mathsf{b} \in \mathcal{D}_\beta$. Finally, we say $\mathcal{A}$ is *standard* if it is a frame structure where $\mathcal{D}$ is a standard frame. Note that these definitions impose restrictions on the domains for function types only.

We call a frame $\mathcal{D}$ *functional* [*full*] if the corresponding frame structure $(\mathcal{D}, @)$ is functional [full].

REMARK 3.1.6. It is easy to see that every frame is functional. Furthermore, an applicative structure is standard iff it is a full frame structure.

DEFINITION 3.1.7 (Homomorphism). Let

$$\mathcal{A}^1 = (\mathcal{D}^1, @^1)$$

and

$$\mathcal{A}^2 = (\mathcal{D}^2, @^2)$$

be applicative structures. A *homomorphism* is a typed function

$$\kappa\colon \mathcal{D}^1 \longrightarrow \mathcal{D}^2$$

such that for all types α and β, all $\mathsf{f} \in \mathcal{D}^1_{\alpha\beta}$, and $\mathsf{b} \in \mathcal{D}^1_\beta$ we have

$$\kappa(\mathsf{f})@^2\kappa(\mathsf{b}) = \kappa(\mathsf{f}@^1\mathsf{b}).$$

The two applicative structures $\mathcal{A}^1$ and $\mathcal{A}^2$ are *isomorphic* if there are homomorphisms $i : \mathcal{A}^1 \longrightarrow \mathcal{A}^2$ and $j : \mathcal{A}^2 \longrightarrow \mathcal{A}^1$ which are mutually inverse at each type.

3.2. Evaluations

Evaluations are applicative structures with a notion of evaluation for well-formed formulas in $\mathit{wff}(\mathcal{S})$ over a signature $\mathcal{S}$.

DEFINITION 3.2.1 (Assignments). Let $\mathcal{D}$ be a typed family of nonempty sets. An *assignment* φ *into* $\mathcal{D}$ is a typed function $\varphi\colon \mathcal{V} \longrightarrow \mathcal{D}$. We denote the set of all assignments into $\mathcal{D}$ by $\mathit{Assignments}(\mathcal{D})$. That is, $\mathit{Assignments}(\mathcal{D})$ is $\mathfrak{F}_\mathcal{T}(\mathcal{V}, \mathcal{D})$.

Given an assignment φ, variable x_α, and value $\mathsf{a} \in \mathcal{D}_\alpha$, we use $\varphi, [\mathsf{a}/x]$ to denote the assignment such that $(\varphi, [\mathsf{a}/x])(x) = \mathsf{a}$ and $(\varphi, [\mathsf{a}/x])(y) = \varphi(y)$ for variables y other than x.

DEFINITION 3.2.2 (Weak Evaluation). Let $\mathcal{A} = (\mathcal{D}, @)$ be an applicative structure and $\mathcal{E}\colon \mathit{Assignments}(\mathcal{D}) \longrightarrow \mathfrak{F}_\mathcal{T}(\mathit{wff}(\mathcal{S}), \mathcal{D})$ be a function. We will consider the argument of $\mathcal{E}$ as a parameter to $\mathcal{E}$ and write $\mathcal{E}_\varphi$ for the typed function $\mathcal{E}_\varphi\colon \mathit{wff}(\mathcal{S}) \longrightarrow \mathcal{D}$ that maps terms in $\mathit{wff}_\alpha(\mathcal{S})$ to objects in $\mathcal{D}_\alpha$ for each type α. $\mathcal{E}$ is called a *weak* $\mathcal{S}$-*evaluation function* for $\mathcal{A}$ if for any assignment φ into $\mathcal{A}$, we have

1. $\mathcal{E}_\varphi\big|_\mathcal{V} = \varphi$
2. $\mathcal{E}_\varphi([\mathbf{F}\,\mathbf{B}]) = \mathcal{E}_\varphi(\mathbf{F})@\mathcal{E}_\varphi(\mathbf{B})$ for any $\mathbf{F}$ in $\mathit{wff}_{\alpha\beta}(\mathcal{S})$ and $\mathbf{B}$ in $\mathit{wff}_\beta(\mathcal{S})$ and types α and β.
3. For any assignments φ and ψ, $\mathcal{E}_\varphi(W) = \mathcal{E}_\psi(W)$ for any parameter W and $\mathcal{E}_\varphi(c) = \mathcal{E}_\psi(c)$ for any constant c.
4. $\mathcal{E}_\varphi([\lambda x_\beta\, \mathbf{A}])@\mathsf{b} = \mathcal{E}_{\varphi,[\mathsf{b}/x]}(\mathbf{A})$ for all $\mathbf{A} \in \mathit{wff}_\alpha(\mathcal{S})$ and $\mathsf{b} \in \mathcal{D}_\beta$.

DEFINITION 3.2.3 (Evaluation). Let $\mathcal{A} = (\mathcal{D}, @)$ be an applicative structure and $\mathcal{E}\colon \mathit{Assignments}(\mathcal{D}) \longrightarrow \mathfrak{F}_\mathcal{T}(\mathit{wff}(\mathcal{S}), \mathcal{D})$ be a function. As above, we write $\mathcal{E}_\varphi$ instead of $\mathcal{E}(\varphi)$. $\mathcal{E}$ is called an $\mathcal{S}$-*evaluation function* for $\mathcal{A}$ if for any assignment φ into $\mathcal{A}$, we have

1. $\mathcal{E}_\varphi\big|_{\mathcal{V}} = \varphi$
2. $\mathcal{E}_\varphi([\mathbf{F}\,\mathbf{B}]) = \mathcal{E}_\varphi(\mathbf{F})@\mathcal{E}_\varphi(\mathbf{B})$ for any $\mathbf{F}$ in $\mathit{wff}_{\alpha\beta}(\mathcal{S})$ and $\mathbf{B}$ in $\mathit{wff}_\beta(\mathcal{S})$ and types α and β.
3. $\mathcal{E}_\varphi(\mathbf{A}) = \mathcal{E}_\psi(\mathbf{A})$ for any type α and $\mathbf{A} \in \mathit{wff}_\alpha(\mathcal{S})$, whenever φ and ψ coincide on $\mathbf{Free}(\mathbf{A})$
4. $\mathcal{E}_\varphi(\mathbf{A}) = \mathcal{E}_\varphi(\mathbf{A}^{\downarrow\beta})$ for all $\mathbf{A} \in \mathit{wff}_\alpha(\mathcal{S})$.

We call $\mathcal{J} := (\mathcal{D}, @, \mathcal{E})$ an *evaluation* if $(\mathcal{D}, @)$ is an applicative structure and $\mathcal{E}$ is an evaluation function for $(\mathcal{D}, @)$. We call $\mathcal{E}_\varphi(\mathbf{A}_\alpha) \in \mathcal{D}_\alpha$ the *denotation* of $\mathbf{A}_\alpha$ in $\mathcal{J}$ for φ. (Note that since $\mathcal{E}$ is a function, the denotation in $\mathcal{J}$ is unique. However, for a given applicative structure $\mathcal{A}$, there may be many possible evaluation functions.)

If $\mathbf{A}$ is a closed formula, then $\mathcal{E}_\varphi(\mathbf{A})$ is independent of φ, since $\mathbf{Free}(\mathbf{A})$ is empty. In these cases we sometimes drop the reference to φ from $\mathcal{E}_\varphi(\mathbf{A})$ and simply write $\mathcal{E}(\mathbf{A})$.

We call an evaluation $\mathcal{J} := (\mathcal{D}, @, \mathcal{E})$ *functional* [*full*, *standard*] if the applicative structure $(\mathcal{D}, @)$ is *functional* [*full*, *standard*]. We say $\mathcal{J}$ is an *evaluation over a frame* if $(\mathcal{D}, @)$ is a frame structure.

Evaluations generalize evaluations over frames, which are the basis for Henkin models, to the nonfunctional case. We cannot in general assume the evaluation function is uniquely determined by its values on constants and parameters as this requires functionality.

REMARK 3.2.4 ($\mathcal{S}$-Evaluations respect β-Equality). Let

$$\mathcal{J} := (\mathcal{D}, @, \mathcal{E})$$

be an $\mathcal{S}$-evaluation and $\mathbf{A} \stackrel{\beta}{=} \mathbf{B}$. For all assignments φ into $(\mathcal{D}, @)$, we have $\mathcal{E}_\varphi(\mathbf{A}) = \mathcal{E}_\varphi(\mathbf{A}^{\downarrow\beta}) = \mathcal{E}_\varphi(\mathbf{B}^{\downarrow\beta}) = \mathcal{E}_\varphi(\mathbf{B})$.

Every evaluation function is a weak evaluation function. When the applicative structure is functional, every weak evaluation function is an evaluation function. On the other hand, there is a weak evaluation function over a nonfunctional applicative structure which is not an evaluation function (cf. Example 4.3.8).

LEMMA 3.2.5. *Let $(\mathcal{D}, @)$ be an applicative structure. Every $\mathcal{S}$-evaluation function $\mathcal{E}$ is a weak $\mathcal{S}$-evaluation function.*

PROOF. The facts that $\mathcal{E}_\varphi(W) = \mathcal{E}_\psi(W)$ and $\mathcal{E}_\varphi(c) = \mathcal{E}_\psi(c)$ for all parameters W and constants c hold since both $\mathbf{Free}(W)$ and $\mathbf{Free}(c)$

are empty. For any $\mathbf{A} \in wff_\alpha(\mathcal{S})$ and $\mathsf{b} \in \mathcal{D}_\beta$, we have

$$\mathcal{E}_\varphi([\lambda x_\beta \mathbf{A}])@\mathsf{b} = \mathcal{E}_{\varphi,[\mathsf{b}/x]}([[\lambda x_\beta \mathbf{A}]\, x]) = \mathcal{E}_{\varphi,[\mathsf{b}/x]}(\mathbf{A})$$

using Remark 3.2.4. □

LEMMA 3.2.6. *Let $(\mathcal{D}, @)$ be a functional applicative structure and $\mathcal{E}$ be a weak $\mathcal{S}$-evaluation function. For every $\mathbf{A} \in wff_\alpha(\mathcal{S})$,*

$$\mathcal{E}_\varphi(\mathbf{A}) = \mathcal{E}_\psi(\mathbf{A})$$

whenever φ and ψ coincide on $\mathbf{Free}(\mathbf{A})$.

PROOF. We prove this by induction on $\mathbf{A}$. If $\mathbf{A}$ is a constant or parameter, then we know this by the definition of weak evaluation functions. If $\mathbf{A}$ is a variable x, then

$$\mathcal{E}_\varphi(\mathbf{A}) = \varphi(x) = \psi(x) = \mathcal{E}_\psi(\mathbf{A})$$

by our assumption on φ and ψ. If $\mathbf{A}$ is an application of the form $[\mathbf{F}\,\mathbf{B}]$, we have

$$\mathcal{E}_\varphi([\mathbf{F}\,\mathbf{B}]) = \mathcal{E}_\varphi(\mathbf{F})@\mathcal{E}_\varphi(\mathbf{B}) = \mathcal{E}_\psi(\mathbf{F})@\mathcal{E}_\psi(\mathbf{B}) = \mathcal{E}_\psi([\mathbf{F}\,\mathbf{B}])$$

by the inductive hypothesis.

Suppose $\mathbf{A}$ is a λ-abstraction of the form $[\lambda x_\beta \mathbf{C}_\gamma]$. Note that for any $\mathsf{b} \in \mathcal{D}_\beta$, $\varphi, [\mathsf{b}/x]$ and $\psi, [\mathsf{b}/x]$ coincide on $\mathbf{Free}(\mathbf{C})$ since φ and ψ coincide on $\mathbf{Free}([\lambda x_\beta \mathbf{C}_\gamma])$. By the inductive hypothesis and the definition of weak evaluation functions, we have

$$\mathcal{E}_\varphi([\lambda x_\beta \mathbf{C}_\gamma])@\mathsf{b} = \mathcal{E}_{\varphi,[\mathsf{b}/x]}(\mathbf{C}_\gamma) = \mathcal{E}_{\psi,[\mathsf{b}/x]}(\mathbf{C}_\gamma) = \mathcal{E}_\psi([\lambda x_\beta \mathbf{C}_\gamma])@\mathsf{b}.$$

By functionality, $\mathcal{E}_\varphi(\mathbf{A}) = \mathcal{E}_\psi(\mathbf{A})$. □

A key property we will need is the *Substitution-Value Lemma* which states that $\mathcal{E}_{\varphi,[\mathcal{E}_\varphi(\mathbf{B})/x]}(\mathbf{A}) = \mathcal{E}_\varphi([\mathbf{B}/x]\mathbf{A})$ for any assignment φ, types α and β, formulas $\mathbf{A} \in wff_\alpha(\mathcal{S})$, $\mathbf{B} \in wff_\beta(\mathcal{S})$, and variable x_β. A simple proof establishes the *Substitution-Value Lemma* for evaluation functions.

LEMMA 3.2.7 (Substitution-Value Lemma). *Let $(\mathcal{D}, @, \mathcal{E})$ be an evaluation. For any types α and β, variable x_β, assignment φ into $\mathcal{D}$, and formulas $\mathbf{A} \in wff_\alpha(\mathcal{S})$ and $\mathbf{B} \in wff_\beta(\mathcal{S})$, we have*

$$\mathcal{E}_{\varphi,[\mathcal{E}_\varphi(\mathbf{B})/x]}(\mathbf{A}) = \mathcal{E}_\varphi([\mathbf{B}/x]\mathbf{A}).$$

PROOF. We can compute

$$\begin{aligned}
\mathcal{E}_{\varphi,[\mathcal{E}_\varphi(\mathbf{B})/x]}(\mathbf{A}) &= \mathcal{E}_{\varphi,[\mathcal{E}_\varphi(\mathbf{B})/x]}([[\lambda x\, \mathbf{A}]\, x]) \\
&= \mathcal{E}_{\varphi,[\mathcal{E}_\varphi(\mathbf{B})/x]}([\lambda x\, \mathbf{A}])@\mathcal{E}_{\varphi,[\mathcal{E}_\varphi(\mathbf{B})/x]}(x) \\
&= \mathcal{E}_\varphi([\lambda x\, \mathbf{A}])@\mathcal{E}_\varphi(\mathbf{B}) \\
&= \mathcal{E}_\varphi([[\lambda x\, \mathbf{A}]\, \mathbf{B}]) \\
&= \mathcal{E}_\varphi([\mathbf{B}/x]\mathbf{A})
\end{aligned}$$

using the fact that $\mathcal{E}$ respects β-equality (cf. Remark 3.2.4) and the other properties of $\mathcal{E}$ (cf. Definition 3.2.3). □

We can also prove the *Substitution-Value Lemma* for weak evaluations over functional applicative structures by induction on $\mathbf{A}$. The *Substitution-Value Lemma* for weak evaluations over functional applicative structures will be used to determine every weak evaluation is an evaluation.

LEMMA 3.2.8 (Functional Substitution-Value Lemma). *Let* $(\mathcal{D}, @)$ *be a functional applicative structure and* $\mathcal{E}$ *be a weak* $\mathcal{S}$*-evaluation function. We have* $\mathcal{E}_{\varphi,[\mathcal{E}_\varphi(\mathbf{B})/x]}(\mathbf{A}) = \mathcal{E}_\varphi([\mathbf{B}/x]\mathbf{A})$ *for all* $\mathbf{A} \in \mathit{wff}_\alpha(\mathcal{S})$, $\mathbf{B} \in \mathit{wff}_\beta(\mathcal{S})$, $x_\beta \in \mathcal{V}_\beta$ *and assignments* φ *into* $\mathcal{D}$.

PROOF. We prove this by induction on $\mathbf{A}$. If $\mathbf{A}$ is the variable x, we immediately have $\mathcal{E}_{\varphi,[\mathcal{E}_\varphi(\mathbf{B})/x]}(x) = \mathcal{E}_\varphi(\mathbf{B}) = \mathcal{E}_\varphi([\mathbf{B}/x]x)$. If $\mathbf{A}$ is a variable y other than x, we have $\mathcal{E}_{\varphi,[\mathcal{E}_\varphi(\mathbf{B})/x]}(y) = \varphi(y) = \mathcal{E}_\varphi([\mathbf{B}/x]y)$. If $\mathbf{A}$ is a constant or parameter, then we can use the definition of weak evaluation functions to conclude

$$\mathcal{E}_{\varphi,[\mathcal{E}_\varphi(\mathbf{B})/x]}(\mathbf{A}) = \mathcal{E}_\varphi(\mathbf{A}) = \mathcal{E}_\varphi([\mathbf{B}/x]\mathbf{A}).$$

Suppose $\mathbf{A}$ is an application of the form $[\mathbf{F}\,\mathbf{C}]$. By induction we have

$$\begin{aligned}
&\mathcal{E}_{\varphi,[\mathcal{E}_\varphi(\mathbf{B})/x]}([\mathbf{F}\,\mathbf{C}]) = \mathcal{E}_{\varphi,[\mathcal{E}_\varphi(\mathbf{B})/x]}(\mathbf{F})@\mathcal{E}_{\varphi,[\mathcal{E}_\varphi(\mathbf{B})/x]}(\mathbf{C}) \\
&= \mathcal{E}_\varphi([\mathbf{B}/x]\mathbf{F})@\mathcal{E}_\varphi([\mathbf{B}/x]\mathbf{C}) = \mathcal{E}_\varphi([\mathbf{B}/x][\mathbf{F}\,\mathbf{C}]).
\end{aligned}$$

Suppose $\mathbf{A}$ is a λ-abstraction of the form $[\lambda y_\gamma\, \mathbf{D}_\delta]$. (By our implicit treatment of α-conversion, we can assume x and y are not the same variable.) For each $\mathsf{c} \in \mathcal{D}_\gamma$, we use induction to compute

$$\begin{aligned}
&\mathcal{E}_{\varphi,[\mathcal{E}_\varphi(\mathbf{B})/x]}([\lambda y_\gamma\, \mathbf{D}])@\mathsf{c} = \mathcal{E}_{\varphi,[\mathcal{E}_\varphi(\mathbf{B})/x],[\mathsf{c}/y]}(\mathbf{D}) \\
&= \mathcal{E}_{\varphi,[\mathsf{c}/y]}([\mathbf{B}/x]\mathbf{D}) = \mathcal{E}_\varphi([\mathbf{B}/x][\lambda y_\gamma\, \mathbf{D}])@\mathsf{c}.
\end{aligned}$$

By functionality, we have $\mathcal{E}_{\varphi,[\mathcal{E}_\varphi(\mathbf{B})/x]}([\lambda y_\gamma\, \mathbf{D}]) = \mathcal{E}_\varphi([\mathbf{B}/x][\lambda y_\gamma\, \mathbf{D}])$. □

LEMMA 3.2.9. *Let $(\mathcal{D}, @)$ be a functional applicative structure and $\mathcal{E}$ be a weak $\mathcal{S}$-evaluation function. If $\mathbf{A} \in wff_\alpha(\mathcal{S})$ β-reduces to $\mathbf{B} \in wff_\alpha(\mathcal{S})$ in one step, then $\mathcal{E}_\varphi(\mathbf{A}) = \mathcal{E}_\varphi(\mathbf{B})$.*

PROOF. The proof is by induction on the position of the β-redex in $\mathbf{A}$. For the base case where $\mathbf{A}$ is the β-redex, we calculate

$$\begin{aligned}\mathcal{E}_\varphi([[\lambda x_\beta\, \mathbf{C}]\, \mathbf{D}]) &= \mathcal{E}_\varphi([\lambda x_\beta\, \mathbf{C}])@\mathcal{E}_\varphi(\mathbf{D}) \\ &= \mathcal{E}_{\varphi,[\mathcal{E}_\varphi(\mathbf{D})/x]}(\mathbf{C}) = \mathcal{E}_\varphi([\mathbf{D}/x]\mathbf{C})\end{aligned}$$

using the definition of weak evaluations and Lemma 3.2.8. □

LEMMA 3.2.10. *Let $(\mathcal{D}, @)$ be a functional applicative structure and $\mathcal{E}$ be a weak $\mathcal{S}$-evaluation function. We have $\mathcal{E}_\varphi(\mathbf{A}) = \mathcal{E}_\varphi(\mathbf{A}^{\downarrow\beta})$ for all $\mathbf{A} \in wff_\alpha(\mathcal{S})$.*

PROOF. This follows by induction on the number of β-reductions from $\mathbf{A}$ to $\mathbf{A}^{\downarrow\beta}$ using Lemma 3.2.9. □

THEOREM 3.2.11. *Let $(\mathcal{D}, @)$ be a functional applicative structure. If $\mathcal{E}$ be a weak $\mathcal{S}$-evaluation function, then $\mathcal{E}$ is an $\mathcal{S}$-evaluation function.*

PROOF. This follows from Lemmas 3.2.6 and 3.2.10. □

Functionality can be factored into two weaker properties.

DEFINITION 3.2.12 (Weakly Functional Evaluations). Let

$$\mathcal{J} = (\mathcal{D}, @, \mathcal{E})$$

be an $\mathcal{S}$-evaluation. We say $\mathcal{J}$ is *η-functional* if

$$\mathcal{E}_\varphi(\mathbf{A}) = \mathcal{E}_\varphi(\mathbf{A}^{\downarrow})$$

for any type α, formula $\mathbf{A} \in wff_\alpha(\mathcal{S})$, and assignment φ. We say $\mathcal{J}$ is *ξ-functional* if for all types $\alpha, \beta \in \mathcal{T}$, $\mathbf{M}, \mathbf{N} \in wff_\beta(\mathcal{S})$, assignments φ, and variables x_α,

$$\mathcal{E}_\varphi(\lambda x_\alpha \mathbf{M}) = \mathcal{E}_\varphi(\lambda x_\alpha \mathbf{N})$$

whenever $\mathcal{E}_{\varphi,[\mathsf{a}/x]}(\mathbf{M}) = \mathcal{E}_{\varphi,[\mathsf{a}/x]}(\mathbf{N})$ for every $\mathsf{a} \in \mathcal{D}_\alpha$.

As proven in [19], an evaluation is functional iff it is both η-functional and ξ-functional. We reprove this result in our context in which the signature $\mathcal{S}$ may not contain the logical constants assumed in [19]. We begin with two lemmas about functional $\mathcal{S}$-evaluations.

LEMMA 3.2.13. *Let $\mathcal{J} := (\mathcal{D}, @, \mathcal{E})$ be a functional $\mathcal{S}$-evaluation. For any assignment φ into $\mathcal{J}$ and $\mathbf{F} \in \mathit{wff}_{\alpha\beta}(\mathcal{S})$ where $x_\beta \notin \mathbf{Free}(\mathbf{F})$, we have $\mathcal{E}_\varphi([\lambda x_\beta \blacksquare \mathbf{F}\, x]) = \mathcal{E}_\varphi(\mathbf{F})$.*

PROOF. Let $\mathsf{b} \in \mathcal{D}_\beta$ be given. Since $\mathcal{E}$ respects β-equality (cf. Remark 3.2.4) and $x_\beta \notin \mathbf{Free}(\mathbf{F})$, we can compute

$$\begin{aligned}\mathcal{E}_\varphi([\lambda x \blacksquare \mathbf{F}\, x])@\mathsf{b} &= \mathcal{E}_{\varphi,[\mathsf{b}/x]}([\lambda x \blacksquare \mathbf{F}\, x])@\mathsf{b} = \mathcal{E}_{\varphi,[\mathsf{b}/x]}([[\lambda x \blacksquare \mathbf{F}\, x]\, x]) \\ &= \mathcal{E}_{\varphi,[\mathsf{b}/x]}([\mathbf{F}\, x]) = \mathcal{E}_{\varphi,[\mathsf{b}/x]}(\mathbf{F})@\mathsf{b} = \mathcal{E}_\varphi(\mathbf{F})@\mathsf{b}.\end{aligned}$$

Generalizing over b, we conclude $\mathcal{E}_\varphi([\lambda x \blacksquare \mathbf{F} x]) = \mathcal{E}_\varphi(\mathbf{F})$ by functionality. □

LEMMA 3.2.14. *Let $\mathcal{J} := (\mathcal{D}, @, \mathcal{E})$ be a functional $\mathcal{S}$-evaluation. If a formula $\mathbf{A}$ η-reduces to $\mathbf{B}$ in one step, then $\mathcal{E}_\varphi(\mathbf{A}) = \mathcal{E}_\varphi(\mathbf{B})$ for any assignment φ into $\mathcal{J}$.*

PROOF. The proof is by induction on the position of the η-redex in $\mathbf{A}$. □

We can now establish that functionality is equivalent to the conjunction of η-functionality and ξ-functionality.

LEMMA 3.2.15. *An $\mathcal{J} := (\mathcal{D}, @, \mathcal{E})$ evaluation is functional iff it is both η-functional and ξ-functional.*

PROOF. Using a simple induction on the number of $\beta\eta$-reduction steps using Lemma 3.2.14 and Remark 3.2.4, we conclude that functionality implies η-functionality.

We next prove functionality implies ξ-functionality. Assume $\mathcal{J}$ is functional. Let two terms $\mathbf{M}, \mathbf{N} \in \mathit{wff}_\beta(\mathcal{S})$, an assignment φ, and a variable x_α be given. Assume $\mathcal{E}_{\varphi,[\mathsf{a}/x]}(\mathbf{M}) = \mathcal{E}_{\varphi,[\mathsf{a}/x]}(\mathbf{N})$ for every $\mathsf{a} \in \mathcal{D}_\alpha$. For each $\mathsf{a} \in \mathcal{D}_\alpha$, we compute

$$\mathcal{E}_\varphi(\lambda x_\alpha \mathbf{M})@\mathsf{a} = \mathcal{E}_{\varphi,[\mathsf{a}/x]}(\mathbf{M}) = \mathcal{E}_{\varphi,[\mathsf{a}/x]}(\mathbf{N}) = \mathcal{E}_\varphi(\lambda x_\alpha \mathbf{N})@\mathsf{a}.$$

By functionality, we have $\mathcal{E}_\varphi(\lambda x_\alpha \mathbf{M}) = \mathcal{E}_\varphi(\lambda x_\alpha \mathbf{N})$ as desired.

Finally, assume $\mathcal{J}$ is both η-functional and ξ-functional. Let $\mathsf{f}, \mathsf{g} \in \mathcal{D}_{\alpha\beta}$ be given. Assume $\mathsf{f}@\mathsf{b} = \mathsf{g}@\mathsf{b}$ for all $\mathsf{b} \in \mathcal{D}_\beta$. Choose distinct variables $f_{\alpha\beta}$, $g_{\alpha\beta}$ and x_β. Let φ be an assignment with $\varphi(f) = \mathsf{f}$ and $\varphi(g) = \mathsf{g}$. For any $\mathsf{b} \in \mathcal{D}_\beta$, we have

$$\mathcal{E}_{\varphi,[\mathsf{b}/x]}(f\,x) = \mathsf{f}@\mathsf{b} = \mathsf{g}@\mathsf{b} = \mathcal{E}_{\varphi,[\mathsf{b}/x]}(g\,x)$$

and so

$$\mathcal{E}_\varphi(\lambda x\, \blacksquare\, f\,x) = \mathcal{E}_\varphi(\lambda x\, \blacksquare\, g\,x)$$

by ξ-functionality. Combining this with η-functionality, we compute

$$\mathsf{f} = \mathcal{E}_\varphi(f) = \mathcal{E}_\varphi(\lambda x\, \blacksquare\, f\,x) = \mathcal{E}_\varphi(\lambda x\, \blacksquare\, g\,x) = \mathcal{E}_\varphi(g) = \mathsf{g}$$

as desired. □

The notion of ξ-functionality will only play an incidental role in this book.

Sometimes we may want to modify an evaluation by changing the value of a parameter. We now prove changing the value of a parameter makes sense.

DEFINITION 3.2.16 (Evaluation Change of Parameter). Let $\mathcal{J}$ be an $\mathcal{S}$-evaluation $(\mathcal{D}, @, \mathcal{E})$, W_α be a parameter and $\mathsf{a} \in \mathcal{D}_\alpha$. For each term $\mathbf{B} \in \mathit{wff}_\beta(\mathcal{S})$ choose a variable $x^{\mathbf{B}} \notin \mathbf{Free}(\mathbf{B})$ of type α. For every assignment φ, define $\mathcal{E}_\varphi^{W \mapsto \mathsf{a}}(\mathbf{B}) := \mathcal{E}_{\varphi,[\mathsf{a}/x^{\mathbf{B}}]}([x^{\mathbf{B}}/W]\mathbf{B})$.

The next lemma verifies that the definition of $\mathcal{E}^{W \mapsto \mathsf{a}}$ does not depend on the choice of variables $x^{\mathbf{B}} \notin \mathbf{Free}(\mathbf{B})$.

LEMMA 3.2.17. *Let $\mathcal{J} := (\mathcal{D}, @, \mathcal{E})$ be an $\mathcal{S}$-evaluation, W_α be a parameter or constant and $\mathsf{a} \in \mathcal{D}_\alpha$. If $\mathbf{B} \in \mathit{wff}_\beta(\mathcal{S})$, $x_\alpha \notin \mathbf{Free}(\mathbf{B})$ and $y_\alpha \notin \mathbf{Free}(\mathbf{B})$, then $\mathcal{E}_{\varphi,[\mathsf{a}/x]}([x/W]\mathbf{B}) = \mathcal{E}_{\varphi,[\mathsf{a}/y]}([y/W]\mathbf{B})$.*

PROOF. This follows from the Substitution-Value Lemma. □

We now verify $\mathcal{E}^{W \mapsto \mathsf{a}}$ is an evaluation function.

THEOREM 3.2.18 (Evaluation Change of Parameter). *Let*

$$\mathcal{J} := (\mathcal{D}, @, \mathcal{E})$$

be an $\mathcal{S}$-evaluation, W_α be a parameter and $\mathsf{a} \in \mathcal{D}_\alpha$. The function $\mathcal{E}^{W \mapsto \mathsf{a}}$ (cf. Definition 3.2.16) is an $\mathcal{S}$-evaluation function. If $\mathcal{J}$

is η-functional [ξ-functional], then $(\mathcal{D}, @, \mathcal{E}^{W\mapsto \mathsf{a}})$ is η-functional [ξ-functional]. Furthermore,

$$\mathcal{E}^{W\mapsto \mathsf{a}}(W) = \mathsf{a} \ \ \textit{and}\ \ \mathcal{E}_{\varphi}^{W\mapsto \mathsf{a}}(\mathbf{B}) = \mathcal{E}_{\varphi}(\mathbf{B})$$

for every assignment φ and $\mathbf{B} \in \mathit{wff}_{\beta}(\mathcal{S})$ with $W \notin \mathbf{Params}(\mathbf{B})$.

PROOF. The proof simply involves checking the conditions. See Appendix A.1. □

We will also have occasion to extend $\mathcal{S}_1$-evaluations to be $\mathcal{S}_2$-evaluations when $\mathcal{S}_1 \subseteq \mathcal{S}_2$. We now demonstrate that such extensions exist. The technique is similar to the one used to change the value of a parameter.

DEFINITION 3.2.19 (Evaluation Extension). Let $\mathcal{S}_1$ and $\mathcal{S}_2$ be signatures with $\mathcal{S}_1 \subseteq \mathcal{S}_2$, $\mathcal{A} = (\mathcal{D}, @)$ be an applicative structure, $\mathcal{E}$ be an $\mathcal{S}_1$-evaluation function for $\mathcal{A}$, and $\mathcal{I}\colon (\mathcal{S}_2 \setminus \mathcal{S}_1) \to \mathcal{D}$ be a typed function. We call an $\mathcal{S}_2$-evaluation function $\mathcal{E}^2$ for $\mathcal{A}$ an *evaluation extension of $\mathcal{E}$ by $\mathcal{I}$* if

1. $\mathcal{E}^2\big|_{\mathit{wff}(\mathcal{S}_1)} = \mathcal{E}$,
2. $\mathcal{E}^2\big|_{\mathcal{S}_2\setminus\mathcal{S}_1} = \mathcal{I}$, and
3. if $\mathcal{E}$ is η-functional, then $\mathcal{E}^2$ is η-functional.

We can prove that evaluation extensions exist. The details are in Appendix A.1.

LEMMA 3.2.20. *Let $\mathcal{J} := (\mathcal{D}, @, \mathcal{E})$ be an $\mathcal{S}$-evaluation, $c_\alpha \notin \mathcal{S}$ be a logical constant and $\mathsf{a} \in \mathcal{D}_\alpha$. There is an evaluation extension $\mathcal{E}'$ of $\mathcal{E}$ by $\mathcal{I} : \{c\} \to \mathcal{D}$ where $\mathcal{I}(c) := \mathsf{a}$.*

PROOF. See Appendix A.1. □

THEOREM 3.2.21 (Evaluation Extension). *Let $\mathcal{S}_1$ and $\mathcal{S}_2$ be signatures with $\mathcal{S}_1 \subseteq \mathcal{S}_2$, $\mathcal{A} = (\mathcal{D}, @)$ be an applicative structure, $\mathcal{E}$ be an $\mathcal{S}_1$-evaluation function for $\mathcal{A}$, and $\mathcal{I}\colon (\mathcal{S}_2 \setminus \mathcal{S}_1) \to \mathcal{D}$ be a typed function. There exists an evaluation extension $\mathcal{E}'$ of $\mathcal{E}$ by $\mathcal{I}$.*

PROOF. See Appendix A.1. □

We can also restrict an evaluation from a signature to a smaller signature.

NOTATION *Suppose $\mathcal{A} = (\mathcal{D}, @)$ is an applicative structure, $\mathcal{E}$ is an $\mathcal{S}$-evaluation function and $\mathcal{W} \subseteq \mathit{wff}(\mathcal{S})$. We use the notation $\mathcal{E}\big|_{\mathcal{W}}$*

to refer to the collection of typed functions $\mathcal{E}_\varphi|_{\mathcal{W}}\colon \mathcal{W} \to \mathcal{D}$ *such that* $\mathcal{E}_\varphi|_{\mathcal{W}}(\mathbf{A}) = \mathcal{E}_\varphi(\mathbf{A})$ *for every* $\mathbf{A}_\alpha \in \mathcal{W}_\alpha$ *and each assignment* φ *into* $\mathcal{A}$.

LEMMA 3.2.22. *Let* $\mathcal{S}_1$ *and* $\mathcal{S}_2$ *be signatures with* $\mathcal{S}_1 \subseteq \mathcal{S}_2$. *Let* $\mathcal{A} = (\mathcal{D}, @)$ *be an applicative structure and* $\mathcal{E}$ *be an* $\mathcal{S}_2$*-evaluation function for* $\mathcal{A}$. *Then* $\mathcal{E}|_{\mathit{wff}(\mathcal{S}_1)}$ *is an* $\mathcal{S}_1$*-evaluation function for* $\mathcal{A}$. *We also have the following:*

1. $(\mathcal{D}, @, \mathcal{E}|_{\mathit{wff}(\mathcal{S}_1)})$ *is* η*-functional iff* $(\mathcal{D}, @, \mathcal{E})$ *is* η*-functional.*
2. $(\mathcal{D}, @, \mathcal{E}|_{\mathit{wff}(\mathcal{S}_1)})$ *is* ξ*-functional iff* $(\mathcal{D}, @, \mathcal{E})$ *is* ξ*-functional.*

PROOF. All the properties which make $\mathcal{E}$ an evaluation function are still true when one restricts to $\mathit{wff}(\mathcal{S}_1) \subseteq \mathit{wff}(\mathcal{S}_2)$. For example, if $\mathbf{A} \in \mathit{wff}(\mathcal{S}_1)$, then

$$\mathcal{E}_\varphi|_{\mathit{wff}(\mathcal{S}_1)}(\mathbf{A}) = \mathcal{E}_\varphi(\mathbf{A}) = \mathcal{E}_\varphi(\mathbf{A}^{\downarrow\beta}) = \mathcal{E}_\varphi|_{\mathit{wff}(\mathcal{S}_1)}(\mathbf{A}^{\downarrow\beta}).$$

Suppose $(\mathcal{D}, @, \mathcal{E})$ is η-functional and $\mathbf{A} \in \mathit{wff}_\alpha(\mathcal{S}_1)$. Then

$$\mathcal{E}_\varphi|_{\mathit{wff}(\mathcal{S}_1)}(\mathbf{A}) = \mathcal{E}_\varphi(\mathbf{A}) = \mathcal{E}_\varphi(\mathbf{A}^{\downarrow}) = \mathcal{E}_\varphi|_{\mathit{wff}(\mathcal{S}_1)}(\mathbf{A}^{\downarrow}).$$

Thus $(\mathcal{D}, @, \mathcal{E}|_{\mathit{wff}(\mathcal{S}_1)})$ is η-functional.

Suppose $(\mathcal{D}, @, \mathcal{E}|_{\mathit{wff}(\mathcal{S}_1)})$ is η-functional and $\mathbf{A} \in \mathit{wff}_\alpha(\mathcal{S}_2)$. Since $\mathbf{Consts}(\mathbf{A})$ is finite, there are constants $c^1_{\beta^1}, \ldots, c^n_{\beta^n} \in \mathcal{S}_2$ such that

$$\{c^1, \ldots, c^n\} = (\mathbf{Consts}(\mathbf{A}) \setminus \mathcal{S}_1).$$

Let $x^1_{\beta^1}, \ldots, x^n_{\beta^n}$ be distinct variables not in $\mathbf{Free}(\mathbf{A})$. Define substitutions $\theta : \mathbf{Free}(\mathbf{A}) \to \mathit{wff}(\mathcal{S}_2)$ and $\sigma : \{c^1, \ldots, c^n\} \to \mathit{wff}(\mathcal{S}_1)$ by $\theta(x^i) := c^i$ and $\sigma(c^i) := x^i$ for each i $(1 \leq i \leq n)$. One can easily verify $\sigma(\mathbf{A}) \in \mathit{wff}(\mathcal{S}_1)$, $\theta(\sigma(\mathbf{A})) = \mathbf{A}$ and $\theta(\sigma(\mathbf{A})^{\downarrow}) = \mathbf{A}^{\downarrow}$. Let φ be an assignment into $\mathcal{D}$. By η-functionality of $\mathcal{E}|_{\mathit{wff}(\mathcal{S}_1)}$, we have

$$\mathcal{E}_\varphi|_{\mathit{wff}(\mathcal{S}_1)}(\sigma(\mathbf{A})) = \mathcal{E}_\varphi|_{\mathit{wff}(\mathcal{S}_1)}(\sigma(\mathbf{A})^{\downarrow}).$$

Applying the Substitution-Value Lemma n times, we have

$$\mathcal{E}_\varphi(\mathbf{A}) = \mathcal{E}_\varphi(\theta(\sigma(\mathbf{A}))) = \mathcal{E}_{\varphi,[\mathcal{E}(c^n)/x^n],\ldots,[\mathcal{E}(c^1)/x^1]}(\sigma(\mathbf{A}))$$

$$= \mathcal{E}_{\varphi,[\mathcal{E}(c^n)/x^n],\ldots,[\mathcal{E}(c^1)/x^1]}(\sigma(\mathbf{A})^{\downarrow}) = \mathcal{E}_\varphi(\theta(\sigma(\mathbf{A})^{\downarrow})) = \mathcal{E}_\varphi(\mathbf{A}^{\downarrow}).$$

Thus $(\mathcal{D}, @, \mathcal{E})$ is η-functional.

A similar argument verifies the result for ξ-functionality. □

Recall $cwff^{\downarrow\beta}$ $[cwff^{\downarrow}]$ is the set of closed β-normal [$\beta\eta$-normal] well-formed formulas. We can use this family of terms to define an evaluation.

DEFINITION 3.2.23 (Term Evaluations). Let $\mathbf{F}_{\alpha\beta}@^{\beta}\mathbf{B}_{\beta}$ be $[\mathbf{F}\,\mathbf{B}]^{\downarrow\beta}$ for each $\mathbf{F}_{\alpha\beta} \in cwff^{\downarrow\beta}$ and $\mathbf{B}_{\beta} \in cwff^{\downarrow\beta}$. For the definition of an evaluation function, let φ be an assignment into $cwff^{\downarrow\beta}$. Note that $\sigma := \varphi\big|_{\mathbf{Free}(\mathbf{A})}$ is a substitution, since $\mathbf{Free}(\mathbf{A})$ is finite. Thus we can choose $\mathcal{E}^{\beta}_{\varphi}(\mathbf{A}) := \sigma(\mathbf{A})^{\downarrow\beta}$. We call $\mathcal{TS}(\mathcal{S})^{\beta} := (cwff^{\downarrow\beta}, @^{\beta}, \mathcal{E}^{\beta})$ the *β-term evaluation.*

Analogously, we can define $\mathcal{TS}(\mathcal{S})^{\beta\eta} := (cwff^{\downarrow}, @^{\beta\eta}, \mathcal{E}^{\beta\eta})$ to be the *$\beta\eta$-term evaluation.*

The name *term evaluation* in the previous definition is justified by the following lemma.

LEMMA 3.2.24. *$\mathcal{TS}_{\mathcal{S}}(\mathcal{S})^{\beta}$ and $\mathcal{TS}_{\mathcal{S}}(\mathcal{S})^{\beta\eta}$ are $\mathcal{S}$-evaluations. Furthermore, $\mathcal{TS}_{\mathcal{S}}(\mathcal{S})^{\beta\eta}$ is η-functional.*

PROOF. The fact that $(cwff^{\downarrow\beta}, @^{\beta})$ is an applicative structure is immediate. For each type α, ${cwff_{\alpha}}^{\downarrow\beta}$ is nonempty (since $\mathcal{P}_{\alpha}$ is nonempty) and $@^{\beta} : {cwff_{\alpha\beta}}^{\downarrow\beta} \times {cwff_{\beta}}^{\downarrow\beta} \to {cwff_{\alpha}}^{\downarrow\beta}$. We next check that $\mathcal{E}^{\beta}$ is an evaluation function.

1. $\mathcal{E}^{\beta}_{\varphi}(x) = \varphi\big|_{\mathbf{Free}(x)}(x) = \varphi(x)$.
2. $\mathcal{E}^{\beta}_{\varphi}$ respects application since
$$\sigma(\mathbf{F}\,\mathbf{A})^{\downarrow\beta} = (\sigma(\mathbf{F})^{\downarrow\beta}\sigma(\mathbf{A})^{\downarrow\beta})^{\downarrow\beta}$$
where $\sigma := \varphi\big|_{\mathbf{Free}(\mathbf{F}\,\mathbf{A})}$.
3. Suppose φ and ψ coincide on $\mathbf{Free}(\mathbf{A})$. That is, $\varphi\big|_{\mathbf{Free}(\mathbf{A})}$ and $\psi\big|_{\mathbf{Free}(\mathbf{A})}$ are the same substitution σ. Thus
$$\mathcal{E}^{\beta}_{\varphi}(\mathbf{A}) = \sigma(\mathbf{A})^{\downarrow\beta} = \mathcal{E}^{\beta}_{\psi}(\mathbf{A}).$$
4. Since $\mathbf{Free}(\mathbf{A}^{\downarrow\beta}) \subseteq \mathbf{Free}(\mathbf{A})$, we know
$$\mathcal{E}^{\beta}_{\varphi}(\mathbf{A}) = \sigma(\mathbf{A})^{\downarrow\beta} = \sigma(\mathbf{A}^{\downarrow\beta})^{\downarrow\beta} = \mathcal{E}^{\beta}_{\varphi}(\mathbf{A}^{\downarrow\beta})$$
where $\sigma := \varphi\big|_{\mathbf{Free}(\mathbf{A})}$.

By a similar argument, $\mathcal{TS}(\mathcal{S})^{\beta\eta}$ is an η-functional $\mathcal{S}$-evaluation. □

prop.	where	holds when			for all
$\mathfrak{L}_{\top}(\mathsf{a})$	$\mathsf{a} \in \mathcal{D}_o$	$\upsilon(\mathsf{a}) = \mathtt{T}$			
$\mathfrak{L}_{\bot}(\mathsf{b})$	$\mathsf{b} \in \mathcal{D}_o$	$\upsilon(\mathsf{b}) = \mathtt{F}$			
$\mathfrak{L}_{\neg}(\mathsf{n})$	$\mathsf{n} \in \mathcal{D}_{oo}$	$\upsilon(\mathsf{n}@\mathsf{a}) = \mathtt{T}$	iff	$\upsilon(\mathsf{a}) = \mathtt{F}$	$\mathsf{a} \in \mathcal{D}_o$
$\mathfrak{L}_{\vee}(\mathsf{d})$	$\mathsf{d} \in \mathcal{D}_{ooo}$	$\upsilon(\mathsf{d}@\mathsf{a}@\mathsf{b}) = \mathtt{T}$	iff	$\upsilon(\mathsf{a}) = \mathtt{T}$ or $\upsilon(\mathsf{b}) = \mathtt{T}$	$\mathsf{a}, \mathsf{b} \in \mathcal{D}_o$
$\mathfrak{L}_{\wedge}(\mathsf{c})$	$\mathsf{c} \in \mathcal{D}_{ooo}$	$\upsilon(\mathsf{c}@\mathsf{a}@\mathsf{b}) = \mathtt{T}$	iff	$\upsilon(\mathsf{a}) = \mathtt{T}$ and $\upsilon(\mathsf{b}) = \mathtt{T}$	$\mathsf{a}, \mathsf{b} \in \mathcal{D}_o$
$\mathfrak{L}_{\supset}(\mathsf{i})$	$\mathsf{i} \in \mathcal{D}_{ooo}$	$\upsilon(\mathsf{i}@\mathsf{a}@\mathsf{b}) = \mathtt{T}$	iff	$\upsilon(\mathsf{a}) = \mathtt{F}$ or $\upsilon(\mathsf{b}) = \mathtt{T}$	$\mathsf{a}, \mathsf{b} \in \mathcal{D}_o$
$\mathfrak{L}_{\equiv}(\mathsf{e})$	$\mathsf{e} \in \mathcal{D}_{ooo}$	$\upsilon(\mathsf{e}@\mathsf{a}@\mathsf{b}) = \mathtt{T}$	iff	$\upsilon(\mathsf{a}) = \upsilon(\mathsf{b})$	$\mathsf{a}, \mathsf{b} \in \mathcal{D}_o$
$\mathfrak{L}_{\Pi^\alpha}(\pi)$	$\pi \in \mathcal{D}_{o(o\alpha)}$	$\upsilon(\pi@\mathsf{f}) = \mathtt{T}$	iff	$\forall \mathsf{a} \in \mathcal{D}_\alpha\ \upsilon(\mathsf{f}@\mathsf{a}) = \mathtt{T}$	$\mathsf{f} \in \mathcal{D}_{o\alpha}$
$\mathfrak{L}_{\Sigma^\alpha}(\sigma)$	$\sigma \in \mathcal{D}_{o(o\alpha)}$	$\upsilon(\sigma@\mathsf{f}) = \mathtt{T}$	iff	$\exists \mathsf{a} \in \mathcal{D}_\alpha\ \upsilon(\mathsf{f}@\mathsf{a}) = \mathtt{T}$	$\mathsf{f} \in \mathcal{D}_{o\alpha}$
$\mathfrak{L}_{=^\alpha}(\mathsf{q})$	$\mathsf{q} \in \mathcal{D}_{o\alpha\alpha}$	$\upsilon(\mathsf{q}@\mathsf{a}@\mathsf{b}) = \mathtt{T}$	iff	$\mathsf{a} = \mathsf{b}$	$\mathsf{a}, \mathsf{b} \in \mathcal{D}_\alpha$

TABLE 3.1. Logical Properties of $\upsilon : \mathcal{D}_o \to \{\mathtt{T}, \mathtt{F}\}$

DEFINITION 3.2.25 (Homomorphism on Evaluations). A *homomorphism* κ from an evaluation $\mathcal{J}^1 = (\mathcal{D}^1, @^1, \mathcal{E}^1)$ to an evaluation $\mathcal{J}^2 = (\mathcal{D}^2, @^2, \mathcal{E}^2)$ is a typed function $\kappa\colon \mathcal{D}^1 \to \mathcal{D}^2$ such that κ is a homomorphism from the applicative structure $(\mathcal{D}^1, @^1)$ to the applicative structure $(\mathcal{D}^2, @^2)$ and $\kappa(\mathcal{E}^1_\varphi(\mathbf{A})) = \mathcal{E}^2_{\kappa\varphi}(\mathbf{A})$ for every $\mathbf{A} \in \mathit{wff}_\alpha(\mathcal{S})$ and assignment φ into $\mathcal{J}^1$. Here, $\kappa\varphi$ is the assignment into $\mathcal{J}^2$ defined by $\kappa\varphi(x_\beta) := \kappa(\varphi(x_\beta))$ for every type β and variable x_β, i.e., the composition of κ and φ.

3.3. Models

The semantic notions so far make sense independent of the set of base types. We now obtain a notion of models by requiring specialized behavior on the type o of truth values. We use the notion of a valuation, which intuitively gives a truth-value interpretation to the domain $\mathcal{D}_o$ of an evaluation which is consistent with the intuitive interpretations of the logical constants. Our most general notion of model will be evaluations that have valuations.

DEFINITION 3.3.1. Fix two values $\mathtt{T} \neq \mathtt{F}$. Let $\mathcal{A} := (\mathcal{D}, @)$ be an applicative structure and $\upsilon\colon \mathcal{D}_o \to \{\mathtt{T}, \mathtt{F}\}$ be a function. For each logical constant c_α and element $\mathsf{a} \in \mathcal{D}_\alpha$, we define the properties $\mathfrak{L}_c(\mathsf{a})$ with respect to υ in Table 3.1.

It is important to note that the properties in Table 3.1 are defined relative to an applicative structure $(\mathcal{D}, @)$ and a function υ from $\mathcal{D}_o$ to $\{\mathtt{T}, \mathtt{F}\}$. In particular, the properties in no way depend upon an evaluation function. We define the condition for υ to be a valuation and

$$\mathcal{M} = (\mathcal{D}, @, \mathcal{E}, \upsilon)$$

to be an $\mathcal{S}$-model. The intuition is that the internal logical constants in $\mathcal{S}$ must be interpreted properly.

DEFINITION 3.3.2 ($\mathcal{S}$-Model). Let $\mathcal{J} := (\mathcal{D}, @, \mathcal{E})$ be an evaluation, then a function $v\colon \mathcal{D}_o \to \{\mathtt{T}, \mathtt{F}\}$ is called an *$\mathcal{S}$-valuation* for $\mathcal{J}$ if for every logical constant $c \in \mathcal{S}$, $\mathfrak{L}_c(\mathcal{E}(c))$ holds with respect to v. Furthermore, $\mathcal{M} := (\mathcal{D}, @, \mathcal{E}, v)$ is called an *$\mathcal{S}$-model.*

We say φ is an assignment into $\mathcal{M}$ if it is an assignment into the underlying applicative structure $(\mathcal{D}, @)$.

A model $\mathcal{M} := (\mathcal{D}, @, \mathcal{E}, v)$ is called *functional* [*full*, *standard*] if the applicative structure $(\mathcal{D}, @)$ is functional [full, standard]. We say $\mathcal{M}$ is a *model over a frame* if $(\mathcal{D}, @)$ is a frame structure. We say $\mathcal{M}$ is *η-functional* [*ξ-functional*] if the evaluation $(\mathcal{D}, @, \mathcal{E})$ is η-functional [*ξ-functional*].

DEFINITION 3.3.3. Let $\mathcal{S}$ be a signature and $\mathcal{M} = (\mathcal{D}, @, \mathcal{E}, v)$ be an $\mathcal{S}$-model. We define when a proposition $\mathbf{M}$ is *satisfied* by an assignment φ in $\mathcal{M}$ (written $\mathcal{M} \models_\varphi \mathbf{M}$) by induction on propositions.

- For $\mathbf{A} \in \mathit{wff}_o(\mathcal{S})$, $\mathcal{M} \models_\varphi \mathbf{A}$ if $v(\mathcal{E}_\varphi(\mathbf{A})) = \mathtt{T}$.
- For $\mathbf{A}, \mathbf{B} \in \mathit{wff}_\alpha(\mathcal{S})$, $\mathcal{M} \models_\varphi [\mathbf{A} \mathrel{\dot{=}}^\alpha \mathbf{B}]$ if $\mathcal{E}_\varphi(\mathbf{A}) = \mathcal{E}_\varphi(\mathbf{B})$.
- $\mathcal{M} \models_\varphi \dot{\top}$.
- $\mathcal{M} \models_\varphi \dot{\neg}\mathbf{M}$ if $\mathcal{M} \not\models_\varphi \mathbf{M}$.
- $\mathcal{M} \models_\varphi [\mathbf{M} \mathrel{\dot{\vee}} \mathbf{N}]$ if $\mathcal{M} \models_\varphi \mathbf{M}$ or $\mathcal{M} \models_\varphi \mathbf{N}$.
- $\mathcal{M} \models_\varphi [\dot{\forall} x_\alpha \mathbf{M}]$ if $\mathcal{M} \models_{\varphi, [\mathsf{a}/x]} \mathbf{M}$ for every $\mathsf{a} \in \mathcal{D}_\alpha$.

We say a proposition $\mathbf{M}$ is *valid* in $\mathcal{M}$ if $\mathcal{M} \models_\varphi \mathbf{M}$ for all assignments φ. When $\mathbf{M}$ is valid in $\mathcal{M}$, we use the notation $\mathcal{M} \models \mathbf{M}$ and say $\mathcal{M}$ is a *model* of $\mathbf{M}$. Finally, we use the notation $\mathcal{M} \models_\varphi \Phi$ for a set of propositions $\Phi \subseteq \mathit{prop}(\mathcal{S})$ if $\mathcal{M} \models_\varphi \mathbf{A}$ for all $\mathbf{A} \in \Phi$, and $\mathcal{M} \models \Phi$ if $\mathcal{M} \models_\varphi \Phi$ for every assignment φ. When $\mathcal{M} \models \Phi$, we say $\mathcal{M}$ is a *model* of Φ.

Since we have introduced abbreviations for other (external) logical constructors on propositions, it is worthwhile to note these satisfy the appropriate properties.

LEMMA 3.3.4. *Let $\mathcal{S}$ be a signature, $\mathcal{M} = (\mathcal{D}, @, \mathcal{E}, v)$ be an $\mathcal{S}$-model and φ be an assignment in $\mathcal{M}$.*

$\underset{\cdot}{\bot}$: $\mathcal{M} \not\models_\varphi \underset{\cdot}{\bot}$.

$\underset{\cdot}{\wedge}$: $\mathcal{M} \models_\varphi [\mathbf{M} \underset{\cdot}{\wedge} \mathbf{N}]$ *iff both* $\mathcal{M} \models_\varphi \mathbf{M}$ *and* $\mathcal{M} \models_\varphi \mathbf{N}$.

$\underset{\cdot}{\supset}$: $\mathcal{M} \models_\varphi [\mathbf{M} \underset{\cdot}{\supset} \mathbf{N}]$ *iff either* $\mathcal{M} \not\models_\varphi \mathbf{M}$ *or* $\mathcal{M} \models_\varphi \mathbf{N}$.

$\underset{\cdot}{\equiv}$: $\mathcal{M} \models_\varphi [\mathbf{M} \underset{\cdot}{\equiv} \mathbf{N}]$ *iff either*

$$\mathcal{M} \models_\varphi \mathbf{M} \text{ and } \mathcal{M} \models_\varphi \mathbf{N},$$

or

$$\mathcal{M} \not\models_\varphi \mathbf{M} \text{ and } \mathcal{M} \not\models_\varphi \mathbf{N}.$$

$\underset{\cdot}{\exists}$: $\mathcal{M} \models_\varphi [\underset{\cdot}{\exists} x_\alpha \mathbf{M}]$ *iff there exists some* $\mathsf{a} \in \mathcal{D}_\alpha$ *such that* $\mathcal{M} \models_{\varphi,[\mathsf{a}/x]} \mathbf{M}$.

PROOF. Each case is easily verified. □

We now turn to semantic properties describing different forms of extensionality. For functional extensionality, we have already defined when a model is functional, η-functional or ξ-functional. We can characterize Boolean extensionality by requiring the valuation to be injective.

DEFINITION 3.3.5 (Properties η, ξ, $\mathfrak{f}$ and $\mathfrak{b}$). Let $\mathcal{M} = (\mathcal{D}, @, \mathcal{E}, \upsilon)$ be a model. We say that $\mathcal{M}$ has *property*

η: if $\mathcal{M}$ is η-functional.
ξ: if $\mathcal{M}$ is ξ-functional.
$\mathfrak{f}$: if $\mathcal{M}$ is functional.
$\mathfrak{b}$: if υ is injective.

In [19], we defined property $\mathfrak{b}$ to hold when $\mathcal{D}_o$ has two elements. However, without assuming enough logical constants are in the signature, $\mathcal{D}_o$ may only contain one element (cf. Examples 3.6.23 and 3.6.24). In such cases, υ is clearly injective. On the other hand, in such signatures one may have a model where $\mathcal{D}_o$ has two elements but υ is not injective since both members of $\mathcal{D}_o$ may map to `T` (or `F`). Under very weak assumptions on the signature, we will know that υ is surjective. In such a case, property $\mathfrak{b}$ (as defined above) is equivalent to $\mathcal{D}_o$ having two elements.

LEMMA 3.3.6. *Suppose $\mathcal{S}$ is a signature and $\mathcal{M} = (\mathcal{D}, @, \mathcal{E}, v)$ is an $\mathcal{S}$-model. If either $\top, \bot \in \mathcal{S}$ or $\neg \in \mathcal{S}$, then v is surjective.*

PROOF. Suppose $\top, \bot \in \mathcal{S}$. By $\mathfrak{L}_\top(\mathcal{E}(\top))$ and $\mathfrak{L}_\bot(\mathcal{E}(\bot))$, we have $v(\mathcal{E}(\top)) = \mathtt{T}$ and $v(\mathcal{E}(\bot)) = \mathtt{F}$. Thus v is surjective.

Suppose $\neg \in \mathcal{S}$. Choose any $\mathsf{a} \in \mathcal{D}_o$. We know $v(\mathcal{E}(\neg)@\mathsf{a}) \neq v(\mathsf{a})$ by $\mathfrak{L}_\neg(\mathcal{E}(\neg))$. Thus v is surjective. □

THEOREM 3.3.7 (Property $\mathfrak{b}$). *Let $\mathcal{S}$ be a signature and*

$$\mathcal{M} = (\mathcal{D}, @, \mathcal{E}, v)$$

be an $\mathcal{S}$-model. Suppose either $\top, \bot \in \mathcal{S}$ or $\neg \in \mathcal{S}$. Then $\mathcal{M}$ satisfies $\mathfrak{b}$ iff $\mathcal{D}_o$ has two elements.

PROOF. By Lemma 3.3.6, we know v is surjective. Thus $\mathcal{M}$ satisfies property $\mathfrak{b}$ iff v is bijective iff $\mathcal{D}_o$ has two elements. □

DEFINITION 3.3.8 (Extensional Models). An *extensional $\mathcal{S}$-model* is an $\mathcal{S}$-model satisfying properties $\mathfrak{f}$ and $\mathfrak{b}$. A *Henkin $\mathcal{S}$-model* is an extensional $\mathcal{S}$-model over a frame with $\mathcal{D}_o = \{\mathtt{T}, \mathtt{F}\}$ and v the identity function.

We now define classes of models depending on which extensionality properties are satisfied.

DEFINITION 3.3.9 (Model Classes). For each $*$ in the set

$$\{\beta, \beta\eta, \beta\xi, \beta\mathfrak{f}, \beta\mathfrak{b}, \beta\eta\mathfrak{b}, \beta\xi\mathfrak{b}, \beta\mathfrak{f}\mathfrak{b}\}$$

we define the model class $\mathfrak{M}_*(\mathcal{S})$ to be the class of all $\mathcal{S}$-models satisfying the extensionality properties listed in $*$.

In this book, we will concentrate on the model classes $\mathfrak{M}_\beta(\mathcal{S})$, $\mathfrak{M}_{\beta\eta}(\mathcal{S})$, and $\mathfrak{M}_{\beta\mathfrak{f}\mathfrak{b}}(\mathcal{S})$. Note that these model classes are defined as follows:

1. $\mathfrak{M}_\beta(\mathcal{S})$ is the class of all $\mathcal{S}$-models.
2. $\mathfrak{M}_{\beta\eta}(\mathcal{S})$ is the class of all $\mathcal{S}$-models which satisfy property η.
3. $\mathfrak{M}_{\beta\mathfrak{f}\mathfrak{b}}(\mathcal{S})$ is the class of all extensional $\mathcal{S}$-models. That is, $\mathfrak{M}_{\beta\mathfrak{f}\mathfrak{b}}(\mathcal{S})$ is the class of all $\mathcal{S}$-models which satisfy properties $\mathfrak{f}$ and $\mathfrak{b}$.

We will establish soundness (cf. Theorems 3.4.6 and 3.4.9) and completeness (cf. Theorem 5.6.6) of the sequent calculi $\mathcal{G}^{\mathcal{S}}_\beta$ and $\mathcal{G}^{\mathcal{S}}_{\beta\eta}$ with respect to $\mathfrak{M}_\beta(\mathcal{S})$ and $\mathfrak{M}_{\beta\eta}(\mathcal{S})$, respectively. By Lemma 3.2.15 we know $\mathfrak{M}_{\beta\mathfrak{fb}}(\mathcal{S}) \subseteq \mathfrak{M}_{\beta\eta}(\mathcal{S}) \subseteq \mathfrak{M}_{\beta}(\mathcal{S})$. We will establish soundness (cf. Theorem 3.4.14) and completeness (cf. Theorem 5.7.19) of the sequent calculus $\mathcal{G}^{\mathcal{S}}_{\beta\mathfrak{fb}}$ with respect to $\mathfrak{M}_{\beta\mathfrak{fb}}(\mathcal{S})$ (extensional $\mathcal{S}$-models).

REMARK 3.3.10 (Henkin Models). The definition of Henkin $\mathcal{S}_{Ch}$-models corresponds to the original definition of Henkin models. Andrews noted in [3], a model over a frame may not satisfy the sentence

$$\underset{\cdot}{\forall} f_{\iota\iota}\, \underset{\cdot}{\forall} g_{\iota\iota} \,\blacksquare\, \underset{\cdot}{\forall} x_{\iota}\, [f\, x \doteq^{\iota} g\, x] \supset f \doteq^{\iota\iota} g.$$

Consequently, a Henkin $\mathcal{S}_{Ch}$-model may not satisfy this sentence. Further consequences of the existence of such a model are discussed in Remark 6.6.3. Andrews defined a "general model" to be a Henkin model which satisfies the condition

(a_0): For each type α, $\mathcal{D}_{o\alpha\alpha}$ contains the identity relation $\mathsf{q}_{o\alpha\alpha}$ on $\mathcal{D}_{\alpha}$.

Such a "general model" corresponds to our Henkin $\mathcal{S}^{=}_{Ch}$-model. (That is, a model where $=^{\alpha}$ is in the signature.)

LEMMA 3.3.11. *Let $\mathcal{M} = (\mathcal{D}, @, \mathcal{E}, \upsilon)$ be an $\mathcal{S}$-model, φ be an assignment and $\mathbf{M}$ be a proposition. $\mathcal{M} \models_{\varphi} \mathbf{M}$ iff $\mathcal{M} \models_{\varphi} \mathbf{M}^{\downarrow\beta}$. Furthermore, if $\mathcal{M}$ satisfies property η, then $\mathcal{M} \models_{\varphi} \mathbf{M}$ iff $\mathcal{M} \models_{\varphi} \mathbf{M}^{\downarrow}$.*

PROOF. The proof is by a simple induction on the proposition $\mathbf{M}$. □

We know $\mathcal{E}_{\varphi,[\mathcal{E}_{\varphi}(\mathbf{B})/x]}(\mathbf{A}) = \mathcal{E}_{\varphi}([\mathbf{B}/x]\mathbf{A})$ by the Substitution-Value Lemma. We can extend this to the level of models and (external) propositions by verifying $\mathcal{M} \models_{\varphi,[\mathcal{E}_{\varphi}(\mathbf{B})/x]} \mathbf{M}$ iff $\mathcal{M} \models_{\varphi} [\mathbf{B}/x]\mathbf{M}$.

LEMMA 3.3.12 (Model Substitution-Value Lemma). *Let*

$$\mathcal{M} = (\mathcal{D}, @, \mathcal{E}, \upsilon)$$

be an $\mathcal{S}$-model. For any type β, variable x_{β}, formula $\mathbf{B} \in \mathit{wff}_{\beta}(\mathcal{S})$, proposition $\mathbf{M} \in \mathit{prop}(\mathcal{S})$ and assignment φ, we have

$$\mathcal{M} \models_{\varphi,[\mathcal{E}_{\varphi}(\mathbf{B})/x]} \mathbf{M} \text{ iff } \mathcal{M} \models_{\varphi} [\mathbf{B}/x]\mathbf{M}.$$

PROOF. The proof is by a simple induction on $\mathbf{M}$ using the Substitution-Value Lemma (cf. Lemma 3.2.7) for the base cases where $\mathbf{M} \in \mathit{wff}_{o}(\mathcal{S})$ or $\mathbf{M}$ is $[\mathbf{A} \doteq^{\alpha} \mathbf{C}]$. □

Next we verify that we can change the interpretation of a parameter at the level of models.

DEFINITION 3.3.13 (Model Change of Parameter). Let

$$\mathcal{M} = (\mathcal{D}, @, \mathcal{E}, v)$$

be an $\mathcal{S}$-model, W_α be a parameter and $\mathsf{a} \in \mathcal{D}_\alpha$. We define

$$\mathcal{M}^{W \mapsto \mathsf{a}} := (\mathcal{D}, @, \mathcal{E}^{W \mapsto \mathsf{a}}, v)$$

where $\mathcal{E}^{W \mapsto \mathsf{a}}$ is defined in Definition 3.2.16.

THEOREM 3.3.14 (Model Change of Parameter). *Let*

$$\mathcal{M} = (\mathcal{D}, @, \mathcal{E}, v)$$

be an $\mathcal{S}$-model, W_α be a parameter and $\mathsf{a} \in \mathcal{D}_\alpha$. Then $\mathcal{M}^{W \mapsto \mathsf{a}}$ is an $\mathcal{S}$-model. Furthermore, we have:

1. *For any $\mathbf{M} \in prop(\mathcal{S})$, $x_\alpha \notin \mathbf{Free}(\mathbf{M})$ and assignment φ into $\mathcal{D}$, $\mathcal{M}^{W \mapsto \mathsf{a}} \models_\varphi \mathbf{M}$ iff $\mathcal{M} \models_{\varphi,[\mathsf{a}/x]} [x/W]\mathbf{M}$.*
2. *If $\mathcal{M}$ satisfies property η, then $\mathcal{M}^{W \mapsto \mathsf{a}}$ satisfies property η.*
3. *If $\mathcal{M}$ satisfies property ξ, then $\mathcal{M}^{W \mapsto \mathsf{a}}$ satisfies property ξ.*
4. *If $\mathcal{M}$ satisfies property $\mathfrak{f}$, then $\mathcal{M}^{W \mapsto \mathsf{a}}$ satisfies property $\mathfrak{f}$.*
5. *If $\mathcal{M}$ satisfies property $\mathfrak{b}$, then $\mathcal{M}^{W \mapsto \mathsf{a}}$ satisfies property $\mathfrak{b}$.*

In particular, if $\mathcal{M} \in \mathfrak{M}_*(\mathcal{S})$, then $\mathcal{M}^{W \mapsto \mathsf{a}} \in \mathfrak{M}_*(\mathcal{S})$ for each $*$ in $\{\beta, \beta\eta, \beta\xi, \beta\mathfrak{f}, \beta\mathfrak{b}, \beta\eta\mathfrak{b}, \beta\xi\mathfrak{b}, \beta\mathfrak{f}\mathfrak{b}\}$.

PROOF. See Appendix A.1. □

Next we verify that we can restrict any $\mathcal{S}_2$-model to obtain an $\mathcal{S}_1$-model when $\mathcal{S}_1 \subseteq \mathcal{S}_2$. Then we will characterize when $\mathcal{S}_1$-models can be extended to an $\mathcal{S}_2$-model. This will be especially useful when we begin to study conservation in Chapter 6.

THEOREM 3.3.15 (Model Restriction). *Let $\mathcal{S}_1$ and $\mathcal{S}_2$ be signatures with $\mathcal{S}_1 \subseteq \mathcal{S}_2$ and $\mathcal{M}_2 = (\mathcal{D}, @, \mathcal{E}, v)$ be an $\mathcal{S}_2$-model. Then*

$$\mathcal{M}_1 := (\mathcal{D}, @, \mathcal{E}|_{wff(\mathcal{S}_1)}, v)$$

is an $\mathcal{S}_1$-model. Furthermore, we have:

1. *For any $\mathbf{M} \in prop(\mathcal{S}_1)$ and assignment φ into $\mathcal{D}$, $\mathcal{M}_1 \models_\varphi \mathbf{M}$ iff $\mathcal{M}_2 \models_\varphi \mathbf{M}$.*
2. *$\mathcal{M}_1$ satisfies property η iff $\mathcal{M}_2$ satisfies property η.*
3. *$\mathcal{M}_1$ satisfies property ξ iff $\mathcal{M}_2$ satisfies property ξ.*
4. *$\mathcal{M}_1$ satisfies property $\mathfrak{f}$ iff $\mathcal{M}_2$ satisfies property $\mathfrak{f}$.*
5. *$\mathcal{M}_1$ satisfies property $\mathfrak{b}$ iff $\mathcal{M}_2$ satisfies property $\mathfrak{b}$.*

In particular, $\mathcal{M}_1 \in \mathfrak{M}_*(\mathcal{S}_1)$ *iff* $\mathcal{M}_2 \in \mathfrak{M}_*(\mathcal{S}_2)$ *for*

$$* \in \{\beta, \beta\eta, \beta\xi, \beta\mathfrak{f}, \beta\mathfrak{b}, \beta\eta\mathfrak{b}, \beta\xi\mathfrak{b}, \beta\mathfrak{f}\mathfrak{b}\}.$$

PROOF. By Lemma 3.2.22, we know $\mathcal{E}\big|_{wff(\mathcal{S}_1)}$ is an evaluation function. To check $\mathcal{M}_1$ is an $\mathcal{S}_1$-model, for each logical constant $c \in \mathcal{S}_1$ we must check $\mathfrak{L}_c(\mathcal{E}\big|_{wff(\mathcal{S}_1)}(c))$ (cf. Table 3.1) holds with respect to υ. For any $c \in \mathcal{S}_1 \subseteq \mathcal{S}_2$ we know $\mathfrak{L}_c(\mathcal{E}\big|_{wff(\mathcal{S}_1)}(c))$ holds since $\mathcal{E}\big|_{wff(\mathcal{S}_1)}(c) = \mathcal{E}(c)$ and $\mathcal{M}_2$ is an $\mathcal{S}_2$-model.

1. We prove $\mathcal{M}_1 \models_\varphi \mathbf{M}$ iff $\mathcal{M}_2 \models_{\varphi,[\mathsf{a}/x]} \mathbf{M}$ by a simple induction on $\mathbf{M} \in prop(\mathcal{S}_1)$. If $\mathbf{M} \in wff_o(\mathcal{S}_1)$, then the equivalence follows from $\upsilon(\mathcal{E}_\varphi(\mathbf{M})) = \upsilon(\mathcal{E}_\varphi\big|_{wff(\mathcal{S}_1)}(\mathbf{M}))$. If $\mathbf{M}$ is of the form $[\mathbf{A} \doteq^\alpha \mathbf{B}]$ where $\mathbf{A}$ and $\mathbf{B}$ are in $wff_\alpha(\mathcal{S}_1)$, then the equivalence follows from

$$\mathcal{E}_\varphi(\mathbf{A}) = \mathcal{E}_\varphi\big|_{wff(\mathcal{S}_1)}(\mathbf{A}) \text{ and } \mathcal{E}_\varphi(\mathbf{B}) = \mathcal{E}_\varphi\big|_{wff(\mathcal{S}_1)}(\mathbf{B}).$$

   If $\mathbf{M}$ is $\dot{\top}$, then the equivalence is trivial.

   If $\mathbf{M}$ is of the form $[\dot{\neg}\mathbf{N}]$, then by the inductive hypothesis we know $\mathcal{M}_1 \models_\varphi \mathbf{M}$ iff $\mathcal{M}_1 \not\models_\varphi \mathbf{N}$ iff $\mathcal{M}_2 \not\models_\varphi \mathbf{N}$ iff $\mathcal{M}_2 \models_\varphi \mathbf{M}$. Similarly, if $\mathbf{M}$ is of the form $[\mathbf{N} \dot{\vee} \mathbf{P}]$, then the equivalence follows by the inductive hypothesis.

   Suppose $\mathbf{M}$ is of the form $[\dot{\forall} x_\alpha \, \mathbf{N}]$. By the inductive hypothesis, for any $\mathsf{a} \in \mathcal{D}_\alpha$ we know

$$\mathcal{M}_1 \models_{\varphi,[\mathsf{a}/x]} \mathbf{N} \text{ iff } \mathcal{M}_2 \models_{\varphi,[\mathsf{a}/x]} \mathbf{N}.$$

   Thus $\mathcal{M}_1 \models_\varphi \mathbf{M}$ iff $\mathcal{M}_2 \models_\varphi \mathbf{M}$.
2. By Lemma 3.2.22, we know $(\mathcal{D}, @, \mathcal{E}\big|_{wff(\mathcal{S}_1)})$ is η-functional iff $(\mathcal{D}, @, \mathcal{E})$ is η-functional. Hence $\mathcal{M}_1$ satisfies property η iff $\mathcal{M}_2$ satisfies property η.
3. The proof for ξ-functionality is analagous to the argument for η-functionality again using Lemma 3.2.22.
4. We know $\mathcal{M}_1$ satisfies property $\mathfrak{f}$ iff $(\mathcal{D}, @)$ is functional iff $\mathcal{M}_2$ satisfies property $\mathfrak{f}$.
5. Finally, $\mathcal{M}_1$ satisfies property $\mathfrak{b}$ iff υ is injective iff $\mathcal{M}_2$ satisfies property $\mathfrak{b}$.

□

THEOREM 3.3.16. *Let $\mathcal{S}_1$ and $\mathcal{S}_2$ be signatures with $\mathcal{S}_1 \subseteq \mathcal{S}_2$ and $\mathcal{M}_1 = (\mathcal{D}, @, \mathcal{E}^1, \upsilon)$ be an $\mathcal{S}_1$-model. Suppose $\mathcal{I} : (\mathcal{S}_2 \setminus \mathcal{S}_1) \rightarrow \mathcal{D}$ is an interpretation of constants such that $\mathfrak{L}_c(\mathcal{I}(c))$ holds with respect to υ for all $c \in \mathcal{S}_2 \setminus \mathcal{S}_1$. Then there is an $\mathcal{S}_2$-model $\mathcal{M}_2 = (\mathcal{D}, @, \mathcal{E}^2, \upsilon)$. where $\mathcal{E}^2$ is an evaluation extension of $\mathcal{E}^1$ by $\mathcal{I}$. Furthermore, we have the following:*

1. *For every $\mathbf{M} \in prop(\mathcal{S}_1)$ and assignment φ into $(\mathcal{D}, @)$, $\mathcal{M}_1 \models_\varphi \mathbf{M}$ iff $\mathcal{M}_2 \models_\varphi \mathbf{M}$.*
2. *If $\mathcal{M}_1$ satisfies property η, then $\mathcal{M}_2$ satisfies property η.*
3. *If $\mathcal{M}_1$ satisfies property ξ, then $\mathcal{M}_2$ satisfies property ξ.*
4. *If $\mathcal{M}_1$ satisfies property $\mathfrak{f}$, then $\mathcal{M}_2$ satisfies property $\mathfrak{f}$.*
5. *If $\mathcal{M}_1$ satisfies property $\mathfrak{b}$, then $\mathcal{M}_2$ satisfies property $\mathfrak{b}$.*

In particular, if $\mathcal{M}_1 \in \mathfrak{M}_*(\mathcal{S}_1)$, then $\mathcal{M}_2 \in \mathfrak{M}_*(\mathcal{S}_2)$ for each

$$* \in \{\beta, \beta\eta, \beta\xi, \beta\mathfrak{f}, \beta\mathfrak{b}, \beta\eta\mathfrak{b}, \beta\xi\mathfrak{b}, \beta\mathfrak{f}\mathfrak{b}\}.$$

PROOF. By Theorem 3.2.21, we know there is an $\mathcal{S}_2$-evaluation extension $\mathcal{E}^2$ of $\mathcal{E}^1$ by $\mathcal{I}$. For each $c \in \mathcal{S}_2 \setminus \mathcal{S}_1$, we have $\mathcal{E}^2(c) = \mathcal{I}(c)$. Since we have assumed $\mathfrak{L}_c(\mathcal{I}(c))$ holds with respect to υ, we know $\mathcal{M}_2 := (\mathcal{D}, @, \mathcal{E}^2, \upsilon)$ is an $\mathcal{S}_2$-model. The remaining assertions follow from Theorem 3.3.15 since $\mathcal{E}^1 = \mathcal{E}^2|_{wff(\mathcal{S}_1)}$. □

There are many examples of signatures $\mathcal{S}_1$ and $\mathcal{S}_2$ with $\mathcal{S}_1 \subseteq \mathcal{S}_2$ where there are $\mathcal{S}_1$-models $\mathcal{M}$ which cannot be extended to an $\mathcal{S}_2$-model. This occurs when there is a logical constant $c_\alpha \in \mathcal{S}_2 \setminus \mathcal{S}_1$ where there is no candidate a in the domain of type α satisfying $\mathfrak{L}_c(\mathsf{a})$. When such candidates *do* exist, we will say the model realizes c. In particular, when such candidates exist we will be able to extend a model $\mathcal{M}$ to be a model for a larger signature without specifying the particular interpretations of the new logical constants.

DEFINITION 3.3.17. Let $\mathcal{S}$ be a signature and $\mathcal{M} = (\mathcal{D}, @, \mathcal{E}, \upsilon)$ be an $\mathcal{S}$-model. For each logical constant $c_\alpha \in \mathcal{S}_{all}$, we say $\mathcal{M}$ *realizes* c (or c *is realized by* $\mathcal{M}$) if there is some $\mathsf{a} \in \mathcal{D}_\alpha$ such that $\mathfrak{L}_c(\mathsf{a})$ holds with respect to υ. We say $\mathcal{M}$ *realizes* a signature $\mathcal{S}_2$ (or $\mathcal{S}_2$ *is realized by* $\mathcal{M}$) if $\mathcal{M}$ realizes every $c_\alpha \in \mathcal{S}_2$. We define the *maximal signature* of $\mathcal{M}$ to be

$$\mathcal{S}^{\mathcal{M}} := \{c \in \mathcal{S}_{all} \mid \mathcal{M} \text{ realizes } c\}.$$

We say a class of models $\mathfrak{M}$ realizes a signature $\mathcal{S}_2$ if every $\mathcal{M} \in \mathfrak{M}$ realizes $\mathcal{S}_2$.

Clearly any $\mathcal{S}$-model realizes every $c_\alpha \in \mathcal{S}$ since $\mathcal{E}(c)$ acts as the witness. Consequently, we know $\mathcal{S} \subseteq \mathcal{S}^{\mathcal{M}}$ for any $\mathcal{S}$-model $\mathcal{M}$. We now verify that we can extend any $\mathcal{S}$-model $\mathcal{M}$ to an $\mathcal{S}^{\mathcal{M}}$-model.

THEOREM 3.3.18. *Let $\mathcal{S}$ be a signature and $\mathcal{M} = (\mathcal{D}, @, \mathcal{E}, \upsilon)$ be in $\mathfrak{M}_*(\mathcal{S})$ where $* \in \{\beta, \beta\eta, \beta\mathfrak{fb}\}$. There is an $\mathcal{S}^{\mathcal{M}}$-evaluation $\mathcal{E}'$ such that*

$$\mathcal{M}' := (\mathcal{D}, @, \mathcal{E}', \upsilon) \in \mathfrak{M}_*(\mathcal{S}^{\mathcal{M}})$$

and $\mathcal{M} \models_\varphi \mathbf{M}$ iff $\mathcal{M}' \models_\varphi \mathbf{M}$ for any $\mathbf{M} \in prop(\mathcal{S})$ and assignment φ into $\mathcal{D}$. Furthermore, for any logical constant c, $\mathcal{M}'$ realizes c iff $c \in \mathcal{S}^{\mathcal{M}}$.

PROOF. This follows from Theorem 3.3.16 if we let $\mathcal{I}(c)$ be some a such that $\mathfrak{L}_c(\mathsf{a})$ holds for each $c \in (\mathcal{S}^{\mathcal{M}} \setminus \mathcal{S})$. (In the cases without extensionality, there may be more than one way to choose $\mathcal{I}$. We will discover in Theorem 3.6.15 that $\mathcal{I}$ is uniquely determined in the extensional case.)

Let c be any logical constant. Since the applicative structure $(\mathcal{D}, @)$ and valuation υ is the same in $\mathcal{M}$ as $\mathcal{M}'$, we know $\mathcal{M}'$ realizes c iff $\mathcal{M}$ realizes c iff $c \in \mathcal{S}^{\mathcal{M}}$. □

COROLLARY 3.3.19. *Let $\mathcal{S}$ be a signature, $* \in \{\beta, \beta\eta, \beta\mathfrak{fb}\}$ and $\mathcal{M} = (\mathcal{D}, @, \mathcal{E}, \upsilon) \in \mathfrak{M}_*(\mathcal{S})$. Suppose $\mathcal{S}'$ is a signature with $\mathcal{S}' \subseteq \mathcal{S}^{\mathcal{M}}$. There is an $\mathcal{S}'$-evaluation $\mathcal{E}'$ such that $\mathcal{M}' := (\mathcal{D}, @, \mathcal{E}', \upsilon) \in \mathfrak{M}_*(\mathcal{S}')$ and $\mathcal{M} \models_\varphi \mathbf{M}$ iff $\mathcal{M}' \models_\varphi \mathbf{M}$ for any $\mathbf{M} \in prop(\mathcal{S} \cap \mathcal{S}')$ and assignment φ into $\mathcal{D}$. Furthermore, for any logical constant c, $\mathcal{M}'$ realizes c iff $c \in \mathcal{S}^{\mathcal{M}}$.*

PROOF. By Theorem 3.3.18, we can extend the evaluation $\mathcal{E}$ to an $\mathcal{S}^{\mathcal{M}}$-evaluation $\mathcal{E}''$ so that $\mathcal{M}'' := (\mathcal{D}, @, \mathcal{E}'', \upsilon)$ is in $\mathfrak{M}_*(\mathcal{S}^{\mathcal{M}})$ and $\mathcal{M} \models_\varphi \mathbf{M}$ iff $\mathcal{M}' \models_\varphi \mathbf{M}$ for any $\mathbf{M} \in prop(\mathcal{S})$ and assignment φ into $\mathcal{D}$.

Applying Theorem 3.3.15, we conclude $\mathcal{E}' := \mathcal{E}''|_{wff(\mathcal{S}')}$ is the desired $\mathcal{S}'$-evaluation and $\mathcal{M}' := (\mathcal{D}, @, \mathcal{E}', \upsilon)$ is the desired $\mathcal{S}'$-model in $\mathfrak{M}_*(\mathcal{S}')$ such that $\mathcal{M} \models_\varphi \mathbf{M}$ iff $\mathcal{M}' \models_\varphi \mathbf{M}$ for any $\mathbf{M} \in prop(\mathcal{S} \cap \mathcal{S}')$ and assignment φ into $\mathcal{D}$.

Finally, note that $\mathcal{M}'$ realizes c iff $\mathcal{M}$ realizes c since the underlying applicative structures and valuations of $\mathcal{M}$ and $\mathcal{M}'$ are the same. □

3.4. Soundness of Sequent Calculi

DEFINITION 3.4.1. We say a model $\mathcal{M}$ *satisfies* a (one-sided) sequent Γ ($\mathcal{M} \models \Gamma$) if there exists some sentence $\mathbf{M} \in \Gamma$ such that $\mathcal{M} \models \mathbf{M}$.

DEFINITION 3.4.2 (Soundness). We say a sequent rule (with $n \geq 0$)

$$\frac{\Gamma_1 \quad \cdots \quad \Gamma_n}{\Gamma}\; Rule$$

is *sound* with respect to a model $\mathcal{M}$ if $\mathcal{M} \models \Gamma$ whenever $\mathcal{M} \models \Gamma_i$ for every i $(1 \leq i \leq n)$. We say the rule is *sound* with respect to a model class $\mathfrak{M}$ if every model $\mathcal{M} \in \mathfrak{M}$ models Γ whenever every model $\mathcal{M} \in \mathfrak{M}$ models Γ_i for every i $(1 \leq i \leq n)$.

LEMMA 3.4.3. *Let $\mathcal{M} = (\mathcal{D}, @, \mathcal{E}, v)$ be an $\mathcal{S}$-model. Every rule of $\mathcal{G}^{\mathcal{S}}_{\beta}$ other than $\mathcal{G}(\dot{\forall}^W)$ is sound with respect to $\mathcal{M}$.*

PROOF.

$\mathcal{G}(Init)$: If $\mathbf{A} \in cwff_o(\mathcal{S})$ and $\mathbf{A}, \dot{\neg}\mathbf{A} \in \Gamma$, then

$$\mathcal{M} \models \mathbf{A} \text{ or } \mathcal{M} \models \dot{\neg}\mathbf{A}.$$

Thus $\mathcal{M} \models \Gamma$.

$\mathcal{G}(Contr)$: Suppose $\mathcal{M} \models \Gamma, \mathbf{M}, \mathbf{M}$. Either $\mathcal{M} \models \Gamma$ or $\mathcal{M} \models \mathbf{M}$. In either case, $\mathcal{M} \models \Gamma, \mathbf{M}$.

$\mathcal{G}(\dot{\top})$: By Definition 3.3.3, $\mathcal{M} \models \dot{\top}$ and so $\mathcal{M} \models \Gamma, \dot{\top}$.

$\mathcal{G}(\dot{\neg}\dot{\neg})$: Suppose $\mathcal{M} \models \Gamma, \mathbf{M}$. Either $\mathcal{M} \models \Gamma$ or $\mathcal{M} \models \mathbf{M}$. If $\mathcal{M} \models \mathbf{M}$, then $\mathcal{M} \models \dot{\neg}\dot{\neg}\mathbf{M}$. Hence $\mathcal{M} \models \Gamma, \dot{\neg}\dot{\neg}\mathbf{M}$.

$\mathcal{G}(\dot{\vee})$: Suppose $\mathcal{M} \models \Gamma, \mathbf{M}, \mathbf{N}$. Either

$$\mathcal{M} \models \Gamma, \mathcal{M} \models \mathbf{M} \text{ or } \mathcal{M} \models \mathbf{N}.$$

Either $\mathcal{M} \models \Gamma$ or $\mathcal{M} \models [\mathbf{M} \dot{\vee} \mathbf{N}]$. Thus $\mathcal{M} \models \Gamma, [\mathbf{M} \dot{\vee} \mathbf{N}]$.

$\mathcal{G}(\dot{\neg}\dot{\vee})$: Suppose $\mathcal{M} \models \Gamma, \dot{\neg}\mathbf{M}$ and $\mathcal{M} \models \Gamma, \dot{\neg}\mathbf{N}$. If $\mathcal{M} \models \Gamma$, then $\mathcal{M} \models \Gamma, \dot{\neg}[\mathbf{M} \dot{\vee} \mathbf{N}]$. Otherwise, both $\mathcal{M} \models \dot{\neg}\mathbf{M}$ and $\mathcal{M} \models \dot{\neg}\mathbf{N}$. Hence $\mathcal{M} \not\models \mathbf{M}$ and $\mathcal{M} \not\models \mathbf{N}$. Thus $\mathcal{M} \not\models [\mathbf{M} \dot{\vee} \mathbf{N}]$ and so $\mathcal{M} \models \dot{\neg}[\mathbf{M} \dot{\vee} \mathbf{N}]$. Therefore, $\mathcal{M} \models \Gamma, \dot{\neg}[\mathbf{M} \dot{\vee} \mathbf{N}]$.

$\mathcal{G}(\dot{\neg}\dot{\forall}, \mathcal{S})$: Suppose $\mathcal{M} \models \Gamma, \dot{\neg}[[\mathbf{C}/x]\mathbf{M}]$ for some $\mathbf{C} \in cwff_\alpha(\mathcal{S})$. If $\mathcal{M} \models \Gamma$, then $\mathcal{M} \models \Gamma, \dot{\neg}[\dot{\forall} x_\alpha \,\blacksquare\, \mathbf{M}]$. Otherwise, we must have $\mathcal{M} \models \dot{\neg}[[\mathbf{C}/x]\mathbf{M}]$. Let φ be any assignment into $\mathcal{M}$. By Lemma 3.3.12, we know $\mathcal{M} \models_{\varphi,[\mathcal{E}(\mathbf{C})/x]} \dot{\neg}\mathbf{M}$. Hence we have

$$\mathcal{M} \not\models_{\varphi,[\mathcal{E}(\mathbf{C})/x]} \mathbf{M} \text{ and so } \mathcal{M} \not\models_{\varphi} [\dot{\forall} x_\alpha \,\blacksquare\, \mathbf{M}].$$

Thus $\mathcal{M} \models_\varphi \dot{\neg}[\dot{\forall} x_\alpha \centerdot \mathbf{M}]$. In this case, $\mathcal{M} \models \Gamma, \dot{\neg}[\dot{\forall} x_\alpha \centerdot \mathbf{M}]$

$\mathcal{G}(\beta)$: Suppose $\mathcal{M} \models \Gamma, \mathbf{M}^{\downarrow\beta}$. If $\mathcal{M} \models \Gamma$, then $\mathcal{M} \models \Gamma, \mathbf{M}$. By Lemma 3.3.11, if $\mathcal{M} \models \mathbf{M}^{\downarrow\beta}$, then $\mathcal{M} \models \mathbf{M}$. In either case, $\mathcal{M} \models \Gamma, \mathbf{M}$.

$\mathcal{G}(\sharp)$,$\mathcal{G}(\dot{\neg}\sharp)$: These rules are proven sound using the appropriate property in Table 3.1. We only prove soundness for the case when $\Pi^\alpha \in \mathcal{S}$ and $\mathbf{A}$ is $[\Pi^\alpha\, \mathbf{F}]$. Recall that $[\Pi^\alpha\, \mathbf{F}]^\sharp$ is $[\dot{\forall} x_\alpha \centerdot \mathbf{F}\, x]$ (cf. Definition 2.1.38). Suppose $\mathcal{M} \models \Gamma, [\dot{\forall} x_\alpha \centerdot \mathbf{F}\, x]$. We must prove $\mathcal{M} \models \Gamma, [\Pi^\alpha\, \mathbf{F}]$. If $\mathcal{M} \models \Gamma$, then we are done. Otherwise, $\mathcal{M} \models [\dot{\forall} x_\alpha \centerdot \mathbf{F}\, x]$. Thus $\mathcal{M} \models_{\varphi,[\mathsf{a}/x]} \mathbf{F}\, x$ for every $\mathsf{a} \in \mathcal{D}_\alpha$. This means $\upsilon(\mathcal{E}(\mathbf{F})@\mathsf{a}) = \mathtt{T}$ for every $\mathsf{a} \in \mathcal{D}_\alpha$. By $\mathfrak{L}_{\Pi^\alpha}(\mathcal{E}(\Pi^\alpha))$, we have $\upsilon(\mathcal{E}(\Pi^\alpha)@\mathcal{E}(\mathbf{F})) = \mathtt{T}$. Therefore, $\mathcal{M} \models [\Pi^\alpha\, \mathbf{F}]$, as desired.

□

The rule $\mathcal{G}(\dot{\forall}^W)$ is not sound with respect to arbitrary models, but is sound with respect to appropriate model classes. To verify soundness of the rule $\mathcal{G}(\dot{\forall}^W)$ of a class $\mathfrak{M}$ of models may involve changing the value of the parameter W in a model, yielding a different model in $\mathfrak{M}$.

LEMMA 3.4.4. *Let W be a parameter. Suppose $\mathfrak{M}$ is a model class such that $\mathcal{M}^{W \mapsto \mathsf{a}} \in \mathfrak{M}$ whenever $\mathcal{M} = (\mathcal{D}, @, \mathcal{E}, \upsilon) \in \mathfrak{M}$ and $\mathsf{a} \in \mathcal{D}_\alpha$. The rule $\mathcal{G}(\dot{\forall}^W)$ is sound with respect to $\mathfrak{M}$.*

PROOF. Suppose $\Gamma, [\dot{\forall} x_\alpha\, \mathbf{M}]$ is a sequent such that W is not in **Params**$(\Gamma, \mathbf{M})$ and $\mathcal{M} \models \Gamma, [W_\alpha/x]\mathbf{M}$ for every $\mathcal{M} \in \mathfrak{M}$. We argue by contradiction. Assume there is some $\mathcal{M} = (\mathcal{D}, @, \mathcal{E}, \upsilon) \in \mathfrak{M}$ such that $\mathcal{M} \not\models \Gamma, [\dot{\forall} x_\alpha\, \mathbf{M}]$. Hence $\mathcal{M} \not\models \Gamma$ and $\mathcal{M} \not\models [\dot{\forall} x_\alpha\, \mathbf{M}]$. Thus there must be some assignment φ and $\mathsf{a} \in \mathcal{D}_\alpha$ such that $\mathcal{M} \not\models_{\varphi,[\mathsf{a}/x]} \mathbf{M}$. By our assumption on the model class $\mathfrak{M}$, $\mathcal{M}^{W \mapsto \mathsf{a}} \in \mathfrak{M}$. By Theorem 3.3.14, for every assignment ψ into $\mathcal{D}$ and $\mathbf{N} \in prop(\mathcal{S})$, $\mathcal{M}^{W \mapsto \mathsf{a}} \models_\psi \mathbf{N}$ iff $\mathcal{M} \models_{\psi,[\mathsf{a}/x]} [x/W]\mathbf{N}$ where $x \notin$ **Free**$(\mathbf{N})$. Hence $\mathcal{M}^{W \mapsto \mathsf{a}} \not\models \Gamma$ since $\mathcal{M} \not\models \Gamma$ and W does not occur in any sentence in Γ. By our assumption, $\mathcal{M}^{W \mapsto \mathsf{a}} \in \mathfrak{M}$ and $\mathcal{M}^{W \mapsto \mathsf{a}} \not\models \Gamma$ imply $\mathcal{M}^{W \mapsto \mathsf{a}} \models [W/x]\mathbf{M}$. Since W does not occur in $\mathbf{M}$, we know $[x/W][W/x]\mathbf{M}$ is $\mathbf{M}$. Thus $\mathcal{M} \models_{\varphi,[\mathsf{a}/x]} \mathbf{M}$, contradicting our choice of φ and a. □

LEMMA 3.4.5. *Every rule of $\mathcal{G}^{\mathcal{S}}_\beta$ is sound with respect to $\mathfrak{M}_\beta(\mathcal{S})$.*

Proof. For every rule except $\mathcal{G}(\dot{\forall}^W)$, soundness follows directly from Lemma 3.4.3. Note that by Theorem 3.3.14, $\mathcal{M}^{W \mapsto \mathsf{a}} \in \mathfrak{M}_\beta(\mathcal{S})$ whenever $\mathcal{M} = (\mathcal{D}, @, \mathcal{E}, v) \in \mathfrak{M}_\beta(\mathcal{S})$ and $\mathsf{a} \in \mathcal{D}_\alpha$. Soundness of the rule $\mathcal{G}(\dot{\forall}^W)$ with respect to $\mathfrak{M}_\beta(\mathcal{S})$ follows from Lemma 3.4.4. □

Theorem 3.4.6 (Soundness for $\mathcal{G}^{\mathcal{S}}_\beta$). *If a sequent Γ is derivable in $\mathcal{G}^{\mathcal{S}}_\beta$ and $\mathcal{M} \in \mathfrak{M}_\beta(\mathcal{S})$, then $\mathcal{M} \models \Gamma$.*

Proof. This follows by induction on the derivation of Γ and the soundness of each rule of $\mathcal{G}^{\mathcal{S}}_\beta$ with respect to $\mathfrak{M}_\beta(\mathcal{S})$ as given by Lemma 3.4.5. □

Lemma 3.4.7. *Let $\mathcal{M}$ be an $\mathcal{S}$-model satisfying property η. The rule $\mathcal{G}(\beta\eta)$ is sound with respect to $\mathcal{M}$.*

Proof. Suppose $\mathcal{M} \models \Gamma, \mathbf{M}^{\downarrow}$. If $\mathcal{M} \models \Gamma$, then $\mathcal{M} \models \Gamma, \mathbf{M}$. By Lemma 3.3.11, if $\mathcal{M} \models \mathbf{M}^{\downarrow}$, then $\mathcal{M} \models \mathbf{M}$. In either case, $\mathcal{M} \models \Gamma, \mathbf{M}$. □

Lemma 3.4.8. *Every rule of $\mathcal{G}^{\mathcal{S}}_{\beta\eta}$ is sound with respect to $\mathfrak{M}_{\beta\eta}(\mathcal{S})$.*

Proof. For every rule other than $\mathcal{G}(\beta\eta)$ and $\mathcal{G}(\dot{\forall}^W)$ soundness follows from Lemma 3.4.3. Soundness of the rule $\mathcal{G}(\beta\eta)$ with respect to $\mathfrak{M}_{\beta\eta}(\mathcal{S})$ follows from Lemma 3.4.7. By Theorem 3.3.14, $\mathcal{M}^{W \mapsto \mathsf{a}} \in \mathfrak{M}_{\beta\eta}(\mathcal{S})$ whenever $\mathcal{M} = (\mathcal{D}, @, \mathcal{E}, v) \in \mathfrak{M}_{\beta\eta}(\mathcal{S})$ and $\mathsf{a} \in \mathcal{D}_\alpha$. Soundness of $\mathcal{G}(\dot{\forall}^W)$ follows from Lemma 3.4.4. □

Theorem 3.4.9 (Soundness for $\mathcal{G}^{\mathcal{S}}_{\beta\eta}$). *If a sequent Γ is derivable in $\mathcal{G}^{\mathcal{S}}_{\beta\eta}$ and $\mathcal{M} \in \mathfrak{M}_{\beta\eta}(\mathcal{S})$, then $\mathcal{M} \models \Gamma$.*

Proof. This follows by induction on the derivation of Γ and the soundness of each rule of $\mathcal{G}^{\mathcal{S}}_{\beta\eta}$ with respect to $\mathfrak{M}_{\beta\eta}(\mathcal{S})$ as given by Lemma 3.4.8. □

Lemma 3.4.10. *Let $\mathcal{M} = (\mathcal{D}, @, \mathcal{E}, v)$ be an $\mathcal{S}$-model. Each of the equality rules in Figure 2.5 is sound with respect to $\mathcal{M}$.*

Proof.

$\mathcal{G}(Init^{\dot{=}})$: Suppose H is a parameter and

$$\mathcal{M} \models \Gamma, [[H\,\overline{\mathbf{A}^n}] \doteq^o [H\,\overline{\mathbf{B}^n}]].$$

If $\mathcal{M} \models \Gamma$, then we are done. Otherwise, $\mathcal{M} \models [\mathbf{A}^i \doteq \mathbf{B}^i]$ and so $\mathcal{E}(\mathbf{A}^i) = \mathcal{E}(\mathbf{B}^i)$ for each i $(1 \leq i \leq n)$. This implies

$$\mathcal{E}([H\,\overline{\mathbf{A}^n}]) = \mathcal{E}(H)@\mathcal{E}(\mathbf{A}^1)@\cdots@\mathcal{E}(\mathbf{A}^n)$$
$$= \mathcal{E}(H)@\mathcal{E}(\mathbf{B}^1)@\cdots@\mathcal{E}(\mathbf{B}^n) = \mathcal{E}([H\,\overline{\mathbf{B}^n}]).$$

If $\mathcal{M} \models [H\,\overline{\mathbf{A}^n}]$, then $\mathcal{M} \models \Gamma, [H\,\overline{\mathbf{A}^n}], \dot{\neg}[H\,\overline{\mathbf{B}^n}]$. Otherwise, we may assume $\mathcal{M} \not\models [H\,\overline{\mathbf{A}^n}]$. Since $v(\mathcal{E}([H\,\overline{\mathbf{A}^n}]))$ equals $v(\mathcal{E}([H\,\overline{\mathbf{B}^n}]))$, we know $\mathcal{M} \not\models [H\,\overline{\mathbf{B}^n}]$ and hence $\mathcal{M} \models \dot{\neg}[H\,\overline{\mathbf{B}^n}]$. Therefore, in any case we have $\mathcal{M} \models \Gamma, [H\,\overline{\mathbf{A}^n}], \dot{\neg}[H\,\overline{\mathbf{B}^n}]$.

$\mathcal{G}(Dec)$**:** Let H be a parameter. Suppose $\mathcal{M} \models \Gamma, [\mathbf{A}^i \doteq \mathbf{B}^i]$ for $1 \leq i \leq n$. If $\mathcal{M} \models \Gamma$, then we are done. Otherwise, $\mathcal{M} \models [\mathbf{A}^i \doteq \mathbf{B}^i]$ for each i $(1 \leq i \leq n)$. That is, $\mathcal{E}(\mathbf{A}^i) = \mathcal{E}(\mathbf{B}^i)$ for each i $(1 \leq i \leq n)$. Thus

$$\mathcal{E}(H\,\overline{\mathbf{A}^n}) = \mathcal{E}(H\,\overline{\mathbf{B}^n})$$

and so

$$\mathcal{M} \models \Gamma, [[H\,\overline{\mathbf{A}^n}] \doteq^{\iota} [H\,\overline{\mathbf{B}^n}]]$$

as desired.

$\mathcal{G}(EUnif_1)$**:** Suppose $\mathcal{M} \models \Gamma, [\mathbf{A} \doteq^{\iota} \mathbf{C}]$ and $\mathcal{M} \models \Gamma, [\mathbf{B} \doteq^{\iota} \mathbf{D}]$. If $\mathcal{M} \models \Gamma$, then we are done. Otherwise, $\mathcal{E}(\mathbf{A}) = \mathcal{E}(\mathbf{C})$ and $\mathcal{E}(\mathbf{B}) = \mathcal{E}(\mathbf{D})$. If $\mathcal{M} \models [\mathbf{C} \doteq^{\iota} \mathbf{D}]$, then we have

$$\mathcal{M} \models \Gamma, \dot{\neg}[\mathbf{A} \doteq^{\iota} \mathbf{B}], [\mathbf{C} \doteq^{\iota} \mathbf{D}].$$

Otherwise, $\mathcal{E}(\mathbf{C}) \neq \mathcal{E}(\mathbf{D})$. This implies $\mathcal{E}(\mathbf{A}) \neq \mathcal{E}(\mathbf{B})$. Thus in this case $\mathcal{M} \models \Gamma, \dot{\neg}[\mathbf{A} \doteq^{\iota} \mathbf{B}], [\mathbf{C} \doteq^{\iota} \mathbf{D}]$ as well.

$\mathcal{G}(EUnif_2)$**:** This case is analogous to $\mathcal{G}(EUnif_1)$.

$\mathcal{G}(\dot{\neg} \doteq^{o})$**:** Suppose $\mathcal{M} \models \Gamma, \mathbf{A}, \mathbf{B}$ and $\mathcal{M} \models \Gamma, \dot{\neg}\mathbf{A}, \dot{\neg}\mathbf{B}$. We argue by contradiction. Assume $\mathcal{M} \not\models \Gamma, \dot{\neg}[\mathbf{A} \doteq^{o} \mathbf{B}]$. Hence $\mathcal{M} \not\models \Gamma$ and $\mathcal{E}(\mathbf{A}) = \mathcal{E}(\mathbf{B})$. If $\mathcal{M} \models \mathbf{A}$, then $\mathcal{M} \models \dot{\neg}\mathbf{B}$ since $\mathcal{M} \models \Gamma, \dot{\neg}\mathbf{A}, \dot{\neg}\mathbf{B}$. This contradicts $v(\mathcal{E}(\mathbf{A})) = v(\mathcal{E}(\mathbf{B}))$. If $\mathcal{M} \not\models \mathbf{A}$, then $\mathcal{M} \models \mathbf{B}$ since $\mathcal{M} \models \Gamma, \mathbf{A}, \mathbf{B}$. This also contradicts $v(\mathcal{E}(\mathbf{A})) = v(\mathcal{E}(\mathbf{B}))$.

$\mathcal{G}(\dot{\neg} \doteq^{\rightarrow}, \mathcal{S})$**:** Suppose $\mathcal{M} \models \Gamma, \dot{\neg}[[\mathbf{G}\,\mathbf{B}] \doteq^{\alpha} [\mathbf{H}\,\mathbf{B}]]$ where $\mathbf{B}$ is in $cwff_{\beta}(\mathcal{S})$. If $\mathcal{M} \models \Gamma$, then we are done. Otherwise,

$$\mathcal{E}(\mathbf{G})@\mathcal{E}(\mathbf{B}) = \mathcal{E}([\mathbf{G}\,\mathbf{B}]) \neq \mathcal{E}([\mathbf{H}\,\mathbf{B}]) = \mathcal{E}(\mathbf{H})@\mathcal{E}(\mathbf{B}).$$

This implies $\mathcal{E}(\mathbf{G}) \neq \mathcal{E}(\mathbf{H})$. Thus $\mathcal{M} \models \dot{\neg}[\mathbf{G} \doteq^{\alpha\beta} \mathbf{H}]$, as desired.

□

LEMMA 3.4.11. *Let $\mathcal{M} = (\mathcal{D}, @, \mathcal{E}, v)$ be an $\mathcal{S}$-model satisfying property $\mathfrak{b}$. The extensional rule $\mathcal{G}(\mathfrak{b})$ is sound with respect to $\mathcal{M}$.*

PROOF. Suppose $\mathcal{M} \models \Gamma, \neg\mathbf{A}, \mathbf{B}$ and $\mathcal{M} \models \Gamma, \neg\mathbf{B}, \mathbf{A}$. We are done if $\mathcal{M} \models \Gamma$. Otherwise,

$$\mathcal{M} \models \neg\mathbf{A}, \mathbf{B} \text{ and } \mathcal{M} \models \neg\mathbf{B}, \mathbf{A}.$$

If $\mathcal{M} \models \mathbf{A}$, then $\mathcal{M} \models \neg\mathbf{A}, \mathbf{B}$ implies $\mathcal{M} \models \mathbf{B}$. In this case, $\mathcal{E}(\mathbf{A})$ equals $\mathcal{E}(\mathbf{B})$ by property $\mathfrak{b}$. If $\mathcal{M} \models \neg\mathbf{A}$, then $\mathcal{M} \models \neg\mathbf{B}, \mathbf{A}$ implies $\mathcal{M} \models \neg\mathbf{B}$. In this case, $\mathcal{E}(\mathbf{A}) = \mathcal{E}(\mathbf{B})$ by property $\mathfrak{b}$. In either case, $\mathcal{M} \models [\mathbf{A} \doteq^{o} \mathbf{B}]$, as desired. □

LEMMA 3.4.12. *Let W_β be a parameter. The rule $\mathcal{G}(\mathfrak{f}^W)$ is sound with respect to $\mathfrak{M}_{\beta\mathfrak{f}\mathfrak{b}}(\mathcal{S})$.*

PROOF. Suppose $\Gamma, [\mathbf{G}_{\alpha\beta} \doteq^{\alpha\beta} \mathbf{H}_{\alpha\beta}]$ is a sequent where

$$W \notin \mathbf{Params}(\Gamma) \cup \mathbf{Params}(\mathbf{G}) \cup \mathbf{Params}(\mathbf{H}).$$

Also, suppose $\mathcal{M} \models \Gamma, [[\mathbf{G}\, W] \doteq^{\alpha} [\mathbf{H}\, W]]$ for every $\mathcal{M} \in \mathfrak{M}_{\beta\mathfrak{f}\mathfrak{b}}(\mathcal{S})$. We argue by contradiction. Assume there is some

$$\mathcal{M} = (\mathcal{D}, @, \mathcal{E}, v) \in \mathfrak{M}_{\beta\mathfrak{f}\mathfrak{b}}(\mathcal{S})$$

such that $\mathcal{M} \not\models \Gamma, [\mathbf{G}_{\alpha\beta} \doteq^{\alpha\beta} \mathbf{H}_{\alpha\beta}]$. Hence $\mathcal{M} \not\models [\mathbf{G}_{\alpha\beta} \doteq^{\alpha\beta} \mathbf{H}_{\alpha\beta}]$. This means $\mathcal{E}(\mathbf{G}) \neq \mathcal{E}(\mathbf{H})$. By functionality, there is some $\mathsf{b} \in \mathcal{D}_\beta$ such that $\mathcal{E}(\mathbf{G})@\mathsf{b} \neq \mathcal{E}(\mathbf{H})@\mathsf{b}$. By Theorem 3.3.14, we know that $\mathcal{M}^{W\mapsto\mathsf{b}} \in \mathfrak{M}_{\beta\mathfrak{f}\mathfrak{b}}(\mathcal{S})$ and for every assignment ψ into $\mathcal{D}$ and proposition $\mathbf{N}$, $\mathcal{M}^{W\mapsto\mathsf{b}} \models_\psi \mathbf{N}$ iff $\mathcal{M} \models_{\psi,[\mathsf{b}/x]} [x/W]\mathbf{N}$ where $x_\beta \notin \mathbf{Free}(\mathbf{N})$. In particular, $\mathcal{M}^{W\mapsto\mathsf{b}} \not\models \Gamma$ since $\mathcal{M} \not\models \Gamma$ and W does not occur in any sentence in Γ. By assumption, $\mathcal{M}^{W\mapsto\mathsf{b}} \in \mathfrak{M}_{\beta\mathfrak{f}\mathfrak{b}}(\mathcal{S})$ and $\mathcal{M}^{W\mapsto\mathsf{b}} \not\models \Gamma$ imply $\mathcal{M}^{W\mapsto\mathsf{b}} \models [[\mathbf{G}\, W] \doteq^{\alpha} [\mathbf{H}\, W]]$. Since

$$W \notin (\mathbf{Params}(\mathbf{G}) \cup \mathbf{Params}(\mathbf{H}))$$

, we know

$$\mathcal{E}^{W\mapsto\mathsf{b}}(\mathbf{G}) = \mathcal{E}(\mathbf{G}) \text{ and } \mathcal{E}^{W\mapsto\mathsf{b}}(\mathbf{H}) = \mathcal{E}(\mathbf{H})$$

by Theorem 3.2.18. Thus

$$\mathcal{E}(\mathbf{G})@\mathsf{b} = \mathcal{E}^{W\mapsto\mathsf{b}}([\mathbf{G}\, W]) = \mathcal{E}^{W\mapsto\mathsf{b}}([\mathbf{H}\, W]) = \mathcal{E}(\mathbf{H})@\mathsf{b}$$

contradicting the choice of b. □

LEMMA 3.4.13. *Every rule of* $\mathcal{G}^{\mathcal{S}}_{\beta\mathfrak{fb}}$ *is sound with respect to* $\mathfrak{M}_{\beta\mathfrak{fb}}(\mathcal{S})$.

PROOF. Every rule in Figure 2.3 except $\mathcal{G}(\dot\forall^W)$ is sound by Lemma 3.4.3. By Theorem 3.3.14, $\mathcal{M}^{W \mapsto \mathsf{a}} \in \mathfrak{M}_{\beta\mathfrak{fb}}(\mathcal{S})$ whenever

$$\mathcal{M} = (\mathcal{D}, @, \mathcal{E}, \upsilon)$$

is in $\mathfrak{M}_{\beta\mathfrak{fb}}(\mathcal{S})$ and $\mathsf{a} \in \mathcal{D}_\alpha$. Soundness of $\mathcal{G}(\dot\forall^W)$ follows from Lemma 3.4.4. The rule $\mathcal{G}(\beta\eta)$ is sound by Lemmas 3.2.15 and 3.4.7. The equational rules in Figure 2.5 are sound by Lemma 3.4.10. The extensional rule $\mathcal{G}(\mathfrak{b})$ is sound by Lemma 3.4.11. The extensional rule $\mathcal{G}(\mathfrak{f}^W)$ is sound by Lemma 3.4.12. □

THEOREM 3.4.14 (Soundness for $\mathcal{G}^{\mathcal{S}}_{\beta\mathfrak{fb}}$). *If a sequent* Γ *is derivable in* $\mathcal{G}^{\mathcal{S}}_{\beta\mathfrak{fb}}$ *and* $\mathcal{M} \in \mathfrak{M}_{\beta\mathfrak{fb}}(\mathcal{S})$, *then* $\mathcal{M} \models \Gamma$.

PROOF. This follows by induction on the derivation of Γ and the soundness of each rule of $\mathcal{G}^{\mathcal{S}}_{\beta\mathfrak{fb}}$ with respect to $\mathfrak{M}_{\beta\mathfrak{fb}}(\mathcal{S})$ as given by Lemma 3.4.13. □

3.5. Model Isomorphisms

We next define the notion of an isomorphism between two models and determine every extensional $\mathcal{S}$-model is isomorphic to a Henkin $\mathcal{S}$-model if either $\top, \bot \in \mathcal{S}$ or $\neg \in \mathcal{S}$. Consequently, the class of such extensional $\mathcal{S}$-models is simply the closure of the class of Henkin $\mathcal{S}$-models under isomorphism.

DEFINITION 3.5.1 (Model Homomorphism). Let

$$\mathcal{M}^1 = (\mathcal{D}^1, @^1, \mathcal{E}^1, \upsilon^1)$$

and

$$\mathcal{M}^2 = (\mathcal{D}^2, @^2, \mathcal{E}^2, \upsilon^2)$$

be models. A *homomorphism* from $\mathcal{M}^1$ to $\mathcal{M}^2$ is a typed function $\kappa : \mathcal{D}^1 \to \mathcal{D}^2$ such that κ is a homomorphism from the evaluation $(\mathcal{D}^1, @^1, \mathcal{E}^1)$ to the evaluation $(\mathcal{D}^2, @^2, \mathcal{E}^2)$ and $\upsilon^1(\mathsf{a}) = \upsilon^2(\kappa(\mathsf{a}))$ for every $\mathsf{a} \in \mathcal{D}^1_o$.

DEFINITION 3.5.2 (Model Isomorphism). Let

$$\mathcal{M}^1 = (\mathcal{D}^1, @^1, \mathcal{E}^1, \upsilon^1)$$

and

$$\mathcal{M}^2 = (\mathcal{D}^2, @^2, \mathcal{E}^2, \upsilon^2)$$

be models. An *isomorphism* from $\mathcal{M}^1$ to $\mathcal{M}^2$ is a homomorphism i from $\mathcal{M}^1$ to $\mathcal{M}^2$ such that there exists a homomorphism j from $\mathcal{M}^2$ to $\mathcal{M}^1$ where $j_\alpha : \mathcal{D}^2_\alpha \to \mathcal{D}^1_\alpha$ is the inverse of $i_\alpha : \mathcal{D}^1_\alpha \to \mathcal{D}^2_\alpha$ at each type α. Two models are said to be *isomorphic* if there is such an isomorphism. (It is clear from the definition that this is an equivalence relation between models.)

The class of Henkin models is not closed under isomorphism of models; neither is the class of standard models. Given a Henkin (or standard) model with domains $\mathcal{D}$, we may choose

$$\mathcal{D}'_{\alpha\beta} := \{(0, f) \mid f \in \mathcal{D}_{\alpha\beta}\}$$

and define @ appropriately. In this manner, we can define an isomorphic model which is not a model over a frame, hence not a Henkin (or standard) model.

LEMMA 3.5.3. *Let*

$$\mathcal{M}^1 = (\mathcal{D}^1, @^1, \mathcal{E}^1, \upsilon^1)$$

and

$$\mathcal{M}^2 = (\mathcal{D}^2, @^2, \mathcal{E}^2, \upsilon^2)$$

be isomorphic models with isomorphism $i : \mathcal{M}^1 \to \mathcal{M}^2$ *and inverse* j. *For any proposition* $\mathbf{M}$ *and assignment* φ *into* $\mathcal{M}^2$, $\mathcal{M}^1 \models_{j\varphi} \Phi$ *iff* $\mathcal{M}^2 \models_\varphi \Phi$.

PROOF. We prove this by induction on propositions $\mathbf{M}$. See Appendix A.1. □

LEMMA 3.5.4. *Let*

$$\mathcal{M}^1 = (\mathcal{D}^1, @^1, \mathcal{E}^1, \upsilon^1)$$

and

$$\mathcal{M}^2 = (\mathcal{D}^2, @^2, \mathcal{E}^2, \upsilon^2)$$

be isomorphic models.

1. *For any set of sentences* Φ, $\mathcal{M}^1 \models \Phi$ *implies* $\mathcal{M}^2 \models \Phi$.
2. *If* $\mathcal{M}^1$ *satisfies property* η, *then* $\mathcal{M}^2$ *satisfies property* η.
3. *If* $\mathcal{M}^1$ *satisfies property* ξ, *then* $\mathcal{M}^2$ *satisfies property* ξ.
4. *If* $\mathcal{M}^1$ *satisfies property* $\mathfrak{f}$, *then* $\mathcal{M}^2$ *satisfies property* $\mathfrak{f}$.

5. *If $\mathcal{M}^1$ satisfies property* $\mathfrak{b}$, *then* $\mathcal{M}^2$ *satisfies property* $\mathfrak{b}$. *In particular, if* $\mathcal{M}^1 \in \mathfrak{M}_*(\mathcal{S})$, *then* $\mathcal{M}^2 \in \mathfrak{M}_*(\mathcal{S})$ *for each*

$$* \in \{\beta, \beta\eta, \beta\xi, \beta\mathfrak{f}, \beta\mathfrak{b}, \beta\eta\mathfrak{b}, \beta\xi\mathfrak{b}, \beta\mathfrak{f}\mathfrak{b}\}.$$

PROOF. See Appendix A.1. □

THEOREM 3.5.5 (Models Over Frames). *Let* $\mathcal{M} = (\mathcal{D}, @, \mathcal{E}, \upsilon)$ *be an extensional model. There is an isomorphic model*

$$\mathcal{M}^h = (\mathcal{D}^h, @^h, \mathcal{E}^h, \upsilon^h)$$

over a frame such that $\mathcal{D}^h_o \subseteq \{\mathtt{T}, \mathtt{F}\}$ *and* υ^h *is the inclusion (or identity) map.*

PROOF. See Appendix A.1. □

COROLLARY 3.5.6 (Henkin Models). *Let* $\mathcal{M} = (\mathcal{D}, @, \mathcal{E}, \upsilon)$ *be an extensional* $\mathcal{S}$*-model such that* $\mathcal{D}_o$ *has two elements. There is an isomorphic Henkin* $\mathcal{S}$*-model* $\mathcal{M}^h$.

PROOF. This follows directly from Theorem 3.5.5. □

COROLLARY 3.5.7 (Henkin Models). *Assume* $\mathcal{S}$ *is a signature such that* $\top, \bot \in \mathcal{S}$ *or* $\neg \in \mathcal{S}$. *Let* $\mathcal{M} = (\mathcal{D}, @, \mathcal{E}, \upsilon)$ *be an extensional* $\mathcal{S}$*-model. There is an isomorphic Henkin* $\mathcal{S}$*-model* $\mathcal{M}^h$.

PROOF. This follows directly from Theorem 3.3.7 and Corollary 3.5.6. □

3.6. Consequences of Functionality

We can simplify the semantics by restricting our attention to functional evaluations and models. Functional evaluations are uniquely determined by the interpretations they give constants and parameters (cf. Theorem 3.6.8). Consequently, we can construct functional evaluations by giving a frame and an interpretation for parameters and constants. Furthermore, the interpretation of the logical constants in an extensional $\mathcal{S}$-model are uniquely determined (cf. Theorem 3.6.15). Therefore, we can construct functional models by giving a frame and an interpretation for parameters (cf. Theorem 3.6.20). Such constructions require the given frame to include possible interpretations for special λ-terms called *combinators* (cf. Definitions 3.6.1 and 3.6.9).

3.6.1. Combinators. Every λ-term can be β-expanded to a λ-term which only have very special λ-abstractions. Restricting our attention to such terms provides a way to define an evaluation function by defining the function only on these special terms.

DEFINITION 3.6.1 (**SK**-Combinatory Formulas). For all types α, β, and γ, we define two families of closed formulas we call *combinators*:

- $\mathbf{K}_{\alpha\beta\alpha} := \lambda x_\alpha \lambda y_\beta\, x$
- $\mathbf{S}_{\gamma\alpha(\beta\alpha)(\gamma\beta\alpha)} := \lambda u_{\gamma\beta\alpha} \lambda v_{\beta\alpha} \lambda w_\alpha \,\blacksquare\, u\, w \,\blacksquare\, v\, w$

We define the set of **SK**-*combinatory formulas* to be the least subset of $\bigcup_\alpha \mathit{wff}_\alpha(\mathcal{S})$ containing every **K** and **S**, every constant, every parameter, every variable, and closed under application.

As proven in [4], every formula can be β-expanded to an **SK**-combinatory formula. (This is true in the untyped case as well. See, for example, [14; 34].) The way to handle λ-abstraction is to define an abstraction operation on **SK**-combinatory formulas. We define this operation following Definition 7.1.5 in [14] (and adopting his $\lambda^* x$ notation for the operation). The same definition is given as Definition 9.20 in [34] (called "weak abstraction") who credit the definition to Curry and Feys [27]. This definition is also implicit in Andrews' proof of Lemma 1 in [4].

DEFINITION 3.6.2 (Weak Abstraction). For each variable x_α, we define an operation $\lambda^* x_\alpha$ taking **SK**-combinatory formulas of type β to **SK**-combinatory formulas of type $(\beta\alpha)$ by induction.

- If $x_\alpha \notin \mathbf{Free}(\mathbf{B}_\beta)$, then let $\lambda^* x_\alpha(\mathbf{B}) := [\mathbf{K}_{\beta\alpha\beta}\, \mathbf{B}]$.
- Let $\lambda^* x_\alpha(x_\alpha) := [\mathbf{S}_{\alpha\alpha(\alpha\alpha\alpha)(\alpha(\alpha\alpha)\alpha)}\, \mathbf{K}_{\alpha(\alpha\alpha)\alpha}\, \mathbf{K}_{\alpha\alpha\alpha}]$.
- Let $\lambda^* x_\alpha([\mathbf{F}_{\beta\delta}\, \mathbf{D}_\delta]) := [\mathbf{S}_{\beta\alpha(\delta\alpha)(\beta\delta\alpha)}\, \lambda^* x_\alpha(\mathbf{F})\, \lambda^* x_\alpha(\mathbf{D})]$.

LEMMA 3.6.3. *For any* **SK**-*combinatory formula* $\mathbf{B}_\beta$ *and variable* x_α, $\mathbf{Free}(\lambda^* x_\alpha(\mathbf{B})) = (\mathbf{Free}(\mathbf{B}) \setminus \{x_\alpha\})$.

PROOF. This follows by an easy induction on **B**. □

LEMMA 3.6.4. *For any* **SK**-*combinatory formula* $\mathbf{B}_\beta$ *and variable* x_α,

$$\lambda^* x_\alpha(\mathbf{B}) \xrightarrow{\beta} [\lambda x_\alpha\, \mathbf{B}].$$

PROOF. We prove this by induction on **B**. If $x_\alpha \notin \mathbf{Free}(\mathbf{B})$, then

$$\lambda^* x_\alpha(\mathbf{B}) = [\mathbf{K}\,\mathbf{B}] \xrightarrow{\beta} [\lambda x_\alpha\, \mathbf{B}]$$

by the definition of **K**. By the definitions of **S** and **K**, we have

$$\lambda^* x_\alpha(x) = [\mathbf{S}\,\mathbf{K}\,\mathbf{K}] \xrightarrow{\beta} [\lambda x_\alpha \centerdot \mathbf{K}\, x \centerdot \mathbf{K}\, x] \xrightarrow{\beta} [\lambda x_\alpha\, x]$$

By the inductive hypothesis and the definition of **S**, we have

$$\lambda^* x_\alpha([\mathbf{F}\,\mathbf{D}]) = \mathbf{S}\,\lambda^* x(\mathbf{F})\,\lambda^* x(\mathbf{D}) \xrightarrow{\beta} [\lambda x_\alpha \centerdot \lambda^* x(\mathbf{F})\, x\, [\lambda^* x(\mathbf{D})\, x]]$$
$$\xrightarrow{\beta} [\lambda x_\alpha \centerdot [\lambda x_\alpha\, \mathbf{F}]\, x \centerdot [\lambda x_\alpha\, \mathbf{D}] x] \xrightarrow{\beta} [\lambda x_\alpha \centerdot \mathbf{F}\,\mathbf{D}]$$

□

We can now define an operation which β-expands any formula to an **SK**-combinatory formula. This definition is implicit in Andrews' proof of Proposition 1 in [4].

DEFINITION 3.6.5. We define an operation taking $\mathbf{A} \in \mathit{wff}_\alpha(\mathcal{S})$ to an **SK**-combinatory formula $\mathbf{A}^\natural$ by induction on **A**.

- Let $c^\natural := c$ for any constant $c \in \mathcal{S}$.
- Let $W^\natural := W$ for any parameter W.
- Let $x^\natural := x$ for any variable x.
- Let $[\mathbf{F}\,\mathbf{B}]^\natural := [\mathbf{F}^\natural\,\mathbf{B}^\natural]$.
- Let $[\lambda x_\alpha\, \mathbf{B}]^\natural := \lambda^* x_\alpha(\mathbf{B}^\natural)$.

LEMMA 3.6.6. *For every* $\mathbf{A} \in \mathit{wff}_\alpha(\mathcal{S})$, $\mathbf{Free}(\mathbf{A}^\natural) = \mathbf{Free}(\mathbf{A})$.

PROOF. This follows by an easy induction on **A** and Lemma 3.6.3.

□

LEMMA 3.6.7. *For every* $\mathbf{A} \in \mathit{wff}_\alpha(\mathcal{S})$, $\mathbf{A}^\natural$ *is an* **SK**-*combinatory formula and* $\mathbf{A}^\natural \xrightarrow{\beta} \mathbf{A}$.

PROOF. The fact that $\mathbf{A}^\natural$ is an **SK**-combinatory formula is obvious by inspecting Definition 3.6.5. We prove $\mathbf{A}^\natural \xrightarrow{\beta} \mathbf{A}$ by induction on **A**. If **A** is a constant, parameter or variable, then $\mathbf{A}^\natural$ is already **A**. If **A** is $[\mathbf{F}\,\mathbf{B}]$, then we have

$$[\mathbf{F}\,\mathbf{B}]^\natural = [\mathbf{F}^\natural\,\mathbf{B}^\natural] \xrightarrow{\beta} [\mathbf{F}\,\mathbf{B}]$$

by the inductive hypothesis. If **A** is $[\lambda y_\beta\, \mathbf{D}]$, then we have

$$[\lambda y_\beta\, \mathbf{D}]^\natural = [\lambda^* y_\beta(\mathbf{D}^\natural)] \xrightarrow{\beta} [\lambda y_\beta\, \mathbf{D}^\natural] \xrightarrow{\beta} [\lambda y_\beta\, \mathbf{D}]$$

by Lemma 3.6.4 and the inductive hypothesis. □

3.6.2. Functional Evaluations and Models. We now prove an evaluation function $\mathcal{E}$ over a functional applicative structure is uniquely determined by the values $\mathcal{E}(W)$ for parameters $W \in \mathcal{P}$ and the values $\mathcal{E}(c)$ for constants $c \in \mathcal{S}$.

THEOREM 3.6.8 (Uniqueness of Functional Evaluations). *Suppose* $(\mathcal{D}, @)$ *is a functional applicative structure and* $\mathcal{S}$ *is a signature. If* $\mathcal{E}^1$ *and* $\mathcal{E}^2$ *are* $\mathcal{S}$*-evaluations for* $(\mathcal{D}, @)$ *such that* $\mathcal{E}^1|_{\mathcal{S}\cup\mathcal{P}} = \mathcal{E}^2|_{\mathcal{S}\cup\mathcal{P}}$, *then* $\mathcal{E}^1 = \mathcal{E}^2$.

PROOF. By induction on $\mathbf{A} \in \mathit{wff}_\alpha(\mathcal{S})$, we prove $\mathcal{E}^1_\varphi(\mathbf{A}) = \mathcal{E}^2_\varphi(\mathbf{A})$ for every assignment φ. If $\mathbf{A}$ is a constant or parameter, then we have assumed $\mathcal{E}^1(\mathbf{A}) = \mathcal{E}^2(\mathbf{A})$. If $\mathbf{A}$ is a variable x_α, we know

$$\mathcal{E}^1_\varphi(x) = \varphi(x) = \mathcal{E}^2_\varphi(x).$$

If $\mathbf{A}$ is an application of the form $[\mathbf{F}\,\mathbf{B}]$, then we know

$$\mathcal{E}^1_\varphi([\mathbf{F}\,\mathbf{B}]) = \mathcal{E}^1_\varphi(\mathbf{F})@\mathcal{E}^1_\varphi(\mathbf{B}) = \mathcal{E}^2_\varphi(\mathbf{F})@\mathcal{E}^2_\varphi(\mathbf{B}) = \mathcal{E}^2_\varphi([\mathbf{F}\,\mathbf{B}])$$

by the inductive hypotheses at $\mathbf{F}$ and $\mathbf{B}$.

For the λ-abstraction step, we need functionality. Suppose $\mathbf{A}$ is of the form $[\lambda x_\beta\, \mathbf{C}_\gamma]$. We know $\mathcal{E}^1_{\varphi,[\mathsf{b}/x]}(\mathbf{C}) = \mathcal{E}^2_{\varphi,[\mathsf{b}/x]}(\mathbf{C})$ for every $\mathsf{b} \in \mathcal{D}_\beta$ by the inductive hypothesis. Hence $\mathcal{E}^1_\varphi(\mathbf{A})@\mathsf{b} = \mathcal{E}^2_\varphi(\mathbf{A})@\mathsf{b}$ for every $\mathsf{b} \in \mathcal{D}_\beta$. By functionality $\mathcal{E}^1_\varphi(\mathbf{A}) = \mathcal{E}^2_\varphi(\mathbf{A})$. □

We could prove a similar result for the nonfunctional case if we further assume $\mathcal{E}^1(\mathbf{K}) = \mathcal{E}^2(\mathbf{K})$ and $\mathcal{E}^1(\mathbf{S}) = \mathcal{E}^2(\mathbf{S})$ for every combinator $\mathbf{K}$ and $\mathbf{S}$.

We next demonstrate that under certain closure conditions, one can always extend an interpretation of constants and parameters in a functional applicative structure to an evaluation.

DEFINITION 3.6.9 (Combinatory). Let $\mathcal{A} = (\mathcal{D}, @)$ be an applicative structure. We say $\mathcal{A}$ is *combinatory* if for all types α, β and γ there exist elements $\mathsf{k} \in \mathcal{D}_{\alpha\beta\alpha}$ and $\mathsf{s} \in \mathcal{D}_{\gamma\alpha(\beta\alpha)(\gamma\beta\alpha)}$ such that

1. $\mathsf{k}@\mathsf{a}@\mathsf{b} = \mathsf{a}$ for every $\mathsf{a} \in \mathcal{D}_\alpha$ and $\mathsf{b} \in \mathcal{D}_\beta$, and
2. $\mathsf{s}@\mathsf{g}@\mathsf{f}@\mathsf{a} = \mathsf{g}@\mathsf{a}@(\mathsf{f}@\mathsf{a})$ for every $\mathsf{a} \in \mathcal{D}_\alpha$, $\mathsf{f} \in \mathcal{D}_{\beta\alpha}$ and $\mathsf{g} \in \mathcal{D}_{\gamma\beta\alpha}$.

If $\mathcal{D}$ is a frame, then $\mathcal{D}$ is *combinatory* if the corresponding frame structure $(\mathcal{D}, @)$ is combinatory.

LEMMA 3.6.10. *Every full applicative structure is combinatory.*

PROOF. Let $\mathcal{A} = (\mathcal{D}, @)$ be a full applicative structure. We must verify the existence of values k and s satisfying the appropriate properties. Fix types α and β. For each a, let $k_{\mathsf{a}} : \mathcal{D}_\beta \to \mathcal{D}_\alpha$ be the constant function $k_{\mathsf{a}}(\mathsf{b}) := \mathsf{a}$. Since $\mathcal{A}$ is full, there is some $\mathsf{k}_{\mathsf{a}} \in \mathcal{D}_{\alpha\beta}$ such that $\mathsf{k}_{\mathsf{a}}@\mathsf{b} = \mathsf{a}$ for every b. Let $k : \mathcal{D}_\alpha \to \mathcal{D}_{\alpha\beta}$ be the function defined by $k(\mathsf{a}) := \mathsf{k}_{\mathsf{a}}$. Using fullness again, there is some $\mathsf{k} \in \mathcal{D}_{\alpha\beta\alpha}$ such that $\mathsf{k}@\mathsf{a} = \mathsf{k}_{\mathsf{a}}$ for every $\mathsf{a} \in \mathcal{D}_\alpha$. Thus $\mathsf{k}@\mathsf{a}@\mathsf{b} = \mathsf{k}_{\mathsf{a}}@\mathsf{b} = \mathsf{a}$ for every $\mathsf{a} \in \mathcal{D}_\alpha$ and $\mathsf{b} \in \mathcal{D}_\beta$.

The same procedure verifies there are appropriate values s. □

We now prove that one can always extend interpretations of constants and parameters in a combinatory functional applicative structure to an evaluation. Theorem 3.6.12 below is similar to Theorem 1 in [4], except that we do not assume the applicative structure is a frame structure and we do not insist that a representative for equality exists in each $\mathcal{D}_{o\alpha\alpha}$.

LEMMA 3.6.11. *Let $(\mathcal{D}, @)$ be a combinatory functional applicative structure and $\mathcal{S}$ be a signature. Further suppose $\mathcal{E}^\natural$ is a function taking any assignment φ and* **SK***-combinatory formula $\mathbf{A}_\alpha$ to a value $\mathcal{E}^\natural_\varphi(\mathbf{A}) \in \mathcal{D}_\alpha$ satisfying the following properties:*

1. *For each variable x, $\mathcal{E}^\natural_\varphi(x) = \varphi(x)$.*
2. *For any* **SK***-combinatory formulas $\mathbf{F}_{\gamma\beta}$ and $\mathbf{D}_\beta$, we have*
$$\mathcal{E}^\natural_\varphi([\mathbf{F}\,\mathbf{D}]) = \mathcal{E}^\natural_\varphi(\mathbf{F})@\mathcal{E}^\natural_\varphi(\mathbf{D}).$$
3. *$\mathcal{E}^\natural_\varphi(\mathbf{A}) = \mathcal{E}^\natural_\psi(\mathbf{A})$ whenever φ and ψ coincide on $\mathbf{Free}(\mathbf{A})$.*
4. *For each combinator $\mathbf{K}_{\alpha\beta\alpha}$, $\mathcal{E}^\natural_\varphi(\mathbf{K})@\mathsf{a}@\mathsf{b} = \mathsf{a}$ for every $\mathsf{a} \in \mathcal{D}_\alpha$ and $\mathsf{b} \in \mathcal{D}_\beta$.*
5. *For each combinator $\mathbf{S}_{\gamma\alpha(\beta\alpha)(\gamma\beta\alpha)}$, $\mathcal{E}^\natural_\varphi(\mathbf{S})@\mathsf{g}@\mathsf{f}@\mathsf{a} = \mathsf{g}@\mathsf{a}@(\mathsf{f}@\mathsf{a})$ for every $\mathsf{a} \in \mathcal{D}_\alpha$, $\mathsf{f} \in \mathcal{D}_{\beta\alpha}$ and $\mathsf{g} \in \mathcal{D}_{\gamma\beta\alpha}$.*

Then for every **SK***-combinatory formula $\mathbf{A}$, variable x_β and value $\mathsf{b} \in \mathcal{D}_\beta$, we have $\mathcal{E}^\natural_\varphi(\lambda^* x_\beta(\mathbf{A}))@\mathsf{b} = \mathcal{E}^\natural_{\varphi,[\mathsf{b}/x]}(\mathbf{A})$.*

PROOF. The proof is by induction on $\mathbf{A}$. See Appendix A.1. □

THEOREM 3.6.12 (Existence of Functional Evaluations). *Suppose $(\mathcal{D}, @)$ is a combinatory functional applicative structure and $\mathcal{S}$ is a signature. For any interpretation of constants and parameters*
$$\mathcal{I} : \mathcal{S} \cup \mathcal{P} \to \mathcal{D},$$

there is a unique $\mathcal{S}$-evaluation $\mathcal{E}$ such that $\mathcal{E}|_{\mathcal{S}\cup\mathcal{P}} = \mathcal{I}$.

PROOF. We only prove existence since uniqueness follows from Theorem 3.6.8. We first define $\mathcal{E}^{\natural}_{\varphi}(\mathbf{A})$ for **SK**-combinatory formulas $\mathbf{A}$ by induction.

- Let $\mathcal{E}^{\natural}_{\varphi}(x) := \varphi(x)$ for every variable x.
- Let $\mathcal{E}^{\natural}_{\varphi}(c) := \mathcal{I}(c)$ for every constant $c \in \mathcal{S}$.
- Let $\mathcal{E}^{\natural}_{\varphi}(W) := \mathcal{I}(W)$ for every parameter W.
- For each $\mathbf{K}_{\alpha\beta\alpha}$ let $\mathcal{E}^{\natural}_{\varphi}(\mathbf{K})$ be some $\mathsf{k} \in \mathcal{D}_{\alpha\beta\alpha}$ such that

$$\mathsf{k}@\mathsf{a}@\mathsf{b} = \mathsf{a}$$

  for every $\mathsf{a} \in \mathcal{D}_{\alpha}$ and $\mathsf{b} \in \mathcal{D}_{\beta}$.
- For each $\mathbf{S}_{\gamma\alpha(\beta\alpha)(\gamma\beta\alpha)}$ let $\mathcal{E}^{\natural}\varphi(\mathbf{S})$ be some $\mathsf{s} \in \mathcal{D}_{\gamma\alpha(\beta\alpha)(\gamma\beta\alpha)}$ such that

$$\mathsf{s}@\mathsf{g}@\mathsf{f}@\mathsf{a} = \mathsf{g}@\mathsf{a}@(\mathsf{f}@\mathsf{a})$$

  for every $\mathsf{a} \in \mathcal{D}_{\alpha}$, $\mathsf{f} \in \mathcal{D}_{\beta\alpha}$ and $\mathsf{g} \in \mathcal{D}_{\gamma\beta\alpha}$.
- Let $\mathcal{E}^{\natural}_{\varphi}([\mathbf{F}\,\mathbf{D}]) := \mathcal{E}^{\natural}_{\varphi}(\mathbf{F})@\mathcal{E}^{\natural}_{\varphi}(\mathbf{D})$ for combinatory terms $\mathbf{F}_{\alpha\delta}$ and $\mathbf{D}_{\delta}$.

An easy induction verifies that $\mathcal{E}^{\natural}_{\varphi}(\mathbf{A}) = \mathcal{E}^{\natural}_{\psi}(\mathbf{A})$ whenever φ and ψ coincide on $\mathbf{Free}(\mathbf{A})$.

Now, for arbitrary $\mathbf{A} \in \mathit{wff}_{\alpha}(\mathcal{S})$, we define $\mathcal{E}_{\varphi}(\mathbf{A}) := \mathcal{E}^{\natural}_{\varphi}(\mathbf{A}^{\natural})$ (cf. Definition 3.6.5). The fact that $\mathcal{E}$ extends $\mathcal{I}$ follows from the definition of $c^{\natural}$ and $W^{\natural}$ and the definition of $\mathcal{E}^{\natural}$. Using these facts, we compute

$$\mathcal{E}_{\varphi}(c) = \mathcal{E}^{\natural}_{\varphi}(c^{\natural}) = \mathcal{E}^{\natural}_{\varphi}(c) = \mathcal{I}(c)$$

for $c \in \mathcal{S}$ and

$$\mathcal{E}_{\varphi}(W) = \mathcal{E}^{\natural}_{\varphi}(W^{\natural}) = \mathcal{E}^{\natural}_{\varphi}(W) = \mathcal{I}(W)$$

for $W \in \mathcal{P}$.

We now check that $\mathcal{E}$ is a weak $\mathcal{S}$-evaluation function (cf. Definition 3.2.2).

1. For any variable x, $\mathcal{E}_{\varphi}(x) = \mathcal{E}^{\natural}_{\varphi}(x^{\natural}) = \mathcal{E}^{\natural}_{\varphi}(x) = \varphi(x)$.
2. For application,

$$\mathcal{E}_{\varphi}([\mathbf{F}\,\mathbf{D}]) = \mathcal{E}^{\natural}_{\varphi}([\mathbf{F}\,\mathbf{D}]^{\natural}) = \mathcal{E}^{\natural}_{\varphi}([\mathbf{F}^{\natural}\,\mathbf{D}^{\natural}])$$
$$= \mathcal{E}^{\natural}_{\varphi}(\mathbf{F}^{\natural})@\mathcal{E}^{\natural}_{\varphi}(\mathbf{D}^{\natural}) = \mathcal{E}_{\varphi}(\mathbf{F})@\mathcal{E}_{\varphi}(\mathbf{D}).$$

3. For any assignments φ and ψ, $\mathcal{E}_\varphi(W) = \mathcal{I}(W) = \mathcal{E}_\psi(W)$ and $\mathcal{E}_\varphi(c) = \mathcal{I}(c) = \mathcal{E}_\psi(c)$ for any parameter W and constant c.
4. By Lemma 3.6.11,

$$\mathcal{E}_\varphi([\lambda x_\beta\, \mathbf{A}])@\mathsf{b} = \mathcal{E}^\natural_\varphi(\lambda^* x_\beta(\mathbf{A}^\natural))@\mathsf{b} = \mathcal{E}^\natural_{\varphi,[\mathsf{b}/x]}(\mathbf{A}^\natural) = \mathcal{E}_{\varphi,[\mathsf{b}/x]}(\mathbf{A}).$$

Thus $\mathcal{E}$ is a weak evaluation function. Therefore, $\mathcal{E}$ is an evaluation function by Theorem 3.2.11. □

COROLLARY 3.6.13. *Let $\mathcal{A} = (\mathcal{D}, @)$ be a full and functional applicative structure and $\mathcal{S}$ be a signature. For any interpretation of constants and parameters $\mathcal{I} : \mathcal{S} \cup \mathcal{P} \to \mathcal{D}$, there is a unique $\mathcal{S}$-evaluation $\mathcal{E}$ such that $\mathcal{E}\big|_{\mathcal{S}\cup\mathcal{P}} = \mathcal{I}$.*

PROOF. We know $(\mathcal{D}, @)$ is combinatory by Lemma 3.6.10. The result follows from Theorem 3.6.12. □

In particular, to give an evaluation over standard frames we need only give the interpretation of constants and parameters.

COROLLARY 3.6.14. *Let $(\mathcal{D}, @)$ be a standard frame structure and $\mathcal{S}$ be a signature. For any interpretation of constants and parameters $\mathcal{I} : \mathcal{S} \cup \mathcal{P} \to \mathcal{D}$, there is a unique $\mathcal{S}$-evaluation $\mathcal{E}$ such that $\mathcal{E}\big|_{\mathcal{S}\cup\mathcal{P}} = \mathcal{I}$.*

PROOF. This follows directly from Corollary 3.6.13. □

Once we add a valuation to obtain extensional models, the interpretation of logical constants must be unique. Hence an evaluation for an extensional model is determined by its values on parameters.

THEOREM 3.6.15 (Uniqueness of Extensional Models). *Let $\mathcal{S}$ be a signature. Suppose $\mathcal{M}_1 = (\mathcal{D}, @, \mathcal{E}_1, \upsilon)$ and $\mathcal{M}_2 = (\mathcal{D}, @, \mathcal{E}_2, \upsilon)$ are extensional $\mathcal{S}$-models. If $\mathcal{E}_1\big|_{\mathcal{P}} = \mathcal{E}_2\big|_{\mathcal{P}}$, then $\mathcal{E}_1 = \mathcal{E}_2$.*

PROOF. By Theorem 3.6.8 we only need to check $\mathcal{E}_1\big|_{\mathcal{S}} = \mathcal{E}_2\big|_{\mathcal{S}}$. Since $\mathcal{M}_1$ and $\mathcal{M}_2$ are extensional, we know $(\mathcal{D}, @)$ is functional and υ is injective. We use these facts to check $\mathcal{E}_1(c) = \mathcal{E}_2(c)$ for each $c \in \mathcal{S}$.

$\top$: By $\mathfrak{L}_\top(\mathcal{E}_1(\top))$ and $\mathfrak{L}_\top(\mathcal{E}_2(\top))$, we know

$$\upsilon(\mathcal{E}_1(\top)) = \mathtt{T} = \upsilon(\mathcal{E}_2(\top)).$$

Since υ is injective, $\mathcal{E}_1(\top) = \mathcal{E}_2(\top)$.

$\bot$: As with the logical constant $\top$, $\upsilon(\mathcal{E}_1(\bot)) = \mathtt{F} = \upsilon(\mathcal{E}_2(\bot))$ implies $\mathcal{E}_1(\bot) = \mathcal{E}_2(\bot)$.

$\neg$: By functionality, it is enough to prove $\mathcal{E}_1(\neg)@\mathsf{a} = \mathcal{E}_2(\neg)@\mathsf{a}$ for each $\mathsf{a} \in \mathcal{D}_o$. By $\mathfrak{L}_\neg(\mathcal{E}_1(\neg))$ and $\mathfrak{L}_\neg(\mathcal{E}_2(\neg))$, if $\upsilon(\mathsf{a}) = \mathtt{F}$, then

$$\upsilon(\mathcal{E}_1(\neg)@\mathsf{a}) = \mathtt{T} = \upsilon(\mathcal{E}_2(\neg)@\mathsf{a})$$

and hence $\mathcal{E}_1(\neg)@\mathsf{a} = \mathcal{E}_2(\neg)@\mathsf{a}$. If $\upsilon(\mathsf{a}) = \mathtt{T}$, then

$$\upsilon(\mathcal{E}_1(\neg)@\mathsf{a}) = \mathtt{F} = \upsilon(\mathcal{E}_2(\neg)@\mathsf{a})$$

and so $\mathcal{E}_1(\neg)@\mathsf{a} = \mathcal{E}_2(\neg)@\mathsf{a}$. In either case, $\mathcal{E}_1(\neg)@\mathsf{a} = \mathcal{E}_2(\neg)@\mathsf{a}$ and by functionality $\mathcal{E}_1(\neg) = \mathcal{E}_2(\neg)$.

$\vee$: Again using functionality, we prove $\mathcal{E}_1(\vee)@\mathsf{a}@\mathsf{b} = \mathcal{E}_2(\vee)@\mathsf{a}@\mathsf{b}$ for each $\mathsf{a}, \mathsf{b} \in \mathcal{D}_o$. Using $\mathfrak{L}_\vee(\mathcal{E}_1(\vee))$ and $\mathfrak{L}_\vee(\mathcal{E}_2(\vee))$, we examine two cases. If $\upsilon(\mathsf{a}) = \mathtt{T}$ or $\upsilon(\mathsf{b}) = \mathtt{T}$, then

$$\upsilon(\mathcal{E}_1(\vee)@\mathsf{a}@\mathsf{b}) = \mathtt{T} = \upsilon(\mathcal{E}_2(\vee)@\mathsf{a}@\mathsf{b}).$$

If $\upsilon(\mathsf{a}) = \mathtt{F}$ and $\upsilon(\mathsf{b}) = \mathtt{F}$, then

$$\upsilon(\mathcal{E}_1(\vee)@\mathsf{a}@\mathsf{b}) = \mathtt{F} = \upsilon(\mathcal{E}_2(\vee)@\mathsf{a}@\mathsf{b}).$$

Hence in either case, $\mathcal{E}_1(\vee)@\mathsf{a}@\mathsf{b} = \mathcal{E}_2(\vee)@\mathsf{a}@\mathsf{b}$ since υ is injective.

$\wedge, \supset, \equiv$: Analogous to $\vee$.

Π^α: We prove $\mathcal{E}_1(\Pi^\alpha)@\mathsf{f} = \mathcal{E}_2(\Pi^\alpha)@\mathsf{f}$ for each $\mathsf{f} \in \mathcal{D}_{o\alpha}$ using $\mathfrak{L}_{\Pi^\alpha}(\mathcal{E}_1(\Pi^\alpha))$ and $\mathfrak{L}_{\Pi^\alpha}(\mathcal{E}_2(\Pi^\alpha))$. If $\upsilon(\mathsf{f}@\mathsf{a}) = \mathtt{T}$ for every $\mathsf{a} \in \mathcal{D}_\alpha$, then

$$\upsilon(\mathcal{E}_1(\Pi^\alpha)@\mathsf{f}) = \mathtt{T} = \upsilon(\mathcal{E}_2(\Pi^\alpha)@\mathsf{f}).$$

Otherwise,

$$\upsilon(\mathcal{E}_1(\Pi^\alpha)@\mathsf{f}) = \mathtt{F} = \upsilon(\mathcal{E}_2(\Pi^\alpha)@\mathsf{f}).$$

In either case, $\mathcal{E}_1(\Pi^\alpha)@\mathsf{f} = \mathcal{E}_2(\Pi^\alpha)@\mathsf{f}$ by injectivity of υ. Thus $\mathcal{E}_1(\Pi^\alpha) = \mathcal{E}_2(\Pi^\alpha)$ by functionality.

Σ^α: Analogous to Π^α.

$=^\alpha$: By $\mathfrak{L}_{=^\alpha}(\mathcal{E}_1(=^\alpha))$ and $\mathfrak{L}_{=^\alpha}(\mathcal{E}_2(=^\alpha))$, we can check

$$\upsilon(\mathcal{E}_1(=^\alpha)@\mathsf{a}@\mathsf{b}) = \upsilon(\mathcal{E}_2(=^\alpha)@\mathsf{a}@\mathsf{b})$$

for every $\mathsf{a}, \mathsf{b} \in \mathcal{D}_\alpha$. By injectivity of υ and functionality, we know $\mathcal{E}_1(=^\alpha) = \mathcal{E}_2(=^\alpha)$.

□

3.6.3. Frames. Using the results above, we can simplify our consideration of extensional models over frames.

DEFINITION 3.6.16. Let $\mathcal{D}$ be a frame, $\mathcal{D}_o \subseteq \{\mathtt{T}, \mathtt{F}\}$ and $\mathcal{I} : \mathcal{P} \to \mathcal{D}$ be a typed function. We call $\mathcal{F} := (\mathcal{D}, \mathcal{I})$ a *frame interpretation.* We say $\mathcal{F}$ is *standard* if $\mathcal{D}$ is a standard frame. We say $\mathcal{F}$ is *combinatory* if $\mathcal{D}$ is combinatory.

Note that frame interpretations do not depend on the signature $\mathcal{S}$. Instead, for each combinatory frame interpretation $\mathcal{F}$, we can determine a maximal signature $\mathcal{S}'$ for which $\mathcal{F}$ determines a unique extensional $\mathcal{S}$-model for any $\mathcal{S} \subseteq \mathcal{S}'$. Furthermore, up to isomorphism every extensional $\mathcal{S}$-model can be constructed in this way.

DEFINITION 3.6.17. Suppose $\mathcal{D}$ is a frame such that $\mathcal{D}_o \subseteq \{\mathtt{T}, \mathtt{F}\}$. Let @ denote function application so $\mathcal{A} := (\mathcal{D}, @)$ is a frame structure. Let $\upsilon : \mathcal{D}_o \to \{\mathtt{T}, \mathtt{F}\}$ be the inclusion map. We say $\mathcal{D}$ *realizes* a logical constant c_α (or c *is realized by* $\mathcal{D}$) if there is some $\mathsf{a} \in \mathcal{D}_\alpha$ such that $\mathfrak{L}_c(\mathsf{a})$ holds with respect to this υ. We say $\mathcal{D}$ *realizes* a signature $\mathcal{S}$ (or $\mathcal{S}$ *is realized by* $\mathcal{D}$) if it realizes every $c \in \mathcal{S}$. We define

$$\mathcal{S}^{\mathcal{D}} := \{c \in \mathcal{S}_{all} \mid \mathcal{D} \text{ realizes } c\}.$$

We now prove that every standard frame $\mathcal{D}$ with $\mathcal{D}_o = \{\mathtt{T}, \mathtt{F}\}$ realizes every logical constant. In Examples 3.6.23 and 3.6.24 we consider degenerate cases where $\mathcal{D}_o$ is a singleton.

LEMMA 3.6.18. *Let $\mathcal{D}$ be a standard frame with $\mathcal{D}_o = \{\mathtt{T}, \mathtt{F}\}$. Then $\mathcal{D}$ realizes $\mathcal{S}_{all}$.*

PROOF. For $\top$ and $\bot$, we have obvious witnesses $\mathtt{T} \in \mathcal{D}_o$ and $\mathtt{F} \in \mathcal{D}_o$. We take the obvious function in each remaining case. □

We now demonstrate how to extend a combinatory frame interpretation to an extensional model.

DEFINITION 3.6.19 (Frame Models). Let $\mathcal{F} = (\mathcal{D}, \mathcal{I})$ be a combinatory frame interpretation and $\mathcal{S}$ be any signature. We say $\mathcal{M}$ is a *frame $\mathcal{S}$-model induced by $\mathcal{F}$* if $\mathcal{M} = (\mathcal{D}, @, \mathcal{E}, \upsilon) \in \mathfrak{M}_{\beta\mathfrak{f}\mathfrak{b}}(\mathcal{S})$ where @ is function application, $\upsilon : \mathcal{D}_o \to \{\mathtt{T}, \mathtt{F}\}$ is the inclusion function and $\mathcal{E}|_{\mathcal{P}} = \mathcal{I}$.

THEOREM 3.6.20 (Frame Models). *Let $\mathcal{F} = (\mathcal{D}, \mathcal{I})$ be a combinatory frame interpretation. For any signature $\mathcal{S} \subseteq \mathcal{S}^{\mathcal{D}}$, there is a unique frame $\mathcal{S}$-model $\mathcal{M}$ induced by $\mathcal{F}$. Furthermore, $\mathcal{S}^{\mathcal{M}} = \mathcal{S}^{\mathcal{D}}$.*

PROOF. Let @ denote function application and $\upsilon : \mathcal{D}_o \to \{\mathtt{T}, \mathtt{F}\}$ be the inclusion function. Since $\mathcal{S} \subseteq \mathcal{S}^{\mathcal{D}}$, for each $c_\alpha \in \mathcal{S}$ we can define $\mathcal{I}'(c)$ to be an element of $\mathcal{D}_\alpha$ such that $\mathfrak{L}_c(\mathcal{I}'(c))$ holds with respect to υ. By Theorem 3.6.12 we can extend $\mathcal{I} \cup \mathcal{I}' : \mathcal{S} \cup \mathcal{P} \to \mathcal{D}$ to an $\mathcal{S}$-evaluation function $\mathcal{E}$ such that $\mathcal{E}|_{\mathcal{P}} = \mathcal{I}$ and $\mathcal{E}|_{\mathcal{S}} = \mathcal{I}'$. Let $\mathcal{M} := (\mathcal{D}, @, \mathcal{E}, \upsilon)$. By the choice of $\mathcal{I}'$, $\mathcal{M}$ is an $\mathcal{S}$-model. Clearly, $\mathcal{M}$ is extensional since υ is injective and $(\mathcal{D}, @)$ is functional as a frame structure. Uniqueness of $\mathcal{M}$ follows from Theorem 3.6.15.

Note that the underlying applicative structure of $\mathcal{M}$ is the frame structure $(\mathcal{D}, @)$. Also, υ is the inclusion function. Hence $\mathcal{M}$ realizes c_α iff $\mathcal{D}$ realizes c_α for every $c_\alpha \in \mathcal{S}_{all}$. Therefore, $\mathcal{S}^{\mathcal{M}} = \mathcal{S}^{\mathcal{D}}$. □

COROLLARY 3.6.21 (Standard Frame Interpretations). *Let*

$$\mathcal{F} = (\mathcal{D}, \mathcal{I})$$

be a standard frame interpretation. If $\mathcal{S} \subseteq \mathcal{S}^{\mathcal{D}}$, then there is a unique frame $\mathcal{S}$-model $\mathcal{M} = (\mathcal{D}, @, \mathcal{E}, \upsilon)$ induced by $\mathcal{F}$.

PROOF. Let @ denote function application. Since $\mathcal{D}$ is standard, we know $(\mathcal{D}, @)$ is full. Thus $\mathcal{D}$ is combinatory by Lemma 3.6.10. The proof is completed by an appeal to Theorem 3.6.20. □

We now give an example of a standard $\mathcal{S}$-model to determine all our model classes are nonempty.

EXAMPLE 3.6.22 (Singleton Model). Let $\mathcal{D}$ be the standard frame with $\mathcal{D}_o := \{\mathtt{T}, \mathtt{F}\}$ and $\mathcal{D}_\iota := \{0\}$. Let $\mathcal{I}(W_\alpha) \in \mathcal{D}_\alpha$ be arbitrary for parameters W. Note that $(\mathcal{D}, \mathcal{I})$ is a standard frame interpretation. By Lemma 3.6.18 we know $\mathcal{S}^{\mathcal{D}} = \mathcal{S}_{all}$. For any signature $\mathcal{S}$, we can use Corollary 3.6.21 to define the standard extensional $\mathcal{S}$-model $\mathcal{M} := (\mathcal{D}, @, \mathcal{E}, id)$ induced by $(\mathcal{D}, \mathcal{I})$. Hence

$$\mathcal{M} \in \mathfrak{M}_{\beta\mathfrak{fb}}(\mathcal{S}) \subseteq \mathfrak{M}_{\beta\eta}(\mathcal{S}) \subseteq \mathfrak{M}_{\beta}(\mathcal{S}).$$

There are two other extensional "singleton" models. We will be able to use these to establish a few independence results (cf. Lemmas 6.4.2 and 6.4.3).

EXAMPLE 3.6.23 (Singleton True Model). Let $\mathcal{D}$ be the standard frame with $\mathcal{D}_o := \{\mathtt{T}\}$ and $\mathcal{D}_\iota := \{0\}$. Clearly, each $\mathcal{D}_\alpha$ is a singleton. Let $\mathcal{I}(W_\alpha)$ be the unique member of $\mathcal{D}_\alpha$ for parameters $W_\alpha \in \mathcal{P}$. Note that $(\mathcal{D}, \mathcal{I})$ is a standard frame interpretation.

Assume $\mathcal{S}$ is a signature such that $\mathcal{S} \cap \{\bot, \neg\} = \emptyset$. Let υ be the inclusion of $\{\mathtt{T}\}$ into $\{\mathtt{T}, \mathtt{F}\}$. We must check that if $c_\alpha \in \mathcal{S}$, then $\mathfrak{L}_c(\mathsf{a})$ holds (with respect to the inclusion υ) for the unique $\mathsf{a} \in \mathcal{D}_\alpha$. Let d be the unique element of $\mathcal{D}_{ooo}$. For each type α, let e^α be the unique element of $\mathcal{D}_{o(o\alpha)}$ and q^α be the unique element of $\mathcal{D}_{o\alpha\alpha}$.

$\top$: Clearly, $\mathfrak{L}_\top(\mathtt{T})$ holds.

$\wedge$, $\vee$, $\supset$, $\equiv$: Each of $\mathfrak{L}_\wedge(\mathsf{d})$, $\mathfrak{L}_\vee(\mathsf{d})$, $\mathfrak{L}_\supset(\mathsf{d})$ and $\mathfrak{L}_\equiv(\mathsf{d})$ hold since $\mathsf{d}(\mathsf{a})(\mathsf{b}) = \mathtt{T}$, $\mathsf{a} = \mathtt{T}$ and $\mathsf{b} = \mathtt{T}$ hold for any $\mathsf{a}, \mathsf{b} \in \mathcal{D}_o$.

Π^α, Σ^α: Both $\mathfrak{L}_{\Pi^\alpha}(\mathsf{e}^\alpha)$ and $\mathfrak{L}_{\Sigma^\alpha}(\mathsf{e}^\alpha)$ hold since for any $\mathsf{f} \in \mathcal{D}_{o\alpha}$ we have $\mathsf{e}^\alpha(\mathsf{f}) = \mathtt{T}$ and $\mathsf{f}(\mathsf{a}) = \mathtt{T}$ (where a is the unique element of $\mathcal{D}_\alpha$).

$=^\alpha$: To check $\mathfrak{L}_{=^\alpha}(\mathsf{q}^\alpha)$ holds, note that for any $\mathsf{a}, \mathsf{b} \in \mathcal{D}_\alpha$ we have $\mathsf{q}^\alpha(\mathsf{a})(\mathsf{b}) = \mathtt{T}$ and $\mathsf{a} = \mathsf{b}$.

By Corollary 3.6.21, $(\mathcal{D}, \mathcal{I})$ induces a unique frame $\mathcal{S}$-model

$$\mathcal{M} = (\mathcal{D}, @, \mathcal{E}, \upsilon).$$

We can also build a singleton model with $\mathcal{D}_o = \{\mathtt{F}\}$.

EXAMPLE 3.6.24 (Singleton False Model). Assume $\mathcal{S}$ is an equality-free signature with $\mathcal{S} \cap \{\top, \neg, \supset, \equiv\} = \emptyset$. Let $\mathcal{D}$ be the standard frame with $\mathcal{D}_o := \{\mathtt{F}\}$ and $\mathcal{D}_\iota := \{0\}$ so that each $\mathcal{D}_\alpha$ is a singleton. Let υ be the inclusion of $\{\mathtt{F}\}$ into $\{\mathtt{T}, \mathtt{F}\}$. As in Example 3.6.23, we rely on Corollary 3.6.21 and only check that $\mathfrak{L}_c(\mathsf{a})$ for each $c_\alpha \in \mathcal{S}$, where a is the unique element of $\mathcal{D}_\alpha$. Let d be the unique element of $\mathcal{D}_{ooo}$. For each type α, let e^α be the unique element of $\mathcal{D}_{o(o\alpha)}$.

$\bot$: Clearly, $\mathfrak{L}_\bot(\mathtt{F})$ holds.

$\wedge$, $\vee$: Both $\mathfrak{L}_\wedge(\mathsf{d})$ and $\mathfrak{L}_\vee(\mathsf{d})$ hold since $\mathsf{d}(\mathsf{a})(\mathsf{b}) = \mathtt{F}$, $\mathsf{a} = \mathtt{F}$ and $\mathsf{b} = \mathtt{F}$ hold for any $\mathsf{a}, \mathsf{b} \in \mathcal{D}_o$.

Π^α, Σ^α: Both $\mathfrak{L}_{\Pi^\alpha}(\mathsf{e}^\alpha)$ and $\mathfrak{L}_{\Sigma^\alpha}(\mathsf{e}^\alpha)$ hold since for any $\mathsf{f} \in \mathcal{D}_{o\alpha}$ we have $\mathsf{e}^\alpha(\mathsf{f}) = \mathtt{F}$ and $\mathsf{f}(\mathsf{a}) = \mathtt{F}$ (where a is the unique element of $\mathcal{D}_\alpha$).

Using Corollary 3.6.21, this gives an $\mathcal{S}$-model $\mathcal{M} := (\mathcal{D}, @, \mathcal{E}, \upsilon)$.

Finally, we prove every extensional $\mathcal{S}$-model is isomorphic to a frame $\mathcal{S}$-model induced by some combinatory frame interpretation.

THEOREM 3.6.25. *Let $\mathcal{S}$ be a signature and $\mathcal{M} = (\mathcal{D}, @, \mathcal{E}, \upsilon)$ be an extensional $\mathcal{S}$-model. Then there exists a frame interpretation $\mathcal{F}^h$ and frame $\mathcal{S}$-model $\mathcal{M}^h$ induced by $\mathcal{F}^h$ such that $\mathcal{M}$ is isomorphic to $\mathcal{M}^h$.*

PROOF. By Theorem 3.5.5 there is an $\mathcal{S}$-model

$$\mathcal{M}^h = (\mathcal{D}^h, @^h, \mathcal{E}^h, \upsilon^h)$$

over a frame isomorphic to $\mathcal{M}$ where $\mathcal{D}^h_o \subseteq \{\mathtt{T}, \mathtt{F}\}$ and υ^h the inclusion map. Define the frame interpretation $\mathcal{F}^h := (\mathcal{D}^h, \mathcal{E}^h|_{\mathcal{P}})$. The frame interpretation $\mathcal{F}^h$ is combinatory since $\mathcal{E}^h(\mathbf{K})$ and $\mathcal{E}^h(\mathbf{S})$ provide the appropriate witnesses. Clearly $\mathcal{M}^h$ is the frame $\mathcal{S}$-model induced by $\mathcal{F}^h$. □

3.6.4. Operations in Frames. The domains of combinatory frames $\mathcal{D}$ are closed under all λ-definable operations. It is useful to record this for particular operations and introduce notations for these operations.

LEMMA 3.6.26. *Let $\mathcal{D}$ be a combinatory frame. For any type α, the identity function $id : \mathcal{D}_\alpha \to \mathcal{D}_\alpha$ is in $\mathcal{D}_{\alpha\alpha}$.*

PROOF. Let $\mathcal{S} := \emptyset$ and $\mathcal{I} : \mathcal{P} \to \mathcal{D}$ be some interpretation of parameters. By Theorem 3.6.12 there is a unique $\mathcal{S}$-evaluation $\mathcal{E}$ such that $\mathcal{E}|_{\mathcal{S}\cup\mathcal{P}} = \mathcal{I}$. Let φ be some assignment into $\mathcal{D}$. For any $\mathsf{a} \in \mathcal{D}_\alpha$, we know

$$\mathcal{E}_\varphi(\lambda x_\alpha x)(\mathsf{a}) = \mathcal{E}_{\varphi,[\mathsf{a}/x]}(x) = \mathsf{a} = id(\mathsf{a}).$$

Thus $id = \mathcal{E}(\lambda x_\alpha x) \in \mathcal{D}_{\alpha\alpha}$. □

Projection functions form another important class of functions.

DEFINITION 3.6.27. Let A, B and F be sets with $F \subseteq A^B$. For each $\mathsf{b} \in B$, the *projection function* $proj_{\mathsf{b}} : F \to A$ *induced by* b is defined by $proj_{\mathsf{b}}(\mathsf{u}) := \mathsf{u}(\mathsf{b})$ for each $\mathsf{u} \in F$. We say some $g : F \to A$ is a *projection function* if there is some $\mathsf{b} \in B$ such that $g = proj_{\mathsf{b}}$.

LEMMA 3.6.28. *Let $\mathcal{D}$ be a combinatory frame and $\alpha, \beta \in \mathcal{T}$ be types. For each $\mathsf{b} \in \mathcal{D}_\beta$, the projection function $proj_{\mathsf{b}} : \mathcal{D}_{\alpha\beta} \to \mathcal{D}_\alpha$ is in $\mathcal{D}_{\alpha(\alpha\beta)}$.*

PROOF. Let $\mathcal{S} := \emptyset$ and $\mathcal{I} : \mathcal{P} \to \mathcal{D}$ be some interpretation of parameters. By Theorem 3.6.12 there is a unique $\mathcal{S}$-evaluation $\mathcal{E}$ such that $\mathcal{E}\big|_{\mathcal{S}\cup\mathcal{P}} = \mathcal{I}$. Let y_β be a variable and φ be some assignment into $\mathcal{D}$ with $\varphi(y_\beta) = \mathsf{b}$. For any $\mathsf{u} \in \mathcal{D}_{\alpha\beta}$, we compute

$$\mathcal{E}_\varphi(\lambda u_{\alpha\beta}[u\, y])(\mathsf{u}) = \mathcal{E}_{\varphi,[\mathsf{u}/u]}(u\, y) = \mathsf{u}(\mathsf{b}) = proj_{\mathsf{b}}(\mathsf{u}).$$

Thus $proj_{\mathsf{b}} = \mathcal{E}_\varphi(\lambda u_{\alpha\beta}[u\, y]) \in \mathcal{D}_{\alpha(\alpha\beta)}$. □

When studying combinatory frames $\mathcal{D}$ which realize $\neg$, $\wedge$ and $\vee$, it is useful to reduce the problem of studying the elements of a domain $\mathcal{D}_{o\alpha}$ to studying certain collections of elements from which the others can be defined. Describing $\mathcal{D}_{o\alpha}$ in this way is analogous to describing a topology by giving a basis.

For convenience, we extend the notation for binary unions, binary intersections and complements to characteristic functions.

DEFINITION 3.6.29. Let A be a set and $\mathsf{f}, \mathsf{g} : A \to \{\mathtt{T}, \mathtt{F}\}$. We define $\overline{\mathsf{f}} : A \to \{\mathtt{T}, \mathtt{F}\}$ to be

$$\overline{\mathsf{f}}(\mathsf{a}) := \begin{cases} \mathtt{F} & \text{if } \mathsf{f}(\mathsf{a}) = \mathtt{T} \\ \mathtt{T} & \text{if } \mathsf{f}(\mathsf{a}) = \mathtt{F} \end{cases}$$

for each $\mathsf{a} \in A$. We define $(\mathsf{f} \cup \mathsf{g}) : A \to \{\mathtt{T}, \mathtt{F}\}$ to be

$$(\mathsf{f} \cup \mathsf{g})(\mathsf{a}) := \begin{cases} \mathtt{T} & \text{if } \mathsf{f}(\mathsf{a}) = \mathtt{T} \text{ or } \mathsf{g}(\mathsf{a}) = \mathtt{T} \\ \mathtt{F} & \text{otherwise} \end{cases}$$

for each $\mathsf{a} \in A$. We define $(\mathsf{f} \cap \mathsf{g}) : A \to \{\mathtt{T}, \mathtt{F}\}$ to be

$$(\mathsf{f} \cap \mathsf{g})(\mathsf{a}) := \begin{cases} \mathtt{T} & \text{if } \mathsf{f}(\mathsf{a}) = \mathtt{T} \text{ and } \mathsf{g}(\mathsf{a}) = \mathtt{T} \\ \mathtt{F} & \text{otherwise} \end{cases}$$

for each $\mathsf{a} \in A$.

The frames of interest are closed under these operations.

LEMMA 3.6.30. *Let $\mathcal{D}$ be a combinatory frame realizing $\neg$, $\wedge$, and $\vee$ where $\mathcal{D}_o = \{\mathtt{T}, \mathtt{F}\}$. Let α be any type. For any $\mathsf{f} \in \mathcal{D}_{o\alpha}$, $\overline{\mathsf{f}} \in \mathcal{D}_{o\alpha}$. For any $\mathsf{f}, \mathsf{g} \in \mathcal{D}_{o\alpha}$, $\mathsf{f} \cup \mathsf{g} \in \mathcal{D}_{o\alpha}$ and $\mathsf{f} \cap \mathsf{g} \in \mathcal{D}_{o\alpha}$.*

PROOF. Let $\mathcal{I} : \mathcal{P} \to \mathcal{D}$ be any interpretation of parameters. Note that $(\mathcal{D}, \mathcal{I})$ is a combinatory frame interpretation. Let

$$\mathcal{S} := \{\neg, \wedge, \vee\} \subseteq \mathcal{S}^{\mathcal{D}}.$$

By Theorem 3.6.20 there is a unique frame $\mathcal{S}$-model $\mathcal{M} = (\mathcal{D}, @, \mathcal{E}, \upsilon)$ induced by $(\mathcal{D}, \mathcal{I})$. Note that υ is the identity function and $@$ is function application. Let $f_{o\alpha}$, $g_{o\alpha}$ and x_α be distinct variables.

Suppose $\mathsf{f} \in \mathcal{D}_{o\alpha}$. Let φ be an assignment with $\varphi(f) = \mathsf{f}$. Using $\mathfrak{L}_\neg(\mathcal{E}(\neg))$, we compute

$$\mathcal{E}_\varphi(\lambda x_\alpha[\neg f\, x])(\mathsf{a}) = \mathcal{E}_{\varphi,[\mathsf{a}/x]}(\neg f\, x) = \overline{\mathsf{f}}(\mathsf{a})$$

for each $\mathsf{a} \in \mathcal{D}_\alpha$. Hence $\overline{\mathsf{f}} = \mathcal{E}_\varphi(\lambda x_\alpha[\neg f\, x]) \in \mathcal{D}_{o\alpha}$.

Suppose $\mathsf{f}, \mathsf{g} \in \mathcal{D}_{o\alpha}$. Let φ be an assignment with $\varphi(f) = \mathsf{f}$ and $\varphi(g) = \mathsf{g}$. Using $\mathfrak{L}_\vee(\mathcal{E}(\vee))$ and $\mathfrak{L}_\wedge(\mathcal{E}(\wedge))$, we compute

$$\mathcal{E}_\varphi(\lambda x_\alpha[f\, x \vee g\, x])(\mathsf{a}) = \mathcal{E}_{\varphi,[\mathsf{a}/x]}(f\, x \vee g\, x) = (\mathsf{f} \cup \mathsf{g})(\mathsf{a})$$

and

$$\mathcal{E}_\varphi(\lambda x_\alpha[f\, x \wedge g\, x])(\mathsf{a}) = \mathcal{E}_{\varphi,[\mathsf{a}/x]}(f\, x \wedge g\, x) = (\mathsf{f} \cap \mathsf{g})(\mathsf{a})$$

for each $\mathsf{a} \in \mathcal{D}_\alpha$. Thus

$$\mathsf{f} \cup \mathsf{g} = \mathcal{E}_\varphi(\lambda x_\alpha[f\, x \vee g\, x]) \in \mathcal{D}_{o\alpha}$$

and

$$\mathsf{f} \cap \mathsf{g} = \mathcal{E}_\varphi(\lambda x_\alpha[f\, x \wedge g\, x]) \in \mathcal{D}_{o\alpha}.$$

□

We next define the closure of a set of functions under these operations.

DEFINITION 3.6.31. Let A be a set and $S \subseteq \{\mathtt{T}, \mathtt{F}\}^A$. Let

$$k_\mathtt{T} : A \longrightarrow \{\mathtt{T}, \mathtt{F}\} \text{ and } k_\mathtt{F} : A \longrightarrow \{\mathtt{T}, \mathtt{F}\}$$

be the constant functions $k_\mathtt{T}(\mathsf{a}) := \mathtt{T}$ and $k_\mathtt{F}(\mathsf{a}) := \mathtt{F}$. We define the *propositional closure* $\overline{S}$ of S to be the least subset of $\{\mathtt{T}, \mathtt{F}\}^A$ such that

- $S \subseteq \overline{S}$,
- $k_\mathtt{T} \in \overline{S}$,
- $k_\mathtt{F} \in \overline{S}$,
- $\overline{\mathsf{f}} \in \overline{S}$ whenever $\mathsf{f} \in \overline{S}$,
- $\mathsf{f} \cup \mathsf{g} \in \overline{S}$ whenever $\mathsf{f}, \mathsf{g} \in \overline{S}$, and
- $\mathsf{f} \cap \mathsf{g} \in \overline{S}$ whenever $\mathsf{f}, \mathsf{g} \in \overline{S}$.

LEMMA 3.6.32. *Let A be a set and $S \subseteq \{\mathtt{T}, \mathtt{F}\}^A$. If S is infinite, then $\overline{S}$ has the same cardinality as S.*

PROOF. Let $S^0 := S \cup \{k_{\mathtt{T}}, k_{\mathtt{F}}\}$ and for each $n \in \mathbb{N}$ let

$$S^{n+1} := S^n \cup \{\bar{\mathsf{f}} \mid \mathsf{f} \in S^n\} \cup \{\mathsf{f} \cup \mathsf{g} \mid \mathsf{f}, \mathsf{g} \in S^n\} \cup \{\mathsf{f} \cap \mathsf{g} \mid \mathsf{f}, \mathsf{g} \in S^n\}.$$

It is easy to check $\overline{S} = \bigcup_{n \in \mathbb{N}} S^n$. If S is infinite, it is easy to prove each S^n has the same cardinality as S. Thus the countable union $\bigcup_{n \in \mathbb{N}} S^n$ has the same cardinality as S. □

THEOREM 3.6.33. *Let $\mathcal{D}$ be a combinatory frame realizing $\neg$, $\wedge$, and $\vee$ where $\mathcal{D}_o = \{\mathtt{T}, \mathtt{F}\}$. Let α be any type and $S \subseteq \mathcal{D}_{o\alpha}$. Then $\overline{S} \subseteq \mathcal{D}_{o\alpha}$. Furthermore, if S is infinite and $\overline{S} = \mathcal{D}_{o\alpha}$, then $\mathcal{D}_{o\alpha}$ has the same cardinality as S.*

PROOF. Let $\mathsf{k} \in \mathcal{D}_{o\alpha o}$ be such that $\mathsf{k}(\mathsf{b})(\mathsf{a}) = \mathsf{b}$ for each $\mathsf{a} \in \mathcal{D}_\alpha$ and $\mathsf{b} \in \mathcal{D}_o$. The constant $\mathtt{T}$ function $\mathsf{k}(\mathtt{T})$ and constant $\mathtt{F}$ function $\mathsf{k}(\mathtt{F})$ are members of $\mathcal{D}_{o\alpha}$. By Lemma 3.6.30 we know $\mathcal{D}_{o\alpha}$ is closed under the operations $\mathsf{f} \mapsto \bar{\mathsf{f}}$, $(\mathsf{f}, \mathsf{g}) \mapsto (\mathsf{f} \cup \mathsf{g})$ and $(\mathsf{f}, \mathsf{g}) \mapsto (\mathsf{f} \cap \mathsf{g})$. Since $\overline{S}$ is the least such set, $\overline{S} \subseteq \mathcal{D}_{o\alpha}$.

Assume S is infinite and $\overline{S} = \mathcal{D}_{o\alpha}$. By Lemma 3.6.32 we know $\overline{S} = \mathcal{D}_{o\alpha}$ has the same cardinality as S. □

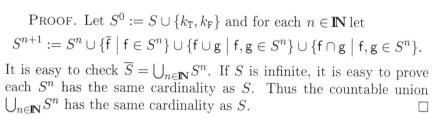

CHAPTER 4

Semantic Constructions

In Chapter 3, we defined semantics for fragments of type theory. We now introduce techniques for constructing various frames, applicative structures, evaluations and models.

In Section 4.1 we consider applicative structures $\langle \mathcal{D}, @ \rangle$ in which each member of the domain $\mathcal{D}_\alpha$ is a pair $\langle \mathbf{A}_\alpha, a \rangle$ where $\mathbf{A}$ is a closed normal term of type α and a is some value. We will call such an a a *possible value* for $\mathbf{A}$ (cf. Definition 4.1.2) and $\langle \mathcal{D}, @ \rangle$ a *possible values structure* (cf. Definition 4.1.1). Our notion of possible values structures provides a general theory which includes Andrews' V-complexes (cf. [2]). We will use possible values structures to construct models in $\mathfrak{M}_\beta(\mathcal{S})$ and $\mathfrak{M}_{\beta\eta}(\mathcal{S})$ for certain sets of sentences (cf. Theorem 5.6.4). The model existence theorem will be used to establish completeness of the equality-free sequent calculi $\mathcal{G}^{\mathcal{S}}_{\beta}$ and $\mathcal{G}^{\mathcal{S}}_{\beta\eta}$ relative to the model classes $\mathfrak{M}_\beta(\mathcal{S})$ and $\mathfrak{M}_{\beta\eta}(\mathcal{S})$ for equality-free signatures $\mathcal{S}$ (cf. Theorem 5.6.6).

In order to establish completeness of the extensional sequent calculus $\mathcal{G}^{\mathcal{S}}_{\beta\mathfrak{fb}}$ relative to the model class $\mathfrak{M}_{\beta\mathfrak{fb}}(\mathcal{S})$ for arbitrary signatures $\mathcal{S}$ (cf. Theorem 5.7.19), we use a possible values structure with a partial equivalence relation on each typed domain (cf. Theorem 5.7.17). In Section 4.2 we demonstrate how to construct quotients of models using such partial equivalence relations.

In Section 4.3 we construct non-extensional models using retractions. Retraction models will be used in Chapter 6 to relate extensional and non-extensional models (cf. Theorem 6.1.3).

In Sections 4.4 and 4.5 we construct combinatorial frames using logical relations. Such frames will be used in Chapter 6 to establish independence results in the presence of extensionality (cf. Theorems 6.7.8 and 6.7.40).

4.1. Possible Values

In Chapter 5, we will prove completeness of the sequent calculi $\mathcal{G}_*^{\mathcal{S}}$ with respect to the model classes $\mathfrak{M}_*(\mathcal{S})$ for each $* \in \{\beta, \beta\eta, \beta\mathfrak{f}\mathfrak{b}\}$ and signature $\mathcal{S} \subseteq \mathcal{S}_*$. Such completeness proofs for cut-free higher-order calculi are notoriously subtle. We know such proofs must require strong techniques since completeness implies cut-elimination and cut-elimination implies the consistency of analysis (cf. [5]). Takahashi and Prawitz introduced semantic techniques in [61] and [56] for proving cut-elimination for "relational" formulations of simple type theory (cf. Remark 2.1.44). Andrews extended the technique to Church-style formulations (with all function types) using V-complexes in [2]. The basic idea of the technique is to define $\mathcal{D}_o$ in a way that determines the truth value of some propositions and delays determining the truth value of others. (Prawitz used the phrase "possible values" in [56].) In this section, we use the same technique to define applicative structures and evaluations appropriate for proving completeness.

For the remainder of this section, let $*$ be β or $\beta\eta$. Likewise, let $\mathbf{A}^{\downarrow *}$ denote $\mathbf{A}^{\downarrow\beta}$ if $*$ is β or $\mathbf{A}^{\downarrow}$ if $*$ is $\beta\eta$.

DEFINITION 4.1.1 (Possible Values Structure). A *possible values structure for* $*$ is an applicative structure $\mathcal{F} = (\mathcal{D}, @)$ satisfying the following:

1. For each type $\alpha \in \mathcal{T}$, $\mathsf{a} \in \mathcal{D}_\alpha$ implies $\mathsf{a} = \langle \mathbf{A}, a\rangle$ for some a and term $\mathbf{A} \in cwff_\alpha(\mathcal{S})$ such that $\mathbf{A}^{\downarrow *} = \mathbf{A}$.
2. At each base type $\alpha \in \{o, \iota\}$, for every $\mathbf{A} \in cwff_\alpha(\mathcal{S})$, there exists some p with $\langle \mathbf{A}^{\downarrow *}, p\rangle \in \mathcal{D}_\alpha$.
3. For each function type $\alpha\beta$, $\langle \mathbf{G}, g\rangle \in \mathcal{D}_{\alpha\beta}$ iff $\mathbf{G} \in cwff_{\alpha\beta}(\mathcal{S})$, $\mathbf{G}^{\downarrow *} = \mathbf{G}$, $g : \mathcal{D}_\beta \longrightarrow \mathcal{D}_\alpha$ and for every $\langle \mathbf{B}, b\rangle \in \mathcal{D}_\beta$ the first component of $g(\langle \mathbf{B}, b\rangle)$ is $[\mathbf{G}\,\mathbf{B}]^{\downarrow *}$.
4. For each $\langle \mathbf{G}, g\rangle \in \mathcal{D}_{\alpha\beta}$ and $\langle \mathbf{B}, b\rangle \in \mathcal{D}_\beta$,
$$\langle \mathbf{G}, g\rangle @ \langle \mathbf{B}, b\rangle = g(\langle \mathbf{B}, b\rangle).$$

DEFINITION 4.1.2 (Possible Value). Let $\mathcal{A} = (\mathcal{D}, @)$ be a possible values structure for $*$. We call p a *possible value* for $\mathbf{A} \in cwff_\alpha(\mathcal{S})$ if $\langle \mathbf{A}^{\downarrow *}, p\rangle \in \mathcal{D}_\alpha$.

LEMMA 4.1.3 (Possible Values Exist). *Let $\mathcal{F}$ be a possible values structure for $*$. For each closed term $\mathbf{A} \in cwff_\alpha(\mathcal{S})$, there is a possible value p for $\mathbf{A}$ in $\mathcal{F}$.*

PROOF. We prove this by induction on the type α as in [2].

$\alpha \in \{o, \iota\}$: By the definition of a possible values structure, we know there is a p with $\langle \mathbf{A}^{\downarrow *}, p\rangle \in \mathcal{D}_\alpha$. The desired possible value is p.

$\gamma\beta$: By the inductive hypothesis, there are possible values $p^{[\mathbf{A}\,\mathbf{B}]}$ for $[\mathbf{A}\,\mathbf{B}]$ for each $\langle \mathbf{B}, b\rangle \in \mathcal{D}_\beta$. Using the axiom of choice, there is a function $p : \mathcal{D}_\beta \longrightarrow \mathcal{D}_\gamma$ such that

$$p(\langle \mathbf{B}, b\rangle) = \langle [\mathbf{A}\,\mathbf{B}]^{\downarrow *}, p^{[\mathbf{A}\,\mathbf{B}]}\rangle.$$

This guarantees p is a possible value for $\mathbf{A}$.

□

Since we are interested in interpreting a term $\mathbf{A} \in cwff_\alpha(\mathcal{S})$ to be a pair of the form $\langle \mathbf{A}^{\downarrow *}, a\rangle$ it is helpful to adopt a notation to reflect this.

DEFINITION 4.1.4. Let $\mathcal{A} = (\mathcal{D}, @)$ be a possible values structure for $*$. We define

$$\mathcal{D}_\alpha^{\mathbf{A}} := \{\langle \mathbf{A}^{\downarrow *}, a\rangle \in \mathcal{D}_\alpha \mid a \text{ is a possible value for } \mathbf{A}\}.$$

for each $\mathbf{A} \in cwff_\alpha$.

DEFINITION 4.1.5. Let $\mathcal{A} = (\mathcal{D}, @)$ be a possible values structure for $*$ and φ be an assignment into $\mathcal{A}$. For any $\mathbf{A} \in wff_\alpha(\mathcal{S})$, we define $\varphi_1(\mathbf{A})$ to be $\theta(\mathbf{A}) \in cwff_\alpha(\mathcal{S})$ where θ is the substitution with $\mathbf{Dom}(\theta) = \mathbf{Free}(\mathbf{A})$ and $\varphi(x_\beta) = \langle \theta(x_\beta), b\rangle \in \mathcal{D}_\beta$ for each variable $x_\beta \in \mathbf{Free}(\mathbf{A})$. We define $\varphi_1^*(\mathbf{A})$ to be $\varphi_1(\mathbf{A})^{\downarrow *}$.

REMARK 4.1.6. Note that the syntactic condition on $\mathcal{D}_{\alpha\beta}$ guarantees that if $\langle \mathbf{G}, g\rangle \in \mathcal{D}_{\alpha\beta}$, then $g : \mathcal{D}_\beta \longrightarrow \mathcal{D}_\alpha$ restricts to a mapping $g : \mathcal{D}_\beta^{\mathbf{B}} \longrightarrow \mathcal{D}_\alpha^{\mathbf{GB}}$ for any $\mathbf{B} \in cwff_\beta(\mathcal{S})$.

DEFINITION 4.1.7 (Possible Values Evaluation). A *possible values evaluation for* $*$ is an evaluation $\mathcal{J} = (\mathcal{D}, @, \mathcal{E})$ where $(\mathcal{D}, @)$ is a possible values structure for $*$ and $\mathcal{E}_\varphi(\mathbf{A}) \in \mathcal{D}_\alpha^{\varphi_1(\mathbf{A})}$ for every $\mathbf{A} \in wff_\alpha(\mathcal{S})$ and assignment φ.

We can always extend an appropriate interpretation of parameters and constants in a possible values structure to obtain a possible values evaluation.

THEOREM 4.1.8. *Let $\mathcal{A} = (\mathcal{D}, @)$ be a possible values structure for $*$ and $\mathcal{I} : \mathcal{P} \cup \mathcal{S} \to \mathcal{D}$ be an interpretation of parameters and constants such that $\mathcal{I}(W_\alpha) \in \mathcal{D}_\alpha^W$ for every parameter $W \in \mathcal{P}$ and $\mathcal{I}(c_\alpha) \in \mathcal{D}_\alpha^c$ for every constant $c \in \mathcal{S}$. There is an evaluation function $\mathcal{E}$ such that $\mathcal{J} := (\mathcal{D}, @, \mathcal{E})$ is a possible values evaluation for $*$, $\mathcal{E}(W_\alpha) = \mathcal{I}(W_\alpha)$ for every parameter W_α and $\mathcal{E}(c_\alpha) = \mathcal{I}(c_\alpha)$ for every constant c_α. Also, if $\mathcal{A}$ is a possible values structure for $\beta\eta$, then $\mathcal{J}$ is η-functional.*

PROOF. The evaluation function $\mathcal{E}$ can be defined by induction on terms. See Appendix A.2. □

4.2. Pers Over Domains

Models constructed using possible values are not extensional. However, we will still use a possible values evaluation to establish completeness of $\mathcal{G}_{\beta\mathfrak{f}\mathfrak{b}}^{\mathcal{S}}$. We construct an extensional model by taking a quotient of a possible values evaluation by a partial equivalence relation (cf. Theorem 5.7.16). The per (partial equivalence relation) will be used to make some elements equal and to throw other elements away (those which are not related to themselves by the per). Taking quotients by pers is more general than taking quotients by equivalence relations. Accordingly, we must require more conditions for the quotient to be well-defined. We now develop the theory of such constructions.

DEFINITION 4.2.1 ((Typed) Per). A (*typed*) *per* $\sim$ over a typed collection of sets $\mathcal{D}$ is a collection of partial equivalence relations (symmetric and transitive relations) $\sim_\alpha$ on each set $\mathcal{D}_\alpha$. For any two assignments φ and ψ into $\mathcal{D}$, we use the notation $\varphi \sim \psi$ to indicate that $\varphi(x) \sim \psi(x)$ for every variable x. We call φ a *$\sim$-assignment* if $\varphi \sim \varphi$. We say $\sim$ is a *per over an applicative structure* $(\mathcal{D}, @)$ or a *per over an evaluation* $(\mathcal{D}, @, \mathcal{E})$ whenever $\sim$ is a per over $\mathcal{D}$.

At each type α, we define $\overline{\mathcal{D}_\alpha^\sim} := \{\mathsf{a} \in \mathcal{D}_\alpha \mid \mathsf{a} \sim \mathsf{a}\}$.

DEFINITION 4.2.2 (Per Properties). Let $\mathcal{D}$ be a typed collection of nonempty sets and $\sim$ be a per over $\mathcal{D}$. We define the following properties for $\sim$ over $\mathcal{D}$.

∂_α^{tot}: The per $\sim$ is an equivalence relation on $\mathcal{D}_\alpha$.

$\partial_\alpha^{\subseteq}$: For all $\mathsf{a}, \mathsf{b} \in \mathcal{D}_\alpha$, if $\mathsf{a} \sim \mathsf{b}$, then $\mathsf{a} = \mathsf{b}$.

$\partial^{\neq\emptyset}$: For every type α, there exists some $\mathsf{a} \in \mathcal{D}_\alpha$ such that $\mathsf{a} \sim \mathsf{a}$.

Let $(\mathcal{D}, @)$ be an applicative structure. We define the following properties for $\sim$ over $(\mathcal{D}, @)$.

$\partial^{@}$: For all $\alpha, \beta \in \mathcal{T}$, $\mathsf{g}, \mathsf{h} \in \mathcal{D}_{\beta\alpha}$ and $\mathsf{a}, \mathsf{b} \in \mathcal{D}_\alpha$, if $\mathsf{g} \sim \mathsf{h}$ and $\mathsf{a} \sim \mathsf{b}$, then $\mathsf{g}@\mathsf{a} \sim \mathsf{h}@\mathsf{b}$.

$\partial^{\mathfrak{f}}$: For all $\alpha, \beta \in \mathcal{T}$, $\mathsf{g}, \mathsf{h} \in \mathcal{D}_{\beta\alpha}$, $\mathsf{g} \sim \mathsf{h}$ whenever for all $\mathsf{a}, \mathsf{b} \in \mathcal{D}_\alpha$ $\mathsf{a} \sim \mathsf{b}$ implies $\mathsf{g}@\mathsf{a} \sim \mathsf{h}@\mathsf{b}$.

Let $\mathcal{E}$ be an $\mathcal{S}$-evaluation function for $(\mathcal{D}, @)$. We define the following properties for $\sim$ over the evaluation $(\mathcal{D}, @, \mathcal{E})$.

$\partial^{\mathcal{P}}$: For every parameter $W_\alpha \in \mathcal{P}$, $\mathcal{E}(W) \sim \mathcal{E}(W)$ in $\mathcal{D}_\alpha$.

$\partial^{\mathcal{S}}$: For every constant $c_\alpha \in \mathcal{S}$, $\mathcal{E}(c) \sim \mathcal{E}(c)$ in $\mathcal{D}_\alpha$.

∂^{sub}: For every type α, every $\mathbf{A} \in \mathit{wff}_\alpha$, and all assignments φ and ψ, if $\varphi \sim \psi$, then $\mathcal{E}_\varphi(\mathbf{A}) \sim \mathcal{E}_\psi(\mathbf{A})$.

Finally, let $\upsilon : \mathcal{D}_o \to \{\mathtt{T}, \mathtt{F}\}$ be a function. We define the following properties for υ and $\sim$ over the applicative structure $(\mathcal{D}, @)$.

∂^{υ}: For all $\mathsf{a}, \mathsf{b} \in \mathcal{D}_o$, $\mathsf{a} \sim \mathsf{b}$ implies $\upsilon(\mathsf{a}) = \upsilon(\mathsf{b})$.

$\partial^{\mathfrak{b}}$: For all $\mathsf{a}, \mathsf{b} \in \mathcal{D}_o$, $\upsilon(\mathsf{a}) = \upsilon(\mathsf{b})$ implies $\mathsf{a} \sim \mathsf{b}$.

We further define the properties $\mathfrak{L}_c^{\partial}(\mathsf{a})$ which characterize logical constants with respect to υ and $\sim$ over the applicative structure $(\mathcal{D}, @)$ in Table 4.1.

DEFINITION 4.2.3 (Per Quotient of Applicative Structures). Let $\sim$ be a per over an applicative structure $\mathcal{A} := (\mathcal{D}, @)$. We say an applicative structure $(\mathcal{D}^{\sim}, @^{\sim})$ is the *per quotient structure of* $\mathcal{A}$ *by* $\sim$ if for each type α

$$\mathcal{D}^{\sim}_\alpha = \{[\![\mathsf{a}]\!]_\sim \mid \mathsf{a} \in \overline{\mathcal{D}^{\sim}_\alpha}\}$$

and each type $(\gamma\beta)$, $\mathsf{f} \in \overline{\mathcal{D}^{\sim}_{\gamma\beta}}$ and $\mathsf{b} \in \overline{\mathcal{D}^{\sim}_\beta}$,

$$[\![\mathsf{f}]\!]_\sim @^{\sim} [\![\mathsf{b}]\!]_\sim = [\![\mathsf{f}@\mathsf{b}]\!]_\sim.$$

It is clear from the definition that there can be at most one per quotient structure of $\mathcal{A}$ by $\sim$. We now demonstrate existence of such a quotient under certain conditions.

prop.	where	holds when			for all
$\mathfrak{L}^{\partial}_{\top}(\mathsf{a})$	$\mathsf{a} \in \mathcal{D}_o$	$\upsilon(\mathsf{a}) = \mathtt{T}$			
$\mathfrak{L}^{\partial}_{\bot}(\mathsf{b})$	$\mathsf{b} \in \mathcal{D}_o$	$\upsilon(\mathsf{b}) = \mathtt{F}$			
$\mathfrak{L}^{\partial}_{\neg}(\mathsf{n})$	$\mathsf{n} \in \mathcal{D}_{oo}$	$\upsilon(\mathsf{n}@\mathsf{a}) = \mathtt{T}$	iff	$\upsilon(\mathsf{a}) = \mathtt{F}$	$\mathsf{a} \in \overline{\mathcal{D}^{\sim}_o}$
$\mathfrak{L}^{\partial}_{\vee}(\mathsf{d})$	$\mathsf{d} \in \mathcal{D}_{ooo}$	$\upsilon(\mathsf{d}@\mathsf{a}@\mathsf{b}) = \mathtt{T}$	iff	$\upsilon(\mathsf{a}) = \mathtt{T}$ or $\upsilon(\mathsf{b}) = \mathtt{T}$	$\mathsf{a}, \mathsf{b} \in \overline{\mathcal{D}^{\sim}_o}$
$\mathfrak{L}^{\partial}_{\wedge}(\mathsf{c})$	$\mathsf{c} \in \mathcal{D}_{ooo}$	$\upsilon(\mathsf{c}@\mathsf{a}@\mathsf{b}) = \mathtt{T}$	iff	$\upsilon(\mathsf{a}) = \mathtt{T}$ and $\upsilon(\mathsf{b}) = \mathtt{T}$	$\mathsf{a}, \mathsf{b} \in \overline{\mathcal{D}^{\sim}_o}$
$\mathfrak{L}^{\partial}_{\supset}(\mathsf{i})$	$\mathsf{i} \in \mathcal{D}_{ooo}$	$\upsilon(\mathsf{i}@\mathsf{a}@\mathsf{b}) = \mathtt{T}$	iff	$\upsilon(\mathsf{a}) = \mathtt{F}$ or $\upsilon(\mathsf{b}) = \mathtt{T}$	$\mathsf{a}, \mathsf{b} \in \overline{\mathcal{D}^{\sim}_o}$
$\mathfrak{L}^{\partial}_{\equiv}(\mathsf{e})$	$\mathsf{e} \in \mathcal{D}_{ooo}$	$\upsilon(\mathsf{e}@\mathsf{a}@\mathsf{b}) = \mathtt{T}$	iff	$\upsilon(\mathsf{a}) = \upsilon(\mathsf{b})$	$\mathsf{a}, \mathsf{b} \in \overline{\mathcal{D}^{\sim}_o}$
$\mathfrak{L}^{\partial}_{\Pi^\alpha}(\pi)$	$\pi \in \mathcal{D}_{o(o\alpha)}$	$\upsilon(\pi@\mathsf{f}) = \mathtt{T}$	iff	$\forall \mathsf{a} \in \overline{\mathcal{D}^{\sim}_\alpha}\ \upsilon(\mathsf{f}@\mathsf{a}) = \mathtt{T}$	$\mathsf{f} \in \overline{\mathcal{D}^{\sim}_{o\alpha}}$
$\mathfrak{L}^{\partial}_{\Sigma^\alpha}(\sigma)$	$\sigma \in \mathcal{D}_{o(o\alpha)}$	$\upsilon(\sigma@\mathsf{f}) = \mathtt{T}$	iff	$\exists \mathsf{a} \in \overline{\mathcal{D}^{\sim}_\alpha}\ \upsilon(\mathsf{f}@\mathsf{a}) = \mathtt{T}$	$\mathsf{f} \in \overline{\mathcal{D}^{\sim}_{o\alpha}}$
$\mathfrak{L}^{\partial}_{=^\alpha}(\mathsf{q})$	$\mathsf{q} \in \mathcal{D}_{o\alpha\alpha}$	$\upsilon(\mathsf{q}@\mathsf{a}@\mathsf{b}) = \mathtt{T}$	iff	$\mathsf{a} \sim \mathsf{b}$	$\mathsf{a}, \mathsf{b} \in \overline{\mathcal{D}^{\sim}_\alpha}$

TABLE 4.1. Logical Properties of $\upsilon : \mathcal{D}_o \to \{\mathtt{T}, \mathtt{F}\}$ and $\sim$

LEMMA 4.2.4 (Per Quotient of Applicative Structures). *Let $\sim$ be a per over an applicative structure $\mathcal{A} := (\mathcal{D}, @)$ satisfying $\partial^{\neq\emptyset}$ and $\partial^{@}$. There is a per quotient structure $(\mathcal{D}^{\sim}, @^{\sim})$ of $\mathcal{A}$ by $\sim$.*

PROOF. We define

$$\mathcal{D}^{\sim}_\alpha := \{[\![\mathsf{a}]\!]_\sim \mid \mathsf{a} \in \overline{\mathcal{D}^{\sim}_\alpha}\}.$$

The fact that $\mathcal{D}^{\sim}_\alpha$ is nonempty for each type α follows directly from $\partial^{\neq\emptyset}$. For each type α and equivalence class $\mathsf{A} \in \mathcal{D}^{\sim}_\alpha$, choose some representative $\mathsf{A}^* \in \mathsf{A} \subseteq \mathcal{D}_\alpha$. For each $\mathsf{A} \in \mathcal{D}^{\sim}_\alpha$ and $\mathsf{a} \in \mathsf{A}$, we know $\mathsf{a} \sim \mathsf{A}^*$. Hence $\mathsf{a} \sim ([\![\mathsf{a}]\!]_\sim)^*$ for each $\mathsf{a} \in \overline{\mathcal{D}^{\sim}_\alpha}$. Define

$$\mathsf{F}@^{\sim}\mathsf{B} := [\![\mathsf{F}^*@\mathsf{B}^*]\!]_\sim$$

for $\mathsf{F} \in \mathcal{D}^{\sim}_{\alpha\beta}$, $\mathsf{B} \in \mathcal{D}^{\sim}_\beta$ and $\alpha, \beta \in \mathcal{T}$. Let $\mathsf{f} \in \overline{\mathcal{D}^{\sim}_{\gamma\beta}}$ and $\mathsf{b} \in \overline{\mathcal{D}^{\sim}_\beta}$ be given. We have

$$(([\![\mathsf{f}]\!]_\sim)^*@([\![\mathsf{b}]\!]_\sim)^*) \sim (\mathsf{f}@\mathsf{b})$$

by $\partial^{@}$ since $([\![\mathsf{f}]\!]_\sim)^* \sim \mathsf{f}$ and $([\![\mathsf{b}]\!]_\sim)^* \sim \mathsf{b}$. Hence

$$[\![\mathsf{f}]\!]_\sim @^{\sim} [\![\mathsf{b}]\!]_\sim = [\![([\![\mathsf{f}]\!]_\sim)^*@([\![\mathsf{b}]\!]_\sim)^*]\!]_\sim = [\![\mathsf{f}@\mathsf{b}]\!]_\sim.$$

Thus $(\mathcal{D}^{\sim}, @^{\sim})$ is the per quotient of $\mathcal{A}$ by $\sim$. □

DEFINITION 4.2.5. Let $\sim$ be a per over an applicative structure $\mathcal{A}$ satisfying $\partial^{\neq\emptyset}$ and $\partial^{@}$ and $\mathcal{A}^{\sim}$ be the per quotient of $\mathcal{A}$ by $\sim$. For any $\sim$-assignment φ, we define an assignment $\varphi^{\sim}$ into $\mathcal{A}^{\sim}$ by $\varphi^{\sim}(x) := [\![\varphi(x)]\!]_\sim$ for each variable x.

When constructing a per over an applicative structure satisfying $\partial^{@}$ and $\partial^{\mathfrak{f}}$, it is enough to specify $\sim$ only on the base types, since this determines $\sim$ on all function types. We will formulate the extension in terms of applicative structures, so we can build such pers before we interpret formulas.

DEFINITION 4.2.6 (Functional Per Extension). Let $(\mathcal{D}, @)$ be an applicative structure, $\sim_o$ be a partial equivalence relation on $\mathcal{D}_o$, and $\sim_\iota$ be a partial equivalence relation on $\mathcal{D}_\iota$. We define the *functional per extension* $\sim$ *of* $\sim_o$ *and* $\sim_\iota$ *over* $(\mathcal{D}, @)$ by induction on types.

o: For $\mathsf{a}, \mathsf{b} \in \mathcal{D}_o$, $\mathsf{a} \sim \mathsf{b}$ if $\mathsf{a} \sim_o \mathsf{b}$.

ι: For $\mathsf{a}, \mathsf{b} \in \mathcal{D}_\iota$, $\mathsf{a} \sim \mathsf{b}$ if $\mathsf{a} \sim_\iota \mathsf{b}$.

$\alpha\beta$: For $\mathsf{g}, \mathsf{h} \in \mathcal{D}_{\alpha\beta}$, $\mathsf{g} \sim \mathsf{h}$ if we have $\mathsf{g}@\mathsf{a} \sim \mathsf{h}@\mathsf{b}$ for every $\mathsf{a} \sim \mathsf{b}$ in $\mathcal{D}_\beta$.

Note that $\sim$ agrees with $\sim_o$ on $\mathcal{D}_o$ and $\sim_\iota$ on $\mathcal{D}_\iota$.

LEMMA 4.2.7. *The functional per extension* $\sim$ *of* $\sim_o$ *and* $\sim_\iota$ *over* $(\mathcal{D}, @)$ *is a typed per over* $(\mathcal{D}, @)$ *satisfying* $\partial^{@}$ *and* $\partial^{\mathfrak{f}}$.

PROOF. The proof is by a simple induction on types. See Appendix A.2. □

LEMMA 4.2.8. *Let* $\sim$ *be a per over an evaluation* $(\mathcal{D}, @, \mathcal{E})$. *Suppose* $\sim$ *satisfies* $\partial^{@}$, $\partial^{\mathfrak{f}}$, $\partial^{\mathcal{P}}$ *and* $\partial^{\mathcal{S}}$. *Then* $\sim$ *satisfies* ∂^{sub}.

PROOF. We prove $\mathcal{E}_\varphi(\mathbf{A}) \sim \mathcal{E}_\psi(\mathbf{A})$ for all assignments φ and ψ such that $\varphi \sim \psi$ and every $\mathbf{A} \in \mathit{wff}_\alpha(\mathcal{S})$ by induction on $\mathbf{A}$. If $\mathbf{A}$ is a parameter or constant, we have $\mathcal{E}_\varphi(\mathbf{A}) \sim \mathcal{E}_\psi(\mathbf{A})$ by $\partial^{\mathcal{P}}$ or $\partial^{\mathcal{S}}$. If $\mathbf{A}$ is a variable x_α, we have $\mathcal{E}_\varphi(x) = \varphi(x) \sim \psi(x) = \mathcal{E}_\psi(x)$. For application, we assume $\mathcal{E}_\varphi(\mathbf{G}) \sim \mathcal{E}_\psi(\mathbf{G})$ and $\mathcal{E}_\varphi(\mathbf{C}) \sim \mathcal{E}_\psi(\mathbf{C})$. By $\partial^{@}$, we know

$$\mathcal{E}_\varphi([\mathbf{G}\,\mathbf{C}]) = \mathcal{E}_\varphi(\mathbf{G})@\mathcal{E}_\varphi(\mathbf{C}) \sim \mathcal{E}_\psi(\mathbf{G})@\mathcal{E}_\psi(\mathbf{C}) = \mathcal{E}_\psi([\mathbf{G}\,\mathbf{C}]).$$

For abstraction we use $\partial^{\mathfrak{f}}$. We must verify that $\mathcal{E}_\varphi([\lambda x_\alpha\, \mathbf{C}_\gamma])$ and $\mathcal{E}_\psi([\lambda x_\alpha\, \mathbf{C}_\gamma])$ send related values to related results. Let $\mathsf{a} \sim \mathsf{b}$ in $\mathcal{D}_\alpha$ be given. By the inductive hypothesis, we have $\mathcal{E}_{\varphi,[\mathsf{a}/x]}(\mathbf{C}) \sim \mathcal{E}_{\psi,[\mathsf{b}/x]}(\mathbf{C})$, so

$$\mathcal{E}_\varphi([\lambda x\, \mathbf{C}])@\mathsf{a} = \mathcal{E}_{\varphi,[\mathsf{a}/x]}(\mathbf{C}) \sim \mathcal{E}_{\psi,[\mathsf{b}/x]}(\mathbf{C}) = \mathcal{E}_\psi([\lambda x\, \mathbf{C}])@\mathsf{b}.$$

By $\partial^{\mathfrak{f}}$, $\mathcal{E}_\varphi([\lambda x\, \mathbf{C}]) \sim \mathcal{E}_\psi([\lambda x\, \mathbf{C}])$. □

We can use evaluations with pers to define a semantics that differs from those defined in Definition 3.3.2.

DEFINITION 4.2.9 (Per Model). Suppose $\sim$ is a per over an $\mathcal{S}$-evaluation $(\mathcal{D}, @, \mathcal{E})$ and $v : \mathcal{D}_o \to \{\mathtt{T}, \mathtt{F}\}$. We call $\mathscr{P} := (\mathcal{D}, @, \mathcal{E}, v, \sim)$ an *$\mathcal{S}$-per model* if $\partial^{\mathcal{P}}$, $\partial^{\mathcal{S}}$, $\partial^{@}$, $\partial^{\mathfrak{f}}$, $\partial^{\mathfrak{b}}$ and ∂^{v} hold and $\mathfrak{L}^{\partial}_{c}(\mathcal{E}(c))$ holds for each $c \in \mathcal{S}$. An *assignment* φ into $\mathscr{P}$ is an assignment φ into $\mathcal{D}$. An assignment φ into $\mathscr{P}$ is a *$\sim$-assignment* if $\varphi \sim \varphi$.

We now define a notion of satisfaction for per models. The treatment of equality and universal quantifiers differs from satisfaction for models (cf. Definition 3.3.3). In particular, universal quantifiers only consider elements in $\overline{\mathcal{D}^{\sim}_{\alpha}}$. The quotient of the per model by the per $\sim$ (cf. Theorem 4.2.11) will not include any elements from $(\mathcal{D}_\alpha \setminus \overline{\mathcal{D}^{\sim}_{\alpha}})$.

DEFINITION 4.2.10. Let $\mathcal{S}$ be a signature and $\mathscr{P} = (\mathcal{D}, @, \mathcal{E}, v, \sim)$ be an $\mathcal{S}$-per model. We define when a proposition $\mathbf{M}$ is *satisfied* by an assignment φ in $\mathscr{P}$ (written $\mathscr{P} \models\!\!\approx_{\varphi} \mathbf{M}$) by induction on propositions.

- For $\mathbf{A} \in \mathit{wff}_o(\mathcal{S})$, $\mathscr{P} \models\!\!\approx_{\varphi} \mathbf{A}$ if $v(\mathcal{E}_\varphi(\mathbf{A})) = \mathtt{T}$.
- For $\mathbf{A}, \mathbf{B} \in \mathit{wff}_\alpha(\mathcal{S})$, $\mathscr{P} \models\!\!\approx_{\varphi} [\mathbf{A} \doteq^{\alpha} \mathbf{B}]$ if $\mathcal{E}_\varphi(\mathbf{A}) \sim \mathcal{E}_\varphi(\mathbf{B})$.
- $\mathscr{P} \models\!\!\approx_{\varphi} \dot{\top}$.
- $\mathscr{P} \models\!\!\approx_{\varphi} \dot{\neg} \mathbf{M}$ if $\mathscr{P} \not\models\!\!\approx_{\varphi} \mathbf{M}$.
- $\mathscr{P} \models\!\!\approx_{\varphi} [\mathbf{M} \dot{\vee} \mathbf{N}]$ if $\mathscr{P} \models\!\!\approx_{\varphi} \mathbf{M}$ or $\mathscr{P} \models\!\!\approx_{\varphi} \mathbf{N}$.
- $\mathscr{P} \models\!\!\approx_{\varphi} [\dot{\forall} x_\alpha\, \mathbf{M}]$ if $\mathscr{P} \models\!\!\approx_{\varphi, [\mathsf{a}/x]} \mathbf{M}$ for every $\mathsf{a} \in \overline{\mathcal{D}^{\sim}_{\alpha}}$.

We write $\mathscr{P} \models\!\!\approx \mathbf{M}$ if $\mathscr{P} \models\!\!\approx_{\varphi} \mathbf{M}$ for every $\sim$-assignment φ. For a set of propositions Φ, we write $\mathscr{P} \models\!\!\approx \Phi$ if $\mathscr{P} \models\!\!\approx \mathbf{M}$ for every $\mathbf{M} \in \Phi$.

Finally, we determine that every $\mathcal{S}$-per model corresponds to an extensional model in $\mathfrak{M}_{\beta\mathfrak{f}\mathfrak{b}}(\mathcal{S})$.

THEOREM 4.2.11. *Suppose $\mathscr{P} = (\mathcal{D}, @, \mathcal{E}, v, \sim)$ is an $\mathcal{S}$-per model. Then there is an extensional $\mathcal{S}$-model $\mathcal{M} \in \mathfrak{M}_{\beta\mathfrak{f}\mathfrak{b}}(\mathcal{S})$ such that $\mathcal{M} \models \mathbf{M}$ iff $\mathscr{P} \models\!\!\approx \mathbf{M}$ for all $\mathbf{M} \in \mathit{prop}(\mathcal{S})$.*

PROOF. The extensional model $\mathcal{M}$ is obtained by taking the per quotient structure of $(\mathcal{D}, @)$ and appropriately defining an evaluation function and valuation. See Appendix A.2. □

4.3. Retraction Models

In this section we construct models using retractions. Such constructions will be used to obtain models in which different forms of extensionality fail.

DEFINITION 4.3.1 (Retractions on Function Types). Let $(\mathcal{D}, @)$ be an applicative structure. A collection of functions

$$\imath^{\alpha\beta} : \mathcal{D}_\alpha^{\mathcal{D}_\beta} \to \mathcal{D}_{\alpha\beta}$$

is called a *retraction on function types* if for all types α and β and functions $f : \mathcal{D}_\beta \to \mathcal{D}_\alpha$, we have $\imath^{\alpha\beta}(f)@\mathsf{b} = f(\mathsf{b})$ for every $\mathsf{b} \in \mathcal{D}_\beta$.

Often a retraction is described in terms of an injection $\imath$ and a one-sided inverse $\jmath$. Here, we should think of the application function $@$ (Curried) as being the one-sided inverse of $\imath$. That is, we can define $\jmath : \mathcal{D}_{\alpha\beta} \to \mathcal{D}_\alpha^{\mathcal{D}_\beta}$ by $\jmath(\mathsf{f})(\mathsf{b}) := \mathsf{f}@\mathsf{b}$. Then $\jmath(\imath(f))(\mathsf{b}) = \imath(f)@\mathsf{b} = f(\mathsf{b})$ and so $\jmath(\imath(f)) = f$ for every $f : \mathcal{D}_\beta \to \mathcal{D}_\alpha$.

LEMMA 4.3.2. *An applicative structure $\mathcal{A} = (\mathcal{D}, @)$ with retraction $\imath$ on function types is full. Furthermore, each $\imath^{\alpha\beta}$ is injective.*

PROOF. Given any $f : \mathcal{D}_\beta \to \mathcal{D}_\alpha$, $\imath(f) \in \mathcal{D}_{\alpha\beta}$ is the witness that $\mathcal{A}$ is full since $\imath(f)@\mathsf{b} = f(\mathsf{b})$ for every $\mathsf{b} \in \mathcal{D}_\beta$. To check $\imath^{\alpha\beta}$ is injective, suppose we have functions $f, g : \mathcal{D}_\beta \to \mathcal{D}_\alpha$ with $\imath(f) = \imath(g)$. Hence

$$f(\mathsf{b}) = \imath(f)@\mathsf{b} = \imath(g)@\mathsf{b} = g(\mathsf{b})$$

for any $\mathsf{b} \in \mathcal{D}_\beta$. Thus $f = g$ (as functions). □

LEMMA 4.3.3. *An applicative structure $\mathcal{A} = (\mathcal{D}, @)$ with retraction $\imath$ on function types is functional iff $\imath^{\alpha\beta} : \mathcal{D}_\alpha^{\mathcal{D}_\beta} \to \mathcal{D}_{\alpha\beta}$ is a surjection for all types α and β.*

PROOF. Suppose $\mathcal{A}$ is functional. For any $\mathsf{f} \in \mathcal{D}_{\alpha\beta}$, we define the function $f : \mathcal{D}_\beta \to \mathcal{D}_\alpha$ by $f(\mathsf{b}) := \mathsf{f}@\mathsf{b}$ for every $\mathsf{b} \in \mathcal{D}_\beta$. Since $\imath$ is a retraction on function types, we know $\imath(f)@\mathsf{b} = \mathsf{f}@\mathsf{b}$ for every b. By functionality, $\imath(f) = \mathsf{f}$ and $\imath$ is surjective.

For the converse, suppose for all types α and β, $\imath^{\alpha\beta}$ is surjective. Suppose we have $\mathsf{f}, \mathsf{g} \in \mathcal{D}_{\alpha\beta}$ such that $\mathsf{f}@\mathsf{b} = \mathsf{g}@\mathsf{b}$ for every $\mathsf{b} \in \mathcal{D}_\beta$. By surjectivity, there exist functions $f, g : \mathcal{D}_\beta \to \mathcal{D}_\alpha$ such that $\imath(f) = \mathsf{f}$ and $\imath(g) = \mathsf{g}$. Since

$$f(\mathsf{b}) = \imath(f)@\mathsf{b} = \mathsf{f}@\mathsf{b} = \mathsf{g}@\mathsf{b} = \imath(g)@\mathsf{b} = g(\mathsf{b})$$

for every $\mathsf{b} \in \mathcal{D}_\beta$, we know $f = g$ (as functions). Thus we have

$$\mathsf{f} = \imath(f) = \imath(g) = \mathsf{g}$$

and $\mathcal{A}$ is functional. □

We can use a retraction on function types to extend any interpretation of constants and parameters to an evaluation.

DEFINITION 4.3.4. Let $(\mathcal{D}, @)$ be an applicative structure with retraction i on function types. Given a signature $\mathcal{S}$ and an interpretation of constants and parameters $\mathcal{I} : \mathcal{S} \cup \mathcal{P} \to \mathcal{D}$, we define the *retraction extension* $\mathcal{E}_\varphi(\mathbf{A})$ *of* $\mathcal{I}$ by induction on $\mathbf{A}$.

- $\mathcal{E}_\varphi(x) := \varphi(x)$ for any variable x.
- $\mathcal{E}_\varphi(W) := \mathcal{I}(W)$ for any parameter W.
- $\mathcal{E}_\varphi(c) := \mathcal{I}(c)$ for any constant c.
- $\mathcal{E}_\varphi([\mathbf{F}\,\mathbf{B}]) := \mathcal{E}_\varphi(\mathbf{F})@\mathcal{E}_\varphi(\mathbf{B})$.
- $\mathcal{E}_\varphi([\lambda x_\beta\, \mathbf{C}_\alpha]) := i(f) \in \mathcal{D}_{\alpha\beta}$ where $f : \mathcal{D}_\beta \to \mathcal{D}_\alpha$ is defined by $f(\mathsf{b}) := \mathcal{E}_{\varphi,[\mathsf{b}/x]}(\mathbf{C})$ for every $\mathsf{b} \in \mathcal{D}_\beta$.

We prove the retraction extension $\mathcal{E}$ of $\mathcal{I}$ is an evaluation function using two lemmas. First, we note that the value of $\mathcal{E}_\varphi(\mathbf{A})$ does not depend on the value of φ on variables $y_\beta \notin \mathbf{Free}(\mathbf{A})$.

LEMMA 4.3.5. *Let $(\mathcal{D}, @)$ be an applicative structure with retraction i on function types. Let $\mathcal{S}$ be a signature, $\mathcal{I} : \mathcal{S} \cup \mathcal{P} \to \mathcal{D}$ be an interpretation of constants and parameters, and $\mathcal{E}$ be the retraction extension of $\mathcal{I}$. If $\mathbf{A} \in wff_\alpha(\mathcal{S})$, then $\mathcal{E}_\varphi(\mathbf{A}) = \mathcal{E}_\psi(\mathbf{A})$ for any φ and ψ are assignments which coincide on $\mathbf{Free}(\mathbf{A})$.*

PROOF. The proof is by an easy induction on $\mathbf{A}$ and the definition of $\mathcal{E}$ which is left to the reader. $\square$

Next, we note that $\mathcal{E}$ respects β-normalization.

LEMMA 4.3.6. *Let $(\mathcal{D}, @)$ be an applicative structure with retraction i on function types. Let $\mathcal{S}$ be a signature, $\mathcal{I} : \mathcal{S} \cup \mathcal{P} \to \mathcal{D}$ be interpretation of constants and parameters, and $\mathcal{E}$ be the retraction extension of $\mathcal{I}$. For any $\mathbf{A}$ and assignment φ, $\mathcal{E}_\varphi(\mathbf{A}) = \mathcal{E}_\varphi(\mathbf{A}^{\downarrow})$.*

PROOF. See Appendix A.2. $\square$

We can verify $\mathcal{E}$ is an evaluation function using the lemmas above.

THEOREM 4.3.7. *Let $(\mathcal{D}, @)$ be an applicative structure with retraction i on function types. For any signature $\mathcal{S}$ and interpretation of constants and parameters $\mathcal{I} : \mathcal{S} \cup \mathcal{P} \to \mathcal{D}$, the retraction extension $\mathcal{E}$ of $\mathcal{I}$ is an $\mathcal{S}$-evaluation function. Furthermore, the following are equivalent:*

1. *For all types α and β, $i^{\alpha\beta}$ is surjective.*
2. *$(\mathcal{D}, @)$ is functional.*
3. *$(\mathcal{D}, @, \mathcal{E})$ is η-functional.*

PROOF. By definition of $\mathcal{E}$, we know $\mathcal{E}_\varphi(x) = \varphi(x)$ for any variable x and $\mathcal{E}_\varphi([\mathbf{F}\,\mathbf{B}]) = \mathcal{E}_\varphi(\mathbf{F})@\mathcal{E}_\varphi(\mathbf{B})$. By Lemma 4.3.5, we know $\mathcal{E}_\varphi(\mathbf{A}) = \mathcal{E}_\psi(\mathbf{A})$ whenever φ and ψ coincide on $\mathbf{Free}(\mathbf{A})$. The fact that $\mathcal{E}_\varphi(\mathbf{A}) = \mathcal{E}_\varphi(\mathbf{A}^{\downarrow\beta})$ follows from Lemma 4.3.6. Thus $\mathcal{E}$ is an $\mathcal{S}$-evaluation function.

We now verify the equivalences regarding functionality.

1. If every $i^{\alpha\beta}$ is surjective, then we know $(\mathcal{D}, @)$ is functional by Lemma 4.3.3.
2. If $(\mathcal{D}, @)$ is functional, then we know $(\mathcal{D}, @, \mathcal{E})$ is η-functional by Lemma 3.2.15.
3. Suppose $(\mathcal{D}, @, \mathcal{E})$ is η-functional. Let $\mathsf{f} \in \mathcal{D}_{\alpha\beta}$ be given. Let $g_{\alpha\beta}$ be a variable and φ be an assignment with $\varphi(g) = \mathsf{f}$. Using the definition of $\mathcal{E}$ (cf. Definition 4.3.4), we have
$$\mathcal{E}_\varphi([\lambda x_\beta \centerdot g\, x]) = i(f)$$
where $f : \mathcal{D}_\beta \to \mathcal{D}_\alpha$ is defined by
$$f(\mathsf{b}) := \mathcal{E}_{\varphi,[\mathsf{b}/x]}(g\, x) = \mathsf{f}@\mathsf{b}$$
for every $\mathsf{b} \in \mathcal{D}_\beta$. By η-functionality, we know
$$i(f) = \mathcal{E}_\varphi([\lambda x_\beta \centerdot g\, x]) = \mathcal{E}_\varphi(g) = \mathsf{f}.$$
Thus $i^{\alpha\beta}$ is surjective for all types α and β.

□

In particular, evaluations built using retractions cannot distinguish between functionality and η-functional. Examples of evaluations which are η-functional but not functional are constructed in [19]. Such examples can be constructed using possible values (cf. Theorem 4.1.8).

We can use retractions to construct a weak evaluation that is not an evaluation. Consequently, one really does need the stronger conditions defining evaluations (cf. Definition 3.2.3) as opposed to weak evaluations (cf. Definition 3.2.2) in the absence of functionality.

EXAMPLE 4.3.8 (Nonfunctional Weak Evaluations). Let I be $\{1, 2\}$, $\mathcal{D}_\iota$ be $\{0\}$, $\mathcal{D}_o$ be $\{\mathtt{T}, \mathtt{F}\}$, and $\mathcal{D}_{\alpha\beta}$ be $I \times \mathcal{D}_\alpha^{\mathcal{D}_\beta}$ for all types

α and β. Let application be defined by $\langle n, f\rangle @\mathsf{b} := f(\mathsf{b})$ for every $\langle n, f\rangle \in \mathcal{D}_{\alpha\beta}$ and $\mathsf{b} \in \mathcal{D}_\beta$. We define a retraction by $i^{\alpha\beta}(f) := \langle 1, f\rangle$. Let $\mathcal{I} : \mathcal{S} \cup \mathcal{P} \to \mathcal{D}$ be arbitrary and let $\mathcal{E}$ be the retraction extension of $\mathcal{I}$. We will modify $\mathcal{E}$ to obtain a weak evaluation $\mathcal{E}'$ which is not an evaluation. It is easy to compute

$$\mathcal{E}_\varphi([\lambda x_\iota \centerdot [\lambda y_\iota\, y_\iota]\, x]) = \langle 1, id\rangle$$

where $id : \mathcal{D}_\iota \to \mathcal{D}_\iota$ is the (unique) function from the singleton $\mathcal{D}_\iota$ into itself. Let $\mathcal{E}'_\varphi([\lambda x_\iota \centerdot [\lambda y_\iota\, y_\iota]\, x]) := \langle 2, id\rangle$ and $\mathcal{E}'_\varphi(\mathbf{A}) := \mathcal{E}_\varphi(\mathbf{A})$ if $\mathbf{A}$ is not (α-equivalent to) $[\lambda x_\iota \centerdot [\lambda y_\iota\, y_\iota]\, x]$.

Since $\mathcal{E}'$ and $\mathcal{E}$ coincide on constants, parameters, variables and applications, we need only check $\mathcal{E}'_\varphi([\lambda x_\beta\, \mathbf{A}])@\mathsf{b} = \mathcal{E}'_{\varphi,[\mathsf{b}/x]}(\mathbf{A})$ to conclude $\mathcal{E}'$ is a weak evaluation. We check

$$\mathcal{E}'_\varphi([\lambda x_\iota \centerdot [\lambda y\, y]\, x])@0 = 0 = \mathcal{E}'_{\varphi,[0/x]}([[\lambda y\, y]\, x]).$$

If $\beta \neq \iota$ or $\mathbf{A}$ is not $[[\lambda y\, y]\, x_\iota]$, we have

$$\mathcal{E}'_\varphi(\lambda x_\beta\, \mathbf{A})@\mathsf{b} = \mathcal{E}_\varphi(\lambda x_\beta\, \mathbf{A})@\mathsf{b} = \mathcal{E}_{\varphi,[\mathsf{b}/x]}(\mathbf{A}) = \mathcal{E}'_{\varphi,[\mathsf{b}/x]}(\mathbf{A}).$$

Thus $\mathcal{E}'$ is a weak evaluation function. However, $\mathcal{E}'$ is not an evaluation function, since

$$\mathcal{E}'([\lambda x_\iota \centerdot [\lambda y\, y]\, x]) = \langle 2, id\rangle \neq \langle 1, id\rangle = \mathcal{E}'([\lambda x_\iota\, x]).$$

DEFINITION 4.3.9 (Retractions on Booleans). Let $(\mathcal{D}, @)$ be an applicative structure. A pair of functions

$$b : \{\mathtt{T}, \mathtt{F}\} \to \mathcal{D}_o$$

and

$$v : \mathcal{D}_o \to \{\mathtt{T}, \mathtt{F}\}$$

is called a *retraction on Booleans* if $v(b(\mathtt{T})) = \mathtt{T}$ and $v(b(\mathtt{F})) = \mathtt{F}$.

DEFINITION 4.3.10. Let $(\mathcal{D}, @)$ be an applicative structure with retraction i on function types and retraction (b, v) on Booleans. Let

$k_\mathtt{T} : \{\mathtt{T}, \mathtt{F}\} \to \{\mathtt{T}, \mathtt{F}\}$	denote the constant true function,
$k_\mathtt{F} : \{\mathtt{T}, \mathtt{F}\} \to \{\mathtt{T}, \mathtt{F}\}$	denote the constant false function,
$n : \{\mathtt{T}, \mathtt{F}\} \to \{\mathtt{T}, \mathtt{F}\}$	denote the negation function, and
$id : \{\mathtt{T}, \mathtt{F}\} \to \{\mathtt{T}, \mathtt{F}\}$	denote the identity function.

For any signature $\mathcal{S}$ of logical constants we define the *retraction of logical constants* $\mathcal{I}^\mathcal{S} : \mathcal{S} \to \mathcal{D}$ as follows.

$\top$: If $\top \in \mathcal{S}$, then $\mathcal{I}^{\mathcal{S}}(\top) := b(\mathtt{T})$.

$\bot$: If $\bot \in \mathcal{S}$, then $\mathcal{I}^{\mathcal{S}}(\bot) := b(\mathtt{F})$.

$\neg$: If $\neg \in \mathcal{S}$, then $\mathcal{I}^{\mathcal{S}}(\neg) := i^{oo}(b \circ n \circ v)$.

$\wedge$: Let $c : \{\mathtt{T}, \mathtt{F}\} \to \mathcal{D}_{oo}$ be the function $c(\mathtt{T}) := i^{oo}(b \circ id \circ v)$ and $c(\mathtt{F}) := i^{oo}(b \circ k_{\mathtt{F}} \circ v)$. If $\wedge \in \mathcal{S}$, then $\mathcal{I}^{\mathcal{S}}(\wedge) := i^{ooo}(c \circ v)$.

$\vee$: Let $d : \{\mathtt{T}, \mathtt{F}\} \to \mathcal{D}_{oo}$ be the function $d(\mathtt{T}) := i^{oo}(b \circ k_{\mathtt{T}} \circ v)$ and $d(\mathtt{F}) := i^{oo}(b \circ id \circ v)$. If $\vee \in \mathcal{S}$, then $\mathcal{I}^{\mathcal{S}}(\vee) := i^{ooo}(d \circ v)$.

$\equiv$: Let $e : \{\mathtt{T}, \mathtt{F}\} \to \mathcal{D}_{oo}$ be the function $e(\mathtt{T}) := i^{oo}(b \circ id \circ v)$ and $e(\mathtt{F}) := i^{oo}(b \circ n \circ v)$. If $\equiv \in \mathcal{S}$, then $\mathcal{I}^{\mathcal{S}}(\equiv) := i^{ooo}(e \circ v)$.

$\supset$: Let $f : \{\mathtt{T}, \mathtt{F}\} \to \mathcal{D}_{oo}$ be the function $f(\mathtt{T}) := i^{oo}(b \circ id \circ v)$ and $f(\mathtt{F}) := i^{oo}(b \circ k_{\mathtt{T}} \circ v)$. If $\supset \in \mathcal{S}$, then $\mathcal{I}^{\mathcal{S}}(\supset) := i^{ooo}(f \circ v)$.

Π^α: Let $\pi^\alpha : \mathcal{D}_{o\alpha} \to \mathcal{D}_o$ be the function

$$\pi^\alpha(\mathsf{f}) := \begin{cases} b(\mathtt{T}) & \text{if } \forall \mathsf{a} \in \mathcal{D}_\alpha\, v(\mathsf{f}@\mathsf{a}) = \mathtt{T} \\ b(\mathtt{F}) & \text{otherwise.} \end{cases}$$

If $\Pi^\alpha \in \mathcal{S}$, then $\mathcal{I}^{\mathcal{S}}(\Pi^\alpha) := i^{o(o\alpha)}(\pi^\alpha)$.

Σ^α: Let $\sigma^\alpha : \mathcal{D}_{o\alpha} \to \mathcal{D}_o$ be the function

$$\sigma^\alpha(\mathsf{f}) := \begin{cases} b(\mathtt{T}) & \text{if } \exists \mathsf{a} \in \mathcal{D}_\alpha\, v(\mathsf{f}@\mathsf{a}) = \mathtt{T} \\ b(\mathtt{F}) & \text{otherwise.} \end{cases}$$

If $\Sigma^\alpha \in \mathcal{S}$, then $\mathcal{I}^{\mathcal{S}}(\Sigma^\alpha) := i^{o(o\alpha)}(\sigma^\alpha)$.

$=^\alpha$: For each $\mathsf{a} \in \mathcal{D}_\alpha$, let $s_{\mathsf{a}} : \mathcal{D}_\alpha \to \mathcal{D}_o$ be the function

$$s_{\mathsf{a}}(\mathsf{b}) := \begin{cases} b(\mathtt{T}) & \text{if } \mathsf{a} = \mathsf{b} \\ b(\mathtt{F}) & \text{otherwise.} \end{cases}$$

Let $q^\alpha : \mathcal{D}_\alpha \to \mathcal{D}_{o\alpha}$ be the function $q^\alpha(\mathsf{a}) := i^{o\alpha}(s_{\mathsf{a}})$. If $=^\alpha \in \mathcal{S}$, then $\mathcal{I}^{\mathcal{S}}(=^\alpha) := i^{o\alpha\alpha}(q^\alpha)$.

DEFINITION 4.3.11. Let $(\mathcal{D}, @)$ be an applicative structure with retraction i on function types and retraction (b, v) on Booleans. Let $\mathcal{S}$ be a signature and $\mathcal{I} : \mathcal{P} \to \mathcal{D}$ be an interpretation of parameters. We define $\mathcal{M} := (\mathcal{D}, @, \mathcal{E}, v)$ to be the *retraction model of* $\mathcal{I}$ where $\mathcal{E}$ is the retraction extension of $\mathcal{I} \cup \mathcal{I}^{\mathcal{S}}$ and $\mathcal{I}^{\mathcal{S}} : \mathcal{S} \to \mathcal{D}$ is the retraction of logical constants.

THEOREM 4.3.12 (Retraction Models). *Let $(\mathcal{D}, @)$ be an applicative structure with retraction i on function types and retraction (b, υ) on Booleans. Let $\mathcal{S}$ be a signature and $\mathcal{I} : \mathcal{P} \to \mathcal{D}$ be an interpretation of parameters. The retraction model $\mathcal{M} := (\mathcal{D}, @, \mathcal{E}, \upsilon)$ of $\mathcal{I}$ is an $\mathcal{S}$-model.*

PROOF. By Theorem 4.3.7 we know $\mathcal{E}$ is an $\mathcal{S}$-evaluation function. Also, we know how the logical constants are interpreted since $\mathcal{E}|_{\mathcal{S}} = \mathcal{I}^{\mathcal{S}}$. To check that $\mathcal{M}$ is an $\mathcal{S}$-model, we must check $\mathfrak{L}_c(\mathcal{I}^{\mathcal{S}}(c))$ holds with respect to υ (cf. Table 3.1) using the definition of $\mathcal{I}^{\mathcal{S}}$ (cf. Definition 4.3.10) for each $c \in \mathcal{S}$.

$\top$: $\upsilon(\mathcal{I}^{\mathcal{S}}(\top)) = \upsilon(b(\mathtt{T})) = \mathtt{T}$.

$\bot$: $\upsilon(\mathcal{I}^{\mathcal{S}}(\bot)) = \upsilon(b(\mathtt{F})) = \mathtt{F}$.

$\neg$: For any $\mathsf{a} \in \mathcal{D}_o$, we compute

$$\upsilon(\mathcal{I}^{\mathcal{S}}(\neg)@\mathsf{a}) = \upsilon(i(b \circ n \circ \upsilon)@\mathsf{a}) = \upsilon(b(n(\upsilon(\mathsf{a})))) = n(\upsilon(\mathsf{a})).$$

Recall that $n : \{\mathtt{T}, \mathtt{F}\} \to \{\mathtt{T}, \mathtt{F}\}$ is the negation function. Hence

$$\upsilon(\mathcal{I}^{\mathcal{S}}(\neg)@\mathsf{a}) = \mathtt{T} \text{ iff } n(\upsilon(\mathsf{a})) = \mathtt{T} \text{ iff } \upsilon(\mathsf{a}) = \mathtt{F}.$$

$\wedge$: For any $\mathsf{a}, \mathsf{b} \in \mathcal{D}_o$, we compute

$$\upsilon(\mathcal{I}^{\mathcal{S}}(\wedge)@\mathsf{a}@\mathsf{b}) = \upsilon(i(c \circ \upsilon)@\mathsf{a}@\mathsf{b}) = \upsilon(c(\upsilon(\mathsf{a}))@\mathsf{b}).$$

Using the definition of $c : \{\mathtt{T}, \mathtt{F}\} \to \mathcal{D}_{oo}$, there are two possibilities. If $\upsilon(\mathsf{a}) = \mathtt{T}$, then

$$\upsilon(c(\upsilon(\mathsf{a}))@\mathsf{b}) = \upsilon(i(b \circ id \circ \upsilon)@\mathsf{b}) = \upsilon(b(id(\upsilon(\mathsf{b})))) = \upsilon(\mathsf{b}).$$

If $\upsilon(\mathsf{a}) = \mathtt{F}$, then

$$\upsilon(c(\upsilon(\mathsf{a}))@\mathsf{b}) = \upsilon(i(b \circ k_{\mathtt{F}} \circ \upsilon)@\mathsf{b}) = \upsilon(b(\mathtt{F})) = \mathtt{F}.$$

From this, we see that $\upsilon(\mathcal{I}^{\mathcal{S}}(\wedge)@\mathsf{a}@\mathsf{b}) = \mathtt{T}$ iff $\upsilon(\mathsf{a}) = \mathtt{T}$ and $\upsilon(\mathsf{b}) = \mathtt{T}$.

$\vee$, $\equiv$, $\supset$: Analogous to $\wedge$.

Π^α: For any $\mathsf{f} \in \mathcal{D}_{o\alpha}$, we compute

$$\upsilon(\mathcal{I}^{\mathcal{S}}(\Pi^\alpha)@\mathsf{f}) = \upsilon(i(\pi^\alpha)@\mathsf{f}) = \upsilon(\pi^\alpha(\mathsf{f})).$$

By the definition of π^α, if $\upsilon(\mathsf{f}@\mathsf{a}) = \mathtt{T}$ for every $\mathsf{a} \in \mathcal{D}_\alpha$, then

$$\upsilon(\pi^\alpha(\mathsf{f})) = \upsilon(b(\mathtt{T})) = \mathtt{T}.$$

Otherwise,

$$\upsilon(\pi^\alpha(\mathsf{f})) = \upsilon(b(\mathtt{F})) = \mathtt{F}$$

as desired.

Σ^α: Analogous to Π^α.

$=^\alpha$: For any $\mathsf{a}, \mathsf{b} \in \mathcal{D}_\alpha$, we compute

$$\upsilon(\mathcal{I}^{\mathcal{S}}(=^\alpha)@\mathsf{a}@\mathsf{b}) = \upsilon(i(q^\alpha)@\mathsf{a}@\mathsf{b}) = \upsilon(q^\alpha(\mathsf{a})@\mathsf{b})$$
$$= \upsilon(i(s_{\mathsf{a}})@\mathsf{b}) = \upsilon(s_{\mathsf{a}}(\mathsf{b})).$$

By the definition of s_{a}, we know

$$\upsilon(s_{\mathsf{a}}(\mathsf{b})) = \upsilon(b(\mathtt{T})) = \mathtt{T}$$

if $\mathsf{a} = \mathsf{b}$ and

$$\upsilon(s_{\mathsf{a}}(\mathsf{b})) = \upsilon(b(\mathtt{F})) = \mathtt{F}$$

otherwise.

□

We can use retraction models to demonstrate that the inclusions

$$\mathfrak{M}_{\beta\mathfrak{f}\mathfrak{b}}(\mathcal{S}) \subseteq \mathfrak{M}_{\beta\eta}(\mathcal{S}) \subseteq \mathfrak{M}_{\beta}(\mathcal{S})$$

are proper. We will also use such models to establish independence results (cf. Theorem 6.1.3).

EXAMPLE 4.3.13 ($\mathfrak{M}_{\beta\eta}(\mathcal{S})$). Let I be $\{1, 2\}$, $\mathcal{D}_o$ be $I \times \{\mathtt{T}, \mathtt{F}\}$, $\mathcal{D}_\iota$ be $\{0\}$, and $\mathcal{D}_{\alpha\beta}$ be $\mathcal{D}_\alpha^{\mathcal{D}_\beta}$. Let @ be function application. Let $i^{\alpha\beta} : \mathcal{D}_\alpha^{\mathcal{D}_\beta} \longrightarrow \mathcal{D}_{\alpha\beta}$ be the identity function. This clearly forms a retraction on function types as $i(f)@\mathsf{b} = f@\mathsf{b} = f(\mathsf{b})$ for any $f : \mathcal{D}_\beta \longrightarrow \mathcal{D}_\alpha$ and $\mathsf{b} \in \mathcal{D}_\beta$. Let $\upsilon : \mathcal{D}_o \longrightarrow \{\mathtt{T}, \mathtt{F}\}$ be $\upsilon(\langle n, \mathsf{a}\rangle) := \mathsf{a}$ and $b : \{\mathtt{T}, \mathtt{F}\} \longrightarrow \mathcal{D}_o$ be $b(\mathsf{a}) := \langle 1, \mathsf{a}\rangle$. Clearly, $\upsilon(b(\mathtt{T})) = \mathtt{T}$ and $\upsilon(b(\mathtt{F})) = \mathtt{F}$. Hence (b, υ) is a retraction on Booleans. Let $\mathcal{I} : \mathcal{P} \longrightarrow \mathcal{D}$ be arbitrary and

$$\mathcal{M} := \mathcal{M}^{\mathcal{I}} := (\mathcal{D}, @, \mathcal{E}, \upsilon)$$

be the retraction model of $\mathcal{I}$. The model $\mathcal{M}$ satisfies property η (by Theorem 4.3.7) since each $i^{\alpha\beta}$ is surjective. However, $\mathcal{M}$ does not satisfy property $\mathfrak{b}$ since υ is not injective (e.g., $\upsilon(\langle 1, \mathtt{T}\rangle) = \mathtt{T} = \upsilon(\langle 2, \mathtt{T}\rangle)$). Thus $\mathcal{M} \in \mathfrak{M}_{\beta\eta}(\mathcal{S}) \setminus \mathfrak{M}_{\beta\mathfrak{f}\mathfrak{b}}(\mathcal{S})$.

EXAMPLE 4.3.14 ($\mathfrak{M}_{\beta}(\mathcal{S})$). We start with the same applicative structure $(\mathcal{D}, @)$ and retraction i on function types as in Example 4.3.8. Since $\mathcal{D}_o := \{\mathtt{T}, \mathtt{F}\}$, we can let $b, \upsilon : \{\mathtt{T}, \mathtt{F}\} \longrightarrow \{\mathtt{T}, \mathtt{F}\}$ both be the identity function. Clearly, (b, υ) is a retraction on Booleans.

For any $\mathcal{I} : \mathcal{P} \to \mathcal{D}$, we can let $\mathcal{M} := \mathcal{M}^{\mathcal{I}} := (\mathcal{D}, @, \mathcal{E}, v)$ be the retraction model of $\mathcal{I}$. By Theorem 4.3.7 we know $\mathcal{M}$ does not satisfy property η since no $\langle 2, f\rangle \in \mathcal{D}_{\alpha\beta}$ is in the image of $i^{\alpha\beta}$. Thus $\mathcal{M} \in \mathfrak{M}_{\beta}(\mathcal{S}) \setminus \mathfrak{M}_{\beta\eta}(\mathcal{S})$. (Note that since v is the identity function, it is injective, and so $\mathcal{M}$ does satisfy property $\mathfrak{b}$.)

4.4. Logical Relation Frames

We introduced n-ary logical relations on terms in Definition 2.3.1. These syntactic logical relations will be used to help prove extensional completeness (cf. Theorem 5.2.11). In addition to being a proof technique for verifying syntactic properties, logical relations also provide a way of studying semantics of type theories. In fact, logical relations were introduced by Plotkin [55] as a way to characterize λ-definable elements of a frame.

Logical relations are defined hereditarily on function types once the relation is given at the base types. We will use the idea of logical relations to define domains of function types given domains of base types and relations on these domains. The construction will also generalize from n-ary relations to A-ary relations where A is an arbitrary nonempty index set. The logical relations construction will give examples of combinatory frames (cf. Definitions 3.1.4 and 3.6.9) realizing some (but not necessarily all) logical constants. Using the results of Section 3.6, we can use such combinatory frames to construct models for fragments of extensional type theory.

Let A be any nonempty set. For any nonempty set $\mathcal{D}$, we can consider a function $f : A \to \mathcal{D}$ to be an A-tuple of elements of $\mathcal{D}$. Similarly, we can consider a set $\mathcal{R} \subseteq \mathcal{D}^A$ to be an A-ary relation (i.e., a collection of A-tuples).

To define a frame, we can start with any nonempty sets $\mathcal{D}_\iota$ and $\mathcal{D}_o$. We also start with any $\mathcal{R}_\iota \subseteq \mathcal{D}_\iota^A$ and $\mathcal{R}_o \subseteq \mathcal{D}_o^A$ which contain all the constant functions. Then we can define the remaining domains by induction on types. Furthermore, when we define $\mathcal{D}_\alpha$, we will also define $\mathcal{R}_\alpha \subseteq \mathcal{D}_\alpha^A$.

It is useful to define polymorphic operations S and K on functions into a frame. These operations behave similarly to the **S** and **K** combinators. The operations will be used to define logical relation frames and to verify that these frames are combinatory.

DEFINITION 4.4.1. Let $\mathcal{D}$ be a frame and A be a nonempty set. For each type α and $\mathsf{a} \in \mathcal{D}_\alpha$, let $K(\mathsf{a}) : A \to \mathcal{D}_\alpha$ be the constant function defined by $K(\mathsf{a})(x) := \mathsf{a}$ for each $x \in A$. For all types α and β and functions $p : A \to \mathcal{D}_{\alpha\beta}$ and $f : A \to \mathcal{D}_\beta$, let $S(p, f) : A \to \mathcal{D}_\alpha$ be defined by $S(p, f)(x) := p(x)(f(x))$ for each $x \in A$.

DEFINITION 4.4.2 (Logical Relation Frame). Let $\mathcal{D}$ be a frame and A be a nonempty set. We say $\mathcal{D}$ is an *A-ary logical relation frame with relation $\mathcal{R}$* if

- $\mathcal{R}_\alpha \subseteq \mathcal{D}_\alpha^A$ for all types α,
- $K(\mathsf{a}) \in \mathcal{R}_\alpha$ for all $\mathsf{a} \in \mathcal{D}_\alpha$ and all types α, and
- for all types α and β

$$\mathcal{D}_{\alpha\beta} = \{h : \mathcal{D}_\beta \to \mathcal{D}_\alpha \mid (h \circ f) \in \mathcal{R}_\alpha \text{ for every } f \in \mathcal{R}_\beta\}$$

and

$$\mathcal{R}_{\alpha\beta} = \{p : A \to \mathcal{D}_{\alpha\beta} \mid S(p, f) \in \mathcal{R}_\alpha \text{ for every } f \in \mathcal{R}_\beta\}.$$

If we start with appropriate values of $\mathcal{D}$ and $\mathcal{R}$ at base types, we can extend this by induction on types to obtain a logical relation frame.

DEFINITION 4.4.3 (Logical Relation Extension). Let A, $\mathcal{D}_\iota$, $\mathcal{D}_o$, $\mathcal{R}_\iota \subseteq \mathcal{D}_\iota^A$ and $\mathcal{R}_o \subseteq \mathcal{D}_o^A$ be nonempty sets. By induction on types, we extend $\mathcal{D}$ and $\mathcal{R}$ to be defined on all types by

$$\mathcal{D}_{\alpha\beta} := \{h : \mathcal{D}_\beta \to \mathcal{D}_\alpha \mid (h \circ f) \in \mathcal{R}_\alpha \text{ for every } f \in \mathcal{R}_\beta\}$$

and

$$\mathcal{R}_{\alpha\beta} := \{p : A \to \mathcal{D}_{\alpha\beta} \mid S(p, f) \in \mathcal{R}_\alpha \text{ for every } f \in \mathcal{R}_\beta\}$$

for all types α and β. We say $(\mathcal{D}, \mathcal{R})$ is the *A-ary logical relation extension of* $(\mathcal{D}_\iota, \mathcal{D}_o, \mathcal{R}_\iota, \mathcal{R}_o)$.

We now give conditions under which logical relation extensions define logical relation frames. First, we need every element to be related to itself by $\mathcal{R}$. That is, we need to know the operator K defined above maps into $\mathcal{R}_\alpha$ for each type α.

LEMMA 4.4.4. *Let A, $\mathcal{D}_\iota$, $\mathcal{D}_o$, $\mathcal{R}_\iota \subseteq \mathcal{D}_\iota^A$ and $\mathcal{R}_o \subseteq \mathcal{D}_o^A$ be nonempty sets. Assume*

- *$K(\mathsf{b}) \in \mathcal{R}_o$ for every $\mathsf{b} \in \mathcal{D}_o$ and*
- *$K(\mathsf{c}) \in \mathcal{R}_\iota$ for every $\mathsf{c} \in \mathcal{D}_\iota$.*

Let $(\mathcal{D}, \mathcal{R})$ be the A-ary logical relation extension of $(\mathcal{D}_\iota, \mathcal{D}_o, \mathcal{R}_\iota, \mathcal{R}_o)$. Then $K(\mathsf{a}) \in \mathcal{R}_\alpha$ for each type α and $\mathsf{a} \in \mathcal{D}_\alpha$.

PROOF. We know $K(\mathsf{a}) \in \mathcal{D}_\alpha$ for each $\mathsf{a} \in \mathcal{D}_\alpha$ by assumption for base types $\alpha \in \{\iota, o\}$. Assume α is a function type $(\beta\gamma)$. Let $h \in \mathcal{D}_{\beta\gamma}$ be given. We must verify $K(h) : A \to \mathcal{D}_{\beta\gamma}$ is in $\mathcal{R}_{\beta\gamma}$. Let $f \in \mathcal{R}_\gamma$ be given. We must check $S(K(h), f) \in \mathcal{R}_\beta$. For each $x \in A$, we have

$$S(K(h), f)(x) = K(h)(x)(f(x)) = h(f(x)).$$

Thus $S(K(h), f) = (h \circ f) \in \mathcal{R}_\beta$ since $h \in \mathcal{D}_{\beta\gamma}$. □

We can now use constant functions to demonstrate that every domain of such a logical relation extension is nonempty.

LEMMA 4.4.5. *Let A, $\mathcal{D}_\iota$, $\mathcal{D}_o$, $\mathcal{R}_\iota \subseteq \mathcal{D}_\iota^A$ and $\mathcal{R}_o \subseteq \mathcal{D}_o^A$ be nonempty sets. Assume*

- *$K(\mathsf{b}) \in \mathcal{R}_o$ for every $\mathsf{b} \in \mathcal{D}_o$ and*
- *$K(\mathsf{c}) \in \mathcal{R}_\iota$ for every $\mathsf{c} \in \mathcal{D}_\iota$.*

Let $(\mathcal{D}, \mathcal{R})$ be the A-ary logical relation extension of $(\mathcal{D}_\iota, \mathcal{D}_o, \mathcal{R}_\iota, \mathcal{R}_o)$. Then $\mathcal{D}_\alpha$ is nonempty for each type α.

PROOF. We prove $\mathcal{D}_\alpha$ is nonempty by induction on the type α. By assumption, $\mathcal{D}_o$ and $\mathcal{D}_\iota$ are nonempty. Assume $\mathcal{D}_\beta$ is nonempty and choose some $\mathsf{b} \in \mathcal{D}_\beta$. Let $k_{\mathsf{b}}^\gamma : \mathcal{D}_\gamma \to \mathcal{D}_\beta$ be the constant function $k_{\mathsf{b}}^\gamma(\mathsf{c}) := \mathsf{b}$ for all $\mathsf{c} \in \mathcal{D}_\gamma$. For any $f \in \mathcal{R}_\beta$, $(k_{\mathsf{b}}^\gamma \circ f) = K(\mathsf{b}) \in \mathcal{R}_\beta$ by Lemma 4.4.4. Hence $k_{\mathsf{b}}^\gamma \in \mathcal{D}_{\beta\gamma}$ and so $\mathcal{D}_{\beta\gamma}$ is nonempty. □

Since the domains are nonempty, the logical relation extension is a frame.

LEMMA 4.4.6. *Let A, $\mathcal{D}_\iota$, $\mathcal{D}_o$, $\mathcal{R}_\iota \subseteq \mathcal{D}_\iota^A$ and $\mathcal{R}_o \subseteq \mathcal{D}_o^A$ be nonempty sets. Assume*

- *$K(\mathsf{b}) \in \mathcal{R}_o$ for every $\mathsf{b} \in \mathcal{D}_o$ and*
- *$K(\mathsf{c}) \in \mathcal{R}_\iota$ for every $\mathsf{c} \in \mathcal{D}_\iota$.*

Let $(\mathcal{D}, \mathcal{R})$ be the A-ary logical relation extension of $(\mathcal{D}_\iota, \mathcal{D}_o, \mathcal{R}_\iota, \mathcal{R}_o)$. Then $\mathcal{D}$ is an A-ary logical relation frame with relation $\mathcal{R}$.

PROOF. By Lemma 4.4.5, we know each $\mathcal{D}_\alpha$ is nonempty. By the definition of $\mathcal{D}$ at function types, we know each $\mathcal{D}_{\alpha\beta} \subseteq \mathcal{D}_\alpha^{\mathcal{D}_\beta}$. Hence $\mathcal{D}$ is a frame.

We know $\mathcal{R}_\alpha \subseteq \mathcal{D}_\alpha^A$ on base types by assumption and every function type by definition. By Lemma 4.4.4, for any $\mathsf{a} \in \mathcal{D}_\alpha$ we know $K(\mathsf{a}) \in \mathcal{R}_\alpha$. Finally, we know

$$\mathcal{D}_{\alpha\beta} = \{h : \mathcal{D}_\beta \to \mathcal{D}_\alpha \mid (h \circ f) \in \mathcal{R}_\alpha \text{ for every } f \in \mathcal{R}_\beta\}$$

and

$$\mathcal{R}_{\alpha\beta} = \{p : A \to \mathcal{D}_{\alpha\beta} \mid S(p, f) \in \mathcal{R}_\alpha \text{ for every } f \in \mathcal{R}_\beta\}.$$

for function types by definition. □

We now prove that logical relation frames are combinatory (cf. Definition 3.6.9).

THEOREM 4.4.7. *Let A be a nonempty set and $\mathcal{D}$ be an A-ary logical relation frame with relation $\mathcal{R}$. Then $\mathcal{D}$ is combinatory.*

PROOF. Let α and β be types. We must verify there is a $\mathsf{k} \in \mathcal{D}_{\alpha\beta\alpha}$ such that $\mathsf{k}(\mathsf{a})(\mathsf{b}) = \mathsf{a}$ for every $\mathsf{a} \in \mathcal{D}_\alpha$ and $\mathsf{b} \in \mathcal{D}_\beta$. This, of course, supplies the only possible definition of k as the function $\mathsf{k} : \mathcal{D}_\alpha \to \mathcal{D}_\alpha^{\mathcal{D}_\beta}$ *defined* by $\mathsf{k}(\mathsf{a})(\mathsf{b}) := \mathsf{a}$. To check $\mathsf{k} \in \mathcal{D}_{\alpha\beta\alpha}$, we first check that k maps from $\mathcal{D}_\alpha$ into $\mathcal{D}_{\alpha\beta}$. For each $\mathsf{a} \in \mathcal{D}_\alpha$, we must verify $\mathsf{k}(\mathsf{a}) \in \mathcal{D}_{\alpha\beta}$. Let $f \in \mathcal{R}_\beta$ be given. Since $(\mathsf{k}(\mathsf{a}) \circ f) : A \to \mathcal{D}_\alpha$ is the constant a function, we know $(\mathsf{k}(\mathsf{a}) \circ f) \in \mathcal{R}_\alpha$ by Lemma 4.4.4. Hence $\mathsf{k}(\mathsf{a}) \in \mathcal{D}_{\alpha\beta}$. To check $\mathsf{k} \in \mathcal{D}_{\alpha\beta\alpha}$, let $g \in \mathcal{R}_\alpha$ be given. In order to check $(\mathsf{k} \circ g) \in \mathcal{R}_{\alpha\beta}$, let $f \in \mathcal{R}_\beta$ be given. Now, we must check that $S(\mathsf{k} \circ g, f) \in \mathcal{R}_\alpha$. For each $x \in A$, we compute

$$S(\mathsf{k} \circ g, f)(x) = (\mathsf{k} \circ g)(x)(f(x)) = \mathsf{k}(g(x))(f(x)) = g(x).$$

Thus $S(\mathsf{k} \circ g, f) = g \in \mathcal{R}_\alpha$.

Checking that the **S** combinators have interpretations is similar. □

Theorem 4.4.7 ensures logical relation frames have domains sufficient to interpret λ-terms (cf. Theorem 3.6.20).

REMARK 4.4.8. Suppose A is a nonempty set and $\mathcal{D}$ is an A-ary logical relation frame with relation $\mathcal{R}$. One can easily check $(\mathcal{R}, S)$ and $(\mathcal{D}^A, S)$ are combinatory applicative structures where

$$S : (\mathcal{D}_{\alpha\beta})^A \times (\mathcal{D}_\beta)^A \to (\mathcal{D}_\alpha)^A$$

is defined by $S(p, f)(x) := p(x)(f(x))$. We know by the definition of $\mathcal{R}_{\alpha\beta}$ that S restricts to a map $S : \mathcal{R}_{\alpha\beta} \times \mathcal{R}_\beta \to \mathcal{R}_\alpha$. Note that $(\mathcal{R}, S)$

is a substructure of $(\mathcal{D}^A, S)$ and that $\mathcal{R}$ is a unary logical relation on $(\mathcal{D}^A, S)$. Suppose $\mathcal{F}$ is a filter on $\mathcal{P}(A)$. The filter induces an equivalence relation $\sim^{\mathcal{F}}$ by $f \sim^{\mathcal{F}} g$ iff $\{i \in A \mid f(i) = g(i)\} \in \mathcal{F}$. The equivalence $\sim^{\mathcal{F}}$ respects the application operator S and hence is a congruence on the applicative structures $(\mathcal{R}, S)$ and $(\mathcal{D}^A, S)$. One can construct the quotient applicative structure of either of these by the congruence $\sim^{\mathcal{F}}$. If $\mathcal{F}$ is a principal ultrafilter, then the quotient will clearly be isomorphic to the frame $\mathcal{D}$. Other filters and ultrafilters may give new combinatory applicative structures (cf. Remark 6.7.42).

REMARK 4.4.9. For readers familiar with category theory (cf. [39; 41; 40; 42]) it is useful to give a categorical description of logical relation frames. Let A be any nonempty set. We can define a category $\mathfrak{C}$ with objects $\langle \mathcal{D}, \mathcal{R} \rangle$ where $\mathcal{R} \subseteq \mathcal{D}^A$ and for every $\mathsf{d} \in \mathcal{D}$ the constant function $K(\mathsf{d})$ is in $\mathcal{R}$. The set of arrows $Hom(\langle \mathcal{D}, \mathcal{R} \rangle, \langle \mathcal{D}', \mathcal{R}' \rangle)$ of $\mathfrak{C}$ from $\langle \mathcal{D}, \mathcal{R} \rangle$ to $\langle \mathcal{D}', \mathcal{R}' \rangle$ consists of functions $f : \mathcal{D} \to \mathcal{D}'$ such that $(f \circ p) \in \mathcal{R}'$ for every $p \in \mathcal{R}$. Composition of arrows is simply functional composition. The category $\mathfrak{C}$ is Cartesian closed. The exponent

$$\langle \mathcal{D}', \mathcal{R}' \rangle^{\langle \mathcal{D}, \mathcal{R} \rangle}$$

of $\langle \mathcal{D}', \mathcal{R}' \rangle$ by $\langle \mathcal{D}, \mathcal{R} \rangle$ is given by $\langle \mathcal{D}'', \mathcal{R}'' \rangle$ where

$$\mathcal{D}'' := \{h : \mathcal{D} \to \mathcal{D}' \mid (h \circ f) \in \mathcal{R}' \text{ for every } f \in \mathcal{R}\}$$

and

$$\mathcal{R}'' := \{p : A \to \mathcal{D}'' \mid S(p, f) \in \mathcal{R}' \text{ for every } f \in \mathcal{R}\}.$$

Note that the definition of the exponent corresponds to our definition of $\mathcal{D}_{\alpha\beta}$ and $\mathcal{R}_{\alpha\beta}$ at function types. We can conclude that any A-ary logical relation frame $\mathcal{D}$ with relation $\mathcal{R}$ is the *standard* interpretation of typed λ-calculus in the Cartesian closed category $\mathfrak{C}$ with given objects $\langle \mathcal{D}_\iota, \mathcal{R}_\iota \rangle$ and $\langle \mathcal{D}_o, \mathcal{R}_o \rangle$ interpreting the base types.

4.5. Models with Specified Sets

Let A be any nonempty set and $\mathcal{B} \subseteq \mathcal{P}(A)$ be a collection of sets such that $\emptyset, A \in \mathcal{B}$. We will now demonstrate how to construct a combinatory frame (cf. Definitions 3.1.4 and 3.6.9) where $\mathcal{D}_\iota = A$, $\mathcal{D}_o = \{\mathtt{T}, \mathtt{F}\}$ and the sets in $\mathcal{B}$ correspond to the sets represented in $\mathcal{D}_{o\iota}$. In other words, we can specify precisely what sets should be in $\mathcal{D}_{o\iota}$ by choosing $\mathcal{B}$. The resulting evaluation induces a model for different signatures depending on properties of $\mathcal{B}$. We can use the fact that some logical constants cannot be interpreted in the corresponding model to determine that these logical constants must be used in (instantiations involved in) any proof of a theorem. We will define the frame as a logical relation frame (cf. Definition 4.4.2).

NOTATION *Let A_1 and A_2 be sets and $f : A_1 \to A_2$ be a function. This function f induces an* inverse image *function*

$$f^{-1} : \mathcal{P}(A_2) \to \mathcal{P}(A_1)$$

defined by

$$f^{-1}(Y) := \{x \in A_1 \mid f(x) \in Y\}.$$

Also, if $y \in A_2$, then we write $f^{-1}(y)$ to denote $f^{-1}(\{y\})$. If $B_1 \subseteq \mathcal{P}(A_1)$ and $B_2 \subseteq \mathcal{P}(A_2)$, then we write $f^{-1} : B_2 \to B_1$ for the restriction of f^{-1} to B_2 if $f^{-1}(Y) \in B_1$ whenever $Y \in B_2$. In particular, given a nonempty set A and $\mathcal{B} \subseteq \mathcal{P}(A)$ as above, we write $f^{-1} : \mathcal{B} \to \mathcal{B}$ to denote this restriction if $\{x \in A \mid f(x) \in X\} \in \mathcal{B}$ for every $X \in \mathcal{B}$.

We define a logical relation frame by defining $\mathcal{D}$ and $\mathcal{R}$ on base types.

DEFINITION 4.5.1. Let A be a nonempty set and $\mathcal{B} \subseteq \mathcal{P}(A)$ be such that $\emptyset, A \in \mathcal{B}$. Let $\mathcal{D}_\iota := A$ and $\mathcal{D}_o := \{\mathtt{T}, \mathtt{F}\}$. Let

$$\mathcal{R}_\iota := \{f : A \to A \mid f^{-1}(X) \in \mathcal{B} \text{ for every } X \in \mathcal{B}\}$$

and

$$\mathcal{R}_o := \{f : A \to \mathcal{D}_o \mid f^{-1}(\mathtt{T}) \in \mathcal{B}\}.$$

Let $(\mathcal{D}, \mathcal{R})$ be the A-ary logical relation extension $(\mathcal{D}_\iota, \mathcal{D}_o, \mathcal{R}_\iota, \mathcal{R}_o)$. We call $\mathcal{D}$ the *$\mathcal{B}$-specified frame with relation $\mathcal{R}$.*

Note that in Definition 4.5.1 for any $f \in \mathcal{R}_\iota$ we know f^{-1} restricts to a function $f^{-1} : \mathcal{B} \to \mathcal{B}$.

We use Lemma 4.4.6 to determine $\mathcal{D}$ is a frame.

LEMMA 4.5.2. *Let A be a nonempty set, $\mathcal{B} \subseteq \mathcal{P}(A)$ be such that $\emptyset, A \in \mathcal{B}$ and $\mathcal{D}$ be the $\mathcal{B}$-specified frame with relation $\mathcal{R}$. Then $\mathcal{D}$ is an A-ary logical relation frame with relation $\mathcal{R}$.*

PROOF. Using Lemma 4.4.6, we need only verify $K(\mathsf{b}) \in \mathcal{R}_o$ for every $\mathsf{b} \in \mathcal{D}_o$ and $K(\mathsf{c}) \in \mathcal{R}_\iota$ for every $\mathsf{c} \in \mathcal{D}_\iota$. Note that

$$(K(\mathtt{T}))^{-1}(\mathtt{T}) = \{x \in A \mid K(\mathtt{T})(x) = \mathtt{T}\} = A \in \mathcal{B}$$

and

$$(K(\mathtt{F}))^{-1}(\mathtt{T}) = \{x \in A \mid K(\mathtt{F})(x) = \mathtt{T}\} = \emptyset \in \mathcal{B}.$$

Hence $K(\mathsf{b}) \in \mathcal{R}_o$ for every $\mathsf{b} \in \mathcal{D}_o$. Let $\mathsf{c} \in \mathcal{D}_\iota$ be given. For any set $X \in \mathcal{B}$, $(K(\mathsf{c}))^{-1}(X)$ is either A (if $\mathsf{c} \in X$) or $\emptyset$ (if $\mathsf{c} \notin X$). In either case, $K(\mathsf{c})^{-1}(X) \in \mathcal{B}$ for every $X \in \mathcal{B}$. Hence $K(\mathsf{c}) \in \mathcal{R}_\iota$. Thus $\mathcal{D}$ is an A-ary logical relation frame with relation $\mathcal{R}$. In particular, $\mathcal{D}$ is a frame. □

Since $\mathcal{B}$-specified frames are logical relation frames, we know they are combinatory. We now obtain further results about $\mathcal{B}$-specified frames.

LEMMA 4.5.3. *Let A be a nonempty set, $\mathcal{B} \subseteq \mathcal{P}(A)$ be such that $\emptyset, A \in \mathcal{B}$ and $\mathcal{D}$ be the $\mathcal{B}$-specified frame with relation $\mathcal{R}$. The identity function on A is in $\mathcal{R}_\iota$.*

PROOF. The identity function $id : A \to A$ satisfies the property that $id^{-1}(X) = X \in \mathcal{B}$ whenever $X \in \mathcal{B}$. Hence $id \in \mathcal{R}_\iota$. □

We will now record the correspondence between $\mathcal{B}$ and the domain $\mathcal{D}_{o\iota}$.

LEMMA 4.5.4. *Let A be a nonempty set, $\mathcal{B} \subseteq \mathcal{P}(A)$ be such that $\emptyset, A \in \mathcal{B}$ and $\mathcal{D}$ be the $\mathcal{B}$-specified frame with relation $\mathcal{R}$. Let χ_X be the function from $\mathcal{D}_\iota$ to $\mathcal{D}_o$ defined by*

$$\chi_X(a) := \begin{cases} \mathtt{T} & \text{if } a \in X \\ \mathtt{F} & \text{if } a \notin X \end{cases}$$

for each $X \in \mathcal{B}$ and $a \in A$. That is, χ_X is the characteristic function of X. Then $\chi_X \in \mathcal{D}_{o\iota}$ for each $X \in \mathcal{B}$ and the map $X \mapsto \chi_X$ is a bijection from $\mathcal{B}$ to $\mathcal{D}_{o\iota}$.

PROOF. Let $X \in \mathcal{B}$ be given. To determine $\chi_X \in \mathcal{D}_{o\iota}$, let $f \in \mathcal{R}_\iota$ be given. To check $(\chi_X \circ f) \in \mathcal{R}_o$, we must verify $(\chi_X \circ f)^{-1}(\mathtt{T}) \in \mathcal{B}$. We compute

$$(\chi_X \circ f)^{-1}(\mathtt{T}) = f^{-1}({\chi_X}^{-1}(\mathtt{T})) = f^{-1}(X) \in \mathcal{B}$$

since $f \in \mathcal{R}_\iota$ and $X \in \mathcal{B}$.

The fact that $X \mapsto \chi_X$ is injective is trivial, since any $X \in \mathcal{B}$ is determined by its characteristic function. To verify surjectivity, let $h \in \mathcal{D}_{o\iota}$ be given. Hence $h : A \to \{\mathtt{T}, \mathtt{F}\}$ and for every $f \in \mathcal{R}_\iota$ we know

$(h \circ f) \in \mathcal{R}_o$. By Lemma 4.5.3, $id \in \mathcal{R}_\iota$ and so $h = (h \circ id) \in \mathcal{R}_o$. By definition of $\mathcal{R}_o$, this means $h^{-1}(\mathtt{T}) \in \mathcal{B}$. Let $X := h^{-1}(\mathtt{T}) \in \mathcal{B}$. Note that $\chi_X(x) = \mathtt{T}$ iff $x \in X$ iff $h(x) = \mathtt{T}$. Thus $h = \chi_X$ and we are done. □

LEMMA 4.5.5. *Let A be a nonempty set, $\mathcal{B} \subseteq \mathcal{P}(A)$ be such that $\emptyset, A \in \mathcal{B}$ and $\mathcal{D}$ be the $\mathcal{B}$-specified frame with relation $\mathcal{R}$. For each type α, $\mathcal{D}_{\alpha\iota} \subseteq \mathcal{R}_\alpha$. Furthermore, $\mathcal{D}_{o\iota} = \mathcal{R}_o$ and $\mathcal{D}_{\iota\iota} = \mathcal{R}_\iota$.*

PROOF. By Lemma 4.5.3, $id \in \mathcal{R}_\iota$. Consequently, if $h \in \mathcal{D}_{\alpha\iota}$, then $h = (h \circ id)$ is in $\mathcal{R}_\alpha$. We conclude the inclusion $\mathcal{D}_{\alpha\iota} \subseteq \mathcal{R}_\alpha$.

At base types, we verify the reverse inclusions.

o: Suppose $f \in \mathcal{R}_o$. Let $X := f^{-1}(\mathtt{T}) \in \mathcal{B}$. Since $f = \chi_X$, we know $f \in \mathcal{D}_{o\iota}$ by Lemma 4.5.4.

ι: Suppose $f \in \mathcal{R}_\iota$. That is, $f : A \to A$ and $f^{-1}(X) \in \mathcal{B}$ for any $X \in \mathcal{B}$. Since $\mathcal{D}_\iota = A$, $f : \mathcal{D}_\iota \to \mathcal{D}_\iota$. To check $f \in \mathcal{D}_{\iota\iota}$, let $g \in \mathcal{R}_\iota$ be given. For any $X \in \mathcal{B}$, $f^{-1}(X) \in \mathcal{B}$ and so

$$(f \circ g)^{-1}(X) = g^{-1}(f^{-1}(X)) \in \mathcal{B}.$$

This establishes $(f \circ g) \in \mathcal{R}_\iota$ and so $f \in \mathcal{D}_{\iota\iota}$.

□

We turn now to the question of which logical constants are realized by such frames. Since we have assumed $\mathcal{D}_o = \{\mathtt{T}, \mathtt{F}\}$, we can always interpret $\top$ and $\bot$. To interpret the logical constants $\neg$, $\vee$, $\wedge$, $\supset$ or $\equiv$, we must assume closure conditions on $\mathcal{B}$.

LEMMA 4.5.6. *Let A be a nonempty set, $\mathcal{B} \subseteq \mathcal{P}(A)$ be such that $\emptyset, A \in \mathcal{B}$ and $\mathcal{D}$ be the $\mathcal{B}$-specified frame with relation $\mathcal{R}$. We have the following:*

$\top, \bot$: *$\mathcal{D}$ realizes $\top$ and $\bot$.*

$\neg$: *$\mathcal{D}$ realizes $\neg$ iff $\mathcal{B}$ is closed under complements relative to A (i.e., $A \setminus X \in \mathcal{B}$ whenever $X \in \mathcal{B}$).*

$\vee$: *$\mathcal{D}$ realizes $\vee$ iff $\mathcal{B}$ is closed under binary unions.*

$\wedge$: *$\mathcal{D}$ realizes $\wedge$ iff $\mathcal{B}$ is closed under binary intersections.*

$\supset$: *$\mathcal{D}$ realizes $\supset$ iff $((A \setminus X) \cup Y) \in \mathcal{B}$ whenever $X, Y \in \mathcal{B}$.*

$\equiv$: *$\mathcal{D}$ realizes $\equiv$ iff $\{x \in A \mid x \in X$ iff $x \in Y\} \in \mathcal{B}$ whenever $X, Y \in \mathcal{B}$.*

PROOF. Clearly $\mathfrak{L}_{\mathsf{T}}(\mathtt{T})$ and $\mathfrak{L}_{\mathsf{T}}(\mathtt{F})$ hold with respect to the identity function $id : \mathcal{D}_o \to \{\mathtt{T}, \mathtt{F}\}$. For each $c_\alpha \in \{\neg, \vee, \wedge, \supset, \equiv\}$, there is only one possible function a which will satisfy the property $\mathfrak{L}_c(\mathsf{a})$ with respect to id. That is, we are forced to interpret each of these logical constant as the function determined by truth tables. We must check in each case that this function is actually in the appropriate domain.

$\neg$: Let $n : \mathcal{D}_o \to \mathcal{D}_o$ be the function defined by $n(\mathtt{T}) := \mathtt{F}$ and $n(\mathtt{F}) := \mathtt{T}$. Suppose $\mathcal{B}$ is closed under complements relative to A. Let $f \in \mathcal{R}_o$ be given. Then $f : A \to \{\mathtt{T}, \mathtt{F}\}$ and $f^{-1}(\mathtt{T}) \in \mathcal{B}$. Hence

$$(n \circ f)^{-1}(\mathtt{T}) = f^{-1}(n^{-1}(\mathtt{T})) = f^{-1}(\mathtt{F}) = (A \setminus f^{-1}(\mathtt{T})) \in \mathcal{B}$$

since $\mathcal{B}$ is closed under complements relative to A. Thus $n \in \mathcal{D}_{oo}$ witnesses that $\mathcal{D}$ realizes $\neg$.

On the other hand, if we know $n \in \mathcal{D}_{oo}$ and $X \in \mathcal{B}$, then $\chi_X \in \mathcal{D}_{o\iota} \subseteq \mathcal{R}_o$ by Lemmas 4.5.4 and 4.5.5. Hence $n \circ \chi_X \in \mathcal{R}_o$ and so $(A \setminus X) = \{a \in A \mid n(\chi_X(a)) = \mathtt{T}\} \in \mathcal{B}$.

$\vee$: Let $d : \mathcal{D}_o \to \mathcal{D}_o^{\mathcal{D}_o}$ be the function defined by

$$d(\mathsf{a})(\mathsf{b}) := \begin{cases} \mathtt{T} & \text{if } \mathsf{a} = \mathtt{T} \text{ or } \mathsf{b} = \mathtt{T} \\ \mathtt{F} & \text{otherwise.} \end{cases}$$

First, note that $d : \mathcal{D}_o \to \mathcal{D}_{oo}$ since $d(\mathtt{T})$ is the constant $\mathtt{T}$ function and $d(\mathtt{F})$ is the identity function (both of which are in $\mathcal{D}_{oo}$ since the frame $\mathcal{D}$ is combinatory).

Suppose $\mathcal{B}$ is closed under binary unions. To check $d \in \mathcal{D}_{ooo}$, we let $f, g \in \mathcal{R}_o$ and verify $S(d \circ f, g) \in \mathcal{R}_o$. For each $x \in A$, we compute

$$S(d \circ f, g)(x) = d(f(x))(g(x)).$$

Hence

$$\begin{aligned} S(d \circ f, g)(x) = \mathtt{T} \quad &\text{iff} \quad d(f(x))(g(x)) = \mathtt{T} \\ &\text{iff} \quad f(x) = \mathtt{T} \text{ or } g(x) = \mathtt{T}. \end{aligned}$$

Thus $(S(d \circ f, g))^{-1}(\mathtt{T}) = f^{-1}(\mathtt{T}) \cup g^{-1}(\mathtt{T}) \in \mathcal{B}$ since $\mathcal{B}$ is closed under binary unions. Therefore, $d \in \mathcal{D}_{ooo}$ and $\mathcal{D}$ realizes $\vee$.

Suppose $d \in \mathcal{D}_{ooo}$ and $X, Y \in \mathcal{B}$. Then $S(d \circ \chi_X, \chi_Y) \in \mathcal{R}_o$. Thus $X \cup Y = \{a \in A \mid d(\chi_X(a))(\chi_Y(a)) = \mathtt{T}\} \in \mathcal{B}$.

$\wedge, \supset, \equiv$: These are all analogous to the $\vee$ case.

□

DEFINITION 4.5.7. Let A be a nonempty set. We call $\mathcal{B} \subseteq \mathcal{P}(A)$ a *field of sets* if $\emptyset, A \in \mathcal{B}$ and $\mathcal{B}$ is closed under complements, finite unions and finite intersections.

Note that any field of sets $\mathcal{B}$ is a Boolean algebra under the subset ordering with bottom element $\emptyset$, top element A, unions as joins, and intersections as meets. In fact, the Stone Representation Theorem implies every Boolean Algebra is isomorphic to some field of sets.

THEOREM 4.5.8. *Let A be a nonempty set, $\mathcal{B} \subseteq \mathcal{P}(A)$ be a field of sets and $\mathcal{D}$ be the $\mathcal{B}$-specified frame with relation $\mathcal{R}$. Then $\mathcal{D}$ realizes $\mathcal{S}_{Bool}$.*

PROOF. This follows immediately from Lemma 4.5.6. □

CHAPTER 5

Completeness

In Chapter 2 we introduced sequent calculi $\mathcal{G}_*^{\mathcal{S}}$ relative to signatures $\mathcal{S}$. In Chapter 3 we defined model classes $\mathfrak{M}_\beta(\mathcal{S})$, $\mathfrak{M}_{\beta\eta}(\mathcal{S})$ and $\mathfrak{M}_{\beta\mathfrak{fb}}(\mathcal{S})$ relative to signatures $\mathcal{S}$. We also established soundness of the sequent calculi with respect to the corresponding model classes (cf. Theorems 3.4.6, 3.4.9 and 3.4.14). Now we turn to the task of establishing completeness of these sequent calculi with respect to the corresponding model classes. In the present chapter we will prove the elementary sequent calculi $\mathcal{G}_\beta^{\mathcal{S}}$ and $\mathcal{G}_{\beta\eta}^{\mathcal{S}}$ are complete with respect to the model classes $\mathfrak{M}_\beta(\mathcal{S})$ and $\mathfrak{M}_{\beta\eta}(\mathcal{S})$, respectively, for any equality-free signature $\mathcal{S}$ (cf. Theorem 5.6.6). The argument for the elementary case uses a straightforward generalization of the V-complex construction in [2]. We will also prove the extensional sequent calculus $\mathcal{G}_{\beta\mathfrak{fb}}^{\mathcal{S}}$ is complete with respect to the model class $\mathfrak{M}_{\beta\mathfrak{fb}}(\mathcal{S})$ for any signature $\mathcal{S}$ (cf. Theorem 5.7.19). The argument for the extensional case makes use of an applicative structure similar to the one used in the elementary case and a partial equivalence relation over this applicative structure. The admissibility of the cut rule for any of these sequent calculi follows from these completeness results (cf. Corollaries 5.6.8 and 5.7.20).

5.1. Abstract Consistency

We now develop an abstract consistency method appropriate for restricted signatures $\mathcal{S}$ in both the elementary and extensional cases.

DEFINITION 5.1.1. Let $\mathscr{C}$ be a class of sets.

1. $\mathscr{C}$ is called *closed under subsets* if for any sets S and T, $S \in \mathscr{C}$ whenever $S \subseteq T$ and $T \in \mathscr{C}$.
2. $\mathscr{C}$ has *finite character* if for every set S we have $S \in \mathscr{C}$ iff every finite subset of S is a member of $\mathscr{C}$.

LEMMA 5.1.2. *If $\mathscr{C}$ has finite character, then $\mathscr{C}$ is closed under subsets.*

PROOF. Suppose $S \subseteq T$ and $T \in \mathscr{C}$. Every finite subset A of S is a finite subset of T, and since $\mathscr{C}$ has finite character we know that $A \in \mathscr{C}$. Thus $S \in \mathscr{C}$. □

We will now introduce a technical side-condition for witness parameters.

DEFINITION 5.1.3 (Sufficiently Pure). Let $\mathcal{S}$ be a signature and Φ be a set of $\mathcal{S}$-sentences. Φ is called *sufficiently pure* if for each type α there is a set $\mathcal{W}_\alpha \subseteq \mathcal{P}_\alpha$ of parameters with equal cardinality to $\mathit{wff}_\alpha(\mathcal{S})$ such that the parameters in $\mathcal{W}$ do not occur in the sentences of Φ.

Recall that we assumed $\mathcal{P}_\alpha$ has a common (infinite) cardinality $\aleph_s$ for every type α. This guarantees that every finite set of sentences is sufficiently pure.

NOTATION *For reasons of legibility we will write $S * a$ for $S \cup \{a\}$, where S is a set. We will use this notation with the convention that $*$ associates to the left.*

DEFINITION 5.1.4 (Properties for Abstract Consistency Classes). Let $\mathscr{C}$ be a class of sets of $\mathcal{S}$-sentences. We define the following conditions for $\mathscr{C}$ where $\Phi \in \mathscr{C}$.

∇_c : If $\mathbf{A} \in \mathit{cwff}_o(\mathcal{S})$ is an atom, then $\mathbf{A} \notin \Phi$ or $\dot\neg\mathbf{A} \notin \Phi$.

∇_β : If $\mathbf{M} \in \Phi$, then $\Phi * \mathbf{M}^{\downarrow\beta} \in \mathscr{C}$.

$\nabla_{\beta\eta}$: If $\mathbf{M} \in \Phi$, then $\Phi * \mathbf{M}^{\downarrow} \in \mathscr{C}$.

$\nabla_\bot$: $\dot\neg\dot\top \notin \Phi$.

$\nabla_\neg$: If $\dot\neg\dot\neg\mathbf{M} \in \Phi$, then $\Phi * \mathbf{M} \in \mathscr{C}$.

$\nabla_\vee$: If $\mathbf{M} \dot\vee \mathbf{N} \in \Phi$, then $\Phi * \mathbf{M} \in \mathscr{C}$ or $\Phi * \mathbf{N} \in \mathscr{C}$.

$\nabla_\wedge$: If $\dot\neg[\mathbf{M} \dot\vee \mathbf{N}] \in \Phi$, then $\Phi * \dot\neg\mathbf{M} * \dot\neg\mathbf{N} \in \mathscr{C}$.

$\nabla_\forall$: If $[\dot\forall x_\alpha \mathbf{P}] \in \Phi$, then $\Phi * [\mathbf{C}/x]\mathbf{P} \in \mathscr{C}$ for each $\mathbf{C} \in \mathit{cwff}_\alpha(\mathcal{S})$.

$\nabla_\exists$: If $\dot\neg[\dot\forall x_\alpha \mathbf{P}] \in \Phi$, then $\Phi * \dot\neg[W/x]\mathbf{P} \in \mathscr{C}$ for any parameter W_α which does not occur in any sentence of Φ.

$\nabla^\sharp$: Suppose $\mathbf{A} \in \mathit{cwff}_o(\mathcal{S})$. If $\mathbf{A}_o \in \Phi$, then $\Phi * \mathbf{A}^\sharp \in \mathscr{C}$. If $\dot\neg\mathbf{A}_o \in \Phi$, then $\Phi * \dot\neg\mathbf{A}^\sharp \in \mathscr{C}$.

∇_m : If H is a parameter and $\{\neg[H\,\overline{\mathbf{A}^n}], [H\,\overline{\mathbf{B}^n}]\} \subseteq \Phi$, then there is an i with $1 \leq i \leq n$ such that $\Phi * \neg[\mathbf{A}^i \doteq \mathbf{B}^i] \in \mathscr{C}$.

∇_{dec}: If H is a parameter and $\neg[[H\,\overline{\mathbf{A}^n}] \doteq^\iota [H\,\overline{\mathbf{B}^n}]] \in \Phi$, then there is an i with $1 \leq i \leq n$ such that $\Phi * \neg[\mathbf{A}^i \doteq \mathbf{B}^i] \in \mathscr{C}$.

$\nabla_{\mathfrak{b}}$: If $\neg[\mathbf{A} \doteq^o \mathbf{B}] \in \Phi$, then

$$\Phi * \mathbf{A} * \neg\mathbf{B} \in \mathscr{C} \text{ or } \Phi * \neg\mathbf{A} * \mathbf{B} \in \mathscr{C}.$$

$\nabla_{\mathfrak{f}}$: If $\neg[\mathbf{G} \doteq^{\alpha\beta} \mathbf{H}] \in \Phi$, then $\Phi * \neg[[\mathbf{G}\,W] =^\alpha [\mathbf{H}\,W]] \in \mathscr{C}$ for any parameter W_β which does not occur in any sentence of Φ.

$\nabla_{=}^{o}$: If $[\mathbf{A}_o \doteq^o \mathbf{B}_o] \in \Phi$, then

$$\Phi * \mathbf{A} * \mathbf{B} \in \mathscr{C} \text{ or } \Phi * \neg\mathbf{A} * \neg\mathbf{B} \in \mathscr{C}.$$

$\nabla_{=}^{\rightarrow}$: If $[\mathbf{G}_{\alpha\beta} \doteq^{\alpha\beta} \mathbf{H}_{\alpha\beta}] \in \Phi$, then $\Phi * [[\mathbf{G}\,\mathbf{B}] \doteq^\alpha [\mathbf{H}\,\mathbf{B}]] \in \mathscr{C}$ for every $\mathbf{B} \in \mathit{cwff}_\beta(\mathcal{S})$.

$\nabla_{=}^{u}$: Suppose $[\mathbf{A}_\iota \doteq^\iota \mathbf{B}_\iota] \in \Phi$ and $\neg[\mathbf{C}_\iota \doteq^\iota \mathbf{D}_\iota] \in \Phi$. Then

$$\Phi * \neg[\mathbf{A} \doteq^\iota \mathbf{C}] \in \mathscr{C} \text{ or } \Phi * \neg[\mathbf{B} \doteq^\iota \mathbf{D}] \in \mathscr{C}.$$

Also,

$$\Phi * \neg[\mathbf{A} \doteq^\iota \mathbf{D}] \in \mathscr{C} \text{ or } \Phi * \neg[\mathbf{B} \doteq^\iota \mathbf{C}] \in \mathscr{C}.$$

DEFINITION 5.1.5 (Elementary Abstract Consistency Classes). Let $\mathcal{S}$ be an equality-free signature and $\mathscr{C}$ be a class of sets of $\mathcal{S}$-sentences.

- We say $\mathscr{C}$ is an *elementary $\mathcal{S}$-abstract consistency class* if every $\Phi \in \mathscr{C}$ is equality-free, $\mathscr{C}$ is closed under subsets and $\mathscr{C}$ satisfies ∇_c, ∇_β, $\nabla_\perp$, $\nabla_\neg$, $\nabla_\vee$, $\nabla_\wedge$, $\nabla_\forall$, $\nabla_\exists$ and $\nabla^\sharp$. We let $\mathfrak{Acc}_\beta(\mathcal{S})$ be the collection of elementary $\mathcal{S}$-abstract consistency classes.
- We say $\mathscr{C}$ is an *elementary $\mathcal{S}$-abstract consistency class with η* if every $\Phi \in \mathscr{C}$ is equality-free, $\mathscr{C}$ is closed under subsets and $\mathscr{C}$ satisfies ∇_c, $\nabla_{\beta\eta}$, $\nabla_\perp$, $\nabla_\neg$, $\nabla_\vee$, $\nabla_\wedge$, $\nabla_\forall$, $\nabla_\exists$ and $\nabla^\sharp$. We let $\mathfrak{Acc}_{\beta\eta}(\mathcal{S})$ be the collection of elementary $\mathcal{S}$-abstract consistency classes with η.

DEFINITION 5.1.6 (Extensional Abstract Consistency Classes). Let $\mathcal{S}$ be a signature and $\mathscr{C}$ be a class of sets of $\mathcal{S}$-sentences. We say $\mathscr{C}$ is an *extensional $\mathcal{S}$-abstract consistency class* if $\mathscr{C}$ is closed under subsets and satisfies $\nabla_{\beta\eta}$, $\nabla_{\bot}$, $\nabla_{\neg}$, $\nabla_{\vee}$, $\nabla_{\wedge}$, $\nabla_{\forall}$, $\nabla_{\exists}$, ∇_m, ∇_{dec}, $\nabla_{\mathfrak{b}}$, $\nabla_{\mathfrak{f}}$, $\nabla_{=}^{o}$, $\nabla_{=}^{\rightarrow}$, $\nabla_{=}^{u}$ and $\nabla^{\sharp}$.[1] We let $\mathfrak{Acc}_{\beta\mathfrak{fb}}(\mathcal{S})$ be the collection of extensional $\mathcal{S}$-abstract consistency classes.

For each $* \in \{\beta, \beta\eta, \beta\mathfrak{fb}\}$ and signature $\mathcal{S} \subseteq \mathcal{S}_*$ (cf. Definition 2.2.5) we have defined a collection $\mathfrak{Acc}_*(\mathcal{S})$ such that $\Phi \subseteq sent_*(\mathcal{S})$ (cf. Definition 2.2.6) for all $\Phi \in \mathscr{C}$ and $\mathscr{C} \in \mathfrak{Acc}_*(\mathcal{S})$. We say $\mathscr{C}$ is an *abstract consistency class* if $\mathscr{C} \in \mathfrak{Acc}_*(\mathcal{S})$ for some $* \in \{\beta, \beta\eta, \beta\mathfrak{fb}\}$ and $\mathcal{S} \subseteq \mathcal{S}_*$. In particular, every abstract consistency class at least satisfies $\nabla_{\neg}$, $\nabla_{\vee}$, $\nabla_{\wedge}$, $\nabla_{\forall}$, $\nabla_{\exists}$, $\nabla^{\sharp}$ and either ∇_{β} or $\nabla_{\beta\eta}$.

REMARK 5.1.7. $\mathfrak{Acc}_{\beta}(\mathcal{S}_{Ch})$ closely corresponds to the abstract consistency class discussed by Andrews in [2]. There are (technical) differences. In [2] α-conversion is handled in the ∇_{β} rule using α-standardized forms. We have included $\dot{\top}$ as a basic proposition and the corresponding abstract consistency condition $\nabla_{\bot}$. Finally, the conditions in [2] for the logical constants in $\mathcal{S}_{Ch}$ follow from $\nabla^{\sharp}$ and the corresponding condition for propositions in Definition 5.1.5.

5.2. Abstract Compatibility

We will use a notion of compatibility of terms in order to prove results in the extensional case. We can define a compatibility relation on closed terms induced by a class of sets of sentences. The compatibility relation will always be symmetric and respect $\beta\eta$-normalization. The relation will be reflexive under certain assumptions.

DEFINITION 5.2.1 ($\mathscr{C}$-Compatibility). Let $\mathscr{C}$ be a class of sets of $\mathcal{S}$-sentences. We say two closed terms $\mathbf{A}, \mathbf{B} \in cwff_{\alpha}(\mathcal{S})$ are $\mathscr{C}$-*compatible* (written $\mathbf{A} \parallel \mathbf{B}$) by induction on the type α.

o: $\mathbf{A}_o \parallel \mathbf{B}_o$ if $\{\mathbf{A}^{\downarrow}, \neg\mathbf{B}^{\downarrow}\} \notin \mathscr{C}$ and $\{\neg\mathbf{A}^{\downarrow}, \mathbf{B}^{\downarrow}\} \notin \mathscr{C}$.

ι: $\mathbf{A}_{\iota} \parallel \mathbf{B}_{\iota}$ if $\{\neg[\mathbf{A}^{\downarrow} \dot{=}^{\iota} \mathbf{B}^{\downarrow}]\} \notin \mathscr{C}$ and $\{\neg[\mathbf{B}^{\downarrow} \dot{=}^{\iota} \mathbf{A}^{\downarrow}]\} \notin \mathscr{C}$.

$\gamma\beta$: $\mathbf{A}_{\gamma\beta} \parallel \mathbf{B}_{\gamma\beta}$ if $[\mathbf{A}\,\mathbf{C}] \parallel [\mathbf{B}\,\mathbf{D}]$ whenever $\mathbf{C}, \mathbf{D} \in cwff_{\beta}(\mathcal{S})$ and $\mathbf{C} \parallel \mathbf{D}$.

Note that $\parallel$ is a binary logical relation on $cwff(\mathcal{S})$ (cf. Definition 2.3.1).

[1] Note that we do not require ∇_{c} in this case. We will determine ∇_{c} follows from the other conditions.

Intuitively, if two terms are $\mathscr{C}$-compatible, then they *might* be equal in an extensional model of some set of sentences $\Phi \in \mathscr{C}$. Notice that $\mathscr{C}$-compatibility is not an equivalence relation as it need not be transitive. We can easily verify compatibility is symmetric and respects $\beta\eta$-normalization.

LEMMA 5.2.2. *Let $\mathscr{C}$ be a class of sets of sentences and $\|$ be the compatibility relation induced by $\mathscr{C}$. For any $\mathbf{A}, \mathbf{B} \in cwff_\alpha(\mathcal{S})$, if $\mathbf{A} \| \mathbf{B}$, then $\mathbf{B} \| \mathbf{A}$.*

PROOF. We prove this by induction on types. At base types the definition of compatibility is symmetric. At a function type $\gamma\beta$, assume $\mathbf{A}_{\gamma\beta} \| \mathbf{B}_{\gamma\beta}$ and $\mathbf{C}_\beta \| \mathbf{D}_\beta$. By the inductive hypothesis at β, we know $\mathbf{D}_\beta \| \mathbf{C}_\beta$. By the definition of compatibility at function types, $[\mathbf{A}\,\mathbf{D}] \| [\mathbf{B}\,\mathbf{C}]$. By the inductive hypothesis at γ, we know $[\mathbf{B}\,\mathbf{C}] \| [\mathbf{A}\,\mathbf{D}]$. Generalizing over $\mathbf{C}$ and $\mathbf{D}$, we know $\mathbf{B} \| \mathbf{A}$. □

A similar induction verifies compatibility respects normal forms.

LEMMA 5.2.3. *Let $\mathscr{C}$ be a class of sets of sentences and $\|$ be the compatibility relation induced by $\mathscr{C}$. For any $\mathbf{A}, \mathbf{B} \in cwff_\alpha(\mathcal{S})$, $\mathbf{A} \| \mathbf{B}$ iff $\mathbf{A}^{\downarrow} \| \mathbf{B}^{\downarrow}$.*

PROOF. This follows by a simple induction on types. For base types $\|$ is defined in terms of normal forms. For function types we can use the definition of $\|$ and the induction hypothesis. □

LEMMA 5.2.4. *Let $\mathscr{C}$ be a class of sets of sentences. The compatibility relation $\|$ induced by $\mathscr{C}$ is an admissible binary logical relation (cf. Definition 2.3.1).*

PROOF. We know $\|$ is a logical relation by the definition of the relation at function types (cf. Definition 5.2.1). To check the logical relation $\|$ is admissible, suppose

- $\mathbf{B}^1, \mathbf{B}^2 \in cwff_\beta(\mathcal{S})$,
- $[\lambda x_\beta\, \mathbf{A}^1], [\lambda x_\beta\, \mathbf{A}^2] \in cwff_{\alpha\beta}(\mathcal{S})$,
- and $([\mathbf{B}^1/x]\mathbf{A}^1) \| ([\mathbf{B}^2/x]\mathbf{A}^2)$.

Lemma 5.2.3 implies $([\mathbf{B}^1/x]\mathbf{A}^1)^{\downarrow} \| ([\mathbf{B}^2/x]\mathbf{A}^2)^{\downarrow}$ and hence

$$[[\lambda x_\beta\, \mathbf{A}^1]\, \mathbf{B}^1]^{\downarrow} \| [[\lambda x_\beta\, \mathbf{A}^2]\, \mathbf{B}^2]^{\downarrow}.$$

Using Lemma 5.2.3 again we conclude $[[\lambda x_\beta\, \mathbf{A}^1]\, \mathbf{B}^1] \| [[\lambda x_\beta\, \mathbf{A}^2]\, \mathbf{B}^2]$. □

The compatibility relation induced by $\mathscr{C}$ is only reflexive if we assume appropriate properties of $\mathscr{C}$. We will need to use a reflexive compatibility relation both for extensional abstract consistency classes and later for extensional Hintikka sets. In order to develop a theory that works for both contexts, we define a notion of an *abstract compatibility class*. This will satisfy many of the same properties as an extensional abstract consistency class, but will insist on the existence of parameter witnesses.

DEFINITION 5.2.5 (Abstract Compatibility Classes). Let $\mathscr{C}$ be a class of sets of $\mathcal{S}$-sentences. We define the following conditions for $\mathscr{C}$ where $\Phi \in \mathscr{C}$.

$\nabla_{\exists}^{w}$: If $\dot{\neg}[\dot{\forall} x_\alpha \mathbf{P}] \in \Phi$, then there exists a parameter W_α such that $\Phi * \dot{\neg}[W/x]\mathbf{P} \in \mathscr{C}$.

$\nabla_{\mathfrak{f}}^{w}$: If $\dot{\neg}[\mathbf{G} \doteq^{\alpha\beta} \mathbf{H}] \in \Phi$, then there exists a parameter W_β such that $\Phi * \dot{\neg}[[\mathbf{G}\,W] \doteq^{\alpha} [\mathbf{H}\,W]] \in \mathscr{C}$.

We call $\mathscr{C}$ an *$\mathcal{S}$-abstract compatibility class* if it is closed under subsets and satisfies $\nabla_{\beta\eta}$, $\nabla_{\bot}$, $\nabla_{\neg}$, $\nabla_{\vee}$, $\nabla_{\wedge}$, $\nabla_{\forall}$, $\nabla_{\exists}^{w}$, ∇_{m}, ∇_{dec}, $\nabla_{\mathfrak{b}}$, $\nabla_{\mathfrak{f}}^{w}$, $\nabla_{=}^{o}$, $\nabla_{=}^{\rightarrow}$, $\nabla_{=}^{u}$ and $\nabla^{\sharp}$.

Since we have assumed there are infinitely many parameters, there is a simple way to transform any extensional $\mathcal{S}$-abstract consistency class into an $\mathcal{S}$-abstract compatibility class.

DEFINITION 5.2.6. For each extensional $\mathcal{S}$-abstract consistency class $\mathscr{C}$, let

$$\mathscr{C}^f := \{\Phi \in \mathscr{C} \mid \Phi \text{ is finite}\}.$$

THEOREM 5.2.7. *If $\mathscr{C}$ is an extensional $\mathcal{S}$-abstract consistency class, then $\mathscr{C}^f$ is an $\mathcal{S}$-abstract compatibility class.*

PROOF. One can easily check all of the conditions $\nabla_{\beta\eta}$, $\nabla_{\bot}$, $\nabla_{\neg}$, $\nabla_{\vee}$, $\nabla_{\wedge}$, $\nabla_{\forall}$, ∇_{m}, ∇_{dec}, $\nabla_{\mathfrak{b}}$, $\nabla_{=}^{o}$, $\nabla_{=}^{\rightarrow}$, $\nabla_{=}^{u}$ and $\nabla^{\sharp}$. The reason is that $\mathscr{C}$ already satisfies these properties and none of the properties insist infinite sets must be in $\mathscr{C}$. Also, $\mathscr{C}^f$ is closed under subsets since $\mathscr{C}$ is closed under subsets.

To check $\nabla_{\exists}^{w}$ holds, suppose $\Phi \in \mathscr{C}^f$ and $\dot{\neg}[\dot{\forall} x_\alpha \mathbf{P}] \in \Phi$. Since Φ is finite, there is some parameter W_α with $W \notin \mathbf{Params}(\Phi)$. By $\nabla_{\exists}$, we must have $\Phi * \dot{\neg}[W/x]\mathbf{P} \in \mathscr{C}$. Since $\Phi * \dot{\neg}[W/x]\mathbf{P}$ is finite, we have $\Phi * \dot{\neg}[W/x]\mathbf{P} \in \mathscr{C}^f$.

The same proof verifies $\nabla_{\mathfrak{f}}^{w}$ holds for $\mathscr{C}^f$ since $\nabla_{\mathfrak{f}}$ holds for $\mathscr{C}$. $\square$

Now we turn to proving lemmas to establish reflexivity of compatibility induced by an abstract compatibility class. First we prove a general result from which we can conclude parameters are compatible with themselves. It is worth explicitly pointing out that much of the work for showing reflexivity of compatibility is contained in the proof of this lemma. For this reason we give the full proof.

LEMMA 5.2.8. *Let $\mathscr{C}$ be an $\mathcal{S}$-abstract compatibility class and $\|$ be the compatibility relation induced by $\mathscr{C}$.*

1. *If $\mathbf{A}_\alpha \| \mathbf{B}_\alpha$, then $\{\dot\neg[\mathbf{A}_\alpha{}^{\downarrow} \doteq^\alpha \mathbf{B}_\alpha{}^{\downarrow}]\} \notin \mathscr{C}$.*
2. *Suppose $H_{\alpha\alpha^n\ldots\alpha^1}$ is a parameter and $\mathbf{C}^i, \mathbf{D}^i \in cwff_{\alpha^i}(\mathcal{S})$ are such that*

$$\{\dot\neg[(\mathbf{C}^i)^{\downarrow} \doteq^{\alpha^i} (\mathbf{D}^i)^{\downarrow}]\} \notin \mathscr{C} \text{ and } \{\dot\neg[(\mathbf{D}^i)^{\downarrow} \doteq^{\alpha^i} (\mathbf{C}^i)^{\downarrow}]\} \notin \mathscr{C}$$

for each i $(1 \leq i \leq n)$. Then $[H\,\overline{\mathbf{C}^n}] \| [H\,\overline{\mathbf{D}^n}]$.

PROOF. We can prove this by mutual induction on the type α.

1.

ι: This is immediate from Definition 5.2.1.

o: If $\{\dot\neg[\mathbf{A}_o{}^{\downarrow} \doteq^o \mathbf{B}_\alpha{}^{\downarrow}]\} \in \mathscr{C}$, then by $\nabla_\mathfrak{b}$ either $\{\mathbf{A}^{\downarrow}, \dot\neg\mathbf{B}^{\downarrow}\} \in \mathscr{C}$ or $\{\dot\neg\mathbf{A}^{\downarrow}, \mathbf{B}^{\downarrow}\} \in \mathscr{C}$. In either case $\mathbf{A}$ and $\mathbf{B}$ are $\mathscr{C}$-incompatible by Definition 5.2.1.

$\gamma\beta$: Assume $\{\dot\neg[\mathbf{A}_{\gamma\beta}{}^{\downarrow} \doteq \mathbf{B}_{\gamma\beta}{}^{\downarrow}]\} \in \mathscr{C}$ and $\mathbf{A} \| \mathbf{B}$. By $\nabla_\mathfrak{f}^w$, $\nabla_{\beta\eta}$ and closure under subsets, there is a parameter W_β with

$$\{\dot\neg[[\mathbf{A}\,W]^{\downarrow} \doteq^\beta [\mathbf{B}\,W]^{\downarrow}]\} \in \mathscr{C}.$$

Applying the inductive hypothesis for part (2) with W (and no arguments) at the lower type β, we have $W \| W$. By Definition 5.2.1, we have $[\mathbf{A}\,W] \| [\mathbf{B}\,W]$ since $\mathbf{A} \| \mathbf{B}$. We can apply the inductive hypothesis for part (1) at type γ to obtain

$$\{\dot\neg[[\mathbf{A}\,W]^{\downarrow} \doteq^\beta [\mathbf{B}\,W]^{\downarrow}]\} \notin \mathscr{C},$$

a contradiction.

2.

ι: If $[H\,\overline{\mathbf{C}^n}] \| [H\,\overline{\mathbf{D}^n}]$ does not hold, then by definition we have

$$\{\dot\neg[[H\,\mathbf{C}^{1\downarrow} \cdots \mathbf{C}^{n\downarrow}] \doteq^\iota [H\,\mathbf{D}^{1\downarrow} \cdots \mathbf{D}^{n\downarrow}]]\} \in \mathscr{C}$$

or

$$\{\dot\neg[[H\,\mathbf{D}^{1\downarrow} \cdots \mathbf{D}^{n\downarrow}] \doteq^\iota [H\,\mathbf{C}^{1\downarrow} \cdots \mathbf{C}^{n\downarrow}]]\} \in \mathscr{C}.$$

In either case we can apply ∇_{dec} to obtain $\neg\{[\mathbf{C}^{i\downarrow} \doteq \mathbf{D}^{i\downarrow}]\} \in \mathscr{C}$ or $\neg\{[\mathbf{D}^{i\downarrow} \doteq \mathbf{C}^{i\downarrow}]\} \in \mathscr{C}$ for some i, contradicting our assumption.

o: If $[H\,\overline{\mathbf{C}^n}] \parallel [H\,\overline{\mathbf{D}^n}]$ does not hold, then by definition we have

$$\{\neg[H\,\mathbf{C}^{1\downarrow}\cdots\mathbf{C}^{n\downarrow}], [H\,\mathbf{D}^{1\downarrow}\cdots\mathbf{D}^{n\downarrow}]\} \in \mathscr{C}$$

or

$$\{[H\,\mathbf{C}^{1\downarrow}\cdots\mathbf{C}^{n\downarrow}], \neg[H\,\mathbf{D}^{1\downarrow}\cdots\mathbf{D}^{n\downarrow}]\} \in \mathscr{C}.$$

Applying ∇_m we conclude that either $\neg\{[\mathbf{C}^{i\downarrow} \doteq \mathbf{D}^{i\downarrow}]\} \in \mathscr{C}$ or $\neg\{[\mathbf{D}^{i\downarrow} \doteq \mathbf{C}^{i\downarrow}]\} \in \mathscr{C}$ for some i, contradicting our assumption.

$\gamma\beta$: To verify $[H\,\mathbf{C}^1\cdots\mathbf{C}^n] \parallel [H\,\mathbf{D}^1\cdots\mathbf{D}^n]$, let $\mathbf{A}, \mathbf{B} \in cwff_\beta(\mathcal{S})$ with $\mathbf{A} \parallel \mathbf{B}$ be given. By symmetry (cf. Lemma 5.2.2), we know $\mathbf{B} \parallel \mathbf{A}$. Applying the inductive hypothesis for part (1) to $\mathbf{A}$ and $\mathbf{B}$ at type β, we know

$$\{\neg[\mathbf{A}^{\downarrow} \doteq^\beta \mathbf{B}^{\downarrow}]\} \notin \mathscr{C} \text{ and } \{\neg[\mathbf{B}^{\downarrow} \doteq^\beta \mathbf{A}^{\downarrow}]\} \notin \mathscr{C}.$$

We can apply the inductive hypothesis for part (2) to the two terms $[H\,\mathbf{C}^1\cdots\mathbf{C}^n\,\mathbf{A}]$ and $[H\,\mathbf{D}^1\cdots\mathbf{D}^n\,\mathbf{B}]$ at type γ to obtain

$$[H\,\mathbf{C}^1\cdots\mathbf{C}^n\,\mathbf{A}] \parallel [H\,\mathbf{D}^1\cdots\mathbf{D}^n\,\mathbf{B}].$$

Generalizing over $\mathbf{A}$ and $\mathbf{B}$ we have

$$[H\,\mathbf{C}^1\cdots\mathbf{C}^n] \parallel [H\,\mathbf{D}^1\cdots\mathbf{D}^n].$$

□

We can also prove a result concerning compatibility of terms in the presence of equations.

LEMMA 5.2.9. *Let $\mathscr{C}$ be an $\mathcal{S}$-abstract compatibility class, $\parallel$ be the compatibility relation induced by $\mathscr{C}$, α be a type and $\mathbf{A}, \mathbf{B}, \mathbf{C}, \mathbf{D}$ be in $cwff_\alpha(\mathcal{S})$. If $\mathbf{A}_\alpha \parallel \mathbf{C}_\alpha$ and $\{[\mathbf{A}^{\downarrow} \doteq^\alpha \mathbf{B}^{\downarrow}], \neg[\mathbf{C}^{\downarrow} \doteq^\alpha \mathbf{D}^{\downarrow}]\} \in \mathscr{C}$, then $\mathbf{B}$ and $\mathbf{D}$ are $\mathscr{C}$-incompatible.*

PROOF. The proof is by induction on α. See Appendix A.3. □

Next we verify logical constants are compatible with themselves.

LEMMA 5.2.10. *Let $\mathscr{C}$ be an $\mathcal{S}$-abstract compatibility class and $\|$ be the compatibility relation induced by $\mathscr{C}$. For every logical constant $c \in \mathcal{S}$, we have $c \| c$.*

PROOF. The proof is by a case analysis on the logical constant c. We use Lemma 5.2.8 if c is Π^α or Σ^α for some type α. We use Lemma 5.2.9 if c is $=^\alpha$ for some type α. See Appendix A.3. □

We can now conclude compatibility is reflexive using the Fundamental Theorem of Logical Relations (cf. Theorem 2.3.2).

THEOREM 5.2.11. *Let $\mathscr{C}$ be an $\mathcal{S}$-abstract compatibility class and $\|$ be the compatibility relation induced by $\mathscr{C}$. Then $\mathbf{A} \| \mathbf{A}$ for every closed term $\mathbf{A} \in cwff_\alpha(\mathcal{S})$.*

PROOF. We already know $\|$ is an admissible binary logical relation on $cwff(\mathcal{S})$ (cf. Lemma 5.2.4). For parameters W we apply Lemma 5.2.8:2 to conclude $W \| W$. Logical constants $c \in \mathcal{S}$ are compatible with themselves by Lemma 5.2.10. Hence $\mathbf{A} \| \mathbf{A}$ for every closed term $\mathbf{A} \in cwff_\alpha(\mathcal{S})$ by Theorem 2.3.2. □

5.3. Nonatomic Consistency

We often refer to property ∇_c as "atomic consistency". In Theorem 5.3.4, we will establish the corresponding property (called "nonatomic consistency") for all sentences. We consider the elementary case in which

$$\mathscr{C} \in \mathfrak{Acc}_\beta(\mathcal{S}) \cup \mathfrak{Acc}_{\beta\eta}(\mathcal{S})$$

separately from the extensional case in which

$$\mathscr{C} \in \mathfrak{Acc}_{\beta\mathfrak{fb}}(\mathcal{S}).$$

LEMMA 5.3.1. *Let $\mathcal{S}$ be an equality-free signature and $\mathscr{C}$ be an elementary $\mathcal{S}$-abstract consistency class [with η]. For all β-normal $\mathbf{A} \in cwff_o(\mathcal{S})$, $\{\mathbf{A}, \neg\mathbf{A}\} \notin \mathscr{C}$.*

PROOF. (We follow a similar argument in [2], Lemma 3.3.3.) We prove this by induction on the number of occurrences of logical constants in $\mathbf{A}$. See Appendix A.3. □

Now we consider the more complicated case where $\mathscr{C}$ is an extensional abstract consistency class. Note that in this case we know $\mathscr{C}$ satisfies ∇_m instead of ∇_c. We use the compatibility relation induced by $\mathscr{C}$ to establish nonatomic consistency.

LEMMA 5.3.2 (Reflexive Consistency). *Let $\mathscr{C}$ be an extensional $\mathcal{S}$-abstract consistency class. For any* $\mathbf{A} \in cwff_\alpha(\mathcal{S})$, $\{\neg[\mathbf{A} \doteq^\alpha \mathbf{A}]\} \notin \mathscr{C}$.

PROOF. Let $\mathscr{C}^f$ be $\{\Phi \in \mathscr{C} \mid \Phi \text{ is finite}\}$ and $\|$ be the compatibility relation induced by $\mathscr{C}^f$. By Theorem 5.2.7, we know $\mathscr{C}^f$ is an $\mathcal{S}$-abstract compatibility class. By Theorem 5.2.11, we have $\mathbf{A} \| \mathbf{A}$. By Lemma 5.2.8:1, we conclude $\{\neg[\mathbf{A} \doteq^\alpha \mathbf{A}]\} \notin \mathscr{C}^f$. Thus $\{\neg[\mathbf{A} \doteq^\alpha \mathbf{A}]\} \notin \mathscr{C}$. □

LEMMA 5.3.3. *Let $\mathscr{C}$ be an extensional $\mathcal{S}$-abstract consistency class. For any* $\mathbf{A} \in cwff_o(\mathcal{S})$, $\{\mathbf{A}, \neg\mathbf{A}\} \notin \mathscr{C}$.

PROOF. Let $\mathscr{C}^f$ be $\{\Phi \in \mathscr{C} \mid \Phi \text{ is finite}\}$ and $\|$ be the compatibility relation induced by $\mathscr{C}^f$. By Theorem 5.2.7, we know $\mathscr{C}^f$ is an $\mathcal{S}$-abstract compatibility class. By Theorem 5.2.11, we have $\mathbf{A} \| \mathbf{A}$. By the definition of compatibility at type o, this implies the finite set $\{\mathbf{A}, \neg\mathbf{A}\}$ is not in $\mathscr{C}$. □

THEOREM 5.3.4 (Nonatomic Consistency). *Let* $* \in \{\beta, \beta\eta, \beta\mathfrak{fb}\}$ *and* $\mathcal{S} \subseteq \mathcal{S}_*$ *be a signature. For any* $\mathscr{C} \in \mathfrak{Acc}_*(\mathcal{S})$, *any* $\Phi \in \mathscr{C}$ *and any $\mathcal{S}$-sentence* $\mathbf{M}$, *either* $\mathbf{M} \notin \Phi$ *or* $\neg\mathbf{M} \notin \Phi$.

PROOF. We prove this by induction on the construction of $\mathbf{M}$. See Appendix A.3. □

5.4. Abstract Consistency for Sequent Calculi

We now define abstract consistency classes corresponding to the sequent calculi $\mathcal{G}_*^{\mathcal{S}}$ for each $* \in \{\beta, \beta\eta, \beta\mathfrak{fb}\}$ and $\mathcal{S} \subseteq \mathcal{S}_*$.

DEFINITION 5.4.1 (Φ-Sequent). For any set of sentences Φ, we call a (one-sided) sequent Γ a *Φ-sequent* if every $\mathbf{M} \in \Gamma$ is of the form $\neg\mathbf{N}$ where $\mathbf{N} \in \Phi$. For each Γ we define $\Phi_\Gamma := \{\neg\mathbf{M} \mid \mathbf{M} \in \Gamma\}$.

DEFINITION 5.4.2 (Classes for Sequent Calculi). For each $*$ in $\{\beta, \beta\eta, \beta\mathfrak{fb}\}$ and signature $\mathcal{S} \subseteq \mathcal{S}_*$, let $\mathscr{C}_*^{\mathcal{S}}$ be the class of all sets $\Phi \subseteq sent_*(\mathcal{S})$ such that no Φ-sequent is $\mathcal{G}_*^{\mathcal{S}}$-derivable.

LEMMA 5.4.3. *Let* $* \in \{\beta, \beta\eta, \beta\mathfrak{fb}\}$, $\mathcal{S} \subseteq \mathcal{S}_*$ *be a signature and* Γ *be an $\mathcal{S}$-sequent such that* $|\Gamma| \subseteq sent_*(\mathcal{S})$. *If* $\Phi_\Gamma \notin \mathscr{C}_*^{\mathcal{S}}$, *then* Γ *is* $\mathcal{G}_*^{\mathcal{S}}$-*derivable.*

PROOF. The sequent Γ is a multiset $\mathbf{M}^1, \ldots, \mathbf{M}^n$. Hence Φ_Γ is the set $\{\neg\mathbf{M}^i \mid 1 \leq i \leq n\}$. Suppose $\Phi_\Gamma \notin \mathcal{C}_*^{\mathcal{S}}$. Then there is a $\mathcal{G}_*^{\mathcal{S}}$-derivable Φ_Γ-sequent Γ_1. By weakening and Lemma 2.2.11, the sequent $\neg\neg\mathbf{M}^1, \ldots, \neg\neg\mathbf{M}^n$ is derivable. Applying Lemma 2.2.12 n times we know Γ is derivable. □

LEMMA 5.4.4. *Let* $* \in \{\beta, \beta\eta, \beta\mathfrak{fb}\}$, $\mathcal{S} \subseteq \mathcal{S}_*$ *be a signature,* $\Phi \subseteq sent_*(\mathcal{S})$ *be a set of* $\mathcal{S}$*-sentences and* $\mathbf{M} \in sent_*(\mathcal{S})$. *If* $\Phi * \mathbf{M}$ *is not in* $\mathcal{C}_*^{\mathcal{S}}$, *then there is a* Φ*-sequent* Γ *such that* $\Gamma, \neg\mathbf{M}$ *is* $\mathcal{G}_*^{\mathcal{S}}$*-derivable.*

PROOF. Since $\Phi * \mathbf{M} \notin \mathcal{C}_*^{\mathcal{S}}$ there must be a $\mathcal{G}_*^{\mathcal{S}}$-derivable $\Phi * \mathbf{M}$-sequent Γ_1. By weakening (Lemma 2.2.10), we can assume $\neg\mathbf{M} \in \Gamma_1$. Let Γ be the Φ-sequent $(|\Gamma_1| \setminus \{\neg\mathbf{M}\})^{ms}$. Clearly, the multisets $\Gamma, \neg\mathbf{M}$ and $|\Gamma_1|^{ms}$ are equal. Thus $\Gamma, \neg\mathbf{M}$ is derivable by Lemma 2.2.11. □

LEMMA 5.4.5. *Let* $* \in \{\beta, \beta\eta, \beta\mathfrak{fb}\}$, $\mathcal{S} \subseteq \mathcal{S}_*$ *be a signature,* $\Phi \subseteq sent_*(\mathcal{S})$ *and* $\mathbf{M}, \mathbf{N} \in sent_*(\mathcal{S})$. *If* $\Phi * \mathbf{M} * \mathbf{N} \notin \mathcal{C}_*^{\mathcal{S}}$, *then there is a* Φ*-sequent* Γ *such that* $\Gamma, \neg\mathbf{M}, \neg\mathbf{N}$ *is* $\mathcal{G}_*^{\mathcal{S}}$*-derivable.*

PROOF. This is analogous to Lemma 5.4.4 using Lemma 2.2.11 and weakening to obtain exactly one copy of $\neg\mathbf{M}$ and one copy of $\neg\mathbf{N}$. □

LEMMA 5.4.6. *Let* $* \in \{\beta, \beta\eta, \beta\mathfrak{fb}\}$, $\mathcal{S} \subseteq \mathcal{S}_*$ *be a signature,* $\Phi \subseteq sent_*(\mathcal{S})$ *and* $\mathbf{M} \in sent_*(\mathcal{S})$. *If* $\Phi * \neg\mathbf{M} \notin \mathcal{C}_*^{\mathcal{S}}$, *then there is a* Φ*-sequent* Γ *such that* $\Gamma, \mathbf{M}$ *is* $\mathcal{G}_*^{\mathcal{S}}$*-derivable.*

PROOF. Since $\Phi * \neg\mathbf{M} \notin \mathcal{C}_*^{\mathcal{S}}$, by Lemma 5.4.4 there is a Φ-sequent Γ such that $\Gamma, \neg\neg\mathbf{M}$ is derivable. By Lemma 2.2.12, there is a derivation of $\Gamma, \mathbf{M}$. □

THEOREM 5.4.7 (Abstract Consistency for Sequent Calculi). *Suppose* $* \in \{\beta, \beta\eta, \beta\mathfrak{fb}\}$ *and* $\mathcal{S} \subseteq \mathcal{S}_*$ *is a signature. Then* $\mathcal{C}_*^{\mathcal{S}} \in \mathfrak{Acc}_*(\mathcal{S})$ *and* $\mathcal{C}_*^{\mathcal{S}}$ *has finite character.*

PROOF. Suppose $\Phi \notin \mathcal{C}_*^{\mathcal{S}}$ and every finite subset $\Phi_0 \in \mathcal{C}_*^{\mathcal{S}}$. If $*$ is β or $\beta\eta$, then $\{\mathbf{M}\} \in \mathcal{C}_*^{\mathcal{S}}$ implies $\mathbf{M}$ is equality-free for every $\mathbf{M} \in \Phi$. Since $\Phi \notin \mathcal{C}_*^{\mathcal{S}}$ we know some Φ-sequent Γ has a $\mathcal{G}_*^{\mathcal{S}}$-derivation. Hence we have a finite subset Φ_Γ of Φ such that $\Phi_\Gamma \notin \mathcal{C}_*^{\mathcal{S}}$. Consequently, each class $\mathcal{C}_*^{\mathcal{S}}$ has finite character and is thus closed under subsets by Lemma 5.1.2.

We prove the other conditions by contradiction. We verify a few of the conditions here and check the remaining conditions in Appendix A.3.

∇_c : We need only check this in case $* \in \{\beta, \beta\eta\}$ (since ∇_c is not required in the definition of extensional abstract consistency classes). Let $\mathbf{A} \in cwff_o(\mathcal{S})$ be an atom. Assume $\Phi \in \mathscr{C}_*^{\mathcal{S}}$ and $\neg\mathbf{A}, \mathbf{A} \in \Phi$. Then $\neg\neg\mathbf{A}, \neg\mathbf{A}$ is a Φ-sequent and has the derivation

$$\dfrac{\dfrac{}{\mathbf{A}, \neg\mathbf{A}}\;\mathcal{G}(Init)}{\neg\neg\mathbf{A}, \neg\mathbf{A}}\;\mathcal{G}(\neg\neg)$$

This is a contradiction.

$\nabla_\vee$: Assume $\Phi \in \mathscr{C}_*^{\mathcal{S}}$, $[\mathbf{M} \vee \mathbf{N}] \in \Phi$,

$$\Phi * \mathbf{M} \notin \mathscr{C}_*^{\mathcal{S}} \text{ and } \Phi * \mathbf{N} \notin \mathscr{C}_*^{\mathcal{S}}.$$

By Lemma 5.4.4, there are Φ-sequents Γ_1 and Γ_2 and $\mathcal{G}_*^{\mathcal{S}}$-derivations $\mathcal{D}$ of $\Gamma_1, \neg\mathbf{M}$ and $\mathcal{E}$ of $\Gamma_2, \neg\mathbf{N}$. Let Γ be the Φ-sequent Γ_1, Γ_2. By weakening (Lemma 2.2.10), there are derivations $\mathcal{D}'$ of $\Gamma, \neg\mathbf{M}$ and $\mathcal{E}'$ of $\Gamma, \neg\mathbf{N}$. This is contradicted by the derivation

$$\dfrac{\begin{matrix}\mathcal{D}' \\ \Gamma, \neg\mathbf{M}\end{matrix} \qquad \begin{matrix}\mathcal{E}' \\ \Gamma, \neg\mathbf{N}\end{matrix}}{\Gamma, \neg[\mathbf{M} \vee \mathbf{N}]}\;\mathcal{G}(\neg\vee)$$

of the Φ-sequent $\Gamma, \neg[\mathbf{M} \vee \mathbf{N}]$.

$\nabla_\wedge$: Assume $\Phi \in \mathscr{C}_*^{\mathcal{S}}$, $\neg[\mathbf{M} \vee \mathbf{N}] \in \Phi$, $\Phi * \neg\mathbf{M} * \neg\mathbf{N} \notin \mathscr{C}_*^{\mathcal{S}}$. Applying Lemma 5.4.5 we know there is a Φ-sequent Γ and a $\mathcal{G}_*^{\mathcal{S}}$-derivation $\mathcal{D}_1$ of $\Gamma, \neg\neg\mathbf{M}, \neg\neg\mathbf{N}$. Applying Lemma 2.2.12 twice we have a derivation $\mathcal{D}_2$ of $\Gamma, \mathbf{M}, \mathbf{N}$. The derivation

$$\dfrac{\dfrac{\begin{matrix}\mathcal{D}_2 \\ \Gamma, \mathbf{M}, \mathbf{N}\end{matrix}}{\Gamma, [\mathbf{M} \vee \mathbf{N}]}\;\mathcal{G}(\vee)}{\Gamma, \neg\neg[\mathbf{M} \vee \mathbf{N}]}\;\mathcal{G}(\neg\neg)$$

of the Φ-sequent $\Gamma, \neg\neg[\mathbf{M} \vee \mathbf{N}]$ contradicts $\Phi \in \mathscr{C}_*^{\mathcal{S}}$.

$\nabla_\exists$: Assume $\Phi \in \mathscr{C}_*^{\mathcal{S}}$, $\neg[\forall x_\alpha \mathbf{M}] \in \Phi$ and $\Phi * \neg[W/x]\mathbf{M} \notin \mathscr{C}_*^{\mathcal{S}}$ where W_α is a parameter $W \notin \mathbf{Params}(\Phi)$. By Lemma 5.4.6, there is a Φ-sequent Γ and a $\mathcal{G}_*^{\mathcal{S}}$-derivation $\mathcal{D}$ of $\Gamma, [W/x]\mathbf{M}$.

Since Γ is a Φ-sequent we know $W \notin \mathbf{Params}(\Gamma)$. We obtain a contradiction by deriving the Φ-sequent $\Gamma, \dot\neg\dot\neg[\dot\forall x_\alpha \mathbf{M}]$ as follows:

$$\dfrac{\dfrac{\begin{matrix}\mathcal{D}\\ \Gamma, [[W/x]\mathbf{M}]\end{matrix}}{\Gamma, [\dot\forall x_\alpha \mathbf{M}]}\ \mathcal{G}(\dot\forall^W)}{\Gamma, \dot\neg\dot\neg[\dot\forall x_\alpha \mathbf{M}]}\ \mathcal{G}(\dot\neg\dot\neg)$$

∇_β : We must check ∇_β when $*$ is β. Suppose $\mathbf{M} \in \Phi$ and $\Phi \in \mathscr{C}_\beta^\mathcal{S}$. Assume $\Phi * \mathbf{M}^{\downarrow\beta} \notin \mathscr{C}_\beta^\mathcal{S}$. By Lemma 5.4.4, there is a Φ-sequent Γ and a $\mathcal{G}_\beta^\mathcal{S}$-derivation $\mathcal{D}$ of $\Gamma, \dot\neg\mathbf{M}^{\downarrow\beta}$. The fact that $\Gamma, \dot\neg\mathbf{M}$ is a Φ-sequent with derivation

$$\dfrac{\begin{matrix}\mathcal{D}\\ \Gamma, \dot\neg\mathbf{M}^{\downarrow\beta}\end{matrix}}{\Gamma, \dot\neg\mathbf{M}}\ \mathcal{G}(\beta)$$

contradicts $\Phi \in \mathscr{C}_\beta^\mathcal{S}$.

When $*$ is $\beta\mathfrak{fb}$, we must check the remaining conditions. We verify ∇_m and $\nabla_\mathfrak{b}$ here and the other conditions for extensional abstract consistency classes in Appendix A.3.

∇_m : Suppose H is a parameter. Assume $\{\dot\neg[H\,\overline{\mathbf{A}^n}], [H\,\overline{\mathbf{B}^n}]\} \subseteq \Phi$, $\Phi \in \mathscr{C}_{\beta\mathfrak{fb}}^\mathcal{S}$ and $\Phi * \dot\neg[\mathbf{A}^i \doteq \mathbf{B}^i] \notin \mathscr{C}_{\beta\mathfrak{fb}}^\mathcal{S}$ for every i $(1 \le i \le n)$. By Lemma 5.4.6 and weakening, there is a Φ-sequent Γ such that there are $\mathcal{G}_{\beta\mathfrak{fb}}^\mathcal{S}$-derivations $\mathcal{D}_i$ of $\Gamma, [\mathbf{A}^i \doteq \mathbf{B}^i]$. The derivation

$$\dfrac{\dfrac{\begin{matrix}\mathcal{D}_1\\ \Gamma, [\mathbf{A}^1 \doteq \mathbf{B}^1]\end{matrix} \quad \cdots \quad \begin{matrix}\mathcal{D}_n\\ \Gamma, [\mathbf{A}^n \doteq \mathbf{B}^n]\end{matrix}}{\Gamma, [H\,\overline{\mathbf{A}^n}], \dot\neg[H\,\overline{\mathbf{B}^n}]}\ \mathcal{G}(Init^{\doteq})}{\Gamma, \dot\neg\dot\neg[[H\,\overline{\mathbf{A}^n}], \dot\neg[H\,\overline{\mathbf{B}^n}]]}\ \mathcal{G}(\dot\neg\dot\neg)$$

of the Φ-sequent $\Gamma, \dot\neg\dot\neg[H\,\overline{\mathbf{A}^n}], \dot\neg[H\,\overline{\mathbf{B}^n}]]$ contradicts $\Phi \in \mathscr{C}_{\beta\mathfrak{fb}}^\mathcal{S}$.

$\nabla_\mathfrak{b}$: Suppose $\Phi \in \mathscr{C}_{\beta\mathfrak{fb}}^\mathcal{S}$ and $\dot\neg[\mathbf{A} \doteq^o \mathbf{B}] \in \Phi$. Assume

$$\Phi * \mathbf{A} * \dot\neg\mathbf{B} \notin \mathscr{C}_{\beta\mathfrak{fb}}^\mathcal{S} \text{ and } \Phi * \dot\neg\mathbf{A} * \mathbf{B} \notin \mathscr{C}_{\beta\mathfrak{fb}}^\mathcal{S}.$$

By Lemma 5.4.5 and weakening, there is a Φ-sequent Γ such that $\Gamma, \dot\neg\mathbf{A}, \dot\neg\dot\neg\mathbf{B}$ and $\Gamma, \dot\neg\dot\neg\mathbf{A}, \dot\neg\mathbf{B}$ are derivable. Using Lemma 2.2.12 there are derivations of $\Gamma, \dot\neg\mathbf{A}, \mathbf{B}$ and $\Gamma, \mathbf{A}, \dot\neg\mathbf{B}$. Using

$\mathcal{G}(\mathfrak{b})$ and $\mathcal{G}(\dot\neg\dot\neg)$ we can derive the Φ-sequent $\Gamma, \dot\neg\dot\neg[\mathbf{A} \doteq^o \mathbf{B}]$, a contradiction.

□

Now we can use these abstract consistency classes to prove admissibility of certain rules.

COROLLARY 5.4.8 (Admissibility of General Init). *Suppose $*$ is in $\{\beta, \beta\eta, \beta\mathfrak{fb}\}$ and $\mathcal{S} \subseteq \mathcal{S}_*$ is a signature. Both of the rules $\mathcal{G}(GInit)$ and $\mathcal{G}(\dot\neg \doteq)$ are admissible in $\mathcal{G}_*^{\mathcal{S}}$.*

PROOF. To check $\mathcal{G}(GInit)$ is admissible, suppose there is no derivation of $\Gamma, \mathbf{M}, \dot\neg\mathbf{M}$ where $\mathbf{M} \in sent_*(\mathcal{S})$. By weakening, there is no derivation of $\mathbf{M}, \dot\neg\mathbf{M}$. By Lemma 5.4.3, $\{\mathbf{M}, \dot\neg\mathbf{M}\} \in \mathscr{C}_*^{\mathcal{S}}$. This contradicts Theorem 5.4.7 and Theorem 5.3.4 (nonatomic consistency).
Admissibility of $\mathcal{G}(\dot\neg \doteq)$ now follows from Lemma 2.2.14. □

COROLLARY 5.4.9 (Admissibility of Reflexivity). *Let $\mathcal{S}$ be a signature. The rule $\mathcal{G}(Refl)$ is admissible in $\mathcal{G}_{\beta\mathfrak{fb}}^{\mathcal{S}}$.*

PROOF. To check $\mathcal{G}(Refl)$ is admissible, suppose there is no derivation of $\Gamma, [\mathbf{A} \doteq^\alpha \mathbf{A}]$. By weakening, there is no derivation of $[\mathbf{A} \doteq^\alpha \mathbf{A}]$. By Lemma 5.4.3, $\{\dot\neg[\mathbf{A} \doteq^\alpha \mathbf{A}]\} \in \mathscr{C}_{\beta\mathfrak{fb}}^{\mathcal{S}}$. This contradicts Theorem 5.4.7 and Lemma 5.3.2. □

5.5. Hintikka Sets

DEFINITION 5.5.1 (Hintikka Properties). Let $\mathcal{H}$ be a set of sentences. We define the following properties which $\mathcal{H}$ may satisfy.

$\vec{\nabla}_c$: $\mathbf{M} \notin \mathcal{H}$ or $\dot\neg\mathbf{M} \notin \mathcal{H}$.

$\vec{\nabla}_\beta$: If $\mathbf{M} \in \mathcal{H}$, then $\mathbf{M}^{\downarrow\beta} \in \mathcal{H}$.

$\vec{\nabla}_{\beta\eta}$: If $\mathbf{M} \in \mathcal{H}$, then $\mathbf{M}^{\downarrow} \in \mathcal{H}$.

$\vec{\nabla}_\bot$: $\dot\neg\dot\top \notin \mathcal{H}$.

$\vec{\nabla}_\neg$: If $\dot\neg\dot\neg\mathbf{M} \in \mathcal{H}$, then $\mathbf{M} \in \mathcal{H}$.

$\vec{\nabla}_\vee$: If $[\mathbf{M} \dot\vee \mathbf{N}] \in \mathcal{H}$, then $\mathbf{M} \in \mathcal{H}$ or $\mathbf{N} \in \mathcal{H}$.

$\vec{\nabla}_\wedge$: If $\dot\neg[\mathbf{M} \dot\vee \mathbf{N}] \in \mathcal{H}$, then $\dot\neg\mathbf{M} \in \mathcal{H}$ and $\dot\neg\mathbf{N} \in \mathcal{H}$.

$\vec{\nabla}_\forall$: If $[\dot\forall x_\alpha \mathbf{M}] \in \mathcal{H}$, then $[\mathbf{A}/x]\mathbf{M} \in \mathcal{H}$ for every $\mathbf{A} \in cwff_\alpha(\mathcal{S})$.

$\vec{\nabla}_{\exists}$: If $\neg[\forall x_\alpha \mathbf{M}] \in \mathcal{H}$, then $\neg[W/x]\mathbf{M} \in \mathcal{H}$ for some parameter W_α.

$\vec{\nabla}^{\sharp}$: If $\mathbf{A}_o \in \mathcal{H}$, then $\mathbf{A}^{\sharp} \in \mathcal{H}$. Also, if $\neg\mathbf{A}_o \in \mathcal{H}$, then $\neg\mathbf{A}^{\sharp} \in \mathcal{H}$.

$\vec{\nabla}_m$: If H is a parameter, $\neg[H\,\overline{\mathbf{A}^n}] \in \mathcal{H}$ and $[H\,\overline{\mathbf{B}^n}] \in \mathcal{H}$, then there is an i with $1 \leq i \leq n$ such that $\neg[\mathbf{A}^i \doteq \mathbf{B}^i] \in \mathcal{H}$.

$\vec{\nabla}_{dec}$: If H is a parameter and $\neg[[H\,\overline{\mathbf{A}^n}] \doteq^{\iota} [H\,\overline{\mathbf{B}^n}]] \in \mathcal{H}$, then there is an i with $1 \leq i \leq n$ such that $\neg[\mathbf{A}^i \doteq \mathbf{B}^i] \in \mathcal{H}$.

$\vec{\nabla}_{\mathfrak{b}}$: If $\neg[\mathbf{A}_o \doteq^o \mathbf{B}_o] \in \mathcal{H}$, then $\{\mathbf{A}, \neg\mathbf{B}\} \subseteq \mathcal{H}$ or $\{\neg\mathbf{A}, \mathbf{B}\} \subseteq \mathcal{H}$.

$\vec{\nabla}_{\mathfrak{f}}$: If $\neg[\mathbf{G}_{\alpha\beta} \doteq^{\alpha\beta} \mathbf{H}_{\alpha\beta}] \in \mathcal{H}$, then there is a parameter W_β such that $\neg[\mathbf{G}\,W \doteq^{\alpha} \mathbf{H}\,W] \in \mathcal{H}$.

$\vec{\nabla}^{o}_{=}$: If $\mathbf{A}_o \doteq^o \mathbf{B}_o \in \mathcal{H}$, then $\{\mathbf{A}, \mathbf{B}\} \subseteq \mathcal{H}$ or $\{\neg\mathbf{A}, \neg\mathbf{B}\} \subseteq \mathcal{H}$.

$\vec{\nabla}^{\rightarrow}_{=}$: If $\mathbf{G}_{\alpha\beta} \doteq^{\alpha\beta} \mathbf{H}_{\alpha\beta} \in \mathcal{H}$, then $[\mathbf{G}\,\mathbf{B} \doteq^{\beta} \mathbf{H}\,\mathbf{B}] \in \mathcal{H}$ for every term $\mathbf{B} \in \mathit{cwff}_\beta(\mathcal{S})$.

$\vec{\nabla}^{r}_{=}$: $\neg[\mathbf{A}_\iota \doteq^{\iota} \mathbf{A}] \notin \mathcal{H}$.

$\vec{\nabla}^{u}_{=}$: Suppose $[\mathbf{A}_\iota \doteq^{\iota} \mathbf{B}_\iota] \in \mathcal{H}$ and $\neg[\mathbf{C}_\iota \doteq^{\iota} \mathbf{D}_\iota] \in \mathcal{H}$. Then

$$\neg[\mathbf{A} \doteq^{\iota} \mathbf{C}] \in \mathcal{H} \text{ or } \neg[\mathbf{B} \doteq^{\iota} \mathbf{D}] \in \mathcal{H}.$$

Also,

$$\neg[\mathbf{A} \doteq^{\iota} \mathbf{D}] \in \mathcal{H} \text{ or } \neg[\mathbf{B} \doteq^{\iota} \mathbf{C}] \in \mathcal{H}.$$

DEFINITION 5.5.2 (Elementary Hintikka Set). Let $\mathcal{S}$ be an equality-free signature and $\mathcal{H} \subseteq \mathit{sent}(\mathcal{S})$ be a set of $\mathcal{S}$-sentences.

- We say $\mathcal{H}$ is an *elementary $\mathcal{S}$-Hintikka set* if $\mathcal{H}$ is equality-free and satisfies $\vec{\nabla}_c$, $\vec{\nabla}_\beta$, $\vec{\nabla}_\bot$, $\vec{\nabla}_\neg$, $\vec{\nabla}_\vee$, $\vec{\nabla}_\wedge$, $\vec{\nabla}_\forall$, $\vec{\nabla}_\exists$ and $\vec{\nabla}^{\sharp}$. We let $\mathfrak{Hint}_\beta(\mathcal{S})$ be the collection of elementary $\mathcal{S}$-Hintikka sets.
- We say $\mathcal{H}$ is an *η-elementary $\mathcal{S}$-Hintikka set* if $\mathcal{H}$ is equality-free and satisfies $\vec{\nabla}_c$, $\vec{\nabla}_{\beta\eta}$, $\vec{\nabla}_\bot$, $\vec{\nabla}_\neg$, $\vec{\nabla}_\vee$, $\vec{\nabla}_\wedge$, $\vec{\nabla}_\forall$, $\vec{\nabla}_\exists$ and $\vec{\nabla}^{\sharp}$. We let $\mathfrak{Hint}_{\beta\eta}(\mathcal{S})$ be the collection of η-elementary $\mathcal{S}$-Hintikka sets.

REMARK 5.5.3 (Semivaluations). Schütte defines a notation of semivaluation (for a relational formulation of simple type theory) in [59]. A *semivaluation* is a partial function assigning terms of type o to either T (true) or F (false) satisfying certain closure properties similar to those for Hintikka sets. In our context a semivaluation could be defined as a partial function V from $\mathit{sent}(\mathcal{S})$ to $\{\mathtt{T}, \mathtt{F}\}$ satisfying closure properties similar to those for elementary $\mathcal{S}$-Hintikka sets. Given any semivaluation V one can define a set

$$\mathcal{H}_V := \{\mathbf{M} \mid V(\mathbf{M}) = \mathtt{T}\}.$$

Conversely, given any elementary $\mathcal{S}$-Hintikka $\mathcal{H}$ one can define a partial function

$$V_{\mathcal{H}}(\mathbf{M}) := \begin{cases} \mathtt{T} & \text{if } \mathbf{M} \in \mathcal{H} \\ \mathtt{F} & \text{if } \neg\mathbf{M} \in \mathcal{H} \\ \text{undefined} & \text{otherwise.} \end{cases}$$

If we define the notion of semivaluation appropriately, then $\mathcal{H}_V$ will be an elementary $\mathcal{S}$-Hintikka set whenever V is a semivaluation, and $V_{\mathcal{H}}$ will be a semivaluation whenever $\mathcal{H}$ is an elementary $\mathcal{S}$-Hintikka set. Thus there is no significant difference between discussing semivaluations and Hintikka sets.

Schütte proves every semivaluation (as defined in [59]) can be extended to a *partial valuation* (satisfying more properties than a semivaluation). Theorem IV of [59] states that every derivable wff is strictly definable iff every partial valuation is extendible to a total valuation. Analogously, one can prove that cut-elimination for the calculus $\mathcal{G}_\beta^{\mathcal{S}}$ holds iff every elementary $\mathcal{S}$-Hintikka set can be extended to a *saturated* $\mathcal{S}$-Hintikka set. (A Hintikka set $\mathcal{H}$ is saturated if for every sentence $\mathbf{M} \in \mathit{sent}(\mathcal{S})$ either $\mathbf{M} \in \mathcal{H}$ or $\neg\mathbf{M} \in \mathcal{H}$.) Schütte does not prove in [59] that every semivaluation can be extended to a total valuation. Prawitz directly constructs such extensions in [56]. (A similar construction in [61] also implies such extensions exist.) Both Takahashi and Prawitz conclude cut-elimination results for relational formulations of type theory. In [2] Andrews constructs V-complexes for a semivaluation V for elementary type theory (using a formulation of higher-order logic with simply typed λ-terms). The V-complex construction implicitly gives a total extension of V by considering the function $\mathscr{V}_\varphi^2$ (cf. [2]) on the terms of type o (where φ is an appropriate assignment of variables). We follow a similar approach to construct a model for an elementary $\mathcal{S}$-Hintikka set $\mathcal{H}$ (cf. Theorem 5.6.4). If we consider a semivaluation V, the model of $\mathcal{H}_V$ constructed in the proof of Theorem 5.6.4 allows us to extend V to be a total valuation.

In summary, the foundation for proving cut-elimination for versions of higher-order logic was given in [59] in terms of total extensions of semivaluations. Such extensions exist as a consequence of [61], [56] and [2]. We are following a very similar approach here while leaving open the possibility of changing the signature $\mathcal{S}$ of logical constants and the amount of extensionality assumed.

DEFINITION 5.5.4 (Extensional Hintikka Set). Let $\mathcal{S}$ be a signature and $\mathcal{H} \subseteq sent(\mathcal{S})$ be a set of $\mathcal{S}$-sentences. We say $\mathcal{H}$ is an *extensional $\mathcal{S}$-Hintikka set* if it satisfies $\vec{\nabla}_c$, $\vec{\nabla}_{\beta\eta}$, $\vec{\nabla}_\bot$, $\vec{\nabla}_\neg$, $\vec{\nabla}_\vee$, $\vec{\nabla}_\wedge$, $\vec{\nabla}_\forall$, $\vec{\nabla}_\exists$, $\vec{\nabla}_m$, $\vec{\nabla}_{dec}$, $\vec{\nabla}_\mathfrak{b}$, $\vec{\nabla}_\mathfrak{f}$, $\vec{\nabla}_=^o$, $\vec{\nabla}_=^\to$, $\vec{\nabla}_=^r$, $\vec{\nabla}_=^u$ and $\vec{\nabla}^\sharp$. We let $\mathfrak{Hint}_{\beta\mathfrak{fb}}(\mathcal{S})$ be the collection of extensional $\mathcal{S}$-Hintikka sets.

For each $* \in \{\beta, \beta\eta, \beta\mathfrak{fb}\}$ and $\mathcal{S} \subseteq \mathcal{S}_*$ we have defined a collection $\mathfrak{Hint}_*(\mathcal{S})$ such that $\mathcal{H} \subseteq sent_*(\mathcal{S})$ for every $\mathcal{H} \in \mathfrak{Hint}_*(\mathcal{S})$. We say $\mathcal{H}$ is a *Hintikka set* if $\mathcal{H} \in \mathfrak{Hint}_*(\mathcal{S})$ for some $* \in \{\beta, \beta\eta, \beta\mathfrak{fb}\}$ and $\mathcal{S} \subseteq \mathcal{S}_*$.

We will use Hintikka sets to obtain models in the appropriate model classes. For example, every extensional $\mathcal{S}$-Hintikka set has an extensional $\mathcal{S}$-model. First we demonstrate how to obtain Hintikka sets.

LEMMA 5.5.5 (Hintikka Lemma). *Let $* \in \{\beta, \beta\eta, \beta\mathfrak{fb}\}$, $\mathcal{S} \subseteq \mathcal{S}_*$ be a signature and $\mathscr{C}$ be an abstract consistency class in $\mathfrak{Acc}_*(\mathcal{S})$. Suppose a set $\mathcal{H} \in \mathscr{C}$ satisfies the following properties:*

1. *$\mathcal{H}$ is subset-maximal in $\mathscr{C}$ (i.e., for each sentence $\mathbf{D} \in cwff_o(\mathcal{S})$ such that $\mathcal{H} * \mathbf{D} \in \mathscr{C}$, we already have $\mathbf{D} \in \mathcal{H}$).*
2. *$\mathcal{H}$ satisfies $\vec{\nabla}_\exists$.*
3. *If $*$ is $\beta\mathfrak{fb}$, then $\vec{\nabla}_\mathfrak{f}$ holds in $\mathcal{H}$.*

Then $\mathcal{H} \in \mathfrak{Hint}_*(\mathcal{S})$.

PROOF. $\mathcal{H}$ satisfies $\vec{\nabla}_\exists$ by assumption. Also, if $*$ is $\beta\mathfrak{fb}$, then we have explicitly assumed $\mathcal{H}$ satisfies $\vec{\nabla}_\mathfrak{f}$. Since $\mathcal{H} \in \mathscr{C}$ we know $\mathcal{H} \subseteq sent_*(\mathcal{S})$. The fact that $\mathcal{H} \in \mathscr{C}$ satisfies $\vec{\nabla}_c$ follows directly from nonatomic consistency (Theorem 5.3.4). If $*$ is $\beta\mathfrak{fb}$, then $\mathcal{H}$ satisfies $\vec{\nabla}_=^r$ by Lemma 5.3.2.

$\mathcal{H}$ satisfies $\vec{\nabla}_\bot$ since $\mathcal{H} \in \mathscr{C}$ and $\mathscr{C}$ satisfies $\nabla_\bot$. Every other $\vec{\nabla}_*$ property follows directly from the corresponding ∇_* property and maximality of $\mathcal{H}$ in $\mathscr{C}$. For example, to check $\vec{\nabla}_\neg$, suppose $\neg\neg\mathbf{M} \in \mathcal{H}$. By $\nabla_\neg$, we know $\mathcal{H} * \mathbf{M} \in \mathscr{C}$. By maximality of $\mathcal{H}$, we have $\mathbf{M} \in \mathcal{H}$. Checking $\vec{\nabla}_\beta$, $\vec{\nabla}_{\beta\eta}$ (if $* \in \{\beta\eta, \beta\mathfrak{fb}\}$), $\vec{\nabla}_\wedge$, $\vec{\nabla}_\forall$ and $\vec{\nabla}^\sharp$ hold for $\mathcal{H}$ follows exactly this same pattern. Also, if $*$ is $\beta\mathfrak{fb}$, $\vec{\nabla}_=^\to$ follows the pattern. To check $\vec{\nabla}_\vee$, given $[\mathbf{M} \mathbin{\dot\vee} \mathbf{N}] \in \mathcal{H}$, $\nabla_\vee$ implies $\mathcal{H} * \mathbf{M} \in \mathscr{C}$ or $\mathcal{H} * \mathbf{N} \in \mathscr{C}$.

Thus either $\mathbf{M} \in \mathcal{H}$ or $\mathbf{N} \in \mathcal{H}$. If $*$ is $\beta\mathfrak{fb}$, then we can check $\vec{\nabla}_m$, $\vec{\nabla}_{dec}$, $\vec{\nabla}_{\mathfrak{b}}$, $\vec{\nabla}^o_=$ and $\vec{\nabla}^u_=$ by doing a similar case analysis. □

LEMMA 5.5.6 (Abstract Extension Lemma). *Let* $* \in \{\beta, \beta\eta, \beta\mathfrak{fb}\}$, $\mathcal{S} \subseteq \mathcal{S}_*$ *be a signature and* $\mathscr{C}$ *be an abstract consistency class of finite character in* $\mathfrak{Acc}_*(\mathcal{S})$ *(cf. Definitions 5.1.1, 5.1.5 and 5.1.6). For every sufficiently pure* $\Phi \in \mathscr{C}$ *there exists a Hintikka set* $\mathcal{H} \in \mathfrak{Hint}_*(\mathcal{S})$ *such that* $\Phi \subseteq \mathcal{H}$.

PROOF. We can use transfinite induction to extend Φ to a set $\mathcal{H}$ in $\mathscr{C}$ with appropriate witnesses which is subset-maximal. This $\mathcal{H}$ will be a Hintikka set by Lemma 5.5.5. The full proof is in Appendix A.3. □

5.6. Completeness of Elementary Sequent Calculi

We will now prove every [η-]elementary $\mathcal{S}$-Hintikka set has a model in $\mathfrak{M}_\beta(\mathcal{S})$ [$\mathfrak{M}_{\beta\eta}(\mathcal{S})$]. We directly follow Andrews V-complex construction [2] (generalizing over the signature) by using a possible values evaluation.

DEFINITION 5.6.1 ($\mathcal{H}$-Possible Booleans). Let $\mathcal{H}$ be an [η-]elementary $\mathcal{S}$-Hintikka set. We define the set $\mathcal{B}^{\mathbf{A}}_{\mathcal{H}}$ of *$\mathcal{H}$-possible booleans* for $\mathbf{A}$ by

- $\mathcal{B}^{\mathbf{A}}_{\mathcal{H}} := \{\mathtt{T}, \mathtt{F}\}$ if $\mathbf{A} \notin \mathcal{H}$ and $\neg\mathbf{A} \notin \mathcal{H}$,
- $\mathcal{B}^{\mathbf{A}}_{\mathcal{H}} := \{\mathtt{T}\}$ if $\mathbf{A} \in \mathcal{H}$, and
- $\mathcal{B}^{\mathbf{A}}_{\mathcal{H}} := \{\mathtt{F}\}$ if $\neg\mathbf{A} \in \mathcal{H}$.

Note that by $\vec{\nabla}_c$ we cannot have $\mathbf{A}, \neg\mathbf{A} \in \mathcal{H}$. Hence $\mathcal{B}^{\mathbf{A}}_{\mathcal{H}}$ is defined (and nonempty) for every $\mathbf{A} \in cwff_o(\mathcal{S})$.

DEFINITION 5.6.2 (Andrews Structure). Let $\mathcal{H}$ be an [η-]elementary $\mathcal{S}$-Hintikka set. A possible values structure

$$\mathcal{A}^{\mathcal{H}} := (\mathcal{D}, @) \quad [\mathcal{A}^{\mathcal{H}}_\eta := (\mathcal{D}, @)]$$

called the [η-]*Andrews Structure for* $\mathcal{H}$ is defined as follows:

- Let $\mathcal{D}_o$ be the set of pairs $\langle \mathbf{A}_o, p \rangle$ where $\mathbf{A} \in cwff_o(\mathcal{S})$ is β-normal [$\beta\eta$-normal] and $p \in \mathcal{B}^{\mathbf{A}}_{\mathcal{H}}$.

- Let $\mathcal{D}_\iota$ be the set of pairs $\langle \mathbf{A}_\iota, \iota\rangle$ where $\mathbf{A} \in cwff_\iota(\mathcal{S})$ is β-normal [$\beta\eta$-normal].
- Let $\mathcal{D}_{\alpha\beta}$ be the set of pairs $\langle \mathbf{G}, g\rangle$ where $\mathbf{G} \in cwff_{\alpha\beta}(\mathcal{S})$ is β-normal [$\beta\eta$-normal] and $g : \mathcal{D}_\beta \to \mathcal{D}_\alpha$ is such that for every $\langle \mathbf{B}, b\rangle \in \mathcal{D}_\alpha$, $g(\langle \mathbf{B}, b\rangle) = \langle \mathbf{A}, a\rangle$ implies $\mathbf{A}$ is the β-normal form [$\beta\eta$-normal form] of $[\mathbf{G}\,\mathbf{B}]$.

We define @ by

$$\langle \mathbf{G}, g\rangle @ \langle \mathbf{B}, b\rangle = g(\langle \mathbf{B}, b\rangle)$$

for each $\langle \mathbf{G}, g\rangle \in \mathcal{D}_{\alpha\beta}$ and $\langle \mathbf{B}, b\rangle \in \mathcal{D}_\beta$.

It is trivial to check the conditions of Definition 4.1.1 so that the Andrews Structure for an elementary $\mathcal{S}$-Hintikka set is a possible values structure for β. It is also easy to check that the η-Andrews Structure for an η-elementary $\mathcal{S}$-Hintikka set is a possible values structure for $\beta\eta$. The next lemma will be used to determine that any model based on a possible values evaluation over the Andrews Structure for $\mathcal{H}$ must satisfy the sentences in $\mathcal{H}$.

LEMMA 5.6.3. *Let $\mathcal{H}$ be an [η-]elementary $\mathcal{S}$-Hintikka set where $\mathcal{S}$ is equality-free and*

$$\mathcal{M} = (\mathcal{D}, @, \mathcal{E}, \upsilon)$$

be an $\mathcal{S}$-model where

$$\mathcal{A}^{\mathcal{H}} = (\mathcal{D}, @)$$

is the [η-]Andrews Structure for $\mathcal{H}$ (cf. Definition 5.6.2),

$$(\mathcal{D}, @, \mathcal{E})$$

is a possible values evaluation for $\beta[\eta]$ (cf. Definition 4.1.7) and $\upsilon : \mathcal{D}_o \to \{\mathtt{T}, \mathtt{F}\}$ is defined by

$$\upsilon(\langle \mathbf{A}_o, a\rangle) := a.$$

If $\mathbf{M}$ is a proposition, φ is an assignment and $\varphi_1^*(\mathbf{M}) \in \mathcal{H}$ (cf. Definition 4.1.5), then $\mathcal{M} \models_\varphi \mathbf{M}$.

PROOF. We prove this by induction on the measure $|\mathbf{M}|_{\mathsf{e}}$ of propositions $\mathbf{M}$ (cf. Definition 2.1.27). Note that $\mathcal{H}$ is equality-free since it is elementary (cf. Definition 5.5.2). We first consider the base cases.

If $\mathbf{M}$ is $\dot{\top}$, then we certainly have $\mathcal{M} \models_\varphi \dot{\top}$. If $\mathbf{M}$ is $\dot{\neg}\dot{\top}$, then we have $\dot{\neg}\dot{\top} \notin \mathcal{H}$ by $\vec{\nabla}_\perp$. If $\mathbf{M}$ is $[\mathbf{A}_\alpha \doteq^\alpha \mathbf{B}_\alpha]$ or $\dot{\neg}[\mathbf{A}_\alpha \doteq^\alpha \mathbf{B}]$, then $\varphi_1^*(\mathbf{M}) \notin \mathcal{H}$ since $\mathcal{H}$ is equality-free.

Suppose $\mathbf{M} \in wff_o(\mathcal{S})$ and $\varphi_1^*(\mathbf{M}) \in \mathcal{H}$. Since $(\mathcal{D}, @, \mathcal{E})$ is a possible values evaluation, we know the first component of $\mathcal{E}_\varphi(\mathbf{M})$ is $\varphi_1^*(\mathbf{M})$. Hence $\mathcal{B}_{\mathcal{H}}^{\varphi_1^*(\mathbf{M})} = \{\mathtt{T}\}$ yielding $\mathcal{E}_\varphi(\mathbf{M}) = \langle \varphi_1^*(\mathbf{M}), \mathtt{T}\rangle$ and so finally $\upsilon(\mathcal{E}_\varphi(\mathbf{M})) = \mathtt{T}$. Thus $\mathcal{M} \models_\varphi \mathbf{M}$.

Suppose $\mathbf{M}$ is $\dot\neg\mathbf{N}$ where $\mathbf{N} \in wff_o(\mathcal{S})$ and $\varphi_1(\dot\neg\mathbf{N}) \in \mathcal{H}$. Since $(\mathcal{D}, @, \mathcal{E})$ is a possible values evaluation, we know the first component of $\mathcal{E}_\varphi(\mathbf{N})$ is $\varphi_1^*(\mathbf{N})$. Since $\dot\neg\varphi_1^*(\mathbf{N}) \in \mathcal{H}$ we know $\mathcal{B}_{\mathcal{H}}^{\varphi_1^*(\mathbf{N})} = \{\mathtt{F}\}$. Thus $\upsilon(\mathcal{E}_\varphi(\mathbf{N})) = \mathtt{F}$ and $\mathcal{M} \models_\varphi \dot\neg\mathbf{N}$.

We next consider the inductive cases in which $\mathbf{M}$ is of the form $\dot\neg\mathbf{N}$.

Suppose $\mathbf{N}$ is $\dot\neg\mathbf{P}$ and $\dot\neg\dot\neg\varphi_1^*(\mathbf{P}) \in \mathcal{H}$. By $\vec\nabla_\neg$, we know $\varphi_1^*(\mathbf{P}) \in \mathcal{H}$. Since $|\mathbf{P}|_\mathbf{e} < |\mathbf{M}|_\mathbf{e}$ we have $\mathcal{M} \models_\varphi \mathbf{P}$ and so $\mathcal{M} \models_\varphi \dot\neg\dot\neg\mathbf{P}$.

Suppose $\mathbf{N}$ is $[\mathbf{P} \dot\vee \mathbf{Q}]$ and $\dot\neg\varphi_1^*([\mathbf{P} \dot\vee \mathbf{Q}]) \in \mathcal{H}$. We know

$$\dot\neg\varphi_1^*(\mathbf{P}) \in \mathcal{H} \text{ and } \dot\neg\varphi_1^*(\mathbf{Q}) \in \mathcal{H}$$

by $\vec\nabla_\wedge$. By Lemma 2.1.28, $|\dot\neg\mathbf{P}|_\mathbf{e} < |\mathbf{M}|_\mathbf{e}$ and $|\dot\neg\mathbf{Q}|_\mathbf{e} < |\mathbf{M}|_\mathbf{e}$. By the inductive hypothesis, we know $\mathcal{M} \models_\varphi \dot\neg\mathbf{P}$ and $\mathcal{M} \models_\varphi \dot\neg\mathbf{Q}$. Hence $\mathcal{M} \not\models_\varphi \mathbf{P}$ and $\mathcal{M} \not\models_\varphi \mathbf{Q}$. Thus $\mathcal{M} \not\models_\varphi [\mathbf{P} \dot\vee \mathbf{Q}]$ and $\mathcal{M} \models_\varphi \mathbf{M}$.

Suppose $\mathbf{N}$ is $[\dot\forall x_\alpha \mathbf{P}]$ and $\dot\neg\varphi_1^*(\mathbf{N}) \in \mathcal{H}$. By $\vec\nabla_\exists$ and $\vec\nabla_\beta$ (or $\vec\nabla_{\beta\eta}$), there is some parameter W_α such that $\dot\neg\varphi_1^*([W/x]\mathbf{P}) \in \mathcal{H}$. By Lemma 2.1.28, $|\dot\neg[W/x]\mathbf{P}|_\mathbf{e} < |\mathbf{M}|_\mathbf{e}$. By the inductive hypothesis, we have $\mathcal{M} \models_\varphi \dot\neg[W/x]\mathbf{P}$. Let $\mathsf{w} := \mathcal{E}(W)$. By Lemma 3.3.12, $\mathcal{M} \not\models_{\varphi,[\mathsf{w}/x]} \mathbf{P}$ and so $\mathcal{M} \not\models_\varphi \dot\forall x_\alpha \mathbf{P}$. Thus $\mathcal{M} \models_\varphi \mathbf{M}$.

Finally, we consider the inductive cases in which $\mathbf{M}$ is not of the form $\dot\neg\mathbf{N}$.

Suppose $\mathbf{M}$ is $[\mathbf{P} \dot\vee \mathbf{Q}]$ and $\varphi_1^*([\mathbf{P} \dot\vee \mathbf{Q}]) \in \mathcal{H}$. By $\vec\nabla_\vee$, $\varphi_1^*(\mathbf{P}) \in \mathcal{H}$ or $\varphi_1^*(\mathbf{Q}) \in \mathcal{H}$. Since $|\mathbf{P}|_\mathbf{e} < |\mathbf{M}|_\mathbf{e}$ and $|\mathbf{Q}|_\mathbf{e} < |\mathbf{M}|_\mathbf{e}$, we know $\mathcal{M} \models_\varphi \mathbf{P}$ or $\mathcal{M} \models_\varphi \mathbf{Q}$. Thus $\mathcal{M} \models_\varphi \mathbf{M}$.

Suppose $\mathbf{M}$ is $[\dot\forall x_\alpha \mathbf{P}]$ and $\varphi_1^*(\mathbf{M}) \in \mathcal{H}$. Let $\langle \mathbf{A}, a\rangle \in \mathcal{D}_\alpha$ be given. Note that $(\varphi, [\langle \mathbf{A}, a\rangle/x])_1(\mathbf{P})^{\downarrow *}$ is $\varphi_1^*([\mathbf{A}/x]\mathbf{P})$ (since $\mathbf{A}$ is closed). By $\vec\nabla_\forall$ and $\vec\nabla_\beta$ (or $\vec\nabla_{\beta\eta}$), $\varphi_1^*([\mathbf{A}/x]\mathbf{P}) \in \mathcal{H}$. Since $|\mathbf{P}|_\mathbf{e} < |\mathbf{M}|_\mathbf{e}$ we know $\mathcal{M} \models_{\varphi,[\langle \mathbf{A},a\rangle/x]} \mathbf{P}$. Thus $\mathcal{M} \models_\varphi \mathbf{M}$. □

THEOREM 5.6.4 (Model Existence for Elementary Hintikka Sets). *Let $\mathcal{S}$ be an equality-free signature and $\mathcal{H}$ be an $[\eta$-$]$elementary $\mathcal{S}$-Hintikka set. There exists a model $\mathcal{M} \in \mathfrak{M}_\beta(\mathcal{S})$ $[\mathcal{M} \in \mathfrak{M}_{\beta\eta}(\mathcal{S})]$ such that $\mathcal{M} \models \mathcal{H}$.*

PROOF. Let $\mathcal{A}^{\mathcal{H}} = (\mathcal{D}, @)$ be the $[\eta$-$]$Andrews Structure (cf. Definition 5.6.2). If we choose interpretations for parameters and constants, then we can use Theorem 4.1.8 to extend these choices to a possible values evaluation for $\beta[\eta]$. For parameters W_α we choose $\mathcal{I}(W) := \langle W, p^W \rangle$ where p^W is any possible value for W which exists by Lemma 4.1.3. To make this an $\mathcal{S}$-model, we need a valuation. We take the obvious choice of $\upsilon(\langle \mathbf{A}_o, a \rangle) := a$ for a valuation $\upsilon : \mathcal{D}_o \to \{\mathtt{T}, \mathtt{F}\}$. For this to be a valuation we must make proper choices for the interpretations of the logical constants in $\mathcal{S}$.

$\top$: If $\top \in \mathcal{S}$, let $\mathcal{I}(\top) := \langle \top, \mathtt{T} \rangle$ so that $\upsilon(\mathcal{I}(\top)) = \mathtt{T}$ and $\mathfrak{L}_\top(\mathcal{I}(\top))$ holds. By $\vec{\nabla}^\sharp$ and $\vec{\nabla}_\bot$, we know $\mathtt{T} \in \mathcal{B}_{\mathcal{H}}^{\top}$.

$\bot$: If $\bot \in \mathcal{S}$, let $\mathcal{I}(\bot) := \langle \bot, \mathtt{F} \rangle$ so that $\upsilon(\mathcal{I}(\bot)) = \mathtt{F}$ and $\mathfrak{L}_\bot(\mathcal{I}(\bot))$ holds. By $\vec{\nabla}^\sharp$ and $\vec{\nabla}_\bot$, we know $\mathtt{F} \in \mathcal{B}_{\mathcal{H}}^{\top}$.

$\neg$: Suppose $\neg \in \mathcal{S}$. We use the notation $\overline{\mathtt{T}} := \mathtt{F}$ and $\overline{\mathtt{F}} := \mathtt{T}$. Let $\mathcal{I}(\neg)$ be $\langle \neg, n \rangle$ where $n : \mathcal{D}_o \to \mathcal{D}_o$ is defined by

$$n(\langle \mathbf{A}, a \rangle) := \langle \neg \mathbf{A}, \overline{a} \rangle.$$

To verify n is well-defined we must check $n(\langle \mathbf{A}, a \rangle) \in \mathcal{D}_o$. Let $\langle \mathbf{A}, a \rangle$ be an arbitrary member of $\mathcal{D}_o$. Suppose $a = \mathtt{T}$ and $\langle \neg \mathbf{A}, \overline{a} \rangle \notin \mathcal{D}_o$. Hence $\neg \mathbf{A} \in \mathcal{H}$ and $\overline{\neg} \mathbf{A} \in \mathcal{H}$ (by $\vec{\nabla}^\sharp$). This implies $\langle \mathbf{A}, \mathtt{T} \rangle \notin \mathcal{D}_o$, a contradiction. Suppose $a = \mathtt{F}$ and $\langle \neg \mathbf{A}, \overline{a} \rangle \notin \mathcal{D}_o$. Thus $\overline{\neg}\neg \mathbf{A} \in \mathcal{H}$ and $\overline{\neg}\overline{\neg} \mathbf{A} \in \mathcal{H}$ (by $\vec{\nabla}^\sharp$) and $\mathbf{A} \in \mathcal{H}$ (by $\vec{\nabla}_\neg$). Consequently, we have $\langle \mathbf{A}, \mathtt{F} \rangle \notin \mathcal{D}_o$, a contradiction.

To check $\mathfrak{L}_\neg(\mathcal{I}(\neg))$ holds for υ, note $\upsilon(\mathcal{I}(\neg) @ \langle \mathbf{A}, a \rangle) = \mathtt{T}$ iff $\overline{a} = \mathtt{T}$ iff $a = \mathtt{F}$ iff $\upsilon(\langle \mathbf{A}, a \rangle) = \mathtt{F}$.

$\vee$: Suppose $\vee \in \mathcal{S}$. For each $\langle \mathbf{A}, \mathtt{F} \rangle \in \mathcal{D}_o$, let $p^{\vee}_{\langle \mathbf{A}, \mathtt{F} \rangle} : \mathcal{D}_o \to \mathcal{D}_o$ be the function defined by

$$p^{\vee}_{\langle \mathbf{A}, \mathtt{F} \rangle}(\langle \mathbf{B}, b \rangle) := \langle [\mathbf{A} \vee \mathbf{B}], b \rangle.$$

For each $\langle \mathbf{A}, \mathtt{T} \rangle \in \mathcal{D}_o$, let $p^{\vee}_{\langle \mathbf{A},\mathtt{T} \rangle} : \mathcal{D}_o \to \mathcal{D}_o$ be the function defined by

$$p^{\vee}_{\langle \mathbf{A},\mathtt{T} \rangle}(\langle \mathbf{B}, b \rangle) := \langle [\mathbf{A} \vee \mathbf{B}], \mathtt{T} \rangle.$$

The properties $\vec{\nabla}^{\sharp}$, $\vec{\nabla}_{\vee}$, $\vec{\nabla}_{\wedge}$ and $\vec{\nabla}_{c}$ of $\mathcal{H}$ guarantees these are well-defined and $\langle [\vee\, \mathbf{A}], p^{\vee}_{\langle \mathbf{A},a \rangle} \rangle \in \mathcal{D}_{oo}$. Let $p^{\vee} : \mathcal{D}_o \to \mathcal{D}_{oo}$ be the function defined by

$$p^{\vee}(\langle \mathbf{A}, a \rangle) := \langle [\vee\, \mathbf{A}], p^{\vee}_{\langle \mathbf{A},a \rangle} \rangle.$$

Let $\mathcal{I}(\vee) := \langle \vee, p^{\vee} \rangle \in \mathcal{D}_{ooo}$. By the definition of $p^{\vee}$, we immediately know $\mathfrak{L}_{\vee}(\mathcal{I}(\vee))$ holds for υ.

$\wedge$,$\supset$,$\equiv$: We define $\mathcal{I}(\wedge)$, $\mathcal{I}(\supset)$ and $\mathcal{I}(\equiv)$ (when any of these are in $\mathcal{S}$) analogously to the $\vee$ case.

Π^{α}: Suppose $\Pi^{\alpha} \in \mathcal{S}$. Let $\pi : \mathcal{D}_{o\alpha} \to \mathcal{D}_o$ be defined by

$$\pi(\langle \mathbf{F}, f \rangle) := \begin{cases} \langle [\Pi^{\alpha}\, \mathbf{F}], \mathtt{T} \rangle & \text{if } \upsilon(f(\mathsf{a})) = \mathtt{T} \text{ for every } \mathsf{a} \in \mathcal{D}_{\alpha} \\ \langle [\Pi^{\alpha}\, \mathbf{F}], \mathtt{F} \rangle & \text{otherwise} \end{cases}$$

for each $\langle \mathbf{F}, f \rangle \in \mathcal{D}_{o\alpha}$. In order to check π is well-defined let $\langle \mathbf{F}, f \rangle \in \mathcal{D}_{o\alpha}$ be arbitrary. Suppose

$$\pi(\langle \mathbf{F}, f \rangle) = \langle [\Pi^{\alpha}\, \mathbf{F}], \mathtt{T} \rangle \notin \mathcal{D}_o.$$

Then $\neg[\Pi^{\alpha}\, \mathbf{F}] \in \mathcal{H}$. By $\vec{\nabla}^{\sharp}$, $\vec{\nabla}_{\exists}$ and $\vec{\nabla}_{\beta}$ $[\vec{\nabla}_{\beta\eta}]$, we know there is some parameter W_{α} such that $\neg[\mathbf{F}\, W]^{\downarrow *} \in \mathcal{H}$. By Lemma 4.1.3, there is some possible value p^{W} for W. Since $\neg[\mathbf{F}\, W]^{\downarrow *} \in \mathcal{H}$ we must have $f(\langle W, p^{W} \rangle) = \langle [\mathbf{F}\, W]^{\downarrow *}, \mathtt{F} \rangle$, contradicting our assumption that $\pi(\langle \mathbf{F}, f \rangle) = \langle [\Pi^{\alpha}\, \mathbf{F}], \mathtt{T} \rangle$. On the other hand, suppose $\pi(\langle \mathbf{F}, f \rangle) = \langle [\Pi^{\alpha}\, \mathbf{F}], \mathtt{F} \rangle$ and $\langle [\Pi^{\alpha}\, \mathbf{F}], \mathtt{F} \rangle \notin \mathcal{D}_o$. We know $[\Pi^{\alpha}\, \mathbf{F}] \in \mathcal{H}$ since $\langle [\Pi^{\alpha}\, \mathbf{F}], \mathtt{F} \rangle \notin \mathcal{D}_o$. Since $\pi(\langle \mathbf{F}, f \rangle) = \langle [\Pi^{\alpha}\, \mathbf{F}], \mathtt{F} \rangle$ we know some $\langle \mathbf{A}, a \rangle \in \mathcal{D}_{\alpha}$ exists such that

$$f(\langle \mathbf{A}, a \rangle) = \langle [\mathbf{F}\, \mathbf{A}]^{\downarrow *}, \mathtt{F} \rangle.$$

Hence $\langle [\mathbf{F}\, \mathbf{A}]^{\downarrow *}, \mathtt{F} \rangle \in \mathcal{D}_o$. By $\vec{\nabla}^{\sharp}$, $\vec{\nabla}_{\forall}$ and $\vec{\nabla}_{\beta}$ $[\vec{\nabla}_{\beta\eta}]$, we conclude $[\mathbf{F}\, \mathbf{A}]^{\downarrow *} \in \mathcal{H}$, contradicting $\langle [\mathbf{F}\, \mathbf{A}]^{\downarrow *}, \mathtt{F} \rangle \in \mathcal{D}_o$. Thus π is well-defined.

We let $\mathcal{I}(\Pi^{\alpha}) := \langle \Pi^{\alpha}, \pi \rangle \in \mathcal{D}_{o(o\alpha)}$. Note that $\mathfrak{L}_{\Pi^{\alpha}}(\mathcal{I}(\Pi^{\alpha}))$ holds for υ by the definition of π.

Σ^α: This is analogous to the Π^α case.

$=^\alpha$: We have assumed $=^\alpha \notin \mathcal{S}$.

We use Theorem 4.1.8 to extend these choices to an evaluation function $\mathcal{E}$ where $(\mathcal{D}, @, \mathcal{E})$ is a possible values evaluation for $\beta[\eta]$.

Let $\mathcal{M} := (\mathcal{D}, @, \mathcal{E}, \upsilon)$. Since we have noted the appropriate property for each choice $\mathcal{I}(c) = \mathcal{E}(c)$ where $c \in \mathcal{S}$, we know $\mathcal{M}$ is an $\mathcal{S}$-model. That is, $\mathcal{M} \in \mathfrak{M}_\beta(\mathcal{S})$. If $\mathcal{H}$ is an η-elementary $\mathcal{S}$-Hintikka set, then $\mathcal{A}^{\mathcal{H}}$ is a possible values structure for $\beta\eta$. Thus Theorem 4.1.8 also implies $(\mathcal{D}, @, \mathcal{E})$ is η-functional. Therefore, $\mathcal{M} \in \mathfrak{M}_{\beta\eta}(\mathcal{S})$.

Finally, we verify $\mathcal{M} \models \mathcal{H}$. Let $\mathbf{M} \in \mathcal{H}$ and any assignment φ be given. Since $\mathbf{M}$ is a sentence we know $\varphi_1^*(\mathbf{M})$ is $\mathbf{M}^{\downarrow *}$. By $\vec{\nabla}_\beta$ $[\vec{\nabla}_{\beta\eta}]$, we have $\mathbf{M}^{\downarrow *} \in \mathcal{H}$. Thus $\mathcal{M} \models_\varphi \mathbf{M}$ by Lemma 5.6.3. Therefore, $\mathcal{M} \models \mathcal{H}$. □

THEOREM 5.6.5 (Model Existence for Abstract Consistency Classes). *Let $\mathcal{S}$ be an equality-free signature, $\mathscr{C} \in \mathfrak{Acc}_\beta(\mathcal{S})$ $[\mathscr{C} \in \mathfrak{Acc}_{\beta\eta}(\mathcal{S})]$ be an elementary $\mathcal{S}$-abstract consistency class [with η] of finite character and $\Phi \in \mathscr{C}$ be a sufficiently pure set of $\mathcal{S}$-sentences. There exists a model $\mathcal{M} \in \mathfrak{M}_\beta(\mathcal{S})$ $[\mathcal{M} \in \mathfrak{M}_{\beta\eta}(\mathcal{S})]$ such that $\mathcal{M} \models \Phi$.*

PROOF. There is a Hintikka set $\mathcal{H}$ in $\mathfrak{Hint}_\beta(\mathcal{S})$ $[\mathfrak{Hint}_{\beta\eta}(\mathcal{S})]$ such that $\Phi \subseteq \mathcal{H}$ by Lemma 5.5.6. By Theorem 5.6.4, there is a model $\mathcal{M}$ in $\mathfrak{M}_\beta(\mathcal{S})$ $[\mathfrak{M}_{\beta\eta}(\mathcal{S})]$ such that $\mathcal{M} \models \mathcal{H}$. Hence $\mathcal{M} \models \Phi$. □

THEOREM 5.6.6 (Completeness of $\mathcal{G}_\beta^{\mathcal{S}}$ and $\mathcal{G}_{\beta\eta}^{\mathcal{S}}$). *Let $\mathcal{S}$ be an equality-free signature. If a sequent Γ is not derivable in $\mathcal{G}_\beta^{\mathcal{S}}$ $[\mathcal{G}_{\beta\eta}^{\mathcal{S}}]$, then there is an $\mathcal{S}$-model $\mathcal{M} \in \mathfrak{M}_\beta(\mathcal{S})$ $[\mathcal{M} \in \mathfrak{M}_{\beta\eta}(\mathcal{S})]$ such that $\mathcal{M} \not\models \Gamma$.*

PROOF. Suppose Γ is not derivable in $\mathcal{G}_\beta^{\mathcal{S}}$ $[\mathcal{G}_{\beta\eta}^{\mathcal{S}}]$. By Lemma 5.4.3, Φ_Γ is in $\mathscr{C}_\beta^{\mathcal{S}}$ $[\mathscr{C}_{\beta\eta}^{\mathcal{S}}]$. Note that $\mathscr{C}_\beta^{\mathcal{S}}$ $[\mathscr{C}_{\beta\eta}^{\mathcal{S}}]$ is an abstract consistency class in $\mathfrak{Acc}_\beta(\mathcal{S})$ $[\mathfrak{Acc}_{\beta\eta}(\mathcal{S})]$ of finite character by Theorem 5.4.7. Also, Φ_Γ is sufficiently pure since it is finite and there are infinitely many parameters of each type. By Theorem 5.6.5, there is a model $\mathcal{M}$ in $\mathfrak{M}_\beta(\mathcal{S})$ $[\mathfrak{M}_{\beta\eta}(\mathcal{S})]$ such that $\mathcal{M} \models \Phi_\Gamma$. In particular, for each $\mathbf{M} \in \Gamma$, $\mathcal{M} \models \neg\mathbf{M}$. Thus $\mathcal{M} \not\models \Gamma$. □

REMARK 5.6.7 (Cut Elimination). We can now give a semantic proof of cut-elimination. One may be tempted to try a Gentzen-style proof-theoretic proof. Note that the measure $|\mathbf{M}|_{\mathbf{e}}$ on propositions $\mathbf{M}$ (cf. Definition 2.1.27) does satisfy $|[\mathbf{A}/x]\mathbf{M}|_{\mathbf{e}} < |[\dot{\forall} x_\alpha \mathbf{M}]|_{\mathbf{e}}$ for any x_α, $\mathbf{A} \in wff_\alpha(\mathcal{S})$ and proposition $\mathbf{M}$. This would permit the (often problematic) induction step on the rule $\mathcal{G}(\dot{\neg}\dot{\forall}, \mathcal{S})$ to succeed. However, we have simply moved the problem by distinguishing between internal and external logical connectives and quantifiers. The inductive proof would fail at $\mathcal{G}(\sharp)$ since, in general, $|\mathbf{A}^\sharp|_{\mathbf{e}}$ may be bigger than $|\mathbf{A}|_{\mathbf{e}}$ for $\mathbf{A} \in wff_o(\mathcal{S})$. (On the other hand, such an inductive proof should be possible for signatures without equality and without quantifiers.)

COROLLARY 5.6.8 (Cut Elimination for Elementary Sequent Calculi). *Let $\mathcal{S}$ be an equality-free signature. Suppose $\Gamma, \mathbf{M}$ and $\Gamma, \dot{\neg}\mathbf{M}$ are both derivable in $\mathcal{G}^{\mathcal{S}}_\beta$ $[\mathcal{G}^{\mathcal{S}}_{\beta\eta}]$. Then Γ is derivable in $\mathcal{G}^{\mathcal{S}}_\beta$ $[\mathcal{G}^{\mathcal{S}}_{\beta\eta}]$. That is, the rule $\mathcal{G}(Cut)$ is admissible in $\mathcal{G}^{\mathcal{S}}_\beta$ $[\mathcal{G}^{\mathcal{S}}_{\beta\eta}]$.*

PROOF. Suppose Γ is not derivable in $\mathcal{G}^{\mathcal{S}}_\beta$ $[\mathcal{G}^{\mathcal{S}}_{\beta\eta}]$. By completeness (Theorem 5.6.6), there is an $\mathcal{S}$-model $\mathcal{M} \in \mathfrak{M}_\beta(\mathcal{S})$ $[\mathcal{M} \in \mathfrak{M}_{\beta\eta}(\mathcal{S})]$ such that $\mathcal{M} \not\models \Gamma$. By soundness (Theorem 3.4.6 or Theorem 3.4.9), $\mathcal{M} \models \Gamma, \mathbf{M}$ and $\mathcal{M} \models \Gamma, \dot{\neg}\mathbf{M}$. Thus $\mathcal{M} \models \mathbf{M}$ and $\mathcal{M} \models \dot{\neg}\mathbf{M}$, a contradiction. □

5.7. Completeness of Extensional Sequent Calculi

We now turn to the construction of extensional $\mathcal{S}$-models for extensional $\mathcal{S}$-Hintikka sets.

THEOREM 5.7.1. *If $\mathcal{H} \in \mathfrak{Hint}_{\beta\mathfrak{f}\mathfrak{b}}(\mathcal{S})$, then the power set $\mathcal{P}(\mathcal{H})$ is an abstract compatibility class.*

PROOF. Each of the properties $\nabla_{\beta\eta}$, $\nabla_\bot$, $\nabla_\neg$, $\nabla_\vee$, $\nabla_\wedge$, $\nabla_\forall$, $\nabla^w_\exists$, ∇_m, ∇_{dec}, $\nabla_\mathfrak{b}$, $\nabla^w_\mathfrak{f}$, $\nabla^o_=$, $\nabla^\to_=$, $\nabla^u_=$ and $\nabla^\sharp$ follows trivially from the corresponding $\vec{\nabla}_*$ property for $\mathcal{H}$. □

Given an extensional Hintikka set $\mathcal{H} \in \mathfrak{Hint}_{\beta\mathfrak{f}\mathfrak{b}}(\mathcal{S})$, we can use the compatibility relation induced by the abstract compatibility class $\mathcal{P}(\mathcal{H})$ to prove the existence of an extensional model for $\mathcal{H}$.

DEFINITION 5.7.2 ($\mathcal{H}$-Compatibility). Let $\mathcal{H} \in \mathfrak{Hint}_{\beta\mathfrak{f}\mathfrak{b}}(\mathcal{S})$ be an extensional $\mathcal{S}$-Hintikka set. We say two closed terms $\mathbf{A}, \mathbf{B} \in cwff_\alpha(\mathcal{S})$ are *$\mathcal{H}$-compatible* when they are $\mathcal{P}(\mathcal{H})$-compatible (cf. Definition 5.2.5). We will also denote this by $\mathbf{A} \parallel \mathbf{B}$. We say a set $S \subseteq cwff_\alpha(\mathcal{S})$

is *$\mathcal{H}$-compatible* if $\mathbf{A} \parallel \mathbf{B}$ for every pair $\mathbf{A}, \mathbf{B} \in S$. A set S is *maximally $\mathcal{H}$-compatible* if it is $\mathcal{H}$-compatible and for any $\mathcal{H}$-compatible set T, $S \subseteq T$ implies $S = T$.

LEMMA 5.7.3. *Let $\mathcal{H}$ be an extensional $\mathcal{S}$-Hintikka set. For any $\mathcal{H}$-compatible set $S \subseteq cwff_\alpha(\mathcal{S})$ there is a maximally $\mathcal{H}$-compatible $S^* \subseteq cwff_\alpha(\mathcal{S})$ such that $S \subseteq S^*$.*

PROOF. Let $\mathcal{C}$ be the set of $\mathcal{H}$-compatible subsets S' of $cwff_\alpha(\mathcal{S})$ with $S \subseteq S'$. Given any chain

$$S_0 \subseteq S_1 \subseteq \cdots \subseteq S_n \subseteq \cdots \subseteq cwff_\alpha(\mathcal{S})$$

of $\mathcal{H}$-compatible sets with $S \subseteq S_0$, it is easy to see the union $\bigcup_{n \in \mathbb{N}} S_n$ is $\mathcal{H}$-compatible. Hence any chain in $\mathcal{C}$ has an upper bound in $\mathcal{C}$. Therefore, there is a maximal $S^* \in \mathcal{C}$ by Zorn's Lemma. □

LEMMA 5.7.4. *Let $\mathcal{H}$ be an extensional $\mathcal{S}$-Hintikka set. For every closed term $\mathbf{A}_\alpha$ there is a maximally $\mathcal{H}$-compatible $S \subseteq cwff_\alpha(\mathcal{S})$ such that $\mathbf{A} \in S$.*

PROOF. By Theorem 5.2.11, we know $\{\mathbf{A}\}$ is $\mathcal{H}$-compatible. Hence there is a maximally $\mathcal{H}$-compatible set S with $\{\mathbf{A}\} \subseteq S$ by Lemma 5.7.3. □

LEMMA 5.7.5. *Let $\mathcal{H}$ be an extensional $\mathcal{S}$-Hintikka set. If either*

$$[\mathbf{A}_\iota \doteq^\iota \mathbf{B}_\iota] \in \mathcal{H} \text{ or } [\mathbf{B}_\iota \doteq^\iota \mathbf{A}_\iota] \in \mathcal{H},$$

then

$$S^{\mathbf{A}} := \{\mathbf{C} \in cwff_\iota(\mathcal{S}) \mid \mathbf{A} \parallel \mathbf{C}\}$$

is the unique maximally $\mathcal{H}$-compatible set containing $\mathbf{A}$. Furthermore, $\mathbf{B} \in S^{\mathbf{A}}$.

PROOF. See Appendix A.3. □

LEMMA 5.7.6. *Let $\mathcal{H}$ be an extensional $\mathcal{S}$-Hintikka set and S and T be maximally $\mathcal{H}$-compatible sets. If $[\mathbf{A}_\iota \doteq^\iota \mathbf{B}_\iota] \in \mathcal{H}$, $\mathbf{A} \in S$ and $\mathbf{B} \in T$, then $S = T$.*

PROOF. Applying Lemma 5.7.5 we have $S = S^{\mathbf{A}} = S^{\mathbf{B}} = T$. □

Fix an extensional $\mathcal{S}$-Hintikka set $\mathcal{H}$. We start by constructing a possible values structure.

DEFINITION 5.7.7 ($\mathcal{H}$-Compatibility Structure). We define a possible values structure we will call the *$\mathcal{H}$-compatibility structure*

$$\mathcal{F}^{\mathcal{H}} := (\mathcal{D}, @)$$

by

- Let $\mathcal{D}_o$ be the set of pairs $\langle \mathbf{A}_o, p\rangle$ where $\mathbf{A} \in cwff_o(\mathcal{S})$ is $\beta\eta$-normal and $p \in \mathcal{B}_{\mathcal{H}}^{\mathbf{A}}$.
- Let $\mathcal{D}_\iota$ be the set of pairs $\langle \mathbf{A}_\iota, S\rangle$ where $\mathbf{A} \in cwff_\iota(\mathcal{S})$ is $\beta\eta$-normal, $\mathbf{A} \in S$ and S is a maximally $\mathcal{H}$-compatible subset of $cwff_\iota(\mathcal{S})$.
- Let $\mathcal{D}_{\alpha\beta}$ be the set of pairs $\langle \mathbf{G}, g\rangle$ where $\mathbf{G} \in cwff_{\alpha\beta}(\mathcal{S})$ is $\beta\eta$-normal and $g : \mathcal{D}_\beta \longrightarrow \mathcal{D}_\alpha$ such that for every $\langle \mathbf{B}, b\rangle \in \mathcal{D}_\beta$, $g(\langle \mathbf{B}, b\rangle) = \langle \mathbf{A}, a\rangle$ implies $\mathbf{A}$ is the $\beta\eta$-normal form of $[\mathbf{G}\,\mathbf{B}]$.

We define @ by

$$\langle \mathbf{G}, g\rangle @ \langle \mathbf{B}, b\rangle := g(\langle \mathbf{B}, b\rangle)$$

for each $\langle \mathbf{G}, g\rangle \in \mathcal{D}_{\alpha\beta}$ and $\langle \mathbf{B}, b\rangle \in \mathcal{D}_\beta$.

LEMMA 5.7.8 (Compatibility Structures). *$\mathcal{F}^{\mathcal{H}}$ is a possible values structure.*

PROOF. It is trivial to check most of the conditions that $\mathcal{F}^{\mathcal{H}}$ is a possible values structure. To determine every $\mathbf{A} \in cwff_\iota(\mathcal{S})$ has a possible value we can use Lemma 5.7.4 to obtain a set S with $\langle \mathbf{A}^{\downarrow}, S\rangle \in \mathcal{D}_\iota$. For every $\mathbf{A} \in cwff_o(\mathcal{S})$ we know $\mathcal{B}_{\mathcal{H}}^{(\mathbf{A}^{\downarrow})}$ is nonempty. In $\mathcal{F}^{\mathcal{H}}$ we have $\langle \mathbf{A}^{\downarrow}, \mathsf{p}\rangle$ for $\mathsf{p} \in \mathcal{B}_{\mathcal{H}}^{(\mathbf{A}^{\downarrow})}$. □

DEFINITION 5.7.9 ($\mathcal{H}$-Compatibility Per). We define the *$\mathcal{H}$-compatibility per* $\sim$ on the $\mathcal{H}$-compatibility structure $\mathcal{F}^{\mathcal{H}}$ to be the functional per extension (cf. Definition 4.2.6) of the equivalence relations defined on base types as follows:

- In $\mathcal{D}_o$, $\langle \mathbf{A}_o, p\rangle \sim \langle \mathbf{B}_o, q\rangle$ if $p = q$.
- In $\mathcal{D}_\iota$, $\langle \mathbf{A}_\iota, S\rangle \sim \langle \mathbf{B}_\iota, R\rangle$ if $S = R$.

By Definition 4.2.6 at function types, $\langle \mathbf{G}_{\alpha\beta}, g\rangle \sim \langle \mathbf{H}_{\alpha\beta}, h\rangle$ if for every $\langle \mathbf{A}_\beta, a\rangle \sim \langle \mathbf{B}_\beta, b\rangle$ in $\mathcal{D}_\beta$, we have $g(\langle \mathbf{A}_\beta, a\rangle) \sim h(\langle \mathbf{B}_\beta, b\rangle)$ in $\mathcal{D}_\alpha$.

NOTATION *As in Definition 4.2.1 at each type α we let*

$$\overline{\mathcal{D}_\alpha^{\sim}} := \{\langle \mathbf{A}, a\rangle \in \mathcal{D}_\alpha \mid \langle \mathbf{A}, a\rangle \sim \langle \mathbf{A}, a\rangle\}$$

Combining this with the notation restricting the first components, we let $\overline{\mathcal{D}_\alpha^{\mathbf{A}}} := \mathcal{D}_\alpha^{\mathbf{A}} \cap \overline{\mathcal{D}_\alpha^{\sim}}$.

Since every $\mathbf{A} \in cwff_\alpha(\mathcal{S})$ has a possible value by Lemma 4.1.3, we can choose a particular one $r^{\mathbf{A}}$ for each $\mathbf{A}$. This will act as a default value when necessary.

In order to prove every closed $\beta\eta$-normal term has a related possible value we first show the more general result contained in Lemma 5.7.10.

LEMMA 5.7.10. *For each type* α, *we have*

1. *If* $\langle \mathbf{A}, a\rangle \sim \langle \mathbf{B}, b\rangle$ *in* $\mathcal{D}_\alpha$, *then* $\mathbf{A} \parallel \mathbf{B}$.
2. *If* $S \subseteq cwff_\alpha(\mathcal{S})$ *is* $\mathcal{H}$*-compatible, then there is a family* $(p^{\mathbf{A}})_{\mathbf{A}\in S}$ *such that* $p^{\mathbf{A}}$ *is a possible value for each* $\mathbf{A} \in S$ *and*

$$\langle \mathbf{A}^{\downarrow}, p^{\mathbf{A}}\rangle \sim \langle \mathbf{B}^{\downarrow}, p^{\mathbf{B}}\rangle$$

for each $\mathbf{A}, \mathbf{B} \in S$.

PROOF. These two statements are proven by a mutual induction on the type α. The cases for part 1 are proven as follows:

- o: $\langle \mathbf{A}, a\rangle \sim \langle \mathbf{B}, b\rangle$ implies $a = b$ in $\{\mathtt{T}, \mathtt{F}\}$. If $\mathbf{A}$ and $\mathbf{B}$ were incompatible, then without loss of generality $\mathbf{A}, \neg\mathbf{B} \in \mathcal{H}$. Since $\mathbf{A} \in \mathcal{H}$ the value a cannot be $\mathtt{F}$ and so $a = \mathtt{T}$. Since $\neg\mathbf{B} \in \mathcal{H}$ the value b cannot be $\mathtt{T}$ and so $b = \mathtt{F}$. Since $\mathtt{T} \neq \mathtt{F}$, we have a contradiction.
- ι: $\langle \mathbf{A}, A\rangle \sim \langle \mathbf{B}, B\rangle$ implies $A = B$. Hence $\mathbf{A} \parallel \mathbf{B}$ as members of the $\mathcal{H}$-compatible set A.
- $\gamma\beta$: Suppose $\langle \mathbf{G}, g\rangle \sim \langle \mathbf{H}, h\rangle$. Let $\mathbf{A} \parallel \mathbf{B}$ in $cwff_\beta(\mathcal{S})$ be given. We can apply the inductive hypothesis for part 2 at type β to the set $S := \{\mathbf{A}, \mathbf{B}\}$ to obtain $p^{\mathbf{A}}$ and $p^{\mathbf{B}}$ with

$$\langle \mathbf{A}^{\downarrow}, p^{\mathbf{A}}\rangle \sim \langle \mathbf{B}^{\downarrow}, p^{\mathbf{B}}\rangle.$$

  Thus $g(\langle \mathbf{A}^{\downarrow}, p^{\mathbf{A}}\rangle) \sim h(\langle \mathbf{B}^{\downarrow}, p^{\mathbf{B}}\rangle)$. The first components of $g(\langle \mathbf{A}^{\downarrow}, p^{\mathbf{A}}\rangle)$ and $h(\langle \mathbf{B}^{\downarrow}, p^{\mathbf{B}}\rangle)$ are $[\mathbf{G}\,\mathbf{A}]^{\downarrow}$ and $[\mathbf{H}\,\mathbf{B}]^{\downarrow}$, respectively. Applying the inductive hypothesis for part 1 to these terms at type γ, we have $[\mathbf{G}\,\mathbf{A}]^{\downarrow} \parallel [\mathbf{H}\,\mathbf{B}]^{\downarrow}$. By Lemma 5.2.3, $[\mathbf{G}\,\mathbf{A}] \parallel [\mathbf{H}\,\mathbf{B}]$. Generalizing over $\mathbf{A}$ and $\mathbf{B}$, we have $\mathbf{G} \parallel \mathbf{H}$.

The cases for part 2 are proven as follows:

- o: We must either be able to let $p^{\mathbf{A}} := \mathtt{T}$ for every $\mathbf{A} \in S$ or let $p^{\mathbf{A}} := \mathtt{F}$ for every $\mathbf{A} \in S$. If neither is the case, then by the definition of $\mathcal{D}_o$ there must be $\mathbf{A}, \mathbf{B} \in S$ with $\mathbf{A}^{\downarrow}, \neg\mathbf{B}^{\downarrow} \in \mathcal{H}$. This contradicts $\mathcal{H}$-compatibility of S.

ι: By Lemma 5.7.3, there is a maximally $\mathcal{H}$-compatible set S^* such that $S \subseteq S^*$. Let $p^{\mathbf{A}} := S^*$ for each $\mathbf{A} \in S$. By definition of $\mathcal{D}_\iota$, $\langle \mathbf{A}^{\downarrow}, S^* \rangle \in \mathcal{D}_\iota$

$\gamma\beta$: Suppose we are given the set $S \subseteq cwff_{\gamma\beta}(\mathcal{S})$.

For $\langle \mathbf{B}, b \rangle \in \mathcal{D}_\beta \setminus \overline{\mathcal{D}_\beta^{\sim}}$ and $\mathbf{G} \in S$, we let $p^{[\mathbf{G}\,\mathbf{B}]}$ be the default possible value $r^{[\mathbf{G}\,\mathbf{B}]}$ for $\mathbf{GB}$.

For each $\langle \mathbf{B}, b \rangle \in \overline{\mathcal{D}_\beta^{\sim}}$, we choose a particular representative $\langle \mathbf{B}^*, b^* \rangle$ in the equivalence class of $\langle \mathbf{B}, b \rangle$ with respect to $\sim$. For a particular $\langle \mathbf{B}^*, b^* \rangle$, let

$$\mathcal{B} := \{\langle \mathbf{B}, b \rangle \mid \langle \mathbf{B}, b \rangle \sim \langle \mathbf{B}^*, b^* \rangle\}$$

and let

$$\mathcal{G}_{\mathcal{B}} := \{[\mathbf{G}\,\mathbf{B}] \mid \mathbf{G} \in S, \langle \mathbf{B}, b \rangle \in \mathcal{B} \text{ for some } b\}.$$

For each $\langle \mathbf{B}, b \rangle, \langle \mathbf{C}, c \rangle \in \mathcal{B}$, we apply the inductive hypothesis for part 1 to $\langle \mathbf{B}, b \rangle \sim \langle \mathbf{C}, c \rangle$ at type β to conclude $\mathbf{B} \parallel \mathbf{C}$. Thus the set $\mathcal{G}_{\mathcal{B}}$ is $\mathcal{H}$-compatible by the definition of $\parallel$ at function types and the fact that S is $\mathcal{H}$-compatible.

By applying the inductive hypothesis for part 2 to $\mathcal{G}_{\mathcal{B}}$ at type γ, we obtain related possible values $p^{[\mathbf{G}\,\mathbf{B}]}$ for each $[\mathbf{G}\,\mathbf{B}] \in \mathcal{G}_{\mathcal{B}}$.

Now, for each $\mathbf{G} \in S$, we can use the axiom of choice (at the metalevel) to define a function $p^{\mathbf{G}} : \mathcal{D}_\beta \to \mathcal{D}_\gamma$ such that

$$p^{\mathbf{G}}(\langle \mathbf{B}, b \rangle) = \langle [\mathbf{G}\,\mathbf{B}]^{\downarrow}, p^{[\mathbf{G}\,\mathbf{B}]} \rangle$$

This $p^{\mathbf{G}}$ does map into $\mathcal{D}_\gamma$ since each $p^{[\mathbf{G}\,\mathbf{B}]}$ is a possible value for $[\mathbf{G}\,\mathbf{B}]$. Note that the choices of $p^{[\mathbf{G}\,\mathbf{B}]}$ imply the functions $p^{\mathbf{G}}$ are related as

$$p^{\mathbf{G}}(\langle \mathbf{B}, b \rangle) = \langle [\mathbf{G}\,\mathbf{B}]^{\downarrow}, p^{[\mathbf{G}\,\mathbf{B}]} \rangle \sim \langle [\mathbf{H}\,\mathbf{C}]^{\downarrow}, p^{[\mathbf{H}\,\mathbf{C}]} \rangle = p^{\mathbf{H}}(\langle \mathbf{C}, c \rangle)$$

whenever $\langle \mathbf{B}, b \rangle \sim \langle \mathbf{C}, c \rangle$ and $\mathbf{G}, \mathbf{H} \in S$. Therefore, for each $\mathbf{G}, \mathbf{H} \in S$, we have $\langle \mathbf{G}, p^{\mathbf{G}} \rangle \sim \langle \mathbf{H}, p^{\mathbf{H}} \rangle$

□

THEOREM 5.7.11 (Related Possible Values Exist). *For every closed term $\mathbf{A} \in cwff_\alpha(\mathcal{S})$ there is a possible value a such that $\langle \mathbf{A}^{\downarrow}, a \rangle \in \overline{\mathcal{D}_\alpha^{\sim}}$. That is, $\overline{\mathcal{D}_\alpha^{\mathbf{A}}}$ is nonempty.*

PROOF. This follows simply by applying Lemma 5.7.10:2 to the singleton set $\{\mathbf{A}\}$ since $\mathbf{A} \parallel \mathbf{A}$ by Theorem 5.2.11. □

LEMMA 5.7.12. *If* $\langle \mathbf{A}, a \rangle, \langle \mathbf{A}, b \rangle \in \mathcal{D}_\alpha$ *and* $[\mathbf{A} \doteq^\alpha \mathbf{B}] \in \mathcal{H}$, *then* $a = b$.

PROOF. This follows by induction on the type α. See Appendix A.3. □

LEMMA 5.7.13. *If* $\langle \mathbf{A}, a \rangle \in \mathcal{D}_\alpha$ *and* $[\mathbf{A} \doteq^\alpha \mathbf{B}] \in \mathcal{H}$, *then* $\langle \mathbf{A}, a \rangle \sim \langle \mathbf{A}, a \rangle$.

PROOF. By Theorem 5.7.11, there is some possible value a_1 for $\mathbf{A}$ with $\langle \mathbf{A}, a_1 \rangle \in \overline{\mathcal{D}_\alpha^\sim}$. By Lemma 5.7.12, we know $a = a_1$ and so $\langle \mathbf{A}, a \rangle \in \overline{\mathcal{D}_\alpha^\sim}$, i.e., $\langle \mathbf{A}, a \rangle \sim \langle \mathbf{A}, a \rangle$. □

LEMMA 5.7.14. *If* $\langle \mathbf{A}, a \rangle, \langle \mathbf{B}, b \rangle \in \mathcal{D}_\alpha$ *and* $[\mathbf{A} =^\alpha \mathbf{B}] \in \mathcal{H}$, *then* $\langle \mathbf{A}, a \rangle \sim \langle \mathbf{B}, b \rangle$.

PROOF. The proof is by induction on the type α. See Appendix A.3. □

We have now constructed the appropriate applicative structure and a per over the structure. Assuming we can use this to build a possible values evaluation for $\beta\eta$, we will have a per model of $\mathcal{H}$.

LEMMA 5.7.15. *Let* $\mathcal{H} \in \mathfrak{Hint}_{\beta\mathfrak{fb}}(\mathcal{S})$ *be an extensional* $\mathcal{S}$*-Hintikka set and* $\mathscr{P} = (\mathcal{D}, @, \mathcal{E}, \upsilon, \sim)$ *be an* $\mathcal{S}$*-per model where* $\mathcal{F}^{\mathcal{H}} = (\mathcal{D}, @)$ *is the* $\mathcal{H}$*-compatibility structure (cf. Definition 5.7.7),* $\sim$ *is the* $\mathcal{H}$*-compatibility per (cf. Definition 5.7.9),* $(\mathcal{D}, @, \mathcal{E})$ *is a possible values evaluation for* $\beta\eta$ *and* $\upsilon : \mathcal{D}_o \to \{\mathtt{T}, \mathtt{F}\}$ *is defined by* $\upsilon(\langle \mathbf{A}_o, a \rangle) := a$. *If* $\mathbf{M}$ *is a proposition,* φ *is a* $\sim$*-assignment and* $\varphi_1^{\beta\eta}(\mathbf{M}) \in \mathcal{H}$ *(cf. Definition 4.1.5), then* $\mathscr{P} \models\!\approx_\varphi \mathbf{M}$.

PROOF. The proof is by induction on $|\mathbf{M}|_\mathbf{e}$ (cf. Definition 2.1.27) and follows the same pattern as Lemma 5.6.3 except now the signature $\mathcal{S}$ and set $\mathcal{H}$ need not be equality-free. We only verify the equation base cases here and check the remaining cases in Appendix A.3.

Suppose $\mathbf{M}$ is $[\mathbf{A}_\alpha \doteq^\alpha \mathbf{B}_\alpha]$ and $\varphi_1^{\beta\eta}([\mathbf{A} \doteq^\alpha \mathbf{B}]) \in \mathcal{H}$. Hence we have $[\varphi_1^{\beta\eta}(\mathbf{A}) \doteq^\alpha \varphi_1^{\beta\eta}(\mathbf{B})] \in \mathcal{H}$. Since

$$\mathcal{E}_\varphi(\mathbf{A}) \in \mathcal{D}_\alpha^{\varphi_1(\mathbf{A})} \text{ and } \mathcal{E}_\varphi(\mathbf{B}) \in \mathcal{D}_\alpha^{\varphi_1(\mathbf{B})}$$

we know $\mathcal{E}_\varphi(\mathbf{A}) \sim \mathcal{E}_\varphi(\mathbf{B})$ by Lemma 5.7.14. Thus $\mathscr{P} \models\!\approx_\varphi [\mathbf{A} \doteq^\alpha \mathbf{B}]$.

Suppose $\mathbf{M}$ is $\neg[\mathbf{A}_\alpha \doteq^\alpha \mathbf{B}_\alpha]$ and $\neg[\varphi_1^{\beta\eta}(\mathbf{A}) \doteq^\alpha \varphi_1^{\beta\eta}(\mathbf{B})] \in \mathcal{H}$. Assume $\mathcal{E}_\varphi(\mathbf{A}) \sim \mathcal{E}_\varphi(\mathbf{B})$. By Lemma 5.7.10:1, we must have

$$\varphi_1^{\beta\eta}(\mathbf{A}) \parallel \varphi_1^{\beta\eta}(\mathbf{B})$$

. By Lemma 5.2.8:1, we conclude $\neg[\varphi_1^{\beta\eta}(\mathbf{A}) \doteq^\alpha \varphi_1^{\beta\eta}(\mathbf{B})] \notin \mathcal{H}$, a contradiction. Hence $\mathcal{E}_\varphi(\mathbf{A}) \not\sim \mathcal{E}_\varphi(\mathbf{B})$. Thus $\mathscr{P} \models\!\approx_\varphi \neg[\mathbf{A}_\alpha \doteq^\alpha \mathbf{B}_\alpha]$. □

We can now use the construction to prove the following per model existence theorem.

THEOREM 5.7.16 (Per Model Existence for Extensional Hintikka Sets). *Let $\mathcal{H}$ be an extensional $\mathcal{S}$-Hintikka set (i.e., $\mathcal{H} \in \mathfrak{Hint}_{\beta\mathfrak{fb}}(\mathcal{S})$). There is an $\mathcal{S}$-per model $\mathscr{P}$ such that $\mathscr{P} \models\!\approx \mathcal{H}$.*

PROOF. Let $(\mathcal{D}, @) := \mathcal{F}^{\mathcal{H}}$ be the $\mathcal{H}$-compatibility structure and let $\sim$ be the $\mathcal{H}$-compatibility per over $(\mathcal{D}, @)$. This is a possible values structure by Lemma 5.7.8. To define an evaluation, we first need to interpret the parameters and constants. For parameters W_α then choose any possible value p^W with $\langle W, p^W \rangle \in \overline{\mathcal{D}_\alpha^\sim}$. Such a possible value exists by Theorem 5.7.11. Let $\mathcal{I}(W) := \langle W, p^W \rangle$.

To interpret the logical constants, we must check that we can interpret each $c \in \mathcal{S}$ in the intended way. First we define a function $\upsilon : \mathcal{D}_o \to \{\mathtt{T}, \mathtt{F}\}$ by $\upsilon(\langle \mathbf{A}, \mathsf{p} \rangle) := \mathsf{p}$. We define the appropriate value and then check that this is an appropriate possible value such that the pair satisfies the corresponding property in Table 4.1 with respect to υ and $\sim$. We demonstrate how to define $\mathcal{I}(c)$ for some logical constants $c \in \mathcal{S}$ here and for other logical constants in Appendix A.3.

$=^\alpha$: For each $\langle \mathbf{A}, a \rangle \in \mathcal{D}_\alpha$ let $s^{\langle \mathbf{A}, a \rangle} : \mathcal{D}_\alpha \to \mathcal{D}_o$ be defined by

$$s^{\langle \mathbf{A}, a \rangle}(\langle \mathbf{B}, b \rangle) := \begin{cases} \langle [\mathbf{A} =^\alpha \mathbf{B}], \mathtt{T} \rangle & \text{if } \langle \mathbf{A}, a \rangle \sim \langle \mathbf{B}, b \rangle \\ \langle [\mathbf{A} =^\alpha \mathbf{B}], \mathtt{F} \rangle & \text{otherwise.} \end{cases}$$

To check this is well-defined, we must verify $s^{\langle \mathbf{A}, a \rangle}(\langle \mathbf{B}, b \rangle) \in \mathcal{D}_o$. There are two possibilities. First suppose

$$s^{\langle \mathbf{A}, a \rangle}(\langle \mathbf{B}, b \rangle) = \langle [\mathbf{A} =^\alpha \mathbf{B}], \mathtt{T} \rangle.$$

By the definition of $s^{\langle \mathbf{A}, a \rangle}$, we have $\langle \mathbf{A}, a \rangle \sim \langle \mathbf{B}, b \rangle$. By Lemma 5.7.10:1, we have $\mathbf{A} \parallel \mathbf{B}$. We know $\neg[\mathbf{A} =^\alpha \mathbf{B}] \notin \mathcal{H}$ by Lemma

5.2.8:1. Hence $\langle[\mathbf{A} =^{\alpha} \mathbf{B}], \mathtt{T}\rangle \in \mathcal{D}_o$. Next suppose

$$s^{\langle \mathbf{A},a\rangle}(\langle \mathbf{B}, b\rangle) = \langle[\mathbf{A} =^{\alpha} \mathbf{B}], \mathtt{F}\rangle.$$

By definition this means $\langle \mathbf{A}, a\rangle \not\sim \langle \mathbf{B}, b\rangle$. By Lemma 5.7.14, we conclude $[\mathbf{A} =^{\alpha} \mathbf{B}] \notin \mathcal{H}$. Thus $\langle[\mathbf{A} =^{\alpha} \mathbf{B}], \mathtt{F}\rangle \in \mathcal{D}_o$.

Let $q^{\alpha} : \mathcal{D}_{\alpha} \longrightarrow \mathcal{D}_{o\alpha}$ be defined by

$$q^{\alpha}(\langle \mathbf{A}, a\rangle) := \langle[=^{\alpha} \mathbf{A}], s^{\langle \mathbf{A},a\rangle}\rangle.$$

Now, let $\mathcal{I}(=^{\alpha}) := \langle =^{\alpha}, q^{\alpha}\rangle$.

To check $\langle =^{\alpha}, q^{\alpha}\rangle \sim \langle =^{\alpha}, q^{\alpha}\rangle$, let $\langle \mathbf{A}, a\rangle \sim \langle \mathbf{C}, c\rangle$ be given. We must verify $q^{\alpha}(\langle \mathbf{A}, a\rangle) \sim q^{\alpha}(\langle \mathbf{C}, c\rangle)$. Let $\langle \mathbf{B}, b\rangle \sim \langle \mathbf{D}, d\rangle$ be given. We can establish $s^{\langle \mathbf{A},a\rangle}(\langle \mathbf{B}, b\rangle) \sim s^{\langle \mathbf{C},c\rangle}(\langle \mathbf{D}, d\rangle)$ by verifying

$$\langle \mathbf{A}, a\rangle \sim \langle \mathbf{B}, b\rangle \text{ is equivalent to } \langle \mathbf{C}, c\rangle \sim \langle \mathbf{D}, d\rangle.$$

If $\langle \mathbf{A}, a\rangle \sim \langle \mathbf{B}, b\rangle$, then

$$\langle \mathbf{C}, c\rangle \sim \langle \mathbf{A}, a\rangle \sim \langle \mathbf{B}, b\rangle \sim \langle \mathbf{D}, d\rangle$$

as desired. If $\langle \mathbf{C}, c\rangle \sim \langle \mathbf{D}, d\rangle$, then

$$\langle \mathbf{A}, a\rangle \sim \langle \mathbf{C}, c\rangle \sim \langle \mathbf{D}, d\rangle \sim \langle \mathbf{B}, b\rangle$$

as desired.

It is easy to check $\mathfrak{L}^{\partial}_{=^{\alpha}}(\mathcal{I}(=^{\alpha}))$ holds using the definition of q^{α}.

$\vee$: For each $\langle \mathbf{A}, a\rangle \in \mathcal{D}_o$ let $d_{\langle \mathbf{A},a\rangle} : \mathcal{D}_o \longrightarrow \mathcal{D}_o$ be defined by

$$d_{\langle \mathbf{A},a\rangle}(\langle \mathbf{B}, b\rangle) := \begin{cases} \langle[\mathbf{A} \vee \mathbf{B}], \mathtt{T}\rangle & a = \mathtt{T} \text{ or } b = \mathtt{T} \\ \langle[\mathbf{A} \vee \mathbf{B}], \mathtt{F}\rangle & \text{otherwise.} \end{cases}$$

Using $\vec{\nabla}^{\sharp}$, $\vec{\nabla}_{\wedge}$, $\vec{\nabla}_{\vee}$, $\vec{\nabla}_{c}$ and the definition of $\sim$ on $\mathcal{D}_o$, we can easily check that each $\langle \mathbf{A}, d_{\langle \mathbf{A},a\rangle}\rangle \in \overline{\mathcal{D}^{\sim}_{oo}}$. Furthermore,

$$\langle \mathbf{A}, d_{\langle \mathbf{A},a\rangle}\rangle \sim \langle \mathbf{A}', d_{\langle \mathbf{A}',a\rangle}\rangle$$

for any other $\langle \mathbf{A}', a\rangle \in \mathcal{D}_o$ by the definition of $\sim$. That is, $\sim$-related values in $\mathcal{D}_o$ give $\sim$-related values d_*. Let $d : \mathcal{D}_o \longrightarrow \mathcal{D}_{oo}$ be defined by $d(\langle \mathbf{A}, a\rangle) := \langle[\vee\, \mathbf{A}], d_{\langle \mathbf{A},a\rangle}\rangle$. We now know

$$\langle \vee, d\rangle \in \overline{\mathcal{D}^{\sim}_{ooo}}.$$

Let $\mathcal{I}(\vee) := \langle \vee, d\rangle$. It is easy to check $\mathfrak{L}^{\partial}_{\vee}(\mathcal{I}(\vee))$ holds using the definitions of d and υ.

$\wedge,\supset,\equiv$: In these cases, we define $\mathcal{I}(c)$ in an analogous way to $\mathcal{I}(\vee)$ and check the appropriate properties.

Now we have an interpretations $\mathcal{I}(W_\alpha) \in \overline{\mathcal{D}_\alpha^W}$ and $\mathcal{I}(c_\alpha) \in \overline{\mathcal{D}_\alpha^c}$ for all parameters W_α and constants c_α. By Theorem 4.1.8, there is an evaluation function $\mathcal{E}$ such that

1. $\mathcal{E}(W_\alpha) = \mathcal{I}(W_\alpha)$ for every parameter W_α,
2. $\mathcal{E}(c_\alpha) = \mathcal{I}(c_\alpha)$ for every constant c_α,
3. $(\mathcal{D}, @, \mathcal{E})$ is a possible values evaluation for $\beta\eta$, hence an evaluation, and
4. $\mathcal{E}_\varphi(\mathbf{A}) \in \mathcal{D}_\alpha^{\varphi_1(\mathbf{A})}$ for each $\mathbf{A} \in \mathit{wff}_\alpha(\mathcal{S})$.

Let $\mathscr{P} := (\mathcal{D}, @, \mathcal{E}, \upsilon, \sim)$. We know $(\mathcal{D}, @, \mathcal{E})$ is an evaluation and $\sim$ is a typed per over this evaluation. Also, for each logical constant $c \in \mathcal{S}$ we noted the property $\mathfrak{L}_c^\partial(\mathcal{I}(c))$ (cf. Table 4.1) holds when we defined $\mathcal{I}(c)$. Since for $\mathcal{E}(c) = \mathcal{I}(c)$ we know $\mathfrak{L}_c^\partial(\mathcal{I}(c))$ holds. We now prove $\mathscr{P}$ is an $\mathcal{S}$-per model by verifying $\partial^\mathcal{P}$, $\partial^\mathcal{S}$, $\partial^@$, $\partial^\mathfrak{f}$, $\partial^\mathfrak{b}$ and ∂^υ.

$\partial^\mathcal{P}$: For each $W \in \mathcal{P}$ we have $\mathcal{E}(W) = \mathcal{I}(W) \sim \mathcal{I}(W) = \mathcal{E}(W)$.

$\partial^\mathcal{S}$: For each constant $c \in \mathcal{S}$ we have $\mathcal{E}(c) = \mathcal{I}(c) \sim \mathcal{I}(c) = \mathcal{E}(c)$.

$\partial^@$, $\partial^\mathfrak{f}$: For each $\mathsf{g}, \mathsf{h} \in \mathcal{D}_{\alpha\beta}$ we know $\mathsf{g} \sim \mathsf{h}$ iff $\mathsf{g}@\mathsf{a} \sim \mathsf{h}@\mathsf{b}$ for every $\mathsf{a}, \mathsf{b} \in \mathcal{D}_\beta$ with $\mathsf{a} \sim \mathsf{b}$ by the definition of $\sim$ on function domains $\mathcal{D}_{\alpha\beta}$.

$\partial^\mathfrak{b}$: If $\upsilon(\langle \mathbf{A}, \mathsf{a}\rangle) = \upsilon(\langle \mathbf{B}, \mathsf{b}\rangle)$, then $\mathsf{a} = \mathsf{b}$ and so $\langle \mathbf{A}, \mathsf{a}\rangle \sim \langle \mathbf{B}, \mathsf{b}\rangle$. Thus $\partial^\mathfrak{b}$ holds.

∂^υ: We have $\mathsf{a} \sim \mathsf{b}$ implies $\upsilon(\mathsf{a}) = \upsilon(\mathsf{b})$ for each $\mathsf{a}, \mathsf{b} \in \mathcal{D}_o$ by the definitions of υ and $\sim$ at type o.

Therefore, $\mathscr{P} := (\mathcal{D}, @, \mathcal{E}, \upsilon, \sim)$ is an $\mathcal{S}$-per model.

We now prove $\mathscr{P} \models\!\!\approx \mathcal{H}$. Let $\mathbf{M} \in \mathcal{H}$ and a $\sim$-assignment φ be given. Since $\mathbf{M}$ is a sentence we know $\varphi_1^{\beta\eta}(\mathbf{M})$ is $\mathbf{M}\!\downarrow$. By $\vec{\nabla}_{\beta\eta}$, we know $\mathbf{M}\!\downarrow \in \mathcal{H}$. Thus $\mathcal{M} \models_\varphi \mathbf{M}$ by Lemma 5.7.15. Therefore, $\mathscr{P} \models \mathcal{H}$, as desired. □

THEOREM 5.7.17 (Model Existence for Extensional Hintikka Sets). *Let $\mathcal{H}$ be an extensional $\mathcal{S}$-Hintikka set (i.e., $\mathcal{H} \in \mathfrak{Hint}_{\beta\mathfrak{fb}}(\mathcal{S})$). There is an extensional $\mathcal{S}$-model $\mathcal{M}$ (i.e., $\mathcal{M} \in \mathfrak{M}_{\beta\mathfrak{fb}}(\mathcal{S})$) such that $\mathcal{M} \models \mathcal{H}$.*

PROOF. By Theorem 5.7.16, there is an $\mathcal{S}$-per model $\mathscr{P}$ such that $\mathscr{P} \models\!\!\approx \mathcal{H}$. We apply Theorem 4.2.11 to obtain an extensional $\mathcal{S}$-model $\mathcal{M}$ such that $\mathcal{M} \models \mathbf{M}$ iff $\mathscr{P} \models \mathbf{M}$ for every proposition $\mathbf{M}$. Thus $\mathcal{M} \models \mathcal{H}$. □

THEOREM 5.7.18 (Model Existence for Extensional Abstract Consistency Classes). *Let* $\mathscr{C} \in \mathfrak{Acc}_{\beta\mathfrak{fb}}(\mathcal{S})$ *be an extensional* $\mathcal{S}$*-abstract consistency class of finite character and* $\Phi \in \mathscr{C}$ *be a sufficiently pure set of* $\mathcal{S}$*-sentences. There exists a model* $\mathcal{M} \in \mathfrak{M}_{\beta\mathfrak{fb}}(\mathcal{S})$ *such that* $\mathcal{M} \models \Phi$.

PROOF. By Lemma 5.5.6, there is a Hintikka set $\mathcal{H}$ in $\mathfrak{Hint}_{\beta\mathfrak{fb}}(\mathcal{S})$ such that $\Phi \subseteq \mathcal{H}$. By Theorem 5.7.17, there is a model $\mathcal{M}$ in $\mathfrak{M}_{\beta\mathfrak{fb}}(\mathcal{S})$ such that $\mathcal{M} \models \mathcal{H}$. Hence $\mathcal{M} \models \Phi$. □

THEOREM 5.7.19 (Completeness of $\mathcal{G}^{\mathcal{S}}_{\beta\mathfrak{fb}}$). *If a sequent* Γ *is not derivable in* $\mathcal{G}^{\mathcal{S}}_{\beta\mathfrak{fb}}$*, then there is an* $\mathcal{S}$*-model* $\mathcal{M} \in \mathfrak{M}_{\beta\mathfrak{fb}}(\mathcal{S})$ *such that* $\mathcal{M} \not\models \Gamma$.

PROOF. Suppose Γ is not derivable in $\mathcal{G}^{\mathcal{S}}_{\beta\mathfrak{fb}}$. By Lemma 5.4.3, the set Φ_Γ is in $\mathscr{C}^{\mathcal{S}}_{\beta\mathfrak{fb}}$. Note that $\mathscr{C}^{\mathcal{S}}_{\beta\mathfrak{fb}}$ is an abstract consistency class in $\mathfrak{Acc}_{\beta\mathfrak{fb}}(\mathcal{S})$ of finite character by Theorem 5.4.7. Also, Φ_Γ is sufficiently pure since it is finite and there are infinitely many parameters of each type. By Theorem 5.7.18, there is a model $\mathcal{M}$ in $\mathfrak{M}_{\beta\mathfrak{fb}}(\mathcal{S})$ such that $\mathcal{M} \models \Phi_\Gamma$. In particular, for each $\mathbf{M} \in \Gamma$ we have $\mathcal{M} \models \neg\mathbf{M}$. Thus $\mathcal{M} \not\models \Gamma$. □

COROLLARY 5.7.20 (Cut Elimination for Extensional Sequent Calculi). *Suppose* $\Gamma, \mathbf{M}$ *and* $\Gamma, \neg\mathbf{M}$ *are both derivable in* $\mathcal{G}^{\mathcal{S}}_{\beta\mathfrak{fb}}$*. Then* Γ *is derivable in* $\mathcal{G}^{\mathcal{S}}_{\beta\mathfrak{fb}}$*. That is, the rule* $\mathcal{G}(Cut)$ *is admissible in* $\mathcal{G}^{\mathcal{S}}_{\beta\mathfrak{fb}}$.

PROOF. Suppose Γ is not derivable in $\mathcal{G}^{\mathcal{S}}_{\beta\mathfrak{fb}}$. By completeness (Theorem 5.7.19), there is an $\mathcal{S}$-model $\mathcal{M} \in \mathfrak{M}_{\beta\mathfrak{fb}}(\mathcal{S})$ such that $\mathcal{M} \not\models \Gamma$. By soundness (Theorem 3.4.14), $\mathcal{M} \models \Gamma, \mathbf{M}$ and $\mathcal{M} \models \Gamma, \neg\mathbf{M}$. Thus $\mathcal{M} \models \mathbf{M}$ and $\mathcal{M} \models \neg\mathbf{M}$, a contradiction. □

CHAPTER 6

Independence and Conservation

We have defined families of sequent calculi $\mathcal{G}_*^{\mathcal{S}}$ that vary with respect to both extensionality $* \in \{\beta, \beta\eta, \beta\mathfrak{fb}\}$ and the signature of logical constants $\mathcal{S} \subseteq \mathcal{S}_*$. We can compare these sequent calculi with respect to derivable sentences $\mathbf{M}$ (thinking of $\mathbf{M}$ as a sequent with the single sentence $\mathbf{M}$). Adding more logical constants to the signature $\mathcal{S}$ corresponds to assuming more set comprehension. Consequently, we expect $\mathcal{G}_*^{\mathcal{S}_2}$ to prove at least as many sentences from $sent_*(\mathcal{S}_1)$ as (and possibly more than) $\mathcal{G}_*^{\mathcal{S}_1}$ whenever $\mathcal{S}_1 \subseteq \mathcal{S}_2 \subseteq \mathcal{S}_*$.

Similarly, with stronger extensionality assumptions we expect to be able to prove more sentences. There are particular sentences which demonstrate this in [19] (with even more forms of extensionality than the three considered here). Below we use two particular sentences (over the empty signature) to demonstrate that adding extensionality is stronger than adding any particular collection of logical constants to the signature (cf. Theorem 6.1.3).

For the remainder of this chapter we compare sequent calculi $\mathcal{G}_*^{\mathcal{S}_1}$ and $\mathcal{G}_*^{\mathcal{S}_2}$ where $\mathcal{S}_1 \subseteq \mathcal{S}_2 \subseteq \mathcal{S}_*$ and the form of extensionality $*$ in $\{\beta, \beta\eta, \beta\mathfrak{fb}\}$ is the same in both sequent calculi. There are some cases where every sentence $\mathbf{M} \in sent_*(\mathcal{S}_1)$ is derivable using the smaller signature $\mathcal{S}_1$ if it is derivable using the larger signature $\mathcal{S}_2$. In these cases we will say $\mathcal{G}_*^{\mathcal{S}_2}$ is conservative over $\mathcal{G}_*^{\mathcal{S}_1}$ (cf. Definition 6.1.2). A semantic characterization of conservation (cf. Theorem 6.1.7) allows us to investigate conservation in terms of models instead of sequent derivations. Using this approach we can establish conservation results involving equalities and quantifiers at different types (cf. Theorem 6.3.6) and logical constants of propositional type (cf. Corollary 6.5.3).

We will further discover certain signatures $\mathcal{S}$ are extensionally complete (cf. Definition 6.6.1) in the sense that $\mathcal{G}_{\beta\mathfrak{fb}}^{\mathcal{S}}$ proves the same sentences as $\mathcal{G}^{\mathcal{S}_{all}}_{\beta\mathfrak{fb}}$ (cf. Theorem 6.6.7). In terms of automated theorem proving this means we can restrict the set of logical constants appearing in proofs to those in an extensionally complete signature $\mathcal{S}$ without losing any theorems we could prove using the full signature $\mathcal{S}_{all}$ of all logical constants.

There are also cases in which there is a sentence $\mathbf{M} \in sent_*(\mathcal{S}_1)$ that is derivable using the larger signature $\mathcal{S}_2$ but is not derivable using the smaller signature $\mathcal{S}_1$. We will establish such independence results by giving signatures which are not strong enough to prove different versions of Cantor's theorem (cf. Corollary 6.7.9 and Theorem 6.7.40).

6.1. Conservation

Suppose $* \in \{\beta, \beta\eta, \beta\mathfrak{fb}\}$ and $\mathcal{S}_1$ and $\mathcal{S}_2$ are signatures with

$$\mathcal{S}_1 \subseteq \mathcal{S}_2 \subseteq \mathcal{S}_*$$

(cf. Definition 2.2.5). Note that $cwff_\alpha(\mathcal{S}_1) \subseteq cwff_\alpha(\mathcal{S}_2)$ for every type α and $sent_*(\mathcal{S}_1) \subseteq sent_*(\mathcal{S}_2)$ (cf. Definition 2.2.6). Every $\mathcal{G}^{\mathcal{S}_1}_*$-derivation can be viewed as a $\mathcal{G}^{\mathcal{S}_2}_*$-derivation.

LEMMA 6.1.1. *Let* $* \in \{\beta, \beta\eta, \beta\mathfrak{fb}\}$, $\mathcal{S}_1 \subseteq \mathcal{S}_2 \subseteq \mathcal{S}_*$ *and* Γ *be an* $\mathcal{S}_1$-*sequent. If* Γ *is* $\mathcal{G}^{\mathcal{S}_1}_*$-*derivable, then* Γ *is* $\mathcal{G}^{\mathcal{S}_2}_*$-*derivable.*

PROOF. The proof is by an easy induction on the $\mathcal{G}^{\mathcal{S}_1}_*$-derivation of Γ. Since $cwff_\alpha(\mathcal{S}_1) \subseteq cwff_\alpha(\mathcal{S}_2)$ we can use the $\mathcal{G}(\dot{\neg}\dot{\forall}, \mathcal{S}_2)$ rule in place of the $\mathcal{G}(\dot{\neg}\dot{\forall}, \mathcal{S}_1)$ rule. Similarly, if $*$ is $\beta\mathfrak{fb}$, then we can use the $\mathcal{G}(\dot{\neg} \doteq^{\rightarrow}, \mathcal{S}_2)$ rule to replace the $\mathcal{G}(\dot{\neg} \doteq^{\rightarrow}, \mathcal{S}_1)$ rule. □

We say $\mathcal{G}^{\mathcal{S}_2}_*$ is conservative over $\mathcal{G}^{\mathcal{S}_1}_*$ if the converse of Lemma 6.1.1 is true for appropriate sequents. In particular, we consider each $\mathbf{M} \in sent_*(\mathcal{S}_1)$ viewed as a singleton sequent.

DEFINITION 6.1.2 (Conservation). Let $* \in \{\beta, \beta\eta, \beta\mathfrak{fb}\}$ and $\mathcal{S}_1$ and $\mathcal{S}_2$ be signatures with $\mathcal{S}_1 \subseteq \mathcal{S}_2 \subseteq \mathcal{S}_*$. We say $\mathcal{G}^{\mathcal{S}_2}_*$ is *conservative over* $\mathcal{G}^{\mathcal{S}_1}_*$ if $\mathbf{M}$ is $\mathcal{G}^{\mathcal{S}_1}_*$-derivable whenever $\mathbf{M}$ is $\mathcal{G}^{\mathcal{S}_2}_*$-derivable for every $\mathbf{M} \in sent_*(\mathcal{S}_1)$. We write

$$\mathcal{S}_1 \trianglelefteq_* \mathcal{S}_2$$

if $\mathcal{G}^{\mathcal{S}_2}_*$ is conservative over $\mathcal{G}^{\mathcal{S}_1}_*$ and write

$$\mathcal{S}_1 \ntrianglelefteq_* \mathcal{S}_2$$

otherwise.

For each $* \in \{\beta, \beta\eta, \beta\mathfrak{fb}\}$ it is easy to check reflexivity and transitivity of $\trianglelefteq_*$ as a binary relation on $\mathcal{P}(\mathcal{S}_*)$. We will use these facts freely below.

One might ask if we can ever obtain a conservation result for $\mathcal{G}_{\beta}^{\mathcal{S}_2}$ over $\mathcal{G}_{\beta\eta}^{\mathcal{S}_1}$ where $\mathcal{S}_1 \subseteq \mathcal{S}_2 \subseteq \mathcal{S}_{all}^{elem}$. Similarly, one could ask if there are conservation results for $\mathcal{G}_{\beta\eta}^{\mathcal{S}_2}$ over $\mathcal{G}_{\beta\mathfrak{fb}}^{\mathcal{S}_1}$. That is, can one capture some level of extensionality by adding logical constants to the signature? We determine in Theorem 6.1.3 that this never happens. In particular, there are sentences in $sent(\emptyset)$ provable in $\mathcal{G}_{\beta\eta}^{\emptyset}$ (where no logical constants are available) but are not provable in $\mathcal{G}_{\beta}^{\mathcal{S}_{all}}$ (where all logical constants are available). Likewise, there are sentences in $sent(\emptyset)$ provable in $\mathcal{G}_{\beta\mathfrak{fb}}^{\emptyset}$ but not in $\mathcal{G}_{\beta\eta}^{\mathcal{S}_{all}}$. Consequently, it only makes sense to compare $\mathcal{G}_{*}^{\mathcal{S}_1}$ and $\mathcal{G}_{*}^{\mathcal{S}_2}$ with the same $* \in \{\beta, \beta\eta, \beta\mathfrak{fb}\}$.

THEOREM 6.1.3 (Extensionality and Conservation). *Let*

$$\mathbf{M} \textit{ be the sentence } \dot{\forall} f_{\iota\iota} \centerdot f \doteq^{\iota\iota} \centerdot \lambda x_\iota \centerdot f\, x$$

and

$$\mathbf{N} \textit{ be the sentence } \dot{\forall} x_o \dot{\forall} y_o \centerdot x \mathrel{\dot{\equiv}} y \mathrel{\dot{\supset}} \centerdot x \doteq^o y.$$

(Note that **M** *and* **N** *are equality-free* $\mathcal{S}$*-sentences for any signature* $\mathcal{S}$*.) We have the following:*

1. **M** *is* $\mathcal{G}_{\beta\eta}^{\emptyset}$*-derivable.*
2. **M** *is not* $\mathcal{G}_{\beta}^{\mathcal{S}_{all}}$*-derivable.*
3. **N** *is* $\mathcal{G}_{\beta\mathfrak{fb}}^{\emptyset}$*-derivable.*
4. **N** *is not* $\mathcal{G}_{\beta\eta}^{\mathcal{S}_{all}}$*-derivable.*

PROOF. There are easy derivations of **M** in $\mathcal{G}_{\beta\eta}^{\emptyset}$ and **N** in $\mathcal{G}_{\beta\mathfrak{fb}}^{\emptyset}$. We can use the retraction models from Examples 4.3.14 and 4.3.13 and soundness to determine **M** is not $\mathcal{G}_{\beta}^{\mathcal{S}_{all}}$-derivable and **N** is not $\mathcal{G}_{\beta\eta}^{\mathcal{S}_{all}}$-derivable. The full proof is in Appendix A.4. □

In spite of Theorem 6.1.3, it would be a mistake to conclude that extensionality and set comprehension are completely independent of one another. If extensionality was unrelated to set comprehension, then we would expect the relations $\mathcal{S}_1 \trianglelefteq_\beta \mathcal{S}_2$, $\mathcal{S}_1 \trianglelefteq_{\beta\eta} \mathcal{S}_2$ and $\mathcal{S}_1 \trianglelefteq_{\beta\mathfrak{fb}} \mathcal{S}_2$ to all be equivalent (with respect to equality-free signatures). We will see below that the relations $\trianglelefteq_\beta$ and $\trianglelefteq_{\beta\eta}$ differ radically from the relation $\trianglelefteq_{\beta\mathfrak{f}\mathfrak{b}}$. In particular, we can have $\mathcal{S}_1 \trianglelefteq_{\beta\mathfrak{f}\mathfrak{b}} \mathcal{S}_2$ without having $\mathcal{S}_1 \trianglelefteq_{\beta\eta} \mathcal{S}_2$. We leave open the possibility that the relations $\trianglelefteq_{\beta}$ and $\trianglelefteq_{\beta\eta}$ are the same (see Conjecture 6.3.16).

We now introduce propositions $\mathbf{L}_c \in prop(\emptyset)$ which characterize when a model realizes a logical constant c (cf. Definition 3.3.17). The idea is that we use connectives at the "external" level of propositions to characterize logical constants at the "internal" level.

DEFINITION 6.1.4. For each $c_\alpha \in \mathcal{S}_{all}$ we define $\mathbf{L}_c$ as follows:

$\top$: Let $\mathbf{L}_\top := a_o$.

$\bot$: Let $\mathbf{L}_\bot := \dot{\neg} a_o$.

$\neg$: Let $\mathbf{L}_\neg := \dot{\forall} p_o \centerdot [a_{oo}\, p] \,\dot{\equiv}\, \dot{\neg} p$.

$\wedge$: Let $\mathbf{L}_\wedge := \dot{\forall} p_o \dot{\forall} q_o \centerdot [a_{ooo}\, p\, q] \,\dot{\equiv} \centerdot p \,\dot{\wedge}\, q$.

$\vee$: Let $\mathbf{L}_\vee := \dot{\forall} p_o \dot{\forall} q_o \centerdot [a_{ooo}\, p\, q] \,\dot{\equiv} \centerdot p \,\dot{\vee}\, q$.

$\equiv$: Let $\mathbf{L}_\equiv := \dot{\forall} p_o \dot{\forall} q_o \centerdot [a_{ooo}\, p\, q] \,\dot{\equiv} \centerdot p \,\dot{\equiv}\, q$.

$\supset$: Let $\mathbf{L}_\supset := \dot{\forall} p_o \dot{\forall} q_o \centerdot [a_{ooo}\, p\, q] \,\dot{\equiv} \centerdot p \,\dot{\supset}\, q$.

Π^α: Let $\mathbf{L}_{\Pi^\alpha} := \dot{\forall} f_{o\alpha} \centerdot [a_{o(o\alpha)}\, f] \,\dot{\equiv} \centerdot \dot{\forall} x_\alpha \centerdot f\, x$.

Σ^α: Let $\mathbf{L}_{\Sigma^\alpha} := \dot{\forall} f_{o\alpha} \centerdot [a_{o(o\alpha)}\, f] \,\dot{\equiv} \centerdot \dot{\exists} x_\alpha \centerdot f\, x$.

$=^\alpha$: Let $\mathbf{L}_{=^\alpha} := \dot{\forall} x_\alpha \dot{\forall} y_\alpha \centerdot [a_{o\alpha\alpha}\, x\, y] \,\dot{\equiv} \centerdot x \,\dot{=}^\alpha\, y$.

Note that each $\mathbf{L}_c$ is a proposition over the empty signature. Furthermore, if c is not an equality constant $=^\alpha$, then $\mathbf{L}_c$ is equality-free. We now prove each of these propositions characterizes when a model realizes a logical constant.

LEMMA 6.1.5. *Let $\mathcal{S}$ be a signature, $\mathcal{M} = (\mathcal{D}, @, \mathcal{E}, \upsilon) \in \mathfrak{M}_*(\mathcal{S})$ and φ be an assignment. For each logical constant c_α and $\mathsf{a} \in \mathcal{D}_\alpha$, $\mathfrak{L}_c(\mathsf{a})$ holds with respect to υ iff $\mathcal{M} \models_{\varphi,[\mathsf{a}/a]} \mathbf{L}_c$.*

PROOF. Each case follows using the definition of $\models$ in Definition 3.3.3 and the properties for abbreviations proven in Lemma 3.3.4. □

LEMMA 6.1.6. *Let $\mathcal{S}$ be a signature and $\mathcal{M} \in \mathfrak{M}_*(\mathcal{S})$. For each logical constant c_α, $\mathcal{M}$ realizes c iff $\mathcal{M} \models \dot{\exists} a_\alpha \mathbf{L}_c$.*

PROOF. Let $\mathcal{M} = (\mathcal{D}, @, \mathcal{E}, v) \in \mathfrak{M}_*(\mathcal{S})$ and a logical constant c_α be given. Suppose $\mathcal{M}$ realizes c. There is some $\mathsf{a} \in \mathcal{D}_\alpha$ such that $\mathfrak{L}_c(\mathsf{a})$ holds. By Lemma 6.1.5, $\mathcal{M} \models_{\varphi,[\mathsf{a}/a]} \mathbf{L}_c$ for any assignment φ. Thus $\mathcal{M} \models \dot{\exists} a_\alpha\, \mathbf{L}_c$.

Likewise, if $\mathcal{M} \models \dot{\exists} a_\alpha\, \mathbf{L}_c$, then we can use Lemma 6.1.5 to conclude there exists some $\mathsf{a} \in \mathcal{D}_\alpha$ such that $\mathfrak{L}_c(\mathsf{a})$ holds. Thus $\mathcal{M}$ realizes c. □

Since we have soundness and completeness of the sequent calculi $\mathcal{G}_*^{\mathcal{S}}$ with respect to the model classes $\mathfrak{M}_*(\mathcal{S})$, we can use these model classes to determine whether $\mathcal{S}_1 \trianglelefteq_* \mathcal{S}_2$ or $\mathcal{S}_1 \ntrianglelefteq_* \mathcal{S}_2$. The next theorem proves the relationship between conservation and models which realize logical constants. We will make use of the sentences $\dot{\exists} a\, \mathbf{L}_c$ in order to verify this relationship.

THEOREM 6.1.7. *Suppose* $* \in \{\beta, \beta\eta, \beta\mathfrak{f}\mathfrak{b}\}$ *and* $\mathcal{S}_1 \subseteq \mathcal{S}_2 \subseteq \mathcal{S}_*$. *The following are equivalent:*

1. $\mathfrak{M}_*(\mathcal{S}_1)$ *realizes* $\mathcal{S}_2$.
2. $\mathfrak{M}_*(\mathcal{S}_1)$ *realizes* $(\mathcal{S}_2 \setminus \mathcal{S}_1)$.
3. $\mathcal{S}_1 \trianglelefteq_* \mathcal{S}_2$. *(That is,* $\mathcal{G}_*^{\mathcal{S}_2}$ *is conservative over* $\mathcal{G}_*^{\mathcal{S}_1}$.*)*

PROOF.

(1)⇒(2): This is obvious since $(\mathcal{S}_2 \setminus \mathcal{S}_1) \subseteq \mathcal{S}_2$.

(2)⇒(3): Assume $\mathfrak{M}_*(\mathcal{S}_1)$ realizes $(\mathcal{S}_2 \setminus \mathcal{S}_1)$ and $\mathcal{S}_1 \ntrianglelefteq_* \mathcal{S}_2$. There is some $\mathbf{M} \in sent_*(\mathcal{S}_1)$ which is $\mathcal{G}_*^{\mathcal{S}_2}$-derivable but not $\mathcal{G}_*^{\mathcal{S}_1}$-derivable. By completeness (cf. Theorems 5.6.6 and 5.7.19), there is some model $\mathcal{M}_1 \in \mathfrak{M}_*(\mathcal{S}_1)$ such that $\mathcal{M}_1 \not\models \mathbf{M}$. Since $\mathfrak{M}_*(\mathcal{S}_1)$ realizes $(\mathcal{S}_2 \setminus \mathcal{S}_1)$ we can define $\mathcal{I} : (\mathcal{S}_2 \setminus \mathcal{S}_1) \to \mathcal{D}$ by $\mathcal{I}(c_\alpha) := \mathsf{a}^c$ where $\mathsf{a}^c \in \mathcal{D}_\alpha$ is some element such that $\mathfrak{L}_c(\mathsf{a}^c)$ holds with respect to v. The existence of $\mathcal{I}$ allows us to apply Theorem 3.3.16 to obtain a model $\mathcal{M}_2 = (\mathcal{D}, @, \mathcal{E}^2, v)$ in $\mathfrak{M}_*(\mathcal{S}_2)$ such that $\mathcal{E}^2|_{wff(\mathcal{S}_1)} = \mathcal{E}^1$. By soundness (cf. Theorems 3.4.6, 3.4.9 and 3.4.14), we must have $\mathcal{M}_2 \models \mathbf{M}$. By Theorem 3.3.15, we have $\mathcal{M}_1 \models \mathbf{M}$, a contradiction.

(3)⇒(1): Let $c \in \mathcal{S}_2$ be given. For any model $\mathcal{M}_2 \in \mathfrak{M}_*(\mathcal{S}_2)$, we know $\mathfrak{L}_c(\mathcal{E}(c))$ holds with respect to v and so $\mathcal{M}_2 \models \dot{\exists} a\, \mathbf{L}_c$ by Lemma 6.1.6. By completeness (cf. Theorems 5.6.6 and

5.7.19), we know $\exists a\, \mathbf{L}_c$ is $\mathcal{G}_*^{\mathcal{S}_2}$-derivable. Since we are assuming $\mathcal{S}_1 \trianglelefteqq_* \mathcal{S}_2$, we know $\exists a\, \mathbf{L}_c$ is $\mathcal{G}_*^{\mathcal{S}_1}$-derivable for every $c \in \mathcal{S}_2$. By soundness (cf. Theorems 3.4.6, 3.4.9 and 3.4.14), we know $\mathcal{M}_1 \models \exists a\, \mathbf{L}_c$ for every $c \in \mathcal{S}_2$ and $\mathcal{M}_1 \in \mathfrak{M}_*(\mathcal{S}_1)$. Thus $\mathcal{M}_1$ realizes $\mathcal{S}_2$ by Lemma 6.1.6. Therefore, $\mathfrak{M}_*(\mathcal{S}_1)$ realizes $\mathcal{S}_2$. □

When proving independence, we will often apply the following corollary.

COROLLARY 6.1.8. *Let* $* \in \{\beta, \beta\eta, \beta\mathfrak{fb}\}$ *and* $\mathcal{S}_1 \subseteq \mathcal{S}_2 \subseteq \mathcal{S}_*$. *Suppose there is a model* $\mathcal{M} \in \mathfrak{M}_*(\mathcal{S}_1)$ *such that for each* $c \in (\mathcal{S}_2 \setminus \mathcal{S}_1)$, $\mathcal{M}$ *does not realize* c. *Then* $\mathcal{S}_1 \trianglelefteqq_* (\mathcal{S}_1 \cup \{c\})$ *for each* $c \in (\mathcal{S}_2 \setminus \mathcal{S}_1)$.

PROOF. For each $c \in (\mathcal{S}_2 \setminus \mathcal{S}_1)$, this follows by applying Theorem 6.1.7 to $\mathcal{S}_1$ and $\mathcal{S}_1 \cup \{c\}$. □

We can now prove that (for equality-free signatures) $\mathcal{S}_1 \trianglelefteqq_\beta \mathcal{S}_2$ implies $\mathcal{S}_1 \trianglelefteqq_{\beta\eta} \mathcal{S}_2$ and that $\mathcal{S}_1 \trianglelefteqq_{\beta\eta} \mathcal{S}_2$ implies $\mathcal{S}_1 \trianglelefteqq_{\beta\mathfrak{fb}} \mathcal{S}_2$.

COROLLARY 6.1.9. *Let* $\mathcal{S}_1$ *and* $\mathcal{S}_2$ *be equality-free signatures such that* $\mathcal{S}_1 \subseteq \mathcal{S}_2$.

- *If* $\mathcal{S}_1 \trianglelefteqq_\beta \mathcal{S}_2$, *then* $\mathcal{S}_1 \trianglelefteqq_{\beta\eta} \mathcal{S}_2$.
- *If* $\mathcal{S}_1 \trianglelefteqq_{\beta\eta} \mathcal{S}_2$, *then* $\mathcal{S}_1 \trianglelefteqq_{\beta\mathfrak{fb}} \mathcal{S}_2$.

PROOF. Suppose $\mathcal{S}_1 \trianglelefteqq_\beta \mathcal{S}_2$. By Theorem 6.1.7, we know $\mathfrak{M}_\beta(\mathcal{S}_1)$ realizes $\mathcal{S}_2$. Since $\mathfrak{M}_{\beta\eta}(\mathcal{S}_1) \subseteq \mathfrak{M}_\beta(\mathcal{S}_1)$ we know $\mathfrak{M}_{\beta\eta}(\mathcal{S}_1)$ realizes $\mathcal{S}_2$. By Theorem 6.1.7, we conclude $\mathcal{S}_1 \trianglelefteqq_{\beta\eta} \mathcal{S}_2$.

Similarly, if $\mathcal{S}_1 \trianglelefteqq_{\beta\eta} \mathcal{S}_2$, then $\mathfrak{M}_{\beta\eta}(\mathcal{S}_1)$ realizes $\mathcal{S}_2$ and so $\mathfrak{M}_{\beta\mathfrak{fb}}(\mathcal{S}_1)$ realizes $\mathcal{S}_2$ and $\mathcal{S}_1 \trianglelefteqq_{\beta\mathfrak{fb}} \mathcal{S}_2$. □

6.2. Equalities and Quantifiers

We next relate equalities and quantifiers at certain types.

LEMMA 6.2.1. *Let* $\mathcal{S}$ *be a signature and* $\mathcal{M} = (\mathcal{D}, @, \mathcal{E}, \upsilon)$ *be an extensional* $\mathcal{S}$*-model. If* $\mathcal{M}$ *realizes* $=^\alpha$ *and* Π^β, *then* $\mathcal{M}$ *realizes* $=^{\alpha\beta}$.

PROOF. Let $\mathsf{q}^\alpha \in \mathcal{D}_{o\alpha\alpha}$ and $\pi^\beta \in \mathcal{D}_{o(o\beta)}$ satisfy $\mathfrak{L}_{=^\alpha}(\mathsf{q}^\alpha)$ and $\mathfrak{L}_{\Pi^\beta}(\pi^\beta)$ with respect to υ. Let $p_{o(o\beta)}$ and $q_{o\alpha\alpha}$ be variables and φ be an assignment with $\varphi(p) := \pi^\beta$ and $\varphi(q) := \mathsf{q}^\alpha$. We define

$$\mathsf{q}^{\alpha\beta} := \mathcal{E}_\varphi(\lambda f_{\alpha\beta} \lambda g_{\alpha\beta}\, [p\, \lambda x_\beta \blacksquare [q\, [f\, x]\, [g\, x]]]).$$

It is easy to check for any $\mathsf{f},\mathsf{g} \in \mathcal{D}_{\alpha\beta}$ we have $\upsilon(\mathsf{q}^{\alpha\beta}@\mathsf{f}@\mathsf{g}) = \mathtt{T}$ iff $\mathsf{f}@\mathsf{b} = \mathsf{g}@\mathsf{b}$ for every $\mathsf{b} \in \mathcal{D}_\beta$. By property $\mathfrak{f}$, this is equivalent to $\mathsf{f} = \mathsf{g}$. Thus $\mathfrak{L}_{=^{\alpha\beta}}(\mathsf{q}^{\alpha\beta})$ holds with respect to υ. □

LEMMA 6.2.2. *Let $\mathcal{M} = (\mathcal{D}, @, \mathcal{E}, \upsilon)$ be an extensional model. If $\mathcal{M}$ realizes $\top$ and $=^{o\alpha}$, then $\mathcal{M}$ realizes Π^α.*

PROOF. Suppose $\mathcal{M}$ realizes $\top$ and $=^{o\alpha}$. Then there exist $\mathsf{p} \in \mathcal{D}_o$ and $\mathsf{q}^{o\alpha} \in \mathcal{D}_{o(o\alpha)(o\alpha)}$ such that $\mathfrak{L}_\top(\mathsf{p})$ and $\mathfrak{L}_{=^{o\alpha}}(\mathsf{q}^{o\alpha})$ hold. Let p_o and $q_{o(o\alpha)(o\alpha)}$ be variables and φ be an assignment with $\varphi(p) := \mathsf{p}$ and $\varphi(q) := \mathsf{q}^{o\alpha}$. We define

$$\pi^\alpha := \mathcal{E}_\varphi(q\,[\lambda x_\alpha\, p_o]).$$

Note that

$$\upsilon(\mathcal{E}_\varphi([\lambda x_\alpha\, p_o])@\mathsf{a}) = \upsilon(\mathcal{E}_\varphi(p)) = \mathtt{T}$$

for any $\mathsf{a} \in \mathcal{D}_\alpha$.

To check $\mathfrak{L}_{\Pi^\alpha}(\pi^\alpha)$ holds, let $\mathsf{f} \in \mathcal{D}_{o\alpha}$ be given. Suppose $\upsilon(\mathsf{f}@\mathsf{a}) = \mathtt{T}$ for every $\mathsf{a} \in \mathcal{D}_\alpha$. By property $\mathfrak{f}$, we must have $\mathsf{f} = \mathcal{E}_\varphi([\lambda x_\alpha\, p_o])$. By $\mathfrak{L}_{=^{o\alpha}}(\mathsf{q}^{o\alpha})$, we have

$$\upsilon(\pi^\alpha@\mathsf{f}) = \upsilon(\mathsf{q}^{o\alpha}@\mathcal{E}_\varphi([\lambda x_\alpha\, p_o])@\mathsf{f}) = \mathtt{T}.$$

Next suppose $\upsilon(\pi^\alpha@\mathsf{f}) = \mathtt{T}$. By $\mathfrak{L}_{=^{o\alpha}}(\mathsf{q}^{o\alpha})$, we must have $\mathsf{f} = \mathcal{E}_\varphi([\lambda x_\alpha\, p_o])$. Thus $\upsilon(\mathsf{f}@\mathsf{a}) = \mathtt{T}$ for every $\mathsf{a} \in \mathcal{D}_\alpha$. □

Of course, if a model realizes negation, then realizing Π^α is equivalent to realizing Σ^α.

LEMMA 6.2.3. *Suppose $\mathcal{M} = (\mathcal{D}, @, \mathcal{E}, \upsilon)$ is a model which realizes $\neg$. For every type α, $\mathcal{M}$ realizes Π^α iff $\mathcal{M}$ realizes Σ^α.*

PROOF. The idea is simply to use $\neg\forall\neg$ for $\exists$ or $\neg\exists\neg$ for $\forall$. □

If an extensional model realizes Π^α for every type α and equalities at base types, then it realizes equalities at all types.

LEMMA 6.2.4. *Let $\mathcal{S}$ be a signature and $\mathcal{M}$ be an extensional $\mathcal{S}$-model. If $\mathcal{M}$ realizes Π^α for every type α and $\mathcal{M}$ realizes $\equiv$ and $=^\iota$, then $\mathcal{M}$ realizes $=^\alpha$ for every type α.*

PROOF. We prove this by induction on the type α. We have assumed $\mathcal{M}$ realizes $=^\iota$. Since $\mathcal{M}$ realizes $\equiv$ and υ is injective, the witness $\mathsf{e} \in \mathcal{D}_{ooo}$ with $\mathfrak{L}_\equiv(\mathsf{e})$ also satisfies $\mathfrak{L}_{=^o}(\mathsf{e})$. Hence $\mathcal{M}$ realizes $=^o$. For function types $(\gamma\beta)$, we know $\mathcal{M}$ realizes $=^{\gamma\beta}$ by Lemma

6.2.1 since $\mathcal{M}$ realizes $=^{\gamma}$ by the inductive hypothesis and Π^{β} by assumption. □

If a domain is finite, then we have quantifiers over that domain under very liberal assumptions.

LEMMA 6.2.5. *Let $\mathcal{S}$ be a signature, $\mathcal{M} = (\mathcal{D}, @, \mathcal{E}, v) \in \mathfrak{M}_*(\mathcal{S})$ and α be a type. Suppose $\mathcal{D}_\alpha$ is finite.*

1. *If $\mathcal{M}$ realizes $\vee$, then $\mathcal{M}$ realizes Σ^α.*
2. *If $\mathcal{M}$ realizes $\wedge$, then $\mathcal{M}$ realizes Π^α.*

PROOF. Enumerate $\mathcal{D}_\alpha$ as $\{\mathsf{a}^1, \ldots, \mathsf{a}^n\}$ where $n > 0$. Choose distinct variables $x^1_\alpha, \ldots, x^n_\alpha$ of type α and choose another variable b_{ooo} not in $\{x^1, \ldots, x^n\}$. Let φ be an assignment with $\varphi(x^i) := \mathsf{a}^i$ for each i $(1 \leq i \leq n)$.

Assume $\mathcal{M}$ realizes $\vee$. Let $\mathsf{d} \in \mathcal{D}_{ooo}$ be an element such that $\mathfrak{L}_\vee(\mathsf{d})$ holds. We can define

$$\sigma^\alpha := \mathcal{E}_{\varphi,[\mathsf{d}/b]}(\lambda p_{o\alpha} \blacksquare [b\,[p\,x^1][b\,[p\,x^2] \;\cdots\; [p\,x^n]\cdots]])$$

and easily check $\mathfrak{L}_{\Sigma^\alpha}(\sigma^\alpha)$ holds.

Similarly, if $\mathcal{M}$ realizes $\wedge$, then we can define

$$\pi^\alpha := \mathcal{E}_{\varphi,[\mathsf{c}/b]}(\lambda p_{o\alpha} \blacksquare [b\,[p\,x^1][b\,[p\,x^2] \;\cdots\; [p\,x^n]\cdots]]).$$

where $\mathfrak{L}_\vee(\mathsf{c})$ holds and check $\mathfrak{L}_{\Pi^\alpha}(\pi^\alpha)$ holds. □

6.3. Retractions on Types

We can also use (internal) injections and surjections between types to determine relationships between equalities and quantifiers.

LEMMA 6.3.1. *Let $* \in \{\beta, \beta\eta, \beta\mathfrak{f}\mathfrak{b}\}$, $\mathcal{S}$ be a signature, α and β be types and $\mathcal{M} = (\mathcal{D}, @, \mathcal{E}, v) \in \mathfrak{M}_*(\mathcal{S})$. Suppose there is some $\mathsf{f} \in \mathcal{D}_{\alpha\beta}$ such that $\mathsf{f}@\mathsf{a} = \mathsf{f}@\mathsf{b}$ implies $\mathsf{a} = \mathsf{b}$ for every $\mathsf{a}, \mathsf{b} \in \mathcal{D}_\beta$. If $\mathcal{M}$ realizes $=^\alpha$, then $\mathcal{M}$ realizes $=^\beta$.*

PROOF. Let $\mathsf{q}^\alpha \in \mathcal{D}_\beta$ satisfy $\mathfrak{L}_{=^\alpha}(\mathsf{q}^\alpha)$, $q_{o\alpha\alpha}$ and $f_{\alpha\beta}$ be variables and φ be an assignment with $\varphi(q) := \mathsf{q}^\alpha$ and $\varphi(f) := \mathsf{f}$. Define

$$\mathsf{q}^\beta := \mathcal{E}_\varphi(\lambda x_\beta \lambda y_\beta \blacksquare q\,[f\,x]\,[f\,y]).$$

One can easily check $\mathfrak{L}_{=^\beta}(\mathsf{q}^\beta)$. □

LEMMA 6.3.2. *Let* $* \in \{\beta, \beta\eta, \beta\mathfrak{fb}\}$, $\mathcal{S}$ *be a signature,* α *and* β *be types and* $\mathcal{M} = (\mathcal{D}, @, \mathcal{E}, v) \in \mathfrak{M}_*(\mathcal{S})$. *Suppose there is some* $\mathsf{g} \in \mathcal{D}_{\beta\alpha}$ *such that for every* $\mathsf{b} \in \mathcal{D}_\beta$ *there is some* $\mathsf{a} \in \mathcal{D}_\alpha$ *such that* $\mathsf{g}@\mathsf{a} = \mathsf{b}$.

1. *If* $\mathcal{M}$ *realizes* Π^α, *then* $\mathcal{M}$ *realizes* Π^β.
2. *If* $\mathcal{M}$ *realizes* Σ^α, *then* $\mathcal{M}$ *realizes* Σ^β.

PROOF. Let $g_{\beta\alpha}$ and $q_{o(o\alpha)}$ be distinct variables and φ be an assignment with such that $\varphi(g) := \mathsf{g}$. Suppose there is some $\pi^\alpha \in \mathcal{D}_{o(o\alpha)}$ such that $\mathfrak{L}_{\Pi^\alpha}(\pi^\alpha)$ holds. Let ψ be $\varphi, [\pi^\alpha/q]$. Define

$$\pi^\beta := \mathcal{E}_\psi(\lambda p_{o\beta}\, [q\, \lambda x_\alpha \centerdot p \centerdot g\, x]).$$

One can easily check $\mathfrak{L}_{\Pi^\beta}(\pi^\beta)$ holds. The case where $\mathcal{M}$ realizes Σ^α is similar. □

We will often apply Lemmas 6.3.1 and 6.3.2 together. It is worthwhile to build a theory of internal retractions to support this.

DEFINITION 6.3.3 (Internal Retractions). Let $(\mathcal{D}, @)$ be an applicative structure and α and β be types. An *internal retraction* from $\mathcal{D}_\alpha$ onto $\mathcal{D}_\beta$ is a pair (g, f) such that $\mathsf{g} \in \mathcal{D}_{\beta\alpha}$, $\mathsf{f} \in \mathcal{D}_{\alpha\beta}$ and $\mathsf{g}@(\mathsf{f}@\mathsf{b}) = \mathsf{b}$ for every $\mathsf{b} \in \mathcal{D}_\beta$.

Intuitively, the existence of an internal retraction from $\mathcal{D}_\alpha$ onto $\mathcal{D}_\beta$ means that $\mathcal{D}_\alpha$ is "at least as big" as $\mathcal{D}_\beta$ and that the model "knows" this. When we have an internal retraction, we can combine Lemmas 6.3.1 and 6.3.2.

LEMMA 6.3.4. *Let* $* \in \{\beta, \beta\eta, \beta\mathfrak{fb}\}$, $\mathcal{S}$ *be a signature,* α *and* β *be types and* $\mathcal{M} = (\mathcal{D}, @, \mathcal{E}, v) \in \mathfrak{M}_*(\mathcal{S})$. *Suppose* (g, f) *is an internal retraction from* $\mathcal{D}_\alpha$ *onto* $\mathcal{D}_\beta$.

1. *If* $\mathcal{M}$ *realizes* $=^\alpha$, *then* $\mathcal{M}$ *realizes* $=^\beta$.
2. *If* $\mathcal{M}$ *realizes* Π^α, *then* $\mathcal{M}$ *realizes* Π^β.
3. *If* $\mathcal{M}$ *realizes* Σ^α, *then* $\mathcal{M}$ *realizes* Σ^β.

PROOF. Note that if $\mathsf{f}@\mathsf{a} = \mathsf{f}@\mathsf{b}$, then

$$\mathsf{a} = \mathsf{g}@(\mathsf{f}@\mathsf{a}) = \mathsf{g}@(\mathsf{f}@\mathsf{b}) = \mathsf{b}$$

for any $\mathsf{a}, \mathsf{b} \in \mathcal{D}_\beta$. Also, given any $\mathsf{b} \in \mathcal{D}_\beta$, we know $\mathsf{f}@\mathsf{b} \in \mathcal{D}_\alpha$ and $\mathsf{g}@(\mathsf{f}@\mathsf{b}) = \mathsf{b}$. The result follows directly from Lemmas 6.3.1 and 6.3.2. □

In certain contexts we will be able to know each model in a class has an internal retraction from a domain of one type onto a domain of another type. In such a case we can use this to establish conservation results. This relationship will induce a partial preorder on simple types (relative to the amount of extensionality and which logical constants are in the signature).

DEFINITION 6.3.5. Let $\mathcal{S}$ be a signature and $* \in \{\beta, \beta\eta, \beta\mathfrak{fb}\}$. We write $\alpha \succeq_*^{\mathcal{S}} \beta$ when every model $\mathcal{M} = (\mathcal{D}, @, \mathcal{E}, \upsilon) \in \mathfrak{M}_*(\mathcal{S})$ has an internal retraction from $\mathcal{D}_\alpha$ onto $\mathcal{D}_\beta$.

We can now combine Lemma 6.3.4 with $\succeq_*^{\mathcal{S}}$ to conclude conservation results.

THEOREM 6.3.6. *Let $* \in \{\beta, \beta\eta, \beta\mathfrak{fb}\}$ and $\mathcal{S}_1 \subseteq \mathcal{S}_2 \subseteq \mathcal{S}_*$. Suppose the following hold:*

- $\mathcal{S}_1 \cap \mathcal{S}_{Bool} = \mathcal{S}_2 \cap \mathcal{S}_{Bool}$.
- *For any $\Pi^\beta \in \mathcal{S}_2$ there is some $\alpha \succeq_*^{\mathcal{S}_1} \beta$ such that $\Pi^\alpha \in \mathcal{S}_1$.*
- *For any $\Sigma^\beta \in \mathcal{S}_2$ there is some $\alpha \succeq_*^{\mathcal{S}_1} \beta$ such that $\Sigma^\alpha \in \mathcal{S}_1$.*
- *For any $=^\beta \in \mathcal{S}_2$ there is some $\alpha \succeq_*^{\mathcal{S}_1} \beta$ such that $=^\alpha \in \mathcal{S}_1$.*[1]

Then $\mathcal{S}_1 \trianglelefteq_* \mathcal{S}_2$.

PROOF. By Theorem 6.1.7, it is enough to prove that $\mathfrak{M}_*(\mathcal{S}_1)$ realizes $(\mathcal{S}_2 \setminus \mathcal{S}_1)$. Since $\mathcal{S}_1 \cap \mathcal{S}_{Bool} = \mathcal{S}_2 \cap \mathcal{S}_{Bool}$, we know every logical constant in $(\mathcal{S}_2 \setminus \mathcal{S}_1)$ is a quantifier or an equality.

Let $\mathcal{M} = (\mathcal{D}, @, \mathcal{E}, \upsilon) \in \mathfrak{M}_*(\mathcal{S}_1)$ be given. For each Π^β $[\Sigma^\beta, =^\beta]$ in $(\mathcal{S}_2 \setminus \mathcal{S}_1)$, let α be a type with $\alpha \succeq_*^{\mathcal{S}_1} \beta$ and Π^α $[\Sigma^\alpha, =^\alpha]$ in $\mathcal{S}_1$. Since $\alpha \succeq_*^{\mathcal{S}_1} \beta$ there is an internal retraction (g, f) from $\mathcal{D}_\alpha$ onto $\mathcal{D}_\beta$. Since $\mathcal{M}$ realizes Π^α $[\Sigma^\alpha, =^\alpha]$ (as an $\mathcal{S}_1$-model), we conclude $\mathcal{M}$ realizes Π^β $[\Sigma^\beta, =^\beta]$ by Lemma 6.3.4. □

We now turn to properties of the relation $\succeq_*^{\mathcal{S}}$.

LEMMA 6.3.7. *Let $* \in \{\beta, \beta\eta, \beta\mathfrak{fb}\}$ and $\mathcal{S}$ be a signature. The relation $\succeq_*^{\mathcal{S}}$ on simple types is a partial preorder. That is, $\succeq_*^{\mathcal{S}}$ is reflexive and transitive (but need not be antisymmetric).*

[1] If $* \in \{\beta, \beta\eta\}$, then this condition is trivial since $\mathcal{S}_2$ is equality-free as a subset of $\mathcal{S}_*$.

PROOF. Let α be a type. In any model $\mathcal{M} = (\mathcal{D}, @, \mathcal{E}, \upsilon)$, we can let $\mathsf{f} := \mathcal{E}(\lambda x_\alpha x) \in \mathcal{D}_{\alpha\alpha}$. Clearly, the pair (f, f) is an internal retraction since $\mathsf{f}@(\mathsf{f}@\mathsf{a}) = \mathsf{f}@\mathsf{a} = \mathsf{a}$ for any $\mathsf{a} \in \mathcal{D}_\alpha$. Thus $\alpha \succeq_*^{\mathcal{S}} \alpha$ and $\succeq_*^{\mathcal{S}}$ is reflexive.

Suppose $\alpha \succeq_*^{\mathcal{S}} \beta$ and $\beta \succeq_*^{\mathcal{S}} \gamma$. Let $\mathcal{M} = (\mathcal{D}, @, \mathcal{E}, \upsilon) \in \mathfrak{M}_*(\mathcal{S})$ be a model. By $\alpha \succeq_*^{\mathcal{S}} \beta$, there is an internal retraction (g, f) with $\mathsf{g} \in \mathcal{D}_{\beta\alpha}$ and $\mathsf{f} \in \mathcal{D}_{\alpha\beta}$. By $\beta \succeq_*^{\mathcal{S}} \gamma$, there is an internal retraction (k, h) with $\mathsf{k} \in \mathcal{D}_{\gamma\beta}$ and $\mathsf{h} \in \mathcal{D}_{\beta\gamma}$. Let $f_{\alpha\beta}$, $g_{\beta\alpha}$, $h_{\beta\gamma}$ and $k_{\gamma\beta}$ be distinct variables. Let φ be an assignment with $\varphi(f) = \mathsf{f}$, $\varphi(g) = \mathsf{g}$, $\varphi(h) = \mathsf{h}$ and $\varphi(k) = \mathsf{k}$. We verify $(\mathcal{E}_\varphi([\lambda x_\alpha \centerdot k \centerdot g\, x]), \mathcal{E}_\varphi([\lambda z_\gamma \centerdot f \centerdot h\, z]))$ is an internal retraction from $\mathcal{D}_\alpha$ onto $\mathcal{D}_\gamma$. This follows from computing

$$\begin{aligned}
&\mathcal{E}_\varphi([\lambda x_\alpha \centerdot k \centerdot g\, x])@(\mathcal{E}_\varphi([\lambda z_\gamma \centerdot f \centerdot h\, z])@\mathsf{c}) \\
&= \mathcal{E}_\varphi([\lambda x_\alpha \centerdot k \centerdot g\, x])@(\mathsf{f}@(\mathsf{h}@\mathsf{c})) \\
&= \mathsf{k}@(\mathsf{g}@(\mathsf{f}@(\mathsf{h}@\mathsf{c}))) = \mathsf{k}@(\mathsf{h}@\mathsf{c}) = \mathsf{c}
\end{aligned}$$

for any $\mathsf{c} \in \mathcal{D}_\gamma$. Thus $\alpha \succeq_*^{\mathcal{S}} \gamma$ and so $\succeq_*^{\mathcal{S}}$ is transitive. □

There is an easy retraction from any domain of a function type $(\alpha\beta)$ onto the domain of the codomain type α. We simply use constant functions and application.

LEMMA 6.3.8. *Let $\mathcal{S}$ be a signature and $* \in \{\beta, \beta\eta, \beta\mathfrak{fb}\}$. For all types α and β, we have $(\alpha\beta) \succeq_*^{\mathcal{S}} \alpha$.*

PROOF. Let $\mathcal{M} = (\mathcal{D}, @, \mathcal{E}, \upsilon) \in \mathfrak{M}_*(\mathcal{S})$ be given. Choose an arbitrary $\mathsf{b} \in \mathcal{D}_\beta$. Let b_β be a variable and let φ an assignment such that $\varphi(b) := \mathsf{b}$. Let $\mathbf{K}_{\alpha\beta\alpha}$ be the $\mathbf{K}$ combinator. Define $\mathsf{f} := \mathcal{E}(\mathbf{K}_{\alpha\beta\alpha})$ and $\mathsf{g} := \mathcal{E}_\varphi([\lambda h_{\alpha\beta} \centerdot h\, b])$. For any $\mathsf{a} \in \mathcal{D}_\alpha$,

$$\mathsf{g}@(\mathsf{f}@\mathsf{a}) = \mathcal{E}_{\varphi,[\mathsf{f}@\mathsf{a}/h]}([h\, b]) = \mathcal{E}(\mathbf{K})@\mathsf{a}@\mathsf{b} = \mathsf{a}.$$

Thus (g, f) is an internal retraction from $\mathcal{D}_{\alpha\beta}$ onto $\mathcal{D}_\alpha$. □

LEMMA 6.3.9. *Let $\mathcal{S}$ be a signature and $* \in \{\beta, \beta\eta, \beta\mathfrak{fb}\}$. For any types $\alpha^0, \ldots, \alpha^n$ where $n \geq 0$, we have $(\alpha^0\alpha^1 \cdots \alpha^n) \succeq_*^{\mathcal{S}} \alpha^0$.*

PROOF. This follows by induction on n. For the base case note that $\alpha^0 \succeq_*^{\mathcal{S}} \alpha^0$ by reflexivity (cf. Lemma 6.3.7). By Lemma 6.3.8, we know

$$(\alpha^0\alpha^1 \cdots \alpha^n\alpha^{n+1}) \succeq_*^{\mathcal{S}} (\alpha^0\alpha^1 \cdots \alpha^n).$$

By the inductive hypothesis, we know $(\alpha^0\alpha^1 \cdots \alpha^n) \succeq_*^{\mathcal{S}} \alpha^0$. Hence we know $(\alpha^0\alpha^1 \cdots \alpha^n\alpha^{n+1}) \succeq_*^{\mathcal{S}} \alpha^0$ by transitivity (cf. Lemma 6.3.7). □

When we assume property $\mathfrak{f}$ we can prove internal retractions at two given types induce internal retractions at higher types.

LEMMA 6.3.10. *Let $\mathcal{S}$ be a signature, $\mathcal{M} = (\mathcal{D}, @, \mathcal{E}, v)$ be a functional $\mathcal{S}$-model and α, β and γ be types. If there is an internal retraction from $\mathcal{D}_\alpha$ onto $\mathcal{D}_\beta$, then there exist internal retractions from $\mathcal{D}_{\alpha\gamma}$ onto $\mathcal{D}_{\beta\gamma}$ and from $\mathcal{D}_{\gamma\alpha}$ onto $\mathcal{D}_{\gamma\beta}$.*

PROOF. The idea of the proof is to use composition to define the retractions on the function types. See Appendix A.4. □

We can now prove the relation $\succeq^{\mathcal{S}}_{\beta\mathfrak{fb}}$ is monotone with respect to function types for each $\mathcal{S}$. (Note that $\succeq^{\mathcal{S}}_{\beta\mathfrak{fb}}$ is *not* contravariant with respect to domain types.)

LEMMA 6.3.11. *Let $\mathcal{S}$ be a signature. If $\alpha \succeq^{\mathcal{S}}_{\beta\mathfrak{fb}} \beta$ and γ is any type, then $(\alpha\gamma) \succeq^{\mathcal{S}}_{\beta\mathfrak{fb}} (\beta\gamma)$ and $(\gamma\alpha) \succeq^{\mathcal{S}}_{\beta\mathfrak{fb}} (\gamma\beta)$.*

PROOF. Suppose $\alpha \succeq^{\mathcal{S}}_{\beta\mathfrak{fb}} \beta$. Let $\mathcal{M} = (\mathcal{D}, @, \mathcal{E}, v) \in \mathfrak{M}_{\beta\mathfrak{fb}}(\mathcal{S})$. There is an internal retraction from $\mathcal{D}_\alpha$ onto $\mathcal{D}_\beta$. By Lemma 6.3.10, there are internal retractions from $\mathcal{D}_{\alpha\gamma}$ onto $\mathcal{D}_{\beta\gamma}$ and $\mathcal{D}_{\gamma\alpha}$ onto $\mathcal{D}_{\gamma\beta}$. Thus $(\alpha\gamma) \succeq^{\mathcal{S}}_{\beta\mathfrak{fb}} (\beta\gamma)$ and $(\gamma\alpha) \succeq^{\mathcal{S}}_{\beta\mathfrak{fb}} (\gamma\beta)$, as desired. □

We can use the results above to establish fairly strong conservation results concerning quantifiers at different types.

LEMMA 6.3.12. *Let $\mathcal{S}$ be a signature and $B \subseteq \mathcal{T}$ be a set of types. Assume there is a type $(o\alpha^n \cdots \alpha^1) \in B$ and a type $(\iota\beta^m \cdots \beta^1) \in B$. Then for any type $\beta \in \mathcal{T}$ there is a type $\alpha \in \mathcal{T}_{gen}(B)$ (cf. Definition 2.1.46) such that $\alpha \succeq^{\mathcal{S}}_{\beta\mathfrak{fb}} \beta$.*

PROOF. We prove this by induction on α.

- o: By assumption $(o\alpha^n \cdots \alpha^1) \in B \subseteq \mathcal{T}_{gen}(B)$. By Lemma 6.3.9 we know $(o\alpha^n \cdots \alpha^1) \succeq^{\mathcal{S}}_{\beta\mathfrak{fb}} o$.
- ι: By assumption and Lemma 6.3.9, we know

$$(\iota\beta^m \cdots \beta^1) \in B \subseteq \mathcal{T}_{gen}(B)$$

and $(\iota\beta^m \cdots \beta^1) \succeq^{\mathcal{S}}_{\beta\mathfrak{fb}} \iota$.
- $(\gamma\delta)$: By the inductive hypothesis, there exist types

$$\gamma^1, \delta^1 \in \mathcal{T}_{gen}(B)$$

such that $\gamma^1 \succeq^{\mathcal{S}}_{\beta\mathfrak{fb}} \gamma$ and $\delta^1 \succeq^{\mathcal{S}}_{\beta\mathfrak{fb}} \delta$. We know $(\gamma^1\delta^1) \succeq^{\mathcal{S}}_{\beta\mathfrak{fb}} (\gamma\delta^1)$ and $(\gamma\delta^1) \succeq^{\mathcal{S}}_{\beta\mathfrak{fb}} (\gamma\delta)$ by Lemma 6.3.11. Hence $(\gamma^1\delta^1) \succeq^{\mathcal{S}}_{\beta\mathfrak{fb}} (\gamma\delta)$ by transitivity (cf. Lemma 6.3.7). Since $(\gamma^1\delta^1) \in \mathcal{T}_{gen}(B)$ we are done.

□

THEOREM 6.3.13. *Let $\mathcal{S}$ be a signature and $B \subseteq \mathcal{T}$ be a set of types. Assume there is a type*

$$(o\alpha^n \cdots \alpha^1) \in B$$

and a type

$$(\iota\beta^m \cdots \beta^1) \in B.$$

Suppose for every $\delta \in \mathcal{T}_{gen}(B)$ we have $\Pi^\delta \in \mathcal{S}$. Then

$$\mathcal{S} \trianglelefteq_{\beta\mathfrak{fb}} (\mathcal{S} \cup \{\Pi^\gamma \mid \gamma \in \mathcal{T}\}).$$

PROOF. By Lemma 6.3.12, for each type β there is an $\alpha \in \mathcal{T}_{gen}(B)$ such that $\alpha \succeq^{\mathcal{S}}_{\beta\mathfrak{fb}} \beta$. Hence for any $\Pi^\beta \in (\mathcal{S} \cup \{\Pi^\gamma \mid \gamma \in \mathcal{T}\})$ there is some $\Pi^\alpha \in \mathcal{S}$ such that $\alpha \succeq^{\mathcal{S}}_{\beta\mathfrak{fb}} \beta$. Thus we can apply Theorem 6.3.6 with $\mathcal{S}$ and $(\mathcal{S} \cup \{\Pi^\gamma \mid \gamma \in \mathcal{T}\})$ to conclude

$$\mathcal{S} \trianglelefteq_{\beta\mathfrak{fb}} (\mathcal{S} \cup \{\Pi^\gamma \mid \gamma \in \mathcal{T}\}).$$

□

In particular, we can apply Theorem 6.3.13 to the set of types generated by $(o\iota)$ and ι.

COROLLARY 6.3.14. *Let $\mathcal{S}$ be a signature such that $\Pi^\alpha \in \mathcal{S}$ for every $\alpha \in \mathcal{T}_{gen}(\{(o\iota), \iota\})$. Then*

$$\mathcal{S} \trianglelefteq_{\beta\mathfrak{fb}} (\mathcal{S} \cup \{\Pi^\gamma \mid \gamma \in \mathcal{T}\}).$$

PROOF. This follows from Theorem 6.3.13 using the types $(o\iota)$ and ι. □

For the next result we must assume η-reduction is preserved.

LEMMA 6.3.15. *Let $\mathcal{S}$ be a signature and $* \in \{\beta\eta, \beta\mathfrak{fb}\}$. For all types α and β, $(\alpha(\alpha(\alpha\beta))) \succeq^{\mathcal{S}}_{*} (\alpha\beta)$.*

PROOF. Let $\mathcal{M} = (\mathcal{D}, @, \mathcal{E}, v) \in \mathfrak{M}_*(\mathcal{S})$ be given. Let $h_{\alpha\beta}$, $k_{\alpha\beta}$, $p_{\alpha(\alpha\beta)}$, $r_{\alpha(\alpha(\alpha\beta))}$ and z_β be distinct variables. Define $\mathsf{f} \in \mathcal{D}_{\alpha(\alpha(\alpha\beta))(\alpha\beta)}$ by

$$\mathsf{f} := \mathcal{E}(\lambda h_{\alpha\beta}\, \lambda p_{\alpha(\alpha\beta)} \blacksquare p\, h).$$

Define $\mathsf{g} \in \mathcal{D}_{\alpha\beta(\alpha(\alpha(\alpha\beta)))}$ by

$$\mathsf{g} := \mathcal{E}(\lambda r_{\alpha(\alpha(\alpha\beta))} \lambda z_\beta \blacksquare r\, [\lambda k_{\alpha\beta} \blacksquare k\, z]).$$

Let $\mathsf{h} \in \mathcal{D}_{\alpha\beta}$ be given. Let φ be an assignment with $\varphi(h) := \mathsf{h}$. We compute

$$\begin{aligned}
\mathsf{g}@(\mathsf{f}@\mathsf{h}) &= \mathsf{g}@\mathcal{E}_\varphi(\lambda p_{\alpha(\alpha\beta)} \blacksquare p\, h) \\
&= \mathcal{E}_\varphi([[\lambda r_{\alpha(\alpha(\alpha\beta))} \lambda z_\beta \blacksquare r\, [\lambda k_{\alpha\beta} \blacksquare k\, z]]\, [\lambda p_{\alpha(\alpha\beta)} \blacksquare p\, h]]) \\
&= \mathcal{E}_\varphi([\lambda z_\beta \blacksquare [\lambda p_{\alpha(\alpha\beta)} \blacksquare p\, h]\, [\lambda k_{\alpha\beta} \blacksquare k\, z]]) \\
&= \mathcal{E}_\varphi([\lambda z_\beta \blacksquare [\lambda k_{\alpha\beta} \blacksquare k\, z]\, h]) \\
&= \mathcal{E}_\varphi([\lambda z_\beta \blacksquare h\, z]).
\end{aligned}$$

By η-functionality of $\mathcal{E}$,

$$\mathcal{E}_\varphi(\lambda z_\beta \blacksquare h\, z) = \mathcal{E}_\varphi(h) = \varphi(h) = \mathsf{h}.$$

Thus (g, f) is an internal retraction from $\mathcal{D}_{\alpha(\alpha(\alpha\beta))}$ onto $\mathcal{D}_{\alpha\beta}$. □

Note in particular that $(o(o(o\alpha))) \succeq^{\mathcal{S}}_{\beta\mathfrak{fb}} (o\alpha)$. From this it follows that

$$\mathcal{S} \cup \{\Pi^{o(o(o\alpha))}\} \trianglelefteq_{\beta\mathfrak{fb}} \mathcal{S} \cup \{\Pi^{o(o(o\alpha))}, \Pi^{o\alpha}\},$$

$$\mathcal{S} \cup \{\Sigma^{o(o(o\alpha))}\} \trianglelefteq_{\beta\mathfrak{fb}} \mathcal{S} \cup \{\Sigma^{o(o(o\alpha))}, \Sigma^{o\alpha}\}$$

and

$$\mathcal{S} \cup \{=^{o(o(o\alpha))}\} \trianglelefteq_{\beta\mathfrak{fb}} \mathcal{S} \cup \{=^{o(o(o\alpha))}, =^{o\alpha}\}$$

for any signature $\mathcal{S}$ and type α.

CONJECTURE 6.3.16. *We conjecture that there is a type α and equality-free signature $\mathcal{S}$ where $\Pi^{o\alpha} \in \mathcal{S}$ such that with η-conversion we have*

$$\mathcal{S} \trianglelefteq_{\beta\eta} \mathcal{S} \cup \{\Pi^{o(o(o\alpha))}\}$$

but without η-conversion we have

$$\mathcal{S} \ntrianglelefteq_{\beta} \mathcal{S} \cup \{\Pi^{o(o(o\alpha))}\}.$$

This would demonstrate necessity of η-conversion in Lemma 6.3.15. Also, if the conjecture is true, then the relations $\trianglelefteq_\beta$ and $\trianglelefteq_{\beta\eta}$ are different.

6.4. Boolean Logical Constants

We now analyze independence and conservation with respect to the constants in $\mathcal{S}_{Bool}$. First we prove the (unsurprising) result that we do not obtain any extra proof strength by adding any of the logical constants $\{\top, \bot, \wedge, \vee, \supset, \equiv\}$ if we already have either $\{\neg, \vee\}$ or $\{\neg, \wedge\}$ in the signature.

LEMMA 6.4.1. *Let* $* \in \{\beta, \beta\eta, \beta\mathfrak{fb}\}$ *and* $\mathcal{S} \subseteq \mathcal{S}_*$ *be a signature. Suppose* $\neg \in \mathcal{S}$ *and either* $\vee \in \mathcal{S}$ *or* $\wedge \in \mathcal{S}$. *Then* $\mathcal{S} \trianglelefteq_{\equiv *} (\mathcal{S} \cup \mathcal{S}_{Bool})$.

PROOF. We can prove this using Theorem 6.1.7 by verifying every $\mathcal{S}$-model realizes $\{\top, \bot, \wedge, \vee, \supset, \equiv\}$. Checking this is straightforward. □

Next we prove some independence results for the constants in $\mathcal{S}_{Bool}$.

LEMMA 6.4.2. *Let* $* \in \{\beta, \beta\eta, \beta\mathfrak{fb}\}$ *and* $\mathcal{S} \subseteq \mathcal{S}_*$ *be a signature such that* $\neg \notin \mathcal{S}$ *and* $\bot \notin \mathcal{S}$. *Then* $\mathcal{S} \ntrianglelefteq_{\not\equiv *} (\mathcal{S} \cup \{\neg\})$ *and* $\mathcal{S} \ntrianglelefteq_{\not\equiv *} (\mathcal{S} \cup \{\bot\})$.

PROOF. Let $\mathcal{M} := (\mathcal{D}, @, \mathcal{E}, id)$ be the singleton true $\mathcal{S}$-model constructed in Example 3.6.23. Obviously $\mathfrak{L}_\bot(\mathtt{T})$ fails. Also, $\mathfrak{L}_\neg(\mathsf{n})$ fails for the unique element in $\mathcal{D}_{oo}$ since $\mathsf{n}(\mathtt{T}) = \mathtt{T}$. That is, $\mathcal{M}$ does not realize $\bot$ and does not realize $\neg$. We know $\mathcal{S} \ntrianglelefteq_{\not\equiv *} \mathcal{S} \cup \{\neg\}$ and $\mathcal{S} \ntrianglelefteq_{\not\equiv *} \mathcal{S} \cup \{\bot\}$ by Corollary 6.1.8.

For a concrete example of a sentence witnessing the lack of conservation, note that $\mathcal{M} \not\models \neg\forall p_o\, p$. The sentence $\neg[\forall p_o\, p]$ is clearly provable if we have $\neg$ or $\bot$ available for instantiations. □

LEMMA 6.4.3. *Let* $* \in \{\beta, \beta\eta, \beta\mathfrak{fb}\}$ *and* $\mathcal{S}$ *be an equality-free signature such that* $\mathcal{S} \cap \{\top, \neg, \supset, \equiv\} = \emptyset$. *Then* $\mathcal{S} \ntrianglelefteq_{\not\equiv *} (\mathcal{S} \cup \{c\})$ *for each* $c \in \{\top, \neg, \supset, \equiv\}$. *Also,* $\mathcal{S} \ntrianglelefteq_{\not\equiv \beta\mathfrak{fb}} (\mathcal{S} \cup \{=^\alpha\})$ *for each type* α.

PROOF. Let $\mathcal{M} := (\mathcal{D}, @, \mathcal{E}, id)$ be the singleton false $\mathcal{S}$-model constructed in Example 3.6.24. We will prove $\mathcal{M}$ does not realize any constant in $\{\top, \neg, \supset, \equiv\} \cup \{=^\alpha \mid \alpha \in \mathcal{T}\}$.

$\top$: Obviously $\mathfrak{L}_\top(\mathtt{F})$ fails.

$\neg$: $\mathfrak{L}_\neg(\mathsf{n})$ fails for the unique element n in $\mathcal{D}_{oo}$ since $\mathsf{n}(\mathtt{F}) = \mathtt{F}$.

$\supset$, $\equiv$: Both $\mathfrak{L}_\supset(\mathsf{d})$ and $\mathfrak{L}_\equiv(\mathsf{d})$ fail for the only $\mathsf{d} \in \mathcal{D}_{ooo}$ since $\mathsf{d}(\mathtt{F})(\mathtt{F}) = \mathtt{F}$.

$=^\alpha$: $\mathfrak{L}_{=^\alpha}(\mathsf{q}^\alpha)$ fails for the unique $\mathsf{q}^\alpha \in \mathcal{D}_{o\alpha\alpha}$ since $\mathsf{q}^\alpha(\mathsf{a})(\mathsf{a}) = \mathtt{F}$ where a is the unique element of $\mathcal{D}_\alpha$.

By Corollary 6.1.8, we are done. □

6.5. Constants of Propositional Type

We now analyze what logical constants must be in the signature to obtain conservation over other logical constants that only involve the type o.

In the extensional case we will prove that we do not obtain any extra proof strength by adding equality or quantifiers over propositional types. The reason is that, in the extensional case, every domain $\mathcal{D}_\alpha$ of a propositional type is finite.

LEMMA 6.5.1. *For any extensional model $\mathcal{M} = (\mathcal{D}, @, \mathcal{E}, \upsilon)$ and propositional type $\alpha \in \mathcal{T}_o$ (cf. Definition 2.1.43), $\mathcal{D}_\alpha$ is finite.*

PROOF. We prove this by induction on α. $\mathcal{D}_o$ must have either one or two elements by property $\mathfrak{b}$. Suppose α is $\beta\gamma$ where β and γ are propositional types. By the inductive hypothesis, we know $\mathcal{D}_\beta$ and $\mathcal{D}_\gamma$ are finite. Consequently, the set of functions $\mathcal{D}_\beta^{\mathcal{D}_\gamma}$ is finite. Let $\epsilon : \mathcal{D}_{\beta\gamma} \to \mathcal{D}_\beta^{\mathcal{D}_\gamma}$ be the Curried version of @, defined by $\epsilon(\mathsf{f})(\mathsf{c}) := \mathsf{f}@\mathsf{c}$. Since $(\mathcal{D}, @)$ is functional, we know ϵ is injective. Thus $\mathcal{D}_{\beta\gamma}$ is also finite. □

We can now collect this into a theorem about quantifiers and equalities at propositional types.

THEOREM 6.5.2. *Let $\mathcal{M} = (\mathcal{D}, @, \mathcal{E}, \upsilon)$ be any extensional model.*

1. *If $\mathcal{M}$ realizes $\vee$, then $\mathcal{M}$ realizes Σ^α for any propositional type α.*
2. *If $\mathcal{M}$ realizes $\wedge$, then $\mathcal{M}$ realizes Π^α for any propositional type α.*
3. *If $\mathcal{M}$ realizes $\wedge$ and $\equiv$, then $\mathcal{M}$ realizes $=^\alpha$ for any propositional type α.*

PROOF. First, by Lemma 6.5.1, we know $\mathcal{D}_\alpha$ is finite for every propositional type α. Consequently, the first two parts of the theorem follow directly from Lemma 6.2.5.

We prove the result about $=^\alpha$ by induction on propositional types. Since $\mathcal{M}$ realizes $\equiv$, there is some $\mathsf{e} \in \mathcal{D}_{ooo}$ such that $\mathfrak{L}_\equiv(\mathsf{e})$ holds. Since υ is injective (by property $\mathfrak{b}$), $\mathfrak{L}_\equiv(\mathsf{e})$ implies $\mathfrak{L}_{=^o}(\mathsf{e})$ and so the base case $\alpha = o$ holds.

If $\alpha = \gamma\beta$, then by the inductive hypothesis $\mathcal{M}$ realizes $=^\gamma$. Also, $\mathcal{M}$ realizes Π^β by Lemma 6.2.5. Therefore, $\mathcal{M}$ realizes $=^{\gamma\beta}$ by Lemma 6.2.1. □

We have the following corollary about (extensional) conservation.

COROLLARY 6.5.3. *Let $\mathcal{S}$ be any signature. If $\neg \in \mathcal{S}$ and either $\wedge \in \mathcal{S}$ or $\vee \in \mathcal{S}$, then*

$$\mathcal{S} \trianglelefteq_{\beta\mathfrak{fb}} (\mathcal{S} \cup \mathcal{S}_{Bool} \cup \{\Pi^\alpha, \Sigma^\alpha, =^\alpha \mid \alpha \in \mathcal{T}_o\}).$$

In particular,

$$\{\neg, \vee\} \trianglelefteq_{\beta\mathfrak{fb}} (\mathcal{S}_{Bool} \cup \{\Pi^\alpha, \Sigma^\alpha, =^\alpha \mid \alpha \in \mathcal{T}_o\}).$$

PROOF. Let $\mathcal{S}_1 := \mathcal{S} \cup \{\top, \bot, \neg, \wedge, \vee, \supset, \equiv\}$. By Lemma 6.4.1, we know $\mathcal{S} \trianglelefteq_{\beta\mathfrak{fb}} \mathcal{S}_1$. By transitivity of $\trianglelefteq_{\beta\mathfrak{fb}}$, it is enough to verify

$$\mathcal{S}_1 \trianglelefteq_{\beta\mathfrak{fb}} (\mathcal{S}_1 \cup \{\Pi^\alpha, \Sigma^\alpha, =^\alpha \mid \alpha \in \mathcal{T}_o\}).$$

Let $\mathcal{M} = (\mathcal{D}, @, \mathcal{E}, \upsilon) \in \mathfrak{M}_{\beta\mathfrak{fb}}(\mathcal{S}_1)$ be given. By Theorem 6.5.2, we know $\mathcal{M}$ realizes

$$\{\Pi^\alpha, \Sigma^\alpha, =^\alpha \mid \alpha \in \mathcal{T}_o\}$$

as desired. By Theorem 6.1.7, we are done. □

6.6. Complete Sets of Logical Constants

In order to restrict the logical constants one needs to use for set instantiations without sacrificing completeness, we determine certain restricted signatures have the same proof strength as the signature $\mathcal{S}_{all}$ with all logical constants.

DEFINITION 6.6.1. We say a signature $\mathcal{S}$ is *extensionally complete* if $\mathcal{S} \trianglelefteq_{\beta\mathfrak{fb}} \mathcal{S}_{all}$.

Clearly, if $\mathcal{S}$ is extensionally complete, then so is any other signature $\mathcal{S}_1$ with $\mathcal{S} \subseteq \mathcal{S}_1 \subseteq \mathcal{S}_{all}$.

For our first example of an extensionally complete signature, we take $\mathcal{S}_{all}^{elem} \cup \{=^\iota\}$. That is, we only need equality at the base type ι to ensure extensional completeness of the signature.

THEOREM 6.6.2. *The signature $\mathcal{S}_{all}^{elem} \cup \{=^\iota\}$ is extensionally complete.*

PROOF. We use Theorem 6.1.7 to prove $\mathcal{S}_{all}^{elem} \cup \{=^{\iota}\} \trianglelefteq_{\beta\mathfrak{fb}} \mathcal{S}_{all}$. We must check each $\mathcal{M} \in \mathfrak{M}_{\beta\mathfrak{fb}}(\mathcal{S}_{all}^{elem} \cup \{=^{\iota}\})$ realizes every logical constant in $\mathcal{S}_{all} \setminus (\mathcal{S}_{all}^{elem} \cup \{=^{\iota}\})$. Since every Π^{α} is in $\mathcal{S}_{all}^{elem}$ and every member of $\mathcal{S}_{all} \setminus (\mathcal{S}_{all}^{elem} \cup \{=^{\iota}\})$ is of the form $=^{\beta}$, this follows from Lemma 6.2.4. □

On the other hand, $\mathcal{S}_{all}^{elem}$ is not extensionally complete. To prove this, it is enough to construct an extensional $\mathcal{S}_{all}^{elem}$-model which does not satisfy some $\mathbf{M} \in cwff_o(\mathcal{S}_{all}^{elem})$ which is a theorem of $\mathcal{G}_{\beta\mathfrak{fb}}^{\mathcal{S}_{all}}$. Andrews constructed such a model (for another purpose) in [3].

REMARK 6.6.3. In [3] Andrews constructs a frame $\mathscr{D}$ with an evaluation function $\mathscr{V}$ such that $\mathscr{D}_o = \{\mathtt{T}, \mathtt{F}\}$, $\mathscr{D}_{\iota} = \{\mathsf{l}, \mathsf{m}, \mathsf{n}\}$, $\mathscr{D}_{o\iota}$ contains only the two constant functions from $\mathscr{D}_{\iota}$ to $\mathscr{D}_o$, and $\mathscr{D}_{\iota\iota}$ contains only the identity function and the three constant functions from $\mathscr{D}_{\iota}$ to $\mathscr{D}_{\iota}$. Andrews proves this is a model in the sense of [31]. In our terms, $\mathscr{D}$ is a combinatory frame realizing $\mathcal{S}_{Ch}$, the signature used in Church's formulation of type theory. Also, $\mathscr{V}$ is an $\mathcal{S}_{Ch}$-evaluation function for $\mathscr{D}$. We can define an $\mathcal{S}_{Ch}$-model $\mathcal{M} := (\mathscr{D}, @, \mathscr{V}, v)$ by taking @ to be function application and v to be the identity. By Lemma 6.2.3, $\mathcal{M}$ realizes Σ^{α} for every type α. By Corollary 6.5.3 and Theorem 6.1.7, we know $\mathcal{M}$ realizes $\mathcal{S}_{Bool}$. Thus $\mathcal{M}$ realizes $\mathcal{S}_{all}^{elem}$.

Andrews proves his model does not satisfy the theorem

$$\dot{\forall} f_{\iota\iota}\, \dot{\forall} g_{\iota\iota} \,\blacksquare\, \dot{\forall} x_{\iota}\, [f\, x \doteq^{\iota} g\, x] \supset f \doteq^{\iota\iota} g.$$

Since the model does not satisfy this theorem (which is a formulation of functional extensionality), Andrews calls this a "nonextensional" model. It is, however, an extensional model (and, in fact, a Henkin model) in the sense of Definition 3.3.8 since it is a model over a frame, $\mathcal{D}_o$ is $\{\mathtt{T}, \mathtt{F}\}$ and v is the identity. Combining this result with extensional completeness of $\mathcal{S}_{all}^{elem} \cup \{=^{\iota}\}$, we know the model in [3] must not realize $=^{\iota}$. Since $\mathscr{D}_{o\iota}$ contains only the constant true and constant false functions, it is easy to check directly that the model does not realize $=^{\iota}$ since otherwise the domain $\mathscr{D}_{o\iota}$ would contain the characteristic functions corresponding to the singletons for the elements in $\mathscr{D}_{\iota}$.

Andrews also proves $\mathscr{V}_{\varphi}([x_{\iota} \doteq^{\iota} y_{\iota}]) = \mathtt{T}$ for any variable assignment φ. However, if $\varphi(x_{\iota}) = \mathsf{l}$ and $\varphi(y_{\iota}) = \mathsf{m}$, then we know

$$\mathscr{V}_{\varphi}([x_{\iota} \doteq^{\iota} y_{\iota}]) = \mathtt{F}$$

since $\mathsf{l} \neq \mathsf{m}$. Thus this model fails to satisfy

$$\dot{\forall} x_\iota \dot{\forall} y_\iota \centerdot x \dot{\doteq}^\iota y \mathbin{\dot{\supset}} x \mathbin{\dot{=}}^\iota y.$$

Consequently, if a model does not realize $=^\iota$, then Leibniz equality at type ι may be different from primitive equality.

REMARK 6.6.4 (Equality in Set Substitutions). Consider the following flawed argument intended to prove $\mathcal{S}_{all}^{elem}$ *is* complete. If we are given any $\mathcal{S}_{all}^{elem} \cup \{=^\iota\}$-derivation $\mathcal{D}$ of a theorem $\mathbf{M} \in cwff_o(\mathcal{S}_{all}^{elem})$, then we simply replace $=^\iota$ in any instantiation in $\mathcal{D}$ with Leibniz equality $\doteq^\iota$ to obtain an $\mathcal{S}_{all}^{elem}$-derivation $\mathcal{E}$ of $\mathbf{M}$. The following example demonstrates why this technique fails in practice. Let $\mathbf{M}$ be the theorem $[A_\iota \dot{\doteq}^\iota B_\iota] \mathbin{\dot{\supset}} [A \mathbin{\dot{=}}^\iota B]$. Let $\mathcal{D}$ be the short $\mathcal{S}_{all}^{elem} \cup \{=^\iota\}$-derivation

$$\dfrac{\dfrac{\dfrac{\dfrac{\dfrac{}{[A \mathbin{\dot{=}} A], [A \mathbin{\dot{=}} B]}\,\mathcal{G}(Refl) \qquad \dfrac{}{\dot{\neg}[A \mathbin{\dot{=}} B], [A \mathbin{\dot{=}} B]}\,\mathcal{G}(GInit)}{\dot{\neg}[A \mathbin{\dot{=}} A \mathbin{\dot{\supset}} A \mathbin{\dot{=}} B], [A \mathbin{\dot{=}} B]}\,\mathcal{G}(\dot{\neg}\dot{\supset})}{\dot{\neg}[[[\lambda x_\iota \centerdot A \mathbin{\dot{=}} x]\, A] \mathbin{\dot{=}} [[\lambda x \centerdot A \mathbin{\dot{=}} x]\, B]], [A \mathbin{\dot{=}} B]}\,\mathcal{G}(\beta\eta)}{\dot{\neg}\dot{\forall} q_{o\iota}\, [q\, A \mathbin{\dot{\supset}} q\, B], A \mathbin{\dot{=}} B}\,\mathcal{G}(\dot{\neg}\dot{\forall})}{[A \dot{\doteq}^\iota B] \mathbin{\dot{\supset}} [A \mathbin{\dot{=}}^\iota B]}\,\mathcal{G}(\dot{\supset})$$

of $\mathbf{M}$ where the formula $[\lambda x_\iota \centerdot A \mathbin{\dot{=}}^\iota x]$ is used in the $\mathcal{G}(\dot{\neg}\dot{\forall})$ rule. Suppose we try to replace $=^\iota$ in this formula with $\doteq^\iota$. We obtain the partial derivation

$$\dfrac{\dfrac{\dfrac{\dfrac{\begin{array}{c}\vdots\\ [A \dot{\doteq} A], [A \mathbin{\dot{=}} B]\end{array} \qquad \begin{array}{c}\vdots\\ \dot{\neg}[A \dot{\doteq} B], [A \mathbin{\dot{=}} B]\end{array}}{\dot{\neg}[A \dot{\doteq} A \mathbin{\dot{\supset}} A \dot{\doteq} B], [A \mathbin{\dot{=}} B]}\,\mathcal{G}(\dot{\neg}\dot{\supset})}{\dot{\neg}[[[\lambda x_\iota \centerdot A \dot{\doteq} x]\, A] \mathbin{\dot{=}} [[\lambda x \centerdot A \dot{\doteq} x]\, B]], [A \mathbin{\dot{=}} B]}\,\mathcal{G}(\beta\eta)}{\dot{\neg}\dot{\forall} q_{o\iota}\, [q\, A \mathbin{\dot{\supset}} q\, B], A \mathbin{\dot{=}} B}\,\mathcal{G}(\dot{\neg}\dot{\forall})}{[A \dot{\doteq}^\iota B] \mathbin{\dot{\supset}} [A \mathbin{\dot{=}}^\iota B]}\,\mathcal{G}(\dot{\supset})$$

This leaves us two goals: $[A \dot{\doteq} A], [A \mathbin{\dot{=}} B]$ and $\dot{\neg}[A \dot{\doteq} B], [A \mathbin{\dot{=}} B]$. Deriving $[A \dot{\doteq} A], [A \mathbin{\dot{=}} B]$ is easy, but the goal $\dot{\neg}[A \dot{\doteq} B], [A \mathbin{\dot{=}} B]$ is essentially the same goal with which we started. The procedure is essentially in a loop.

A similar loop occurs when trying to prove

$$\dot{\forall} f_{\iota\iota}\, \dot{\forall} g_{\iota\iota} \,\blacksquare\, \dot{\forall} x_\iota\, [f\, x \doteq^\iota g\, x] \mathbin{\dot{\supset}} f \doteq^{\iota\iota} g$$

without using $=^\iota$.

Andrews' system $\mathcal{Q}_0$ [9] is a formulation of higher-order logic using equality at every type (and a description operator). This suggests the signature $\mathcal{S}^=$ is extensionally complete. However, this is not quite true. Andrews' **Axiom 1** guarantees $\mathcal{D}_o$ has two elements. This allows one to define $\bot$ in terms of equality as $[\lambda x_o \blacksquare \top] =^{oo} [\lambda x_o \blacksquare x]$ (after defining $\top$ in terms of equality). However, we have no such explicit axiom in $\mathcal{G}^{\mathcal{S}^=}_{\beta\mathfrak{fb}}$. Indeed, the singleton true model in Example 3.6.23 realizes equality at every type but does not realize $\bot$ or $\neg$. To obtain an extensionally complete signature, we must explicitly include $\bot$ or $\neg$ in the signature.

LEMMA 6.6.5. *Let $\mathcal{M} = (\mathcal{D}, @, \mathcal{E}, v)$ be an extensional $\mathcal{S}^=$-model. If $\mathcal{M}$ realizes $\neg$ or $\bot$, then $\mathcal{M}$ realizes $\mathcal{S}_{all}$.*

PROOF. The remaining logical constants in $\mathcal{S}_{all}$ are defined in a manner similar to the definitions given in [9]. See Appendix A.4. □

THEOREM 6.6.6. *The signatures $\mathcal{S}^= \cup \{\bot\}$ and $\mathcal{S}^= \cup \{\neg\}$ are extensionally complete.*

PROOF. Suppose $\mathcal{M} = (\mathcal{D}, @, \mathcal{E}, v) \in \mathfrak{M}_{\beta\mathfrak{fb}}(\mathcal{S}^= \cup \{\bot\})$. Then

$$\mathcal{M}' = (\mathcal{D}, @, \mathcal{E}\big|_{\mathit{wff}(\mathcal{S}^=)}, @)$$

is an extensional $\mathcal{S}^=$-model realizing $\bot$. By Lemma 6.6.5, $\mathcal{M}'$ realizes $\mathcal{S}_{all}$. Hence $\mathcal{M}$ realizes $\mathcal{S}_{all}$. By Theorem 6.1.7, we know $\mathcal{S}^= \cup \{\bot\}$ is extensionally complete.

Similarly, if $\mathcal{M} = (\mathcal{D}, @, \mathcal{E}, v) \in \mathfrak{M}_{\beta\mathfrak{fb}}(\mathcal{S}^= \cup \{\neg\})$, then we can apply Lemma 6.6.5 and Theorem 6.1.7 to conclude $\mathcal{S}^= \cup \{\neg\}$ is extensionally complete. □

Finally, we consider extensionally complete signatures which may not include quantifiers over propositional types or over types of the form (ιo). In particular, we only require quantifiers for types in $\mathcal{T}_{gen}(\{(o\iota), \iota\})$ (cf. Definition 2.1.46).

THEOREM 6.6.7. *Any signature $\mathcal{S}$ such that*

$$\{\neg, \vee, =^\iota\} \cup \{\Pi^\alpha \mid \alpha \in \mathcal{T}_{gen}(\{(o\iota), \iota\})\} \subseteq \mathcal{S}$$

is extensionally complete.

PROOF. Let $\mathcal{M}$ be an extensional $\mathcal{S}$-model. By Lemma 6.4.1, we know $\mathcal{S} \trianglelefteq_{\beta\mathfrak{fb}} \mathcal{S} \cup \mathcal{S}_{Bool}$. Combining this with Corollary 6.3.14 we conclude

$$\mathcal{S} \trianglelefteq_{\beta\mathfrak{fb}} \mathcal{S} \cup \mathcal{S}_{Bool} \cup \{\Pi^{\alpha} \mid \alpha \in \mathcal{T}\}.$$

By Theorem 6.1.7, we know $\mathcal{M}$ realizes

$$\mathcal{S} \cup \mathcal{S}_{Bool} \cup \{\Pi^{\alpha} \mid \alpha \in \mathcal{T}\}.$$

In particular, $\mathcal{M}$ realizes Π^{α} for every type α. By Lemma 6.2.3, $\mathcal{M}$ realizes Σ^{α} for every type α. By Lemma 6.2.4, $\mathcal{M}$ realizes $=^{\alpha}$ for every type α. Hence $\mathcal{M}$ realizes $\mathcal{S}_{all}$. By Theorem 6.1.7, we conclude $\mathcal{S}$ is extensionally complete. □

6.7. Cantor's Theorem

Cantor's theorem intuitively means that the power set of a given set A must be "bigger" than A. Consider two versions of Cantor's theorem which can be expressed in higher-order logic.

X5304$_{\alpha}$: $\neg \exists g_{o\alpha\alpha} \forall f_{o\alpha} \exists j_{\alpha} \centerdot g\, j \doteq^{o\alpha} f$
X5309: $\neg \exists h_{\iota(o\iota)} \forall p_{o\iota} \forall q_{o\iota} \centerdot h\, p \doteq^{\iota} h\, q \supset p \doteq^{o\iota} q$

X5304 is a *surjective* version of Cantor's theorem (for each type α). It states that there does not exist a surjection from a set onto its power set. **X5309** is an *injective* version of Cantor's theorem (discussed in [11]), stating that there does not exist an injection from the power set into the original set.

TPS can easily prove **X5304** automatically using a set substitution that involves a negation. A natural deduction proof for an "internal" version of **X5304**$_{\iota}$ was shown earlier in Figure 2.1. This proof suggests **X5304** is derivable in $\mathcal{G}^{\{\neg\}}_{\beta\mathfrak{fb}}$. We verify this is true by giving an explicit derivation.

LEMMA 6.7.1. *For each type α there is a $\mathcal{G}^{\{\neg\}}_{\beta\mathfrak{fb}}$-derivation of* **X5304**$_{\alpha}$.

PROOF. In Figure 6.1 we show a derivation of **X5304**$_{\alpha}$ (using admissible and derived rules) in the calculus $\mathcal{G}^{\{\neg\}}_{\beta\mathfrak{fb}}$. □

$$
\dfrac{
\dfrac{
\dfrac{
\dfrac{
\dfrac{
\dfrac{
\dfrac{\overbrace{\underset{\cdot}{\neg}[G\,J\,J],\underset{\cdot}{\neg}\underset{\cdot}{\neg}\,[G_{o\alpha\alpha}\,J\,J]}^{\mathcal{G}(GInit)}}{\underset{\cdot}{\neg}[G\,J\,J],\underset{\cdot}{\neg}\neg\,[G_{o\alpha\alpha}\,J\,J]}\,\mathcal{G}(\underset{\cdot}{\neg}\sharp)
\qquad
\dfrac{\overbrace{[G\,J\,J],\underset{\cdot}{\neg}\,[G_{o\alpha\alpha}\,J\,J]}^{\mathcal{G}(GInit)}}{[G\,J\,J],\neg\,[G_{o\alpha\alpha}\,J\,J]}\,\mathcal{G}(\sharp)
}{\underset{\cdot}{\neg}\blacksquare[G\,J\,J]\ \underset{\cdot}{=}^{o}\ \neg\,[G_{o\alpha\alpha}\,J\,J]}\,\mathcal{G}(\underset{\cdot}{\neg}\,\underset{\cdot}{=}^{o})
}{\underset{\cdot}{\neg}\blacksquare[G\,J\,J]\ \underset{\cdot}{=}^{o}\ [[\lambda x_\alpha.\neg\,G_{o\alpha\alpha}\,x\,x]\,J]}\,\mathcal{G}(\beta\eta)
}{\underset{\cdot}{\neg}\,G\,J\ \underset{\cdot}{=}^{o\alpha}\ [\lambda x_\alpha.\neg\,G_{o\alpha\alpha}\,x\,x]}\,\mathcal{G}(\underset{\cdot}{\neg}\,\underset{\cdot}{=}^{\rightarrow})
}{\underset{\cdot}{\neg}\underset{\cdot}{\exists}\,j_\alpha.G\,j\ \underset{\cdot}{=}^{o\alpha}\ [\lambda x_\alpha.\neg\,G_{o\alpha\alpha}\,x\,x]}\,\mathcal{G}(\underset{\cdot}{\neg}\underset{\cdot}{\exists}^{J})
}{\underset{\cdot}{\neg}\underset{\cdot}{\forall}\,f_{o\alpha}\underset{\cdot}{\exists}\,j_\alpha.G\,j\ \underset{\cdot}{=}\ f}\,\mathcal{G}(\underset{\cdot}{\neg}\underset{\cdot}{\forall},\{\neg\})
}{\underset{\cdot}{\neg}\underset{\cdot}{\exists}\,g_{o\alpha\alpha}\underset{\cdot}{\forall}\,f_{o\alpha}\underset{\cdot}{\exists}\,j_\alpha.g\,j\ \underset{\cdot}{=}\ f}\,\mathcal{G}(\underset{\cdot}{\neg}\underset{\cdot}{\exists}^{G})
$$

FIGURE 6.1. Proof of **X5304**

While TPS cannot currently automatically prove the injective version **X5309**, we do have proofs constructed interactively. One of these proofs is shown in Figures 6.2 and 6.3.

The set instantiation involved in line (5) of the proof of **X5309** involves $\neg$, $\wedge$, and a quantifier $\Sigma^{o\iota}$ and an equality $=^{\iota}$. This suggests **X5309** has a proof in $\mathcal{G}_{\beta\mathfrak{fb}}^{\{\neg,\wedge,=^{\iota},\Sigma^{o\iota}\}}$. We prove there is such a derivation by a semantic argument (relying on completeness).

LEMMA 6.7.2. *There is a* $\mathcal{G}_{\beta\mathfrak{fb}}^{\{\neg,\wedge,=^{\iota},\Sigma^{o\iota}\}}$*-derivation of* **X5309**.

PROOF. Suppose there is no such derivation. Then by completeness (cf. Theorem 5.7.19) there is an extensional $\{\neg,\wedge,=^{\iota},\Sigma^{o\iota}\}$-model $\mathcal{M}$ such that $\mathcal{M}\not\models$ **X5309**. By Corollary 3.5.7, we can assume $\mathcal{M}$ is a model $(\mathcal{D},@,\mathcal{E},\upsilon)$ over a frame, $\mathcal{D}_o=\{\mathtt{T},\mathtt{F}\}$ and υ is the identity function. Since @ is simply function application, we will use the usual notation for function application and suppress @.

Since $\mathcal{M}\models\underset{\cdot}{\exists}\,h_{\iota(o\iota)}\underset{\cdot}{\forall}\,p_{o\iota}\underset{\cdot}{\forall}\,q_{o\iota}.h\,p\ \underset{\cdot}{=}^{\iota}\ h\,q\ \underset{\cdot}{\supset}\ p\ \underset{\cdot}{=}^{o\iota}\ q$ there is some element $\mathsf{h}\in\mathcal{D}_{\iota(o\iota)}$ such that $\mathsf{p}=\mathsf{q}$ whenever $\mathsf{h}(\mathsf{p})=\mathsf{h}(\mathsf{q})$ for $\mathsf{p},\mathsf{q}\in\mathcal{D}_{o\iota}$.

(1)	1	$\vdash$	$\neg\neg\exists h_{\iota(o\iota)}\forall p_{o\iota}\forall q_{o\iota}.h\,p = h\,q \supset p = q$	Assume negation
(2)	1	$\vdash$	$\exists h_{\iota(o\iota)}\forall p_{o\iota}\forall q_{o\iota}.h\,p = h\,q \supset p = q$	Neg: 1
(3)	1,3	$\vdash$	$\forall p_{o\iota}\forall q_{o\iota}.h_{\iota(o\iota)}\,p = h\,q \supset p = q$	Choose: $h_{\iota(o\iota)}$ 2
(4)		$\vdash$	$\lambda y_\iota\exists w_{o\iota}[h_{\iota(o\iota)}\,w = y \wedge \neg w\,y] = \lambda y\exists w.h\,w = y \wedge \neg w\,y$	Assert REFL=
(5)		$\vdash$	$\exists z_{o\iota}.z = \lambda y_\iota\exists w_{o\iota}.h_{\iota(o\iota)}\,w = y \wedge \neg w\,y$	EGen: $[\lambda y_\iota\exists w_{o\iota}.h_{\iota(o\iota)}\,w = y \wedge \neg w\,y]$ 4
(6)	6	$\vdash$	$z_{o\iota} = \lambda y_\iota\exists w_{o\iota}.h_{\iota(o\iota)}\,w = y \wedge \neg w\,y$	Hyp
(7)	7	$\vdash$	$z_{o\iota}.h_{\iota(o\iota)}\,z$	Hyp
(8)	6	$\vdash$	$[\lambda f_{o\iota}\lambda g_{o\iota}\forall x_\iota.f\,x = g\,x]\,z_{o\iota}.\lambda y_\iota\exists w_{o\iota}.\,h_{\iota(o\iota)}\,w = y \wedge\neg w\,y$	Equiv-eq: 6
(9)	6	$\vdash$	$\forall x_\iota.z_{o\iota}\,x = \exists w_{o\iota}.h_{\iota(o\iota)}\,w = x \wedge \neg w\,x$	Lambda: 8
(10)	6	$\vdash$	$z_{o\iota}[h_{\iota(o\iota)}\,z] = \exists w_{o\iota}.h\,w = h\,z \wedge \neg w.h\,z$	UI: $[h_{\iota(o\iota)}\,z_{o\iota}]$ 9
(11)	6	$\vdash$	$z_{o\iota}[h_{\iota(o\iota)}\,z] \equiv \exists w_{o\iota}.h\,w = h\,z \wedge \neg w.h\,z$	Equiv-eq: 10
(12)	6,7	$\vdash$	$\exists w_{o\iota}.h_{\iota(o\iota)}\,w = h\,z_{o\iota} \wedge \neg w.h\,z$	RuleP: 7 11
(13)	6,7,13	$\vdash$	$h_{\iota(o\iota)}\,w_{o\iota} = h\,z_{o\iota} \wedge \neg w.h\,z$	Choose: $w_{o\iota}$ 12
(14)	1,3	$\vdash$	$\forall q_{o\iota}.h_{\iota(o\iota)}\,w_{o\iota} = h\,q \supset w = q$	UI: $w_{o\iota}$ 3
(15)	1,3	$\vdash$	$h_{\iota(o\iota)}\,w_{o\iota} = h\,z_{o\iota} \supset w = z$	UI: $z_{o\iota}$ 14
(16)	6,7,13	$\vdash$	$h_{\iota(o\iota)}\,w_{o\iota} = h\,z_{o\iota}$	Conj: 13
(17)	6,7,13	$\vdash$	$\neg w_{o\iota}.h_{\iota(o\iota)}\,z_{o\iota}$	Conj: 13
(18)	6,7,13,1,3	$\vdash$	$w_{o\iota} = z_{o\iota}$	MP: 16 15
(19)	6,7,13,1,3	$\vdash$	$[\lambda f_{o\iota}\lambda g_{o\iota}\forall x_\iota.f\,x = g\,x]\,w_{o\iota}\,z_{o\iota}$	Equiv-eq: 18

FIGURE 6.2. Proof of **X5309** Lines 1-19

(20)	6,7,13,1,3	$\vdash$	$\forall x_\iota \,.\, w_{o\iota}\, x = z_{o\iota}\, x$	Lambda: 19
(21)	6,7,13,1,3	$\vdash$	$w_{o\iota}[h_{\iota(o\iota)}\, z_{o\iota}] = z \,.\, h\, z$	UI: $[h_{\iota(o\iota)}\, z_{o\iota}]$ 20
(22)	6,7,13,1,3	$\vdash$	$w_{o\iota}[h_{\iota(o\iota)}\, z_{o\iota}] \equiv z \,.\, h\, z$	Equiv-eq: 21
(23)	1,3,6,7,13	$\vdash$	$\neg\, z_{o\iota} \,.\, h_{\iota(o\iota)}\, z$	RuleP: 17 22
(24)	1,3,6,7	$\vdash$	$\neg\, z_{o\iota} \,.\, h_{\iota(o\iota)}\, z$	RuleC: 12 23
(25)	1,3,6	$\vdash$	$z_{o\iota}[h_{\iota(o\iota)}\, z] \supset \neg\, z \,.\, h\, z$	Deduct: 24
(26)	26	$\vdash$	$\neg\, z_{o\iota} \,.\, h_{\iota(o\iota)}\, z$	Hyp
(27)	6,26	$\vdash$	$\neg\exists\, w_{o\iota} \,.\, h_{\iota(o\iota)}\, w = h\, z_{o\iota} \wedge \neg\, w \,.\, h\, z$	RuleP: 11 26
(28)	6,26	$\vdash$	$\forall\, w_{o\iota} .\neg .\, h_{\iota(o\iota)}\, w = h\, z_{o\iota} \wedge \neg\, w \,.\, h\, z$	Neg: 27
(29)	6,26	$\vdash$	$\neg .\, h_{\iota(o\iota)}\, z_{o\iota} = h\, z \wedge \neg\, z \,.\, h\, z$	UI: $z_{o\iota}$ 28
(30)	6,26	$\vdash$	$\neg[h_{\iota(o\iota)}\, z_{o\iota} = h\, z] \vee z \,.\, h\, z$	Neg: 29
(31)	6,26	$\vdash$	$h_{\iota(o\iota)}\, z_{o\iota} = h\, z \supset z \,.\, h\, z$	Disj-Imp: 30
(32)		$\vdash$	$h_{\iota(o\iota)}\, z_{o\iota} = h\, z$	Assert REFL=
(33)	6,26	$\vdash$	$z_{o\iota} \,.\, h_{\iota(o\iota)}\, z$	MP: 32 31
(34)	6	$\vdash$	$\neg\, z_{o\iota}[h_{\iota(o\iota)}\, z] \supset z \,.\, h\, z$	Deduct: 33
(35)	1,3,6	$\vdash$	$[z_{o\iota}[h_{\iota(o\iota)}\, z] \supset \neg\, z \,.\, h\, z] \wedge .\neg\, z[h\, z] \supset z \,.\, h\, z$	Conj: 25 34
(36)	1,3,6	$\vdash$	$z_{o\iota}[h_{\iota(o\iota)}\, z] \equiv \neg\, z \,.\, h\, z$	ImpEquiv: 35
(37)	1,3,6	$\vdash$	$\bot$	RuleP: 36
(38)	1,3	$\vdash$	$\bot$	RuleC: 5 37
(39)	1	$\vdash$	$\bot$	RuleC: 2 38
(40)		$\vdash$	$\neg\exists\, h_{\iota(o\iota)} \forall\, p_{o\iota} \forall\, q_{o\iota} \,.\, h\, p = h\, q \supset p = q$	Indirect: 39

FIGURE 6.3. Proof of **X5309** Lines 20-40

Let $h_{o\iota}$ be a variable and φ be an assignment with $\varphi(h) = \mathsf{h}$. We define a "diagonal set"

$$\mathsf{d} := \mathcal{E}_\varphi([\lambda y_\iota \,\blacksquare\, \exists w_{o\iota} \,\blacksquare\, h\, w =^\iota y \wedge \neg[w\, y]]) \in \mathcal{D}_{o\iota}.$$

Note that

$$\mathsf{d}(\mathsf{h}(\mathsf{d})) = \mathcal{E}_{\varphi,[\mathsf{h}(\mathsf{d})/y]}(\exists w_{o\iota} \,\blacksquare\, h\, w =^\iota y \wedge \neg[w\, y]).$$

Assume $\mathsf{d}(\mathsf{h}(\mathsf{d})) = \mathtt{F}$. Then

$$\mathcal{E}_{\varphi,[\mathsf{h}(\mathsf{d})/y],[\mathsf{d}/w]}(\neg[w\,y]) = (\mathcal{E}(\neg)@\mathsf{d}(\mathsf{h}(\mathsf{d}))) = \mathtt{T}$$

by $\mathfrak{L}_\neg(\mathcal{E}(\neg))$. By $\mathfrak{L}_{=^\iota}(\mathcal{E}(=^\iota))$, we know $\mathcal{E}_{\varphi,[\mathsf{h}(\mathsf{d})/y],[\mathsf{d}/w]}(h\,w =^\iota y) = \mathtt{T}$. By $\mathfrak{L}_\wedge(\mathcal{E}(\wedge))$, we know

$$\mathcal{E}_{\varphi,[\mathsf{h}(\mathsf{d})/y],[\mathsf{d}/w]}([h\,w =^\iota y \wedge \neg[w\,y]]) = \mathtt{T}.$$

Finally, by $\mathfrak{L}_{\Sigma^{o\iota}}(\mathcal{E}(\Sigma^{o\iota}))$ we conclude

$$\mathsf{d}(\mathsf{h}(\mathsf{d})) = \mathcal{E}_{\varphi,[\mathsf{h}(\mathsf{d})/y]}([\exists w_{o\iota} \centerdot h\,w =^\iota y \wedge \neg[w\,y]]) = \mathtt{T}$$

which contradicts our assumption.

Hence we must have $\mathsf{d}(\mathsf{h}(\mathsf{d})) = \mathtt{T}$. By $\mathfrak{L}_{\Sigma^{o\iota}}(\mathcal{E}(\Sigma^{o\iota}))$, there must exist some $\mathsf{w} \in \mathcal{D}_{o\iota}$ such that

$$\mathcal{E}_{\varphi,[\mathsf{h}(\mathsf{d})/y],[\mathsf{w}/w]}([h\,w =^\iota y \wedge \neg[w\,y]]) = \mathtt{T}.$$

This implies $\mathcal{E}_{\varphi,[\mathsf{h}(\mathsf{d})/y],[\mathsf{w}/w]}(\neg[w\,y]) = \mathtt{T}$ and

$$\mathsf{h}(\mathsf{w}) = \mathcal{E}_{\varphi,[\mathsf{h}(\mathsf{d})/y],[\mathsf{w}/w]}(h\,w) = \mathcal{E}_{\varphi,[\mathsf{h}(\mathsf{d})/y],[\mathsf{w}/w]}(y) = \mathsf{h}(\mathsf{d}).$$

Since h is an injective function, $\mathsf{w} = \mathsf{d}$. Thus $\mathsf{w}(\mathsf{h}(\mathsf{d})) = \mathsf{d}(\mathsf{h}(\mathsf{d})) = \mathtt{T}$. This contradicts $\mathcal{E}_{\varphi,[\mathsf{h}(\mathsf{d})/y],[\mathsf{w}/w]}(\neg[w\,y]) = \mathtt{T}$, verifying there is no such model $\mathcal{M}$. □

We can use these two versions of Cantor's theorem to analyze independence of signatures. That is, if we can find an extensional $\mathcal{S}$-model $\mathcal{M}$ such that $\mathcal{M} \not\models$ **X5304**, then we know $\mathcal{S} \ntrianglelefteq_{\beta\mathfrak{fb}} (\mathcal{S} \cup \{\neg\})$. Likewise, whenever there is an extensional $\mathcal{S}$-model $\mathcal{M}$ such that $\mathcal{M} \not\models$ **X5309**, then we know $\mathcal{S} \ntrianglelefteq_{\beta\mathfrak{fb}} (\mathcal{S} \cup \{\neg, \wedge, =^\iota, \Sigma^{o\iota}\})$.

It can easily be verified that the singleton true model $\mathcal{M}$ defined in Example 3.6.23 fails to model either **X5304** or **X5309**. Consequently, there is no $\mathcal{G}^{\mathcal{S}}_{\beta\mathfrak{fb}}$-derivation of either **X5304** or **X5309** if $\mathcal{S} \cap \{\bot, \neg\} = \emptyset$. (We already know $\mathcal{S} \ntrianglelefteq_* \mathcal{S} \cup \{\neg\}$ for such $\mathcal{S}$ by Lemma 6.4.2.)

6.7.1. A Model of a Fragment of Extensional Type Theory in which Cantor's Theorems Fail. We will use a $\mathcal{B}$-specified frame to construct an extensional model (without negation) in which both **X5304** and **X5309** fail.

Let A be the real interval $(-1, 1)$ and

$$\mathcal{B} := \{(a, 1) \mid -1 \leq a \leq 1\} \subseteq \mathcal{P}(A).$$

The emptyset is in $\mathcal{B}$ as the (trivial) interval $(1, 1)$ and A is in $\mathcal{B}$ as the interval $(-1, 1)$. The set $\mathcal{B}$ is closed under unions and intersections, but is not closed under complements. Let $\mathcal{D}$ be the $\mathcal{B}$-specified frame with relation $\mathcal{R}$ (cf. Definition 4.5.1). By Lemma 4.5.6, we know $\mathcal{D}$ realizes $\top$, $\bot$, $\vee$ and $\wedge$, but does not realize any of $\neg$, $\supset$ and $\equiv$.

DEFINITION 6.7.3. Let C and D be sets of real numbers. A function $f : C \to D$ is *left continuous* if for any $a \in C$,

$$\lim_{x \to a^-} f(x) = f(a).$$

LEMMA 6.7.4. *Let $f : (-1, 1) \to (-1, 1)$ be any function. $f \in \mathcal{R}_\iota$ iff f is nondecreasing and left continuous.*

PROOF. See Appendix A.4. □

Note that $\mathcal{B}$ is a topology on $(-1, 1)$. By the definition of $\mathcal{R}_\iota$, we know $f \in \mathcal{R}_\iota$ iff f is continuous with respect to this topology. Lemma 6.7.4 establishes continuity of $f : (-1, 1) \to (-1, 1)$ with respect to the topology $\mathcal{B}$ corresponds to f being nondecreasing and left continuous with respect to the usual topology on $(-1, 1)$.

Recall (cf. Definition 4.4.1) for any $\mathsf{a} \in \mathcal{D}_\alpha$ we use the notation $K(\mathsf{a})$ to denote the constant function $K(\mathsf{a}) : (-1, 1) \to \mathcal{D}_\alpha$ defined by $K(\mathsf{a})(x) := \mathsf{a}$. Since $\mathcal{D}$ is an A-ary logical relation frame with relation $\mathcal{R}$ (cf. Definition 4.4.2), we know $K(\mathsf{a}) \in \mathcal{R}_\alpha$. Also, for any $p : (-1, 1) \to \mathcal{D}_{\alpha\beta}$ and $f : (-1, 1) \to \mathcal{D}_\beta$ we use the notation $S(p, f) : (-1, 1) \to \mathcal{D}_\alpha$ to denote the function

$$S(p, f)(x) := p(x)(f(x)).$$

Since $\mathcal{D}$ is an A-ary logical relation frame with relation $\mathcal{R}$, we know $S(p, f) \in \mathcal{R}_\alpha$ whenever $p \in \mathcal{R}_{\alpha\beta}$ and $f \in \mathcal{R}_\beta$.

LEMMA 6.7.5. *Suppose $g \in \mathcal{R}_{o\iota\iota}$, $x, y \in (-1, 1)$ and $g(x)(y) = \mathtt{T}$. For any $y' \in (y, 1)$, we have $g(x)(y') = \mathtt{T}$. Also, for any $x' \in (x, 1)$, we have $g(x')(y) = \mathtt{T}$.*

PROOF. See Appendix A.4. □

We can now characterize the members of $\mathcal{D}_{o\iota\iota}$ and $\mathcal{R}_{o\iota\iota}$.

LEMMA 6.7.6. *Let $g : (-1, 1) \to \{\mathtt{T}, \mathtt{F}\}^{(-1,1)}$ be any function. The following are equivalent:*

1. $g \in \mathcal{D}_{o\iota\iota}$.

2. $g \in \mathcal{R}_{o\iota}$.
3. *There exists a nonincreasing left continuous* $l : (-1, 1) \to \Re$ *(the reals) such that* $g(x)(y) = \mathtt{T}$ *iff* $l(x) < y$ *for every* $x, y \in (-1, 1)$.

PROOF. See Appendix A.4. □

LEMMA 6.7.7. *There is an internal retraction* (g, h) *from* $\mathcal{D}_\iota$ *onto* $\mathcal{D}_{o\iota}$. *In particular,* $\mathsf{g} \in \mathcal{D}_{o\iota\iota}$ *is a surjection and* $\mathsf{h} \in \mathcal{D}_{\iota(o\iota)}$ *is an injection.*

PROOF. Let $\mathsf{g} : (-1, 1) \to \{\mathtt{T}, \mathtt{F}\}^{(-1,1)}$ be the function

$$\mathsf{g}(x)(y) := \begin{cases} \mathtt{T} & \text{if } -2x < y \\ \mathtt{F} & \text{otherwise} \end{cases}$$

Lemma 6.7.6 proves $\mathsf{g} \in \mathcal{D}_{o\iota\iota}$ using the function $l : (-1, 1) \to \Re$ defined by $l(x) := -2x$. By definition, $\mathsf{g}(x)(y) = \mathtt{T}$ iff $l(x) < y$. Clearly l is nonincreasing and left continuous. The graph of this function g is shown as a binary relation in Figure 6.4.

Define $\mathsf{h} : \mathcal{D}_{o\iota} \to \mathcal{D}_\iota$ by

$$\mathsf{h}(\chi_{(a,1)}) := \frac{-a}{2}$$

(using Lemma 4.5.4). To check $\mathsf{h} \in \mathcal{D}_{\iota(o\iota)}$, let $p \in \mathcal{R}_{o\iota}$ be given. By Lemma 6.7.6, there is a nonincreasing left continuous $q : (-1, 1) \to \Re$ such that $p(x)(y) = \mathtt{T}$ iff $q(x) < y$ for every $x, y \in (-1, 1)$. That is, $p(x) = \chi_{(q(x),1)}$ for every $x \in (-1, 1)$. We must check $(\mathsf{h} \circ p) \in \mathcal{R}_\iota$. Let $(b, 1) \in \mathcal{B}$ be given. We must ensure

$$\{x \in (-1, 1) \mid \mathsf{h}(p(x)) > b\} \in \mathcal{B}.$$

We compute

$$\{x \in (-1, 1) \mid \mathsf{h}(p(x)) > b\} = \{x \in (-1, 1) \mid \mathsf{h}(\chi_{(q(x),1)}) > b\}$$
$$= \{x \in (-1, 1) \mid \tfrac{-q(x)}{2} > b\} = \{x \in (-1, 1) \mid q(x) < -2b\}.$$

Let $c := \inf\{y \in (-1, 1) \mid q(y) < -2b\}$. We will determine

$$\{x \in (-1, 1) \mid q(x) < -2b\} = (c, 1) \in \mathcal{B}.$$

First, suppose $x \in (c, 1)$. By the choice of c, there is a $y \in (c, x)$ such that $q(y) < -2b$. Since q is nonincreasing and $x > y$, we know $q(x) \leq q(y) < -2b$. Hence

$$(c, 1) \subseteq \{x \in (-1, 1) \mid q(x) < -2b\} \subseteq (-1, 1).$$

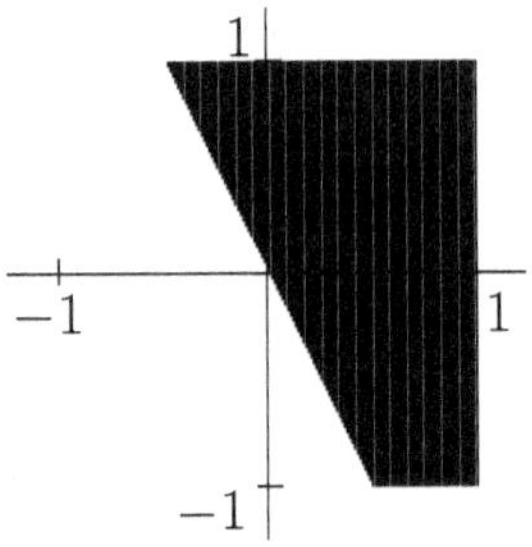

FIGURE 6.4. The Surjection g as a Binary Relation.

Next, suppose $x \in (-1, c]$ and $q(x) < -2b$. Since q is left continuous, there is some $y \in (-1, x)$ such that $q(y) < -2b$. This $y < x \leq c$ contradicts the choice of c as the infimum of such values. Thus

$$\{x \in (-1,1) \mid q(x) < -2b\} = (c, 1) \in \mathcal{B}.$$

Therefore, $(\mathsf{h} \circ p) \in \mathcal{R}_\iota$ and so $\mathsf{h} \in \mathcal{D}_{\iota(o\iota)}$.

Finally, we check that these form a retraction. For $\chi_{(a,1)} \in \mathcal{D}_{o\iota}$, we compute

$$\mathsf{g}(\mathsf{h}(\chi_{(a,1)}))(y) = \mathsf{g}(\frac{-a}{2})(y) = \mathtt{T}$$

iff $-2(\frac{-a}{2}) < y$ iff $a < y$. Thus $\mathsf{g}(\mathsf{h}(\chi_{(a,1)})) = \chi_{(a,1)}$. In particular, g is surjective and h is injective. □

We now use the $\mathcal{B}$-specified frame to induce an extensional model. Let $\mathcal{S}_1 := \{\top, \bot, \wedge, \vee\}$ and

$$\mathcal{S}_2 := \mathcal{S}_1 \cup \{\Pi^\alpha, \Sigma^\alpha \mid \alpha \in \mathcal{T}_o\}.$$

By Lemma 4.5.6, the frame $\mathcal{D}$ realizes $\mathcal{S}_1$. That is, $\mathcal{S}_1 \subseteq \mathcal{S}^{\mathcal{D}}$. Let $\mathcal{I} : \mathcal{P} \to \mathcal{D}$ be arbitrary. By Theorem 3.6.20, there is a unique frame $\mathcal{S}_1$-model $\mathcal{M}_1 = (\mathcal{D}, @, \mathcal{E}^1, \upsilon)$ induced by $(\mathcal{D}, \mathcal{I})$. By Lemma 6.5.1, we know $\mathcal{D}_\alpha$ is finite for every propositional type $\alpha \in \mathcal{T}_o$. Thus we can use Lemma 6.2.5 to conclude $\mathcal{M}_1$ realizes

$$\{\Pi^\alpha, \Sigma^\alpha \mid \alpha \in \mathcal{T}_o\}$$

and so $\mathcal{S}_2 \subseteq \mathcal{S}^{\mathcal{M}_1}$. By Corollary 3.3.19, we can extend $\mathcal{M}_1$ to be an extensional $\mathcal{S}_2$-model $\mathcal{M} = (\mathcal{D}, @, \mathcal{E}^2, \upsilon)$.

THEOREM 6.7.8. *Let*

$$\mathcal{S}_2 := \{\top, \bot, \wedge, \vee\} \cup \{\Pi^\alpha, \Sigma^\alpha \mid \alpha \in \mathcal{T}_o\}$$

*and $\mathcal{S}$ be a signature such that $\mathcal{S} \subseteq \mathcal{S}_2$. There is a model $\mathcal{M} \in \mathfrak{M}_{\beta\mathfrak{f}\mathfrak{b}}(\mathcal{S})$ such that $\mathcal{M} \not\models$ **X5304**$_\iota$ and $\mathcal{M} \not\models$ **X5309**.*

PROOF. By Theorem 3.3.15, $\mathcal{M} := (\mathcal{D}, @, \mathcal{E}^2|_{\mathit{wff}(\mathcal{S})}, \upsilon)$ is an extensional $\mathcal{S}$-model.

We know $\mathcal{M} \not\models$ **X5304**$_\iota$ and $\mathcal{M} \not\models$ **X5309** using the surjective witness $\mathsf{g} \in \mathcal{D}_{o\iota\iota}$ and the injective witness $\mathsf{h} \in \mathcal{D}_{\iota(o\iota)}$ from Lemma 6.7.7. □

COROLLARY 6.7.9. *Let*

$$\mathcal{S}_2 := \{\top, \bot, \wedge, \vee\} \cup \{\Pi^\alpha, \Sigma^\alpha \mid \alpha \in \mathcal{T}_o\}$$

and $\mathcal{S}$ be a signature such that $\mathcal{S} \subseteq \mathcal{S}_2$. Neither **X5304**$_\iota$ *nor* **X5309** *is derivable in $\mathcal{G}^{\mathcal{S}}_{\beta\mathfrak{f}\mathfrak{b}}$.*

PROOF. This follows directly from Theorem 6.7.8 and soundness (cf. Theorem 3.4.14). □

6.7.2. A Model of a Fragment of Extensional Type Theory in which the Injective Cantor Theorem Fails. We have constructed a model in which the injective Cantor theorem fails due to the model lacking negation (cf. Theorem 6.7.8). A more interesting problem is to find an extensional model which realizes $\mathcal{S}_{Bool}$ (and more) but fails to satisfy **X5309**. This must rely on the model failing to realize $\Sigma^{o\iota}$ or $=^\iota$. We construct such a model here which realizes neither $\Sigma^{o\iota}$ nor $=^\iota$ and fails to satisfy **X5309**.

Let $\mathcal{B} \subseteq \mathcal{P}(\mathbb{N})$ be the collection of sets X of natural numbers where X is finite or cofinite. $\mathcal{B}$ is clearly a field of sets. Let $\mathcal{D}$ be the $\mathcal{B}$-specified frame with relation $\mathcal{R}$ (cf. Definition 4.5.1). We will use this frame to construct a model in which the injective Cantor theorem **X5309** fails to hold.

By Definition 4.5.1, we know $\mathcal{D}_\iota := \mathbb{N}$, so $\mathcal{D}_\iota$ is countably infinite. Since there are countably many finite and cofinite subsets of $\mathbb{N}$, $\mathcal{B}$ is also countably infinite. By Lemma 4.5.4, the function taking each $X \in \mathcal{B}$ to a characteristic function χ_X in $\mathcal{D}_{o\iota}$ is a bijection. Hence $\mathcal{D}_{o\iota}$ is also countably infinite.

Since $\mathcal{D}_{o\iota}$ and $\mathcal{D}_\iota$ are countably infinite, there is an injection $h : \mathcal{D}_{o\iota} \to \mathcal{D}_\iota$. The existence of such an h certainly does *not* contradict the injective Cantor theorem. The existence of such an h is often referred to as *Skolem's Paradox.* As is well-known, Skolem's Paradox is not really a paradox, since the injection h exists at the meta-level. This is far different from saying there is such an $h \in \mathcal{D}_{\iota(o\iota)}$. That is, we know the injection h exists, but we do not know it exists *in the model.* We will prove that for the $\mathcal{B}$-specified model, any such injection h *is* in $\mathcal{D}_{\iota(o\iota)}$. This means that with respect to this model, Skolem's Paradox really does provide a counterexample to this version of Cantor's "theorem". This is still not a paradox. It is merely a model demonstrating that the injective Cantor "theorem" is not a theorem if the signature does not include enough quantifiers or equalities.

DEFINITION 6.7.10. Let C be a set and $(a_n)_{n\in\mathbb{N}}$ be a sequence with $a_n \in C$. We say the sequence $(a_n)_{n\in\mathbb{N}}$ *converges to* $a \in C$ (written $a_n \to_n a$) if there exists an $N \in \mathbb{N}$ such that $a_n = a$ for every $n > N$. We say a function $g : \mathbb{N} \to C$ is *eventually constant* if there exists some $a \in C$ such that $g(n) \to_n a$. For each type α, let

$$\mathcal{K}_\alpha := \{g : \mathbb{N} \to \mathcal{D}_\alpha \mid g \text{ is eventually constant}\}.$$

Let C and D be sets and $(f_n)_{n\in\mathbb{N}}$ be a sequence of functions $f_n : D \to C$. We say the sequence $(f_n)_{n\in\mathbb{N}}$ *converges pointwise to* a function $f : D \to C$ (written $f_n \to_n^p f$) if $f_n(d) \to_n f(d)$ for every $d \in D$. For all types α and β, let

$$\mathcal{K}^p_{\alpha\beta} := \{g : \mathbb{N} \to \mathcal{D}_{\alpha\beta} \mid g \text{ converges pointwise to some } f : \mathcal{D}_\beta \to \mathcal{D}_\alpha\}.$$

We start by examining the domains of base type.

LEMMA 6.7.11. $\mathcal{D}_{o\iota} = \mathcal{R}_o = \mathcal{K}_o$.

PROOF. By Lemma 4.5.5, we know $\mathcal{R}_o = \mathcal{D}_{o\iota}$.

To verify $\mathcal{D}_{o\iota} \subseteq \mathcal{K}_o$, let $f \in \mathcal{D}_{o\iota}$ be given. By Lemma 4.5.4, $f = \chi_X$ for some $X \in \mathcal{B}$. If X is finite, then $f(n) = \mathtt{F}$ for every $n > N$ where $N \in \mathbb{N}$ is such that $X \subseteq \{0, \ldots, N\}$. If X is cofinite, then $f(n) = \mathtt{T}$ for every $n > N$ where $N \in \mathbb{N}$ is such that $(\mathbb{N} \setminus X) \subseteq \{0, \ldots, N\}$.

Next suppose $f \in \mathcal{K}_o$. That is, there is some $N \in \mathbb{N}$ and $\mathsf{b} \in \{\mathtt{T}, \mathtt{F}\}$ such that $f(n) = \mathsf{b}$ for every $n > N$. If $\mathsf{b} = \mathtt{F}$, then $f^{-1}(\mathtt{T})$ is finite since

$$f^{-1}(X) \subseteq \{0, \ldots, N\}.$$

If $\mathsf{b} = \mathtt{T}$, then $f^{-1}(\mathtt{T})$ is cofinite since

$$(\mathbb{N} \setminus f^{-1}(\mathtt{T})) \subseteq \{0, \ldots, N\}.$$

In either case $f \in \mathcal{R}_o$. □

DEFINITION 6.7.12. We say a function $f : \mathbb{N} \to \mathbb{N}$ is *uniformly unbounded* if for every $j \in \mathbb{N}$ there is some $I \in \mathbb{N}$ such that $f(i) > j$ for every $i > I$.

LEMMA 6.7.13. *Let $f : \mathbb{N} \to \mathbb{N}$ be any function. The function $f \in \mathcal{R}_\iota$ iff either f is eventually constant or f is uniformly unbounded.*

PROOF. See Appendix A.4. □

We can use this to prove that $\mathcal{D}$ does not realize $=^\iota$, in spite of the fact that $\mathcal{B}$ contains every singleton $\{n\}$ where $n \in \mathbb{N}$.

LEMMA 6.7.14. *The frame $\mathcal{D}$ does not realize $=^\iota$.*

PROOF. Assume $\mathsf{q} \in \mathcal{D}_{o\iota\iota}$ satisfies $\mathfrak{L}_{=^\iota}(\mathsf{q})$. That is, $\mathsf{q}(m)(n) = \mathtt{T}$ iff $n = m$ for every $m, n \in \mathcal{D}_\iota$. Let $id : \mathbb{N} \to \mathcal{D}_\iota$ be the identity function and $f : \mathbb{N} \to \mathcal{D}_\iota$ be defined by $f(2n) := 2n$ and $f(2n+1) := 2n$. Clearly, both f and id are uniformly unbounded. Hence $f, id \in \mathcal{R}_\iota$ by Lemma 6.7.13. Thus $\mathsf{q} \circ f \in \mathcal{R}_{o\iota}$ and $S(\mathsf{q} \circ f, id) \in \mathcal{R}_o$. However,

$$S(\mathsf{q} \circ f, id)^{-1}(\mathtt{T}) = \{n \in \mathbb{N} \mid \mathsf{q}(f(n))(n) = \mathtt{T}\}$$

$$= \{n \in \mathbb{N} \mid n \text{ is even}\} \notin \mathcal{B}$$

contradicting $S(\mathsf{q} \circ f, id) \in \mathcal{R}_o$. □

We next prove $\mathcal{R}_\alpha$ is closed under the operation of making finitely many changes to functions.

DEFINITION 6.7.15. Let C be a set and $f, g : \mathbb{N} \to C$ be functions. We say f and g are *eventually equal* if there exists some $N \in \mathbb{N}$ such that $f(n) = g(n)$ for every $n > N$.

LEMMA 6.7.16. *Suppose $f \in \mathcal{R}_\alpha$ and $g : \mathbb{N} \to \mathcal{D}_\alpha$ are eventually equal. Then $g \in \mathcal{R}_\alpha$.*

PROOF. We prove this by induction on the type α.

o: By Lemma 6.7.11, the function f is eventually constant. Hence g is also eventually constant and $g \in \mathcal{R}_o$.

ι: By Lemma 6.7.13, the function f is eventually constant or uniformly unbounded. If f is eventually constant, then so is g and hence $g \in \mathcal{R}_\iota$. Similarly, if f is uniformly unbounded, then so is g and hence $g \in \mathcal{R}_\iota$.

$\gamma\beta$: Let $h \in \mathcal{R}_\beta$ be given. We must check $S(g, h) \in \mathcal{R}_\gamma$. Since $f \in \mathcal{R}_{\gamma\beta}$, we know $S(f, h) \in \mathcal{R}_\beta$. Since f and g are eventually equal, there is some N such that $f(n) = g(n)$ for every $n > N$. Thus

$$S(f, h)(n) = f(n)(h(n)) = g(n)(h(n)) = S(g, h)(n).$$

That is, $S(f, h)$ and $S(g, h)$ are also eventually equal. By the inductive hypothesis, $S(g, h) \in \mathcal{R}_\gamma$ and hence $g \in \mathcal{R}_{\gamma\beta}$.

□

We can conclude the sets $\mathcal{K}_\alpha$ are lower bounds for the sets $\mathcal{R}_\alpha$.

LEMMA 6.7.17. *For every type α, $\mathcal{K}_\alpha \subseteq \mathcal{R}_\alpha$.*

PROOF. Let $f \in \mathcal{K}_\alpha$ be given. By definition, there is some $\mathsf{a} \in \mathcal{D}_\alpha$ such that $f(n) \to_n \mathsf{a}$. By Lemma 4.4.4, we know $K(\mathsf{a}) \in \mathcal{R}_\alpha$. Since $f(n) \to_n \mathsf{a}$, f and $K(\mathsf{a})$ are eventually equal. Thus $f \in \mathcal{R}_\alpha$ by Lemma 6.7.16. □

We already know $\mathcal{K}_o = \mathcal{R}_o$. On the other hand, $\mathcal{K}_\iota \neq \mathcal{R}_\iota$ since the identity function is uniformly unbounded, hence in $\mathcal{R}_\iota$ but not in $\mathcal{K}_\iota$. When $\mathcal{K}_\beta = \mathcal{R}_\beta$ for a type β we can prove the domains at a higher type are large.

LEMMA 6.7.18. *Let $\alpha, \beta \in \mathcal{T}$. If $\mathcal{R}_\beta \subseteq \mathcal{K}_\beta$, then $\mathcal{D}_{\alpha\beta} = \mathcal{D}_\alpha^{\mathcal{D}_\beta}$.*

PROOF. We know $\mathcal{D}_{\alpha\beta} \subseteq \mathcal{D}_\alpha^{\mathcal{D}_\beta}$ since $\mathcal{D}$ is a frame. To check the reverse inclusion, let $g : \mathcal{D}_\beta \to \mathcal{D}_\alpha$ be given. To check $g \in \mathcal{D}_{\alpha\beta}$, let $h \in \mathcal{R}_\beta \subseteq \mathcal{K}_\beta$ be given. There is some $N \in \mathbb{N}$ and $\mathsf{b} \in \mathcal{D}_\beta$ such that $h(n) = \mathsf{b}$ for every $n > N$. Hence $g(h(n)) = g(\mathsf{b})$ for every $n > N$. That is, $g \circ h$ is eventually constant and hence $g \circ h \in \mathcal{R}_\alpha$ by Lemma 6.7.17. Thus $g \in \mathcal{D}_{\alpha\beta}$, as desired. □

We can use Lemma 6.7.18 to prove certain domains $\mathcal{D}_{o(o\alpha)}$ are the full function space and hence the frame realizes Π^α and Σ^α.

THEOREM 6.7.19. *Suppose $\mathcal{R}_{o\alpha} \subseteq \mathcal{K}_{o\alpha}$. Then the frame $\mathcal{D}$ realizes Π^α and Σ^α.*

PROOF. We define $\pi^\alpha : \mathcal{D}_{o\alpha} \to \mathcal{D}_o$ and $\sigma^\alpha : \mathcal{D}_{o\alpha} \to \mathcal{D}_o$ by

$$\pi^\alpha(f) := \begin{cases} \mathtt{T} & \text{if } \forall \mathsf{a} \in \mathcal{D}_\alpha f(\mathsf{a}) = \mathtt{T} \\ \mathtt{F} & \text{otherwise} \end{cases}$$

and

$$\sigma^\alpha(f) := \begin{cases} \mathtt{T} & \text{if } \exists \mathsf{a} \in \mathcal{D}_\alpha f(\mathsf{a}) = \mathtt{T} \\ \mathtt{F} & \text{otherwise} \end{cases}$$

By Lemma 6.7.18, we know $\pi^\alpha \in \mathcal{D}_{o(o\alpha)}$ and $\sigma^\alpha \in \mathcal{D}_{o(o\alpha)}$ Clearly, $\mathfrak{L}_{\Pi^\alpha}(\pi^\alpha)$ and $\mathfrak{L}_{\Sigma^\alpha}(\sigma^\alpha)$ hold and hence $\mathcal{D}$ realizes Π^α and Σ^α. □

At the moment we can only apply Lemma 6.7.18 to the type o. This does not allow us to yet use Theorem 6.7.19 to conclude certain quantifiers are realized by the frame. Theorem 6.7.19 only provides the motivation for determining types α for which $\mathcal{R}_{o\alpha} \subseteq \mathcal{K}_{o\alpha}$ holds. We now prove $\mathcal{R}_{o\iota} \subseteq \mathcal{K}_{o\iota}$ holds, allowing us to conclude that the frame realizes Π^ι and Σ^ι.

LEMMA 6.7.20. *For every $p \in \mathcal{R}_{o\iota}$ there exist $M, N \in \mathbb{N}$ such that $p(m)(n) = p(M)(N)$ for every $m > M$ and $n > N$.*

PROOF. See Appendix A.4. □

LEMMA 6.7.21. *Every $p \in \mathcal{R}_{o\iota}$ is eventually constant.*

PROOF. See Appendix A.4. □

From Lemmas 6.7.21 and 6.7.17 we know $\mathcal{R}_{o\iota} = \mathcal{K}_{o\iota}$. We can use this to give a simpler characterization of $\mathcal{R}_{o\iota}$. Let $p : \mathbb{N} \to \mathcal{D}_o^{\mathbb{N}}$ be any function. By Lemma 6.7.11, we know $p : \mathbb{N} \to \mathcal{D}_{o\iota}$ iff $p(n)$ is eventually constant for each $n \in \mathbb{N}$. Consequently, p is in $\mathcal{R}_{o\iota}$ iff $p(n)$ is eventually constant for each $n \in \mathbb{N}$ and p is itself eventually constant. The picture in Figure 6.5 shows a typical example of a $p : \mathbb{N} \to (\mathcal{D}_o)^{\mathbb{N}}$ satisfying these conditions. (The painted in portions of the picture indicate values of n and m for which $p(n)(m) = \mathtt{T}$.)

We can now prove $\mathcal{D}_{\iota(o\iota)}$ is the full function space. This guarantees that an "external" injection $h : \mathcal{D}_{o\iota} \to \mathcal{D}_\iota$ is also "internal" in the sense that $h \in \mathcal{D}_{\iota(o\iota)}$. Hence the frame will induce a model in which **X5309** fails.

LEMMA 6.7.22. *For every function $h : \mathcal{D}_{o\iota} \to \mathcal{D}_\iota$, $h \in \mathcal{D}_{\iota(o\iota)}$.*

PROOF. By Lemma 6.7.21, we know $\mathcal{R}_{o\iota} \subseteq \mathcal{K}_{o\iota}$. The result follows from Lemma 6.7.18. □

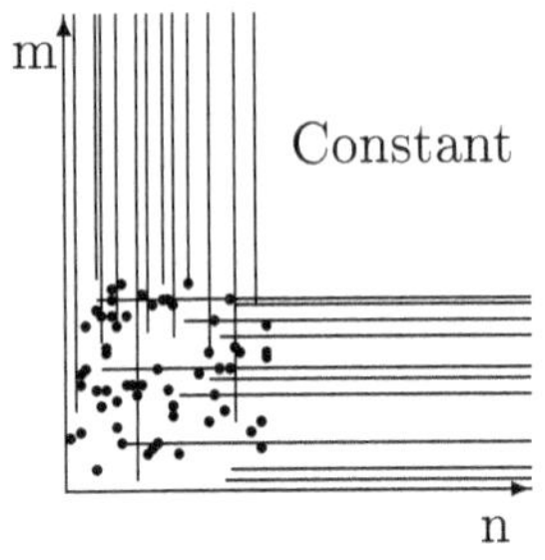

FIGURE 6.5. Members of $\mathcal{R}_{o\iota}$

We can also conclude the frame realizes quantifiers over individuals.

LEMMA 6.7.23. *$\mathcal{D}$ realizes Π^{ι} and Σ^{ι}.*

PROOF. By Lemma 6.7.21, we know $\mathcal{R}_{o\iota} \subseteq \mathcal{K}_{o\iota}$. The result follows from Theorem 6.7.19. □

We now turn this $\mathcal{B}$-specified frame into an extensional model over a signature. Let $\mathcal{S}^{\mathcal{D}}$ be the signature of logical constants realized by $\mathcal{D}$. Let $\mathcal{I} : \mathcal{P} \to \mathcal{D}$ be arbitrary. By Theorem 3.6.20, there is a unique frame $\mathcal{S}^{\mathcal{D}}$-model $\mathcal{M}^{\mathcal{D}} = (\mathcal{D}, @, \mathcal{E}, \upsilon)$ induced by $(\mathcal{D}, \mathcal{I})$. Also, $\mathcal{S}^{\mathcal{M}^{\mathcal{D}}} = \mathcal{S}^{\mathcal{D}}$.

THEOREM 6.7.24. *$\mathcal{M}^{\mathcal{D}} \not\models$* **X5309**.

PROOF. The set $\mathcal{B}$ of finite and cofinite subsets of $\mathbb{N}$ is countably infinite. By Lemma 4.5.4, we know $\mathcal{D}_{o\iota}$ is also countably infinite. Hence both $\mathcal{D}_{\iota}$ and $\mathcal{D}_{o\iota}$ are both countably infinite. Let $h : \mathcal{D}_{o\iota} \to \mathcal{D}_{\iota}$ be an injection. By Lemma 6.7.22, we know $h \in \mathcal{D}_{\iota(o\iota)}$. Thus $\mathcal{M}^{\mathcal{D}} \not\models$ **X5309**. □

COROLLARY 6.7.25. *Let $\mathcal{S}$ be a signature such that $\mathcal{S} \subseteq \mathcal{S}^{\mathcal{D}}$. There is no proof of* **X5309** *in $\mathcal{G}^{\mathcal{S}}_{\beta\mathfrak{fb}}$.*

PROOF. This follows from Theorem 6.7.24, Lemma 6.7.2 and soundness (Theorem 3.4.14) of $\mathcal{G}^{\mathcal{S}}_{\beta\mathfrak{fb}}$ with respect to the model class $\mathfrak{M}_{\beta\mathfrak{fb}}(\mathcal{S})$. □

That is, **X5309** is not provable in the fragment of extensional type theory if we are restricted to quantifiers and equalities realized by $\mathcal{D}$. With more work, we can prove the frame $\mathcal{D}$ realizes additional quantifiers and equalities. We turn to this analysis now.

Lemma 6.7.17 establishes $\mathcal{K}_\alpha$ as a lower bound for $\mathcal{R}_\alpha$. For some function types we can also relate $\mathcal{K}^p_{\alpha\beta}$ to $\mathcal{R}_{\alpha\beta}$.

LEMMA 6.7.26. *Let α and β be types.*

1. *If $\mathcal{R}_\alpha \subseteq \mathcal{K}_\alpha$, then $\mathcal{R}_{\alpha\beta} \subseteq \mathcal{K}^p_{\alpha\beta}$.*
2. *$\mathcal{R}_{o\beta} \subseteq \mathcal{K}^p_{o\beta}$.*
3. *If $\mathcal{R}_\beta \subseteq \mathcal{K}_\beta$, then $\mathcal{K}^p_{\alpha\beta} \subseteq \mathcal{R}_{\alpha\beta}$.*

PROOF. See Appendix A.4. □

We can use Lemma 6.7.26 to establish a kind of continuity result for sequences of certain types.

LEMMA 6.7.27. *Let α, β and γ be types. Suppose $\mathcal{R}_\beta \subseteq \mathcal{K}_\beta$ and $\mathcal{R}_\gamma \subseteq \mathcal{K}_\gamma$. Let $u \in \mathcal{D}_{\alpha\beta}$ be a function and $(u^n)_{n\in\mathbb{N}}$ be a sequence of functions $u^n \in \mathcal{D}_{\alpha\beta}$ such that $u^n \to^p_n u$. Let $m^0 < m^1 < \cdots$ be an increasing sequence of natural numbers. For each $p \in \mathcal{R}_{\gamma(\alpha\beta)}$ there is some $\mathsf{c} \in \mathcal{R}_\gamma$ such that $p(m^n)(u) \to_n \mathsf{c}$ and $p(m^n)(u^n) \to_n \mathsf{c}$.*

PROOF. See Appendix A.4. □

LEMMA 6.7.28. *Let α, β and γ be types. Suppose $\mathcal{R}_\beta \subseteq \mathcal{K}_\beta$ and $\mathcal{R}_\gamma \subseteq \mathcal{K}_\gamma$. Let $u \in \mathcal{D}_{\alpha\beta}$ be a function and $(u^n)_{n\in\mathbb{N}}$ be a sequence of functions $u^n \in \mathcal{D}_{\alpha\beta}$ such that $u^n \to^p_n u$. For each $g \in \mathcal{D}_{\gamma(\alpha\beta)}$, we have $g(u^n) \to_n g(u)$.*

PROOF. Suppose $g \in \mathcal{D}_{\gamma(\alpha\beta)}$. Applying Lemma 6.7.27 to

$$K(g) \in \mathcal{R}_{\gamma(\alpha\beta)}$$

(cf. Lemma 4.4.4), the sequence $(u^n)_{n\in\mathbb{N}}$ and the increasing sequence $(n)_{n\in\mathbb{N}}$ of natural numbers, there is a $\mathsf{c} \in \mathcal{D}_\gamma$ such that

$$g(u^n) = K(g)(n)(u^n) \to_n \mathsf{c} \text{ and } g(u) = K(g)(n)(u) \to_n \mathsf{c}.$$

Hence $g(u) = \mathsf{c}$ and $g(u^n) \to_n g(u)$. □

We now turn to the problem of proving the frame $\mathcal{D}$ realizes quantifiers at higher types such as $o(o(o\iota))$. In order to argue generally, we concentrate on studying domains $\mathcal{D}_{o(o\beta)}$ where $\mathcal{R}_\beta \subseteq \mathcal{K}_\beta$ and $\mathcal{D}_\beta$ is countably infinite. (Note that we already know $\mathcal{R}_{o\iota} \subseteq \mathcal{K}_{o\iota}$ and $\mathcal{D}_{o\iota}$ is countably infinite.) We will determine that every element of such a domain $\mathcal{D}_{o(o\beta)}$ is in the propositional closure of the set of projection functions. This implies all the elements of such $\mathcal{D}_{o(o\beta)}$ are definable by terms of the form

$$[\lambda u_{o\beta}[[u\, y_1^1 \wedge \cdots \wedge u\, y_1^{m_1} \wedge \neg u\, z_1^1 \wedge \cdots \wedge \neg u\, z_1^{k_1}] \vee \cdots \vee [u\, y_n^1 \wedge \cdots \wedge u\, y_n^{m_n} \wedge \neg u\, z_n^1 \wedge \cdots \wedge \neg u\, z_n^{k_n}]]].$$

We can consider $\mathcal{D}_o$ as a topological space with the discrete topology (every set is open). For any type β we can give the function space $\mathcal{D}_o^{\mathcal{D}_\beta}$ the product topology. The product topology is the smallest topology such that for each $\mathsf{b} \in \mathcal{D}_\beta$ the projection function

$$proj_{\mathsf{b}} : (\mathcal{D}_o^{\mathcal{D}_\beta}) \to \mathcal{D}_o$$

(cf. Definition 3.6.27) is continuous. As described in [47] the set

$$\mathscr{S}_\beta := \{proj_{\mathsf{b}}^{-1}(U) \mid \mathsf{b} \in \mathcal{D}_\beta, U \text{ open in } \mathcal{D}_o\}$$

is a subbasis for the product topology. Since $\mathcal{D}_o$ has the discrete topology, $\mathscr{S}_\beta$ is the set

$$\{proj_{\mathsf{b}}^{-1}(\mathsf{a}) \mid \mathsf{b} \in \mathcal{D}_\beta, \mathsf{a} \in \mathcal{D}_o\}.$$

Note that for any $\mathsf{b} \in \beta$ the set

$$\{w \in \mathcal{D}_o^{\mathcal{D}_\beta} \mid w(\mathsf{b}) = \mathtt{T}\}$$

is open as a member of the subbasis $\mathscr{S}_\beta$ (since $w(\mathsf{b}) = proj_{\mathsf{b}}(w)$). The complement of this set is

$$\{w \in \mathcal{D}_o^{\mathcal{D}_\beta} \mid w(\mathsf{b}) = \mathtt{F}\}$$

which is also open as a member of the subbasis $\mathscr{S}_\beta$. Hence each member of $\mathscr{S}_\beta$ is both closed and open. Such sets are called *clopen*.

The basis $\mathscr{B}_\beta$ generated by $\mathscr{S}_\beta$ is obtained by taking finite intersections of the sets in $\mathscr{S}_\beta$. Note that finite intersections of clopen sets are also clopen. Hence each basic set in $\mathscr{B}_\beta$ is clopen. One can easily verify the following characterization of basic sets.

LEMMA 6.7.29. *Let β be a type. For every set $B \subseteq \mathcal{D}_o^{\mathcal{D}_\beta}$ the following are equivalent:*

1. *$B \in \mathscr{B}_\beta$.*
2. *For every $w \in B$ there is a finite set $X \subseteq \mathcal{D}_\beta$ such that for all $u \in \mathcal{D}_o^{\mathcal{D}_\beta}$ we have $u \in B$ iff $u|_X = w|_X$.*

PROOF. See Appendix A.4. □

LEMMA 6.7.30. *Let* β *be a type such that* $\mathcal{R}_\beta \subseteq \mathcal{K}_\beta$ *and* $\mathcal{D}_\beta$ *is countably infinite. Suppose* $p \in \mathcal{R}_{o(o\beta)}$ *and* $w \in \mathcal{D}_{o\beta}$. *There exists a finite set* $X \subseteq \mathcal{D}_\beta$ *and an* $N \in \mathbb{N}$ *such that for every* $u \in \mathcal{D}_{o\beta}$ *and* $n > N$ *if* $u\big|_X = w\big|_X$, *then* $p(n)(u) = p(n)(w)$.

PROOF. See Appendix A.4. □

LEMMA 6.7.31. *Let* β *be a type such that* $\mathcal{R}_\beta \subseteq \mathcal{K}_\beta$ *and* $\mathcal{D}_\beta$ *is countably infinite. Suppose* $p \in \mathcal{R}_{o(o\beta)}$. *For every* $w \in \mathcal{D}_{o\beta}$ *there exist a basic neighborhood* $B \in \mathscr{B}_\beta$ *of* w *and an* $N \in \mathbb{N}$ *such that* $p(n)(u) = p(n)(w)$ *for every* $u \in B$ *and* $n > N$.

PROOF. This follows easily from Lemmas 6.7.30 and 6.7.29. See Appendix A.4. □

Using the Tychonoff Theorem (cf. [47]) we know the topological space $\mathcal{D}_o^{\mathcal{D}_\beta}$ is compact. We can conclude $\mathcal{R}_{o(o\beta)} \subseteq \mathcal{K}_{o(o\beta)}$ and $\mathcal{D}_{o(o\beta)}$ is countable whenever $\mathcal{R}_\beta \subseteq \mathcal{K}_\beta$ and $\mathcal{D}_\beta$ is countably infinite using compactness. When $\mathcal{D}_\beta$ is countably infinite, $\mathcal{D}_o^{\mathcal{D}_\beta}$ is a Cantor space (cf. [17]).

LEMMA 6.7.32. *Let* β *be a type such that* $\mathcal{R}_\beta \subseteq \mathcal{K}_\beta$ *and* $\mathcal{D}_\beta$ *is countably infinite. Suppose* $p \in \mathcal{R}_{o(o\beta)}$. *Then there exists a finite set* $X \subseteq \mathcal{D}_\beta$ *and an* $N \in \mathbb{N}$ *such that for every* $u, v \in \mathcal{D}_{o\beta}$ *and* $n > N$ *if* $u\big|_X = v\big|_X$, *then* $p(n)(u) = p(n)(v)$.

PROOF. We apply Lemma 6.7.31 to obtain a cover of basic open neighborhoods and then use compactness to reduce this to a finite subcover of basic open sets. See Appendix A.4. □

We can now determine $\mathcal{R}_{o(o\beta)} \subseteq \mathcal{K}_{o(o\beta)}$ whenever $\mathcal{R}_\beta \subseteq \mathcal{K}_\beta$ and $\mathcal{D}_\beta$ is countably infinite.

LEMMA 6.7.33. *Suppose* $\mathcal{R}_\beta \subseteq \mathcal{K}_\beta$ *and* $\mathcal{D}_\beta$ *is countably infinite. Then* $\mathcal{R}_{o(o\beta)} \subseteq \mathcal{K}_{o(o\beta)}$.

PROOF. See Appendix A.4. □

In order to study the cardinality of $\mathcal{D}_{o(o\beta)}$, we first prove a consequence of Lemma 6.7.32.

LEMMA 6.7.34. *Suppose $\mathcal{R}_\beta \subseteq \mathcal{K}_\beta$, $\mathcal{D}_\beta$ is countably infinite and $g \in \mathcal{D}_{o(o\beta)}$. There exists a finite set $X \subseteq \mathcal{D}_\beta$ such that for every $u, v \in \mathcal{D}_\beta$, if $u\big|_X = v\big|_X$, then $g(u) = g(v)$.*

PROOF. Applying Lemma 6.7.32 to $K(g) \in \mathcal{R}_{o(o\beta)}$ (cf. Lemma 4.4.4) we obtain a finite set $X \subseteq \mathcal{D}_\beta$ and an $N \in \mathbb{N}$ such that for every $u, v \in \mathcal{D}_\beta$ and $n > N$ if $u\big|_X = v\big|_X$, then $K(g)(n)(u) = K(g)(n)(v)$. Hence for every $u, v \in \mathcal{D}_\beta$, if $u\big|_X = v\big|_X$, then

$$g(u) = K(g)(N+1)(u) = K(g)(N+1)(v) = g(v).$$

□

Assuming $\mathcal{R}_\beta \subseteq \mathcal{K}_\beta$ and $\mathcal{D}_\beta$ is countably infinite, Lemma 6.7.34 implies every $g \in \mathcal{D}_{o(o\beta)}$ is determined by a finite set $X \subseteq \mathcal{D}_\beta$ and the values $g(\chi_Y)$ for each $Y \subseteq X$. This is enough to conclude $\mathcal{D}_{o(o\beta)}$ is countably infinite. In fact, the functions in $\mathcal{D}_{o(o\beta)}$ are precisely the characteristic functions of clopen sets. We can use the set of projection functions for types o and β to describe $\mathcal{D}_{o(o\beta)}$ in such a case. Define

$$\mathfrak{Proj}_\beta := \{proj_{\mathsf{b}} : \mathcal{D}_{o\beta} \to \mathcal{D}_o \mid \mathsf{b} \in \mathcal{D}_\beta\}$$

(cf. Definition 3.6.27).

LEMMA 6.7.35. *Suppose $\mathcal{R}_\beta \subseteq \mathcal{K}_\beta$ and $\mathcal{D}_\beta$ is countably infinite. Then $\mathcal{D}_{o(o\beta)}$ is the propositional closure (cf. Definition 3.6.31) of $\mathfrak{Proj}_\beta$.*

PROOF. See Appendix A.4. □

We know every element of such domains $\mathcal{D}_{o(o\beta)}$ is definable as a consequence of Lemma 6.7.35. We can now prove such domains are countably infinite.

LEMMA 6.7.36. *Suppose $\mathcal{R}_\beta \subseteq \mathcal{K}_\beta$ and $\mathcal{D}_\beta$ is countably infinite. Then $\mathcal{D}_{o(o\beta)}$ is countably infinite.*

PROOF. Since $\mathcal{D}_\beta$ is countably infinite, the set $\mathfrak{Proj}_\beta$ is countable. To prove it is infinite, we must verify $proj_{\mathsf{b}} \neq proj_{\mathsf{c}}$ whenever $\mathsf{b} \neq \mathsf{c}$. Suppose $\mathsf{b} \neq \mathsf{c}$. Define $u^{\mathsf{b}} : \mathcal{D}_\beta \to \mathcal{D}_o$ to be the characteristic function for the unit set $\{\mathsf{b}\}$. That is, $u^{\mathsf{b}}(\mathsf{b}) := \mathtt{T}$ and $u^{\mathsf{b}}(\mathsf{a}) := \mathtt{F}$ for all $\mathsf{a} \neq \mathsf{b}$. By Lemma 6.7.18, we know $u^{\mathsf{b}} \in \mathcal{D}_{o\beta}$. We compute

$$proj_{\mathsf{b}}(u^{\mathsf{b}}) = u^{\mathsf{b}}(\mathsf{b}) = \mathtt{T} \neq \mathtt{F} = u^{\mathsf{b}}(\mathsf{c}) = proj_{\mathsf{c}}(u^{\mathsf{b}}).$$

Hence $proj_{\mathsf{b}} \neq proj_{\mathsf{c}}$. Thus $\mathcal{P}\mathfrak{roj}_\beta$ is infinite.

The set $\mathcal{D}_{o(o\beta)}$ is the propositional closure of $\mathcal{P}\mathfrak{roj}_\beta$ by Lemma 6.7.35. By Theorem 3.6.33, we conclude $\mathcal{D}_{o(o\beta)}$ is countably infinite. □

We can now use induction to prove that the power types alternate between being countably infinite and being the full collection of (characteristic) functions. Likewise, we have an alternation with respect to realization of quantifiers at power types.

LEMMA 6.7.37. *For each $n \in \mathbb{N}$,*

$$\mathcal{R}_{\mathcal{P}^{2n+1}(\iota)} \subseteq \mathcal{K}_{\mathcal{P}^{2n+1}(\iota)}$$

and $\mathcal{D}_{\mathcal{P}^{2n+1}(\iota)}$ is countably infinite.

PROOF. The proof is by induction on $n \in \mathbb{N}$. We know $\mathcal{R}_{o\iota} \subseteq \mathcal{K}_{o\iota}$ by Lemma 6.7.21. Also, the fact that $\mathcal{B}$ is countably infinite implies $\mathcal{D}_{o\iota}$ is countably infinite by Lemma 4.5.4. Thus the result holds for $n = 0$.

Assume $\mathcal{R}_{\mathcal{P}^{2n+1}(\iota)} \subseteq \mathcal{K}_{\mathcal{P}^{2n+1}(\iota)}$ and $\mathcal{D}_{\mathcal{P}^{2n+1}(\iota)}$ is countably infinite. By Lemma 6.7.33, we know $\mathcal{R}_{\mathcal{P}^{2n+3}(\iota)} \subseteq \mathcal{K}_{\mathcal{P}^{2n+3}(\iota)}$. By Lemma 6.7.36, we know $\mathcal{D}_{\mathcal{P}^{2n+3}(\iota)}$ is countably infinite. Thus the induction step also holds. □

THEOREM 6.7.38. *For each $n \in \mathbb{N}$, $\mathcal{D}$ realizes $\Pi^{\mathcal{P}^{2n}(\iota)}$ and $\Sigma^{\mathcal{P}^{2n}(\iota)}$.*

PROOF. By Lemma 6.7.37, we know $\mathcal{R}_{o\mathcal{P}^{2n}(\iota)} \subseteq \mathcal{K}_{o\mathcal{P}^{2n}(\iota)}$. By Theorem 6.7.19, we conclude $\mathcal{D}$ realizes $\Pi^{\mathcal{P}^{2n}(\iota)}$ and $\Sigma^{\mathcal{P}^{2n}(\iota)}$. □

We can use Lemma 6.7.37 to prove the frame $\mathcal{D}$ does not realize $\Pi^{\mathcal{P}^{2n+1}(\iota)}$ and does not realize $\Sigma^{\mathcal{P}^{2n+1}(\iota)}$ for any n.

THEOREM 6.7.39. *For each $n \in \mathbb{N}$, $\mathcal{D}$ realizes neither $\Pi^{\mathcal{P}^{2n+1}(\iota)}$ nor $\Sigma^{\mathcal{P}^{2n+1}(\iota)}$.*

PROOF. Suppose $\pi^n \in \mathcal{D}_{o(o\mathcal{P}^{2n+1}(\iota))}$ satisfies $\mathfrak{L}_{\Pi^{\mathcal{P}^{2n+1}(\iota)}}(\pi^n)$. Let $k^n_{\mathtt{T}}$ be the constant function $k^n_{\mathtt{T}}(\mathsf{b}) := \mathtt{T}$ from $\mathcal{D}_{\mathcal{P}^{2n+1}(\iota)}$ to $\mathcal{D}_o$. Then we have $\pi^n(f) = \mathtt{T}$ iff $f = k^n_{\mathtt{T}}$ for every $f \in \mathcal{D}_{o\mathcal{P}^{2n+1}(\iota)}$.

By Lemma 6.7.37, we know $\mathcal{R}_{\mathcal{P}^{2n+1}(\iota)} \subseteq \mathcal{K}_{\mathcal{P}^{2n+1}(\iota)}$ and $\mathcal{D}_{\mathcal{P}^{2n+1}(\iota)}$ is countably infinite. Since $\mathcal{D}_{\mathcal{P}^{2n+1}(\iota)}$ is countably infinite, we can enumerate it as $\{\mathsf{b}_0, \mathsf{b}_1, \mathsf{b}_2, \ldots\}$ and define $u^m : \mathcal{D}_{\mathcal{P}^{2n+1}(\iota)} \to \mathcal{D}_o$ by

$$u^m(\mathsf{b}_i) = \begin{cases} \mathtt{T} & \text{if } i < m \\ \mathtt{F} & \text{otherwise} \end{cases}$$

for each $m \in \mathbb{N}$ and $\mathsf{b}_i \in \mathcal{D}_{\mathcal{P}^{2n+1}(\iota)}$. Clearly, $u^m \to_m^p k_{\mathtt{T}}^n$. Since each $u^m \neq k_{\mathtt{T}}^n$, we know $\pi^n(u^m) = \mathtt{F}$. On the other hand, we have

$$\pi^n(u^m) \to_m \pi^n(k_{\mathtt{T}}^n)$$

by Lemma 6.7.28 since $\mathcal{R}_{\mathcal{P}^{2n+1}(\iota)} \subseteq \mathcal{K}_{\mathcal{P}^{2n+1}(\iota)}$ and $\mathcal{R}_o \subseteq \mathcal{K}_o$. This contradicts $\pi^n(k_{\mathtt{T}}^n) = \mathtt{T}$.

A similar argument proves $\mathcal{D}$ does not realize $\Sigma^{\mathcal{P}^{2n+1}(\iota)}$. □

Finally, we conclude with independence results.

THEOREM 6.7.40. *Let*

$$\begin{aligned}\mathcal{S}_1 \quad := \quad & \mathcal{S}_{Bool} \cup \{\Pi^{\mathcal{P}^{2n}(\iota)}, \Sigma^{\mathcal{P}^{2n}(\iota)}, =^{\mathcal{P}^{2n+1}(\iota)} \mid n \in \mathbb{N}\} \\ & \cup \{\Pi^{\alpha}, \Sigma^{\alpha}, =^{\alpha} \mid \alpha \in \mathcal{T}_o\}\end{aligned}$$

and $\mathcal{S} \subseteq \mathcal{S}_1$. *For any* $n \in \mathbb{N}$,

$$\mathcal{S} \ntrianglelefteq_{\beta\mathfrak{fb}} \mathcal{S} \cup \{\Pi^{\mathcal{P}^{2n+1}(\iota)}\},$$

$$\mathcal{S} \ntrianglelefteq_{\beta\mathfrak{fb}} \mathcal{S} \cup \{\Sigma^{\mathcal{P}^{2n+1}(\iota)}\},$$

and

$$\mathcal{S} \ntrianglelefteq_{\beta\mathfrak{fb}} \mathcal{S} \cup \{=^{\mathcal{P}^{2n}(\iota)}\}.$$

PROOF. We know $\mathcal{M}^{\mathcal{D}}$ realizes exactly the constants in $\mathcal{S}^{\mathcal{D}}$. By Theorem 4.5.8, we know $\mathcal{S}_{Bool} \subseteq \mathcal{S}^{\mathcal{D}}$. By Theorem 6.7.38, we know

$$\{\Pi^{\mathcal{P}^{2n}(\iota)}, \Sigma^{\mathcal{P}^{2n}(\iota)} \mid n \in \mathbb{N}\} \subseteq \mathcal{S}^{\mathcal{D}}.$$

By Theorem 6.5.2, we know $\mathcal{M}^{\mathcal{D}}$ realizes

$$\{\Pi^{\alpha}, \Sigma^{\alpha}, =^{\alpha} \mid \alpha \in \mathcal{T}_o\}.$$

For each $n \in \mathbb{N}$, since $\mathcal{M}^{\mathcal{D}}$ realizes $\Pi^{\mathcal{P}^{2n}(\iota)}$ and $=^o$, it realizes $=^{\mathcal{P}^{2n+1}(\iota)}$ by Lemma 6.2.1. Thus $\mathcal{M}^{\mathcal{D}}$ realizes $\mathcal{S}_1$.

Let $n \in \mathbb{N}$ be given. Theorem 6.7.39 implies $\Pi^{\mathcal{P}^{2n+1}(\iota)} \notin \mathcal{S}^{\mathcal{D}}$ and $\Sigma^{\mathcal{P}^{2n+1}(\iota)} \notin \mathcal{S}^{\mathcal{D}}$. Applying Corollary 6.1.8 using the model $\mathcal{M}^{\mathcal{D}}$, we know

$$\mathcal{S} \ntrianglelefteq_{\beta\mathfrak{fb}} \mathcal{S} \cup \{\Pi^{\mathcal{P}^{2n+1}(\iota)}\}$$

and

$$\mathcal{S} \ntrianglelefteq_{\beta\mathfrak{fb}} \mathcal{S} \cup \{\Sigma^{\mathcal{P}^{2n+1}(\iota)}\}.$$

To verify

$$\mathcal{S} \ntrianglelefteq_{\beta\mathfrak{fb}} \mathcal{S} \cup \{=^{\mathcal{P}^{2n}(\iota)}\}$$

we must prove $\mathcal{M}^{\mathcal{D}}$ does not realize $=^{\mathcal{P}^{2n}(\iota)}$. For $n = 0$ we know $=^{\iota} \notin \mathcal{S}^{\mathcal{D}}$ by Lemma 6.7.14. Assume $n > 0$. Assume $\mathcal{M}^{\mathcal{D}}$ realizes $=^{\mathcal{P}^{2n}(\iota)}$. By Lemma 6.2.2, the model $\mathcal{M}^{\mathcal{D}}$ must realize $\Pi^{\mathcal{P}^{2n-1}(\iota)}$, a contradiction. □

Since the model $\mathcal{M}^{\mathcal{D}}$ realizes $\neg$, we know $\mathcal{M}^{\mathcal{D}} \models$ **X5304**. Consequently, while there is an internal injection from $\mathcal{D}_{o\iota}$ to $\mathcal{D}_{\iota}$, there is no internal surjection from $\mathcal{D}_{\iota}$ onto $\mathcal{D}_{o\iota}$. We conclude that not every internal injection is part of an internal retraction (cf. Definition 6.3.3).

REMARK 6.7.41. One may find it surprising that realization of a "third-order" quantifier $\Pi^{o(o\iota)}$ does not imply that a model realizes the "second-order" quantifier $\Pi^{o\iota}$. One might hope to simulate the second-order quantifier using the third-order quantifier and singletons. This looks especially hopeful since we have the quantifier $\Pi^{o(o\iota)}$ and equality $=^{o\iota}$ available. However, this must fail (as the model above demonstrates). Consider two possible definitions of $\Pi^{o\iota}$ in terms of $\Pi^{o(o\iota)}$. First consider

$$\lambda x_{o(o\iota)}\, \forall y_{o(o\iota)}\, [y \subseteq^{o(o\iota)} x]$$

where $\subseteq^{o(o\iota)}$ (written in infix) should be interpreted as the subset relation. This definition only works if there is an interpretation for $\subseteq^{o(o\iota)}$ in $\mathcal{D}_{o(o(o\iota))(o(o\iota))}$. Next consider

$$\lambda x_{o(o\iota)}\, \forall y_{o(o\iota)}\, [x\, [\iota^{o\iota}\, y]]$$

where $\iota^{o\iota}_{o\iota(o(o\iota))}$ is a description operator (taking singletons to their unique element). Such a definition only works if the model realizes such a description operator. Consequently, we can use the fact that the model above does not realize $\Pi^{o\iota}$ to conclude that the model does not have an element of $\mathcal{D}_{o(o(o\iota))(o(o\iota))}$ which acts as the subset relation and does not have an element of $\mathcal{D}_{o\iota(o(o\iota))}$ which acts as a description operator (in spite of the fact that the model does realize equality at type $o\iota$).

REMARK 6.7.42. Let $\mathcal{U}$ be a nonprincipal ultrafilter on $\mathcal{P}(\mathbb{N})$. As described in Remark 4.4.8 we can form the quotients of the applicative structures $(\mathcal{R}, S)$ and $(\mathcal{D}^{A}, S)$ by $\sim^{\mathcal{U}}$. The quotient of $(\mathcal{R}, S)$ by $\sim^{\mathcal{U}}$ will only have two truth values since every function in $\mathcal{R}_o$ is eventually constant. On the other hand, the quotient of $(\mathcal{D}^{A}, S)$ by

$\sim^{\mathcal{U}}$ will contain many truth values. Consider the elements of the domain of type ι in the quotient of $(\mathcal{R}, S)$. The set $\mathcal{R}_\iota$ consists of the eventually constant functions $f : \mathbb{N} \to \mathbb{N}$ and uniformly unbounded functions $g : \mathbb{N} \to \mathbb{N}$. The congruence class of any eventually constant $f : \mathbb{N} \to \mathbb{N}$ will correspond to a natural number (the eventual constant value of f). However, the congruence class of a uniformly unbounded function does not correspond to a natural number. Hence the quotient of $(\mathcal{R}, S)$ by $\sim^{\mathcal{U}}$ contains many nonstandard elements and is not isomorphic to the frame $\mathcal{D}$.

Part II

Automated Reasoning

CHAPTER 7

Extensional Expansion Proofs

Expansion proofs provide a compact representation of proofs. In particular, an expansion proof contains the essential information regarding instantiations and which atoms are used in initial sequents without recording all the information about the order of sequent rule applications.

In [43] and [44] Dale Miller defined expansion proofs for theorems of elementary type theory. Such proofs consist of expansion trees with an acyclic dependence relation and a complete mating.

Frank Pfenning defined a notion of extensional expansion proofs in his thesis [53]. These extensional expansion proofs include nodes which introduce extensionality axioms (in terms of Leibniz equality) for functions and relations.

The notion of extensional expansion proofs presented here will not explicitly introduce extensionality axioms. Instead, extensionality will be handled by including nodes for dealing with functional and boolean equations and explicitly including nodes representing connections. Due to the existence of nodes for connections, the extensional expansion proof is not formed using a tree, but a directed acyclic graph (dag).

We will also generalize expansion trees appropriately for restricted signatures by only allowing expansion terms to be chosen from terms constructed using the restricted signature. Also, we will eagerly apply $\beta\eta$-normalization and include nodes for applying the $\mathbf{A}^\sharp$ operation when appropriate.

Before giving technical definitions and proofs, let us consider examples which demonstrate the differences between expansion trees and extensional expansion dags.

Expansion trees are constructed based on the syntax of the formula with extra information associated with quantifiers. A very simple example of an expansion tree for the proposition $[\underset{\cdot}{\neg} A_o \underset{\cdot}{\vee} A_o]$ is

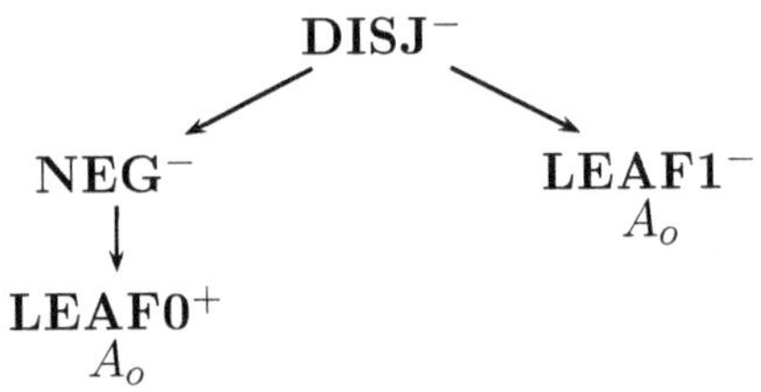

Associated with each node is a polarity + or −. The root node is negative since we think of an expansion tree as providing the basis for a refutation. In this example, we can provide a *complete mating* (cf. [7]) by connecting positive node **LEAF0** with negative node **LEAF1**. In general, we would also need to unify the corresponding formulas, but in this case they are both A. We refer to *mating search* as the search for a complete mating.

In the extensional case unification is not independent from the mating search. That is, we may need to add more connections in order to prove two terms are equal. For this reason, it is useful to build nodes for mates directly into the representation. This is the central reason why we use dags instead of trees. An example of such a dag for $[\underset{\cdot}{\neg}\mathbf{A} \underset{\cdot}{\vee} \mathbf{A}]$ would be of the form

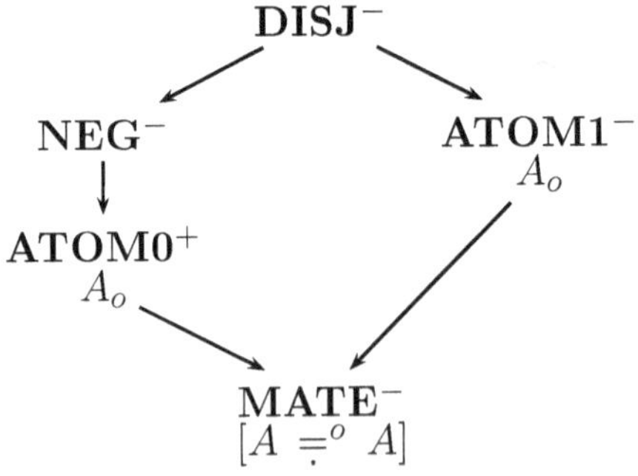

We call the nodes **ATOM0** and **ATOM1** in the dag above *atomic nodes* instead of *leaves* since they are clearly *not* leaves of the structure. In this case, the node **MATE** is a leaf in the sense that it has no children.

THM617 (cf. Definition 2.2.8)

$$[\neg A_o \mathbin{\underset{\cdot}{\vee}} \neg B_o \mathbin{\underset{\cdot}{\vee}} \neg[P_{oo}\, A] \mathbin{\underset{\cdot}{\vee}} [P\, B]].$$

is an example which requires extensionality. An expansion tree for this formula would be

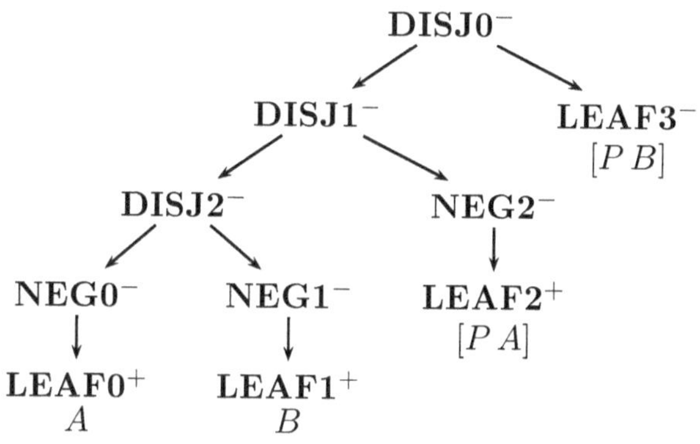

We could try to connect **LEAF2** to **LEAF3**, but we will never be able to syntactically unify $[P\,A]$ and $[P\,B]$. Extensionally, we can prove these are equal since the arguments A_o and B_o are both "true" in the sense that there are positive nodes **LEAF0** and **LEAF1** with associated formulas A and B, respectively. An extensional expansion dag which would provide a proof of **THM617** is shown in Figure 7.1. One can see in this figure that more connections (**MATE1** and **MATE2**) result from the attempt to prove $[P\,A]$ and $[P\,B]$ are equal (**MATE0**). This extensional expansion dag should be compared to the sequent calculus derivation shown in Figure 2.7.

Another simple example is **THM618** (cf. Definition 2.2.8)

$$[\neg[C_\iota \mathrel{\underset{\cdot}{=}^{\iota}} D_\iota] \mathbin{\underset{\cdot}{\vee}} \neg[Q_{o\iota}\, C] \mathbin{\underset{\cdot}{\vee}} [Q\, D]].$$

One can prove this using expansion trees if equality is interpreted as Leibniz equality (requiring instantiation of a set variable). Using primitive equality, we would like to connect two equations between

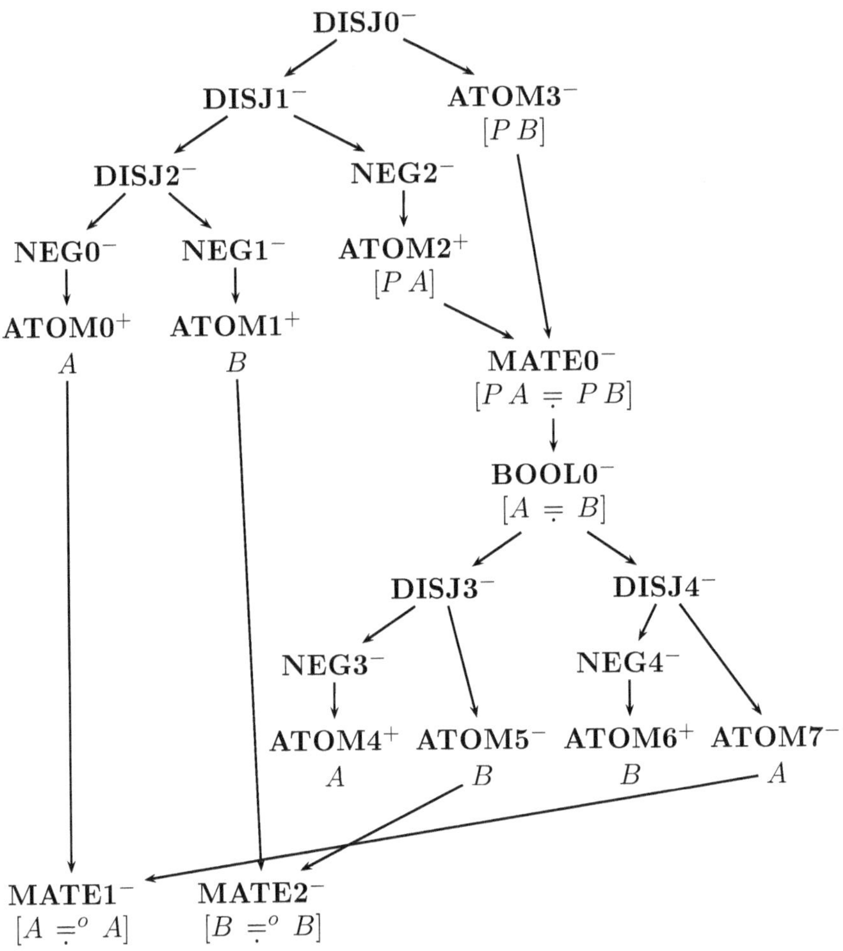

FIGURE 7.1. Extensional Expansion Dag for **THM617**

terms of type ι. This is handled differently from mating atoms, primarily because we must handle equality up to symmetry. An extensional expansion dag which would provide a proof of **THM618** is shown in Figure 7.2. There is a connection in this dag between the positive node **EQN0** and negative node **EQNGOAL1** yielding a

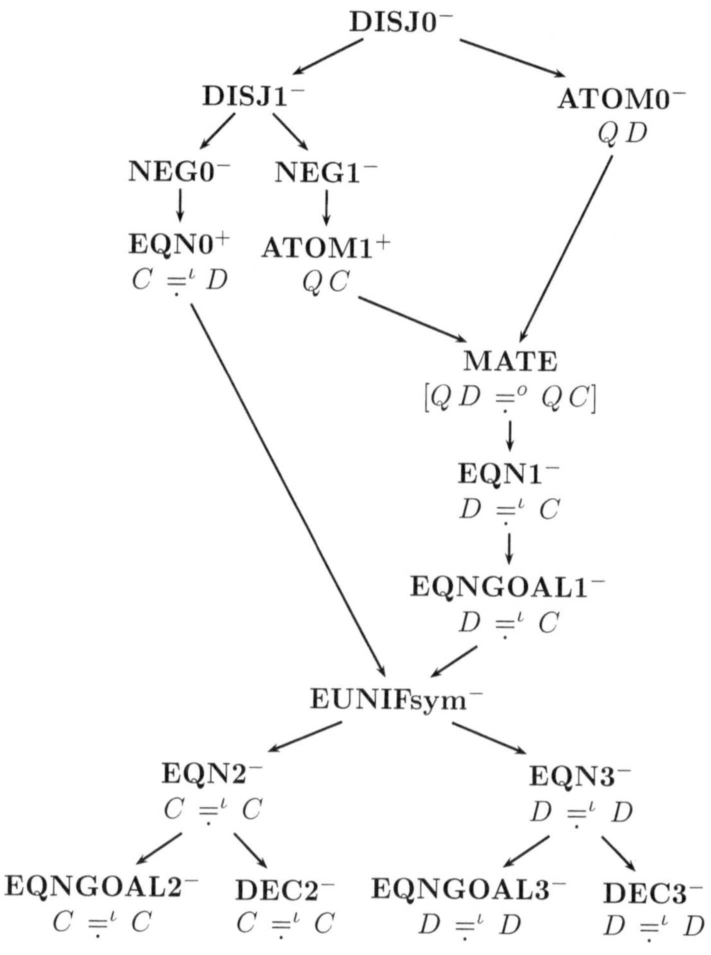

FIGURE 7.2. Extensional Expansion Dag for **THM618**

negative node **EUNIFsym**. We call the node **EUNIFsym** a *symmetric E-unification node.* Such nodes have two children which represent two equations to be proven. This extensional expansion dag should be compared to the sequent calculus derivation shown in Figure 2.8.

A few comments about the negative nodes associated with equations in Figure 7.2 are in order. First we explain the nature of the negative nodes associated with equations and then we explain why the different nodes are needed. The node **EQN1** has one child **EQNGOAL1** associated with the same equation. This child, **EQNGOAL1**, is connected with **EQN0**. On the other hand, **EQN2** and **EQN3** each have two children. The node **EQN2** has children **DEC2** and **EQNGOAL2**. The nodes **EQNGOAL1**, **EQNGOAL2** and **EQNGOAL3** are *equation goal nodes.* The purpose of such nodes is to connect with positive equations. The nodes **DEC2** and **DEC3** are called *decomposition nodes.* Such nodes are only included if the two sides of the equation have the same parameter at the head, e.g. $[H\,\overline{\mathbf{A}^n} \doteq H\,\overline{\mathbf{B}^n}]$ where H is a parameter. In general, the number of children of decomposition nodes depends on the arity of the parameter at the head. In particular, **DEC2** and **DEC3** have no children because C_ι and D_ι both have arity 0. The equation goal child **EQNGOAL1** of **EQN1** plays a vital role in the proof. On the other hand, the decomposition children of **EQN2** and **EQN3** are relevant to the proof while the equation goal nodes **EQNGOAL2** and **EQNGOAL3** are irrelevant. There are examples where both children are important to form the proof. A particular example in which both children are crucial is **THM615**:

$$H_{\iota o}\,[H\,\top =^\iota H\,\bot] \doteq^\iota H\,\bot$$

To understand the complications arising from negative nodes corresponding to equations between terms of base type ι, reconsider the following three rules

$$\frac{\Gamma, [\mathbf{A}^1 \doteq \mathbf{B}^1] \quad \cdots \quad \Gamma, [\mathbf{A}^n \doteq \mathbf{B}^n] \quad H_{\iota\alpha^n\cdots\alpha^1} \in \mathcal{P}}{\Gamma, [[H\,\overline{\mathbf{A}^n}] \doteq^\iota [H\,\overline{\mathbf{B}^n}]]}\,\mathcal{G}(Dec)$$

$$\frac{\Gamma, [\mathbf{A} \doteq^\iota \mathbf{C}] \quad \Gamma, [\mathbf{B} \doteq^\iota \mathbf{D}]}{\Gamma, \neg[\mathbf{A} \doteq^\iota \mathbf{B}], [\mathbf{C} \doteq^\iota \mathbf{D}]}\,\mathcal{G}(EUnif_1)$$

and

$$\frac{\Gamma, [\mathbf{A} \doteq^\iota \mathbf{D}] \quad \Gamma, [\mathbf{B} \doteq^\iota \mathbf{C}]}{\Gamma, \neg[\mathbf{A} \doteq^\iota \mathbf{B}], [\mathbf{C} \doteq^\iota \mathbf{D}]}\,\mathcal{G}(EUnif_2)$$

of the sequent calculus $\mathcal{G}^{\mathcal{S}}_{\beta\mathfrak{fb}}$. If we are attempting to derive a sequent $\Gamma, [[H\,\overline{\mathbf{A}^n}] \doteq^\iota [H\,\overline{\mathbf{B}^n}]]$ using a rule with $[[H\,\overline{\mathbf{A}^n}] \doteq^\iota [H\,\overline{\mathbf{B}^n}]]$ as a principal formula, then we may apply any of these three rules. (We

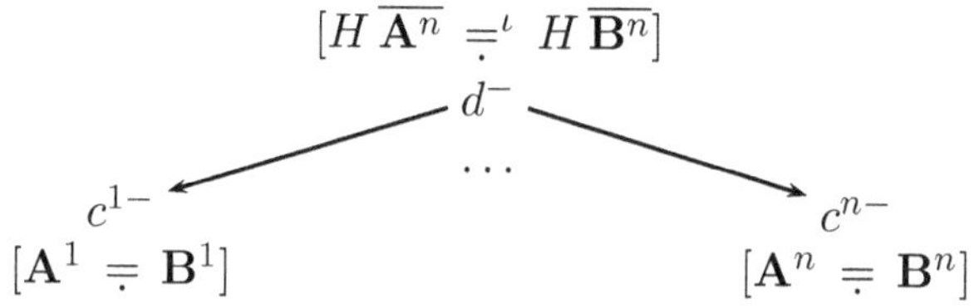

FIGURE 7.3. Decomposition Nodes

may also apply the contraction rule $\mathcal{G}(Contr)$.) Note that the rule $\mathcal{G}(Dec)$ has n different premises (where n is the arity of the parameter H at the head of both sides of the equation) while the rules $\mathcal{G}(EUnif_1)$ and $\mathcal{G}(EUnif_2)$ always have two premises. Furthermore, the rule $\mathcal{G}(Dec)$ has one principal formula while the rules $\mathcal{G}(EUnif_1)$ and $\mathcal{G}(EUnif_2)$ have two principal formulas. In these two ways, the character of the rule $\mathcal{G}(Dec)$ differs fundamentally from the character of the rules $\mathcal{G}(EUnif_1)$ and $\mathcal{G}(EUnif_2)$. There will be a natural correspondence between nodes in an extensional expansion proof and the structure of a sequent calculus derivation. (The correspondence will not be one-to-one since an extensional expansion proof will compile away certain details of a sequent derivation.)

Suppose we wish to represent an application of the $\mathcal{G}(Dec)$ rule. There must be a negative node d in an extensional expansion dag with shallow formula $[[H\,\overline{\mathbf{A}^n}] \underset{\cdot}{=}^{\iota} [H\,\overline{\mathbf{B}^n}]]$ corresponding to the principal formula. To obtain nodes corresponding to the premises of the $\mathcal{G}(Dec)$ rule, we extend the node d to have n (ordered) children $c^1, \ldots, c^n$ where c^i is negative and has shallow formula $[\mathbf{A}^i \underset{\cdot}{=}^{\iota} \mathbf{B}^i]$ for each $i \in \{1, \ldots, n\}$. (In the special case where n is 0, the node d will have no children.) Each new node c^i will have d as the unique parent since $[[H\,\overline{\mathbf{A}^n}] \underset{\cdot}{=}^{\iota} [H\,\overline{\mathbf{B}^n}]]$ is the only principal formula of the rule $\mathcal{G}(Dec)$. This situation is shown in Figure 7.3.

Next suppose we wish to represent an application of either the $\mathcal{G}(EUnif_1)$ or $\mathcal{G}(EUnif_2)$ rule. We only consider the rule $\mathcal{G}(EUnif_1)$ since the rule $\mathcal{G}(EUnif_2)$ is symmetric. Since $\mathcal{G}(EUnif_1)$ has two principal formulas (one of which is negated), there must be a negative node g with shallow formula $[\mathbf{C} \underset{\cdot}{=}^{\iota} \mathbf{D}]$ and a positive node b with shallow formula $[\mathbf{A} \underset{\cdot}{=}^{\iota} \mathbf{B}]$ corresponding to these principal formulas. To obtain nodes corresponding to the premises of the $\mathcal{G}(EUnif_1)$ rule,

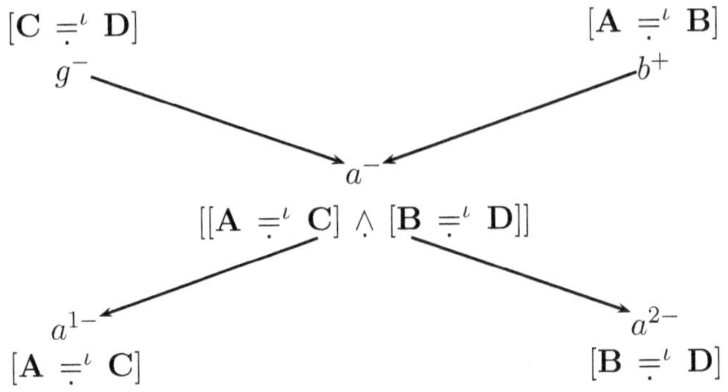

FIGURE 7.4. E-Unification Connections

we need two negative nodes a^1 and a^2 with shallow formulas $[\mathbf{A} \underset{\cdot}{=}^{\iota} \mathbf{C}]$ and $[\mathbf{B} \underset{\cdot}{=}^{\iota} \mathbf{D}]$, respectively. We would like a^1 and a^2 to be children of the nodes g and b. We accomplish this in a natural way by including a connection node a. The node a is a negative node with shallow formula

$$[[\mathbf{A} \underset{\cdot}{=}^{\iota} \mathbf{C}] \underset{\cdot}{\land} [\mathbf{B} \underset{\cdot}{=}^{\iota} \mathbf{D}]]$$

two parents g and b, and two (ordered) children a^1 and a^2. This situation is shown in Figure 7.4.

Now consider a negative node e in an extensional expansion dag with shallow formula $[[H\,\overline{\mathbf{A}^n}] \underset{\cdot}{=}^{\iota} [H\,\overline{\mathbf{B}^n}]]$. Suppose we do not yet know if this should correspond to an application of $\mathcal{G}(Dec)$, $\mathcal{G}(EUnif_1)$ or $\mathcal{G}(EUnif_2)$. Attempting to allow the node e to correspond to a principal formula in any of these rules, we might be persuaded to allow ordered children of e corresponding to a decomposition and also children of e corresponding to E-unification connections. A simpler method for representing such situations is to split the two different uses for negative equations into two different kinds of nodes as in Figure 7.5. The node e will be a *primary node*, the node d will be a *decomposition node* and the node g will be a *equation goal node*. In terms of the sequent calculus, this corresponds to forcing an application of contraction before applying either $\mathcal{G}(Dec)$ or one of the rules $\mathcal{G}(EUnif_1)$ or $\mathcal{G}(EUnif_2)$.

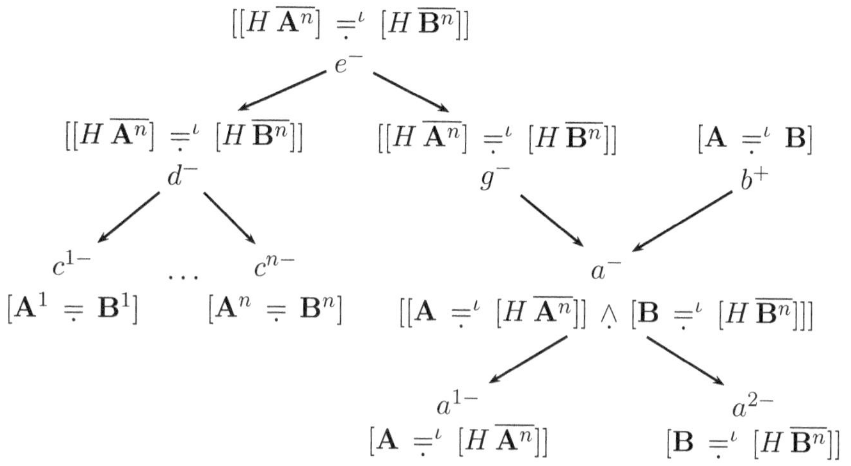

FIGURE 7.5. Decomposition and E-Unification

We now turn from this general discussion to the definitions of extensional expansion dags and extensional expansion proofs. In this chapter we construct extensional expansion dags with certain properties. We will also establish soundness and completeness of extensional expansion proofs by translating back and forth from sequent derivations. In Chapter 8 we will prove a lifting result which forms the basis for establishing completeness of automated search procedures.

Throughout this chapter we assume a fixed (but arbitrary) signature $\mathcal{S}$ except when explicitly noted otherwise.

7.1. Directed Acyclic Graphs

Before defining expansion dags, we first define dags (directed acyclic graphs).

In general, a graph is simply a set of nodes and edges. In a directed graph, the edges have a source node and a target node.

DEFINITION 7.1.1 (Directed Graph). A (finite) *directed graph* $\mathcal{G}$ is a finite set of nodes $\mathcal{N}$ and a binary relation $\rightharpoondown$ on $\mathcal{N}$.

Next we define the notions of parents, children, roots, walks, etc.

DEFINITION 7.1.2. Let $\mathcal{G} = \langle \mathcal{N}, \rightharpoondown \rangle$ be a directed graph and $a, b \in \mathcal{N}$ be nodes. If $a \rightharpoondown b$, then we say a is a *parent* of b (or b is a *child* of a).

We say a is a *leaf node* of $\mathcal{G}$ if $a \in \mathcal{N}$ has no children. We say a is a *root node* of $\mathcal{G}$ if $a \in \mathcal{N}$ has no parents. We define $\mathbf{Roots}(\mathcal{G}) \subseteq \mathcal{N}$ to be the set of root nodes of $\mathcal{G}$.

A *walk* (from b^0 to b^m) in $\mathcal{G}$ is an $(m+1)$-tuple $\langle b^0, \ldots, b^m \rangle$ where $m \geq 0$, $b^0, \ldots, b^m \in \mathcal{N}$ and $b^j \rightharpoondown b^{j+1}$ for each j $(0 \leq j \leq m-1)$. We define $\mathbf{Walks}^{\mathcal{G}}(c, d)$ to be the set of walks from c to d in $\mathcal{G}$. Also, given any set of nodes $A \subseteq \mathcal{N}$, we define

$$\mathbf{Walks}^{\mathcal{G}}(A, d) := \bigcup_{a \in A} \mathbf{Walks}^{\mathcal{G}}(a, d)$$

and

$$\mathbf{Walks}^{\mathcal{G}}(c, A) := \bigcup_{a \in A} \mathbf{Walks}^{\mathcal{G}}(c, a).$$

In particular, let

$$\mathbf{RootWalks}^{\mathcal{G}}(d) := \mathbf{Walks}^{\mathcal{G}}(\mathbf{Roots}(\mathcal{G}), d)$$

be the set of walks from a root node of $\mathcal{G}$ to d.

We say a node b^0 *dominates* a node b^m if there is a walk $\langle b^0, \ldots, b^m \rangle$ in $\mathcal{G}$ with $m \geq 0$.

We say b^0 is an *ancestor* of b^m (or b^m is a *descendant* of b^0) if there is a walk $\langle b^0, \ldots, b^m \rangle$ in $\mathcal{G}$ with $m \geq 1$.

Note that every node dominates itself. On the other hand, nodes are generally not ancestors of themselves.

DEFINITION 7.1.3 (Directed Acyclic Graphs (dags)). A *dag* (directed acyclic graph) is a directed graph $\mathcal{G} = \langle \mathcal{N}, \rightharpoondown \rangle$ such that no node $a \in \mathcal{N}$ is an ancestor of itself.

Acyclicity forces there to be only finitely many walks. Also, we can prove every node has a walk from a root node.

LEMMA 7.1.4. *Let $\mathcal{G} = \langle \mathcal{N}, \rightharpoondown \rangle$ be a dag. The number of walks in $\mathcal{G}$ is finite.*

PROOF. Let n be the number of nodes in $\mathcal{G}$. By acyclicity, no node can be repeated in a walk. Thus there cannot be more than $n!$ walks in $\mathcal{G}$. $\square$

This result allows us to define the depth of a node in a dag. This will be useful for inductive proofs.

DEFINITION 7.1.5. Let $\mathcal{G} = \langle \mathcal{N}, \rightharpoondown \rangle$ be a dag. For each $a \in \mathcal{N}$, let the *depth* $|a|_d^{\mathcal{G}}$ *of* a *in* $\mathcal{G}$ be the largest $n \geq 0$ such that there exists a walk $\langle a^0, \ldots, a^n \rangle \in \mathbf{Walks}^{\mathcal{G}}(\mathcal{N}, a)$.[1] When the dag $\mathcal{G}$ is clear in context, we write $|a|_d$ for $|a|_d^{\mathcal{G}}$.

We also define the height of a node.

DEFINITION 7.1.6. Let $\mathcal{G} = \langle \mathcal{N}, \rightharpoondown \rangle$ be a dag. For each $a \in \mathcal{N}$, let the *height* $|a|_h^{\mathcal{G}}$ *of* a *in* $\mathcal{G}$ be the largest $n \geq 0$ such that there exists a walk $\langle a^0, \ldots, a^n \rangle \in \mathbf{Walks}^{\mathcal{G}}(a, \mathcal{N})$. When the dag $\mathcal{G}$ is clear in context, we write $|a|_h$ for $|a|_h^{\mathcal{G}}$. We further define the *height* $|\mathcal{G}|_h$ of the dag $\mathcal{G}$ to be the largest $n \geq 0$ such that there exists a walk $\langle a^0, \ldots, a^n \rangle \in \mathbf{Walks}^{\mathcal{G}}(\mathcal{N}, \mathcal{N})$.

The depth of any parent of a node is less than the depth of the node. Likewise, the height of any child of a node is less than the height of a node.

LEMMA 7.1.7. *Let* $\mathcal{G} = \langle \mathcal{N}, \rightharpoondown \rangle$ *be a dag and* $a, b \in \mathcal{N}$. *If* $a \rightharpoondown b$, *then* $|a|_d < |b|_d$ *and* $|a|_h > |b|_h$

PROOF. Let m be $|a|_d$. Consequently, there is a walk $\langle c^0, \ldots, c^m \rangle$ in $\mathbf{Walks}^{\mathcal{G}}(\mathcal{N}, a)$ where $c^m = a$. Since $\langle c^0, \ldots, c^m, b \rangle$ is a walk in $\mathbf{Walks}^{\mathcal{G}}(\mathcal{N}, a)$ we have $|a|_d \geq m + 1 > m$. To verify $|a|_h > |b|_h$, we simply note that any walk $\langle b, d^1, \ldots, d^n \rangle$ can be extended to the walk $\langle a, b, d^1, \ldots, d^n \rangle$. □

We can use the depth measure to prove every node in a dag has a walk from a root node. In particular, if there is any node, then there must be at least one root node.

LEMMA 7.1.8. *Let* $\mathcal{G} = \langle \mathcal{N}, \rightharpoondown \rangle$ *be a dag. For every* $a \in \mathcal{N}$, *there exists a walk from a root node of* $\mathcal{G}$ *to* a.

PROOF. We prove this by induction on $|a|_d$. First note that if a is a root node of $\mathcal{G}$, then $\langle a \rangle$ is the only walk from a root to a (since a has no parents). Otherwise, a has some parent b. By the inductive hypothesis there is a walk $\langle c^1, \ldots, c^m, b \rangle \in \mathbf{RootWalks}^{\mathcal{G}}(b)$

[1] $\mathbf{Walks}^{\mathcal{G}}(\mathcal{N}, a)$ is nonempty since $\langle a \rangle \in \mathbf{Walks}^{\mathcal{G}}(\mathcal{N}, a)$.

since $|b|_d < |a|_d$. Thus $\langle c^1, \ldots, c^m, b, a\rangle$ is a walk from a root node of $\mathcal{G}$ to a. □

LEMMA 7.1.9. *Let $\mathcal{G} = \langle \mathcal{N}, \rightharpoondown\rangle$ be a dag. If $\mathcal{N}$ is nonempty, then there is a root node of $\mathcal{G}$ (i.e.,* **Roots**$(\mathcal{G})$ *is nonempty).*

PROOF. Choose $a \in \mathcal{N}$. A walk $\langle c^1, \ldots, c^m, a\rangle \in \mathbf{RootWalks}^{\mathcal{G}}(a)$ exists by Lemma 7.1.8. Thus $c^1 \in \mathbf{Roots}(\mathcal{G})$. □

In order to make it easier to check acyclicity, we sometimes use subgraphs.

DEFINITION 7.1.10 (Subgraphs). Let

$$\mathcal{G} = \langle \mathcal{N}, \rightharpoondown\rangle \text{ and } \mathcal{G}' = \langle \mathcal{N}', \rightharpoondown'\rangle$$

be directed graphs. We say $\mathcal{G}$ is a *subgraph* of $\mathcal{G}'$ if $\mathcal{N} \subseteq \mathcal{N}'$ and for every $a, b \in \mathcal{N}$, if $a \rightharpoondown b$, then $a \rightharpoondown' b$.

LEMMA 7.1.11. *Let $\mathcal{G} = \langle \mathcal{N}, \rightharpoondown\rangle$ and $\mathcal{G}' = \langle \mathcal{N}', \rightharpoondown'\rangle$ be directed graphs. If $\mathcal{G}'$ is a dag and $\mathcal{G}$ is a subgraph of $\mathcal{G}'$, then $\mathcal{G}$ is a dag.*

PROOF. If $\langle a^1, \ldots, a^n, a^1\rangle$ is a cyclic walk in $\mathcal{G}$, then it is also a cyclic walk in $\mathcal{G}'$. □

LEMMA 7.1.12. *Let $\mathcal{G} = \langle \mathcal{N}, \rightharpoondown\rangle$ and $\mathcal{G}' = \langle \mathcal{N}', \rightharpoondown'\rangle$ be directed graphs. If $\mathcal{G}$ is a dag, $\mathcal{G}$ is a subgraph of $\mathcal{G}'$, and every node $a \in \mathcal{N}' \setminus \mathcal{N}$ is a leaf node in $\mathcal{G}'$, then $\mathcal{G}'$ is a dag.*

PROOF. Suppose $\langle a^1, \ldots, a^n, a^1\rangle$ is a cyclic walk in $\mathcal{G}'$. Let a^{n+1} be a^1. Since $\mathcal{G}$ is a dag, there must be some $a^i \in \mathcal{N}' \setminus \mathcal{N}$ where $1 \le i \le n$. By assumption, a^i is a leaf node of $\mathcal{G}'$, contradicting $a^i \rightharpoondown' a^{i+1}$. Thus $\mathcal{G}'$ is a dag. □

7.2. Expansion Structures

Before defining extensional expansion dags we define expansion structures. An expansion structure is essentially a directed graph with annotated nodes. Expansion structures include labels for nodes. These labels include information such as the associated kind of node, polarity and proposition (shallow formula). We will partition the labels into several disjoint sets so the kind of node can be determined. We define a finite number of *kinds* for this purpose.

DEFINITION 7.2.1 (Kinds). Choose distinct values **primary** (the *primary kind*), **dec** (the *decomposition kind*), **eqngoal** (the *equation goal kind*), **mate** (the *mate kind*), **eunif** (the *E-unification kind*) and $\mathbf{eunif}^{sym}$ (the *symmetric E-unification kind*).

We call these values *kinds*.

DEFINITION 7.2.2 (Labels). Let $\mathcal{S}$ be a signature of logical constants. We define mutually disjoint sets $\mathbf{Primary}^{\mathcal{S}}$, $\mathbf{Dec}^{\mathcal{S}}$, $\mathbf{EqnGoal}^{\mathcal{S}}$, $\mathbf{Mate}^{\mathcal{S}}$, $\mathbf{EUnif}^{\mathcal{S}}$ and $\mathbf{EUnif}^{\mathcal{S}}_{sym}$ as follows:

$$
\begin{array}{lll}
\mathbf{Primary}^{\mathcal{S}} & := & \{\langle \mathbf{primary}, p, \mathbf{M}\rangle \mid p \in \{-1,1\}, \mathbf{M} \in prop(\mathcal{S})^{\downarrow}\} \\
\mathbf{Dec}^{\mathcal{S}} & := & \{\langle \mathbf{dec}, -1, [[H\,\mathbf{A}^1 \cdots \mathbf{A}^n] \doteq^{\iota} [H\,\mathbf{B}^1 \cdots \mathbf{B}^n]]\rangle \mid \\
 & & \qquad n \geq 0,\ H_{\iota\alpha^n\cdots\alpha^1} \in \mathcal{P}_{\iota\alpha^n\cdots\alpha^1}, \\
 & & \qquad \mathbf{A}^i, \mathbf{B}^i \in \mathit{wff}_{\alpha^i}(\mathcal{S})^{\downarrow} \text{ for } 1 \leq i \leq n\} \\
\mathbf{EqnGoal}^{\mathcal{S}} & := & \{\langle \mathbf{eqngoal}, -1, [\mathbf{A} \doteq^{\iota} \mathbf{B}]\rangle \mid \mathbf{A}, \mathbf{B} \in \mathit{wff}_{\iota}(\mathcal{S})^{\downarrow}\} \\
\mathbf{Mate}^{\mathcal{S}} & := & \{\langle \mathbf{mate}, -1, [[P\,\mathbf{A}^1 \cdots \mathbf{A}^n] \doteq^{o} [P\,\mathbf{B}^1 \cdots \mathbf{B}^n]]\rangle \mid \\
 & & \qquad n \geq 0,\ P_{o\alpha^n\cdots\alpha^1} \in \mathcal{P}_{o\alpha^n\cdots\alpha^1}, \\
 & & \qquad \mathbf{A}^i, \mathbf{B}^i \in \mathit{wff}_{\alpha^i}(\mathcal{S})^{\downarrow} \text{ for } 1 \leq i \leq n\} \\
\mathbf{EUnif}^{\mathcal{S}} & := & \{\langle \mathbf{eunif}, -1, [[\mathbf{A} \doteq^{\iota} \mathbf{C}] \mathbin{\dot\wedge} [\mathbf{B} \doteq^{\iota} \mathbf{D}]]\rangle \mid \\
 & & \qquad \mathbf{A}, \mathbf{B}, \mathbf{C}, \mathbf{D} \in \mathit{wff}_{\iota}(\mathcal{S})^{\downarrow}\} \\
\mathbf{EUnif}^{\mathcal{S}}_{sym} & := & \{\langle \mathbf{eunif}^{sym}, -1, [[\mathbf{A} \doteq^{\iota} \mathbf{C}] \mathbin{\dot\wedge} [\mathbf{B} \doteq^{\iota} \mathbf{D}]]\rangle \mid \\
 & & \qquad \mathbf{A}, \mathbf{B}, \mathbf{C}, \mathbf{D} \in \mathit{wff}_{\iota}(\mathcal{S})^{\downarrow}\}
\end{array}
$$

Finally, we define $\mathbf{Labels}^{\mathcal{S}}$ to be the union

$$
\mathbf{Primary}^{\mathcal{S}} \cup \mathbf{Dec}^{\mathcal{S}} \cup \mathbf{EqnGoal}^{\mathcal{S}}
$$
$$
\cup \mathbf{Mate}^{\mathcal{S}} \cup \mathbf{EUnif}^{\mathcal{S}} \cup \mathbf{EUnif}^{\mathcal{S}}_{sym}.
$$

We often leave the signature implicit and write **Labels**, **Primary**, **Dec**, **EqnGoal**, **Mate**, **EUnif** and $\mathbf{EUnif}_{sym}$ for $\mathbf{Labels}^{\mathcal{S}}$, $\mathbf{Primary}^{\mathcal{S}}$, $\mathbf{Dec}^{\mathcal{S}}$, $\mathbf{EqnGoal}^{\mathcal{S}}$, $\mathbf{Mate}^{\mathcal{S}}$, $\mathbf{EUnif}^{\mathcal{S}}$ and $\mathbf{EUnif}^{\mathcal{S}}_{sym}$, respectively.

We include a distinct kind with the polarity and formula to ensure mutual disjointness of the sets defined in Definition 7.2.2. Without this extra information the sets $\mathbf{Primary}^{\mathcal{S}}$, $\mathbf{EUnif}^{\mathcal{S}}$ and $\mathbf{EUnif}^{\mathcal{S}}_{sym}$ would intersect. Also, the sets $\mathbf{Primary}^{\mathcal{S}}$, $\mathbf{Dec}^{\mathcal{S}}$ and $\mathbf{EqnGoal}^{\mathcal{S}}$ would intersect, as would $\mathbf{Primary}^{\mathcal{S}}$ and $\mathbf{Mate}^{\mathcal{S}}$.

DEFINITION 7.2.3 (Expansion Structure). Let $\mathcal{S}$ be a signature of logical constants. An *$\mathcal{S}$-expansion structure* is a tuple

$$\langle \mathcal{N}, succ, label, sel, exps, conns, decs \rangle$$

where

- $\mathcal{N}$ is a finite set,
- $succ \subseteq \mathcal{N} \times (\bigcup_{n \geq 0} \mathcal{N}^n)$ is a partial function,
- $label : \mathcal{N} \to \mathbf{Labels}^{\mathcal{S}}$,
- $sel \subseteq \mathcal{N} \times \mathcal{P}$ is a partial function,
- $exps \subseteq \mathcal{N} \times \bigcup_{\alpha \in \mathcal{T}} \mathit{wff}_{\alpha}(\mathcal{S}) \times \mathcal{N}$ is a finite set,
- $conns \subseteq \mathcal{N}^3$ and
- $decs \subseteq \mathcal{N}^2$.

Given such an $\mathcal{S}$-expansion structure

$$\mathcal{Q} = \langle \mathcal{N}, succ, label, sel, exps, conns, decs \rangle$$

we use the following notation and terminology:

- $\mathcal{N}^{\mathcal{Q}} := \mathcal{N}$ is the set of *nodes*.
- $succ^{\mathcal{Q}} := succ$ is the *successor function*.
- $label^{\mathcal{Q}} := label$ is the *labeling function*.
- $sel^{\mathcal{Q}} := sel$ is the *selected parameter function*.
- $exps^{\mathcal{Q}} := exps$ is the set of *expansion arcs*.
- $conns^{\mathcal{Q}} := conns$ is the set of *connections*.
- $decs^{\mathcal{Q}} := decs$ is the set of *decomposition arcs*.

We often refer to $\mathcal{S}$-expansion structures simply as *expansion structures* and leave the signature $\mathcal{S}$ implicit.

Each node in an expansion structure has an associated kind, polarity and shallow formula.

DEFINITION 7.2.4 (Kind, Polarity, Shallow Formula). Let $\mathcal{Q}$ be an expansion structure. For each node $a \in \mathcal{N}^{\mathcal{Q}}$ we know

$$label^{\mathcal{Q}}(a) = \langle k, p, \mathbf{M} \rangle$$

for some kind k, $p \in \{-1, 1\}$ and $\mathbf{M} \in prop(\mathcal{S})^{\downarrow}$. We define the *kind* of a (denoted $kind^{\mathcal{Q}}(a)$) to be k. We define the *polarity* of a (denoted $pol^{\mathcal{Q}}(a)$) to be $p \in \{-1, 1\}$. We say the node is *positive* if $p = 1$ and say the node is *negative* if $p = -1$. We define the *shallow formula* of a (denoted $sh^{\mathcal{Q}}(a)$) to be $\mathbf{M} \in prop(\mathcal{S})$.

When the expansion structure $\mathcal{Q}$ is clear in context, we write $kind(a)$ for $kind^{\mathcal{Q}}(a)$, $pol(a)$ for $pol^{\mathcal{Q}}(a)$ and $sh(a)$ for $sh^{\mathcal{Q}}(a)$.

Sometimes we wish to combine the polarity with the shallow formula by adding a negation. We define two choices.

DEFINITION 7.2.5 (Signed and Dual Shallow Formulas). Let $\mathcal{Q}$ be an expansion structure and a be a node in $\mathcal{Q}$. We define the *signed shallow formula* $\underline{sh}^{\mathcal{Q}}(a)$ of a to be $sh^{\mathcal{Q}}(a)$ if a is positive and $\dot{\neg} sh^{\mathcal{Q}}(a)$ if a is negative. We define the *dual shallow formula* $\overline{sh}^{\mathcal{Q}}(a)$ of a to be $\dot{\neg} sh^{\mathcal{Q}}(a)$ if a is positive and $sh^{\mathcal{Q}}(a)$ if a is negative.

When the expansion structure $\mathcal{Q}$ is clear in context, we write $\underline{sh}(a)$ for $\underline{sh}^{\mathcal{Q}}(a)$ and $\overline{sh}(a)$ for $\overline{sh}^{\mathcal{Q}}(a)$.

We say a node is *extended* if the node is in the domain of the successor function. Note, however, that the successor function may have value $\langle\rangle$. In such a case the node is extended, but the successor function lists zero children. The successor function does not necessarily list *all* the children of a node (cf. Definition 7.2.11 and Figure 7.6). Instead, the successor function provides a list of the *ordered* children of a node. For example, an extended node a with shallow formula $[\mathbf{M} \dot{\vee} \mathbf{N}]$ should have two ordered children b and c such that $succ(a) = \langle b, c \rangle$, b has shallow formula $\mathbf{M}$ and c has shallow formula $\mathbf{N}$. On the other hand, a positive node e with shallow formula $[\dot{\forall} x_\alpha \mathbf{M}]$ may have several children $c^1, \ldots, c^n$ with shallow formulas $([\mathbf{A}^1/x]\mathbf{M})^{\downarrow}, \ldots, ([\mathbf{A}^n/x]\mathbf{M})^{\downarrow}$. Since the order of these n children is unimportant, they are represented by elements of the set of expansion arcs and are not listed by the successor function. Similarly, certain nodes may have children induced by connections and certain other nodes may have children induced by the set of decomposition arcs. We will see that a node a may be extended, $succ(a) = \langle\rangle$ and yet a has a child b (e.g., if there is a decomposition arc $\langle a, b\rangle \in decs$).

DEFINITION 7.2.6. Let $\mathcal{Q}$ be an expansion structure and $b \in \mathcal{N}^{\mathcal{Q}}$ be a node in $\mathcal{Q}$. We say b is *extended in* $\mathcal{Q}$ if $b \in \mathbf{Dom}(succ^{\mathcal{Q}})$.

One can classify nodes b based on the value of $label(b)$. For linguistic convenience we define adjectives for nodes to indicate their classification.

DEFINITION 7.2.7. Let $\mathcal{Q}$ be an expansion structure and $b \in \mathcal{N}^{\mathcal{Q}}$ be a node in $\mathcal{Q}$.

- We say b is a *primary node* if $kind^{\mathcal{Q}}(b) = \mathbf{primary}$.
- We say b is a *true node* if b is a positive primary node with shallow formula $\dot{\top}$.
- We say b is a *false node* if b is a negative primary node with shallow formula $\dot{\top}$.
- We say b is a *negation node* if b is a primary node with shallow formula $\dot{\neg}\mathbf{M}$ for some $\mathbf{M} \in prop(\mathcal{S})^{\downarrow}$.
- We say b is a *disjunction node* if b is a positive primary node with shallow formula $[\mathbf{M} \dot{\vee} \mathbf{N}]$ for some $\mathbf{M}, \mathbf{N} \in prop(\mathcal{S})^{\downarrow}$.
- We say b is a *conjunction node* if b is a negative primary node with shallow formula $[\mathbf{M} \dot{\vee} \mathbf{N}]$ for some $\mathbf{M}, \mathbf{N} \in prop(\mathcal{S})^{\downarrow}$.
- We say b is a *selection node* if b is a negative primary node with shallow formula $[\dot{\forall} x_\alpha \mathbf{M}]$ for some variable x_α and proposition $\mathbf{M} \in prop(\mathcal{S})^{\downarrow}$.
- We say b is an *expansion node* if b is a positive primary node with shallow formula $[\dot{\forall} x_\alpha \mathbf{M}]$ for some variable $x_\alpha \in \mathcal{V}$ and proposition $\mathbf{M} \in prop(\mathcal{S})^{\downarrow}$.
- We say b is an *atomic node* if b is a primary node with shallow formula $[P\,\overline{\mathbf{A}^n}] \in wff_o{}^{\downarrow}$ for some parameter $P \in \mathcal{P}_{o\alpha^n\cdots\alpha^1}$ and some terms $\mathbf{A}^i \in wff_{\alpha^i}{}^{\downarrow}$ for each i $(1 \leq i \leq n)$.

- We say b is an *flexible node* if b is a primary node with shallow formula $[x\,\overline{\mathbf{A}^n}] \in wff_o{\downarrow}$ for some variable $x \in \mathcal{V}_{o\alpha^n\cdots\alpha^1}$ and some terms $\mathbf{A}^i \in wff_{\alpha^i}{\downarrow}$ for each i $(1 \leq i \leq n)$.

- We say b is an *deepening node* if b is a primary node with shallow formula $[c\,\overline{\mathbf{A}^n}] \in wff_o{\downarrow}$ for some constant $c \in \mathcal{S}_{o\alpha^n\cdots\alpha^1}$ and some terms $\mathbf{A}^i \in wff_{\alpha^i}{\downarrow}$ for each i $(1 \leq i \leq n)$.

- We say b is an *equation node* if b is a primary node with shallow formula $[\mathbf{C} \doteq^{\iota} \mathbf{D}]$ for some $\mathbf{C}, \mathbf{D} \in wff_\iota{\downarrow}$.

- We say b is a *functional node* if b is a primary node with shallow formula $[\mathbf{F} \doteq^{\alpha\beta} \mathbf{G}]$ for some $\mathbf{F}, \mathbf{G} \in wff_{\alpha\beta}{\downarrow}$.

- We say b is a *Boolean node* if b is a primary node with shallow formula $[\mathbf{A} \doteq^{o} \mathbf{B}]$ for some $\mathbf{A}, \mathbf{B} \in wff_o{\downarrow}$.

- We say b is a *decomposition node* if $kind^{\mathcal{Q}}(b) = \mathbf{dec}$.

- We say b is an *equation goal node* if $kind^{\mathcal{Q}}(b) = \mathbf{eqngoal}$.

- We say b is a *decomposable node* if it is a negative equation node with shallow formula $[H\,\overline{\mathbf{A}^n} \doteq^{\iota} H\,\overline{\mathbf{B}^n}]$ where $n \geq 0$, $H_{\iota\alpha^n\cdots\alpha^1} \in \mathcal{P}$ and $\mathbf{A}^i, \mathbf{B}^i \in wff_{\alpha^i}{\downarrow}$ for each i $(1 \leq i \leq n)$.

- We say b is a *mate node* if $kind^{\mathcal{Q}}(b) = \mathbf{mate}$.

- We say b is an *E-unification node* if $kind^{\mathcal{Q}}(b) = \mathbf{eunif}$.

- We say b is a *symmetric E-unification node* if $kind^{\mathcal{Q}}(b)$ equals $\mathbf{eunif}^{sym}$.

- We say b is a *connection node* if it is a mate node, an E-unification node, or a symmetric E-unification node.

- We say b is a *potentially connecting node* if it is an atomic node, an equation goal node or a positive equation node.

Note that true nodes have dual shallow formula $\dot{\neg}\dot{\top}$, false nodes have dual shallow formula $\dot{\top}$, disjunction nodes have dual shallow formula $\dot{\neg}[\mathbf{M} \dot{\vee} \mathbf{N}]$ and conjunction nodes have dual shallow formula $[\mathbf{M} \dot{\vee} \mathbf{N}]$. While this may be confusing, it is intentional. We name them true, false, disjunction and conjunction nodes to indicate the role they play in refutations. The *dual* formula plays the role of the formula in the (proof-oriented) sequent calculus. One could reverse the role of polarities in extensional expansion dags to avoid this confusion, but this would diverge significantly from the implementation in TPS and would complicate the discussion of results in Chapter 10.

REMARK 7.2.8 (Primary Nodes). Nodes which are not primary must either be a decomposition node, an equation goal node or a connection node. Since each such node is negative we can conclude that every positive node is primary. The fact that every positive node is primary is more an accident of this definition than a fundamental property of primary nodes. Consequently, we will still usually explicitly point out when positive nodes are primary.

One can sometimes determine the classification from the dual shallow formula of the node.

LEMMA 7.2.9. *Let $\mathcal{Q}$ be an expansion structure and $b \in \mathcal{N}^{\mathcal{Q}}$ be a node in $\mathcal{Q}$.*

1. *If $\overline{sh}(b) = \dot{\top}$, then b is a false node.*
2. *If $\overline{sh}(b) = [\mathbf{M} \dot{\vee} \mathbf{N}]$, then b is a conjunction node.*
3. *If $\overline{sh}(b) = [\dot{\forall} x_\alpha \mathbf{M}]$, then b is a selection node.*
4. *If $\overline{sh}(b) = [\mathbf{A} \dot{=}^\iota \mathbf{B}]$ and b is primary, then b is a negative equation node.*
5. *If $\overline{sh}(b) = [\mathbf{F} \dot{=}^{\alpha\beta} \mathbf{G}]$, then b is a negative functional node.*
6. *If $\overline{sh}(b) = [\mathbf{A} \dot{=}^o \mathbf{B}]$ and b is primary, then b is a negative Boolean node.*

PROOF. In each case the node b is negative since the dual shallow formula is not a negation. Also, the node b is primary either by assumption or by inspecting Definition 7.2.2. The results follow by inspecting Definitions 7.2.4 and 7.2.7. □

In order to do some inductive proofs on expansion structures, we will use a measure counting the number of nodes, number of expansion arcs and number of connections.

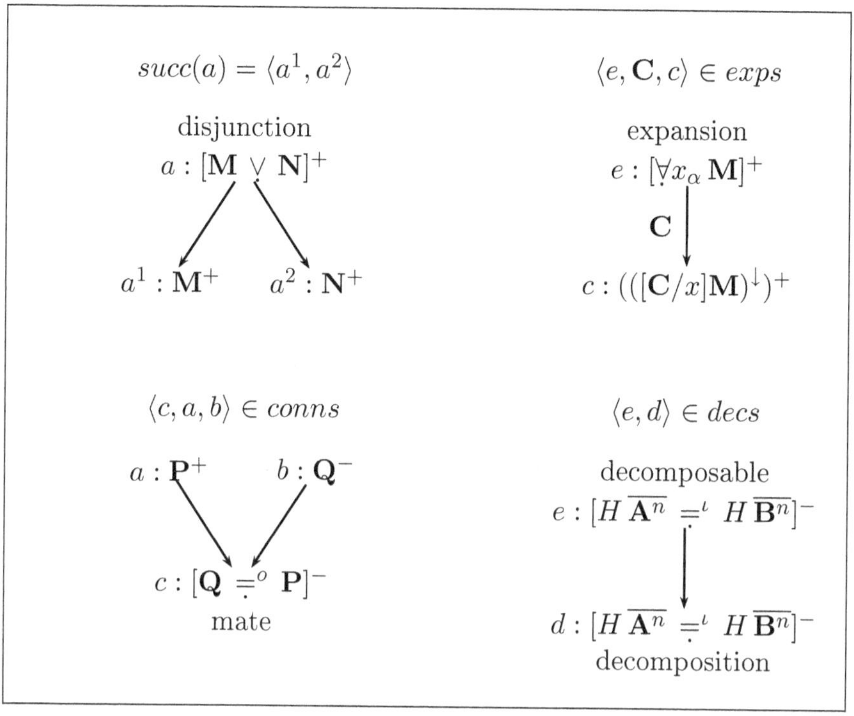

FIGURE 7.6. Edges in Expansion Structures

DEFINITION 7.2.10. Let $\mathcal{Q}$ be an expansion structure. We define $|\mathcal{Q}|$ to be

$$|\mathcal{N}^{\mathcal{Q}}| + |exps^{\mathcal{Q}}| + |conns^{\mathcal{Q}}|.$$

Every expansion structure induces a directed graph. The edges are induced by the successor function, the expansion arcs, the connections and the decompositions arcs. We will later define several properties (cf. Definition 7.3.1) which expansion structures must satisfy in order to be extensional expansion dags. However, we can define the edge relation for expansion structures without using any of these extra properties.

The successor function indicates children in a particular order. For example, disjunction nodes in an extensional expansion dag should have two successors corresponding to the left and right disjunct (cf. $\clubsuit_\vee$ in Definition 7.3.1). Expansion arcs will correspond to children of expansion nodes in extensional expansion dags (cf. $\clubsuit_{exp}$ in Definition 7.3.1). Edges induced by expansion arcs are annotated with a term representing the *expansion term* of the expansion arc. The order of children induced by expansion arcs is not important and is not represented in the expansion structure. Expansion nodes in an extensional expansion dag will have no (ordered) children induced by the successor function (cf. $\clubsuit_{ns}$ in Definition 7.3.1). Connections are triples $\langle c, a, b\rangle$ and will induce two edges, one from a to c and another from b to c. In an extensional expansion dag, the node c will always be a connection node while a and b will be potentially connecting nodes (cf. $\clubsuit_c$, $\clubsuit_m^p$, $\clubsuit_{eu}^p$ and $\clubsuit_{eu}^{sp}$ in Definition 7.3.1). Decomposition arcs in an extensional expansion dag will correspond to an edge from a decomposable node to a decomposition node with the same shallow formula (cf. $\clubsuit_{dec}^p$ in Definition 7.3.1). Examples of each of these situations is shown in Figure 7.6.

DEFINITION 7.2.11. Let $\mathcal{Q}$ be an expansion structure. We define the *edge relation* $\overset{\mathcal{Q}}{\rightarrow}$ *of* $\mathcal{Q}$ to be the binary relation on $\mathcal{N}^{\mathcal{Q}}$ such that $a \overset{\mathcal{Q}}{\rightarrow} b$ holds if either

1. $a \in \mathbf{Dom}(succ^{\mathcal{Q}})$, $succ^{\mathcal{Q}}(a) = \langle a^1, \ldots, a^n\rangle$ and
$$b \in \{a^1, \ldots, a^n\},$$
2. $\langle a, \mathbf{B}_\alpha, b\rangle \in exps^{\mathcal{Q}}$ for some $\mathbf{B}_\alpha \in wff_\alpha$,
3. $\langle b, a, d\rangle \in conns^{\mathcal{Q}}$ or $\langle b, d, a\rangle \in conns^{\mathcal{Q}}$ for some $d \in \mathcal{N}^{\mathcal{Q}}$, or
4. $\langle a, b\rangle \in decs^{\mathcal{Q}}$.

We call $\langle \mathcal{N}^{\mathcal{Q}}, \overset{\mathcal{Q}}{\rightarrow}\rangle$ the *underlying graph of* $\mathcal{Q}$. We extend all the terminology for directed graphs to apply to $\mathcal{Q}$ using the underlying graph of $\mathcal{Q}$. When $\langle \mathcal{N}^{\mathcal{Q}}, \overset{\mathcal{Q}}{\rightarrow}\rangle$ is a dag, we extend all the terminology for dags to apply to $\mathcal{Q}$.

We next define multisets (and sequents) associated with expansion structures.

DEFINITION 7.2.12. Let $\mathcal{Q}$ be an expansion structure. For any set of nodes $\mathcal{R} \subseteq \mathcal{N}^{\mathcal{Q}}$, we define $\Gamma^{\mathcal{Q}}(\mathcal{R})$ to be the finite multiset of propositions containing the dual shallow formula of each of the nodes in $\mathcal{R}$ with multiplicity of $\mathbf{M}$ in $\Gamma^{\mathcal{Q}}(\mathcal{R})$ determined by the number of nodes in $\mathcal{R}$ with dual shallow formula $\mathbf{M}$. Technically, we can follow Definition 2.2.1 and define $\Gamma^{\mathcal{Q}}(\mathcal{R})$ to be the function G from $prop(\mathcal{S})$ to $\mathbb{N}$ as follows:

$$G(\mathbf{M}) := |\{a \in \mathcal{R} \mid \overline{sh}^{\mathcal{Q}}(a) = \mathbf{M}\}|.$$

(Note that $G(\mathbf{M})$ is 0 unless $\overline{sh}^{\mathcal{Q}}(a) = \mathbf{M}$ for some $a \in \mathcal{R}$.) When the expansion structure is clear in context, we write $\Gamma(\mathcal{R})$ for $\Gamma^{\mathcal{Q}}(\mathcal{R})$. Note that $\Gamma(\mathcal{R})$ is an $\mathcal{S}$-sequent if the shallow formula of every node in $\mathcal{R}$ is closed.

DEFINITION 7.2.13. Let $\mathcal{Q}$ be an expansion structure. For any set of nodes $\mathcal{R} \subseteq \mathcal{N}^{\mathcal{Q}}$, we define $\mathbf{Params}^{\mathcal{Q}}(\mathcal{R})$ to be the set

$$\bigcup_{a \in \mathcal{R}} \mathbf{Params}(sh^{\mathcal{Q}}(a))$$

of parameters which occur in the shallow formula of some node in $\mathcal{R}$. We define $\mathbf{Free}^{\mathcal{Q}}(\mathcal{R})$ to be the set

$$\bigcup_{a \in \mathcal{R}} \mathbf{Free}(sh^{\mathcal{Q}}(a))$$

of variables which occur free in the shallow formula of some node in $\mathcal{R}$.

We define $\mathbf{RootParams}(\mathcal{Q})$ to be the set $\mathbf{Params}^{\mathcal{Q}}(\mathbf{Roots}(\mathcal{Q}))$ of parameters which occur in the shallow formula of some root node.

As usual, we write $\mathbf{Params}(\mathcal{R})$ for $\mathbf{Params}^{\mathcal{Q}}(\mathcal{R})$ and $\mathbf{Free}(\mathcal{R})$ for $\mathbf{Free}^{\mathcal{Q}}(\mathcal{R})$ when the expansion structure $\mathcal{Q}$ is clear in context.

Of course, since the shallow formula, signed shallow formula and dual shallow formula of a node can only differ by a negation, they all contain the same parameters and free variables. Hence $\mathbf{Params}(\mathcal{R})$ and $\mathbf{Free}(\mathcal{R})$ are also the sets of parameters and free variables, respectively, which occur in some member of the multiset $\Gamma(\mathcal{R})$.

DEFINITION 7.2.14 (Selected Parameters). Let $\mathcal{Q}$ be an expansion structure. For each node $a \in \mathbf{Dom}(sel^{\mathcal{Q}})$ we know $sel^{\mathcal{Q}}(a) = A_\alpha$ for some parameter A_α. We say A is the *selected parameter* of a.[2] When the expansion structure $\mathcal{Q}$ is clear in context, we use $sel(a)$ to denote $sel^{\mathcal{Q}}(a)$.

[2] We use selected parameters as in [53] rather than selected variables as in [43].

We say A is a *selected parameter* of $\mathcal{Q}$ if it is the selected parameter of some node $a \in \mathcal{N}$. We define $\mathbf{SelPars}(\mathcal{Q})$ to be the (finite) set of all selected parameters of $\mathcal{Q}$.

For any $\mathcal{R} \subseteq \mathcal{N}^{\mathcal{Q}}$, we define $\mathbf{SelPars}^{\mathcal{Q}}(\mathcal{R})$ to be the set

$$\mathbf{SelPars}(\mathcal{Q}) \cap \mathbf{Params}^{\mathcal{Q}}(\mathcal{R})$$

of selected parameters which occur in the shallow formulas of the nodes in $\mathcal{R}$. We define $\mathbf{RootSels}(\mathcal{Q})$ to be the set

$$\mathbf{SelPars}^{\mathcal{Q}}(\mathbf{Roots}(\mathcal{Q}))$$

of selected parameters which occur in the shallow formulas of some root node.

When the expansion structure $\mathcal{Q}$ is clear in context, we write $\mathbf{SelPars}(\mathcal{R})$ for $\mathbf{SelPars}^{\mathcal{Q}}(\mathcal{R})$.

DEFINITION 7.2.15 (Expansion Arcs). Let $\mathcal{Q}$ be an expansion structure. We say $\langle e, \mathbf{C}_\alpha, c\rangle \in exps^{\mathcal{Q}}$ is an *expansion arc* of $\mathcal{Q}$ with *expansion term* $\mathbf{C}_\alpha$ and *expansion child* c.

For any expansion node e in $\mathcal{Q}$, we define the set of *expansion instances* $\mathbf{ExpInsts}^{\mathcal{Q}}(e)$ of e to be

$$\{sh^{\mathcal{Q}}(c) \mid \langle e, \mathbf{C}, c\rangle \in exps^{\mathcal{Q}}\}.$$

When the expansion structure $\mathcal{Q}$ is clear in context, we will write $\mathbf{ExpInsts}(e)$ for $\mathbf{ExpInsts}^{\mathcal{Q}}(e)$.

We say $\mathbf{C}$ is an *expansion term* of $\mathcal{Q}$ if there is some expansion arc of $\mathcal{Q}$ with expansion term $\mathbf{C}$. We define $\mathbf{ExpTerms}(\mathcal{Q})$ to be the set of all expansion terms of $\mathcal{Q}$.

We say a variable x_α is an *expansion variable* if $x_\alpha \in \mathbf{Free}(\mathbf{C})$ for some expansion term $\mathbf{C}$ of $\mathcal{Q}$. We say a parameter B_β is an *expansion parameter* if $B_\beta \in \mathbf{Params}(\mathbf{C})$ for some expansion term $\mathbf{C}$ of $\mathcal{Q}$. We define $\mathbf{ExpVars}(\mathcal{Q})$ to be the set of all expansion variables of $\mathcal{Q}$ and $\mathbf{ExpParams}(\mathcal{Q})$ to be the set of all expansion parameters of $\mathcal{Q}$. That is,

$$\mathbf{ExpVars}(\mathcal{Q}) := \bigcup_{\mathbf{C} \in \mathbf{ExpTerms}(\mathcal{Q})} \mathbf{Free}(\mathbf{C})$$

and

$$\mathbf{ExpParams}(\mathcal{Q}) := \bigcup_{\mathbf{C} \in \mathbf{ExpTerms}(\mathcal{Q})} \mathbf{Params}(\mathbf{C}).$$

Next we define the sets of free variables and parameters of expansion structures. Free variables and parameters may occur in the shallow formula of some node or in an expansion term. Parameters may also occur only as a selected parameter (if, e.g., the corresponding quantifier is vacuous).

DEFINITION 7.2.16. Let $\mathcal{Q}$ be an expansion structure. We define the set of free variables $\mathbf{Free}(\mathcal{Q})$ of $\mathcal{Q}$ to be

$$\mathbf{Free}^{\mathcal{Q}}(\mathcal{N}^{\mathcal{Q}}) \cup \mathbf{ExpVars}(\mathcal{Q}).$$

We say $\mathcal{Q}$ is *closed* if $\mathbf{Free}(\mathcal{Q}) = \emptyset$.

We define the set of parameters $\mathbf{Params}(\mathcal{Q})$ of $\mathcal{Q}$ to be

$$\mathbf{Params}^{\mathcal{Q}}(\mathcal{N}^{\mathcal{Q}}) \cup \mathbf{ExpParams}(\mathcal{Q}) \cup \mathbf{SelPars}(\mathcal{Q}).$$

Note that the set of parameters of $\mathcal{Q}$ always includes the selected parameters of $\mathcal{Q}$.

7.3. Extensional Expansion Dags

For an expansion structure to be an extensional expansion dag, we require several additional properties to hold. These properties are described in Figures 7.7, 7.8 and 7.9. The properties relate parents and children with respect to their kinds, polarities and shallow formulas.

In most cases the properties relate children to a single parent. The exceptions are the properties $\clubsuit_c$, $\clubsuit^p_m$, $\clubsuit^p_{eu}$ and $\clubsuit^{sp}_{eu}$ dealing with connections. A connection is a triple $\langle c, a, b\rangle$ of nodes where a and b are both parents of the node c. The node c may be a mate node, an E-unification node or a symmetric E-unification node. The idea in each case is that a and b have opposite polarities and the connection node c introduces the goal of proving the shallow formulas of a and b are equal. In particular, consider $\clubsuit^p_{eu}$. In this case a is an extended positive equation node with shallow formula $[\mathbf{A} \doteq^\iota \mathbf{B}]$ and b is an extended (negative) equation goal node with shallow formula $[\mathbf{C} \doteq^\iota \mathbf{D}]$. We can prove the formulas $[\mathbf{A} \doteq^\iota \mathbf{B}]$ and $[\mathbf{C} \doteq^\iota \mathbf{D}]$ are equal by proving $\mathbf{A}$ equals $\mathbf{C}$ and $\mathbf{B}$ equals $\mathbf{D}$. The shallow formula of the (negative) connection node c is

$$[[\mathbf{A} \doteq^\iota \mathbf{C}] \mathbin{\dot\wedge} [\mathbf{B} \doteq^\iota \mathbf{D}]]$$

as c corresponds to proving both of these equations.

We now define the properties for expansion structures.

DEFINITION 7.3.1 (Properties of Expansion Structures). Let

$$\mathcal{Q} = \langle \mathcal{N}, succ, label, sel, exps, conns, decs \rangle$$

be an expansion structure. We define the following property for each expansion arc $\langle e, \mathbf{C}_\alpha, c \rangle$ in *exps*:

$\clubsuit_{exp}$: There exists some variable x_α and proposition $\mathbf{M}$ such that e is an extended expansion node with shallow formula $[\underset{\cdot}{\forall} x_\alpha \mathbf{M}]$ and c is a positive primary node with shallow formula $([\mathbf{C}/x]\mathbf{M})^{\downarrow}$.

We define the following properties for each connection $\langle c, a, b \rangle$ in *conns*:

$\clubsuit_c$: The node c is a connection node.

$\clubsuit_m^p$: If c is a mate node with shallow formula

$$[[P\, \mathbf{B}^1_{\alpha^i} \cdots \mathbf{B}^n_{\alpha^n}] \underset{\cdot}{=}^o [P\, \mathbf{A}^1_{\alpha^1} \cdots \mathbf{A}^n_{\alpha^n}]],$$

then a is an extended positive atomic node with shallow formula $[P\, \mathbf{A}^1 \cdots \mathbf{A}^n]$ and b is an extended negative atomic node with shallow formula $[P\, \mathbf{B}^1 \cdots \mathbf{B}^n]$.

$\clubsuit_{eu}^p$: If c is an E-unification node with shallow formula

$$[[\mathbf{A} \underset{\cdot}{=}^\iota \mathbf{C}] \underset{\cdot}{\wedge} [\mathbf{B} \underset{\cdot}{=}^\iota \mathbf{D}]],$$

then a is an extended positive equation node with shallow formula $[\mathbf{A} \underset{\cdot}{=}^\iota \mathbf{B}]$ and b is an extended equation goal node with shallow formula $[\mathbf{C} \underset{\cdot}{=}^\iota \mathbf{D}]$.

$\clubsuit_{eu}^{sp}$: If c is a symmetric E-unification node with shallow formula

$$[[\mathbf{A} \underset{\cdot}{=}^\iota \mathbf{D}] \underset{\cdot}{\wedge} [\mathbf{B} \underset{\cdot}{=}^\iota \mathbf{C}]],$$

then a is an extended positive equation node with shallow formula $[\mathbf{A} \underset{\cdot}{=}^\iota \mathbf{B}]$ and b is an extended equation goal node with shallow formula $[\mathbf{C} \underset{\cdot}{=}^\iota \mathbf{D}]$.

We define the following property for each decomposition arc $\langle e, d \rangle$ in *decs*:

$\clubsuit^p_{dec}$: The node e is an extended negative equation node, the node d is a decomposition node and the shallow formulas of e and d are the same.

We define the following properties for each node a in $\mathcal{N}$:

$\clubsuit_{ns}$: If a is a true node, false node, expansion node or potentially connecting node, then $succ(a) = \langle\rangle$.[3]

$\clubsuit_f$: The node a is not flexible.

$\clubsuit_\neg$: If a is a negation node with shallow formula $\dot{\neg}\mathbf{M}$ and polarity p, then there exists a primary node $a^1 \in \mathcal{N}$ such that $succ(a) = \langle a^1\rangle$, $sh(a^1) = \mathbf{M}$ and $pol(a^1) = -p$.

$\clubsuit_\vee$: If a is a disjunction or conjunction node with shallow formula $[\mathbf{M} \dot{\vee} \mathbf{N}]$ and polarity p, then there exist primary nodes $a^1 \in \mathcal{N}$ and $a^2 \in \mathcal{N}$ such that $succ(a) = \langle a^1, a^2\rangle$, $sh(a^1) = \mathbf{M}$, $sh(a^2) = \mathbf{N}$ and $pol(a^1) = pol(a^2) = p$.

$\clubsuit_{sel}$: If a is a selection node with shallow formula $[\dot{\forall} x_\alpha \mathbf{M}]$, then a is in the domain of the partial function sel and there exists a negative primary node a^1 such that $succ(a) = \langle a^1\rangle$ and $sh(a^1) = [sel(a)/x]\mathbf{M}$.

$\clubsuit_\sharp$: If a is a deepening node with shallow formula $[c\,\overline{\mathbf{A}^n}]$ and polarity p where c is a logical constant, then there is a primary node $a^1 \in \mathcal{N}$ such that $succ(a) = \langle a^1\rangle$, $sh(a^1) = ([c\,\overline{\mathbf{A}^n}]^\sharp)^\downarrow$ and a^1 has polarity p.

$\clubsuit_{dec}$: Suppose a is a decomposition node with shallow formula

$$[[H\,\mathbf{A}^1_{\alpha^1} \cdots \mathbf{A}^n_{\alpha^n}] \dot{=}^\iota [H\,\mathbf{B}^1_{\alpha^1} \cdots \mathbf{B}^n_{\alpha^n}]].$$

Then there exist negative primary nodes $a^1, \ldots, a^n \in \mathcal{N}$ such that $succ(a) = \langle a^1, \ldots, a^n\rangle$ and $sh(a^i) = [\mathbf{A}^i \dot{=}^{\alpha^i} \mathbf{B}^i]$ for each

[3]This does not necessarily make a a leaf node since there may be edges other than those induced by $succ$.

i where $1 \leq i \leq n$.

$\clubsuit_{=}^{-}$: If a is a negative equation node, then there exists an equation goal node $a^1 \in \mathcal{N}$ such that $succ(a) = \langle a^1 \rangle$ and $sh(a^1)$ is $sh(a)$.

$\clubsuit_{=}^{\rightarrow}$: If a is a functional node with shallow formula $[\mathbf{F} \doteq^{\alpha\beta} \mathbf{G}]$ and polarity p, then there exists a primary node $a^1 \in \mathcal{N}$ and a variable $y_\beta \in \mathcal{V}_\beta$ such that $y \notin \mathbf{Free}(\mathbf{F}) \cup \mathbf{Free}(\mathbf{G})$, $succ(a)$ is $\langle a^1 \rangle$, $sh(a^1) = [\dot{\forall} y_\beta \centerdot \mathbf{F}\, y \doteq^{\alpha} \mathbf{G}\, y]^{\downarrow}$ and a^1 has polarity p.

$\clubsuit_{=}^{-o}$: If a is a negative Boolean node with shallow formula $[\mathbf{A} \doteq^{o} \mathbf{B}]$, then there exist negative primary nodes $a^1, a^2 \in \mathcal{N}$ such that $succ(a) = \langle a^1, a^2 \rangle$,

$$sh(a^1) = [\dot{\neg}\mathbf{A} \dot{\vee} \mathbf{B}] \text{ and } sh(a^2) = [\dot{\neg}\mathbf{B} \dot{\vee} \mathbf{A}].$$

$\clubsuit_{=}^{+o}$: If a is a positive Boolean node with $sh(a) = [\mathbf{A} \doteq^{o} \mathbf{B}]$, then there exist *negative*[4] primary nodes $a^1, a^2 \in \mathcal{N}$ such that $succ(a) = \langle a^1, a^2 \rangle$,

$$sh(a^1) = [\mathbf{A} \dot{\vee} \mathbf{B}] \text{ and } sh(a^2) = [\dot{\neg}\mathbf{A} \dot{\vee} \dot{\neg}\mathbf{B}].$$

$\clubsuit_{m}$: If a is a mate node with shallow formula

$$[[P\, \mathbf{A}^1_{\alpha^1} \cdots \mathbf{A}^n_{\alpha^n}] \doteq^{o} [P\, \mathbf{B}^1_{\alpha^1} \cdots \mathbf{B}^n_{\alpha^n}]],$$

then there exist negative primary nodes $a^1, \ldots, a^n \in \mathcal{N}$ such that $succ(a) = \langle a^1, \ldots, a^n \rangle$ and $sh(a^i) = [\mathbf{A}^i \doteq^{\alpha^i} \mathbf{B}^i]$ for each i where $1 \leq i \leq n$.

$\clubsuit_{eu}$: If a is either an E-unification node or a symmetric E-unification node with shallow formula

$$[[\mathbf{A} \doteq^{\iota} \mathbf{C}] \dot{\wedge} [\mathbf{B} \doteq^{\iota} \mathbf{D}]],$$

[4] Notice the change in polarity.

then there exist negative primary nodes $a^1, a^2 \in \mathcal{N}$ such that $succ(a) = \langle a^1, a^2 \rangle$, $sh(a^1) = [\mathbf{A} \doteq^\iota \mathbf{C}]$ and $sh(a^2) = [\mathbf{B} \doteq^\iota \mathbf{D}]$.

DEFINITION 7.3.2 (Extensional Expansion Dag). Let $\mathcal{S}$ be a signature of logical constants and

$$\mathcal{Q} = \langle \mathcal{N}, succ, label, sel, exps, conns, decs \rangle$$

be an $\mathcal{S}$-expansion structure. We say $\mathcal{Q}$ is an *$\mathcal{S}$-extensional expansion dag* if the following hold:

1. The underlying graph $\langle \mathcal{N}, \stackrel{\mathcal{Q}}{\rightarrow} \rangle$ of $\mathcal{Q}$ is a dag.
2. Every node $a \in \mathbf{Dom}(sel)$ is a selection node of $\mathcal{Q}$.
3. The partial function sel is injective.
4. Every expansion arc in $exps$ satisfies $\clubsuit_{exp}$.
5. Every connection in $conns$ satisfies $\clubsuit_c$, $\clubsuit_m^p$, $\clubsuit_{eu}^p$ and $\clubsuit_{eu}^{sp}$.
6. Every decomposition arc in $decs$ satisfies $\clubsuit_{dec}^p$.
7. For each extended node $a \in \mathbf{Dom}(succ)$ the properties $\clubsuit_{ns}$, $\clubsuit_f$, $\clubsuit_\neg$, $\clubsuit_\vee$, $\clubsuit_{sel}$, $\clubsuit_\sharp$, $\clubsuit_{dec}$, $\clubsuit_=^-$, $\clubsuit_=^\rightarrow$, $\clubsuit_=^{-o}$, $\clubsuit_=^{+o}$, $\clubsuit_m$ and $\clubsuit_{eu}$ hold.

We often refer to $\mathcal{S}$-extensional expansion dags as *extensional expansion dags* and leave the signature $\mathcal{S}$ implicit.

REMARK 7.3.3. There are alternative conditions for $\clubsuit_=^{-o}$ and $\clubsuit_=^{+o}$. The choices made here minimize the number of *u-parts* (cf. Definition 7.4.4).

Since every selected parameter of an extensional expansion dag is selected by a unique selection node, we can define a function taking selected parameters to the corresponding node.

DEFINITION 7.3.4. Let $\mathcal{Q}$ be an extensional expansion dag. For every selected parameter A of $\mathcal{Q}$, let $\mathbf{SelNode}^{\mathcal{Q}}(A)$ be the unique selection node of $\mathcal{Q}$ such that $sel^{\mathcal{Q}}(\mathbf{SelNode}^{\mathcal{Q}}(A)) = A$. When the extensional expansion dag is clear in context, we write $\mathbf{SelNode}(A)$ for $\mathbf{SelNode}^{\mathcal{Q}}(A)$.

$\clubsuit_{exp}$:

$$e : [\dot{\forall} x_\alpha \, \mathbf{M}]^+ \xrightarrow{\ \mathbf{C}\ } c : (([\mathbf{C}/x]\mathbf{M})^{\downarrow})^+$$

$\clubsuit_\neg$:

$$a : \dot{\neg}\mathbf{M}^{\pm} \longrightarrow a^1 : \mathbf{M}^{\mp}$$

$\clubsuit_\vee$:

$$a : [\mathbf{M} \dot{\vee} \mathbf{N}]^{\pm} \longrightarrow a^1 : \mathbf{M}^{\pm} \quad a^2 : \mathbf{N}^{\pm}$$

$\clubsuit_{sel}$:

$$a : [\dot{\forall} x_\alpha \, \mathbf{M}]^- \xrightarrow{\ A\ } a^1 : [A/x]\mathbf{M}^-$$

$\clubsuit_\sharp$:

$$a : [c\, \overline{\mathbf{A}^n}]^{\pm} \longrightarrow a^1 : (([c\, \overline{\mathbf{A}^n}]^{\sharp})^{\downarrow})^{\pm}$$

$\clubsuit_{\dot{=}}^-$:

negative equation node

$$a : [\mathbf{A} \doteq^{\iota} \mathbf{B}]^- \longrightarrow a^1 : [\mathbf{A} \doteq^{\iota} \mathbf{B}]^-$$

equation goal

$\clubsuit_{dec}^p$:

negative equation

$$e : [H\, \overline{\mathbf{A}^n} \doteq^{\iota} H\, \overline{\mathbf{B}^n}]^- \longrightarrow d : [H\, \overline{\mathbf{A}^n} \doteq^{\iota} H\, \overline{\mathbf{B}^n}]^-$$

decomposition

FIGURE 7.7. Properties $\clubsuit_{exp}$, $\clubsuit_\neg$, $\clubsuit_\vee$, $\clubsuit_{sel}$, $\clubsuit_\sharp$, $\clubsuit_{=}^-$ and $\clubsuit_{dec}^p$

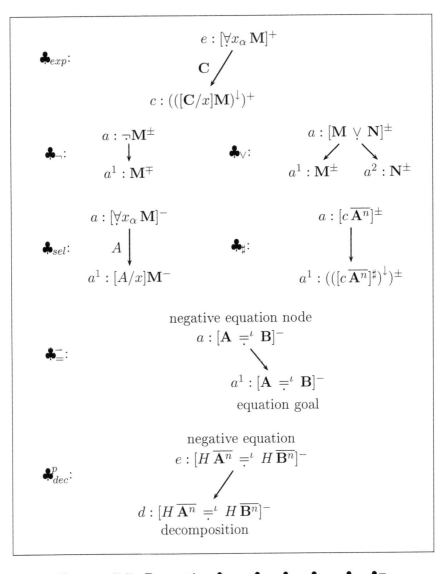

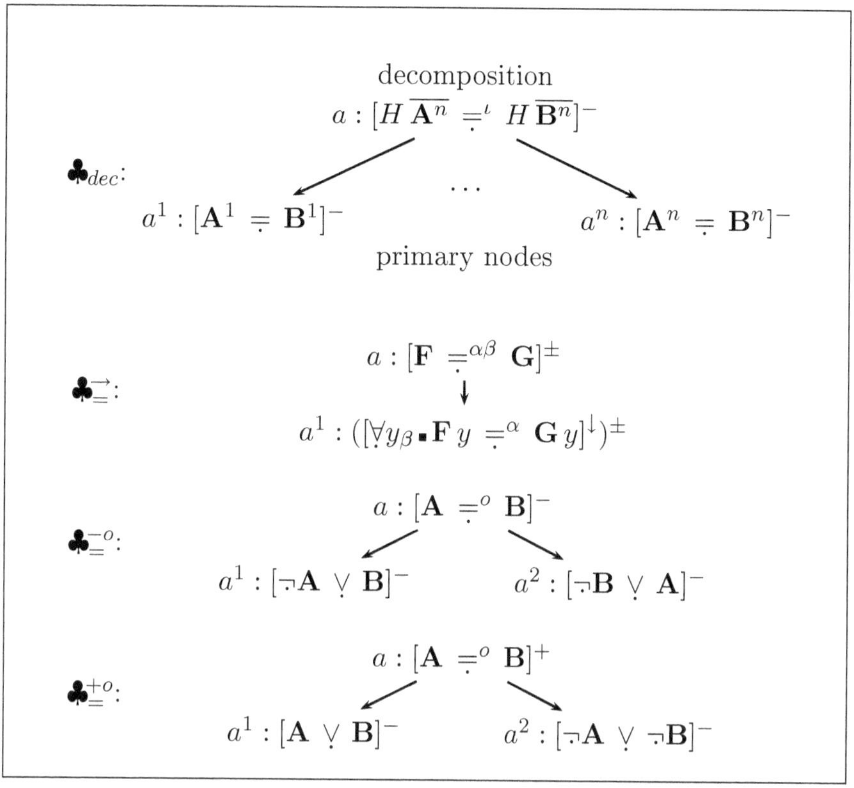

FIGURE 7.8. Properties $\clubsuit_{dec}$, $\clubsuit_{\doteq}^{\rightarrow}$, $\clubsuit_{\doteq}^{-o}$ and $\clubsuit_{\doteq}^{+o}$

The properties $\clubsuit_c$, $\clubsuit_m^p$, $\clubsuit_{eu}^p$ and $\clubsuit_{eu}^{sp}$ guarantee that connections always have a certain form.

LEMMA 7.3.5. *Let $\mathcal{Q}$ be an extensional expansion dag. For any connection $\langle c, a, b\rangle \in conns^{\mathcal{Q}}$, we know c is a connection node, a is an extended, positive potentially connecting node and b is an extended, negative potentially connecting node.*

PROOF. We know c is a connection node by $\clubsuit_c$. Thus c is either a mate node, an E-unification node or a symmetric E-unification node. If c is a mate node, then a is positive, b is negative and both a and b are extended potentially connecting nodes by $\clubsuit_m^p$. If c is an E-unification node (or a symmetric E-unification node), then a is positive, b is

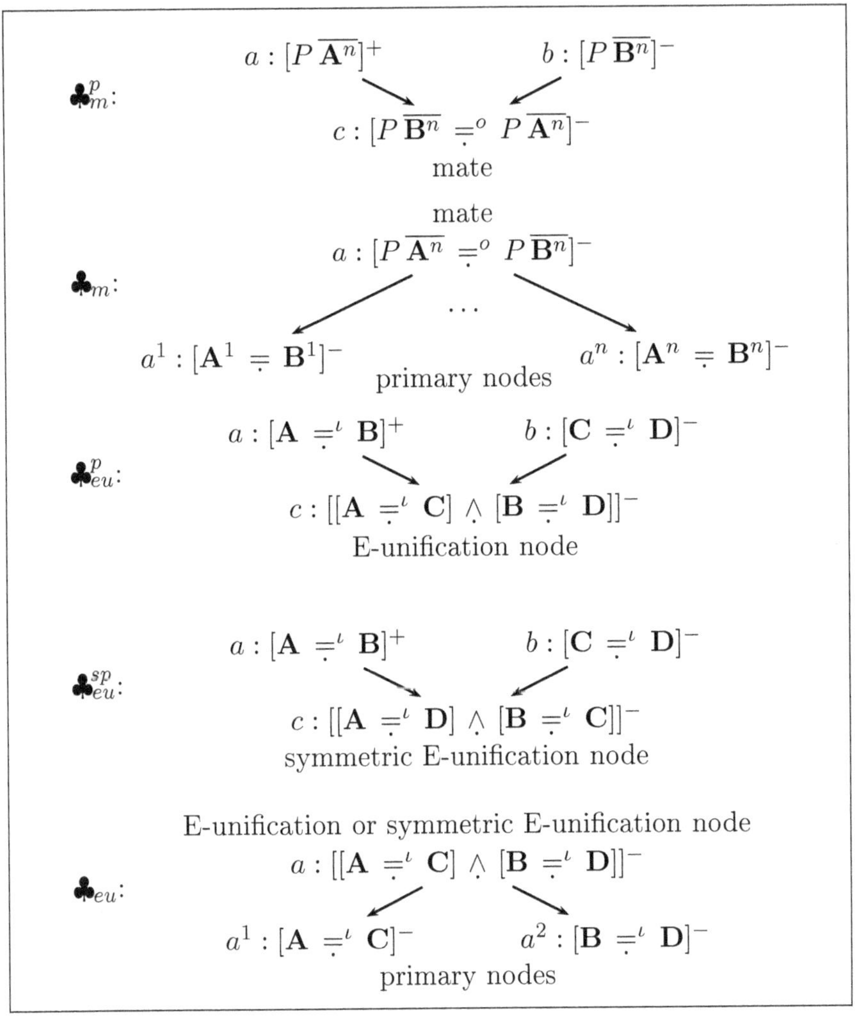

FIGURE 7.9. Properties $\clubsuit^p_m$, $\clubsuit_m$, $\clubsuit^p_{eu}$, $\clubsuit^{sp}_{eu}$ and $\clubsuit_{eu}$

negative and both a and b are extended potentially connecting nodes by $\clubsuit^p_{eu}$ (or $\clubsuit^{sp}_{eu}$). □

Every node of an extensional expansion dag is either an extended node or a leaf node.

LEMMA 7.3.6. *Let $\mathcal{Q}$ be an extensional expansion dag and $a \in \mathcal{N}^{\mathcal{Q}}$ be a node of $\mathcal{Q}$. Either a is extended in $\mathcal{Q}$ or a is a leaf node of $\mathcal{Q}$.*

PROOF. Assume a is not a leaf node of $\mathcal{Q}$. There is some $b \in \mathcal{N}^{\mathcal{Q}}$ such that $a \xrightarrow{\mathcal{Q}} b$. We consider each case in Definition 7.2.11. If $a \in \mathbf{Dom}(succ^{\mathcal{Q}})$, $succ^{\mathcal{Q}}(a) = \langle a^1, \cdots, a^n \rangle$ and $b \in \{a^1, \ldots, a^n\}$, then a is extended. If $\langle a, \mathbf{A}, b \rangle \in exps^{\mathcal{Q}}$ for some $\mathbf{A}$, then a is extended by $\clubsuit_{exp}$. If $\langle a, b \rangle \in decs^{\mathcal{Q}}$, then a is extended by by $\clubsuit^{p}_{dec}$. If either $\langle b, a, d \rangle \in conns^{\mathcal{Q}}$ or $\langle b, d, a \rangle \in conns^{\mathcal{Q}}$, then a is extended by Lemma 7.3.5. Thus a is extended. □

Consequently, flexible nodes in extensional expansion dags must be leaves.

LEMMA 7.3.7. *Let $\mathcal{Q}$ be an extensional expansion dag and $a \in \mathcal{N}^{\mathcal{Q}}$ be a flexible node of $\mathcal{Q}$. Then a is a leaf node of $\mathcal{Q}$.*

PROOF. Suppose $a \xrightarrow{\mathcal{Q}} b$. By Lemma 7.3.6, the flexible node a must be extended, contradicting $\clubsuit_f$. □

All children of potentially connecting nodes in an extensional expansion dag are connection nodes induced by the set of connections. Likewise the children of expansion nodes are those induced by expansion arcs. The children of negative equation nodes are induced by both the *succ* function and the set of decomposition arcs. For all other nodes the children are determined only by the *succ* function.

LEMMA 7.3.8. *Let $\mathcal{Q}$ be an extensional expansion dag. If $a \in \mathcal{N}^{\mathcal{Q}}$ is a potentially connecting node and $a \xrightarrow{\mathcal{Q}} c$, then a is extended, c is a connection node and there exists some b such that $\langle c, a, b \rangle \in conns^{\mathcal{Q}}$ or $\langle c, b, a \rangle \in conns^{\mathcal{Q}}$. Furthermore, we have the following:*

1. *If a is a positive atomic node, then $\langle c, a, b \rangle \in conns^{\mathcal{Q}}$ and c is a mate node.*
2. *If a is a negative atomic node, then $\langle c, b, a \rangle \in conns^{\mathcal{Q}}$ and c is a mate node.*
3. *If a is a positive equation node, then $\langle c, a, b \rangle \in conns^{\mathcal{Q}}$ and c is either an E-unification or a symmetric E-unification node.*
4. *If a is an equation goal node, then $\langle c, b, a \rangle \in conns^{\mathcal{Q}}$ and c is either an E-unification or a symmetric E-unification node.*

PROOF. We can conclude $a \stackrel{\mathcal{Q}}{\rightharpoondown} c$ is induced by a connection using $\clubsuit_{ns}$, $\clubsuit_{exp}$ and $\clubsuit^{p}_{dec}$. The rest follows easily from $\clubsuit_{c}$, $\clubsuit^{p}_{m}$, $\clubsuit^{p}_{eu}$ and $\clubsuit^{sp}_{eu}$. □

Sometimes we wish to consider nodes which are connected without making specific reference to the connection node. Since there are three different kinds of connection nodes, we define three different ways nodes can be connected.

DEFINITION 7.3.9. Let $\mathcal{Q}$ be an extensional expansion dag,

$$k \in \{\mathbf{mate}, \mathbf{eunif}, \mathbf{eunif}^{sym}\}$$

be a kind and $a, b \in \mathcal{N}^{\mathcal{Q}}$ be nodes of $\mathcal{Q}$. We say a pair $\langle a, b\rangle$ is *k-connected in* $\mathcal{Q}$ if there exists a node $c \in \mathcal{N}^{\mathcal{Q}}$ of kind k such that $\langle c, a, b\rangle \in conns^{\mathcal{Q}}$.

Similarly, we say a decomposable node e is decomposed if there is a decomposition arc from e to a node d.

DEFINITION 7.3.10. Let $\mathcal{Q}$ be an extensional expansion dag. We say a node $e \in \mathcal{N}^{\mathcal{Q}}$ is *decomposed* if there exists a node $d \in \mathcal{N}^{\mathcal{Q}}$ such that $\langle e, d\rangle \in decs^{\mathcal{Q}}$. We say $\mathcal{Q}$ is *decomposed* if every decomposable node is decomposed.

Next we consider children of expansion nodes.

LEMMA 7.3.11. *Let $\mathcal{Q}$ be an extensional expansion dag. If $e \in \mathcal{N}^{\mathcal{Q}}$ is an expansion node and $e \stackrel{\mathcal{Q}}{\rightharpoondown} c$, then e is extended and*

$$\langle e, \mathbf{C}, c\rangle \in exps^{\mathcal{Q}}$$

for some term $\mathbf{C}$.

PROOF. Suppose $e \in \mathcal{N}^{\mathcal{Q}}$ is an expansion node and $e \stackrel{\mathcal{Q}}{\rightharpoondown} c$. We know e is extended (by Lemma 7.3.6), $succ^{\mathcal{Q}}(e) = \langle\rangle$ (by $\clubsuit_{ns}$) and $\langle e, c\rangle \notin decs^{\mathcal{Q}}$ (by $\clubsuit^{p}_{dec}$ since e is not a negative equation node). Since e is not a potentially connecting node, we cannot have $\langle c, e, b\rangle$ or $\langle c, b, e\rangle$ in $conns^{\mathcal{Q}}$ for any b by Lemma 7.3.5. Thus $\langle e, \mathbf{C}, c\rangle \in exps^{\mathcal{Q}}$ for some term $\mathbf{C}_{\alpha}$. □

Children of negative equation nodes may be equation goal nodes or decomposition nodes.

LEMMA 7.3.12. *Let $\mathcal{Q}$ be an extensional expansion dag. If $e \in \mathcal{N}^{\mathcal{Q}}$ is a negative equation node with shallow formula* $\mathbf{M}$ *and* $e \overset{\mathcal{Q}}{\rightarrow} c$, *then* e *is extended and there is some equation goal node* a^1 *with shallow formula* $\mathbf{M}$ *such that* $succ(e) = \langle a^1 \rangle$ *and either* c *is* a^1 *or* c *is a decomposition node with shallow formula* $\mathbf{M}$ *and* $\langle e, c \rangle \in decs^{\mathcal{Q}}$.

PROOF. Suppose $e \in \mathcal{N}^{\mathcal{Q}}$ is a negative equation node and $e \overset{\mathcal{Q}}{\rightarrow} c$. We know e is extended by Lemma 7.3.6. By $\clubsuit_{=}^{-}$, there is some equation goal node a^1 such that $succ^{\mathcal{Q}}(e) = \langle a^1 \rangle$. Assume $c \neq a^1$. Since e is not an expansion node, we cannot have $\langle e, \mathbf{C}_\alpha, c \rangle$ in $exps^{\mathcal{Q}}$ for any $\mathbf{C}_\alpha$ by $\clubsuit_{exp}$. Since e is not a potentially connecting node, we cannot have $\langle c, e, b \rangle$ or $\langle c, b, e \rangle$ in $conns^{\mathcal{Q}}$ for any b by Lemma 7.3.5. Hence we must have $\langle e, c \rangle \in decs^{\mathcal{Q}}$. By $\clubsuit_{dec}^{p}$, c is a decomposition node with the same shallow formula as e. □

For nodes other than expansion nodes, potentially connecting nodes and negative equation nodes, the children are listed by the successor function.

LEMMA 7.3.13. *Let $\mathcal{Q}$ be an extensional expansion dag. Suppose* $a \in \mathcal{N}^{\mathcal{Q}}$ *is not an expansion node, not a potentially connecting node and not a negative equation node. If* $a \overset{\mathcal{Q}}{\rightarrow} c$, *then* a *is extended and there exist nodes* $a^1, \ldots, a^n \in \mathcal{N}^{\mathcal{Q}}$ *such that* $succ^{\mathcal{Q}}(a) = \langle a^1, \ldots, a^n \rangle$ *and* $c \in \{a^1, \ldots, a^n\}$.

PROOF. Suppose $a \overset{\mathcal{Q}}{\rightarrow} c$ and a is not an expansion node, not a potentially connecting node and not a negative equation node. We know a is extended by Lemma 7.3.6. By $\clubsuit_{exp}$, $\clubsuit_{dec}^{p}$ and Lemma 7.3.5, the edge $a \overset{\mathcal{Q}}{\rightarrow} c$ cannot be induced by an expansion arc, a decomposition arc or a connection. Hence the edge from a to c must be induced by the successor function. Thus $succ^{\mathcal{Q}}(a) = \langle a^1, \ldots, a^n \rangle$ and $c \in \{a^1, \ldots, a^n\}$ for some $a^1, \ldots, a^n \in \mathcal{N}^{\mathcal{Q}}$. □

Children of nodes listed by the successor function are usually primary.

LEMMA 7.3.14. *Let $\mathcal{Q}$ be an extensional expansion dag,* a *be an extended node with* $a \in \mathbf{Dom}(succ^{\mathcal{Q}})$, *and* $succ^{\mathcal{Q}}(a) = \langle a^1, \ldots, a^n \rangle$. *If* a *is not a negative equation node, then* a^i *is a primary node for each* i $(1 \leq i \leq n)$.

PROOF. This follows from a case analysis on $label^{\mathcal{Q}}(a)$ using $\clubsuit_*$ properties. □

The parents of connection nodes are always parents due to connections.

LEMMA 7.3.15. *Let $\mathcal{Q}$ be an extensional expansion dag. If $c \in \mathcal{N}^{\mathcal{Q}}$ is a connection node, $a \in \mathcal{N}^{\mathcal{Q}}$ and $a \xrightarrow{\mathcal{Q}} c$, then there is some $b \in \mathcal{N}^{\mathcal{Q}}$ such that $\langle c, a, b\rangle \in conns^{\mathcal{Q}}$ or $\langle c, b, a\rangle \in conns^{\mathcal{Q}}$.*

PROOF. The $\clubsuit_*$ properties ensure the edge $a \xrightarrow{\mathcal{Q}} c$ is not induced by an expansion arc, a decomposition arc or the successor function. □

The $\clubsuit_*$ properties guarantee that the shallow formulas of the children of most nodes do not add parameters to the shallow formula of the parents.

LEMMA 7.3.16. *Let $\mathcal{Q}$ be an extensional expansion dag. Suppose $a \in \mathcal{N}^{\mathcal{Q}}$, a is not a selection node, a is extended and*

$$succ^{\mathcal{Q}}(a) = \langle a^1, \ldots, a^n\rangle.$$

Then

$$(\mathbf{Params}(sh(a^1)) \cup \cdots \cup \mathbf{Params}(sh(a^n))) \subseteq \mathbf{Params}(sh(a)).$$

PROOF. This follows by a case analysis on the node a using the appropriate $\clubsuit_*$ properties. □

LEMMA 7.3.17. *Let $\mathcal{Q}$ be an extensional expansion dag and a, b be nodes in $\mathcal{N}^{\mathcal{Q}}$ such that $a \xrightarrow{\mathcal{Q}} b$. Suppose a is not an expansion node, not a selection node and not a potentially connecting node. Then*

$$\mathbf{Params}(sh(b)) \subseteq \mathbf{Params}(sh(a)).$$

PROOF. We must consider the different ways we may have $a \xrightarrow{\mathcal{Q}} b$. If $a \in \mathbf{Dom}(succ^{\mathcal{Q}})$ and $b \in \{a^1, \ldots, a^n\}$ where

$$succ^{\mathcal{Q}}(a) = \langle a^1, \ldots, a^n\rangle,$$

then this follows from Lemma 7.3.16. We cannot have $\langle a, \mathbf{A}_\alpha, b\rangle$ in $exps$ by $\clubsuit_{exp}$ since a is not an expansion node. Since a is not a potentially connecting node, we cannot have $\langle b, a, d\rangle \in conns^{\mathcal{Q}}$ or $\langle b, d, a\rangle \in conns^{\mathcal{Q}}$ by Lemma 7.3.5. Finally, suppose $\langle a, b\rangle \in decs^{\mathcal{Q}}$.

By $\clubsuit^p_{dec}$ the shallow formula of b is the same as the shallow formula of a. Hence we must have $\mathbf{Params}(sh(b)) \subseteq \mathbf{Params}(sh(a))$. □

LEMMA 7.3.18. *Let $\mathcal{Q}$ be an extensional expansion dag. For any connection $\langle c, a, b\rangle \in conns^{\mathcal{Q}}$ we have*

$$\mathbf{Params}(sh(c)) = \mathbf{Params}(sh(a)) \cup \mathbf{Params}(sh(b))$$

and

$$\mathbf{Free}(sh(c)) = \mathbf{Free}(sh(a)) \cup \mathbf{Free}(sh(b)).$$

PROOF. This follows from $\clubsuit_c$, $\clubsuit^p_m$, $\clubsuit^p_{eu}$ and $\clubsuit^{sp}_{eu}$. □

7.4. Extensional Expansion Proofs

We next consider two more conditions required for an extensional expansion dag to be part of an extensional expansion proof. One of the conditions is propositional in nature (cf. Definition 7.4.9) while the other involves the respecting the order of quantifiers (cf. Definition 7.4.17).

DEFINITION 7.4.1 (Conjunctive, Disjunctive, Neutral Nodes). Let $\mathcal{Q}$ be an extensional expansion dag and a be a node in $\mathcal{Q}$.

- We say a is *conjunctive* if a is extended and a is a conjunction node, a true node, an expansion node or a negative equation node.
- We say a is *disjunctive* if a is extended and a is a disjunction node, a false node, a decomposition node, a Boolean node or a connection node.
- We say a is *neutral* if a is extended and a is a negation node, a selection node, a deepening node or a functional node.

The fact that disjunction nodes and false nodes (thinking of false as an empty disjunction) are classified as disjunctive should not be surprising. We have also classified decomposition nodes, Boolean nodes and connection nodes as being disjunctive. Suppose d is an extended decomposition node with shallow formula

$$[F_{\iota\iota\iota}\,\mathbf{A}_\iota\,\mathbf{B}_\iota] \doteq^\iota [F_{\iota\iota\iota}\,\mathbf{C}_\iota\,\mathbf{D}_\iota]. \qquad (1)$$

Decomposition nodes are always negative (cf. Definition 7.2.2). Thinking in terms of refutations, d corresponds to the equation (1) failing to hold. Since the heads of both sides of the equation are the same, either the arguments $\mathbf{A}$ and $\mathbf{C}$ must differ or the arguments $\mathbf{B}$ and $\mathbf{D}$ must differ. That is, if the equation (1) fails, then one of the equations $\mathbf{A} \doteq^{\iota} \mathbf{C}$ and $\mathbf{B} \doteq^{\iota} \mathbf{D}$ must fail. By $\clubsuit_{dec}$, $succ(d) = \langle e^1, e^2\rangle$ where e^1 and e^2 are negative primary nodes with shallow formulas $\mathbf{A} \doteq^{\iota} \mathbf{C}$ and $\mathbf{B} \doteq^{\iota} \mathbf{D}$, respectively. The nodes e^1 and e^2 are disjunctively related in the sense that the proposition

$$\underline{sh}(d) \mathbin{\dot\supset} [\underline{sh}(e^1) \mathbin{\dot\vee} \underline{sh}(e^2)]$$

is true. That is,

$$[\dot\neg[F_{\iota\iota\iota}\,\mathbf{A}_\iota\,\mathbf{B}_\iota] \doteq^{\iota} [F_{\iota\iota\iota}\,\mathbf{C}_\iota\,\mathbf{D}_\iota]] \mathbin{\dot\supset} \centerdot\dot\neg[\mathbf{A} \doteq^{\iota} \mathbf{C}] \mathbin{\dot\vee} \dot\neg[\mathbf{B} \doteq^{\iota} \mathbf{D}]$$

is true. A similar analysis applies to Boolean nodes and connection nodes (i.e., mate nodes, E-unification nodes and symmetric E-unification nodes).

Neutral nodes a only have one child and this child is listed in $succ(a)$.

LEMMA 7.4.2. *Let $\mathcal{Q}$ be an extensional expansion dag and $a \in \mathcal{N}^{\mathcal{Q}}$ be a node. If a is neutral, then it has exactly one child a^1 and $succ^{\mathcal{Q}}(a) = \langle a^1\rangle$.*

PROOF. We know a is extended since a is neutral. By $\clubsuit_{\neg}$, $\clubsuit_{sel}$, $\clubsuit_{\sharp}$ or $\clubsuit_{\doteq}^{\rightarrow}$, we know there exists some node a^1 such that $succ^{\mathcal{Q}}(a) = \langle a^1\rangle$. Hence a has a child a^1. Suppose b is a child of a. Since a is not an expansion node, not a potentially connecting node and not a negative equation node, we must have $b \in \{a^1\}$ by Lemma 7.3.13. That is, $b = a^1$. Thus a^1 is the only child of a. □

All the children of disjunctive nodes d are listed in $succ(d)$.

LEMMA 7.4.3. *Let $\mathcal{Q}$ be an extensional expansion dag, $d \in \mathcal{N}^{\mathcal{Q}}$ be a disjunctive node and $succ^{\mathcal{Q}}(d) = \langle d^1, \ldots, d^n\rangle$. Every child of d is in $\{d^1, \ldots, d^n\}$.*

PROOF. We know a is extended since a is disjunctive. Since d is a disjunctive node, it is not an expansion node, not a potentially connecting node and not a negative equation node. Thus every child of d is in $\{d^1, \ldots, d^n\}$ by Lemma 7.3.13. □

We next define *u-parts*. These could also be called *conjunctive parts* since they contain sets of conjunctively related nodes (thinking in terms of refutations).

DEFINITION 7.4.4 (U-Parts). Let $\mathcal{Q}$ be an extensional expansion dag. We say $\mathfrak{p} \subseteq \mathcal{N}^{\mathcal{Q}}$ is a *u-part of* $\mathcal{Q}$ (or, *unsolved part of* $\mathcal{Q}$) if the following properties hold.

1. If $a \in \mathfrak{p}$ is conjunctive and $a \xrightarrow{\mathcal{Q}} b$, then $b \in \mathfrak{p}$. (That is, all children of a conjunctive node in the u-part must be in the u-part.)
2. If $a \in \mathfrak{p}$ is disjunctive, then there exists some $b \in \mathfrak{p}$ such that $a \xrightarrow{\mathcal{Q}} b$. (That is, some child of a disjunctive node in the u-part must be in the u-part.)
3. For each neutral node $a \in \mathfrak{p}$ and node $a^1 \in \mathcal{N}^{\mathcal{Q}}$ where
$$succ^{\mathcal{Q}}(a) = \langle a^1 \rangle,$$
we have $a^1 \in \mathfrak{p}$. (That is, the unique child of a neutral node in the u-part must be in the u-part.)
4. If $\langle c, a, b \rangle \in conns^{\mathcal{Q}}$ and $a, b \in \mathfrak{p}$, then $c \in \mathfrak{p}$. (That is, if two connected nodes are in the u-part, then the corresponding connection node must be in the u-part.)

For each $\mathcal{R} \subseteq \mathcal{N}^{\mathcal{Q}}$ we say a u-part $\mathfrak{p}$ *includes* $\mathcal{R}$ (or, $\mathfrak{p}$ *is a u-part including* $\mathcal{R}$) if $\mathcal{R} \subseteq \mathfrak{p}$. We define $\boldsymbol{Uparts}(\mathcal{Q})$ to be the set of u-parts of $\mathcal{Q}$. For any $\mathcal{R} \subseteq \mathcal{N}^{\mathcal{Q}}$ we define $\boldsymbol{Uparts}(\mathcal{Q}; \mathcal{R})$ to be the set of u-parts of $\mathcal{Q}$ including $\mathcal{R}$.

Consider condition (2) in the definition of u-parts. A consequence of this condition is that a u-part cannot contain a disjunctive leaf node. Examples of disjunctive leaf nodes are extended false nodes, extended decomposition nodes with shallow formula $[A_\iota \mathrel{\dot{=}^\iota} A_\iota]$ where A_ι is a parameter and extended mate nodes with shallow formula $[P_o \mathrel{\dot{=}^o} P_o]$ where P_o is a parameter. We can think of any u-part as being *unsolved* in the sense that we need to grow the extensional expansion dag until any possible extension of this u-part to the larger extensional expansion dag would necessarily contain a disjunctive leaf node. We now consider a simple example.

Extensional expansion dags are generalizations of expansion trees. The *deep formula* of an expansion tree (cf. [7]) can be viewed as a junctive-form, or jform (cf. [37; 38]), where conjunctions are viewed

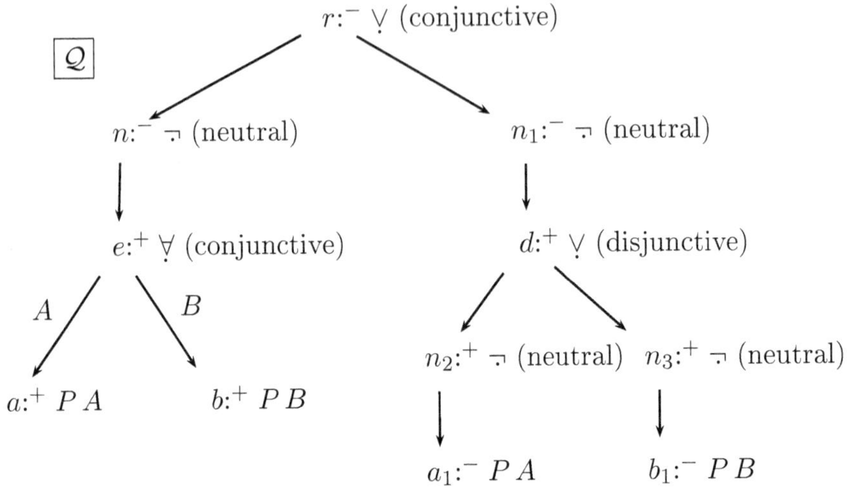

FIGURE 7.10. Extensional Expansion Dag

$$\left[\begin{array}{c} \boxed{1}\,P\,A \\ \boxed{2}\,P\,B \\ \left[\boxed{3}\,\neg P\,A \vee \boxed{4}\,\neg P\,B\right] \end{array}\right]$$

FIGURE 7.11. Jform

vertically and disjunctions are viewed horizontally. Consider the simple theorem

$$[[\dot{\forall} x_\iota \centerdot P_{o\iota}\, x] \;\dot{\supset}\; [P\,A_\iota \;\dot{\wedge}\; P\,B_\iota]]$$

and the extensional expansion dag $\mathcal{Q}$ shown in Figure 7.10. We assume all the nodes in $\mathcal{Q}$ are extended. The children of each node is specified by $succ^{\mathcal{Q}}$ except for the expansion node e. Instead, $succ^{\mathcal{Q}}(e)$ is $\langle\rangle$ and the two children a_1 and b_1 are induced by expansion arcs $\langle e, A, a_1\rangle, \langle e, B, b_1\rangle \in exps^{\mathcal{Q}}$. Since $\mathcal{Q}$ has no connections, $\langle \mathcal{Q}, \overset{\mathcal{Q}}{\rightarrow}\rangle$ is a tree. Viewing $\mathcal{Q}$ as an expansion tree, we can naturally consider

$$[[P_{o\iota}\, A_\iota \;\dot{\wedge}\; P\,B_\iota] \;\dot{\supset}\; [P\,A \;\dot{\wedge}\; P\,B]]$$

to be the deep formula of $\mathcal{Q}$. The corresponding jform is shown in Figure 7.11. There are two vertical paths in this jform, namely $\boxed{1} - \boxed{2} - \boxed{3}$ and $\boxed{1} - \boxed{2} - \boxed{4}$. The notion of a u-part of an extensional expansion dag is a generalization of the notion of a vertical path of such jforms. Among the nodes of $\mathcal{Q}$, the root r is a conjunction node, e is an expansion node and d is a disjunction node. Hence r and e are conjunctive while d is disjunctive. The negation nodes n, n_1, n_2 and n_3 are neutral. Suppose $\mathfrak{p}$ is a u-part of $\mathcal{Q}$ including $\{r\}$. Since r is conjunctive, we must have $\{n, n_1\} \subseteq \mathfrak{p}$ (cf. Definition 7.4.4:1). Since n and n_1 are neutral, we must have $\{e, d\} \subseteq \mathfrak{p}$ (cf. Definition 7.4.4:3). Since e is conjunctive, we must have $\{a, b\} \subseteq \mathfrak{p}$. Thus any u-part of $\mathcal{Q}$ including $\{r\}$ must include $\{a, b, d\}$. For any u-part $\mathfrak{p} \in \mathit{Uparts}(\mathcal{Q})$, $d \in \mathfrak{p}$ implies either $n_2 \in \mathfrak{p}$ or $n_3 \in \mathfrak{p}$ (cf. Definition 7.4.4:2). Likewise, $n_2 \in \mathfrak{p}$ implies $a_1 \in \mathfrak{p}$ for any u-part $\mathfrak{p} \in \mathit{Uparts}(\mathcal{Q})$ (cf. Definition 7.4.4:3). Also, $n_3 \in \mathfrak{p}$ implies $b_1 \in \mathfrak{p}$ for any u-part $\mathfrak{p} \in \mathit{Uparts}(\mathcal{Q})$. The u-part

$$\{r, n, a, b, n_1, d, n_2, a_1\}$$

corresponds to the vertical path $\boxed{1} - \boxed{2} - \boxed{3}$ in the jform in Figure 7.11. The u-part

$$\{r, n, a, b, n_1, d, n_3, b_1\}$$

corresponds to the vertical path $\boxed{1} - \boxed{2} - \boxed{4}$ in the jform in Figure 7.11.

When searching for a proof using jforms, we try to find connections between literals which span the vertical paths (cf. [7]). In terms of the jform in Figure 7.11, the connection $\boxed{1} - \boxed{3}$ spans $\boxed{1} - \boxed{2} - \boxed{3}$ and the connection $\boxed{2} - \boxed{4}$ spans $\boxed{1} - \boxed{2} - \boxed{4}$. When searching for a proof using extensional expansion dags, we add connection nodes to the expansion structure. The connection $\boxed{1} - \boxed{3}$ in the jform corresponds to adding a connection between a and a_1 in $\mathcal{Q}$. An extensional expansion dag $\mathcal{Q}'$ with a connection $\langle c, a, a_1 \rangle$ is shown in Figure 7.12. The node c in $\mathcal{Q}'$ is a mate node. If c were extended (i.e., $c \in \mathbf{Dom}(succ^{\mathcal{Q}'})$), then c would be a disjunctive node with a single child with shallow formula $[A \doteq^{\iota} A]$. However, we assume c is not extended in $\mathcal{Q}'$ and so c is not disjunctive (cf. Definition 7.4.1).

The u-part

$$\{r, n, a, b, n_1, d, n_2, a_1\}$$

of $\mathcal{Q}$ can be extended to obtain a u-part of $\mathcal{Q}'$. By Definition 7.4.4:4, such a u-part $\mathfrak{p}'$ of $\mathcal{Q}'$ must include the connection node c since $\langle c, a, a_1 \rangle \in conns^{\mathcal{Q}'}$, $a \in \mathfrak{p}'$ and $a_1 \in \mathfrak{p}'$. We obtain the u-part

$$\{r, n, a, b, n_1, d, n_2, a_1, c\}$$

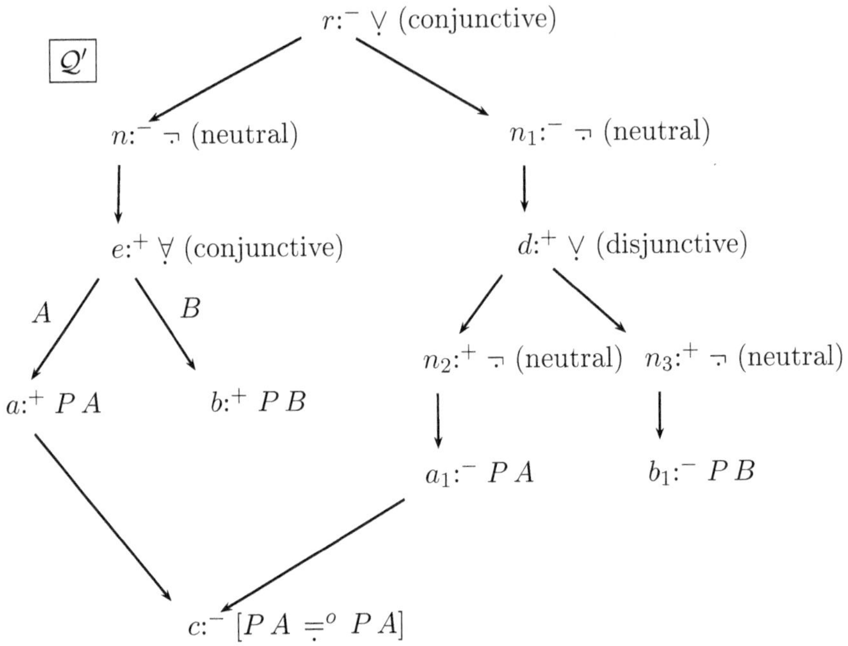

FIGURE 7.12. Dag with Connection

of $\mathcal{Q}'$ by simply adding the new connection node c to the u-part of $\mathcal{Q}$.

In Figure 7.13, we consider an extension $\mathcal{Q}''$ of the extensional expansion dag $\mathcal{Q}'$ by extending c to have a child e_1 (an equation node), extending e_1 to have an equation goal child g, adding a decomposition node d_1, adding a decomposition arc $\langle e_1, d_1\rangle$ and extending d_1 such that $succ^{\mathcal{Q}''}(d_1) = \langle\rangle$. The extended connection node c is a disjunctive node in $\mathcal{Q}''$. Likewise, e_1 is conjunctive and d_1 is disjunctive in $\mathcal{Q}''$. There is no way to extend the u-part

$$\{r, n, a, b, n_1, d, n_2, a_1, c\}$$

of $\mathcal{Q}'$ to be a u-part of $\mathcal{Q}''$. Suppose $\mathfrak{p}''$ is a u-part of $\mathcal{Q}''$ such that

$$\{r, n, a, b, n_1, d, n_2, a_1, c\} \subseteq \mathfrak{p}''.$$

Since e_1 is the only child of the disjunctive node c, we must have $e_1 \in \mathfrak{p}''$. Since e_1 is conjunctive, we must have $\{d_1, g\} \subseteq \mathfrak{p}''$. Since d_1 is disjunctive, some child of d_1 must be in $\mathfrak{p}''$ (cf. Definition 7.4.4:2). However, d_1 is a leaf node. Hence such a u-part $\mathfrak{p}''$ does not exist. In

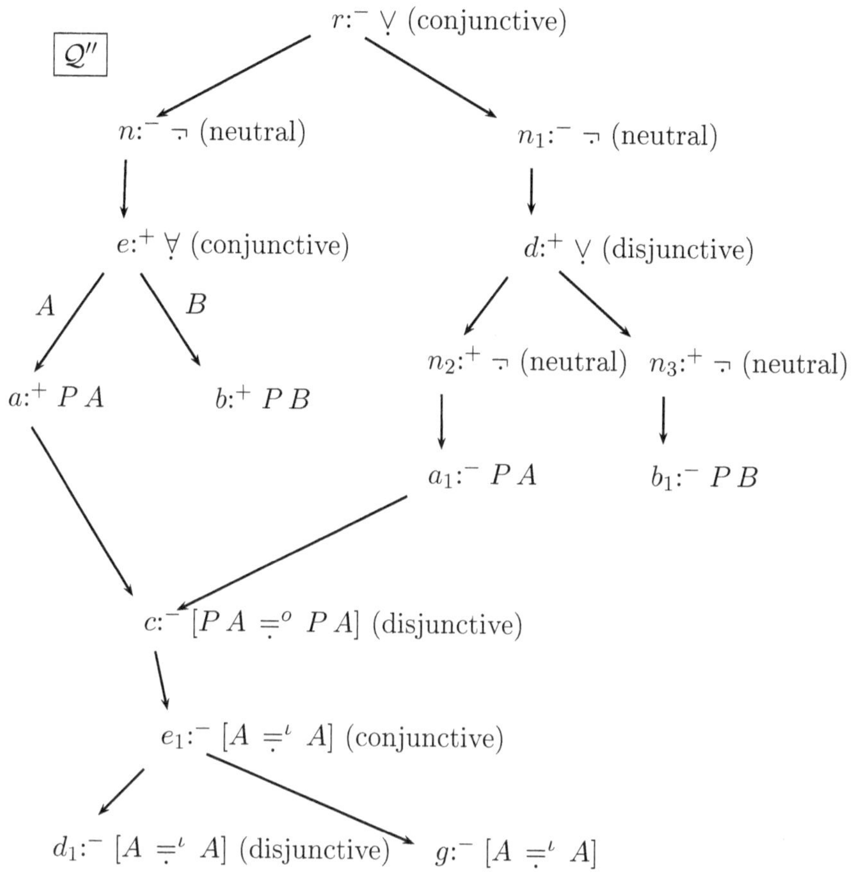

FIGURE 7.13. Dag with Extended Connection

this manner, growing extensional expansion dags using connections leads to the removal of u-parts.

We will not formally consider deep formulas or jforms corresponding to extensional expansion dags. However, jforms for extensional expansion dags can be computed using a technique similar to dissolution (cf. [48; 38]) to handle connections (cf. Remark 9.6.1). Vertical paths of such jforms correspond to u-parts.

REMARK 7.4.5. In traditional mating search methods for theorem proving (cf. [6; 7; 21]), one tries to find a unifiable set of connections so that every vertical path is spanned by a connection. Searching for proofs using extensional expansion dags is a very similar process. The differences primarily result from the fact that equational reasoning requires more than unification. In particular, extensional reasoning requires the process of solving equations and adding connections to be intertwined. We have included explicit connection nodes for this purpose and modified the notion of a vertical path to be a u-part accordingly. We reach a solution by growing extensional expansion dags until there are no remaining u-parts.

LEMMA 7.4.6. *Let $\mathcal{Q}$ be an extensional expansion dag, $\mathfrak{p}$ be a u-part of $\mathcal{Q}$, $d \in \mathfrak{p}$ be a disjunctive node and $succ^{\mathcal{Q}}(d) = \langle d^1, \ldots, d^n \rangle$. There is some i $(1 \leq i \leq n)$ such that $d^i \in \mathfrak{p}$.*

PROOF. By Definition 7.4.4:2, we know there is some $b \in \mathfrak{p}$ such that $d \xrightarrow{\mathcal{Q}} b$. By Lemma 7.4.3, $b \in \{d^1, \ldots, d^n\}$. □

Notice that we have not made any requirement that *only* one child of a disjunctive node is in the u-part. Consequently, a u-part may contain nodes which are disjunctively related. The conditions only force a u-part $\mathfrak{p}$ to contain *enough* nodes.

LEMMA 7.4.7. *Let $\mathcal{Q}$ be an extensional expansion dag and a be a root node of $\mathcal{Q}$. If $\mathfrak{p}$ is a u-part of $\mathcal{Q}$, then $\mathfrak{p} \setminus \{a\}$ is a u-part of $\mathcal{Q}$.*

PROOF. Suppose $\mathfrak{p}$ is a u-part of $\mathcal{Q}$. For any conjunctive or neutral $c \in \mathfrak{p} \setminus \{a\}$, every child of c is in $\mathfrak{p}$, and hence in $\mathfrak{p} \setminus \{a\}$ (since a is a root node). Likewise, for any disjunctive $d \in \mathfrak{p} \setminus \{a\}$, some child of d is in $\mathfrak{p}$, hence in $\mathfrak{p} \setminus \{a\}$. If $\langle c, d, b \rangle$ is a connection in $\mathcal{Q}$ and $d, b \in \mathfrak{p} \setminus \{a\} \subseteq \mathfrak{p}$, then $c \in \mathfrak{p}$ and hence $c \in \mathfrak{p} \setminus \{a\}$ (since a is a root node and c is not). □

LEMMA 7.4.8. *Let $\mathcal{Q}$ be an extensional expansion dag and $a \in \mathcal{N}^{\mathcal{Q}}$. Suppose $\mathfrak{p}$ is a u-part of $\mathcal{Q}$. Assume the following:*

1. *If a is conjunctive, then every child of a is in $\mathfrak{p}$.*
2. *If a is disjunctive, then some child of a is in $\mathfrak{p}$.*
3. *If a is neutral, then the unique child of a is in $\mathfrak{p}$.*
4. *If a is a potentially connecting node, then for every $b \in \mathfrak{p}$ where $\langle c, a, b \rangle \in conns^{\mathcal{Q}}$ or $\langle c, b, a \rangle \in conns^{\mathcal{Q}}$ we have $c \in \mathfrak{p}$.*

Then $\mathfrak{p} \cup \{a\}$ is a u-part of $\mathcal{Q}$.

PROOF. Suppose $\mathfrak{p}$ is a u-part of $\mathcal{Q}$. For each conjunctive $c \in \mathfrak{p}$, every child of c is in $\mathfrak{p} \subseteq \mathfrak{p} \cup \{a\}$. Also, for any disjunctive $d \in \mathfrak{p}$, some child of d is in $\mathfrak{p} \subseteq \mathfrak{p} \cup \{a\}$. For each neutral node $c \in \mathfrak{p}$ with $succ^{\mathcal{Q}}(c) = \langle c^1 \rangle$, we know c^1 is in $\mathfrak{p} \subseteq \mathfrak{p} \cup \{a\}$. For the node a, we have assumed the appropriate children are in $\mathfrak{p}$ when a is conjunctive, disjunctive or neutral.

Suppose $\langle c, d, b \rangle \in conns^{\mathcal{Q}}$ with $d, b \in \mathfrak{p} \cup \{a\}$. Unless $a \in \{d, b\}$, we know $c \in \mathfrak{p} \subseteq \mathfrak{p} \cup \{a\}$ since $\mathfrak{p}$ is a u-part. If $a \in \{d, b\}$, then we have assumed $c \in \mathfrak{p} \subseteq \mathfrak{p} \cup \{a\}$. □

One of the conditions of being an extensional expansion proof is for the extensional expansion dag to have no u-parts which include a set of nodes. We define such a set of nodes to be a *sufficient* set.

DEFINITION 7.4.9. Let $\mathcal{Q}$ be an extensional expansion dag. We say a set $\mathcal{R} \subseteq \mathcal{N}^{\mathcal{Q}}$ of nodes is *sufficient* if no u-part of $\mathcal{Q}$ includes $\mathcal{R}$.

Any superset of a sufficient set is also sufficient.

LEMMA 7.4.10. *Let $\mathcal{Q}$ be an extensional expansion dag. If $\mathcal{R}_1$ is sufficient and $\mathcal{R}_1 \subseteq \mathcal{R}_2 \subseteq \mathcal{N}^{\mathcal{Q}}$, then $\mathcal{R}_2$ is sufficient.*

PROOF. This is trivial since any u-part $\mathfrak{p}$ including $\mathcal{R}_2$ would be a u-part including $\mathcal{R}_1$. □

It follows that if any set of root nodes is sufficient, then the set of all root nodes is also sufficient. On the other hand, the empty set is never sufficient.

LEMMA 7.4.11. *Let $\mathcal{Q}$ be an extensional expansion dag. If $\mathcal{R} \subseteq \mathcal{N}^{\mathcal{Q}}$ is sufficient, then $\mathcal{R} \neq \emptyset$.*

PROOF. The conditions in the definition of u-parts (cf. Definition 7.4.4) are vacuous for the empty set. Thus $\emptyset$ is a u-part including $\emptyset$. □

A node which is not extended can always be deleted from a sufficient set.

LEMMA 7.4.12. *Let $\mathcal{Q}$ be an extensional expansion dag. If $\mathcal{R} \subseteq \mathcal{N}^{\mathcal{Q}}$ is sufficient and $a \in \mathcal{R} \setminus \mathbf{Dom}(succ^{\mathcal{Q}})$, then $\mathcal{R} \setminus \{a\}$ is sufficient.*

PROOF. Suppose $\mathfrak{p}$ is a u-part of $\mathcal{Q}$ and $\mathfrak{p}$ includes $\mathcal{R} \setminus \{a\}$. The assumptions of Lemma 7.4.8 hold for a since a is a leaf (cf. Lemma 7.3.6) and a is not conjunctive, not disjunctive and not neutral. Thus $\mathfrak{p} \cup \{a\}$ is a u-part of $\mathcal{Q}$ including $\mathcal{R}$, contradicting sufficiency of $\mathcal{R}$. □

Under appropriate assumptions we can replace the children of a node a in a sufficient set $\mathcal{R}$ by the node a.

LEMMA 7.4.13. *Let $\mathcal{Q}$ be an extensional expansion dag, $a \in \mathcal{N}^{\mathcal{Q}}$ be a conjunctive node of $\mathcal{Q}$, $\mathcal{K}$ be $\{b \in \mathcal{N}^{\mathcal{Q}} \mid a \xrightarrow{\mathcal{Q}} b\}$ and $\mathcal{R} \subseteq \mathcal{N}^{\mathcal{Q}}$. If $\mathcal{R} \cup \mathcal{K}$ is a sufficient set for $\mathcal{Q}$, then $\mathcal{R} \cup \{a\}$ is a sufficient set for $\mathcal{Q}$.*

PROOF. Suppose there is a u-part $\mathfrak{p}$ which includes $\mathcal{R} \cup \{a\}$. Hence $a \in \mathfrak{p}$ and so $\mathcal{K} \subseteq \mathfrak{p}$ since a is conjunctive. Thus $\mathfrak{p}$ is a u-part including $\mathcal{R} \cup \mathcal{K}$, contradicting sufficiency of $\mathcal{R} \cup \mathcal{K}$. □

LEMMA 7.4.14. *Let $\mathcal{Q}$ be an extensional expansion dag, $a \in \mathcal{N}^{\mathcal{Q}}$ be a disjunctive node of $\mathcal{Q}$, $\mathcal{K}$ be $\{b \in \mathcal{N}^{\mathcal{Q}} \mid a \xrightarrow{\mathcal{Q}} b\}$ and $\mathcal{R} \subseteq \mathcal{N}^{\mathcal{Q}}$. If $\mathcal{R} \cup \{b\}$ is a sufficient set for $\mathcal{Q}$ for each $b \in \mathcal{K}$, then $\mathcal{R} \cup \{a\}$ is a sufficient set for $\mathcal{Q}$.*

PROOF. Suppose there is a u-part $\mathfrak{p}$ which includes $\mathcal{R} \cup \{a\}$. Hence $a \in \mathfrak{p}$. Since a is disjunctive, there is some $b \in \mathcal{K}$ such that $b \in \mathfrak{p}$. Thus $\mathfrak{p}$ is a u-part including $\mathcal{R} \cup \{b\}$, contradicting sufficiency of $\mathcal{R} \cup \{b\}$. □

LEMMA 7.4.15. *Let $\mathcal{Q}$ be an extensional expansion dag, $a \in \mathcal{N}^{\mathcal{Q}}$ be a neutral node of $\mathcal{Q}$ with child b and $\mathcal{R} \subseteq \mathcal{N}^{\mathcal{Q}}$. If $\mathcal{R} \cup \{b\}$ is a sufficient set for $\mathcal{Q}$, then $\mathcal{R} \cup \{a\}$ is a sufficient set for $\mathcal{Q}$.*

PROOF. Suppose there is a u-part $\mathfrak{p}$ which includes $\mathcal{R} \cup \{a\}$. Hence $a \in \mathfrak{p}$. By Lemma 7.4.2, we know $succ^{\mathcal{Q}}(a) = \langle b \rangle$. Hence $b \in \mathfrak{p}$. Thus $\mathfrak{p}$ is a u-part including $\mathcal{R} \cup \{b\}$, contradicting sufficiency of $\mathcal{R} \cup \{b\}$. □

LEMMA 7.4.16. *Let $\mathcal{Q}$ be an extensional expansion dag, $\langle c, a, b \rangle$ be a connection in $conns^{\mathcal{Q}}$ and $\mathcal{R} \subseteq \mathcal{N}^{\mathcal{Q}}$. If $\mathcal{R} \cup \{c\}$ is a sufficient set for $\mathcal{Q}$, then $\mathcal{R} \cup \{a, b\}$ is a sufficient set for $\mathcal{Q}$.*

PROOF. Suppose there is a u-part $\mathfrak{p}$ which includes $\mathcal{R} \cup \{a, b\}$. Hence $a, b \in \mathfrak{p}$ and so $c \in \mathfrak{p}$. Thus $\mathfrak{p}$ is a u-part including $\mathcal{R} \cup \{c\}$, contradicting sufficiency of $\mathcal{R} \cup \{c\}$. □

Another condition for extensional expansion proofs will be acyclicity of an embedding relation on selected parameters.

DEFINITION 7.4.17 (Embedding Relation). Let $\mathcal{Q}$ be an extensional expansion dag. We define the embedding relation $\prec_0^{\mathcal{Q}}$ on selected parameters of $\mathcal{Q}$ as follows. For selected parameters A, B

in **SelPars**($\mathcal{Q}$) we say $A \prec_0^{\mathcal{Q}} B$ holds if there exists an expansion arc $\langle c, \mathbf{C}, d\rangle \in exps^{\mathcal{Q}}$ such that $A \in \mathbf{Params}(\mathbf{C})$ and d dominates **SelNode**(B). We say the embedding relation $\prec_0^{\mathcal{Q}}$ is *acyclic* if there are no selected parameters $A^1, \ldots, A^m$ of $\mathcal{Q}$ with $m \geq 1$ such that $A^i \prec_0^{\mathcal{Q}} A^{i+1}$ for each i $(1 \leq i \leq (m-1))$ and $A^m \prec_0^{\mathcal{Q}} A^1$.

Using sufficient sets and acyclicity of the embedding relation, we can define extensional expansion proofs.

DEFINITION 7.4.18 (Extensional Expansion Proofs). Let $\mathcal{S}$ be a signature of logical constants, $\mathcal{Q}$ be an $\mathcal{S}$-extensional expansion dag and $\mathcal{R} \subseteq \mathcal{N}^{\mathcal{Q}}$ be a set of nodes of $\mathcal{Q}$. We say $\langle \mathcal{Q}, \mathcal{R}\rangle$ is an *$\mathcal{S}$-extensional expansion proof* if $\mathcal{Q}$ is closed, the embedding relation $\prec_0^{\mathcal{Q}}$ is acyclic, $\mathcal{R}$ is a sufficient set for $\mathcal{Q}$ and $\mathbf{SelPars}^{\mathcal{Q}}(\mathcal{R})$ is empty. We often refer to $\mathcal{S}$-extensional expansion proofs as *extensional expansion proofs* and leave the signature $\mathcal{S}$ implicit.

EXAMPLE 7.4.19. Let A_ι be a parameter and $\mathcal{N} := \{*\}$. Let $\mathcal{Q}$ be the expansion structure

$$\langle \mathcal{N}, succ, label, \emptyset, \emptyset, \emptyset, \emptyset\rangle$$

where $succ(*) := \langle\rangle$ and $label(*) := \langle \mathbf{dec}, -1, [A \doteq^\iota A]\rangle \in \mathbf{Dec}$. It is easy to check $\mathcal{Q}$ is an extensional expansion dag. Since there are no selected parameters, the embedding relation $\prec_0^{\mathcal{Q}}$ is vacuously acyclic. The only potential u-part including $\{*\}$ is $\{*\}$. However, this is *not* a u-part since $*$ is a disjunctive node (as a decomposition node) with no children. Thus $\langle \mathcal{Q}, \{*\}\rangle$ is an extensional expansion proof of $[A \doteq^\iota A]$.

EXAMPLE 7.4.20. Consider the extensional expansion dag $\mathcal{Q}$ in Figure 7.14. Assume every node in $\mathcal{Q}$ is extended. Note that $\mathcal{Q}$ has four root nodes: a, b, c and **DIS**. The shallow formula of **DIS** is

$$[\dot{\neg}[A \dot{\vee} B] \dot{\vee} \dot{\neg} C].$$

Note that the mate nodes $\mathbf{MATE}^a$, $\mathbf{MATE}^b$ and $\mathbf{MATE}^c$ are disjunctive leaf nodes. Hence no unsolved parts may contain any of these nodes. Using this fact, we can verify that the set of all root nodes $\mathcal{R} := \{a, b, c, \mathbf{DIS}\}$ is sufficient. Thus $\langle \mathcal{Q}, \mathcal{R}\rangle$ provides a proof of the sequent

$$\dot{\neg} A,\ \dot{\neg} B,\ \dot{\neg} C,\ \dot{\neg}[\dot{\neg}[A \dot{\vee} B] \dot{\vee} \dot{\neg} C].$$

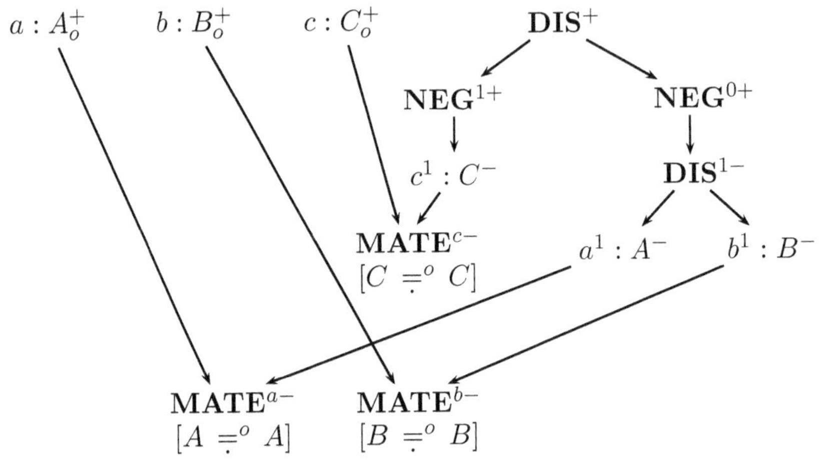

FIGURE 7.14. Dag with Several Sufficient Sets

Furthermore, the sets $\mathcal{R}_a := \{a, c, \mathbf{DIS}\}$ and $\mathcal{R}_b := \{b, c, \mathbf{DIS}\}$ are sufficient. Thus $\langle \mathcal{Q}, \mathcal{R}_a \rangle$ is an extensional expansion proof of the sequent

$$\dot\neg A,\ \dot\neg C,\ \dot\neg[\dot\neg[A \dot\vee B] \dot\vee \dot\neg C]$$

and $\langle \mathcal{Q}, \mathcal{R}_b \rangle$ is an extensional expansion proof of the sequent

$$\dot\neg B,\ \dot\neg C,\ \dot\neg[\dot\neg[A \dot\vee B] \dot\vee \dot\neg C].$$

Other sufficient sets worth noting are $\{a, a^1\}$, $\{\mathbf{MATE}^a\}$, $\{b, b^1\}$, $\{\mathbf{MATE}^b\}$, $\{c, c^1\}$ and $\{\mathbf{MATE}^c\}$.

EXAMPLE 7.4.21. Consider the extensional expansion dag $\mathcal{Q}$ in Figure 7.15 with $\mathbf{Roots}(\mathcal{Q}) = \{\mathbf{EXP}, \mathbf{DIS}\}$. Assume every node in $\mathcal{Q}$ is extended. The decomposition nodes $\mathbf{DEC}^a$ and $\mathbf{DEC}^b$ are disjunctive leaf nodes. Hence no unsolved part may contain either of these nodes. Using this fact, we can determine $\mathbf{Roots}(\mathcal{Q})$ is sufficient and $\langle \mathcal{Q}, \mathbf{Roots}(\mathcal{Q}) \rangle$ is an extensional expansion proof of the sequent

$$\dot\neg[\dot\forall x_\iota\, P\, x],\ [P\, A \dot\wedge P\, B].$$

Note that there is an expansion arc $\langle \mathbf{EXP}, C, c \rangle$ which is not vital to the proof. The fact that this expansion arc is not vital corresponds to the fact that the set $\{a, b, \mathbf{DIS}\}$ is sufficient.

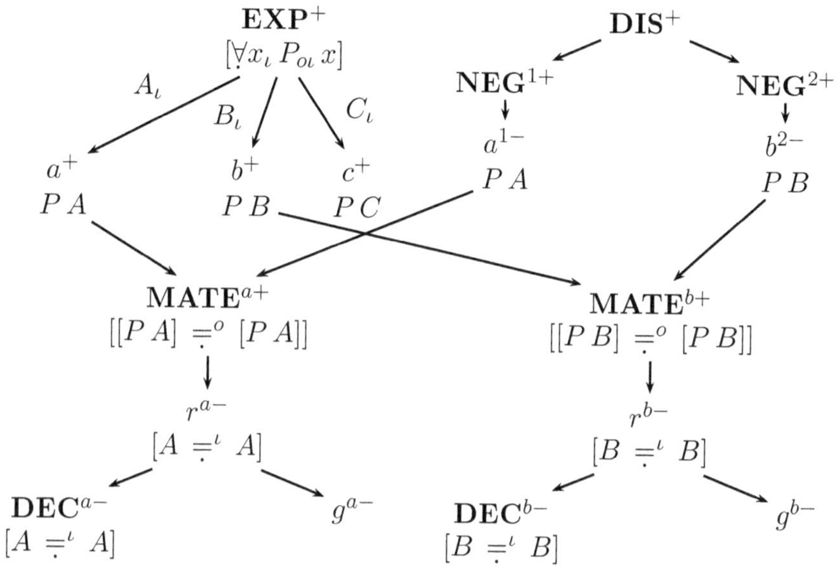

FIGURE 7.15. Dag with an Unnecessary Expansion Arc

We now consider an example demonstrating why the embedding relation is required to be acyclic.

EXAMPLE 7.4.22. Consider the extensional expansion dag $\mathcal{Q}$ in Figure 7.16. The unique root node **NEG** is negative and has shallow formula

$$\dot{\exists} x_\iota \dot{\forall} y_\iota \,\blacksquare\, x \doteq^\iota y.$$

Of course, this is not a theorem. The extensional expansion dag has no u-parts including {**NEG**} since any such u-part must include the decomposition node **DEC** (a disjunctive node with no children). However, this cannot be used to form an extensional expansion proof because the embedding relation is not acyclic. In particular, $A \prec_0^{\mathcal{Q}} A$.

7.5. Further Properties of Extensional Expansion Dags

We now consider additional properties extensional expansion dags and extensional expansion proofs may satisfy.

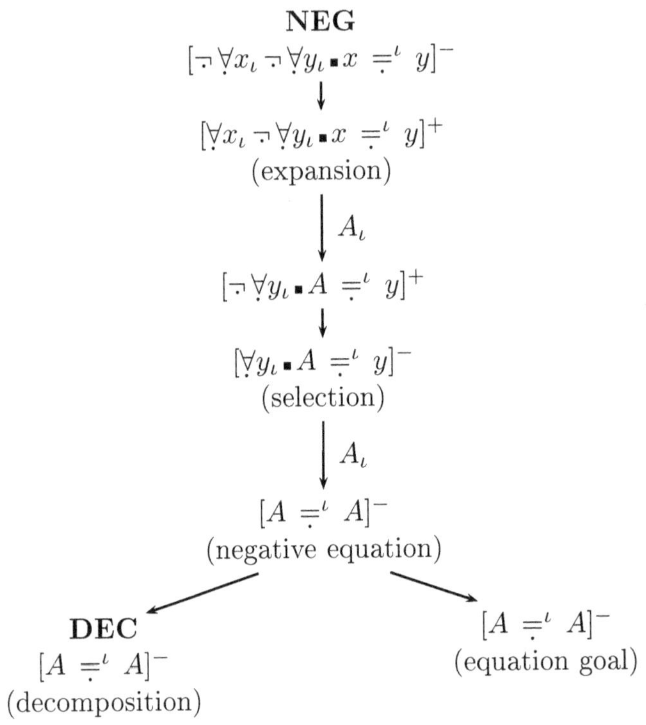

FIGURE 7.16. Dag with Cyclic Embedding Relation

7.5.1. Restricted Extensional Expansion Dags. We defined $\mathcal{W}$-restricted derivations (cf. Definition 2.2.23) by restricting the rules $\mathcal{G}(\underset{\cdot}{\neg}\underset{\cdot}{\forall}, \mathcal{S})$ and $\mathcal{G}(\underset{\cdot}{\neg} \underset{\cdot}{=}^{\rightarrow}, \mathcal{S})$ to terms in $\mathcal{W}$. We similarly define a notion of $\mathcal{W}$-restricted extensional expansion dags.

DEFINITION 7.5.1 (Restricted Extensional Expansion Dags). Let $\mathcal{S}$ be a signature of logical constants, $\mathcal{Q}$ be an $\mathcal{S}$-extensional expansion dag and $\mathcal{W} \subseteq \mathit{wff}(\mathcal{S})$ be a set of formulas. We say $\mathcal{Q}$ is *$\mathcal{W}$-restricted* if $\mathbf{ExpTerms}(\mathcal{Q}) \subseteq \mathcal{W}$. We say an $\mathcal{S}$-extensional expansion proof $\langle \mathcal{Q}, \mathcal{R} \rangle$ is *$\mathcal{W}$-restricted* if $\mathcal{Q}$ is a $\mathcal{W}$-restricted $\mathcal{S}$-extensional expansion dag.

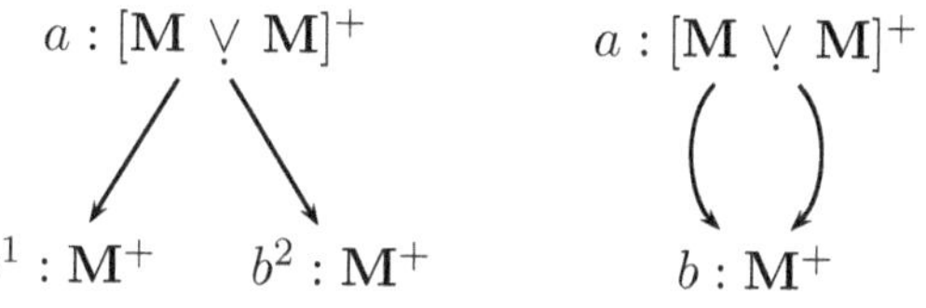

FIGURE 7.17. Structure Sharing

7.5.2. Separated Extensional Expansion Dags. The definition of extensional expansion dags allows structure sharing. For example, consider an extensional expansion dag with an extended disjunction node a with shallow formula $[\mathbf{M} \underset{\cdot}{\vee} \mathbf{M}]$. By $\clubsuit_{\vee}$, we must have $succ(a) = \langle b^1, b^2 \rangle$ where b^1 and b^2 are positive primary nodes which both have shallow formula $\mathbf{M}$. There is no condition which prevents b^1 and b^2 from being the same node b (see Figure 7.17). In general, structure sharing can be useful if one is concerned with implementation issues. Consequently, we have defined extensional expansion dags and extensional expansion proofs in a way which allows structure sharing. Furthermore, we will prove extensional expansion proofs which may contain structure sharing are sound (cf. Theorem 7.9.1).

On the other hand, we will need to disallow such structure sharing to prove lifting results in Chapter 8. For this purpose we define *separated* extensional expansion dags to be extensional expansion dags which do not contain structure sharing. We will prove completeness relative to separated extensional expansion dags (cf. Theorem 7.10.12).

DEFINITION 7.5.2 (Separated Extensional Expansion Dags). Let $\mathcal{Q}$ be an extensional expansion dag. We say $\mathcal{Q}$ is *separated* if the following conditions hold:

1. For every $a \in \mathbf{Dom}(succ^{\mathcal{Q}})$ with $succ^{\mathcal{Q}}(a) = \langle a^1, \ldots, a^n \rangle$ the nodes $a^1, \ldots, a^n$ are distinct.
2. For every node $b \in \mathcal{N}^{\mathcal{Q}}$ if b is not a connection node, then there is at most one node $a \in \mathcal{N}^{\mathcal{Q}}$ such that $a \overset{\mathcal{Q}}{\rightarrow} b$.
3. For every connection node $c \in \mathcal{N}^{\mathcal{Q}}$ there is at most one pair $\langle a, b \rangle$ of nodes such that $\langle c, a, b \rangle \in conns^{\mathcal{Q}}$.

We say an extensional expansion proof $\langle \mathcal{Q}, \mathcal{R} \rangle$ is *separated* if $\mathcal{Q}$ is separated.

7.5.3. Merged Extensional Expansion Dags. As noted earlier, expansion proofs remove unnecessary information (such as the order of application of rules) present in sequent calculus derivations. However, some redundant information may remain. For example, expansion nodes e may have more than one arc with the same expansion term. Likewise, two potentially connecting nodes may share two connection node children of the same kind. We define the notion of a *merged* extensional expansion dag in order to remove this kind of redundancy.

DEFINITION 7.5.3. Let

$$\mathcal{Q} = \langle \mathcal{N}, succ, label, sel, exps, conns, decs \rangle$$

be an extensional expansion dag.

1. We say $\mathcal{Q}$ has *merged expansions* if for any two distinct expansion arcs
$$\langle e, \mathbf{A}^1_\alpha, c^1 \rangle, \langle e, \mathbf{A}^2_\alpha, c^2 \rangle \in exps$$
we have $sh(c^1) \neq sh(c^2)$.
2. We say $\mathcal{Q}$ has *merged connections* if for any two distinct connections
$$\langle c^1, a, b \rangle, \langle c^2, a, b \rangle \in conns$$
we have $kind(c^1) \neq kind(c^2)$.
3. We say $\mathcal{Q}$ has *merged decompositions* if for every node $e \in \mathcal{N}$ there is at most one decomposition arc $\langle e, d \rangle \in decs$.

We say $\mathcal{Q}$ is *merged* if $\mathcal{Q}$ has merged expansions, merged connections and merged decompositions. We say an extensional expansion proof $\langle \mathcal{Q}, \mathcal{R} \rangle$ is *merged* if $\mathcal{Q}$ is merged.

For an extensional expansion dag to be merged, we have not quite required that any pair of potentially connecting nodes have at most one common child. This is because a positive equation node and an equation goal node may have two children in a merged extensional expansion dag. In such a case, one must be an E-unification node while the other must be a symmetric E-unification node. It is true that any pair of atomic nodes must have at most one common child.

LEMMA 7.5.4. *Let $\mathcal{Q}$ be an extensional expansion dag with merged connections, $a \in \mathcal{N}^{\mathcal{Q}}$ be a positive atomic node and $b \in \mathcal{N}^{\mathcal{Q}}$ be a negative atomic node. If $\langle c^1, a, b\rangle, \langle c^2, a, b\rangle \in conns^{\mathcal{Q}}$, then $c^1 = c^2$.*

PROOF. The nodes c^1 and c^2 are both mate nodes by Lemma 7.3.8. Hence $c^1 = c^2$ since $\mathcal{Q}$ has merged connections. □

LEMMA 7.5.5. *Let $\mathcal{Q}$ be an extensional expansion dag with merged connections, $a \in \mathcal{N}^{\mathcal{Q}}$ be a positive equation node and $b \in \mathcal{N}^{\mathcal{Q}}$ be an equation goal node. Suppose $\langle c^1, a, b\rangle, \langle c^2, a, b\rangle \in conns^{\mathcal{Q}}$ and $c^1 \neq c^2$. Then either c^1 is an E-unification node and c^2 is a symmetric E-unification node or c^2 is an E-unification node and c^1 is a symmetric E-unification node. Also, $c \in \{c^1, c^2\}$ for any $\langle c, a, b\rangle \in conns^{\mathcal{Q}}$.*

PROOF. This follows easily from Lemma 7.3.8 and the assumption that $\mathcal{Q}$ has merged connections. □

7.6. Extensional Expansion Subdags

DEFINITION 7.6.1. Let

$$\mathcal{Q} = \langle \mathcal{N}, succ, label, sel, exps, conns, decs\rangle$$

and

$$\mathcal{Q}' = \langle \mathcal{N}', succ', label', sel', exps', conns', decs'\rangle$$

be expansion structures. We say $\mathcal{Q}$ is a *substructure* of $\mathcal{Q}'$ if

$$\mathcal{N} \subseteq \mathcal{N}',$$
$$\mathbf{Dom}(succ) \subseteq \mathbf{Dom}(succ'),$$
$$succ'|_{\mathbf{Dom}(succ)} = succ,$$
$$label'|_{\mathcal{N}} = label,$$
$$\mathbf{Dom}(sel) \subseteq \mathbf{Dom}(sel'),$$
$$sel'|_{\mathbf{Dom}(sel)} = sel,$$
$$exps \subseteq exps',$$
$$conns \subseteq conns'$$

and

$$decs \subseteq decs'.$$

We write $\mathcal{Q} \trianglelefteq \mathcal{Q}'$ when $\mathcal{Q}$ is a substructure of $\mathcal{Q}'$.

We further say $\mathcal{Q}$ is a *root-preserving substructure* of $\mathcal{Q}'$ if $\mathcal{Q} \trianglelefteq \mathcal{Q}'$ and $\mathbf{Roots}(\mathcal{Q}) = \mathbf{Roots}(\mathcal{Q}')$. We write $\mathcal{Q} \trianglelefteq_r \mathcal{Q}'$ when $\mathcal{Q}$ is a root-preserving substructure of $\mathcal{Q}'$.

It is easy to see that the $\trianglelefteq$ and $\trianglelefteq_r$ relations are partial orders on expansion structures. We can use these relations to verify many properties of one expansion structure using another.

LEMMA 7.6.2. *Let $\mathcal{Q}$ and $\mathcal{Q}'$ be expansion structures. If $\mathcal{Q} \trianglelefteq \mathcal{Q}'$, then $\langle \mathcal{N}^{\mathcal{Q}}, \stackrel{\mathcal{Q}}{\rightarrow} \rangle$ is a subgraph of $\langle \mathcal{N}^{\mathcal{Q}'}, \stackrel{\mathcal{Q}'}{\rightarrow} \rangle$*

PROOF. This follows by a simple case analysis on Definition 7.2.11. □

REMARK 7.6.3. When considering two expansion structures $\mathcal{Q}$ and $\mathcal{Q}'$ where $\mathcal{Q} \trianglelefteq \mathcal{Q}'$ we need not distinguish between $\Gamma^{\mathcal{Q}}(\mathcal{R})$ and $\Gamma^{\mathcal{Q}'}(\mathcal{R})$ for $\mathcal{R} \subseteq \mathcal{N}^{\mathcal{Q}}$ since nodes in $\mathcal{Q}$ have the same dual shallow formula with respect to both $\mathcal{Q}$ and $\mathcal{Q}'$. In fact, for any set of nodes $\mathcal{R} \subseteq \mathcal{N}^{\mathcal{Q}'}$ we can write $\Gamma(\mathcal{R})$ for $\Gamma^{\mathcal{Q}'}(\mathcal{R})$ without ambiguity since this will equal $\Gamma^{\mathcal{Q}}(\mathcal{R})$ whenever $\mathcal{R} \subseteq \mathcal{N}^{\mathcal{Q}}$.

LEMMA 7.6.4. *Let $\mathcal{Q}$ and $\mathcal{Q}'$ be expansion structures such that $\mathcal{Q} \trianglelefteq \mathcal{Q}'$,*

$$\mathbf{Dom}(succ^{\mathcal{Q}}) = (\mathbf{Dom}(succ^{\mathcal{Q}'}) \cap \mathcal{N}^{\mathcal{Q}})$$

and

$$\mathbf{Dom}(sel^{\mathcal{Q}}) = (\mathbf{Dom}(sel^{\mathcal{Q}'}) \cap \mathcal{N}^{\mathcal{Q}}).$$

If $\mathcal{Q}'$ is an extensional expansion dag, then $\mathcal{Q}$ is an extensional expansion dag.

PROOF. See Appendix A.5. □

We can also use substructures to verify many $\clubsuit_*$ properties of the larger structure.

LEMMA 7.6.5. *Let $\mathcal{Q}$ be an extensional expansion dag and $\mathcal{Q}'$ be an expansion structure. Suppose $\mathcal{Q} \trianglelefteq \mathcal{Q}'$. Then we have the following:*

1. *Every expansion arc $\langle a, \mathbf{C}, c \rangle \in exps^{\mathcal{Q}}$ satisfies $\clubsuit_{exp}$ with respect to $\mathcal{Q}'$.*
2. *Every connection in $\langle c, a, b \rangle \in conns^{\mathcal{Q}}$ satisfies $\clubsuit_c$, $\clubsuit^p_m$, $\clubsuit^p_{eu}$ and $\clubsuit^{sp}_{eu}$ with respect to $\mathcal{Q}'$.*
3. *Every decomposition arc $\langle e, d \rangle \in decs^{\mathcal{Q}}$ satisfies $\clubsuit^p_{dec}$ with respect to $\mathcal{Q}'$.*

4. *Every* $a \in \mathbf{Dom}(succ^{\mathcal{Q}})$ *satisfies the properties* $\clubsuit_{ns}$, $\clubsuit_{f}$, $\clubsuit_{\neg}$, $\clubsuit_{\vee}$, $\clubsuit_{sel}$, $\clubsuit_{\sharp}$, $\clubsuit_{dec}$, $\clubsuit_{\overline{=}}$, $\clubsuit_{\overline{=}}^{\rightarrow}$, $\clubsuit_{\overline{=}}^{-o}$, $\clubsuit_{\overline{=}}^{+o}$, $\clubsuit_{m}$ *and* $\clubsuit_{eu}$ *with respect to* $\mathcal{Q}'$.

PROOF. See Appendix A.5. □

LEMMA 7.6.6. *Let* $\mathcal{Q}$ *and* $\mathcal{Q}'$ *be extensional expansion dags with* $\mathcal{Q} \trianglelefteq \mathcal{Q}'$. *If* $\mathfrak{p}'$ *is a u-part of* $\mathcal{Q}'$, *then* $\mathfrak{p}' \cap \mathcal{N}^{\mathcal{Q}}$ *is a u-part of* $\mathcal{Q}$.

PROOF. See Appendix A.5. □

LEMMA 7.6.7. *Let* $\mathcal{Q}$ *and* $\mathcal{Q}'$ *be extensional expansion dags with* $\mathcal{Q} \trianglelefteq \mathcal{Q}'$. *If* $\mathcal{R} \subseteq \mathcal{N}^{\mathcal{Q}}$ *is sufficient for* $\mathcal{Q}$, *then* $\mathcal{R}$ *is sufficient for* $\mathcal{Q}'$.

PROOF. Let $\mathcal{N}$ be $\mathcal{N}^{\mathcal{Q}}$. Suppose $\mathfrak{p}$ is a u-part of $\mathcal{Q}'$ which includes $\mathcal{R}$. By Lemma 7.6.6 and $\mathcal{R} \subseteq \mathcal{N}$, we know $\mathfrak{p} \cap \mathcal{N}$ is a u-part of $\mathcal{Q}$ including $\mathcal{R}$, contradicting sufficiency of $\mathcal{R}$ for $\mathcal{Q}$. Hence there is no u-part of $\mathcal{Q}'$ which includes $\mathcal{R}$. □

If we assume more, we can prove a u-part of a substructure is a u-part of the larger structure.

LEMMA 7.6.8. *Let* $\mathcal{Q}$ *and* $\mathcal{Q}'$ *be extensional expansion dags such that* $\mathcal{Q} \trianglelefteq \mathcal{Q}'$ *and let* $\mathcal{N}$ *be* $\mathcal{N}^{\mathcal{Q}}$. *Suppose* $\mathfrak{p}$ *is a u-part of* $\mathcal{Q}$. *Assume the following:*

1. *For any* $a \in \mathfrak{p}$, *if* a *is extended in* $\mathcal{Q}'$, *then* a *is extended in* $\mathcal{Q}$.
2. *For any* $a \in \mathfrak{p}$ *and* $\langle a, \mathbf{D}, c' \rangle \in exps^{\mathcal{Q}'}$, *we have* $c' \in \mathcal{N}$ *and* $a \overset{\mathcal{Q}}{\rightarrow} c'$.
3. *For any* $a, b \in \mathfrak{p}$ *and* $\langle c', a, b \rangle \in conns^{\mathcal{Q}'}$, *we have* $c' \in \mathcal{N}$ *and* $\langle c', a, b \rangle \in conns^{\mathcal{Q}}$.
4. *For any* $a \in \mathfrak{p}$ *and* $\langle a, c' \rangle \in decs^{\mathcal{Q}'}$, *we have* $c' \in \mathcal{N}$ *and* $a \overset{\mathcal{Q}}{\rightarrow} c'$.

Then $\mathfrak{p}$ *is a u-part of* $\mathcal{Q}'$.

PROOF. See Appendix A.5. □

We can also use the substructure relation to aid in checking acyclicity of embedding relations.

LEMMA 7.6.9. *Let* $\mathcal{Q}$ *and* $\mathcal{Q}'$ *be extensional expansion dags such that* $\mathcal{Q} \trianglelefteq \mathcal{Q}'$. *For any* $A, B \in \mathbf{SelPars}(\mathcal{Q})$, *if* $A \prec_0^{\mathcal{Q}} B$, *then* $A \prec_0^{\mathcal{Q}'} B$.

PROOF. Suppose $A \prec_0^{\mathcal{Q}} B$. Let b be $\mathbf{SelNode}^{\mathcal{Q}}(B)$. Then there is some expansion arc $\langle c, \mathbf{D}, d \rangle \in exps^{\mathcal{Q}} \subseteq exps^{\mathcal{Q}'}$ such that

$A \in \mathbf{Params}(\mathbf{D})$ and d dominates b in $\mathcal{Q}$. By Lemma 7.6.2, d dominates b in $\mathcal{Q}'$. Since b is also the selection node for B in $\mathcal{Q}'$, we have $A \prec_0^{\mathcal{Q}'} B$. □

LEMMA 7.6.10. *Let $\mathcal{Q}$ and $\mathcal{Q}'$ be extensional expansion dags such that $\mathcal{Q} \trianglelefteq \mathcal{Q}'$. If $\prec_0^{\mathcal{Q}'}$ is acyclic, then $\prec_0^{\mathcal{Q}}$ is acyclic.*

PROOF. Suppose there is a cycle $A^1 \prec_0^{\mathcal{Q}} \cdots \prec_0^{\mathcal{Q}} A^n \prec_0^{\mathcal{Q}} A^1$. Then we have a cycle $A^1 \prec_0^{\mathcal{Q}'} \cdots \prec_0^{\mathcal{Q}'} A^n \prec_0^{\mathcal{Q}'} A^1$ by Lemma 7.6.9. □

We will also consider particular cases where $\mathcal{Q}'$ is obtained from $\mathcal{Q}$ by adding new children to the nodes in $\mathcal{Q}$.

DEFINITION 7.6.11. Let $\mathcal{Q}$ and $\mathcal{Q}'$ be expansion structures. We say $\mathcal{Q}'$ *augments* $\mathcal{Q}$ if $\mathcal{Q} \trianglelefteq \mathcal{Q}'$, $\mathbf{Roots}(\mathcal{Q}') \subseteq \mathcal{N}^{\mathcal{Q}}$ and for all $a, b \in \mathcal{N}^{\mathcal{Q}'}$ if $a \xrightarrow{\mathcal{Q}'} b$, then $a \in \mathcal{N}^{\mathcal{Q}}$ and either $b \notin \mathcal{N}^{\mathcal{Q}}$ or both $b \in \mathcal{N}^{\mathcal{Q}}$ and $a \xrightarrow{\mathcal{Q}} b$

Intuitively, if $\mathcal{Q}'$ augments $\mathcal{Q}$, then all the new nodes in $\mathcal{Q}'$ are leaves which are children of nodes already in $\mathcal{Q}$. The augmentation relation is not transitive.

LEMMA 7.6.12. *Let $\mathcal{Q}$ and $\mathcal{Q}'$ be expansion structures such that $\mathcal{Q}'$ augments $\mathcal{Q}$. Every node $a \in (\mathcal{N}^{\mathcal{Q}'} \setminus \mathcal{N}^{\mathcal{Q}})$ is a leaf node of $\mathcal{Q}'$.*

PROOF. Suppose $a \in (\mathcal{N}^{\mathcal{Q}'} \setminus \mathcal{N}^{\mathcal{Q}})$ is not a leaf. Then there is some $b \in \mathcal{N}^{\mathcal{Q}'}$ such that $a \xrightarrow{\mathcal{Q}'} b$. Since $\mathcal{Q}'$ augments $\mathcal{Q}$, we must have $a \in \mathcal{N}^{\mathcal{Q}}$, a contradiction. □

LEMMA 7.6.13. *Let $\mathcal{Q}$ and $\mathcal{Q}'$ be expansion structures. If $\mathcal{Q}'$ augments $\mathcal{Q}$, then $\mathcal{Q} \trianglelefteq_r \mathcal{Q}'$.*

PROOF. See Appendix A.5. □

An advantage of assuming $\mathcal{Q}'$ augments $\mathcal{Q}$ is that we can often pull walks back from $\mathcal{Q}'$ to $\mathcal{Q}$.

LEMMA 7.6.14. *Let $\mathcal{Q}$ and $\mathcal{Q}'$ be expansion structures such that $\mathcal{Q}'$ augments $\mathcal{Q}$. If $\langle a^1, \ldots, a^n \rangle$ is a walk in $\mathcal{Q}'$ and $a^n \in \mathcal{N}^{\mathcal{Q}}$, then $\langle a^1, \ldots, a^n \rangle$ is a walk in $\mathcal{Q}$.*

PROOF. We prove this by induction on n. If $n = 1$, then this is trivial since we are assuming $a^n \in \mathcal{N}^{\mathcal{Q}}$. Assume $\langle a^1, \ldots, a^{n+1} \rangle$ is a walk in $\mathcal{Q}'$ and $a^{n+1} \in \mathcal{N}^{\mathcal{Q}}$. Since $\mathcal{Q}'$ augments $\mathcal{Q}$, $a^{n+1} \in \mathcal{N}^{\mathcal{Q}}$

and $a^n \xrightarrow{\mathcal{Q}'} a^{n+1}$, we know $a^n \in \mathcal{N}^{\mathcal{Q}}$ and $a^n \xrightarrow{\mathcal{Q}} a^{n+1}$. By the inductive hypothesis, $\langle a^1, \ldots, a^n \rangle$ is a walk in $\mathcal{Q}$. Thus $\langle a^1, \ldots, a^n, a^{n+1} \rangle$ is a walk in $\mathcal{Q}$. □

We can sometimes prove acyclicity of the embedding relation of an augmented extensional expansion dag using acyclicity of the extensional expansion dag it augments.

LEMMA 7.6.15. *Let $\mathcal{Q}$ and $\mathcal{Q}'$ be extensional expansion dags such that $\mathcal{Q}'$ augments $\mathcal{Q}$. Assume for every*

$$A \in \mathbf{SelPars}(\mathcal{Q}') \setminus \mathbf{SelPars}(\mathcal{Q})$$

and expansion arc $\langle e, \mathbf{C}, c \rangle \in exps^{\mathcal{Q}'}$ we have $A \notin \mathbf{Params}(\mathbf{C})$. If $\prec_0^{\mathcal{Q}}$ is acyclic, then $\prec_0^{\mathcal{Q}'}$ is acyclic.

PROOF. See Appendix A.5. □

Finally, we prove a general lemma which will allow us to conclude augmentations are extensional expansion dags.

LEMMA 7.6.16. *Let*

$$\mathcal{Q} = \langle \mathcal{N}, succ, label, sel, exps, conns, decs \rangle$$

be an extensional expansion dag and

$$\mathcal{Q}' = \langle \mathcal{N}', succ', label', sel', exps', conns', decs' \rangle$$

be an expansion structure such that $\mathcal{Q}'$ augments $\mathcal{Q}$. Assume the following:

1. *Every node $a \in \mathbf{Dom}(sel') \setminus \mathbf{Dom}(sel)$ is a selection node in $\mathcal{Q}'$.*

2. *For all nodes $a \in \mathbf{Dom}(sel') \setminus \mathbf{Dom}(sel)$ and $b \in \mathbf{Dom}(sel')$, if $sel'(a) = sel'(b)$, then $a = b$.*

3. *Every expansion arc in $exps' \setminus exps$ satisfies $\clubsuit_{exp}$ with respect to $\mathcal{Q}'$.*

4. *Every connection in $conns' \setminus conns$ satisfies $\clubsuit_c$, $\clubsuit_m^p$, $\clubsuit_{eu}^p$ and $\clubsuit_{eu}^{sp}$ with respect to $\mathcal{Q}'$.*

5. *Every decomposition arc in $decs' \setminus decs$ satisfies $\clubsuit_{dec}^p$ with respect to $\mathcal{Q}'$.*

6. *Every node* $a \in \mathbf{Dom}(succ') \setminus \mathbf{Dom}(succ)$ *satisfies* $\clubsuit_{ns}$, $\clubsuit_{f}$, $\clubsuit_{\neg}$, $\clubsuit_{\vee}$, $\clubsuit_{sel}$, $\clubsuit_{\sharp}$, $\clubsuit_{dec}$, $\clubsuit_{=}^{-}$, $\clubsuit_{=}^{\rightarrow}$, $\clubsuit_{=}^{-o}$, $\clubsuit_{=}^{+o}$, $\clubsuit_{m}$ *and* $\clubsuit_{eu}$ *with respect to* $\mathcal{Q}'$.

Then $\mathcal{Q}'$ *is an extensional expansion dag.*

PROOF. Most of the conditions we need are either assumed or follow from Lemma 7.6.5. See Appendix A.5. □

LEMMA 7.6.17. *Let* $\mathcal{Q}$ *be a separated extensional expansion dag and* $\mathcal{Q}'$ *be an extensional expansion dag such that* $\mathcal{Q}'$ *augments* $\mathcal{Q}$. *Assume the following:*

1. *For every* $a \in succ^{\mathcal{Q}'} \setminus succ^{\mathcal{Q}}$ *with* $succ^{\mathcal{Q}'}(a) = \langle a^1, \ldots, a^n \rangle$ *the nodes* $a^1, \ldots, a^n$ *are distinct.*
2. *For every* $b \in \mathcal{N}^{\mathcal{Q}'} \setminus \mathcal{N}^{\mathcal{Q}}$ *if* b *is not a connection node in* $\mathcal{Q}'$, *then there is at most one node* $a \in \mathcal{N}^{\mathcal{Q}}$ *such that* $a \overset{\mathcal{Q}'}{\rightarrow} b$.
3. *For every connection node* $c \in \mathcal{N}^{\mathcal{Q}'} \setminus \mathcal{N}^{\mathcal{Q}}$ *there is at most one pair* $\langle a, b \rangle \in \mathcal{N}^{\mathcal{Q}} \times \mathcal{N}^{\mathcal{Q}}$ *such that* $\langle c, a, b \rangle \in conns^{\mathcal{Q}'}$.

Then $\mathcal{Q}'$ *is separated.*

PROOF. See Appendix A.5. □

7.7. Construction of Extensional Expansion Dags

Next we make definitions to ease the construction of extensional expansion dags and extensional expansion proofs. These constructions will be used to establish completeness of extensional expansion proofs.

REMARK 7.7.1. In order to make the definitions uniform, we introduce functions giving new parameters and nodes. For any type α and finite set $\mathcal{A}$ of parameters there is always a parameter

$$A_\alpha \in (\mathcal{P}_\alpha \setminus \mathcal{A})$$

since we have assumed the set $\mathcal{P}_\alpha$ is infinite. We will choose a definite such parameter denoted by $\mathsf{A}_\alpha^{\mathcal{A}}$ for each type α and finite set of parameters. Likewise, for any finite set $\mathcal{N}$ and natural number n

there is always an n-tuple $\langle a^1, \ldots, a^n \rangle$ where $a^1, \ldots, a^n$ are n distinct objects (e.g., natural numbers) and

$$\{a^1, \ldots, a^n\} \cap \mathcal{N} = \emptyset.$$

We choose a definite such n-tuple denoted by $\mathsf{n}_n^{\mathcal{N}}$ for each finite set $\mathcal{N}$ and natural number n.

7.7.1. Initial Extensional Expansion Dags.

DEFINITION 7.7.2. For any $n \geq 0$ and $lab^1, \ldots, lab^n \in \mathbf{Labels}$ we define the expansion structure

$$\mathbf{EInit}(\langle lab^1, \ldots, lab^n \rangle) := \langle \mathcal{N}, \emptyset, label, \emptyset, \emptyset, \emptyset, \emptyset \rangle$$

where $\mathsf{n}_n^{\emptyset} = \langle a^1, \ldots, a^n \rangle$ (cf. Remark 7.7.1), $\mathcal{N} := \{a^1, \ldots, a^n\}$ and $label(a^i) := lab^i$ for each i $(1 \leq i \leq n)$.

LEMMA 7.7.3. *Let $n \geq 0$ and $lab^1, \ldots, lab^n \in \mathbf{Labels}$ and $\mathcal{Q}$ be the expansion structure $\mathbf{EInit}(\langle lab^1, \ldots, lab^n \rangle)$. Then $\mathcal{Q}$ is a merged, separated extensional expansion dag such that*

$$\mathbf{Roots}(\mathcal{Q}) = \mathcal{N}^{\mathcal{Q}} = \{a^1, \ldots, a^n\}.$$

PROOF. Almost every condition one needs to check is vacuous. See Appendix A.5. □

7.7.2. Extending Nodes.

DEFINITION 7.7.4. Let $\mathcal{S}$ be a signature of logical constants,

$$\mathcal{Q} = \langle \mathcal{N}, succ, label, sel, exps, conns, decs \rangle$$

be an $\mathcal{S}$-expansion structure and $a \in \mathcal{N}$ be a node which is neither extended nor flexible. We define an $\mathcal{S}$-expansion structure

$$\mathbf{Extend}(\mathcal{Q}, a) := \langle \mathcal{N}', succ', label', sel', exps, conns, decs \rangle$$

by cases below.

1. If $label(a) = \langle \mathbf{primary}, p, \top \rangle$, then $\mathcal{N}' := \mathcal{N}$, $label' := label$, $sel' := sel$ and $succ'$ is defined by

$$succ'(b) := \begin{cases} succ(b) & \text{if } b \in \mathbf{Dom}(succ) \\ \langle\rangle & \text{if } b = a. \end{cases}$$

2. If $label(a) = \langle \mathbf{primary}, p, [\dot{\neg}\mathbf{M}] \rangle$, then $\mathcal{N}' := \mathcal{N} \cup \{a^1\}$,

$$sel' := sel,$$

$$succ'(b) := \begin{cases} succ(b) & \text{if } b \in \mathbf{Dom}(succ) \\ \langle a^1 \rangle & \text{if } b = a \end{cases}$$

and

$$label'(b) := \begin{cases} label(b) & \text{if } b \in \mathcal{N} \\ \langle \mathbf{primary}, -p, \mathbf{M} \rangle & \text{if } b = a^1 \end{cases}$$

where $\mathsf{n}_1^{\mathcal{N}} = \langle a^1 \rangle$ (cf. Remark 7.7.1).

3. If $label(a) = \langle \mathbf{primary}, p, [\mathbf{M} \dot{\vee} \mathbf{N}] \rangle$, then $\mathcal{N}' := \mathcal{N} \cup \{a^1, a^2\}$, $sel' := sel$,

$$succ'(b) := \begin{cases} succ(b) & \text{if } b \in \mathbf{Dom}(succ) \\ \langle a^1, a^2 \rangle & \text{if } b = a \end{cases}$$

and

$$label'(b) := \begin{cases} label(b) & \text{if } b \in \mathcal{N} \\ \langle \mathbf{primary}, p, \mathbf{M} \rangle & \text{if } b = a^1 \\ \langle \mathbf{primary}, p, \mathbf{N} \rangle & \text{if } b = a^2 \end{cases}$$

where $\mathsf{n}_2^{\mathcal{N}} = \langle a^1, a^2 \rangle$.

4. If $label(a) = \langle \mathbf{primary}, -1, [\dot{\forall} x_\alpha \mathbf{M}] \rangle$, then $\mathcal{N}' := \mathcal{N} \cup \{a^1\}$,

$$succ'(b) := \begin{cases} succ(b) & \text{if } b \in \mathbf{Dom}(succ) \\ \langle a^1 \rangle & \text{if } b = a \end{cases}$$

$$label'(b) := \begin{cases} label(b) & \text{if } b \in \mathcal{N} \\ \langle \mathbf{primary}, -1, [A_\alpha / x_\alpha]\mathbf{M} \rangle & \text{if } b = a^1 \end{cases}$$

and

$$sel'(b) := \begin{cases} sel(b) & \text{if } b \in \mathbf{Dom}(sel) \\ A_\alpha & \text{if } b = a \end{cases}$$

where $\mathsf{n}_1^{\mathcal{N}} = \langle a^1 \rangle$ and A_α is $\mathsf{A}_\alpha^{\mathbf{Params}(\mathcal{Q})}$ (cf. Remark 7.7.1).

5. If $label(a) = \langle \mathbf{primary}, 1, [\dot{\forall} x_\alpha \mathbf{M}] \rangle$, then $\mathcal{N}' := \mathcal{N}$,

$$label' := label, \; sel' := sel$$

and $succ'$ is defined by

$$succ'(b) := \begin{cases} succ(b) & \text{if } b \in \mathbf{Dom}(succ) \\ \langle \rangle & \text{if } b = a. \end{cases}$$

6. If a is an atomic node, then $\mathcal{N}' := \mathcal{N}$,

$$label' := label,\ sel' := sel$$

and $succ'$ is defined by

$$succ'(b) := \begin{cases} succ(b) & \text{if } b \in \mathbf{Dom}(succ) \\ \langle\rangle & \text{if } b = a. \end{cases}$$

7. If $label(a) = \langle \mathbf{primary}, p, [c\,\overline{\mathbf{A}^m}]\rangle$ and $c \in \mathcal{S}$, then

$$\mathcal{N}' := \mathcal{N} \cup \{a^1\},\ sel' := sel,$$

$$succ'(b) := \begin{cases} succ(b) & \text{if } b \in \mathbf{Dom}(succ) \\ \langle a^1\rangle & \text{if } b = a \end{cases}$$

and

$$label'(b) := \begin{cases} label(b) & \text{if } b \in \mathcal{N} \\ \langle \mathbf{primary}, p, ([c\,\overline{\mathbf{A}^m}]^\sharp)^{\downarrow}\rangle & \text{if } b = a^1 \end{cases}$$

where $\mathsf{n}_1^{\mathcal{N}} = \langle a^1\rangle$.

8. If a is a positive equation node, then $\mathcal{N}' := \mathcal{N}$, $label' := label$, $sel' := sel$ and $succ'$ is defined by

$$succ'(b) := \begin{cases} succ(b) & \text{if } b \in \mathbf{Dom}(succ) \\ \langle\rangle & \text{if } b = a. \end{cases}$$

9. If $label(a) = \langle \mathbf{primary}, -1, [\mathbf{C} \doteq^{\iota} \mathbf{D}]\rangle$, then $\mathcal{N}' := \mathcal{N} \cup \{a^1\}$, $sel' := sel$,

$$succ'(b) := \begin{cases} succ(b) & \text{if } b \in \mathbf{Dom}(succ) \\ \langle a^1\rangle & \text{if } b = a \end{cases}$$

and

$$label'(b) := \begin{cases} label(b) & \text{if } b \in \mathcal{N} \\ \langle \mathbf{eqngoal}, -1, [\mathbf{C} \doteq^{\iota} \mathbf{D}]\rangle & \text{if } b = a^1 \end{cases}$$

where $\mathsf{n}_1^{\mathcal{N}} = \langle a^1\rangle$.

10. If a is an equation goal node, then $\mathcal{N}' := \mathcal{N}$, $label' := label$, $sel' := sel$ and $succ'$ is defined by

$$succ'(b) := \begin{cases} succ(b) & \text{if } b \in \mathbf{Dom}(succ) \\ \langle\rangle & \text{if } b = a. \end{cases}$$

11. If $label(a) = \langle \mathbf{primary}, p, [\mathbf{F} \mathrel{\dot{=}}^{\alpha\beta} \mathbf{G}] \rangle$, then $\mathcal{N}' := \mathcal{N}$,
$$sel' := sel$$
$$succ'(b) := \begin{cases} succ(b) & \text{if } b \in \mathbf{Dom}(succ) \\ \langle a^1 \rangle & \text{if } b = a \end{cases}$$
and
$$label'(b) := \begin{cases} label(b) & \text{if } b \in \mathcal{N} \\ \langle \mathbf{primary}, p, [\dot{\forall} y_\beta \centerdot [\mathbf{F}\, y] \mathrel{\dot{=}}^{\alpha\beta} [\mathbf{G}\, y]]^{\downarrow} \rangle & \text{if } b = a^1 \end{cases}$$
where $\mathsf{n}_1^{\mathcal{N}} = \langle a^1 \rangle$ and $y_\beta \notin (\mathbf{Free}(\mathbf{F}) \cup \mathbf{Free}(\mathbf{G}))$.
12. If $label(a) = \langle \mathbf{primary}, -1, [\mathbf{A} \mathrel{\dot{=}}^{o} \mathbf{B}] \rangle$, then
$$\mathcal{N}' := \mathcal{N} \cup \{a^1, a^2\},\ sel' := sel,$$
$$succ'(b) := \begin{cases} succ(b) & \text{if } b \in \mathbf{Dom}(succ) \\ \langle a^1, a^2 \rangle & \text{if } b = a \end{cases}$$
and
$$label'(b) := \begin{cases} label(b) & \text{if } b \in \mathcal{N} \\ \langle \mathbf{primary}, -1, [\dot{\neg}\mathbf{A} \mathrel{\dot{\vee}} \mathbf{B}] \rangle & \text{if } b = a^1 \\ \langle \mathbf{primary}, -1, [\dot{\neg}\mathbf{B} \mathrel{\dot{\vee}} \mathbf{A}] \rangle & \text{if } b = a^2 \end{cases}$$
where $\mathsf{n}_2^{\mathcal{N}} = \langle a^1, a^2 \rangle$.
13. If $label(a) = \langle \mathbf{primary}, 1, [\mathbf{A} \mathrel{\dot{=}}^{o} \mathbf{B}] \rangle$, then
$$\mathcal{N}' := \mathcal{N} \cup \{a^1, a^2\},\ sel' := sel,$$
$$succ'(b) := \begin{cases} succ(b) & \text{if } b \in \mathbf{Dom}(succ) \\ \langle a^1, a^2 \rangle & \text{if } b = a \end{cases}$$
and
$$label'(b) := \begin{cases} label(b) & \text{if } b \in \mathcal{N} \\ \langle \mathbf{primary}, -1, [\mathbf{A} \mathrel{\dot{\vee}} \mathbf{B}] \rangle & \text{if } b = a^1 \\ \langle \mathbf{primary}, -1, [\dot{\neg}\mathbf{A} \mathrel{\dot{\vee}} \dot{\neg}\mathbf{B}] \rangle & \text{if } b = a^2 \end{cases}$$
where $\mathsf{n}_2^{\mathcal{N}} = \langle a^1, a^2 \rangle$.
14. If $label(a) = \langle \mathbf{dec}, -1, [[H\, \mathbf{A}^1 \cdots \mathbf{A}^n] \mathrel{\dot{=}}^{\iota} [H\, \mathbf{B}^1 \cdots \mathbf{B}^n]] \rangle$ where H is a parameter of type $(\iota\alpha^n \cdots \alpha^1)$, then
$$\mathcal{N}' := \mathcal{N} \cup \{a^1, \ldots, a^n\},$$
$$sel' := sel,$$
$$succ'(b) := \begin{cases} succ(b) & \text{if } b \in \mathbf{Dom}(succ) \\ \langle a^1, \ldots, a^n \rangle & \text{if } b = a \end{cases}$$

and

$$label'(b) := \begin{cases} label(b) & \text{if } b \in \mathcal{N} \\ \langle \textbf{primary}, -1, [\mathbf{A}^i \doteq^{\alpha^i} \mathbf{B}^i]\rangle & \text{if } b = a^i \text{ for } 1 \leq i \leq n \end{cases}$$

where $\mathsf{n}_n^{\mathcal{N}} = \langle a^1, \ldots, a^n \rangle$.

15. If $label(a) = \langle \textbf{mate}, -1, [[P\,\mathbf{A}^1 \cdots \mathbf{A}^n] \doteq^o [P\,\mathbf{B}^1 \cdots \mathbf{B}^n]]\rangle$ where P is a parameter of type $(o\alpha^n \cdots \alpha^1)$, then

$$\mathcal{N}' := \mathcal{N} \cup \{a^1, \ldots, a^n\},$$

$$sel' := sel,$$

$$succ'(b) := \begin{cases} succ(b) & \text{if } b \in \mathbf{Dom}(succ) \\ \langle a^1, \ldots, a^n \rangle & \text{if } b = a \end{cases}$$

and

$$label'(b) := \begin{cases} label(b) & \text{if } b \in \mathcal{N} \\ \langle \textbf{primary}, -1, [\mathbf{A}^i \doteq^{\alpha^i} \mathbf{B}^i]\rangle & \text{if } b = a^i \text{ for } 1 \leq i \leq n \end{cases}$$

where $\mathsf{n}_n^{\mathcal{N}} = \langle a^1, \ldots, a^n \rangle$.

16. If $label(a) = \langle k, -1, [[\mathbf{A} \doteq^{\iota} \mathbf{C}] \mathbin{\dot\wedge} [\mathbf{B} \doteq^{\iota} \mathbf{D}]]\rangle$ where k is either **eunif** or $\textbf{eunif}^{sym}$, then $\mathcal{N}' := \mathcal{N} \cup \{a^1, a^2\}$, $sel' := sel$,

$$succ'(b) := \begin{cases} succ(b) & \text{if } b \in \mathbf{Dom}(succ) \\ \langle a^1, a^2 \rangle & \text{if } b = a \end{cases}$$

and

$$label'(b) := \begin{cases} label(b) & \text{if } b \in \mathcal{N} \\ \langle \textbf{primary}, -1, [\mathbf{A} \doteq^{\iota} \mathbf{C}]\rangle & \text{if } b = a^1 \\ \langle \textbf{primary}, -1, [\mathbf{B} \doteq^{\iota} \mathbf{D}]\rangle & \text{if } b = a^2 \end{cases}$$

where $\mathsf{n}_2^{\mathcal{N}} = \langle a^1, a^2 \rangle$.

LEMMA 7.7.5. *Let $\mathcal{Q}$ be an extensional expansion dag, $a \in \mathcal{N}^{\mathcal{Q}}$ be a node which is neither extended nor flexible and $\mathcal{Q}'$ be* $\mathbf{Extend}(\mathcal{Q}, a)$. *For any $c, d \in \mathcal{N}^{\mathcal{Q}'}$, if $c \xrightarrow{\mathcal{Q}'} d$, then either $c, d \in \mathcal{N}^{\mathcal{Q}}$ and $c \xrightarrow{\mathcal{Q}} d$ or $c = a$ and $d \notin \mathcal{N}^{\mathcal{Q}}$.*

PROOF. See Appendix A.5. □

LEMMA 7.7.6. *Let $\mathcal{Q}$ be an extensional expansion dag, $a \in \mathcal{N}^{\mathcal{Q}}$ be a node which is neither extended nor flexible and $\mathcal{Q}'$ be* $\mathbf{Extend}(\mathcal{Q}, a)$.

Then the node a satisfies $\clubsuit_{ns}$, $\clubsuit_{f}$, $\clubsuit_{\neg}$, $\clubsuit_{\vee}$, $\clubsuit_{sel}$, $\clubsuit_{\sharp}$, $\clubsuit_{dec}$, $\clubsuit_{=}^{-}$, $\clubsuit_{=}^{\rightarrow}$, $\clubsuit_{=}^{-o}$, $\clubsuit_{=}^{+o}$, $\clubsuit_{m}$ *and* $\clubsuit_{eu}$ *with respect to* $\mathcal{Q}'$ *and we have*

$$\mathbf{Free}(sh^{\mathcal{Q}'}(b)) \subseteq \mathbf{Free}(sh^{\mathcal{Q}}(a))$$

for every $b \in \mathcal{N}^{\mathcal{Q}'} \setminus \mathcal{N}^{\mathcal{Q}}$.

PROOF. This follows by a case analysis on Definition 7.7.4. □

LEMMA 7.7.7. *Let* $\mathcal{Q}$ *be an extensional expansion dag,* $a \in \mathcal{N}^{\mathcal{Q}}$ *be a node which is neither extended nor flexible and* $\mathcal{Q}'$ *be* $\mathbf{Extend}(\mathcal{Q}, a)$*. Then* $\mathcal{Q}'$ *is an extensional expansion dag,* $\mathcal{Q}'$ *augments* $\mathcal{Q}$*,* $\mathcal{Q} \trianglelefteq_r \mathcal{Q}'$*, a is an extended node of* $\mathcal{Q}'$*,*

$$\mathbf{ExpTerms}(\mathcal{Q}) = \mathbf{ExpTerms}(\mathcal{Q}')$$

and

$$\mathbf{RootSels}(\mathcal{Q}) = \mathbf{RootSels}(\mathcal{Q}').$$

Furthermore, we have the following:

1. *If* $\mathcal{Q}$ *is closed, then* $\mathcal{Q}'$ *is closed.*
2. *If* $\mathcal{Q}$ *is separated, then* $\mathcal{Q}'$ *is separated.*
3. *If* $\mathcal{Q}$ *has merged expansions [connections,decompositions], then* $\mathcal{Q}'$ *has merged expansions [connections,decompositions].*
4. *If* $\mathcal{Q}$ *is merged, then* $\mathcal{Q}'$ *is merged.*
5. *If* $\prec_0^{\mathcal{Q}}$ *is acyclic, then* $\prec_0^{\mathcal{Q}'}$ *is acyclic.*

PROOF. See Appendix A.5. □

7.7.3. Adding Expansion Arcs. We next define a constructor for adding an expansion arc to an expansion node.

DEFINITION 7.7.8. Let

$$\mathcal{Q} = \langle \mathcal{N}, succ, label, sel, exps, conns, decs \rangle$$

be an expansion structure. For any term $\mathbf{C} \in \mathit{wff}_\alpha$ and expansion node $e \in \mathcal{N}$ with shallow formula $[\forall x_\alpha \mathbf{M}]$, we define $\mathbf{EExpand}(\mathcal{Q}, e, \mathbf{C})$ to be the expansion structure

$$\langle \mathcal{N} \cup \{c\}, succ, label', sel, exps \cup \{\langle e, \mathbf{C}, c\rangle\}, conns, decs \rangle$$

where $\mathsf{n}_1^{\mathcal{N}} = \langle c \rangle$ (cf. Remark 7.7.1) and

$$label'(a) := \begin{cases} label(a) & \text{if } a \in \mathcal{N} \\ \langle \mathbf{primary}, 1, ([\mathbf{C}/x]\mathbf{M})^{\downarrow} \rangle & \text{if } a = c. \end{cases}$$

LEMMA 7.7.9. *Let $\mathcal{Q}$ be an expansion structure, $e \in \mathcal{N}^{\mathcal{Q}}$ be an expansion node with shallow formula $[\forall x_\alpha \mathbf{M}]$, $\mathbf{C}_\alpha$ be a term, c be such that $\mathsf{n}_1^{\mathcal{N}^{\mathcal{Q}}} = \langle c \rangle$ and $\mathcal{Q}'$ be $\mathbf{EExpand}(\mathcal{Q}, e, \mathbf{C})$. For all $a, b \in \mathcal{N}^{\mathcal{Q}'}$, if $a \xrightarrow{\mathcal{Q}'} b$, then $a \in \mathcal{N}^{\mathcal{Q}}$ and either $b \in \mathcal{N}^{\mathcal{Q}}$ and $a \xrightarrow{\mathcal{Q}} b$ or $\langle a, b \rangle = \langle e, c \rangle$.*

PROOF. This follows by a simple case analysis on $a \xrightarrow{\mathcal{Q}} b$. □

LEMMA 7.7.10. *Let $\mathcal{Q}$ be an extensional expansion dag, $e \in \mathcal{N}^{\mathcal{Q}}$ be an extended expansion node with shallow formula $[\forall x_\alpha \mathbf{M}]$, $\mathbf{C}_\alpha$ be a term and $\mathcal{Q}'$ be $\mathbf{EExpand}(\mathcal{Q}, e, \mathbf{C})$. Then $\mathcal{Q}'$ is an extensional expansion dag, $\mathcal{Q}'$ augments $\mathcal{Q}$, $\mathcal{Q} \trianglelefteq_r \mathcal{Q}'$, $([\mathbf{C}/x]\mathbf{M})^{\downarrow} \in \mathbf{ExpInsts}^{\mathcal{Q}'}(e)$,*

$$\mathbf{ExpTerms}(\mathcal{Q}') = \mathbf{ExpTerms}(\mathcal{Q}) \cup \{\mathbf{C}\}$$

and

$$\mathbf{RootSels}(\mathcal{Q}') = \mathbf{RootSels}(\mathcal{Q}).$$

Furthermore, we have the following:

1. *If $\mathcal{Q}$ is closed and $\mathbf{C}$ is closed, then $\mathcal{Q}'$ is closed.*
2. *If $\mathcal{Q}$ is separated, then $\mathcal{Q}'$ is separated.*
3. *If $\mathcal{Q}$ has merged connections [decompositions], then $\mathcal{Q}'$ has merged connections [decompositions].*
4. *If $\mathcal{Q}$ has merged expansions and $([\mathbf{C}/x]\mathbf{M})^{\downarrow} \notin \mathbf{ExpInsts}^{\mathcal{Q}}(e)$, then $\mathcal{Q}'$ has merged expansions.*
5. *If $\mathcal{Q}$ is merged and $([\mathbf{C}/x]\mathbf{M})^{\downarrow} \notin \mathbf{ExpInsts}^{\mathcal{Q}}(e)$, then $\mathcal{Q}'$ is merged.*
6. *If $\prec_0^{\mathcal{Q}}$ is acyclic, then $\prec_0^{\mathcal{Q}'}$ is acyclic.*

PROOF. See Appendix A.5. □

7.7.4. Adding Connections. We now give methods for adding connections.

DEFINITION 7.7.11. Let

$$\mathcal{Q} = \langle \mathcal{N}, succ, label, sel, exps, conns, decs \rangle$$

be an expansion structure and $P \in \mathcal{P}$. For any positive atomic node $a \in \mathcal{N}$ with shallow formula $[P\,\overline{\mathbf{A}^n}]$ and negative atomic node $b \in \mathcal{N}$ with shallow formula $[P\,\overline{\mathbf{B}^n}]$, we define $\mathbf{Mate}(\mathcal{Q}, a, b)$ to be the expansion structure

$$\langle \mathcal{N} \cup \{c\}, succ, label', sel, exps, conns \cup \{\langle c, a, b \rangle\}, decs \rangle$$

where $\mathsf{n}_1^{\mathcal{N}} = \langle c \rangle$ (cf. Remark 7.7.1) and

$$label'(d) := \begin{cases} label(d) & \text{if } d \in \mathcal{N} \\ \langle \mathbf{mate}, -1, [[P\,\overline{\mathbf{B}^n}] \doteq^o [P\,\overline{\mathbf{A}^n}]] \rangle & \text{if } d = c. \end{cases}$$

DEFINITION 7.7.12. Let

$$\mathcal{Q} = \langle \mathcal{N}, succ, label, sel, exps, conns, decs \rangle$$

be an expansion structure. For any positive equation node $a \in \mathcal{N}$ with shallow formula $[\mathbf{A} \doteq^\iota \mathbf{B}]$ and equation goal node $b \in \mathcal{N}$ with shallow formula $[\mathbf{C} \doteq^\iota \mathbf{D}]$, we define $\mathbf{EUnif}(\mathbf{eunif}, \mathcal{Q}, a, b)$ to be the expansion structure

$$\langle \mathcal{N} \cup \{c\}, succ, label', sel, exps, conns \cup \{\langle c, a, b \rangle\}, decs \rangle$$

and $\mathbf{EUnif}(\mathbf{eunif}^{sym}, \mathcal{Q}, a, b)$ to be the expansion structure

$$\langle \mathcal{N} \cup \{c\}, succ, label'', sel, exps, conns \cup \{\langle c, a, b \rangle\}, decs \rangle$$

where $\mathsf{n}_1^{\mathcal{N}} = \langle c \rangle$,

$$label'(d) := \begin{cases} label(d) & \text{if } d \in \mathcal{N} \\ \langle \mathbf{eunif}, -1, [[\mathbf{A} \doteq^\iota \mathbf{C}] \mathbin{\dot\wedge} [\mathbf{B} \doteq^\iota \mathbf{D}]] \rangle & \text{if } d = c \end{cases}$$

and

$$label''(d) := \begin{cases} label(d) & \text{if } d \in \mathcal{N} \\ \langle \mathbf{eunif}^{sym}, -1, [[\mathbf{A} \doteq^\iota \mathbf{D}] \mathbin{\dot\wedge} [\mathbf{B} \doteq^\iota \mathbf{C}]] \rangle & \text{if } d = c. \end{cases}$$

LEMMA 7.7.13. *Let*

$$\mathcal{Q} = \langle \mathcal{N}, succ, label, sel, exps, conns, decs \rangle$$

and

$$\mathcal{Q}' = \langle \mathcal{N}', succ', label', sel', exps', conns', decs' \rangle$$

be expansion structures, $a, b \in \mathcal{N}$ be nodes and $c \in \mathcal{N}'$ be a node such that $\mathcal{N}' = \mathcal{N} \cup \{c\}$, $succ' = succ$, $exps' = exps$,

$$conns' = conns \cup \{\langle c, a, b \rangle\}$$

and $decs' = decs$. For every $e, d \in \mathcal{N}'$, if $e \overset{\mathcal{Q}'}{\rightharpoondown} d$, then $e \in \mathcal{N}$ and either $d \in \mathcal{N}$ and $e \overset{\mathcal{Q}}{\rightharpoondown} d$ or $d = c$ and $e \in \{a, b\}$.

PROOF. The only new edges are induced by $\langle c, a, b \rangle$. □

LEMMA 7.7.14. *Let $\mathcal{Q}$ be an extensional expansion dag, $P \in \mathcal{P}$ be a parameter, $a \in \mathcal{N}^{\mathcal{Q}}$ be an extended positive atomic node with shallow formula $[P\,\overline{\mathbf{A}^n}]$, $b \in \mathcal{N}^{\mathcal{Q}}$ be an extended negative atomic node with shallow formula $[P\,\overline{\mathbf{B}^n}]$ and $\mathcal{Q}'$ be* $\mathbf{Mate}(\mathcal{Q}, a, b)$. *Then $\mathcal{Q}'$ is an extensional expansion dag, $\mathcal{Q}'$ augments $\mathcal{Q}$, $\mathcal{Q} \trianglelefteq_r \mathcal{Q}'$, the pair $\langle a, b\rangle$ is* **mate***-connected in $\mathcal{Q}'$,*

$$\mathbf{ExpTerms}(\mathcal{Q}') = \mathbf{ExpTerms}(\mathcal{Q})$$

and

$$\mathbf{RootSels}(\mathcal{Q}') = \mathbf{RootSels}(\mathcal{Q}).$$

Furthermore, we have the following:

1. *If $\mathcal{Q}$ is closed, then $\mathcal{Q}'$ is closed.*
2. *If $\mathcal{Q}$ is separated, then $\mathcal{Q}'$ is separated.*
3. *If $\mathcal{Q}$ has merged expansions [decompositions], then $\mathcal{Q}'$ has merged expansions [decompositions].*
4. *If $\mathcal{Q}$ has merged connections and the pair $\langle a, b\rangle$ is not* **mate***-connected in $\mathcal{Q}$, then $\mathcal{Q}'$ has merged connections.*
5. *If $\mathcal{Q}$ is merged and the pair $\langle a, b\rangle$ is not* **mate***-connected in $\mathcal{Q}$, then $\mathcal{Q}'$ is merged.*
6. *If $\prec_0^{\mathcal{Q}}$ is acyclic, then $\prec_0^{\mathcal{Q}'}$ is acyclic.*

PROOF. See Appendix A.5. □

LEMMA 7.7.15. *Let $\mathcal{Q}$ be an extensional expansion dag, $a \in \mathcal{N}^{\mathcal{Q}}$ be an extended positive equation node, $b \in \mathcal{N}^{\mathcal{Q}}$ be an extended equation goal node, $k \in \{\mathbf{eunif}, \mathbf{eunif}^{sym}\}$ be a kind and*

$$\mathcal{Q}' = \mathbf{EUnif}(k, \mathcal{Q}, a, b).$$

Then $\mathcal{Q}'$ is an extensional expansion dag, $\mathcal{Q}'$ augments $\mathcal{Q}$, $\mathcal{Q} \trianglelefteq_r \mathcal{Q}'$, the pair $\langle a, b\rangle$ is k-connected in $\mathcal{Q}'$,

$$\mathbf{ExpTerms}(\mathcal{Q}') = \mathbf{ExpTerms}(\mathcal{Q})$$

and

$$\mathbf{RootSels}(\mathcal{Q}') = \mathbf{RootSels}(\mathcal{Q}).$$

Furthermore, we have the following:

1. *If $\mathcal{Q}$ is closed, then $\mathcal{Q}'$ is closed.*
2. *If $\mathcal{Q}$ is separated, then $\mathcal{Q}'$ is separated.*
3. *If $\mathcal{Q}$ has merged expansions [decompositions], then $\mathcal{Q}'$ has merged expansions [decompositions].*

4. *If $\mathcal{Q}$ has merged connections and the pair $\langle a, b\rangle$ is not k-connected in $\mathcal{Q}$, then $\mathcal{Q}'$ has merged connections.*
5. *If $\mathcal{Q}$ is merged and the pair $\langle a, b\rangle$ is not k-connected in $\mathcal{Q}$, then $\mathcal{Q}'$ is merged.*
6. *If $\prec_0^{\mathcal{Q}}$ is acyclic, then $\prec_0^{\mathcal{Q}'}$ is acyclic.*

PROOF. Analogous to the proof of Lemma 7.7.14. □

7.7.5. Adding Decomposition Arcs. Finally, we define a way to add decomposition arcs.

DEFINITION 7.7.16. Let

$$\mathcal{Q} = \langle \mathcal{N}, succ, label, sel, exps, conns, decs\rangle$$

be an expansion structure. For any decomposable node $e \in \mathcal{N}$, we define $\mathbf{Dec}(\mathcal{Q}, e)$ to be the expansion structure

$$\langle \mathcal{N} \cup \{d\}, succ, label', sel, exps, conns, decs \cup \{\langle e, d\rangle\}\rangle$$

where $\mathsf{n}_1^{\mathcal{N}} = \langle d\rangle$ (cf. Remark 7.7.1) and

$$label'(a) := \begin{cases} label(a) & \text{if } a \in \mathcal{N} \\ \langle \mathbf{dec}, -1, sh^{\mathcal{Q}}(e)\rangle & \text{if } a = d. \end{cases}$$

LEMMA 7.7.17. *Let $\mathcal{Q}$ be an expansion structure, $e \in \mathcal{N}^{\mathcal{Q}}$ be a decomposable node, d be such that $\mathsf{n}_1^{\mathcal{N}^{\mathcal{Q}}} = \langle d\rangle$ and $\mathcal{Q}'$ be $\mathbf{Dec}(\mathcal{Q}, e)$. For all $a, b \in \mathcal{N}^{\mathcal{Q}'}$, if $a \overset{\mathcal{Q}'}{\rightharpoondown} b$, then $a \in \mathcal{N}^{\mathcal{Q}}$ and either $b \in \mathcal{N}^{\mathcal{Q}}$ and $a \overset{\mathcal{Q}}{\rightharpoondown} b$ or $\langle a, b\rangle = \langle e, d\rangle$.*

PROOF. The only new edge is induced by $\langle e, d\rangle$. □

LEMMA 7.7.18. *Let $\mathcal{Q}$ be an extensional expansion dag, $e \in \mathcal{N}^{\mathcal{Q}}$ be an extended decomposable node and $\mathcal{Q}'$ be $\mathbf{Dec}(\mathcal{Q}, e)$. Then $\mathcal{Q}'$ is an extensional expansion dag, $\mathcal{Q}'$ augments $\mathcal{Q}$, $\mathcal{Q} \trianglelefteq_r \mathcal{Q}'$, e is decomposed in $\mathcal{Q}'$,*

$$\mathbf{ExpTerms}(\mathcal{Q}') = \mathbf{ExpTerms}(\mathcal{Q})$$

and

$$\mathbf{RootSels}(\mathcal{Q}') = \mathbf{RootSels}(\mathcal{Q}).$$

Furthermore, we have the following:

1. *If $\mathcal{Q}$ is closed, then $\mathcal{Q}'$ is closed.*
2. *If $\mathcal{Q}$ is separated, then $\mathcal{Q}'$ is separated.*

3. *If $\mathcal{Q}$ has merged expansions [connections], then $\mathcal{Q}'$ has merged expansions [connections].*
4. *If $\mathcal{Q}$ has merged decompositions and e is not decomposed in $\mathcal{Q}$, then $\mathcal{Q}'$ has merged decompositions.*
5. *If $\mathcal{Q}$ is merged and e is not decomposed in $\mathcal{Q}$, then $\mathcal{Q}'$ is merged.*
6. *If $\prec_0^{\mathcal{Q}}$ is acyclic, then $\prec_0^{\mathcal{Q}'}$ is acyclic.*

PROOF. See Appendix A.5. □

7.8. Deconstruction of Extensional Expansion Dags

In order to prove soundness of extensional expansion proofs, we need methods for deconstructing (removing nodes and arcs from) extensional expansion dags. When proving soundness we will remove nodes and arcs when certain accessibility conditions hold. For example, we will remove certain root nodes called *accessible nodes.*

DEFINITION 7.8.1 (Accessible Nodes). Let $\mathcal{Q}$ be an extensional expansion dag. A node $a \in \mathcal{N}^{\mathcal{Q}}$ is an *accessible node* if $a \in \mathbf{Roots}(\mathcal{Q})$, a is extended and a is either a true node, false node, negation node, disjunction node, conjunction node, selection node, deepening node, decomposition node, negative equation node, functional node, Boolean node or connection node.

In other words, a is an accessible node unless a is either not extended or a is an expansion node, flexible node or a potentially connecting node (atomic node, equation goal node or positive equation node).

An expansion arc is an *accessible expansion* if the corresponding expansion term avoids certain selected parameters. When proving soundness we will avoid removing an expansion arc $\langle e, \mathbf{C}, c\rangle$ until we have removed every selection node a with selected parameter A where $A \in \mathbf{Params}(\mathbf{C})$.

DEFINITION 7.8.2 (Accessible Expansions). Let $\mathcal{Q}$ be an extensional expansion dag. An *accessible expansion* is an expansion arc $\langle e, \mathbf{C}, c\rangle \in exps^{\mathcal{Q}}$ where $e \in \mathbf{Roots}(\mathcal{Q})$ and

$$(\mathbf{Params}(\mathbf{C}) \cap \mathbf{SelPars}(\mathcal{Q})) \subseteq \emptyset.$$

We will remove connections when the parents are root nodes.

DEFINITION 7.8.3 (Accessible Connections). Let $\mathcal{Q}$ be an extensional expansion dag. An *accessible connection* is a connection

$$\langle c, a, b\rangle \in conns^{\mathcal{Q}}$$

where $a, b \in \mathbf{Roots}(\mathcal{Q})$.

We prove in Lemma 7.8.7 that every extensional expansion proof contains an accessible node, accessible expansion or accessible connection. Consequently, we will always be able to make progress when translating from an extensional expansion proof to a sequent calculus derivation. We first prove preliminary results.

LEMMA 7.8.4. *Let $\mathcal{Q}$ be an extensional expansion dag with no accessible connections. For every node $b \in \mathcal{N}^{\mathcal{Q}} \setminus \mathbf{Roots}(\mathcal{Q})$ there exists some extended root node $a \in \mathbf{Roots}(\mathcal{Q})$ such that a is an ancestor of b and a is either an expansion node or an accessible node.*

PROOF. We prove this by induction on the depth $|b|_d$ of b (cf. Definition 7.1.5). Since $b \notin \mathbf{Roots}(\mathcal{Q})$, b has at least one parent.

First suppose b has a parent $c \notin \mathbf{Roots}(\mathcal{Q})$. Then by the inductive hypothesis, there is some extended $a \in \mathbf{Roots}(\mathcal{Q})$ such that a is an ancestor of c (hence of b) and a is either an expansion node or an accessible node.

Next suppose every parent of b is in $\mathbf{Roots}(\mathcal{Q})$. Since b is not a root node, there is some $a \in \mathbf{Roots}(\mathcal{Q})$ such that $a \overset{\mathcal{Q}}{\rightarrow} b$. Since a has a child, a cannot be a flexible node by Lemma 7.3.7 and must be extended by Lemma 7.3.6.

Assume a is a potentially connecting node. By Lemma 7.3.8, there is some node d such that $\langle b, a, d\rangle \in conns^{\mathcal{Q}}$ or $\langle b, d, a\rangle \in conns^{\mathcal{Q}}$. We must have $d \in \mathbf{Roots}(\mathcal{Q})$ since every parent of b is a root node, contradicting our assumption that there are no accessible connections.

Hence a must be an accessible node or an expansion node. □

When the $\prec_0^{\mathcal{Q}}$ is acyclic there must be a minimal selected parameter in $\mathcal{Q}$ with respect to $\prec_0^{\mathcal{Q}}$.

LEMMA 7.8.5. *Let $\mathcal{Q}$ be an extensional expansion dag where the embedding relation $\prec_0^{\mathcal{Q}}$ is acyclic. If there are any selected parameters in $\mathcal{Q}$, then there is a selected parameter B which is minimal with respect to $\prec_0^{\mathcal{Q}}$ in the sense that there are no selected parameters A with $A \prec_0^{\mathcal{Q}} B$.*

PROOF. Assume there is no such minimal selected parameter B. Choose some $B^1 \in \mathbf{SelPars}(\mathcal{Q})$. Since B^1 is not minimal, there must be some selected parameter B^2 with $B^2 \prec_0^{\mathcal{Q}} B^1$. In this manner, we inductively define a descending chain

$$\cdots \prec_0^{\mathcal{Q}} B^n \prec_0^{\mathcal{Q}} \cdots \prec_0^{\mathcal{Q}} B^2 \prec_0^{\mathcal{Q}} B^1.$$

Since there are only finitely many selected parameters in $\mathcal{Q}$, there must exist m and k with $m > k$ such that $B^m = B^k$, contradicting acyclicity of $\prec_0^{\mathcal{Q}}$. □

If $\mathcal{Q}$ is an extensional expansion dag, $\prec_0^{\mathcal{Q}}$ is acyclic and some node of $\mathcal{Q}$ is not a root node, there must be an accessible node, accessible connection or accessible expansion.

LEMMA 7.8.6. *Let $\mathcal{Q}$ be an extensional expansion dag where $\prec_0^{\mathcal{Q}}$ is acyclic. At least one of the following holds:*

1. *There is an accessible node.*
2. *There is an accessible connection.*
3. *There is an accessible expansion.*
4. *Every node is a root node, i.e., $\mathcal{N}^{\mathcal{Q}} \subseteq \mathbf{Roots}(\mathcal{Q})$.*

PROOF. Assume there are no accessible nodes, no accessible connections and there is some non-root node $z \in \mathcal{N}^{\mathcal{Q}} \setminus \mathbf{Roots}(\mathcal{Q})$. Hence every extended root node $a \in \mathbf{Roots}(\mathcal{Q})$ must be either an expansion node or a potentially connecting node. (Flexible nodes are never extended by $\clubsuit_f$.) We will prove there is an accessible expansion. By Lemma 7.8.4, there is an ancestor $e \in \mathbf{Roots}(\mathcal{Q})$ of z such that e is an extended expansion node. In particular, some expansion node in $\mathbf{Roots}(\mathcal{Q})$ has at least one child.

Assume $\mathcal{Q}$ has no selected parameters. Let $e \in \mathbf{Roots}(\mathcal{Q})$ be any extended expansion node with at least one child c. By Lemma 7.3.11, there is some term $\mathbf{C}$ such that $\langle e, \mathbf{C}, c\rangle \in exps^{\mathcal{Q}}$. Since there are no selected parameters, $\langle e, \mathbf{C}, c\rangle$ is an accessible expansion.

Assume $\mathcal{Q}$ has a selected parameter. By Lemma 7.8.5, there is a selected parameter B which is minimal with respect to the embedding relation of $\mathcal{Q}$. Let $b \in \mathcal{N}$ be the selection node with selected parameter B. Note that $b \notin \mathbf{Roots}(\mathcal{Q})$ since there are no accessible nodes. By Lemma 7.8.4, there is an ancestor $e \in \mathbf{Roots}(\mathcal{Q})$ of b where e is an extended expansion node (since there are no accessible nodes). Since

b is not a root node, there must be some $c \in \mathcal{N}^{\mathcal{Q}}$ such that $e \xrightarrow{\mathcal{Q}} c$ and c dominates b. By Lemma 7.3.11, there is some term $\mathbf{C}$ such that $\langle e, \mathbf{C}, c\rangle \in exps^{\mathcal{Q}}$. Since B is minimal with respect to the embedding relation, $\mathbf{Params}(\mathbf{C})$ does not contain any selected parameters. Thus $\langle e, \mathbf{C}, c\rangle$ is an accessible expansion. □

LEMMA 7.8.7. *Let $\langle \mathcal{Q}, \mathcal{R}\rangle$ be an extensional expansion proof. At least one of the following holds:*

1. *There is an accessible node of $\mathcal{Q}$.*
2. *There is an accessible connection of $\mathcal{Q}$.*
3. *There is an accessible expansion of $\mathcal{Q}$.*

PROOF. Assume none of these hold. By Lemma 7.8.6, every node is a root node and each extended node is a potentially connecting node or an expansion node with no children (since any other extended root node with no children would be an accessible node). In this case, $\mathbf{Roots}(\mathcal{Q})$ is a u-part which includes $\mathcal{R}$, contradicting the assumption that $\langle \mathcal{Q}, \mathcal{R}\rangle$ is an extensional expansion proof. □

7.8.1. Removing Root Nodes. We define a way to remove a single root node from an expansion structure.

DEFINITION 7.8.8. Let

$$\mathcal{Q} = \langle \mathcal{N}, succ, label, sel, exps, conns, decs\rangle$$

be an expansion structure and $a \in \mathbf{Roots}(\mathcal{Q})$ be a root node. We define

$$\mathcal{Q}_a^{-\mathbf{r}} := \langle \mathcal{N}^-, succ^-, label^-, sel^-, exps^-, conns^-, decs^-\rangle$$

where

- $\mathcal{N}^- := \mathcal{N} \setminus \{a\}$,
- $succ^- := succ|_{\mathbf{Dom}(succ)\cap\mathcal{N}^-}$,
- $label^- := label|_{\mathcal{N}^-}$,
- $sel^- := sel|_{\mathbf{Dom}(sel)\cap\mathcal{N}^-}$,
- $exps^- := \{\langle e, \mathbf{C}, c\rangle \in exps \mid e, c \in \mathcal{N}^-\}$,
- $conns^- := \{\langle c, b, d\rangle \in conns \mid b, c, d \in \mathcal{N}^-\}$ and

- $decs^- := \{\langle e, d\rangle \in decs \mid e, d \in \mathcal{N}^-\}$.

If $\mathcal{Q}$ is an extensional expansion dag and a is a root node of $\mathcal{Q}$, then $\mathcal{Q}_a^{-\mathbf{r}}$ is an extensional expansion dag such that $|\mathcal{Q}_a^{-\mathbf{r}}| < |\mathcal{Q}|$ (cf. Definition 7.2.10).

LEMMA 7.8.9. *Let $\mathcal{Q}$ be an extensional expansion dag and $a \in \mathbf{Roots}(\mathcal{Q})$ be a root node. Then $\mathcal{Q}_a^{-\mathbf{r}}$ is an extensional expansion dag and $\mathcal{Q}_a^{-\mathbf{r}} \trianglelefteq \mathcal{Q}$. Furthermore, we have the following:*

1. $|\mathcal{Q}_a^{-\mathbf{r}}| < |\mathcal{Q}|$.
2. *If the embedding relation $\prec_0^{\mathcal{Q}}$ is acyclic, then the embedding relation $\prec_0^{\mathcal{Q}_a^{-\mathbf{r}}}$ is acyclic.*
3. *If $\mathfrak{p}$ is a u-part of $\mathcal{Q}_a^{-\mathbf{r}}$, then $\mathfrak{p}$ is a u-part of $\mathcal{Q}$.*

PROOF. It is easy to check $\mathcal{Q}_a^{-\mathbf{r}} \trianglelefteq \mathcal{Q}$. Note also that

$$\mathbf{Dom}(succ^{\mathcal{Q}_a^{-\mathbf{r}}}) = (\mathbf{Dom}(succ^{\mathcal{Q}}) \cap \mathcal{N}^{\mathcal{Q}_a^{-\mathbf{r}}})$$

and

$$\mathbf{Dom}(sel^{\mathcal{Q}_a^{-\mathbf{r}}}) = (\mathbf{Dom}(sel^{\mathcal{Q}}) \cap \mathcal{N}^{\mathcal{Q}_a^{-\mathbf{r}}}).$$

By Lemma 7.6.4, $\mathcal{Q}_a^{-\mathbf{r}}$ is an extensional expansion dag. Since we have removed the node a (and possibly some expansion arcs and connections), we have $|\mathcal{Q}_a^{-\mathbf{r}}| < |\mathcal{Q}|$. The embedding relation $\prec_0^{\mathcal{Q}_a^{-\mathbf{r}}}$ is acyclic whenever $\prec_0^{\mathcal{Q}}$ is acyclic by Lemma 7.6.10. For any u-part $\mathfrak{p}$ of $\mathcal{Q}_a^{-\mathbf{r}}$, we can use Lemma 7.6.8 to conclude $\mathfrak{p}$ is also a u-part of $\mathcal{Q}$. The assumptions of Lemma 7.6.8 hold since the removed node a was a root node of $\mathcal{Q}$. □

If a root node is not part of the sufficient set of an extensional expansion proof, then we can always remove the node.

LEMMA 7.8.10. *Let $\langle \mathcal{Q}, \mathcal{R}\rangle$ be an extensional expansion proof and let $a \in (\mathbf{Roots}(\mathcal{Q}) \setminus \mathcal{R})$ be a root node. Then $\langle \mathcal{Q}_a^{-\mathbf{r}}, \mathcal{R}\rangle$ is also an extensional expansion proof.*

PROOF. See Appendix A.5. □

We can remove a neutral root node from an extensional expansion proof if we replace the neutral node in the sufficient set with its unique child.

LEMMA 7.8.11. *Let $\langle \mathcal{Q}, \mathcal{R} \rangle$ be an extensional expansion proof. Suppose $a \in \mathcal{R} \cap \mathbf{Roots}(\mathcal{Q})$ is a neutral node with child b. Then*

$$\langle \mathcal{Q}_a^{-\mathbf{r}}, ((\mathcal{R} \setminus \{a\}) \cup \{b\}) \rangle$$

is also an extensional expansion proof.

PROOF. See Appendix A.5. □

We can remove a conjunctive accessible node from an extensional expansion proof if we replace the node in the sufficient set with all of the children of the node.

LEMMA 7.8.12. *Let $\langle \mathcal{Q}, \mathcal{R} \rangle$ be an extensional expansion proof. Suppose $a \in \mathcal{R} \cap \mathbf{Roots}(\mathcal{Q})$ is a conjunctive accessible node. Then*

$$\langle \mathcal{Q}_a^{-\mathbf{r}}, ((\mathcal{R} \setminus \{a\}) \cup \{b \in \mathcal{N}^{\mathcal{Q}} \mid a \overset{\mathcal{Q}}{\rightarrow} b\}) \rangle$$

is an extensional expansion proof.

PROOF. See Appendix A.5. □

A special case of Lemma 7.8.12 is when the conjunctive accessible node is not a negative equation node. In such cases the children are listed by the successor function.

LEMMA 7.8.13. *Let $\langle \mathcal{Q}, \mathcal{R} \rangle$ be an extensional expansion proof. Suppose $a \in \mathcal{R} \cap \mathbf{Roots}(\mathcal{Q})$ is a conjunctive accessible node,*

$$succ^{\mathcal{Q}}(a) = \langle a^1, \ldots, a^n \rangle$$

and a is not a negative equation node. Then

$$\langle \mathcal{Q}_a^{-\mathbf{r}}, ((\mathcal{R} \setminus \{a\}) \cup \{a^1, \ldots, a^n\}) \rangle$$

is an extensional expansion proof.

PROOF. By Lemma 7.3.13, $a^1, \ldots, a^n$ are the only children of a. The result follows directly from Lemma 7.8.12. □

We can remove a disjunctive node from an extensional expansion proof if we replace the node in the sufficient set with some child of the node.

LEMMA 7.8.14. *Let $\langle \mathcal{Q}, \mathcal{R} \rangle$ be an extensional expansion proof. Suppose $a \in \mathcal{R} \cap \mathbf{Roots}(\mathcal{Q})$ is a disjunctive node and*

$$succ^{\mathcal{Q}}(a) = \langle a^1, \ldots, a^n \rangle$$

where $n \geq 1$. For each i $(1 \leq i \leq n)$,

$$\langle \mathcal{Q}_a^{-\mathbf{r}}, ((\mathcal{R} \setminus \{a\}) \cup \{a^i\}) \rangle$$

is an extensional expansion proof.

PROOF. See Appendix A.5. □

7.8.2. Removing Expansion Arcs. We next develop a way to remove expansion arcs from root expansion nodes.

DEFINITION 7.8.15. Let

$$\mathcal{Q} = \langle \mathcal{N}, succ, label, sel, exps, conns, decs \rangle$$

be an expansion structure. For each $\langle e, \mathbf{C}, c \rangle \in exps$, we define $\mathcal{Q}^{-\mathbf{e}}_{\langle e, \mathbf{C}, c \rangle}$ to be the expansion structure

$$\langle \mathcal{N}, succ, label, sel, exps^-, conns, decs \rangle$$

where $exps^- := exps \setminus \{\langle e, \mathbf{C}, c \rangle\}$.

If we remove an expansion arc from an extensional expansion dag, the resulting expansion structure is an extensional expansion dag.

LEMMA 7.8.16. *Let $\mathcal{Q}$ be an extensional expansion dag and $\langle e, \mathbf{C}, c \rangle$ be an expansion arc in $\mathcal{Q}$. Then $\mathcal{Q}^{-\mathbf{e}}_{\langle e, \mathbf{C}, c \rangle}$ is an extensional expansion dag such that $\mathcal{Q}^{-\mathbf{e}}_{\langle e, \mathbf{C}, c \rangle} \trianglelefteq \mathcal{Q}$. Furthermore, we have the following:*

1. $|\mathcal{Q}^{-\mathbf{e}}_{\langle e, \mathbf{C}, c \rangle}| < |\mathcal{Q}|$
2. *If the embedding relation $\prec_0^{\mathcal{Q}}$ is acyclic, then the embedding relation $\prec_0^{\mathcal{Q}^{-\mathbf{e}}_{\langle e, \mathbf{C}, c \rangle}}$ is acyclic.*
3. *If $\mathfrak{p}$ is a u-part of $\mathcal{Q}^{-\mathbf{e}}_{\langle e, \mathbf{C}, c \rangle}$ and $e \notin \mathfrak{p}$, then $\mathfrak{p}$ is a u-part of $\mathcal{Q}$.*

PROOF. Let $\mathcal{Q}^-$ be $\mathcal{Q}^{-\mathbf{e}}_{\langle e, \mathbf{C}, c \rangle}$. Clearly, $\mathcal{Q}^- \trianglelefteq \mathcal{Q}$. Note also that

$$\mathbf{Dom}(succ^{\mathcal{Q}^-}) = \mathbf{Dom}(succ^{\mathcal{Q}}) = (\mathbf{Dom}(succ^{\mathcal{Q}}) \cap \mathcal{N}^{\mathcal{Q}^-})$$

and

$$\mathbf{Dom}(sel^{\mathcal{Q}^-}) = \mathbf{Dom}(sel^{\mathcal{Q}}) = (\mathbf{Dom}(sel^{\mathcal{Q}}) \cap \mathcal{N}^{\mathcal{Q}^-}).$$

We conclude $\mathcal{Q}^-$ is an extensional expansion dag by Lemma 7.6.4. Since we have deleted a single expansion arc, $|\mathcal{Q}^-| = (|\mathcal{Q}| - 1) < |\mathcal{Q}|$. The result concerning the embedding relation follows from Lemma 7.6.10 and the result for u-parts follows from Lemma 7.6.8. □

We can delete an accessible expansion from an extensional expansion proof (when the expansion node is in the sufficient set) if we add the expansion child to the sufficient set. If we tried to delete an expansion arc which were not accessible, a selected parameter would occur in the shallow formula of the expansion child (unless the quantifier corresponding to the expansion node were vacuous).

LEMMA 7.8.17. *Let* $\langle \mathcal{Q}, \mathcal{R} \rangle$ *be an extensional expansion proof. Suppose* $\langle e, \mathbf{C}_\alpha, c \rangle$ *is an accessible expansion in* $\mathcal{Q}$. *If* $e \in \mathcal{R}$, *then*

$$\langle \mathcal{Q}^{-\mathbf{e}}_{\langle e, \mathbf{C}, c \rangle}, (\mathcal{R} \cup \{c\}) \rangle$$

is an extensional expansion proof.

PROOF. See Appendix A.5. □

7.8.3. Removing Connections. Finally, we turn to the task of removing connection nodes.

DEFINITION 7.8.18. Let

$$\mathcal{Q} = \langle \mathcal{N}, succ, label, sel, exps, conns, decs \rangle$$

be an expansion structure. For any $\langle c, a, b \rangle \in conns$, we define $\mathcal{Q}^{-\mathbf{c}}_{\langle c,a,b \rangle}$ to be the expansion structure

$$\langle \mathcal{N}, succ, label, sel, exps, conns^-, decs \rangle$$

where $conns^- := conns \setminus \{\langle c, a, b \rangle\}$.

If we remove a connection from an extensional expansion dag, the resulting expansion structure is an extensional expansion dag.

LEMMA 7.8.19. *Let* $\mathcal{Q}$ *be an extensional expansion dag and* $\langle c, a, b \rangle$ *be a connection in* $\mathcal{Q}$. *Then* $\mathcal{Q}^{-\mathbf{c}}_{\langle c,a,b \rangle}$ *is an extensional expansion dag such that* $\mathcal{Q}^{-\mathbf{c}}_{\langle c,a,b \rangle} \trianglelefteq \mathcal{Q}$. *Furthermore, we have the following:*

1. $|\mathcal{Q}^{-\mathbf{c}}_{\langle c,a,b \rangle}| < |\mathcal{Q}|$
2. *If the embedding relation* $\prec_0^{\mathcal{Q}}$ *is acyclic, then the embedding relation* $\prec_0^{\mathcal{Q}^{-\mathbf{c}}_{\langle c,a,b \rangle}}$ *is acyclic.*
3. *If* $\mathfrak{p}$ *is a u-part of* $\mathcal{Q}^{-\mathbf{c}}_{\langle c,a,b \rangle}$ *and either* $a \notin \mathfrak{p}$ *or* $b \notin \mathfrak{p}$, *then* $\mathfrak{p}$ *is a u-part of* $\mathcal{Q}$.

PROOF. Let $\mathcal{Q}^-$ be $\mathcal{Q}^{-\mathbf{c}}_{\langle c,a,b\rangle}$. Clearly, $\mathcal{Q}^- \trianglelefteq \mathcal{Q}$. Also,

$$\mathbf{Dom}(succ^{\mathcal{Q}^-}) = \mathbf{Dom}(succ^{\mathcal{Q}}) = (\mathbf{Dom}(succ^{\mathcal{Q}}) \cap \mathcal{N}^{\mathcal{Q}^-})$$

and

$$\mathbf{Dom}(sel^{\mathcal{Q}^-}) = \mathbf{Dom}(sel^{\mathcal{Q}}) = (\mathbf{Dom}(sel^{\mathcal{Q}}) \cap \mathcal{N}^{\mathcal{Q}^-}).$$

Consequently, we know $\mathcal{Q}^-$ is an extensional expansion dag by Lemma 7.6.4. Since we have deleted a single connection, we have

$$|\mathcal{Q}^-| = (|\mathcal{Q}| - 1) < |\mathcal{Q}|.$$

The result concerning the embedding relation follows from Lemma 7.6.10.

Suppose $\mathfrak{p}$ is a u-part of $\mathcal{Q}^-$ and either $a \notin \mathfrak{p}$ or $b \notin \mathfrak{p}$. If $d^1, d^2 \in \mathfrak{p}$ and $\langle e, d^1, d^2\rangle \in conns$, then $\langle e, d^1, d^2\rangle \neq \langle c, a, b\rangle$ (since $\{d^1, d^2\} \subseteq \mathfrak{p}$ and $\{a, b\} \not\subseteq \mathfrak{p}$) and so $\langle e, d^1, d^2\rangle \in conns^-$. Thus we can apply Lemma 7.6.8 to conclude $\mathfrak{p}$ is a u-part of $\mathcal{Q}$. □

When the potentially connecting nodes of an accessible connection are in the sufficient set of an extensional expansion proof we can delete the connection from the extensional expansion dag and add the connection node to the sufficient set.

LEMMA 7.8.20. *Let $\langle \mathcal{Q}, \mathcal{R}\rangle$ be an extensional expansion proof. Suppose $\langle c, a, b\rangle$ is an accessible connection in $\mathcal{Q}$. If $a, b \in \mathcal{R}$, then $\langle \mathcal{Q}^{-\mathbf{c}}_{\langle c,a,b\rangle}, \mathcal{R} \cup \{c\}\rangle$. an extensional expansion proof.*

PROOF. See Appendix A.5. □

7.9. Soundness of Extensional Expansion Proofs

In order to establish soundness of extensional expansion proofs, we will construct $\mathcal{G}^{\mathcal{S}}_{\beta\mathfrak{fb}}$ derivations of $\Gamma(\mathcal{R})$ from extensional expansion proofs $\langle \mathcal{Q}, \mathcal{R}\rangle$. The proof of Theorem 7.9.1 provides an algorithm for translating extensional expansion proofs into $\mathcal{G}^{\mathcal{S}}_{\beta\mathfrak{fb}}$-derivations.

THEOREM 7.9.1 (Soundness of Extensional Expansion Proofs). *Let $\mathcal{S}$ be a signature of logical constants and $\langle \mathcal{Q}, \mathcal{R}\rangle$ be an $\mathcal{S}$-extensional expansion proof. There is a $\mathcal{G}^{\mathcal{S}}_{\beta\mathfrak{fb}}$-derivation $\mathcal{D}$ of $\Gamma^{\mathcal{Q}}(\mathcal{R})$.*

PROOF. This is proven by induction on the measure $|\mathcal{Q}|$ (cf. Definition 7.2.10). First assume there is some $a \in \mathbf{Roots}(\mathcal{Q}) \setminus \mathcal{R}$. In this case we can simply apply Lemma 7.8.10 to conclude $\langle \mathcal{Q}_a^{-\mathbf{r}}, \mathcal{R}\rangle$ is an extensional expansion proof. Also, $\mathcal{Q}_a^{-\mathbf{r}} \trianglelefteq \mathcal{Q}$ and $|\mathcal{Q}_a^{-\mathbf{r}}| < |\mathcal{Q}|$ by Lemma 7.8.9. Applying the inductive hypothesis, we have a derivation $\mathcal{D}$ of $\Gamma(\mathcal{R})$ (cf. Remark 7.6.3). For the remainder of the proof we assume $\mathbf{Roots}(\mathcal{Q}) \subseteq \mathcal{R}$.

By Lemma 7.8.7, there is either an accessible node, accessible connection or accessible expansion. We consider the three cases separately. Certain subcases are argued in Appendix A.5.

I. Assume there is an accessible node $a \in \mathbf{Roots}(\mathcal{Q})$. Note that, as an accessible node, a is extended, not a potentially connecting node and not an expansion node. (There are no flexible nodes in $\mathcal{Q}$ since $\mathcal{Q}$ is closed.) In each case below, we will delete (at least) the node a from $\mathcal{Q}$ before applying the inductive hypothesis. By Lemma 7.8.9, we know $\mathcal{Q}_a^{-\mathbf{r}}$ is an extensional expansion dag, $\mathcal{Q}_a^{-\mathbf{r}} \trianglelefteq \mathcal{Q}$ and $|\mathcal{Q}_a^{-\mathbf{r}}| < |\mathcal{Q}|$.

Note that $a \in \mathbf{Roots}(\mathcal{Q}) \subseteq \mathcal{R}$. Let $\mathbf{M}$ be the dual shallow formula of a. Hence $\Gamma(\mathcal{R})$ is $\Gamma', \mathbf{M}$ where Γ' is $\Gamma(\mathcal{R} \setminus \{a\})$ (cf. Remark 7.6.3).

- Suppose a is a false node. In this case, $\mathbf{M}$ is $\dot{\top}$. Consequently, we can take $\mathcal{D}$ to be the derivation

$$\frac{}{\Gamma', \dot{\top}}\ \mathcal{G}(\dot{\neg}\dot{\top})$$

- Suppose a is a true node. Hence $\mathbf{M}$ is $\dot{\neg}\dot{\top}$. By Lemma 7.8.13 and $\clubsuit_{ns}$, $\langle \mathcal{Q}_a^{-\mathbf{r}}, \mathcal{R} \setminus \{a\}\rangle$ is an extensional expansion proof. By the inductive hypothesis, there is a derivation $\mathcal{D}'$ of Γ'. By Lemma 2.2.10 (Weakening), there is a derivation $\mathcal{D}$ of $\Gamma', \dot{\neg}\dot{\top}$.
- Suppose a is a conjunction node. Consequently, $\mathbf{M}$ is of the form $[\mathbf{N} \dot{\vee} \mathbf{P}]$. By $\clubsuit_{\vee}$, we know there exist negative nodes b and c with $succ^{\mathcal{Q}}(a) = \langle b, c\rangle$ where b and c have shallow formulas $\mathbf{N}$ and $\mathbf{P}$, respectively. By Lemma 7.8.13, $\langle \mathcal{Q}_a^{-\mathbf{r}}, (\mathcal{R} \setminus \{a\}) \cup \{b, c\}\rangle$ is an extensional expansion proof. By the inductive hypothesis, there is a derivation $\mathcal{D}_1$ of the

sequent $\Gamma', \mathbf{N}, \mathbf{P}$. We complete the derivation $\mathcal{D}$ by

$$\frac{\begin{array}{c}\mathcal{D}_1 \\ \Gamma', \mathbf{N}, \mathbf{P}\end{array}}{\Gamma', [\mathbf{N} \dot{\vee} \mathbf{P}]} \mathcal{G}(\dot{\vee})$$

- Suppose a is a disjunction node. In this case $\mathbf{M}$ is $\dot{\neg}[\mathbf{N} \dot{\vee} \mathbf{P}]$ for some $\mathbf{N}$ and $\mathbf{P}$. By $\clubsuit_{\vee}$, there are positive nodes b and c with $succ^{\mathcal{Q}}(a) = \langle b, c\rangle$ where b and c have dual shallow formulas $\dot{\neg}\mathbf{N}$ and $\dot{\neg}\mathbf{P}$, respectively. By Lemma 7.8.14, both $\langle \mathcal{Q}_a^{-\mathbf{r}}, (\mathcal{R} \setminus \{a\}) \cup \{b\}\rangle$ and $\langle \mathcal{Q}_a^{-\mathbf{r}}, (\mathcal{R} \setminus \{a\}) \cup \{c\}\rangle$ are extensional expansion proofs. Applying the inductive hypothesis to $\mathcal{Q}_a^{-\mathbf{r}}$ with both sufficient sets, we obtain derivations $\mathcal{D}_1$ of $\Gamma, \dot{\neg}\mathbf{N}$ and $\mathcal{D}_2$ of $\Gamma, \dot{\neg}\mathbf{P}$. We complete the derivation $\mathcal{D}$ by

$$\frac{\begin{array}{c}\mathcal{D}_1 \\ \Gamma', \dot{\neg}\mathbf{N}\end{array} \qquad \begin{array}{c}\mathcal{D}_2 \\ \Gamma', \dot{\neg}\mathbf{P}\end{array}}{\Gamma', \dot{\neg}[\mathbf{N} \dot{\vee} \mathbf{P}]} \mathcal{G}(\dot{\neg}\dot{\vee})$$

- Suppose a is a selection node with selected parameter A_α. In this case $\mathbf{M}$ is of the form $[\dot{\forall} x_\alpha \mathbf{N}]$. Since $\langle \mathcal{Q}, \mathcal{R}\rangle$ is an extensional expansion proof, $A \notin \mathbf{Params}(\mathcal{R})$. Hence A does not occur in Γ' and does not occur in $\mathbf{M}$. By $\clubsuit_{sel}$, there is a negative node b where $succ^{\mathcal{Q}}(a) = \langle b\rangle$ and b has dual shallow formula $[A/x]\mathbf{N}$. By Lemma 7.8.11, $\langle \mathcal{Q}_a^{-\mathbf{r}}, (\mathcal{R} \setminus \{a\}) \cup \{b\}\rangle$. is an extensional expansion proof. By the inductive hypothesis, there is a derivation $\mathcal{D}_1$ of $\Gamma', [A/x]\mathbf{N}$. We complete the derivation $\mathcal{D}$ by

$$\frac{\begin{array}{c}\mathcal{D}_1 \\ \Gamma', [A/x]\mathbf{N}\end{array}}{\Gamma', [\dot{\forall} x_\alpha \mathbf{N}]} \mathcal{G}(\dot{\forall}^A)$$

- Suppose a is a decomposition node. Then there is some $n \geq 0$ and parameter H where $\mathbf{M}$ is of the form

$$[[H\, \mathbf{A}^1_{\alpha^1} \cdots \mathbf{A}^n_{\alpha^n}] \doteq^\iota [H\, \mathbf{B}^1_{\alpha^1} \cdots \mathbf{B}^n_{\alpha^n}]].$$

By $\clubsuit_{dec}$, there are negative nodes $a^1, \ldots, a^n$ such that

$$succ^{\mathcal{Q}}(a) = \langle a^1, \ldots, a^n\rangle$$

and for each $i \in \{1, \ldots, n\}$ the shallow formula of a^i must be $[\mathbf{A}^i \mathrel{\dot{=}}^{\alpha^i} \mathbf{B}^i]$.
First consider the case where $n = 0$. In this case we can apply the $\mathcal{G}(Dec)$ for H_ι with no premises. That is, we take $\mathcal{D}$ to be

$$\frac{}{\Gamma', [H \mathrel{\dot{=}}^{\iota} H]}\ \mathcal{G}(Dec)$$

(This is a base case for the induction.)
Next assume $n \geq 1$. By Lemma 7.8.14, for each i $(1 \leq i \leq n)$ we know $\langle \mathcal{Q}_a^{-\mathbf{r}}, (\mathcal{R} \setminus \{a\}) \cup \{a^i\}\rangle$ is an extensional expansion proof. Applying the inductive hypothesis with these sufficient sets, we obtain derivations $\mathcal{D}_i$ of $\Gamma', [\mathbf{A}^i \mathrel{\dot{=}}^{\alpha^i} \mathbf{B}^i]$. We complete the derivation $\mathcal{D}$ by applying the decomposition rule $\mathcal{G}(Dec)$.

- Suppose a is a functional node with shallow formula of the form $[\mathbf{F} \mathrel{\dot{=}}^{\alpha\beta} \mathbf{G}]$ for some $\beta\eta$-normal $\mathbf{F}$ and $\mathbf{G}$. By $\clubsuit_{\dot{=}}^{\rightarrow}$, there is a node b such that $succ^{\mathcal{Q}}(a) = \langle b \rangle$ where b has shallow formula $[\dot{\forall} y_\beta \blacksquare \mathbf{F}\, y \mathrel{\dot{=}}^{\alpha} \mathbf{G}\, y]^{\downarrow}$ and the same polarity as a. By Lemma 7.8.11, $\langle \mathcal{Q}_a^{-\mathbf{r}}, (\mathcal{R} \setminus \{a\}) \cup \{b\}\rangle$ is an extensional expansion proof. By the inductive hypothesis, there is a derivation $\mathcal{D}_1$ of $\Gamma', \overline{sh(b)}$.
If a is negative, then we use the (constructively) admissible rule $\mathcal{G}(\mathfrak{f}\dot{\forall})$ (cf. Lemma 2.2.20) to construct $\mathcal{D}$ as follows:

$$\frac{\begin{array}{c}\mathcal{D}_1\\ \Gamma', [\dot{\forall} y_\beta \blacksquare \mathbf{F}\, y \mathrel{\dot{=}}^{\alpha} \mathbf{G}\, y]^{\downarrow}\end{array}}{\Gamma', [\mathbf{F} \mathrel{\dot{=}}^{\alpha\beta} \mathbf{G}]}\ \mathcal{G}(\mathfrak{f}\dot{\forall})$$

If a is positive, then we use the (constructively) admissible rule $\mathcal{G}(\dot{\neg} \mathrel{\dot{=}}^{\rightarrow} \dot{\forall})$ (cf. Lemma 2.2.21) to complete the derivation $\mathcal{D}$ as follows:

$$\frac{\begin{array}{c}\mathcal{D}_1\\ \Gamma', \dot{\neg}[\dot{\forall} y_\beta \blacksquare \mathbf{F}\, y \mathrel{\dot{=}}^{\alpha} \mathbf{G}\, y]^{\downarrow}\end{array}}{\Gamma', \dot{\neg}[\mathbf{F} \mathrel{\dot{=}}^{\alpha\beta} \mathbf{G}]}\ \mathcal{G}(\dot{\neg} \mathrel{\dot{=}}^{\rightarrow} \dot{\forall})$$

- Suppose a is a negative Boolean node. In this case $\mathbf{M}$ is of the form $[\mathbf{A} \mathrel{\dot{=}}^{o} \mathbf{B}]$. By $\clubsuit_{\dot{=}}^{o}$, there are negative primary nodes b and c where $succ^{\mathcal{Q}}(a) = \langle b, c \rangle$, $sh(b) = [\dot{\neg}\mathbf{A} \mathrel{\dot{\vee}} \mathbf{B}]$ and $sh(c) = [\dot{\neg}\mathbf{B} \mathrel{\dot{\vee}} \mathbf{A}]$.

Since a is disjunctive, Lemma 7.8.14 implies that both

$$\langle \mathcal{Q}_a^{-\mathbf{r}}, (\mathcal{R} \setminus \{a\}) \cup \{b\}\rangle \text{ and } \langle \mathcal{Q}_a^{-\mathbf{r}}, (\mathcal{R} \setminus \{a\}) \cup \{c\}\rangle$$

are extensional expansion proofs.

Assume either b or c is not extended in $\mathcal{Q}$ and call this node $d \in \{b, c\}$. Note that d is also not extended in $\mathcal{Q}_a^{-\mathbf{r}}$. In this case $\mathcal{R} \setminus \{a\}$ is a sufficient set for $\mathcal{Q}_a^{-\mathbf{r}}$ by Lemma 7.4.12. Also,

$$\mathbf{SelPars}^{\mathcal{Q}_a^{-\mathbf{r}}}(\mathcal{R} \setminus \{a\}) \subseteq \mathbf{SelPars}^{\mathcal{Q}_a^{-\mathbf{r}}}((\mathcal{R} \setminus \{a\}) \cup \{d\}) \subseteq \emptyset$$

and so $\langle \mathcal{Q}_a^{-\mathbf{r}}, (\mathcal{R} \setminus \{a\}\rangle$ is an extensional expansion proof. Applying the inductive hypothesis, we have a derivation $\mathcal{D}_0$ of Γ'. By Lemma 2.2.10 (Weakening), there is a derivation $\mathcal{D}$ of $\Gamma(\mathcal{R})$ as desired.

Assume both b and c are extended. By Definition 7.2.7, both b and c are conjunction nodes. By $\clubsuit_\vee$, there exist negative nodes b^1 and b^2 with shallow formulas $\neg\mathbf{A}$ and $\mathbf{B}$, respectively, such that $succ^{\mathcal{Q}}(b) = \langle b^1, b^2\rangle$. Also, $succ^{\mathcal{Q}}(c) = \langle c^1, c^2\rangle$ for some negative nodes c^1 and c^2 with shallow formulas $\neg\mathbf{B}$ and $\mathbf{A}$, respectively.

Applying Lemma 7.8.13 with b and $(\mathcal{R} \setminus \{a\}) \cup \{b\}$, we know

$$\langle (\mathcal{Q}_a^{-\mathbf{r}})_b^{-\mathbf{r}}, (\mathcal{R} \setminus \{a\}) \cup \{b^1, b^2\}\rangle$$

is an extensional expansion proof. Applying Lemma 7.8.13 with c and $(\mathcal{R} \setminus \{a\}) \cup \{c\}$, we know

$$\langle (\mathcal{Q}_a^{-\mathbf{r}})_c^{-\mathbf{r}}, (\mathcal{R} \setminus \{a\}) \cup \{c^1, c^2\}\rangle$$

is an extensional expansion proof.

By Lemma 7.8.9, we know

$$|(\mathcal{Q}_a^{-\mathbf{r}})_b^{-\mathbf{r}}| < |\mathcal{Q}_a^{-\mathbf{r}}| < |\mathcal{Q}|$$

and

$$|(\mathcal{Q}_a^{-\mathbf{r}})_c^{-\mathbf{r}}| < |\mathcal{Q}_a^{-\mathbf{r}}| < |\mathcal{Q}|.$$

Applying the inductive hypothesis to

$$(\mathcal{Q}_a^{-\mathbf{r}})_b^{-\mathbf{r}} \text{ and } (\mathcal{R} \setminus \{a\}) \cup \{b^1, b^2\},$$

we have a derivation $\mathcal{D}_1$ of $\Gamma', \neg\mathbf{A}, \mathbf{B}$. Applying the inductive hypothesis to $(\mathcal{Q}_a^{-\mathbf{r}})_c^{-\mathbf{r}}$ and $(\mathcal{R} \setminus \{a\}) \cup \{c^1, c^2\}$, we have a derivation $\mathcal{D}_2$ of $\Gamma', \neg\mathbf{B}, \mathbf{A}$. We complete the derivation $\mathcal{D}$ by

$$\dfrac{\begin{array}{cc}\mathcal{D}_1 & \mathcal{D}_2 \\ \Gamma', \neg\mathbf{A}, \mathbf{B} & \Gamma', \neg\mathbf{B}, \mathbf{A}\end{array}}{\Gamma', [\mathbf{A} \doteq^o \mathbf{B}]}\ \mathcal{G}(\mathfrak{b})$$

- Suppose a is a mate node. In this case $\mathbf{M}$ is of the form

$$[P\,\overline{\mathbf{A}^n} \doteq^o P\,\overline{\mathbf{B}^n}]$$

for some $n \geq 0$ and parameter $P_{o\alpha^n\cdots\alpha^1}$. By $\clubsuit_m$, there are $n \geq 0$ negative nodes $a^1, \ldots, a^n$ where $succ^{\mathcal{Q}}(a) = \langle a^1, \ldots, a^n\rangle$ and a^i has shallow formula $[\mathbf{A}^i \doteq \mathbf{B}^i]$ for each i $(1 \leq i \leq n)$. First assume $n = 0$. In this case we apply the rule $\mathcal{G}(Init^{\doteq})$ with no premises to the propositional parameter P_o and $\mathcal{G}(\mathfrak{b})$ in order to conclude $[P \doteq^o P]$.

$$\dfrac{\dfrac{}{\Gamma', \neg P, P}\ \mathcal{G}(Init^{\doteq}) \qquad \dfrac{}{\Gamma', \neg P, P}\ \mathcal{G}(Init^{\doteq})}{\Gamma', [P \doteq^o P]}\ \mathcal{G}(\mathfrak{b})$$

Next assume $n \geq 1$. Since a is a disjunctive node, we can apply Lemma 7.8.14 to conclude

$$\langle \mathcal{Q}_a^{-\mathbf{r}}, ((\mathcal{R} \setminus \{a\}) \cup \{a^i\})\rangle$$

is an extensional expansion proof for each i $(1 \leq i \leq n)$. By the inductive hypothesis, there are derivations $\mathcal{D}_i$ of

$$\Gamma, [\mathbf{A}^i \doteq \mathbf{B}^i]$$

for each i $(1 \leq i \leq n)$. Let $\mathcal{D}'$ be the derivation

$$\dfrac{\begin{array}{ccc}\mathcal{D}_1 & & \mathcal{D}_n \\ \Gamma', [\mathbf{A}^1 \doteq \mathbf{B}^1] & \cdots & \Gamma', [\mathbf{A}^n \doteq \mathbf{B}^n]\end{array}}{\Gamma', \neg[P\,\overline{\mathbf{B}^n}], [P\,\overline{\mathbf{A}^n}]}\ \mathcal{G}(Init^{\doteq})$$

We make use of the (constructively) admissible rule $\mathcal{G}(Sym)$ (cf. Lemma 2.2.19) to let $\mathcal{E}'$ be the derivation

$$\dfrac{\dfrac{\begin{array}{c}\mathcal{D}_1 \\ \Gamma', [\mathbf{A}^1 \doteq \mathbf{B}^1]\end{array}}{\Gamma', [\mathbf{B}^1 \doteq \mathbf{A}^1]}\ \mathcal{G}(Sym) \quad \cdots \quad \dfrac{\begin{array}{c}\mathcal{D}_n \\ \Gamma', [\mathbf{A}^n \doteq \mathbf{B}^n]\end{array}}{\Gamma', [\mathbf{B}^n \doteq \mathbf{A}^n]}\ \mathcal{G}(Sym)}{\Gamma', \neg[P\,\overline{\mathbf{A}^n}], [P\,\overline{\mathbf{B}^n}]}\ \mathcal{G}(Init^{\doteq})$$

We complete the derivation $\mathcal{D}$ as follows:

$$\frac{\begin{array}{cc}\mathcal{E}' & \mathcal{D}' \\ \Gamma', \neg[P\,\overline{\mathbf{A}^n}], [P\,\overline{\mathbf{B}^n}] & \Gamma', \neg[P\,\overline{\mathbf{B}^n}], [P\,\overline{\mathbf{A}^n}]\end{array}}{\Gamma', [[P\,\overline{\mathbf{A}^n}] \doteq^o [P\,\overline{\mathbf{B}^n}]]}\,\mathcal{G}(\mathfrak{b})$$

- Suppose a is an E-unification node or a symmetric E-unification node. By $\clubsuit_{eu}$, $succ^{\mathcal{Q}}(a) = \langle a^1, a^2\rangle$, $\mathbf{M}$ is of the form

$$[[\mathbf{A} \doteq^\iota \mathbf{C}] \mathbin{\dot\wedge} [\mathbf{B} \doteq^\iota \mathbf{D}]],$$

$\overline{sh}(a^1) = [\mathbf{A} \doteq^\iota \mathbf{C}]$ and $\overline{sh}(a^2) = [\mathbf{B} \doteq^\iota \mathbf{D}]$. Since a is a disjunctive node, we can apply Lemma 7.8.14 to conclude

$$\langle \mathcal{Q}_a^{-\mathbf{r}}, ((\mathcal{R} \setminus \{a\}) \cup \{a^1\})\rangle$$

and

$$\langle \mathcal{Q}_a^{-\mathbf{r}}, ((\mathcal{R} \setminus \{a\}) \cup \{a^2\})\rangle$$

are both extensional expansion proofs. By the inductive hypothesis, there are derivations $\mathcal{D}_1$ and $\mathcal{D}_2$ of $\Gamma', [\mathbf{A} \doteq^\iota \mathbf{C}]$ and $\Gamma', [\mathbf{B} \doteq^\iota \mathbf{D}]$, respectively. We complete the derivation $\mathcal{D}$ using the derived rule $\mathcal{G}(\dot\wedge)$ as follows:

$$\frac{\begin{array}{cc}\mathcal{D}_1 & \mathcal{D}_2 \\ \Gamma', [\mathbf{A} \doteq^\iota \mathbf{C}] & \Gamma', [\mathbf{B} \doteq^\iota \mathbf{D}]\end{array}}{\Gamma', [[\mathbf{A} \doteq^\iota \mathbf{C}] \mathbin{\dot\wedge} [\mathbf{B} \doteq^\iota \mathbf{D}]]}\,\mathcal{G}(\dot\wedge)$$

- The cases where a is a negation node, deepening node, negative equation node or positive Boolean node are checked in Appendix A.5.

II. Assume there is an accessible connection $\langle c, a, b\rangle$. We know $\mathcal{Q}^{-\mathbf{c}}_{\langle c,a,b\rangle}$ is an extensional expansion dag with $\mathcal{Q}^{-\mathbf{c}}_{\langle c,a,b\rangle} \trianglelefteq \mathcal{Q}$ and $|\mathcal{Q}^{-\mathbf{c}}_{\langle c,a,b\rangle}| < |\mathcal{Q}|$ by Lemma 7.8.19. We know $\{a, b\} \subseteq \mathbf{Roots}(\mathcal{Q}) \subseteq \mathcal{R}$. By Lemma 7.8.20, we know $\langle \mathcal{Q}^{-\mathbf{c}}_{\langle c,a,b\rangle}, \mathcal{R} \cup \{c\}\rangle$ is an extensional expansion proof.

Assume c is not extended in $\mathcal{Q}$. Since $\mathcal{Q}^{-\mathbf{c}}_{\langle c,a,b\rangle} \trianglelefteq \mathcal{Q}$ we know c is not extended in $\mathcal{Q}^{-\mathbf{c}}_{\langle c,a,b\rangle}$. In this case $\mathcal{R}$ is a sufficient set for $\mathcal{Q}^{-\mathbf{c}}_{\langle c,a,b\rangle}$ by Lemma 7.4.12. Also,

$$\mathbf{SelPars}^{\mathcal{Q}^{-\mathbf{c}}_{\langle c,a,b\rangle}}(\mathcal{R}) \subseteq \mathbf{SelPars}^{\mathcal{Q}^{-\mathbf{c}}_{\langle c,a,b\rangle}}(\mathcal{R} \cup \{c\}) \subseteq \emptyset$$

and so $\langle \mathcal{Q}^{-\mathbf{c}}_{\langle c,a,b\rangle}, \mathcal{R}\rangle$ is an extensional expansion proof. Applying the inductive hypothesis, we have a derivation $\mathcal{D}$ of $\Gamma(\mathcal{R})$ (cf. Remark 7.6.3), as desired.

Assume c is extended and let $c^1, \ldots, c^n$ be the nodes such that

$$succ^{\mathcal{Q}}(c) = \langle c^1, \ldots, c^n\rangle.$$

Suppose $n = 0$. In this case c must be a mate node with shallow formula $[P_o \overset{.}{=}^o P_o]$ for some parameter P_o. We can apply the rule $\mathcal{G}(Init^{\overset{.}{=}})$ with no premises to the propositional parameter P_o:

$$\frac{}{\Gamma(\mathcal{R} \setminus \{a, b\}), \dot\neg P, P}\ \mathcal{G}(Init^{\overset{.}{=}})$$

(This is a base case of the induction.)

Now, suppose $n \geq 1$. As a connection node, c is disjunctive. By Lemma 7.8.14, for each i $(1 \leq i \leq n)$, $\langle (\mathcal{Q}^{-\mathbf{c}}_{\langle c,a,b\rangle})^{-\mathbf{r}}_{c}, \mathcal{R} \cup \{c^i\}\rangle$ is an extensional expansion proof. Let Γ' be $\Gamma(\mathcal{R} \setminus \{a, b\})$ and Γ'' be $\Gamma(\mathcal{R})$. By the inductive hypothesis, there are derivations $\mathcal{D}_i$ of $\Gamma'', \overline{sh}(c^i)$. We consider three possibilities for the node a.

1. Suppose c is a mate node. In this case $\clubsuit_m$ and $\clubsuit^p_m$ imply that $\mathbf{M}$ is of the form $[P\,\overline{\mathbf{B}^n} \overset{.}{=}^o P\,\overline{\mathbf{A}^n}]$, $\overline{sh}(a) = \dot\neg[P\,\overline{\mathbf{A}^n}]$, $\overline{sh}(b) = [P\,\overline{\mathbf{B}^n}]$, and $\overline{sh}(c^i) = [\mathbf{B}^i \overset{.}{=} \mathbf{A}^i]$ for each i $(1 \leq i \leq n)$. For each i $(1 \leq i \leq n)$, $\mathcal{D}_i$ is a derivation of $\Gamma'', [\mathbf{B}^i \overset{.}{=} \mathbf{A}^i]$. We complete the derivation $\mathcal{D}$ as follows:

$$\cfrac{\cfrac{\cfrac{\begin{matrix}\mathcal{D}_1 & & \mathcal{D}_n\\ \Gamma'', [\mathbf{B}^1 \overset{.}{=} \mathbf{A}^1] & \cdots & \Gamma'', [\mathbf{B}^n \overset{.}{=} \mathbf{A}^n]\end{matrix}}{\Gamma', [P\,\overline{\mathbf{B}^n}], \dot\neg[P\,\overline{\mathbf{A}^n}], [P\,\overline{\mathbf{B}^n}], \dot\neg[P\,\overline{\mathbf{A}^n}]}\ \mathcal{G}(Init^{\overset{.}{=}})}{\Gamma', \dot\neg[P\,\overline{\mathbf{A}^n}], [P\,\overline{\mathbf{B}^n}], \dot\neg[P\,\overline{\mathbf{A}^n}]}\ \mathcal{G}(Contr)}{\Gamma', [P\,\overline{\mathbf{B}^n}], \dot\neg[P\,\overline{\mathbf{A}^n}]}\ \mathcal{G}(Contr)$$

2. Suppose c is an E-unification node. By $\clubsuit_{eu}$ and $\clubsuit^p_{eu}$, $n = 2$, $\mathbf{M}$ is of the form

$$[[\mathbf{A} \overset{.}{=}^\iota \mathbf{C}] \dot\wedge [\mathbf{B} \overset{.}{=}^\iota \mathbf{D}]],$$

$\overline{sh}(c^1) = [\mathbf{A} \overset{.}{=}^\iota \mathbf{C}]$, $\overline{sh}(c^2) = [\mathbf{B} \overset{.}{=}^\iota \mathbf{D}]$, $\overline{sh}(a) = \dot\neg[\mathbf{A} \overset{.}{=}^\iota \mathbf{B}]$ and $\overline{sh}(b) = [\mathbf{C} \overset{.}{=}^\iota \mathbf{D}]$. Hence

$$\mathcal{D}_1 \text{ is a derivation of } \Gamma'', [\mathbf{A} \overset{.}{=}^\iota \mathbf{C}]$$

and

$$\mathcal{D}_2 \text{ is a derivation of } \Gamma'', [\mathbf{B} \dot{=}^{\iota} \mathbf{D}].$$

We complete the derivation $\mathcal{D}$ as follows:

$$\dfrac{\dfrac{\dfrac{\begin{array}{cc}\mathcal{D}_1 & \mathcal{D}_2 \\ \Gamma'', [\mathbf{A} \dot{=}^{\iota} \mathbf{C}] & \Gamma'', [\mathbf{B} \dot{=}^{\iota} \mathbf{D}]\end{array}}{\Gamma', \dot{\neg}[\mathbf{A} \dot{=}^{\iota} \mathbf{B}], [\mathbf{C} \dot{=}^{\iota} \mathbf{D}], \dot{\neg}[\mathbf{A} \dot{=}^{\iota} \mathbf{B}], [\mathbf{C} \dot{=}^{\iota} \mathbf{D}]}\,\mathcal{G}(EUnif_1)}{\Gamma', [\mathbf{C} \dot{=}^{\iota} \mathbf{D}], \dot{\neg}[\mathbf{A} \dot{=}^{\iota} \mathbf{B}], [\mathbf{C} \dot{=}^{\iota} \mathbf{D}]}\,\mathcal{G}(Contr)}{\Gamma', \dot{\neg}[\mathbf{A} \dot{=}^{\iota} \mathbf{B}], [\mathbf{C} \dot{=}^{\iota} \mathbf{D}]}\,\mathcal{G}(Contr)$$

3. The case where c is a symmetric E-unification node is argued in Appendix A.5.

III. Assume there is an accessible expansion $\langle e, \mathbf{C}_\alpha, c\rangle$. By $\clubsuit_{exp}$, there is some variable x_α and proposition $\mathbf{M}$ where the dual shallow formula of e is $\dot{\neg}[\dot{\forall} x_\alpha \mathbf{M}]$ and the dual shallow formula of c is $\dot{\neg}[[\mathbf{C}/x]\mathbf{M}]^{\downarrow}$ By Lemma 7.8.16, $\mathcal{Q}^{-\mathbf{e}}_{\langle e,\mathbf{C},c\rangle}$ is an extensional expansion dag and $|\mathcal{Q}^{-\mathbf{e}}_{\langle e,\mathbf{C},c\rangle}| < |\mathcal{Q}|$.

Note that $a \in \mathbf{Roots}(\mathcal{Q}) \subseteq \mathcal{R}$. Hence $\Gamma(\mathcal{R})$ is of the form $\Gamma', \dot{\neg}[\dot{\forall} x_\alpha \mathbf{M}]$. By Lemma 7.8.17, $\langle \mathcal{Q}^{-\mathbf{e}}_{\langle e,\mathbf{C},c\rangle}, \mathcal{R} \cup \{c\}\rangle$ is an extensional expansion proof. By the inductive hypothesis, there is a derivation $\mathcal{D}_1$ of

$$\Gamma', \dot{\neg}[\dot{\forall} x_\alpha \mathbf{M}], \dot{\neg}[[\mathbf{C}/x]\mathbf{M}]^{\downarrow}.$$

We complete the derivation $\mathcal{D}$ as follows:

$$\dfrac{\dfrac{\dfrac{\begin{array}{c}\mathcal{D}_1 \\ \Gamma', \dot{\neg}[\dot{\forall} x_\alpha \mathbf{M}], \dot{\neg}[[\mathbf{C}/x]\mathbf{M}]^{\downarrow}\end{array}}{\Gamma', \dot{\neg}[\dot{\forall} x_\alpha \mathbf{M}], \dot{\neg}[[\mathbf{C}/x]\mathbf{M}]}\,\mathcal{G}(\beta\eta)}{\Gamma', \dot{\neg}[\dot{\forall} x_\alpha \mathbf{M}], \dot{\neg}[\dot{\forall} x_\alpha \mathbf{M}]}\,\mathcal{G}(\dot{\neg}\dot{\forall})}{\Gamma', \dot{\neg}[\dot{\forall} x_\alpha \mathbf{M}]}\,\mathcal{G}(Contr)$$

□

REMARK 7.9.2 (Translation Algorithm). The proof of Theorem 7.9.1 provides an algorithm for constructing a derivation of $\Gamma(\mathcal{R})$ given an extensional expansion proof $\langle \mathcal{Q}, \mathcal{R}\rangle$. The proof does make use of the admissible rules $\mathcal{G}(Sym)$, $\mathcal{G}(\mathfrak{f}\dot{\forall})$ and $\mathcal{G}(\neg \dot{=}^{\rightarrow} \dot{\forall})$. However, as noted in Remark 2.2.22, the proofs of admissibility of these rules are constructive. The proof also makes use of Weakening (cf. Lemma 2.2.10) which is constructive. In TPS a similar algorithm which translates from extensional expansion proofs to natural deduction proofs is implemented.

7.10. Completeness of Extensional Expansion Proofs

We next turn to the task of constructing an extensional expansion proof $\langle \mathcal{Q}, \mathcal{R} \rangle$ with $\Gamma(\mathcal{R}) = \Gamma$ whenever there is a proof of Γ in $\mathcal{G}^{\mathcal{S}}_{\beta\mathfrak{f}\mathfrak{b}}$. The completeness proof of Theorem 7.10.12 provides an algorithm for translating $\mathcal{G}^{\mathcal{S}}_{\beta\mathfrak{f}\mathfrak{b}}$-derivations into merged, separated extensional expansion proofs. (The heart of the algorithm is the inductive proof of Lemma 7.10.11.) Furthermore, we can map $\mathcal{W}$-restricted extensional expansion proofs into $\mathcal{W}$-restricted extensional expansion proofs. We start by giving a few more definitions and lemmas to ease the proof.

We will prove completeness by induction on $\mathcal{W}$-restricted derivations of sequents. At each step, we will construct a closed, merged, separated extensional expansion dag $\mathcal{Q}$ where **RootSels**($\mathcal{Q}$) is empty. We will also ensure $\mathcal{Q}$ has an acyclic embedding relation and expansion terms restricted to $\mathcal{W}$. In order to prevent needing to state all these conditions several times, we define the set of extensional expansion dags for which these properties hold.

DEFINITION 7.10.1. Let $\mathcal{W} \subseteq cwff(\mathcal{S})$ be a set of closed formulas. We define $\mathfrak{Nice}_{\mathcal{W}}$ to be the class of all closed, merged, separated extensional expansion dags $\mathcal{Q}$ such that **ExpTerms**($\mathcal{Q}$) $\subseteq \mathcal{W}$, $\prec^{\mathcal{Q}}_{0}$ is acyclic and **RootSels**($\mathcal{Q}$) is empty.

LEMMA 7.10.2. *Let $\mathcal{W} \subseteq cwff(\mathcal{S})$ be a set of closed formulas, $\mathcal{Q} \in \mathfrak{Nice}_{\mathcal{W}}$ and $a \in \mathcal{N}^{\mathcal{Q}}$ be a node which is not flexible. There is some $\mathcal{Q}' \in \mathfrak{Nice}_{\mathcal{W}}$ such that $\mathcal{Q} \trianglelefteq_r \mathcal{Q}'$ and a is extended in $\mathcal{Q}'$.*

PROOF. If a is already extended in $\mathcal{Q}$, then we have $\mathcal{Q} \in \mathfrak{Nice}_{\mathcal{W}}$, $\mathcal{Q} \trianglelefteq_r \mathcal{Q}$ and a is extended in $\mathcal{Q}$.

Assume a is not extended in $\mathcal{Q}$. Let $\mathcal{Q}'$ be **Extend**($\mathcal{Q}, a$). We know $\mathcal{Q}'$ is a closed, merged, separated extensional expansion dag such that $\mathcal{Q} \trianglelefteq_r \mathcal{Q}'$, a is extended in $\mathcal{Q}'$,

$$\mathbf{ExpTerms}(\mathcal{Q}') = \mathbf{ExpTerms}(\mathcal{Q}) \subseteq \mathcal{W},$$

$$\mathbf{RootSels}(\mathcal{Q}') = \mathbf{RootSels}(\mathcal{Q}) = \emptyset$$

and the embedding relation $\prec_0^{\mathcal{Q}'}$ is acyclic by Lemma 7.7.7. Consequently, $\mathcal{Q}' \in \mathfrak{Nice}_{\mathcal{W}}$. □

There are times when we work with a primary node a which has dual shallow formula $\dot{\neg}\mathbf{M}$. Sometimes a will be a positive node with shallow formula $\mathbf{M}$ and other times a will be a negative negation node with shallow formula $\dot{\neg}\mathbf{M}$. In order to prevent having to distinguish between these cases, we define a *positive counterpart of a*.

DEFINITION 7.10.3. Let $\mathcal{Q}$ be an extensional expansion dag and $a, b \in \mathcal{N}^{\mathcal{Q}}$ be nodes. We say b is a *positive counterpart* of a with respect to $\mathcal{Q}$ if either a is positive and $b = a$ or a is an extended negative negation node and $succ^{\mathcal{Q}}(a) = \langle b \rangle$.

LEMMA 7.10.4. *Let $\mathcal{W} \subseteq cwff(\mathcal{S})$ be a set of closed formulas, $\mathcal{Q} \in \mathfrak{Nice}_{\mathcal{W}}$ and $a \in \mathcal{N}^{\mathcal{Q}}$ be a primary node with dual shallow formula $\dot{\neg}\mathbf{M}$. There is some $\mathcal{Q}' \in \mathfrak{Nice}_{\mathcal{W}}$ and some node $b \in \mathcal{N}^{\mathcal{Q}'}$ such that $\mathcal{Q} \trianglelefteq_r \mathcal{Q}'$ and b is a positive counterpart of a with respect to $\mathcal{Q}'$. Furthermore, b is a positive primary node of $\mathcal{Q}'$ with shallow formula $\mathbf{M}$.*

PROOF. First assume a is positive. In this case, we take b to be the positive primary node a and note that the shallow formula of a must be $\mathbf{M}$.

Next assume a is negative. By Lemma 7.10.2, there is some $\mathcal{Q}' \in \mathfrak{Nice}_{\mathcal{W}}$ such that $\mathcal{Q} \trianglelefteq_r \mathcal{Q}'$ and a is extended in $\mathcal{Q}'$. By $\clubsuit_{\neg}$ (in $\mathcal{Q}'$) there is some positive primary node b with shallow formula $\mathbf{M}$ such that $succ^{\mathcal{Q}'}(a) = \langle b \rangle$. The node b is the desired positive counterpart of a with respect to $\mathcal{Q}'$. □

LEMMA 7.10.5. *Let $\mathcal{Q}$ be an extensional expansion dag, $a, b \in \mathcal{N}^{\mathcal{Q}}$ be nodes and $\mathcal{R} \subseteq \mathcal{N}^{\mathcal{Q}}$. If b is a positive counterpart of a and $\mathcal{R} \cup \{b\}$ is sufficient for $\mathcal{Q}$, then $\mathcal{R} \cup \{a\}$ is sufficient for $\mathcal{Q}$.*

PROOF. If a is positive, then this is trivial since $a = b$. Otherwise, a is an extended negative negation node and $succ^{\mathcal{Q}}(a) = \langle b \rangle$. Hence a is a neutral node and b is the unique child of a. If $\mathcal{R} \cup \{b\}$ is sufficient, then $\mathcal{R} \cup \{a\}$ is also sufficient by Lemma 7.4.15. □

LEMMA 7.10.6. *Let $\mathcal{Q}$ and $\mathcal{Q}'$ be extensional expansion dags with $\mathcal{Q} \trianglelefteq \mathcal{Q}'$ and $a, b \in \mathcal{N}^{\mathcal{Q}}$ be nodes of $\mathcal{Q}$. If b is a positive counterpart of*

a with respect to $\mathcal{Q}$, then b is a positive counterpart of a with respect to $\mathcal{Q}'$.

PROOF. Note that $label^{\mathcal{Q}'}(a) = label^{\mathcal{Q}}(a)$. If a is positive in $\mathcal{Q}$ and $b = a$, then a is positive in $\mathcal{Q}'$ since $label^{\mathcal{Q}'}(a) = label^{\mathcal{Q}}(a)$. Also, if a is an extended negative negation node in $\mathcal{Q}$ and $succ^{\mathcal{Q}}(a) = \langle b \rangle$, then a is an extended negative negation node in $\mathcal{Q}'$ and

$$succ^{\mathcal{Q}'}(a) = succ^{\mathcal{Q}}(a) = \langle b \rangle.$$

□

Sometimes we wish to add an expansion arc to an extensional expansion dag without losing the property of being merged.

LEMMA 7.10.7. *Let $\mathcal{W} \subseteq cwff(\mathcal{S})$ be a set of closed formulas. Suppose $\mathcal{Q} \in \mathfrak{Nice}_{\mathcal{W}}$, $a \in \mathcal{N}^{\mathcal{Q}}$ is an expansion node with shallow formula $[\forall x_\alpha \mathbf{M}]$ and $\mathbf{C}_\alpha \in \mathcal{W}$. There is some $\mathcal{Q}' \in \mathfrak{Nice}_{\mathcal{W}}$, term $\mathbf{D}_\alpha \in \mathcal{W}$ and $d \in \mathcal{N}^{\mathcal{Q}'}$ such that $\mathcal{Q} \trianglelefteq_r \mathcal{Q}'$, $\langle a, \mathbf{D}, d \rangle \in exps^{\mathcal{Q}'}$ and $sh^{\mathcal{Q}'}(d)$ is $[[\mathbf{C}/x]\mathbf{M}]^{\downarrow}$.*

PROOF. Let $\mathbf{N}$ be $[[\mathbf{C}/x]\mathbf{M}]^{\downarrow}$. If $\mathbf{N} \in \mathbf{ExpInsts}^{\mathcal{Q}}(a)$, then there is already some $d \in \mathcal{N}^{\mathcal{Q}}$ and $\mathbf{D}_\alpha \in \mathcal{W}$ such that $\langle a, \mathbf{D}, d \rangle \in exps^{\mathcal{Q}}$ and $sh^{\mathcal{Q}}(d)$ is $\mathbf{N}$. Assume $\mathbf{N} \notin \mathbf{ExpInsts}^{\mathcal{Q}}(a)$.

Let $\mathcal{Q}'$ be $\mathbf{EExpand}(\mathcal{Q}, a, \mathbf{C})$. By Lemma 7.7.10, $\mathcal{Q}'$ is a closed, merged, separated extensional expansion dag, $\mathcal{Q} \trianglelefteq_r \mathcal{Q}'$,

$$\mathbf{N} \in \mathbf{ExpInsts}^{\mathcal{Q}'}(a),$$

$$\mathbf{ExpTerms}(\mathcal{Q}') = \mathbf{ExpTerms}(\mathcal{Q}) \cup \{\mathbf{C}\} \subseteq \mathcal{W},$$

$$\mathbf{RootSels}(\mathcal{Q}') = \mathbf{RootSels}(\mathcal{Q}) = \emptyset$$

and the embedding relation $\prec_0^{\mathcal{Q}'}$ is acyclic. Hence $\mathcal{Q}' \in \mathfrak{Nice}_{\mathcal{W}}$ and there exists some $\mathbf{D} \in \mathbf{ExpTerms}(\mathcal{Q}') \subseteq \mathcal{W}$ and $d \in \mathcal{N}^{\mathcal{Q}'}$ such that $\langle a, \mathbf{D}, d \rangle$ is in $exps^{\mathcal{Q}'}$. □

LEMMA 7.10.8. *Let $\mathcal{W} \subseteq cwff(\mathcal{S})$ be a set of closed formulas. Suppose $\mathcal{Q} \in \mathfrak{Nice}_{\mathcal{W}}$, $P_{o\alpha^n \cdots \alpha^1}$ is a parameter, $a \in \mathcal{N}^{\mathcal{Q}}$ is a positive atomic node with shallow formula $[P\,\overline{\mathbf{A}^m}]$ and $b \in \mathcal{N}^{\mathcal{Q}}$ is a negative atomic node with shallow formula $[P\,\overline{\mathbf{B}^m}]$. There is some $\mathcal{Q}' \in \mathfrak{Nice}_{\mathcal{W}}$ and mate node $c \in \mathcal{N}^{\mathcal{Q}'}$ such that $\mathcal{Q} \trianglelefteq_r \mathcal{Q}'$ and $\langle c, a, b \rangle \in conns^{\mathcal{Q}'}$.*

PROOF. Applying Lemma 7.10.2 twice using the nodes a and b, there is some $\mathcal{Q}^0 \in \mathfrak{Nice}_{\mathcal{W}}$ such that $\mathcal{Q} \trianglelefteq_r \mathcal{Q}^0$ and both a and b are extended in $\mathcal{Q}^0$.

If $\langle a, b\rangle$ is already **mate**-connected in $\mathcal{Q}^0$, then there is some mate node c in $\mathcal{Q}^0$ such that $\langle c, a, b\rangle \in conns^{\mathcal{Q}^0}$. Assume $\langle a, b\rangle$ is not **mate**-connected in $\mathcal{Q}^0$.

Let $\mathcal{Q}'$ be $\mathbf{Mate}(\mathcal{Q}^0, a, b)$. By Lemma 7.7.14, $\mathcal{Q}'$ is a closed, merged, separated extensional expansion dag such that $\mathcal{Q}^0 \trianglelefteq_r \mathcal{Q}'$, $\langle a, b\rangle$ is **mate**-connected in $\mathcal{Q}'$,

$$\mathbf{ExpTerms}(\mathcal{Q}') = \mathbf{ExpTerms}(\mathcal{Q}^0) \subseteq \mathcal{W},$$

$$\mathbf{RootSels}(\mathcal{Q}') = \mathbf{RootSels}(\mathcal{Q}^0) = \emptyset$$

and $\prec_0^{\mathcal{Q}'}$ is acyclic. Hence $\mathcal{Q}' \in \mathfrak{Nice}_{\mathcal{W}}$ and $\mathcal{Q} \trianglelefteq_r \mathcal{Q}'$. Since $\langle a, b\rangle$ is **mate**-connected in $\mathcal{Q}'$, there is a mate node c such that $\langle c, a, b$ is in$conns^{\mathcal{Q}'}$. □

LEMMA 7.10.9. *Let $\mathcal{W} \subseteq cwff(\mathcal{S})$ be a set of closed formulas. Suppose $\mathcal{Q} \in \mathfrak{Nice}_{\mathcal{W}}$, $a \in \mathcal{N}^{\mathcal{Q}}$ is a positive equation node and $b \in \mathcal{N}^{\mathcal{Q}}$ is an equation goal node and $k \in \{\mathbf{eunif}, \mathbf{eunif}^{sym}\}$. There is some $\mathcal{Q}' \in \mathfrak{Nice}_{\mathcal{W}}$ and node $c \in \mathcal{N}^{\mathcal{Q}'}$ such that $\mathcal{Q} \trianglelefteq_r \mathcal{Q}'$, $\langle c, a, b\rangle \in conns^{\mathcal{Q}'}$ and $kind^{\mathcal{Q}'}(c) = k$.*

PROOF. The proof is analogous to the proof of Lemma 7.10.8, except we must distinguish between E-unification nodes and symmetric E-unification nodes. □

LEMMA 7.10.10. *Let $\mathcal{W} \subseteq cwff(\mathcal{S})$ be a set of closed formulas. Suppose $\mathcal{Q} \in \mathfrak{Nice}_{\mathcal{W}}$ and $e \in \mathcal{N}^{\mathcal{Q}}$ is a decomposable node. There is some $\mathcal{Q}' \in \mathfrak{Nice}_{\mathcal{W}}$ and decomposition node $d \in \mathcal{N}^{\mathcal{Q}'}$ such that $\mathcal{Q} \trianglelefteq_r \mathcal{Q}'$ and $\langle e, d\rangle \in decs^{\mathcal{Q}'}$.*

PROOF. This proof is similar to the proof of Lemma 7.10.8 except we create a new decomposition node d if necessary and use Definition 7.7.16 and Lemma 7.7.18 to add the new decomposition arc $\langle e, d\rangle$. □

We now perform an induction argument on derivations to grow a given extensional expansion dag to an extensional expansion dag with a sufficient set.

LEMMA 7.10.11. *Let $\mathcal{W} \subseteq cwff(\mathcal{S})$ be a set of closed formulas which is closed under parameter substitutions (cf. Definition 2.1.10).*

Suppose $\mathcal{D}$ is a $\mathcal{W}$-restricted derivation of $\mathbf{M}^1, \ldots, \mathbf{M}^n$. *For any parameter substitution θ, $\mathcal{Q} \in \mathfrak{Nice}_{\mathcal{W}}$, and primary nodes $a^1, \ldots, a^n$ in $\mathcal{N}^{\mathcal{Q}}$, where $\overline{sh}^{\mathcal{Q}}(a^i) = \theta((\mathbf{M}^i)^{\downarrow})$ for each i $(1 \leq i \leq n)$, there exists some $\mathcal{Q}' \in \mathfrak{Nice}_{\mathcal{W}}$ such that $\mathcal{Q} \trianglelefteq_r \mathcal{Q}'$ and $\{a^i \mid 1 \leq i \leq n\}$ is a sufficient set for $\mathcal{Q}'$.*

PROOF. We can prove this by induction on $\mathcal{D}$ using a case analysis on the last rule application of $\mathcal{D}$. Some cases are checked here and the remaining cases are checked in Appendix A.5.

We start by assuming the last rule application of $\mathcal{D}$ is not $\mathcal{G}(Init^{\dot{=}})$, not $\mathcal{G}(EUnif_1)$ and not $\mathcal{G}(EUnif_2)$. Consequently, the last rule application has a single principal formula. Let r be such that $1 \leq r \leq n$ and $\mathbf{M}^r$ is the principal formula of this rule. Define $\mathbf{M} := \mathbf{M}^r$ and $a := a^r$. For each $i \in \{1, \ldots, r-1\}$, define $\mathbf{N}^i := \mathbf{M}^i$ and $b^i := a^i$. For each $i \in \{r, \ldots, n-1\}$, define $\mathbf{N}^i := \mathbf{M}^{i+1}$ and $b^i := a^{i+1}$. Let Γ be $\mathbf{N}^1, \ldots, \mathbf{N}^{n-1}$. Then $\mathcal{D}$ is a derivation of $\Gamma, \mathbf{M}$ where $\mathbf{M}$ is the principal formula of the last rule application of $\mathcal{D}$. Also, notice b^i is primary for each i $(1 \leq i \leq n-1)$ and the dual shallow formula of b^i is $\theta((\mathbf{N}^i)^{\downarrow})$ for each i $(1 \leq i \leq n-1)$.

Suppose $\mathcal{D}$ is of the form

$$\cfrac{\begin{matrix}\mathcal{D}_1 \\ \Gamma, \mathbf{M}, \mathbf{M}\end{matrix}}{\Gamma, \mathbf{M}}\ \mathcal{G}(Contr)$$

Let $\mathbf{N}^n$ and $\mathbf{N}^{n+1}$ be $\mathbf{M}$ and b^n and b^{n+1} be a. We apply the inductive hypothesis to $\mathcal{D}_1$ with θ, $\mathcal{Q}$, $\mathbf{N}^1, \ldots, \mathbf{N}^n, \mathbf{N}^{n+1}$ and $b^1, \ldots, b^n, b^{n+1}$. By the inductive hypothesis, there is some $\mathcal{Q}' \in \mathfrak{Nice}_{\mathcal{W}}$ such that $\mathcal{Q} \trianglelefteq_r \mathcal{Q}'$ and $\{b^i \mid 1 \leq i \leq n+1\}$ is a sufficient set for $\mathcal{Q}'$. Since $\{b^i \mid 1 \leq i \leq n+1\}$ is the same is as $\{a^i \mid 1 \leq i \leq n\}$, we are done.

Suppose $\mathcal{D}$ is of the form

$$\cfrac{\begin{matrix}\mathcal{D}_1 \\ \Gamma, \mathbf{M}^{\downarrow}\end{matrix}}{\Gamma, \mathbf{M}}\ \mathcal{G}(\beta\eta)$$

Let $\mathbf{N}^n$ be $\mathbf{M}^{\downarrow}$ and b^n be a^n. We apply the inductive hypothesis to $\mathcal{D}_1$ with θ, $\mathcal{Q}$, $\mathbf{N}^1, \ldots, \mathbf{N}^n$ and $b^1, \ldots, b^n$. By the inductive hypothesis, there is some $\mathcal{Q}' \in \mathfrak{Nice}_{\mathcal{W}}$ such that $\mathcal{Q} \trianglelefteq_r \mathcal{Q}'$ and $\{b^i \mid 1 \leq i \leq n\}$

is a sufficient set for $\mathcal{Q}'$. Since $\{b^i \mid 1 \leq i \leq n\}$ is the same is as $\{a^i \mid 1 \leq i \leq n\}$, we are done.

Suppose $\mathcal{D}$ is of the form

$$\frac{}{\Gamma, \dot{\top}} \mathcal{G}(\dot{\top})$$

where $\mathbf{M}$ is $\dot{\top}$. By assumption $\overline{sh}^{\mathcal{Q}}(a)$ is $\dot{\top}$. By Lemma 7.2.9:1, we know a is a false node. By Lemma 7.10.2, there is some $\mathcal{Q}' \in \mathfrak{Nice}_{\mathcal{W}}$ such that $\mathcal{Q} \trianglelefteq_r \mathcal{Q}'$ and a is extended in $\mathcal{Q}'$. Hence a is a disjunctive node in $\mathcal{Q}'$ with no children (by $\clubsuit_{ns}$ and Lemma 7.3.13) and so $\{a\}$ is sufficient for $\mathcal{Q}'$. By Lemma 7.4.10, we conclude $\{a^i \mid 1 \leq i \leq n\}$ is sufficient for $\mathcal{Q}'$, as desired.

Suppose $\mathcal{D}$ is of the form

$$\frac{\begin{array}{c}\mathcal{D}_1 \\ \Gamma, \mathbf{P}, \mathbf{Q}\end{array}}{\Gamma, [\mathbf{P} \dot{\vee} \mathbf{Q}]} \mathcal{G}(\dot{\vee})$$

where $\mathbf{M}$ is $[\mathbf{P} \dot{\vee} \mathbf{Q}]$. Let $\mathbf{N}^n$ be $\mathbf{P}$ and $\mathbf{N}^{n+1}$ be $\mathbf{Q}$. The dual shallow formula of a is $\theta([\mathbf{P} \dot{\vee} \mathbf{Q}]^{\downarrow})$. By Lemma 7.2.9:2, we know a is a conjunction node. By Lemma 7.10.2, there is some $\mathcal{Q}^0 \in \mathfrak{Nice}_{\mathcal{W}}$ such that $\mathcal{Q} \trianglelefteq_r \mathcal{Q}^0$ and a is extended in $\mathcal{Q}^0$. By $\clubsuit_{\vee}$, there exist negative primary nodes b^n and b^{n+1} such that $succ^{\mathcal{Q}^0}(a) = \langle b^n, b^{n+1}\rangle$, $sh^{\mathcal{Q}^0}(b^n)$ is $\theta(\mathbf{P}^{\downarrow})$ and $sh^{\mathcal{Q}^0}(b^{n+1})$ is $\theta(\mathbf{Q}^{\downarrow})$. The dual shallow formulas of b^n and b^{n+1} are $\theta(\mathbf{P}^{\downarrow})$ and $\theta(\mathbf{Q}^{\downarrow})$, respectively. We apply the inductive hypothesis to $\mathcal{D}_1$ with θ, $\mathcal{Q}^0$, $\mathbf{N}^1, \ldots, \mathbf{N}^n, \mathbf{N}^{n+1}$ and $b^1, \ldots, b^n, b^{n+1}$. By the inductive hypothesis, there is some $\mathcal{Q}' \in \mathfrak{Nice}_{\mathcal{W}}$ such that $\mathcal{Q}^0 \trianglelefteq_r \mathcal{Q}'$ and $\{b^i \mid 1 \leq i \leq n+1\}$ is a sufficient set for $\mathcal{Q}'$. Note that $\mathcal{Q} \trianglelefteq_r \mathcal{Q}^0 \trianglelefteq_r \mathcal{Q}'$. Since $\mathcal{Q}^0 \trianglelefteq_r \mathcal{Q}'$, we know $a \in \mathcal{N}^{\mathcal{Q}'}$ is a conjunction node of $\mathcal{Q}'$ and $succ^{\mathcal{Q}'}(a) = \langle b^n, b^{n+1}\rangle$. By Lemma 7.3.13, b^n and b^{n+1} are the only children of a in $\mathcal{Q}'$. By Lemma 7.4.13, we can conclude $\{b^1, \ldots, b^{n-1}\} \cup \{a\}$ is sufficient for $\mathcal{Q}'$. That is, $\{a^1, \ldots, a^n\}$ is sufficient for $\mathcal{Q}'$.

Suppose $\mathcal{D}$ is of the form

$$\frac{\begin{array}{c}\mathcal{D}_1 \\ \Gamma, [[W/x]\mathbf{P}]\end{array}}{\Gamma, [\dot{\forall} x_\alpha \mathbf{P}]} \mathcal{G}(\dot{\forall}^W)$$

where $\mathbf{M}$ is $[\dot{\forall} x_\alpha \mathbf{P}]$ and W_α is a parameter which does not occur in any proposition in Γ and does not occur in $\mathbf{P}$. Consequently, we know W does not occur in $\mathbf{N}^i$ for each i $(1 \leq i \leq n-1)$. Let $\mathbf{N}^n$ be $[W/x]\mathbf{P}$. Since the dual shallow formula of a is $[\dot{\forall} x_\alpha\, \theta(\mathbf{P}^\downarrow)]$, we know a is a selection node by Lemma 7.2.9:3. By Lemma 7.10.2, there is some $\mathcal{Q}^0 \in \mathfrak{Nice}_{\mathcal{W}}$ such that $\mathcal{Q} \trianglelefteq_r \mathcal{Q}^0$ and a is extended in $\mathcal{Q}^0$. By $\clubsuit_{sel}$ (in $\mathcal{Q}^0$), $a \in \mathbf{Dom}(sel^{\mathcal{Q}^0})$ and there exists a negative primary node $b^n \in \mathcal{N}^{\mathcal{Q}^0}$ with shallow formula $[A/x]\theta(\mathbf{P}^\downarrow)$ such that $succ^{\mathcal{Q}^0}(a) = \langle b^n \rangle$ where $A = sel^{\mathcal{Q}^0}(a)$. Let ψ be the parameter substitution $\theta, [A/W]$. Since W does not occur in $\mathbf{P}$, we can compute

$$[A/x]\theta(\mathbf{P}^\downarrow) = \psi(([W/x]\mathbf{P})^\downarrow).$$

Hence the dual shallow formula of the negative node b^n can be written as $\psi(([W/x]\mathbf{P})^\downarrow)$. Since W does not occur in $(\mathbf{N}^i)^\downarrow$, we know $sh^{\mathcal{Q}}(b^i)$ is $\psi((\mathbf{N}^i)^\downarrow)$ for each i $(1 \leq i \leq n-1)$. By the inductive hypothesis applied to $\mathcal{D}_1$ with ψ, $\mathcal{Q}^0$, $\mathbf{N}^1, \ldots, \mathbf{N}^n$ and $b^1, \ldots, b^n$, there is some $\mathcal{Q}' \in \mathfrak{Nice}_{\mathcal{W}}$ such that $\mathcal{Q}^0 \trianglelefteq_r \mathcal{Q}'$ and $\{b^i \mid 1 \leq i \leq n\}$ is a sufficient set for $\mathcal{Q}'$. By Lemma 7.4.15, $\{a^i \mid 1 \leq i \leq n\}$ is a sufficient set for $\mathcal{Q}'$. Also, $\mathcal{Q} \trianglelefteq_r \mathcal{Q}^0 \trianglelefteq_r \mathcal{Q}'$.

In Appendix A.5 we check the case where $\mathcal{D}$ has the form

$$\cfrac{\begin{array}{c}\mathcal{D}_1 \\ \Gamma, [[\mathbf{G}\, W] \doteq^\alpha [\mathbf{H}\, W]]\end{array}}{\Gamma, [\mathbf{G} \doteq^{\alpha\beta} \mathbf{H}]}\ \mathcal{G}(\mathfrak{f}^W)$$

$\mathbf{M}$ is $[\mathbf{G} \doteq^{\alpha\beta} \mathbf{H}]$ and W_β is a parameter such that W does not occur in any proposition in Γ and does not occur in $\mathbf{M}$.

The case where $\mathcal{D}$ is of the form

$$\cfrac{\begin{array}{c}\mathcal{D}_1 \\ \Gamma, \mathbf{P}\end{array}}{\Gamma, \dot{\neg}\dot{\neg}\mathbf{P}}\ \mathcal{G}(\dot{\neg}\dot{\neg})$$

and $\mathbf{M}$ is $\dot{\neg}\dot{\neg}\mathbf{P}$ is argued in Appendix A.5.

Suppose $\mathcal{D}$ is of the form

$$\cfrac{\begin{array}{cc}\mathcal{D}_1 & \mathcal{D}_2 \\ \Gamma, \dot{\neg}\mathbf{P} & \Gamma, \dot{\neg}\mathbf{Q}\end{array}}{\Gamma, \dot{\neg}[\mathbf{P} \dot{\vee} \mathbf{Q}]}\ \mathcal{G}(\dot{\neg}\dot{\vee})$$

where $\mathbf{M}$ is $\neg[\mathbf{P} \vee \mathbf{Q}]$. Hence the dual shallow formula of a must be $\neg\theta([\mathbf{P} \vee \mathbf{Q}]^{\downarrow})$. By Lemma 7.10.4, there is some $\mathcal{Q}^0 \in \mathfrak{Nice}_{\mathcal{W}}$ and a positive primary node $c \in \mathcal{N}^{\mathcal{Q}^0}$ such that $\mathcal{Q} \trianglelefteq_r \mathcal{Q}^0$, c is a positive counterpart of a with respect to $\mathcal{Q}^0$ and c has shallow formula $\theta([\mathbf{P} \vee \mathbf{Q}]^{\downarrow})$. By Definition 7.2.7, the node c is a disjunction node. By Lemma 7.10.2, there is some $\mathcal{Q}^1 \in \mathfrak{Nice}_{\mathcal{W}}$ such that $\mathcal{Q}^0 \trianglelefteq_r \mathcal{Q}^1$ and c is extended in $\mathcal{Q}^1$. By $\clubsuit_{\vee}$ (in $\mathcal{Q}^1$), there exist positive primary nodes d and e such that $succ^{\mathcal{Q}^1}(c) = \langle d, e\rangle$, $sh^{\mathcal{Q}^1}(d)$ is $\theta(\mathbf{P}^{\downarrow})$ and $sh^{\mathcal{Q}^1}(e)$ is $\theta(\mathbf{Q}^{\downarrow})$. Consequently, the dual shallow formulas of d and e are $\theta([\neg\mathbf{P}]^{\downarrow})$ and $\theta([\neg\mathbf{Q}]^{\downarrow})$, respectively. Applying the inductive hypothesis to $\mathcal{D}_1$ with θ, $\mathcal{Q}^1$, $\mathbf{N}^1, \ldots \mathbf{N}^{n-1}, \neg\mathbf{P}$ and $b^1, \ldots, b^{n-1}, d$, we obtain some $\mathcal{Q}^2 \in \mathfrak{Nice}_{\mathcal{W}}$ such that $\mathcal{Q}^1 \trianglelefteq_r \mathcal{Q}^2$ and $\{b^1, \ldots, b^{n-1}, d\}$ is sufficient for $\mathcal{Q}^2$. Applying the inductive hypothesis to $\mathcal{D}_1$ with θ, $\mathcal{Q}^2$, $\mathbf{N}^1, \ldots \mathbf{N}^{n-1}, \neg\mathbf{Q}$ and $b^1, \ldots, b^{n-1}, e$, we obtain some $\mathcal{Q}' \in \mathfrak{Nice}_{\mathcal{W}}$ such that $\mathcal{Q}^2 \trianglelefteq_r \mathcal{Q}'$ and $\{b^1, \ldots, b^{n-1}, e\}$ is sufficient for $\mathcal{Q}'$. Hence

$$\mathcal{Q} \trianglelefteq_r \mathcal{Q}^0 \trianglelefteq_r \mathcal{Q}^1 \trianglelefteq_r \mathcal{Q}^2 \trianglelefteq_r \mathcal{Q}'.$$

By Lemma 7.6.7, $\{b^1, \ldots, b^{n-1}, d\}$ is sufficient for $\mathcal{Q}'$. Since

$$\mathcal{Q} \trianglelefteq_r \mathcal{Q}^0 \trianglelefteq_r \mathcal{Q}',$$

c is a disjunction node of $\mathcal{Q}'$ and $succ^{\mathcal{Q}'}(c) = \langle d, e\rangle$. By Lemma 7.4.3, d and e are the only children of c in $\mathcal{Q}'$. By Lemma 7.4.14, $\{b^1, \ldots, b^{n-1}, c\}$ is sufficient for $\mathcal{Q}'$. By Lemma 7.10.5,

$$\{b^1, \ldots, b^{n-1}, a\}, \text{ i.e., } \{a^1, \ldots, a^n\},$$

is sufficient for $\mathcal{Q}'$, as desired.

Suppose $\mathcal{D}$ is of the form

$$\cfrac{\begin{array}{c}\mathcal{D}_1\\ \Gamma, \mathbf{M}^{\sharp}\end{array}}{\Gamma, \mathbf{M}}\ \mathcal{G}(\sharp)$$

where $\mathbf{M} \in cwff_o$. If $\mathbf{M}$ is not of the form $[c\,\overline{\mathbf{A}^m}]$ for some $c \in \mathcal{S}$, then $\mathbf{M}^{\sharp}$ is the same as $\mathbf{M}$ and we can apply the inductive hypothesis to $\mathcal{D}_1$ with θ, $\mathcal{Q}$, $\mathbf{M}^1, \ldots, \mathbf{M}^n$ and $a^1, \ldots, a^n$ to obtain the desired $\mathcal{Q}'$. Assume $\mathbf{M}$ is of the form $[c\,\overline{\mathbf{A}^m}]$ for some $c \in \mathcal{S}$. Since the dual shallow formula of a is $\theta([c\,\overline{\mathbf{A}^m}]^{\downarrow})$, a must be negative. Since $c \notin \mathbf{Dom}(\theta)$ and θ is a parameter substitution, we know $\theta([c\,\overline{\mathbf{A}^m}]^{\downarrow})$ is the same as $[c\,\theta(\mathbf{A}^1)\cdots\theta(\mathbf{A}^m)]^{\downarrow}$. Hence a must be a deepening node by Definition 7.2.7. By Lemma 7.10.2, there is some $\mathcal{Q}^0 \in \mathfrak{Nice}_{\mathcal{W}}$ such that $\mathcal{Q} \trianglelefteq_r \mathcal{Q}^0$ and a is extended in $\mathcal{Q}^0$. By $\clubsuit_{\sharp}$, there is a negative primary node $b^n \in \mathcal{N}^{\mathcal{Q}^0}$ with shallow formula $((\theta(\mathbf{M}^{\downarrow}))^{\sharp})^{\downarrow}$ such that $succ^{\mathcal{Q}^0}(a) = \langle b^n \rangle$. Since θ is a parameter substitution, $\theta(\mathbf{M}^{\downarrow})$ is $(\theta(\mathbf{M}))^{\downarrow}$. By Lemma 2.1.40,

$$(((\theta(\mathbf{M}))^{\downarrow})^{\sharp})^{\downarrow} = ((\theta(\mathbf{M}))^{\sharp})^{\downarrow}.$$

Since $c \notin \mathbf{Dom}(\theta)$, $(\theta(\mathbf{M}))^{\sharp}$ is the same as $\theta(\mathbf{M}^{\sharp})$. Consequently, b^n has dual shallow formula

$$((\theta(\mathbf{M}^{\downarrow}))^{\sharp})^{\downarrow} = (((\theta(\mathbf{M}))^{\downarrow})^{\sharp})^{\downarrow} = ((\theta(\mathbf{M}))^{\sharp})^{\downarrow} = (\theta(\mathbf{M}^{\sharp}))^{\downarrow} = \theta((\mathbf{M}^{\sharp})^{\downarrow}).$$

We apply the inductive hypothesis to $\mathcal{D}_1$ with θ, $\mathcal{Q}^0$,

$$\mathbf{N}^1, \ldots, \mathbf{N}^{n-1}, \mathbf{M}^{\sharp} \text{ and } b^1, \ldots, b^n$$

to obtain some $\mathcal{Q}' \in \mathfrak{Nice}_{\mathcal{W}}$ such that $\mathcal{Q}^0 \trianglelefteq_r \mathcal{Q}'$ and

$$\{b^i \mid 1 \leq i \leq n\}$$

is a sufficient set for $\mathcal{Q}'$. Hence

$$\mathcal{Q} \trianglelefteq_r \mathcal{Q}^0 \trianglelefteq_r \mathcal{Q}'$$

and $\{a^i \mid 1 \leq i \leq n\}$ is a sufficient set for $\mathcal{Q}'$ by Lemma 7.4.15.

The case where $\mathcal{D}$ is of the form

$$\dfrac{\begin{array}{c}\mathcal{D}_1\\ \Gamma, \neg\mathbf{P}^{\sharp}\end{array}}{\Gamma, \neg\mathbf{P}}\,\mathcal{G}(\neg\sharp)$$

$\mathbf{M}$ is $\neg\mathbf{P}$ and $\mathbf{P} \in cwff_o$ is argued in Appendix A.5.

Suppose $\mathcal{D}$ is of the form

$$\dfrac{\begin{array}{c}\mathcal{D}_1\\ \Gamma, \neg[[\mathbf{C}/x]\mathbf{P}]\end{array} \qquad \mathbf{C} \in \mathcal{W}_{\alpha}}{\Gamma, \neg[\forall x_{\alpha}\mathbf{P}]}\,\mathcal{G}(\neg\forall_{re}, \mathcal{W})$$

where $\mathbf{M}$ is $\neg[\forall x_{\alpha}\mathbf{P}]$. Hence a has dual shallow formula $\neg\theta([\forall x_{\alpha}\,\mathbf{P}]^{\downarrow})$. By Lemma 7.10.4, there is some $\mathcal{Q}^0 \in \mathfrak{Nice}_{\mathcal{W}}$ and a positive primary node $e \in \mathcal{N}^{\mathcal{Q}^0}$ such that $\mathcal{Q} \trianglelefteq_r \mathcal{Q}^0$, e is a positive counterpart of a with respect to $\mathcal{Q}^0$ and e has shallow formula $\theta([\forall x_{\alpha}\,\mathbf{P}]^{\downarrow})$. We know e is an expansion node by Definition 7.2.7. We have $\theta(\mathbf{C}) \in \mathcal{W}$ since $\mathcal{W}$

is closed under parameter substitutions. Hence we can apply Lemma 7.10.7 to obtain some $\mathcal{Q}^1 \in \mathfrak{Nice}_{\mathcal{W}}$, term $\mathbf{D} \in \mathcal{W}$ and $d \in \mathcal{N}^{\mathcal{Q}^0}$ such that $\mathcal{Q}^0 \trianglelefteq_r \mathcal{Q}^1$, $\langle e, \mathbf{D}, d\rangle \in exps^{\mathcal{Q}^1}$ and $sh^{\mathcal{Q}^1}(d)$ is $[[\theta(\mathbf{C})/x]\theta(\mathbf{P})]^{\downarrow})$, i.e., $\theta(([\mathbf{C}/x]\mathbf{P})^{\downarrow})$. Let $\mathbf{N}^n$ be $[\dot{\neg}\, [\mathbf{C}/x]\mathbf{P}]$. By $\clubsuit_{exp}$, d is a positive primary node. Thus the dual shallow formula of d in $\mathcal{Q}^1$ is $\theta((\mathbf{N}^n)^{\downarrow})$. Applying the inductive hypothesis to $\mathcal{D}_1$ with θ, $\mathcal{Q}^1$, $\mathbf{N}^1, \ldots, \mathbf{N}^{n-1}, \mathbf{N}^n$ and $b^1, \ldots, b^{n-1}, d$, there is some $\mathcal{Q}' \in \mathfrak{Nice}_{\mathcal{W}}$ such that $\mathcal{Q}^1 \trianglelefteq_r \mathcal{Q}'$ and $\{b^1, \ldots, b^{n-1}, d\}$ is sufficient for $\mathcal{Q}'$. Hence

$$\mathcal{Q} \trianglelefteq_r \mathcal{Q}^0 \trianglelefteq_r \mathcal{Q}^1 \trianglelefteq_r \mathcal{Q}'.$$

Let

$$\mathcal{K} := \{z \in \mathcal{N}^{\mathcal{Q}'} \mid e \overset{\mathcal{Q}'}{\rightarrow} z\}.$$

Since $e \overset{\mathcal{Q}^1}{\rightarrow} d$, we know $e \overset{\mathcal{Q}'}{\rightarrow} d$ and hence $d \in \mathcal{K}$ by Lemma 7.6.2. By Lemma 7.4.10, $\{b^1, \ldots, b^{n-1}\} \cup \mathcal{K}$ is sufficient for $\mathcal{Q}'$. Since e is conjunctive, $\{b^1, \ldots, b^{n-1}, e\}$ is sufficient for $\mathcal{Q}'$ by Lemma 7.4.13. By Lemma 7.10.5, $\{a^i \mid 1 \leq i \leq n\}$ is a sufficient set for $\mathcal{Q}'$.

The case where $\mathcal{D}$ is of the form

$$\frac{\Gamma, \dot{\neg}[[\mathbf{G}\,\mathbf{B}] \mathrel{\dot{=}^{\alpha}} [\mathbf{H}\,\mathbf{B}]] \quad \mathbf{B} \in \mathcal{W}_\beta}{\Gamma, \dot{\neg}[\mathbf{G} \mathrel{\dot{=}^{\alpha\beta}} \mathbf{H}]} \mathcal{G}(\dot{\neg} \overrightarrow{\dot{=}}_{re}, \mathcal{W})$$

and $\mathbf{M}$ is $\dot{\neg}\theta([\mathbf{G} \mathrel{\dot{=}^{\alpha\beta}} \mathbf{H}]^{\downarrow})$ is argued in Appendix A.5.

Suppose $\mathcal{D}$ is of the form

$$\frac{\begin{array}{cc} \mathcal{D}_1 & \mathcal{D}_2 \\ \Gamma, \dot{\neg}\mathbf{A}, \mathbf{B} & \Gamma, \dot{\neg}\mathbf{B}, \mathbf{A} \end{array}}{\Gamma, [\mathbf{A} \mathrel{\dot{=}^{o}} \mathbf{B}]} \mathcal{G}(\mathfrak{b})$$

where $\mathbf{M}$ is $[\mathbf{A} \mathrel{\dot{=}^{o}} \mathbf{B}]$ and $\mathbf{A}, \mathbf{B} \in cwff_o$. Hence the dual shallow formula of a is $[\theta(\mathbf{A}^{\downarrow}) \mathrel{\dot{=}^{o}} \theta(\mathbf{B}^{\downarrow})]$. By Lemma 7.2.9:6, we know a is a negative Boolean node. By Lemma 7.10.2, there is some $\mathcal{Q}^0 \in \mathfrak{Nice}_{\mathcal{W}}$ such that $\mathcal{Q} \trianglelefteq_r \mathcal{Q}^0$ and a is extended in $\mathcal{Q}^0$. By $\clubsuit_{\dot{=}}^{-o}$, there are negative primary nodes c and d such that $succ^{\mathcal{Q}^0}(a) = \langle c, d\rangle$, $sh^{\mathcal{Q}}(c)$ is $[[\dot{\neg}\theta(\mathbf{A}^{\downarrow})] \mathrel{\dot{\vee}} \theta(\mathbf{B}^{\downarrow})]$ and $sh^{\mathcal{Q}}(d)$ is $[[\dot{\neg}\theta(\mathbf{B}^{\downarrow})] \mathrel{\dot{\vee}} \theta(\mathbf{A}^{\downarrow})]$. By Definition 7.2.7, we know c and d are conjunction nodes. Applying Lemma 7.10.2 twice to c and d, there is some $\mathcal{Q}^1 \in \mathfrak{Nice}_{\mathcal{W}}$ such that $\mathcal{Q}^0 \trianglelefteq_r \mathcal{Q}^1$ and the nodes c and d are extended in $\mathcal{Q}^1$. By $\clubsuit_{\vee}$, there exist negative primary nodes c^1, c^2, d^1 and d^2 such that $succ^{\mathcal{Q}^1}(c) = \langle c^1, c^2\rangle$,

$succ^{\mathcal{Q}^1}(d) = \langle d^1, d^2 \rangle$,

$$sh^{\mathcal{Q}^1}(c^1) = [\neg\theta(\mathbf{A}^{\downarrow})],\ sh^{\mathcal{Q}^1}(c^2) = \theta(\mathbf{B}^{\downarrow}),$$

$$sh^{\mathcal{Q}^1}(d^1) = [\neg\theta(\mathbf{B}^{\downarrow})] \text{ and } sh^{\mathcal{Q}^1}(d^2) = \theta(\mathbf{A}^{\downarrow}).$$

We first apply the inductive hypothesis to $\mathcal{D}_1$ with θ, $\mathcal{Q}^1$,

$$\mathbf{N}^1, \ldots, \mathbf{N}^{n-1}, \neg\mathbf{A}, \mathbf{B}$$

and $b^1, \ldots, b^{n-1}, c^1, c^2$ to obtain some $\mathcal{Q}^2 \in \mathfrak{Nice}_{\mathcal{W}}$ such that $\mathcal{Q}^1 \trianglelefteq_r \mathcal{Q}^2$ and $\{b^1, \ldots, b^{n-1}, c^1, c^2\}$ is sufficient for $\mathcal{Q}^2$. Since $\mathcal{Q}^1 \trianglelefteq_r \mathcal{Q}^2$, we know c and d are conjunction nodes of $\mathcal{Q}^2$,

$$succ^{\mathcal{Q}^2}(c) = \langle c^1, c^2 \rangle,\ succ^{\mathcal{Q}^2}(d) = \langle d^1, d^2 \rangle,$$

$$\overline{sh}^{\mathcal{Q}^2}(d^1) = \theta([\neg\mathbf{B}]^{\downarrow}) \text{ and } \overline{sh}^{\mathcal{Q}^2}(d^2) = \theta(\mathbf{A}^{\downarrow}).$$

Note that

$$\mathcal{Q} \trianglelefteq_r \mathcal{Q}^0 \trianglelefteq_r \mathcal{Q}^1 \trianglelefteq_r \mathcal{Q}^2.$$

Since $\mathcal{Q} \trianglelefteq_r \mathcal{Q}^2$ we know $b^i \in \mathcal{N}^{\mathcal{Q}^2}$ and $\overline{sh}^{\mathcal{Q}^2}(b^i)$ is $\theta((\mathbf{N}^i)^{\downarrow})$ for each i $(1 \leq i \leq n-1)$. By Lemma 7.3.13, c^1 and c^2 are the only children of c in $\mathcal{Q}^2$. By Lemma 7.4.13, we can conclude $\{b^1, \ldots, b^{n-1}, c\}$ is sufficient for $\mathcal{Q}^2$. We apply the inductive hypothesis to $\mathcal{D}_2$ with θ, $\mathcal{Q}^2$, $\mathbf{N}^1, \ldots, \mathbf{N}^{n-1}, \neg\mathbf{B}, \mathbf{A}$ and $b^1, \ldots, b^{n-1}, d^1, d^2$ to obtain some $\mathcal{Q}' \in \mathfrak{Nice}_{\mathcal{W}}$ such that $\mathcal{Q}^2 \trianglelefteq_r \mathcal{Q}'$ and $\{b^1, \ldots, b^{n-1}, d^1, d^2\}$ is sufficient for $\mathcal{Q}'$. Hence

$$\mathcal{Q} \trianglelefteq_r \mathcal{Q}^0 \trianglelefteq_r \mathcal{Q}^1 \trianglelefteq_r \mathcal{Q}^2 \trianglelefteq_r \mathcal{Q}'.$$

Since $\mathcal{Q}^2 \trianglelefteq_r \mathcal{Q}'$, we know d is a conjunction node of $\mathcal{Q}'$ with $succ^{\mathcal{Q}'}(d) = \langle d^1, d^2 \rangle$. By Lemma 7.3.13, d^1 and d^2 are the only children of d in $\mathcal{Q}'$. Thus $\{b^1, \ldots, b^n, d\}$ is sufficient for $\mathcal{Q}'$. Since $\mathcal{Q}^2 \trianglelefteq_r \mathcal{Q}'$, we know $\{b^1, \ldots, b^{n-1}, c\}$ is sufficient for $\mathcal{Q}'$ by Lemma 7.6.7. Also, $succ^{\mathcal{Q}'}(a) = \langle c, d \rangle$ since $\mathcal{Q}^0 \trianglelefteq_r \mathcal{Q}'$. By Lemma 7.3.13, c and d are the only children of a in $\mathcal{Q}'$. By Lemma 7.4.14, $\{a^1, \ldots, a^n\}$ is sufficient for $\mathcal{Q}'$, as desired.

The case where $\mathcal{D}$ has the form

$$\frac{\begin{array}{cc} \mathcal{D}_1 & \mathcal{D}_2 \\ \Gamma, \mathbf{A}, \mathbf{B} & \Gamma, \neg\mathbf{A}, \neg\mathbf{B} \end{array}}{\Gamma, \neg[\mathbf{A} \doteq^o \mathbf{B}]}\ \mathcal{G}(\neg \doteq^o)$$

$\mathbf{M}$ is $\neg[\mathbf{A} \doteq^o \mathbf{B}]$ and $\mathbf{A}, \mathbf{B} \in cwff_o$ is argued in Appendix A.5.

Suppose $\mathcal{D}$ is of the form

$$\frac{\begin{array}{ccc}\mathcal{D}_1 & & \mathcal{D}_m \\ \Gamma, [\mathbf{A}^1 \doteq \mathbf{B}^1] & \cdots & \Gamma, \ [\mathbf{A}^m \doteq \mathbf{B}^m]\end{array}}{\Gamma, [[H\,\overline{\mathbf{A}^m}] \doteq^{\iota} [H\,\overline{\mathbf{B}^m}]]}\,\mathcal{G}(Dec)$$

where $\mathbf{M}$ is $[[H\,\overline{\mathbf{A}^m}] \doteq^{\iota} [H\,\overline{\mathbf{B}^m}]]$ and $H_{\iota\alpha^n\cdots\alpha^1}$ is a parameter. Hence a has dual shallow formula

$$[[\theta(H)\,\theta((\mathbf{A}^1)^{\downarrow})\cdots\theta((\mathbf{A}^m)^{\downarrow})] \doteq^{\iota} [\theta(H)\,\theta((\mathbf{B}^1)^{\downarrow})\cdots\theta((\mathbf{B}^m)^{\downarrow})]].$$

By Lemma 7.2.9:4, we know the primary node a must be a negative equation node. Since θ is a parameter substitution, $\theta(H)$ is a parameter and so a is a decomposable node. Applying Lemma 7.10.10 with a, there is some $\mathcal{Q}^* \in \mathfrak{Nice}_{\mathcal{W}}$ and decomposition node d of $\mathcal{Q}^*$ such that $\mathcal{Q} \trianglelefteq_r \mathcal{Q}^*$ and $\langle a, d\rangle \in decs^{\mathcal{Q}^*}$. By $\clubsuit^p_{dec}$, the shallow formula of d in $\mathcal{Q}^*$ is

$$[[\theta(H)\,\theta((\mathbf{A}^1)^{\downarrow})\cdots\theta((\mathbf{A}^m)^{\downarrow})] \doteq^{\iota} [\theta(H)\,\theta((\mathbf{B}^1)^{\downarrow})\cdots\theta((\mathbf{B}^m)^{\downarrow})]].$$

By Lemma 7.10.2, there is some $\mathcal{Q}^0 \in \mathfrak{Nice}_{\mathcal{W}}$ such that $\mathcal{Q}^* \trianglelefteq_r \mathcal{Q}^0$ and d is extended in $\mathcal{Q}^0$. By $\clubsuit_{dec}$, there exist negative primary nodes $d^1, \ldots, d^m$ of $\mathcal{Q}^0$ such that $succ^{\mathcal{Q}^0} = \langle d^1, \ldots, d^m\rangle$ and d^j has shallow formula $\theta([\mathbf{A}^j \doteq \mathbf{B}^j]^{\downarrow})$ for each j $(1 \leq j \leq m)$. Let $\mathbf{P}^j$ be $[\mathbf{A}^j \doteq \mathbf{B}^j]$ for each j $(1 \leq j \leq m)$. Applying the inductive hypothesis m times to the derivations $\mathcal{D}_1, \ldots, \mathcal{D}_m$, we obtain a chain

$$\mathcal{Q}^0 \trianglelefteq_r \cdots \trianglelefteq_r \mathcal{Q}^m$$

such that $\mathcal{Q}^i \in \mathfrak{Nice}_{\mathcal{W}}$ and $\{b^1, \ldots, b^{n-1}, d^i\}$ is a sufficient set for $\mathcal{Q}^i$ for each i $(1 \leq i \leq m)$. Let $\mathcal{Q}'$ be $\mathcal{Q}^m$. Note that

$$\mathcal{Q} \trianglelefteq_r \mathcal{Q}^* \trianglelefteq_r \mathcal{Q}^0 \trianglelefteq_r \cdots \trianglelefteq_r \mathcal{Q}^m = \mathcal{Q}'.$$

By Lemma 7.6.7, $\{b^1, \ldots, b^{n-1}, d^i\}$ is a sufficient set for $\mathcal{Q}'$ for each i such that $1 \leq i \leq m$. By Lemma 7.4.3, $d^1, \ldots, d^m$ are all the children of the decomposition node d in $\mathcal{Q}'$. Thus $\{b^1, \ldots, b^{n-1}, d\}$ is sufficient for $\mathcal{Q}'$ by Lemma 7.4.14. Let $\mathcal{K}$ be the set $\{z \in \mathcal{N}^{\mathcal{Q}'} \mid a \overset{\mathcal{Q}'}{\rightarrow} z\}$. Note that $d \in \mathcal{K}$, so $\{b^1, \ldots, b^{n-1}\} \cup \mathcal{K}$ is sufficient for $\mathcal{Q}'$ by Lemma 7.4.10. Since a is not a leaf node of $\mathcal{Q}'$, it must be extended by Lemma 7.3.6. Hence a is a conjunctive node and so $\{b^1, \ldots, b^{n-1}, a\}$ is sufficient for $\mathcal{Q}'$ by Lemma 7.4.13. Since $\{b^1, \ldots, b^{n-1}, a\}$ is the same as $\{a^1, \ldots, a^n\}$, we are done.

The cases for the rules $\mathcal{G}(Init^{\dot{=}})$, $\mathcal{G}(EUnif_1)$ and $\mathcal{G}(EUnif_2)$ remain. In these cases, there are two principal formulas, exactly one of which is of the form $\dot{\neg}\mathbf{R}$ for some $\mathbf{R}$. Choose $r, s \in \{1, \ldots, n\}$ such that $\mathbf{M}^r$ is the principal formula of the form $\dot{\neg}\mathbf{R}$ and $\mathbf{M}^s$ is the other principal formula. Choose $k^1, \ldots, k^{n-2} \in \{1, \ldots, n\}$ such that

$$1 \leq k^1 < k^2 < \cdots < k^{n-2} \leq n$$

and

$$\{k^1, \ldots, k^{n-2}\} = \{1, \ldots, n\} \setminus \{r, s\}.$$

For each i $(1 \leq i \leq n-2)$, let $b^i := a^{k^i}$ and $\mathbf{N}^i := \mathbf{M}^{k^i}$. Let a' be a^r, b be a^s, $\mathbf{M}$ be $\mathbf{M}^r$ and $\mathbf{Q}$ be $\mathbf{M}^s$. Thus

$$\{a^1, \ldots, a^n\} = \{b^1, \ldots, b^{n-2}, a', b\}$$

and each b^i is a primary node with dual shallow formula $\theta((\mathbf{N}^i)^{\downarrow})$ for each i $(1 \leq i \leq n-2)$.

Suppose $\mathcal{D}$ is of the form

$$\dfrac{\begin{array}{ccc} \mathcal{D}_1 & & \mathcal{D}_m \\ \Gamma, [\mathbf{B}^1 \dot{=} \mathbf{A}^1] & \cdots & \Gamma, [\mathbf{B}^m \dot{=} \mathbf{A}^m] \end{array}}{\Gamma, \dot{\neg}[P\, \overline{\mathbf{A}^n}], [P\, \overline{\mathbf{B}^m}]}\ \mathcal{G}(Init^{\dot{=}})$$

where $P_{o\alpha^m\cdots\alpha^1}$ is a parameter, $\mathbf{A}^i, \mathbf{B}^i \in cwff_{\alpha^i}$ for each i $(1 \leq i \leq m)$, $\mathbf{M}$ is $\dot{\neg}[P\, \overline{\mathbf{A}^m}]$ and $\mathbf{Q}$ is $[P\, \overline{\mathbf{B}^m}]$. Consequently, the dual shallow formulas of a' and b are $\dot{\neg}\theta([P\, \overline{\mathbf{A}^m}]^{\downarrow})$ and $\theta([P\, \overline{\mathbf{B}^m}]^{\downarrow})$, respectively. By Lemma 7.10.4, there is some $\mathcal{Q}^* \in \mathfrak{Nice}_{\mathcal{W}}$ and a positive primary node $a \in \mathcal{N}^{\mathcal{Q}^*}$ such that $\mathcal{Q} \trianglelefteq_r \mathcal{Q}^*$, a is a positive counterpart of a' with respect to $\mathcal{Q}^*$ and a has shallow formula $\theta([P\, \overline{\mathbf{A}^m}]^{\downarrow})$. Since θ is a parameter substitution, $\theta(P)$ is a parameter. By Definition 7.2.7, we know a is a positive atomic node and b is a negative atomic node. By Lemma 7.10.8 (using $\theta(P)$), there is some $\mathcal{Q}^{**} \in \mathfrak{Nice}_{\mathcal{W}}$ and mate node $c \in \mathcal{N}^{\mathcal{Q}^{**}}$ such that $\mathcal{Q}^* \trianglelefteq_r \mathcal{Q}^{**}$ and $\langle c, a, b\rangle \in conns^{\mathcal{Q}^{**}}$. Using $\clubsuit^p_m$, the shallow formula of c must be

$$\theta([[P\, \mathbf{B}^1_{\alpha^i} \cdots \mathbf{B}^m_{\alpha^m}] \dot{=}^o [P\, \mathbf{A}^1_{\alpha^1} \cdots \mathbf{A}^m_{\alpha^m}]]^{\downarrow}).$$

By Lemma 7.10.2, there is some $\mathcal{Q}^0 \in \mathfrak{Nice}_{\mathcal{W}}$ such that $\mathcal{Q}^{**} \trianglelefteq_r \mathcal{Q}^0$ and c is extended in $\mathcal{Q}^0$. By $\clubsuit_m$, there are negative primary nodes $d^1, \ldots, d^m \in \mathcal{N}^{\mathcal{Q}^0}$ such that $succ^{\mathcal{Q}^0}(c) = \langle d^1, \ldots, d^m\rangle$ and for each

$i \in \{1, \ldots, m\}$, $sh(d^i) = \theta([\mathbf{B}^i \doteq^{\alpha^i} \mathbf{A}^i]^\downarrow)$. Applying the inductive hypothesis m times to each derivation $\mathcal{D}^j$ of $\mathbf{N}^1, \ldots, \mathbf{N}^{n-2}, [\mathbf{B}^i \doteq^{\alpha^i} \mathbf{A}^i]$ and using the nodes $b^1, \ldots, b^{n-2}, d^j$, we obtain a chain

$$\mathcal{Q}^0 \trianglelefteq_r \cdots \trianglelefteq_r \mathcal{Q}^m$$

such that $\mathcal{Q}^i \in \mathfrak{Nice}_{\mathcal{W}}$ and $\{b^1, \ldots, b^{n-2}, d^i\}$ is a sufficient set for $\mathcal{Q}^i$ for each i $(1 \leq i \leq m)$. Let $\mathcal{Q}'$ be $\mathcal{Q}^m$. Note that

$$\mathcal{Q} \trianglelefteq_r \mathcal{Q}^* \trianglelefteq_r \mathcal{Q}^{**} \trianglelefteq_r \mathcal{Q}^0 \trianglelefteq_r \cdots \trianglelefteq_r \mathcal{Q}^m = \mathcal{Q}'.$$

By Lemma 7.6.7, $\{b^1, \ldots, b^{n-2}, d^i\}$ is a sufficient set for $\mathcal{Q}'$ for each i such that $1 \leq i \leq m$. By Lemma 7.4.3, $d^1, \ldots, d^m$ are all the children of the mate node c in $\mathcal{Q}'$. Thus $\{b^1, \ldots, b^{n-2}, c\}$ is sufficient by Lemma 7.4.14, $\{b^1, \ldots, b^{n-2}, a, b\}$ is sufficient by Lemma 7.4.16, and so $\{b^1, \ldots, b^{n-2}, a', b\}$ is sufficient by Lemma 7.10.5.

Suppose $\mathcal{D}$ is of the form

$$\frac{\begin{array}{cc}\mathcal{D}_1 & \mathcal{D}_2 \\ \Gamma, [\mathbf{A} \doteq^\iota \mathbf{C}] & \Gamma, [\mathbf{B} \doteq^\iota \mathbf{D}]\end{array}}{\Gamma, \neg[\mathbf{A} \doteq^\iota \mathbf{B}], [\mathbf{C} \doteq^\iota \mathbf{D}]}\ \mathcal{G}(EUnif_1)$$

where $\mathbf{A}, \mathbf{B}, \mathbf{C}, \mathbf{D} \in cwff_\iota$, $\mathbf{M}$ is $\neg[\mathbf{A} \doteq^\iota \mathbf{B}]$ and $\mathbf{Q}$ is $[\mathbf{C} \doteq^\iota \mathbf{D}]$. Consequently, the dual shallow formulas of a' and b are $\neg\theta([\mathbf{A} \doteq^\iota \mathbf{B}]^\downarrow)$ and $\theta([\mathbf{C} \doteq^\iota \mathbf{D}]^\downarrow)$, respectively. By Lemma 7.10.4, there is some $\mathcal{Q}^0 \in \mathfrak{Nice}_{\mathcal{W}}$ and a positive primary node $a \in \mathcal{N}^{\mathcal{Q}^0}$ such that $\mathcal{Q} \trianglelefteq_r \mathcal{Q}^0$, a is a positive counterpart of a' with respect to $\mathcal{Q}^0$ and a has shallow formula $\theta([\mathbf{A} \doteq^\iota \mathbf{B}]^\downarrow)$. By Definition 7.2.7, the primary node a is a positive equation node. By Lemma 7.2.9:4, the primary node b is a negative equation node. There is some $\mathcal{Q}^1 \in \mathfrak{Nice}_{\mathcal{W}}$ such that $\mathcal{Q}^0 \trianglelefteq_r \mathcal{Q}^1$ and b is extended in $\mathcal{Q}^1$ by Lemma 7.10.2. By $\clubsuit_{=}^{-}$, there is an equation goal node e such that $succ^{\mathcal{Q}^1}(b)$ is $\langle e \rangle$ and $sh^{\mathcal{Q}^1}(e)$ is the same as $sh^{\mathcal{Q}^1}(b)$. By Lemma 7.10.9, there is some $\mathcal{Q}^2 \in \mathfrak{Nice}_{\mathcal{W}}$ and E-unification node $c \in \mathcal{N}^{\mathcal{Q}^2}$ such that $\mathcal{Q}^1 \trianglelefteq_r \mathcal{Q}^2$ and $\langle c, a, e \rangle \in conns^{\mathcal{Q}^2}$. By $\clubsuit_{eu}^{p}$, the shallow formula of c must be

$$\theta([[\mathbf{A} \doteq^\iota \mathbf{C}] \dot\wedge [\mathbf{B} \doteq^\iota \mathbf{D}]]^\downarrow).$$

By Lemma 7.10.2, there is some $\mathcal{Q}^3 \in \mathfrak{Nice}_{\mathcal{W}}$ such that $\mathcal{Q}^2 \trianglelefteq_r \mathcal{Q}^3$ and c is extended in $\mathcal{Q}^3$. By $\clubsuit_{eu}$, there are negative equation nodes $c^1, c^2 \in \mathcal{N}^{\mathcal{Q}^3}$ such that $succ^{\mathcal{Q}^3}(c) = \langle c^1, c^2 \rangle$, $sh^{\mathcal{Q}^3}(c^1)$ is $\theta([\mathbf{A} \doteq^\iota \mathbf{C}]^\downarrow)$

and $sh^{\mathcal{Q}^3}(c^2)$ is $\theta([\mathbf{B} \doteq^\iota \mathbf{D}]^\downarrow)$. Consequently, we can apply the inductive hypothesis to $\mathcal{D}_1$ with θ, $\mathcal{Q}^3$, $\mathbf{N}^1, \ldots, \mathbf{N}^{n-2}, [\mathbf{A} \doteq^\iota \mathbf{C}]$ and $b^1, \ldots, b^{n-2}, c^1$ to obtain some $\mathcal{Q}^4 \in \mathfrak{Nice}_{\mathcal{W}}$ such that $\mathcal{Q}^3 \trianglelefteq_r \mathcal{Q}^4$ and $\{b^1, \ldots, b^{n-2}, c^1\}$ is sufficient for $\mathcal{Q}^4$. By the inductive hypothesis with $\mathcal{D}_2$, θ, $\mathcal{Q}^4$, $\mathbf{N}^1, \ldots, \mathbf{N}^{n-2}, [\mathbf{B} \doteq^\iota \mathbf{D}]$ and $b^1, \ldots, b^{n-2}, c^2$, we obtain some $\mathcal{Q}' \in \mathfrak{Nice}_{\mathcal{W}}$ such that $\mathcal{Q}^4 \trianglelefteq_r \mathcal{Q}'$ and the set $\{b^1, \ldots, b^{n-2}, c^2\}$ is sufficient for $\mathcal{Q}'$. Note that

$$\mathcal{Q} \trianglelefteq_r \mathcal{Q}^0 \trianglelefteq_r \mathcal{Q}^1 \trianglelefteq_r \mathcal{Q}^2 \trianglelefteq_r \mathcal{Q}^3 \trianglelefteq_r \mathcal{Q}^4 \trianglelefteq_r \mathcal{Q}'.$$

By Lemma 7.6.7, $\{b^1, \ldots, b^{n-2}, c^1\}$ is sufficient for $\mathcal{Q}'$. Since c is a disjunctive node, $\{b^1, \ldots, b^{n-2}, c\}$ is sufficient for $\mathcal{Q}'$ by Lemmas 7.4.14 and 7.4.3. By Lemma 7.4.16, $\{b^1, \ldots, b^{n-2}, a, e\}$ is sufficient for $\mathcal{Q}'$. Since b is a conjunctive node and $b \overset{\mathcal{Q}'}{\rightharpoondown} e$, $\{b^1, \ldots, b^{n-2}, a, b\}$ is sufficient for $\mathcal{Q}'$ by Lemmas 7.4.13 and 7.4.10. Finally, $\{b^1, \ldots, b^{n-2}, a', b\}$ is sufficient by Lemma 7.10.5.

The case where $\mathcal{D}$ has the form

$$\frac{\begin{array}{cc} \mathcal{D}_1 & \mathcal{D}_2 \\ \Gamma, [\mathbf{A} \doteq^\iota \mathbf{D}] & \Gamma, [\mathbf{B} \doteq^\iota \mathbf{C}] \end{array}}{\Gamma, \neg[\mathbf{A} \doteq^\iota \mathbf{B}], [\mathbf{C} \doteq^\iota \mathbf{D}]} \mathcal{G}(EUnif_2)$$

$\mathbf{A}, \mathbf{B}, \mathbf{C}, \mathbf{D} \in cwff_\iota$, $\mathbf{M}$ is $\neg[\mathbf{A} \doteq^\iota \mathbf{B}]$ and $\mathbf{Q}$ is $[\mathbf{C} \doteq^\iota \mathbf{D}]$ is argued in Appendix A.5. □

THEOREM 7.10.12 (Completeness of Extensional Expansion Proofs). *Let $\mathcal{S}$ be a signature of logical constants and $\mathcal{W} \subseteq cwff(\mathcal{S})$ be a set of closed formulas which is closed under parameter substitutions (cf. Definition 2.1.10). Suppose $\mathcal{D}$ is a $\mathcal{W}$-restricted $\mathcal{G}^{\mathcal{S}}_{\beta\mathfrak{fb}}$-derivation of Γ (cf. Definition 2.2.23). There is a merged, separated $\mathcal{W}$-restricted $\mathcal{S}$-extensional expansion proof $\langle \mathcal{Q}, \mathbf{Roots}(\mathcal{Q})\rangle$ such that $\Gamma(\mathbf{Roots}(\mathcal{Q})) = \Gamma^\downarrow$ and every node in $\mathbf{Roots}(\mathcal{Q})$ is a negative primary node.*

PROOF. The sequent $\Gamma^\downarrow$ is of the form $\mathbf{M}^1, \ldots, \mathbf{M}^n$ for some $n \geq 1$ and sentences $\mathbf{M}^1, \ldots, \mathbf{M}^n \in sent(\mathcal{S})^\downarrow$. Let $\mathcal{Q}^0$ be

$$\mathbf{EInit}(\langle \mathbf{primary}, -1, \mathbf{M}^1\rangle, \ldots, \langle \mathbf{primary}, -1, \mathbf{M}^n\rangle).$$

By Definition 7.7.2, the nodes $a^1, \ldots, a^n$ are all negative primary nodes, $\mathcal{Q}^0$ has no selected parameters and no expansion arcs. Hence

$\prec_0^{\mathcal{Q}^0}$ is acyclic (as an empty relation),

$$\mathbf{RootSels}(\mathcal{Q}^0) \subseteq \mathbf{SelPars}(\mathcal{Q}^0) = \emptyset$$

and

$$\mathbf{ExpTerms}(\mathcal{Q}^0) = \emptyset \subseteq \mathcal{W}.$$

Also, $\mathcal{Q}^0$ is closed since each $\mathbf{M}^i$ is closed and $\mathcal{Q}^0$ has no expansion terms. By Lemma 7.7.3, $\mathcal{Q}^0$ is a merged, separated extensional expansion dag such that

$$\mathbf{Roots}(\mathcal{Q}^0) = \mathcal{N}^{\mathcal{Q}^0} = \{a^1, \ldots, a^n\}.$$

Hence $\mathcal{Q}^0 \in \mathfrak{Nice}_{\mathcal{W}}$.

Applying Lemma 7.10.11 with the derivation $\mathcal{D}$, the empty parameter substitution, $\mathcal{Q}^0$ and the primary nodes $a^1, \ldots, a^n$, we obtain some $\mathcal{Q} \in \mathfrak{Nice}_{\mathcal{W}}$ such that $\mathcal{Q}^0 \trianglelefteq_r \mathcal{Q}$ and $\{a^1, \ldots, a^n\}$ is a sufficient set for $\mathcal{Q}$. That is, $\mathcal{Q}$ is a closed, merged, separated extensional expansion dag such that

$$\mathbf{Roots}(\mathcal{Q}) = \{a^1, \ldots, a^n\},$$

$\mathbf{ExpTerms}(\mathcal{Q}) \subseteq \mathcal{W}$, $\prec_0^{\mathcal{Q}}$ is acyclic and $\mathbf{RootSels}(\mathcal{Q}) = \emptyset$. Since $\mathbf{Roots}(\mathcal{Q})$ is a sufficient set for $\mathcal{Q}$, we know $\langle \mathcal{Q}, \mathbf{Roots}(\mathcal{Q})\rangle$ is an extensional expansion proof. Since $\mathbf{Roots}(\mathcal{Q}) = \{a^1, \ldots, a^n\}$, every root node of $\mathcal{Q}$ is a negative primary node and $\Gamma(\mathbf{Roots}(\mathcal{Q})) = \Gamma^{\downarrow}$.

□

COROLLARY 7.10.13 (Completeness of Extensional Expansion Proofs). *Let $\mathcal{S}$ be a signature of logical constants and $\mathcal{D}$ be a $\mathcal{G}^{\mathcal{S}}_{\beta\mathfrak{fb}}$-derivation of Γ. There is a merged, separated $\mathcal{S}$-extensional expansion proof $\langle \mathcal{Q}, \mathbf{Roots}(\mathcal{Q})\rangle$ such that $\Gamma(\mathbf{Roots}(\mathcal{Q})) = \Gamma^{\downarrow}$ and every node in $\mathbf{Roots}(\mathcal{Q})$ is a negative primary node.*

PROOF. This follows directly from Theorem 7.10.12 since every $\mathcal{G}^{\mathcal{S}}_{\beta\mathfrak{fb}}$-derivation $\mathcal{D}$ is a *cwff*($\mathcal{S}$)-restricted derivation. □

REMARK 7.10.14 (Translation Algorithm). The proofs of Lemma 7.10.11 and Theorem 7.10.12 provide an algorithm for constructing an extensional expansion proof $\langle \mathcal{Q}, \mathbf{Roots}(\mathcal{Q})\rangle$ from a derivation of $\Gamma(\mathbf{Roots}(\mathcal{Q}))$. Such an algorithm is implemented in TPS.

CHAPTER 8

Lifting

We now demonstrate how one can lift an extensional expansion proof to a sequence of extensional expansion dags with expansion variables. Suppose $\langle \mathcal{Q}^*, \{r^*\}\rangle$ is an extensional expansion proof. We will prove that there is a finite sequence

$$\mathcal{Q}^0, \ldots, \mathcal{Q}^n$$

where each $\mathcal{Q}^i$ is an extensional expansion dag related to $\mathcal{Q}^*$. The first extensional expansion dag $\mathcal{Q}^0$ will simply be a root node corresponding to the sentence we wish to prove. For each i $(0 \leq i \leq n-1)$, we obtain $\mathcal{Q}^{i+1}$ by performing a step in $\mathcal{Q}^i$ (e.g., extending a node, connecting two nodes or making an instantiation). Each $\mathcal{Q}^{i+1}$ is, in some sense, closer to $\mathcal{Q}^*$. The final extensional expansion dag $\mathcal{Q}^n$ in the sequence will induce an extensional expansion proof.

Once we know such sequences exist, we will have a basis for complete (non-deterministic) search procedures. During the search a procedure should choose some step leading to a new extensional expansion dag. Since we know there is an extensional expansion proof, the lifting result means there is some sequence of choices which generates a sequence

$$\mathcal{Q}^0, \ldots, \mathcal{Q}^n$$

where $\mathcal{Q}^n$ induces an extensional expansion proof. We will refine the non-determinism by distinguishing between *don't know* non-determinism and *don't care* non-determinism.

8.1. Developed Extensional Expansion Dags

We start by considering steps such as extending nodes, decomposing decomposable nodes and adding expansion arcs to leaf expansion nodes. We will refer to this process as *developing* an extensional expansion dag. This process will terminate. During search, we can always develop an extensional expansion dag before considering other options without losing completeness. Thus the order in which we develop nodes is not imporant. This will be a form of *don't care* non-determinism.

DEFINITION 8.1.1. Let $\mathcal{Q}$ be an expansion structure. We say a node $a \in \mathcal{N}^{\mathcal{Q}}$ is *developed* if the following conditions hold:

1. Either a is flexible or a is extended.
2. If a is a decomposable node, then a is decomposed.
3. If a is an expansion node, then a is not a leaf node.

We say a node $a \in \mathcal{N}^{\mathcal{Q}}$ is *undeveloped* if a is not developed. Let $\mathbf{Dev}(\mathcal{Q})$ be the set of developed nodes of $\mathcal{Q}$ and $\mathbf{UnDev}(\mathcal{Q})$ be the set of undeveloped nodes of $\mathcal{Q}$. We say $\mathcal{Q}$ is *developed* if every node in $\mathcal{Q}$ is developed.

Note that flexible nodes are always developed.

REMARK 8.1.2. In order to simplify the presentation, assume for each type α there is a well-ordering on variables $\mathcal{V}_\alpha$ of type α. Thus for any finite set of variables $\mathcal{A}$ we can freely refer to the *first* variable w_α of type α such that $w_\alpha \notin \mathcal{A}$.

We will define two methods for developing nodes. One provides for *open developments* which introduce new free variables. The other provides for *closed developments* which introduce new parameters, but no new free variables.

DEFINITION 8.1.3. Let $\mathcal{Q}$ be an expansion structure and a be an undeveloped node of $\mathcal{Q}$ (i.e., $a \in \mathbf{UnDev}(\mathcal{Q})$). We define expansion structures $\mathbf{OpDevelop}(\mathcal{Q}, a)$ and $\mathbf{ClDevelop}(\mathcal{Q}, a)$ by cases:

- If a is not extended, then both $\mathbf{OpDevelop}(\mathcal{Q}, a)$ and $\mathbf{ClDevelop}(\mathcal{Q}, a)$ are defined to be $\mathbf{Extend}(\mathcal{Q}, a)$ (cf. Definition 7.7.4).
- If a is an extended decomposable node which is not decomposed, then both $\mathbf{OpDevelop}(\mathcal{Q}, a)$ and $\mathbf{ClDevelop}(\mathcal{Q}, a)$ are defined to be $\mathbf{Dec}(\mathcal{Q}, a)$ (cf. Definition 7.7.16).
- Suppose a is an extended leaf expansion node with shallow formula $[\dot{\forall} x_\alpha \mathbf{M}]$. We define $\mathbf{OpDevelop}(\mathcal{Q}, a)$ to be

$$\mathbf{EExpand}(\mathcal{Q}, e, y_\alpha)$$

(cf. Definition 7.7.8) where y_α is the first variable not in $\mathbf{Free}(\mathcal{Q})$ (cf. Remark 8.1.2). We define $\mathbf{ClDevelop}(\mathcal{Q}, a)$ to be $\mathbf{EExpand}(\mathcal{Q}, e, A_\alpha)$ where A_α is $\mathsf{A}_\alpha^{\mathbf{Params}(\mathcal{Q})} \notin \mathbf{Params}(\mathcal{Q})$ (cf. Remark 7.7.1).

DEFINITION 8.1.4. Let $\mathcal{Q}$ be an expansion structure. We define the following sets and natural numbers:

- Let $\mathbf{UnDev}_1(\mathcal{Q})$ be the set
$$\{a \in \mathbf{UnDev}(\mathcal{Q}) \mid kind^{\mathcal{Q}}(a) \neq \mathbf{eqngoal}\}.$$
- Let $\mathbf{UnDev}_{dec}(\mathcal{Q})$ be the set
$$\{a \in \mathbf{UnDev}(\mathcal{Q}) \mid a \text{ is decomposable}\}.$$
- Let $\mathbf{UnExt}_{=}^{-}(\mathcal{Q})$ be the set
$$\{a \in \mathcal{N}^{\mathcal{Q}} \setminus \mathbf{Dom}(succ^{\mathcal{Q}}) \mid a \text{ is a negative equation node}\}.$$
- Let $\mathbf{UnExt}_{ns}(\mathcal{Q})$ be the set
$$\{a \in \mathcal{N}^{\mathcal{Q}} \setminus \mathbf{Dom}(succ^{\mathcal{Q}}) \mid \begin{array}{r} a \text{ is an expansion node, a true node,} \\ \text{a false node or} \\ \text{a potentially connecting node}\}. \end{array}$$
- Let $|\mathcal{Q}|_{\mathbf{i}}$ be the sum
$$\sum_{a \in \mathbf{UnDev}_1(\mathcal{Q})} |sh^{\mathcal{Q}}(a)|_{\mathbf{i}}$$
(cf. Definition 2.1.31).
- Let $|\mathcal{Q}|_{\mathbf{e}}$ be the sum
$$\sum_{a \in \mathbf{UnDev}_1(\mathcal{Q})} |sh^{\mathcal{Q}}(a)|_{\mathbf{e}}.$$
(cf. Definition 2.1.27).
- Let $|\mathcal{Q}|_{\mathbf{x}}$ be the sum
$$2|\mathbf{UnExt}_{=}^{-}(\mathcal{Q})| + |\mathbf{UnExt}_{ns}(\mathcal{Q})| + |\mathbf{UnDev}_{dec}(\mathcal{Q})|.$$

Finally, let $|\mathcal{Q}|_{\mathbf{D}}$ be the ordinal
$$\omega^2 \cdot |\mathcal{Q}|_{\mathbf{i}} + \omega \cdot |\mathcal{Q}|_{\mathbf{e}} + |\mathcal{Q}|_{\mathbf{x}}.$$

LEMMA 8.1.5. *Let $\mathcal{Q}$ and $\mathcal{Q}'$ be extensional expansion dags such that $\mathcal{Q} \trianglelefteq \mathcal{Q}'$. We have the following:*

1. $\mathbf{UnDev}(\mathcal{Q}') \subseteq \mathbf{UnDev}(\mathcal{Q}) \cup (\mathcal{N}^{\mathcal{Q}'} \setminus \mathcal{N}^{\mathcal{Q}})$.
2. $\mathbf{UnDev}_1(\mathcal{Q}') \subseteq \mathbf{UnDev}_1(\mathcal{Q}) \cup (\mathcal{N}^{\mathcal{Q}'} \setminus \mathcal{N}^{\mathcal{Q}})$.
3. $\mathbf{UnDev}_{dec}(\mathcal{Q}') \subseteq \mathbf{UnDev}_{dec}(\mathcal{Q}) \cup (\mathcal{N}^{\mathcal{Q}'} \setminus \mathcal{N}^{\mathcal{Q}})$.
4. $\mathbf{UnExt}_{\equiv}^{-}(\mathcal{Q}') \subseteq \mathbf{UnExt}_{\equiv}^{-}(\mathcal{Q}) \cup (\mathcal{N}^{\mathcal{Q}'} \setminus \mathcal{N}^{\mathcal{Q}})$.
5. $\mathbf{UnExt}_{ns}(\mathcal{Q}') \subseteq \mathbf{UnExt}_{ns}(\mathcal{Q}) \cup (\mathcal{N}^{\mathcal{Q}'} \setminus \mathcal{N}^{\mathcal{Q}})$.

PROOF. The first result follows from the fact that any node in $\mathcal{Q}$ is developed in $\mathcal{Q}'$ if it is developed in $\mathcal{Q}$. The next two results follow from the first. The remaining two results follow from the fact that any node in $\mathcal{Q}$ is extended in $\mathcal{Q}'$ if it is extended in $\mathcal{Q}$. □

LEMMA 8.1.6. *Let $\mathcal{Q}$ be an extensional expansion dag and $a \in \mathcal{N}^{\mathcal{Q}}$ be an extended node with $succ^{\mathcal{Q}}(a) = \langle a^1, \ldots, a^n \rangle$. If a is a decomposition node, a mate node or a deepening node, then*

$$\sum_{i=1}^{n} |sh^{\mathcal{Q}}(a^i)|_{\mathbf{i}} < |sh^{\mathcal{Q}}(a)|_{\mathbf{i}}$$

PROOF. If a is a mate node or decomposition node, then the result follows from Lemma 2.1.33. If a is a deepening node, then the result follows from Lemma 2.1.42. □

LEMMA 8.1.7. *Let $\mathcal{Q}$ be an extensional expansion dag and $a \in \mathcal{N}^{\mathcal{Q}}$ be an extended node with $succ^{\mathcal{Q}}(a) = \langle a^1, \ldots, a^n \rangle$. If a is a negation node, a disjunction node, a conjunction node, a selection node, a functional node, a Boolean node, a positive equation node, an E-unification node or a symmetric E-unification node, then*

$$\sum_{i=1}^{n} |sh^{\mathcal{Q}}(a^i)|_{\mathbf{i}} \leq |sh^{\mathcal{Q}}(a)|_{\mathbf{i}}$$

and

$$\sum_{i=1}^{n} |sh^{\mathcal{Q}}(a^i)|_{\mathbf{e}} < |sh^{\mathcal{Q}}(a)|_{\mathbf{e}}.$$

PROOF. The proof is by a case analysis on the node a. In each case we use the appropriate $\clubsuit_*$ property to determine n and the shallow formula of a^i for each i $(1 \leq i \leq n)$. The inequalities can then be computed using Definitions 2.1.31 and 2.1.27 and Lemmas 2.1.28, 2.1.30, 2.1.35, 2.1.36 and 2.1.37. □

LEMMA 8.1.8. *Let $\mathcal{Q}$ be an extensional expansion dag, $a \in \mathcal{N}^{\mathcal{Q}}$ be a node which is neither extended nor flexible and $\mathcal{Q}'$ be* $\mathbf{Extend}(\mathcal{Q}, a)$. *Then $|\mathcal{Q}'|_{\mathbf{D}} < |\mathcal{Q}|_{\mathbf{D}}$.*

PROOF. By Definition 7.7.4,

$$(\mathcal{N}^{\mathcal{Q}'} \setminus \mathcal{N}^{\mathcal{Q}}) = \{a^1, \ldots, a^n\}$$

where $a^1, \ldots, a^n$ are distinct nodes and $succ^{\mathcal{Q}'}(a) = \langle a^1, \ldots, a^n \rangle$. We can use Lemma 8.1.5 to bound the sets used in the definition of $|\mathcal{Q}'|_{\mathbf{D}}$ in terms of those sets used in the definition of $|\mathcal{Q}|_{\mathbf{D}}$. If a is a decomposition node, a mate node or a deepening node, then we use Lemma 8.1.6 to conclude $|\mathcal{Q}'|_{\mathbf{i}} < |\mathcal{Q}|_{\mathbf{i}}$. If a is a negation node, a disjunction node, a conjunction node, a selection node, a functional node, a Boolean node, a positive equation node, an E-unification node or a symmetric E-unification node, then we use Lemma 8.1.7 to conclude $|\mathcal{Q}'|_{\mathbf{i}} \leq |\mathcal{Q}|_{\mathbf{i}}$ and $|\mathcal{Q}'|_{\mathbf{e}} < |\mathcal{Q}|_{\mathbf{e}}$. In the remaining cases, $|\mathcal{Q}'|_{\mathbf{i}} \leq |\mathcal{Q}|_{\mathbf{i}}$, $|\mathcal{Q}'|_{\mathbf{e}} \leq |\mathcal{Q}|_{\mathbf{e}}$ and $|\mathcal{Q}'|_{\mathbf{x}} < |\mathcal{Q}|_{\mathbf{x}}$. If a is an expansion node, a true node, a false node or a potentially connecting node, then $|\mathcal{Q}'|_{\mathbf{x}} < |\mathcal{Q}|_{\mathbf{x}}$ since $\mathbf{UnExt}_{ns}(\mathcal{Q}') \subseteq (\mathbf{UnExt}_{ns}(\mathcal{Q}) \setminus \{a\})$. If a is a negative equation node, then $|\mathcal{Q}'|_{\mathbf{x}} < |\mathcal{Q}|_{\mathbf{x}}$ since $|\mathbf{UnExt}_{=}^{-}(\mathcal{Q}')| \leq |\mathbf{UnExt}_{=}^{-}(\mathcal{Q})| - 1$ and $|\mathbf{UnExt}_{ns}(\mathcal{Q}')| \leq |\mathbf{UnExt}_{ns}(\mathcal{Q})| + 1$. □

LEMMA 8.1.9. *Let $\mathcal{Q}$ be an extensional expansion dag and $a \in \mathbf{UnDev}(\mathcal{Q})$ be an extended expansion node with shallow formula $[\dot{\forall} x_\alpha \mathbf{M}]$ such that a is a leaf node of $\mathcal{Q}$. Let $\mathcal{Q}'$ be* $\mathbf{EExpand}(\mathcal{Q}, a, u)$ *where $u_\alpha \in \mathcal{V} \cup \mathcal{P}$. Then $|\mathcal{Q}'|_{\mathbf{i}} \leq |\mathcal{Q}|_{\mathbf{i}}$ and $|\mathcal{Q}'|_{\mathbf{e}} < |\mathcal{Q}|_{\mathbf{e}}$.*

PROOF. This follows from Lemmas 8.1.5, 2.1.35 and 2.1.30. □

LEMMA 8.1.10. *Let $\mathcal{Q}$ be an extensional expansion dag, $a \in \mathbf{UnDev}(\mathcal{Q})$ be an extended decomposable node which is not decomposed and $\mathcal{Q}'$ be* $\mathbf{Dec}(\mathcal{Q}, a)$. *Then $|\mathcal{Q}'|_{\mathbf{i}} \leq |\mathcal{Q}|_{\mathbf{i}}$, $|\mathcal{Q}'|_{\mathbf{e}} \leq |\mathcal{Q}|_{\mathbf{e}}$ and $|\mathcal{Q}'|_{\mathbf{x}} < |\mathcal{Q}|_{\mathbf{x}}$.*

PROOF. This follows from Lemma 8.1.5, $a \in \mathbf{UnDev}_{dec}(\mathcal{Q})$ and

$$\mathbf{UnDev}_{dec}(\mathcal{Q}') \subseteq (\mathbf{UnDev}_{dec}(\mathcal{Q}) \setminus \{a\}).$$

□

LEMMA 8.1.11. *Let $\mathcal{Q}$ be an extensional expansion dag, $a \in \mathbf{UnDev}(\mathcal{Q})$ be an undeveloped node and $\mathcal{Q}'$ be $\mathbf{OpDevelop}(\mathcal{Q}, a)$ or $\mathbf{ClDevelop}(\mathcal{Q}, a)$. Then $\mathcal{Q}'$ is an extensional expansion dag, $\mathcal{Q}'$ augments $\mathcal{Q}$ and $|\mathcal{Q}'|_{\mathbf{D}} < |\mathcal{Q}|_{\mathbf{D}}$. Furthermore, we have the following:*

1. *If $\mathbf{RootSels}(\mathcal{Q})$ is empty, then $\mathbf{RootSels}(\mathcal{Q}')$ is empty.*
2. *If $\mathcal{Q}$ is separated, then $\mathcal{Q}'$ is separated.*
3. *If $\mathcal{Q}$ has merged connections [decompositions, expansions], then $\mathcal{Q}'$ has merged connections [decompositions, expansions].*
4. *If $\prec_0^{\mathcal{Q}}$ is acyclic, then $\prec_0^{\mathcal{Q}'}$ is acyclic.*
5. *If $\mathcal{Q}$ is closed and $\mathcal{Q}'$ is $\mathbf{ClDevelop}(\mathcal{Q}, a)$, then $\mathcal{Q}'$ is closed.*

PROOF. We consider the three cases in Definition 8.1.3. Suppose a is neither extended nor flexible and $\mathcal{Q}'$ is $\mathbf{Extend}(\mathcal{Q}, a)$. By Lemma 7.7.7, $\mathcal{Q}'$ is an extensional expansion dag, $\mathcal{Q}'$ augments $\mathcal{Q}$ and

$$\mathbf{RootSels}(\mathcal{Q}') = \mathbf{RootSels}(\mathcal{Q}).$$

Hence $\mathbf{RootSels}(\mathcal{Q}')$ is empty if $\mathbf{RootSels}(\mathcal{Q})$ is empty. By Lemma 7.7.7, we also know $\mathcal{Q}'$ is separated if $\mathcal{Q}$ is separated, $\mathcal{Q}'$ has merged connections [decompositions,expansions] if $\mathcal{Q}$ has merged connections [decompositions, expansions], $\prec_0^{\mathcal{Q}'}$ is acyclic if $\prec_0^{\mathcal{Q}}$ is acyclic and $\mathcal{Q}'$ is closed if $\mathcal{Q}$ is closed. We know $|\mathcal{Q}'|_{\mathbf{D}} < |\mathcal{Q}|_{\mathbf{D}}$ by Lemma 8.1.8.

Suppose a is an extended decomposable node which is not decomposed and $\mathcal{Q}'$ is $\mathbf{Dec}(\mathcal{Q}, a)$. By Lemma 7.7.18, we know $\mathcal{Q}'$ is an extensional expansion dag, $\mathcal{Q}'$ augments $\mathcal{Q}$ and

$$\mathbf{RootSels}(\mathcal{Q}') = \mathbf{RootSels}(\mathcal{Q}).$$

Hence $\mathbf{RootSels}(\mathcal{Q}')$ is empty if $\mathbf{RootSels}(\mathcal{Q})$ is empty. By Lemma 7.7.18, we also know $\mathcal{Q}'$ is separated if $\mathcal{Q}$ is separated, $\mathcal{Q}'$ has merged connections [expansions] if $\mathcal{Q}$ has merged connections [expansions], $\mathcal{Q}'$ has merged decompositions if $\mathcal{Q}$ has merged decompositions (since a is not decomposed in $\mathcal{Q}$), $\prec_0^{\mathcal{Q}'}$ is acyclic if $\prec_0^{\mathcal{Q}}$ is acyclic and $\mathcal{Q}'$ is closed if $\mathcal{Q}$ is closed. We know $|\mathcal{Q}'|_{\mathbf{i}} \leq |\mathcal{Q}|_{\mathbf{i}}$, $|\mathcal{Q}'|_{\mathbf{e}} \leq |\mathcal{Q}|_{\mathbf{e}}$ and $|\mathcal{Q}'|_{\mathbf{x}} < |\mathcal{Q}|_{\mathbf{x}}$ by Lemma 8.1.10. Therefore,

$$|\mathcal{Q}'|_{\mathbf{D}} = \omega^2 \cdot |\mathcal{Q}'|_{\mathbf{i}} + \omega \cdot |\mathcal{Q}'|_{\mathbf{e}} + |\mathcal{Q}'|_{\mathbf{x}} < \omega^2 \cdot |\mathcal{Q}|_{\mathbf{i}} + \omega \cdot |\mathcal{Q}|_{\mathbf{e}} + |\mathcal{Q}|_{\mathbf{x}} = |\mathcal{Q}|_{\mathbf{D}}$$

as desired.

Suppose a is an extended leaf expansion node with shallow formula $[\forall x_\alpha \mathbf{M}]$. In this case either $\mathcal{Q}'$ is $\mathbf{EExpand}(\mathcal{Q}, a, y)$ where y_α is the first variable not in $\mathbf{Free}(\mathcal{Q})$ (cf. Remark 8.1.2), or $\mathcal{Q}'$ is

EExpand$(\mathcal{Q}, a, A_\alpha)$ where A_α is $\mathsf{A}_\alpha^{\mathbf{Params}(\mathcal{Q})} \notin \mathbf{Params}(\mathcal{Q})$ (cf. Remark 7.7.1). In either case, $\mathcal{Q}'$ is **EExpand**$(\mathcal{Q}, a, u)$ where $u \in \mathcal{V} \cup \mathcal{P}$ and $u \notin (\mathbf{Free}(\mathcal{Q}) \cup \mathbf{Params}(\mathcal{Q}))$. By Lemma 7.7.10, we know $\mathcal{Q}'$ is an extensional expansion dag, $\mathcal{Q}'$ augments $\mathcal{Q}$ and

$$\mathbf{RootSels}(\mathcal{Q}') = \mathbf{RootSels}(\mathcal{Q}).$$

Hence $\mathbf{RootSels}(\mathcal{Q}')$ is empty if $\mathbf{RootSels}(\mathcal{Q})$ is empty. By Lemma 7.7.10, we also know $\mathcal{Q}'$ is separated if $\mathcal{Q}$ is separated, $\mathcal{Q}'$ has merged connections [decompositions] if $\mathcal{Q}$ has merged connections [decompositions] and $\prec_0^{\mathcal{Q}'}$ is acyclic if $\prec_0^{\mathcal{Q}}$ is acyclic. Since a is a leaf node of $\mathcal{Q}$, $\mathbf{ExpInsts}^{\mathcal{Q}}(a)$ is empty (cf. Definition 7.2.15). Hence $([u/x]\mathbf{M})^{\downarrow} \notin \mathbf{ExpInsts}^{\mathcal{Q}}(a)$ and so $\mathcal{Q}'$ has merged expansions if $\mathcal{Q}$ has merged expansions by Lemma 7.7.10. If $\mathcal{Q}'$ is $\mathbf{ClDevelop}(\mathcal{Q}, a)$ and $\mathcal{Q}$ is closed, then $u \in \mathcal{P}$ is closed and so $\mathcal{Q}' = \mathbf{EExpand}(\mathcal{Q}, a, u)$ is closed by Lemma 7.7.10. (If $\mathcal{Q}'$ is $\mathbf{OpDevelop}(\mathcal{Q}, a)$, then $u \in \mathbf{Free}(\mathcal{Q}')$ and so $\mathcal{Q}'$ will definitely not be closed in this case.) We know $|\mathcal{Q}'|_{\mathbf{i}} \leq |\mathcal{Q}|_{\mathbf{i}}$ and $|\mathcal{Q}'|_{\mathbf{e}} < |\mathcal{Q}|_{\mathbf{e}}$ by Lemma 8.1.9. Therefore,

$$|\mathcal{Q}'|_{\mathbf{D}} = \omega^2 \cdot |\mathcal{Q}'|_{\mathbf{i}} + \omega \cdot |\mathcal{Q}'|_{\mathbf{e}} + |\mathcal{Q}'|_{\mathbf{x}} < \omega^2 \cdot |\mathcal{Q}|_{\mathbf{i}} + \omega \cdot |\mathcal{Q}|_{\mathbf{e}} + |\mathcal{Q}|_{\mathbf{x}} = |\mathcal{Q}|_{\mathbf{D}}$$

as desired. □

LEMMA 8.1.12. *There is no infinite sequence of extensional expansion dags* $(\mathcal{Q}^n)_{n \in \mathbb{N}}$ *such that for every* $n \geq 0$ *the extensional expansion dag* $\mathcal{Q}^{n+1}$ *is* $\mathbf{OpDevelop}(\mathcal{Q}^n, a^n)$ $[\mathbf{ClDevelop}(\mathcal{Q}^n, a^n)]$ *for some* $a^n \in \mathbf{UnDev}(\mathcal{Q}^n)$.

PROOF. This follows directly from Lemma 8.1.11 since such an infinite sequence would give an infinite descending chain

$$\cdots < |\mathcal{Q}^n|_{\mathbf{D}} < \cdots < |\mathcal{Q}^0|_{\mathbf{D}} < \omega^3$$

of ordinals. □

LEMMA 8.1.13. *Let* $\mathcal{Q}$ *be an extensional expansion dag. There exists a developed extensional expansion dag* $\mathcal{Q}^*$ *such that* $\mathcal{Q} \trianglelefteq_r \mathcal{Q}^*$ *and the following properties hold:*

1. *If* $\mathcal{Q}$ *is closed, then* $\mathcal{Q}^*$ *is closed.*
2. *If* $\mathbf{RootSels}(\mathcal{Q})$ *is empty, then* $\mathbf{RootSels}(\mathcal{Q}^*)$ *is empty.*
3. *If* $\mathcal{Q}$ *is separated, then* $\mathcal{Q}^*$ *is separated.*
4. *If* $\mathcal{Q}$ *has merged connections [decompositions, expansions], then* $\mathcal{Q}^*$ *has merged connections [decompositions, expansions].*

5. *If* $\prec_0^{\mathcal{Q}}$ *is acyclic, then* $\prec_0^{\mathcal{Q}^*}$ *is acyclic.*

PROOF. We prove this by induction on the ordinal $|\mathcal{Q}|_{\mathbf{D}}$ (cf. Definition 8.1.4). If $\mathbf{UnDev}(\mathcal{Q})$ is empty, then $\mathcal{Q}$ is already a developed extensional expansion dag with the required properties. Otherwise, choose some $a \in \mathbf{UnDev}(\mathcal{Q})$ and let $\mathcal{Q}'$ be $\mathbf{ClDevelop}(\mathcal{Q}, a)$. By Lemma 8.1.11, $\mathcal{Q}'$ is an extensional expansion dag, $\mathcal{Q}'$ augments $\mathcal{Q}$, $|\mathcal{Q}'|_{\mathbf{D}} < |\mathcal{Q}|_{\mathbf{D}}$ and the required properties are preserved. Note that $\mathcal{Q} \trianglelefteq_r \mathcal{Q}'$ by Lemma 7.6.13. By the inductive hypothesis, there is a developed extensional expansion dag $\mathcal{Q}^*$ with the required properties preserved such that $\mathcal{Q} \trianglelefteq_r \mathcal{Q}' \trianglelefteq_r \mathcal{Q}^*$. □

8.2. Substitutions

We define substitution on expansion structures and prove that certain substitutions will preserve the property of being an extensional expansion dag.

LEMMA 8.2.1. *Let* θ *be a substitution such that* $\mathbf{Dom}(\theta) \subset \mathcal{V}$. *For any* $\langle k, p, \mathbf{M}\rangle \in \mathbf{Labels}$ *we have* $\langle k, p, (\theta(\mathbf{M}))^{\downarrow}\rangle \in \mathbf{Labels}$.

PROOF. The proof is by a simple case analysis using the definition of **Labels** (cf. Definition 7.2.2). □

DEFINITION 8.2.2. Let

$$\mathcal{Q} = \langle \mathcal{N}, succ, label, sel, exps, conns, decs\rangle$$

be an expansion structure and θ be a substitution such that $\mathbf{Dom}(\theta) \subset \mathcal{V}$. We define $\theta(\mathcal{Q})$ to be the expansion structure

$$\theta(\mathcal{Q}) := \langle \mathcal{N}, succ, label', sel, exps', conns, decs\rangle$$

where

$$label'(a) := \langle kind^{\mathcal{Q}}(a), pol^{\mathcal{Q}}(a), (\theta(sh^{\mathcal{Q}}(a)))^{\downarrow}\rangle$$

for every $a \in \mathcal{N}$ and

$$exps' := \{\langle e, \theta(\mathbf{C}), c\rangle \mid \langle e, \mathbf{C}, c\rangle \in exps\}.$$

Note that $\theta(\mathcal{Q})$ is not defined if θ substitutes for parameters or logical constants. Potential problems would arise with labels for non-primary nodes (e.g., decomposition nodes) if θ sent a parameter to a term other than a parameter. Also, problems would arise if θ conflated two selected parameters.

LEMMA 8.2.3. *Let $\mathcal{Q}$ be an expansion structure, $\mathbf{A}_\alpha$ be a term and $x_\alpha \in \mathbf{Free}(\mathcal{Q})$ be a variable. Then*

$$\mathbf{Free}([\mathbf{A}/x]\mathcal{Q}) \subseteq (\mathbf{Free}(\mathcal{Q}) \setminus \{x\}) \cup \mathbf{Free}(\mathbf{A})$$

and

$$\mathbf{Params}([\mathbf{A}/x]\mathcal{Q}) \subseteq \mathbf{Params}(\mathcal{Q}) \cup \mathbf{Params}(\mathbf{A}).$$

PROOF. See Appendix A.6. □

Often we can determine properties of a node a in an extensional expansion dag $\mathcal{Q}$ from properties of a in the expansion structure $\theta(\mathcal{Q})$.

LEMMA 8.2.4. *Let $\mathcal{Q}$ be an extensional expansion dag, θ be a substitution such that $\mathbf{Dom}(\theta) \subset \mathcal{V}$, $\mathcal{Q}'$ be the expansion structure $\theta(\mathcal{Q})$ and $a \in \mathcal{N}^{\mathcal{Q}'}$ be a node. Then we have the following:*

1. *If a is a true node in $\mathcal{Q}'$, then a is a true node in $\mathcal{Q}$.*
2. *If a is a false node in $\mathcal{Q}'$, then a is a false node in $\mathcal{Q}$.*
3. *If a is an expansion node in $\mathcal{Q}'$, then a is an expansion node in $\mathcal{Q}$.*
4. *If a is an extended atomic node in $\mathcal{Q}'$, then a is an atomic node in $\mathcal{Q}$.*
5. *If a is an extended deepening node in $\mathcal{Q}'$, then a is a deepening node in $\mathcal{Q}$.*
6. *If a is an equation node in $\mathcal{Q}'$, then a is an equation node in $\mathcal{Q}$.*
7. *If a is an extended potentially connecting node in $\mathcal{Q}'$, then a is a potentially connecting node in $\mathcal{Q}$.*

PROOF. Each case follows easily from definitions. In parts (4), (5) and (7) we assume a is extended in $\mathcal{Q}'$ to ensure a is not a flexible node in $\mathcal{Q}$. □

A substitution does not change the underlying graph of an expansion structure.

LEMMA 8.2.5. *Let $\mathcal{Q}$ be an expansion structure, θ be a substitution such that $\mathbf{Dom}(\theta) \subset \mathcal{V}$ and $\mathcal{Q}'$ be the expansion structure $\theta(\mathcal{Q})$. The graphs $\langle \mathcal{N}^{\mathcal{Q}}, \xrightarrow{\mathcal{Q}} \rangle$ and $\langle \mathcal{N}^{\mathcal{Q}'}, \xrightarrow{\mathcal{Q}'} \rangle$ are the same.*

PROOF. By Definition 8.2.2, the set of nodes $\mathcal{N}^{\mathcal{Q}'}$ is the same as $\mathcal{N}^{\mathcal{Q}}$. Let $\mathcal{N}$ be $\mathcal{N}^{\mathcal{Q}}$ and let $a, b \in \mathcal{N}$ be given. We prove $a \xrightarrow{\mathcal{Q}} b$ iff

$a \overset{\mathcal{Q}'}{\rightarrow} b$. If the edge is induced by the successor function, the set of connections or the set of decomposition arcs, then the equivalence is true since $\mathcal{Q}$ and $\mathcal{Q}'$ share the same successor function, set of connections and set of decomposition arcs. We only need to consider edges induced by expansion arcs. If $\langle a, \mathbf{B}, b\rangle \in exps^{\mathcal{Q}}$ for some $\mathbf{B}$, then $\langle a, \theta(\mathbf{B}), b\rangle \in exps^{\mathcal{Q}'}$. If $\langle a, \mathbf{B}', b\rangle \in exps^{\mathcal{Q}'}$ for some $\mathbf{B}'$, then there is some $\mathbf{B}$ such that $\mathbf{B}'$ is $\theta(\mathbf{B})$ and $\langle a, \mathbf{B}, b\rangle \in exps^{\mathcal{Q}}$. □

We can now prove that substituting into an extensional expansion dag gives an extensional expansion dag. Note that flexible nodes in $\mathcal{Q}$ may correspond to atomic nodes, deepening nodes or flexible nodes in $\theta(\mathcal{Q})$. The choice to allow the successor function to be partial plays an important role here. By $\clubsuit_f$, flexible nodes in an extensional expansion dag $\mathcal{Q}$ are not extended. Since $\mathcal{Q}$ and $\theta(\mathcal{Q})$ share the same successor function, the images of such flexible nodes will not be extended in $\theta(\mathcal{Q})$. Hence when a flexible node is instantiated resulting in an atomic node [deepening node], $\clubsuit_{ns}$ [$\clubsuit_\sharp$] will not hold, but this presents no problem since we have only required these properties to hold for extended nodes (cf. Definition 7.3.2:7).

LEMMA 8.2.6. *Let $\mathcal{Q}$ be an extensional expansion dag and θ be a substitution such that $\mathbf{Dom}(\theta) \subset \mathcal{V}$. Then $\theta(\mathcal{Q})$ is an extensional expansion dag. Furthermore,*

1. $\mathbf{Roots}(\mathcal{Q}) = \mathbf{Roots}(\theta(\mathcal{Q}))$.
2. $\mathbf{SelPars}(\mathcal{Q}) = \mathbf{SelPars}(\theta(\mathcal{Q}))$.
3. *If θ is a ground substitution and $\mathbf{Free}(\mathcal{Q}) \subseteq \mathbf{Dom}(\theta)$, then $\theta(\mathcal{Q})$ is closed.*

PROOF. See Appendix A.6. □

LEMMA 8.2.7. *Let $\mathcal{Q}$ be an extensional expansion dag, $a \in \mathcal{N}^{\mathcal{Q}}$ be a node and θ be a substitution such that $\mathbf{Dom}(\theta) \subset \mathcal{V}$.*

1. *If a is conjunctive in $\mathcal{Q}$, then a is conjunctive in $\theta(\mathcal{Q})$.*
2. *If a is disjunctive in $\mathcal{Q}$, then a is disjunctive in $\theta(\mathcal{Q})$.*
3. *If a is neutral in $\mathcal{Q}$, then a is neutral in $\theta(\mathcal{Q})$.*

PROOF. Each result follows by a simple case analysis using the definitions of conjunctive, disjunctive and neutral. □

LEMMA 8.2.8. *Let $\mathcal{Q}$ be an extensional expansion dag and θ be a substitution such that* $\mathbf{Dom}(\theta) \subset \mathcal{V}$. *For any* $\mathfrak{p} \subseteq \mathcal{N}^{\mathcal{Q}}$, *if* $\mathfrak{p}$ *is a u-part of* $\theta(\mathcal{Q})$, *then* $\mathfrak{p}$ *is a u-part of* $\mathcal{Q}$.

PROOF. This follows easily from Lemma 8.2.7. □

If we apply a substitution θ which does not contain any selected parameter of $\mathcal{Q}$, then we can maintain acyclicity of the embedding relation.

LEMMA 8.2.9. *Let $\mathcal{Q}$ be an extensional expansion dag and θ be a substitution such that* $\mathbf{Dom}(\theta) \subset \mathcal{V}$. *Suppose*

$$\mathbf{Params}(\theta(x)) \cap \mathbf{SelPars}(\mathcal{Q}) = \emptyset$$

for every $x \in \mathbf{Dom}(\theta)$. *For any* $A, B \in \mathbf{SelPars}(\mathcal{Q})$, *if* $A \prec_0^{\theta(\mathcal{Q})} B$, *then* $A \prec_0^{\mathcal{Q}} B$. *Furthermore, if* $\prec_0^{\mathcal{Q}}$ *is acyclic, then* $\prec_0^{\theta(\mathcal{Q})}$ *is acyclic.*

PROOF. Let $\mathcal{Q}'$ be $\theta(\mathcal{Q})$. Note that $\mathcal{Q}'$ is an extensional expansion dag by Lemma 8.2.6. Also, $\mathcal{Q}$ and $\mathcal{Q}'$ have the same selected parameters since $sel^{\mathcal{Q}} = sel^{\mathcal{Q}'}$. Suppose $A, B \in \mathbf{SelPars}(\mathcal{Q})$ and $A \prec_0^{\mathcal{Q}'} B$. Then there exists an expansion arc $\langle e, \mathbf{C}, c\rangle \in exps^{\mathcal{Q}}$ such that $A \in \mathbf{Params}(\theta(\mathbf{C}))$ and c dominates b in $\mathcal{Q}'$ where b is $\mathbf{SelNode}^{\mathcal{Q}'}(B)$. Since $sel^{\mathcal{Q}} = sel^{\mathcal{Q}'}$, b is also $\mathbf{SelNode}^{\mathcal{Q}}(B)$. By Lemma 8.2.5, c dominates b in $\mathcal{Q}$. Since $A \in \mathbf{Params}(\theta(\mathbf{C}))$ we must either have $A \in \mathbf{Params}(\mathbf{C})$ or $A \in \mathbf{Params}(\theta(x))$ for some $x \in \mathbf{Free}(\mathbf{C}) \cap \mathbf{Dom}(\theta)$. However, by assumption $A \notin \mathbf{Params}(\theta(x))$ for any $x \in \mathbf{Dom}(\theta)$. Hence we must have $A \in \mathbf{Params}(\mathbf{C})$ and so $A \prec_0^{\mathcal{Q}} B$.

Applying this result, we know any cycle

$$A^1 \prec_0^{\mathcal{Q}'} \cdots \prec_0^{\mathcal{Q}'} A^n \prec_0^{\mathcal{Q}'} A^1$$

gives a cycle

$$A^1 \prec_0^{\mathcal{Q}} \cdots \prec_0^{\mathcal{Q}} A^n \prec_0^{\mathcal{Q}} A^1$$

where $A^1, \ldots, A^n \in \mathbf{SelPars}(\mathcal{Q}') = \mathbf{SelPars}(\mathcal{Q})$. □

During search we will sometimes need to instantiate an expansion variable with a term containing a selected variable. We define when a parameter is banned for a variable in an extensional expansion dag in order to maintain acyclicity of the embedding relation.

DEFINITION 8.2.10 (Banned). Let $\mathcal{Q}$ be an extensional expansion dag, B be a parameter and x be a variable. We say a B is *banned for x in $\mathcal{Q}$* if there exist selected parameters $A^1, \ldots, A^n \in \mathbf{SelPars}(\mathcal{Q})$ (where $n \geq 1$) and an expansion arc $\langle e, \mathbf{C}, c\rangle \in exps^{\mathcal{Q}}$ such that $x \in \mathbf{Free}(\mathbf{C})$, $A^n = B$,

$$A^1 \prec_0^{\mathcal{Q}} \cdots \prec_0^{\mathcal{Q}} A^n$$

and c dominates $\mathbf{SelNode}^{\mathcal{Q}}(A^1)$.

LEMMA 8.2.11. *Let $\mathcal{Q}$ be an extensional expansion dag, $\mathbf{A}_\alpha$ be a term and x_α be a variable. For every $B, D \in \mathbf{SelPars}([\mathbf{A}/x]\mathcal{Q})$, if $B \prec_0^{([\mathbf{A}/x]\mathcal{Q})} D$ and $B \not\prec_0^{\mathcal{Q}} D$, then $B \in \mathbf{Params}(\mathbf{A})$ and there is an expansion arc $\langle e, \mathbf{C}, c\rangle$ in $\mathcal{Q}$ such that $x \in \mathbf{Free}(\mathbf{C})$ and c dominates $\mathbf{SelNode}^{\mathcal{Q}}(D)$ in $\mathcal{Q}$.*

PROOF. The expansion arc in $[\mathbf{A}/x]\mathcal{Q}$ witnessing $B \prec_0^{([\mathbf{A}/x]\mathcal{Q})} D$ is induced by an expansion arc $\langle e, \mathbf{C}, c\rangle$ in $\mathcal{Q}$. The expansion arc $\langle e, \mathbf{C}, c\rangle$ either witnesses $B \prec_0^{\mathcal{Q}} D$ or we have $B \in \mathbf{Params}(\mathbf{A})$, $x \in \mathbf{Free}(\mathbf{C})$ and c dominates $\mathbf{SelNode}^{\mathcal{Q}}(D)$ in $\mathcal{Q}$. □

LEMMA 8.2.12. *Let $\mathcal{Q}$ be an extensional expansion dag, $\mathbf{A}_\alpha$ be a term, B_β be a parameter and x_α be a variable. Suppose*

$$\mathbf{Params}(\mathbf{A}) \subseteq \{B\} \textit{ and } B \textit{ is not banned for } x.$$

If $\prec_0^{\mathcal{Q}}$ is acyclic, then $\prec_0^{([\mathbf{A}/x]\mathcal{Q})}$ is acyclic.

PROOF. A cycle with respect to the embedding relation $\prec_0^{([\mathbf{A}/x]\mathcal{Q})}$ either induces a cycle with respect to $\prec_0^{\mathcal{Q}}$ or (using Lemma 8.2.11) implies B is banned for x. The full proof is in Appendix A.6. □

8.3. Pre-Solved Parts

We defined the set of u-parts $\mathcal{U}parts(\mathcal{Q})$ of an extensional expansion dag $\mathcal{Q}$ in Definition 7.4.4. Certain u-parts need not be considered during search. Suppose a u-part contains a negative equation node with shallow formula of the form $[[f\,\overline{\mathbf{A}^n}] \doteq^{\iota} [g\,\overline{\mathbf{B}^m}]]$ where f and g are variables. Note that we can solve this equation by substituting $[\lambda \overline{x^n}\, C_\iota]$ and $[\lambda \overline{y^m}\, C_\iota]$ for f and g, respectively. If every u-part contains such a node, then we can construct the extensional expansion proof by simultaneously solving all such equations in this manner.

(The observation that such flex-flex pairs can always be solved simultaneously dates at least to [36].)

DEFINITION 8.3.1. Let $\mathcal{Q}$ be an expansion structure. We say a node $a \in \mathcal{N}^{\mathcal{Q}}$ is a *flex-flex node* of $\mathcal{Q}$ if a is a negative equation node with shallow formula $[\mathbf{A} \stackrel{\iota}{=} \mathbf{B}]$ where $\mathbf{A}$ and $\mathbf{B}$ are flexible terms (i.e., both heads are variables, cf. Definition 2.1.14).

We will also distinguish u-parts which contain negative flexible nodes.

DEFINITION 8.3.2. Let $\mathcal{Q}$ be an extensional expansion dag and $\mathfrak{p}$ be a u-part of $\mathcal{Q}$. We say $\mathfrak{p}$ is *flex-flex* if there exists a flex-flex node $a \in \mathfrak{p}$. We say $\mathfrak{p}$ is *pre-solved* if there is either a flex-flex node in $\mathfrak{p}$ or a negative flexible node in $\mathfrak{p}$. We define $\mathcal{U}parts^{-ps}(\mathcal{Q})$ to be the set of u-parts of $\mathcal{Q}$ which are not pre-solved. That is, $\mathcal{U}parts^{-ps}(\mathcal{Q})$ is

$$\{\mathfrak{p} \in \mathcal{U}parts(\mathcal{Q}) \mid \mathfrak{p} \text{ is not a pre-solved u-part of } \mathcal{Q}\}.$$

For any $\mathcal{R} \subseteq \mathcal{N}^{\mathcal{Q}}$, we define the set

$$\mathcal{U}parts^{-ps}(\mathcal{Q}; \mathcal{R}) := \mathcal{U}parts^{-ps}(\mathcal{Q}) \cap \mathcal{U}parts(\mathcal{Q}; \mathcal{R}).$$

We will demonstrate that we can obtain an extensional expansion proof if we are given an appropriate extensional expansion dag such that every u-part including the roots is pre-solved.

THEOREM 8.3.3. *Let $\mathcal{S}$ be a signature of logical constants and $\mathcal{Q}$ be an $\mathcal{S}$-extensional expansion dag. Assume the following:*

1. $\mathbf{Free}^{\mathcal{Q}}(\mathbf{Roots}(\mathcal{Q}))$ *is empty.*
2. $\mathbf{RootSels}(\mathcal{Q})$ *is empty.*
3. $\prec_0^{\mathcal{Q}}$ *is acyclic.*
4. $\mathcal{U}parts^{-ps}(\mathcal{Q}; \mathbf{Roots}(\mathcal{Q}))$ *is empty. That is, every u-part of $\mathcal{Q}$ which includes $\mathbf{Roots}(\mathcal{Q})$ is pre-solved.*

There is an $(\mathcal{S} \cup \{\top\})$-extensional expansion dag $\mathcal{Q}'$ such that

$$\mathbf{Roots}(\mathcal{Q}') = \mathbf{Roots}(\mathcal{Q}),$$

$\langle \mathcal{Q}', \mathbf{Roots}(\mathcal{Q}) \rangle$ *is an $(\mathcal{S} \cup \{\top\})$-extensional expansion proof and*

$$label^{\mathcal{Q}'}(r) = label^{\mathcal{Q}}(r)$$

for every $r \in \mathbf{Roots}(\mathcal{Q})$.

PROOF. We can remove the remaining pre-solved u-parts by choosing a parameter C_ι (not a selected parameter in $\mathcal{Q}$) and instantiating variables $x_{\iota\alpha^n\cdots\alpha^1}$ with $[\lambda\overline{z^n}\,C_\iota]$ and variables $x_{o\alpha^n\cdots\alpha^1}$ with $[\lambda\overline{z^n}\,\top]$. This substitution will send flex-flex nodes to decomposable nodes (which are not yet decomposed) with shallow formula $[C \stackrel{\cdot}{=}^\iota C]$ and send flexible nodes to deepening nodes (which are not yet extended) with shallow formula $\top$. Then we develop the instantiated extensional expansion dag (using Lemma 8.1.13) to remove all u-parts. The full proof is in Appendix A.6. □

Instead of searching for an extensional expansion proof, we can use Theorem 8.3.3 to justify searching for extensional expansion dags where every u-part (including the roots) is pre-solved. Consequently, we can essentially ignore pre-solved u-parts while determining operations to apply during search.

REMARK 8.3.4. Benzmüller discusses the problem of clauses with flex-flex pairs in his extensional higher-order resolution calculi (cf. page 125 of [18]). In Section 2 of Chapter 4 of [18], a *FlexFlex* unification rule is added to his resolution calculus $\mathcal{ER}$ to obtain an enriched calculus $\mathcal{ER}_f$. This rule is used to prove the Lifting Lemma for $\mathcal{ER}_f$ (Lemma 4.12 of [18]). In terms of search, the LEO resolution theorem prover (cf. [20; 18]) does not apply the (highly branching) *FlexFlex* rule. Benzmüller conjectured that the *FlexFlex* rule is not necessary for completeness of the search. We establish a corresponding result here for search using extensional expansion dags. That is, we never consider flex-flex nodes during search by always delaying pre-solved u-parts. This allows us to prove lifting without using a rule corresponding to the *FlexFlex* rule of [18]. We obtain a further restriction by delaying consideration of negative flexible nodes. If we did not delay negative (or positive) flexible nodes, then we might need to consider mating two flexible nodes during search. Since we never consider pre-solved u-parts, we limit ourselves to mating either two atoms (a *rigid* **mate**-*connection*) or a positive flexible node with an atom (a *flex-rigid* **mate**-*connection*).

8.4. Lifting Maps

Lifting maps injectively map expansion structures into expansion structures preserving the graph structure and labels (up to a substitution). These maps also allow us to measure how close an extensional expansion dag is to a target extensional expansion proof.

DEFINITION 8.4.1. Let $\mathcal{Q}$ be an expansion structure and φ be a substitution. We say φ is a *lifting substitution for* $\mathcal{Q}$ if

- $\mathbf{Dom}(\varphi) \subset \mathcal{V} \cup \mathcal{P}$,
- φ respects parameters,
- $\varphi(x)$ is a closed long $\beta\eta$-form for every $x \in \mathbf{Free}(\mathcal{Q})$ and
- $\varphi(A) = \varphi(B)$ implies $A = B$ for every $A, B \in \mathbf{Params}(\mathcal{Q})$ (i.e., φ is injective on the parameters of $\mathcal{Q}$).

At times we will distinguish the behavior of a lifting substitution φ on parameters. We let $\varphi\big|_{\mathcal{P}}$ denote the parameter substitution with domain $\mathbf{Dom}(\varphi) \cap \mathcal{P}$ determined by $\varphi\big|_{\mathcal{P}}(A) := \varphi(A)$ for $A \in \mathbf{Dom}(\varphi) \cap \mathcal{P}$.

Since each $\varphi(x)$ is assumed to be a long $\beta\eta$-form for $x \in \mathbf{Free}(\mathcal{Q})$, we can define a measure on lifting substitutions using the length (cf. Definition 2.1.31) of each $\varphi(x)$.

DEFINITION 8.4.2. Let $\mathcal{Q}$ be an expansion structure and φ be a lifting substitution for $\mathcal{Q}$. We define $|\varphi|_l^{\mathcal{Q}}$ to be the sum

$$\sum_{x \in \mathbf{Free}(\mathcal{Q})} |\varphi(x)|_l.$$

A lifting map consists of a lifting substitution and an injection on nodes.

DEFINITION 8.4.3 (Lifting Map). Let

$$\mathcal{Q} = \langle \mathcal{N}, succ, label, sel, exps, conns, decs \rangle$$

and

$$\mathcal{Q}^* = \langle \mathcal{N}^*, succ^*, label^*, sel^*, exps^*, conns^*, decs^* \rangle$$

be expansion structures. We say $\langle \varphi, f \rangle$ is a *lifting map from* $\mathcal{Q}$ *to* $\mathcal{Q}^*$ if φ is a lifting substitution for $\mathcal{Q}$ (cf. Definition 8.4.1) and $f : \mathcal{N} \to \mathcal{N}^*$ is an injective function satisfying the following:

1. For every $a \in \mathcal{N}$, if $label(a) = \langle k, p, \mathbf{M} \rangle$, then
$$label^*(f(a)) = \langle k, p, (\varphi(\mathbf{M}))^{\downarrow} \rangle.$$
2. For every $a \in \mathcal{N}$, if $a \in \mathbf{Dom}(succ)$, then $f(a) \in \mathbf{Dom}(succ^*)$.
3. For every $a \in \mathbf{Dom}(succ)$, if $succ(a) = \langle a^1, \ldots, a^n \rangle$, then
$$succ^*(f(a)) = \langle f(a^1), \ldots, f(a^n) \rangle.$$
4. For every $a \in \mathbf{Dom}(sel)$, $f(a) \in \mathbf{Dom}(sel^*)$ and
$$sel^*(f(a)) = \varphi(sel(a)).$$
5. For every $A \in \mathbf{Params}(\mathcal{Q})$ and $a^* \in \mathbf{Dom}(sel^*)$, if
$$\varphi(A) = sel^*(a^*),$$
then there is some $a \in \mathbf{Dom}(sel)$ such that $f(a) = a^*$ and $sel(a) = A$.
6. For all nodes $e, c \in \mathcal{N}$ and every term $\mathbf{C}$, if $\langle e, \mathbf{C}, c \rangle \in exps$, then
$$\langle f(e), \mathbf{C}^*, f(c) \rangle \in exps^*$$
for some term $\mathbf{C}^*$ such that $\varphi(\mathbf{C}) \stackrel{\beta\eta}{=} \mathbf{C}^*$.
7. For all nodes $e, c \in \mathcal{N}$, every term $\mathbf{C}$ and parameter A, if $\langle e, \mathbf{C}, c \rangle \in exps$, $\langle f(e), \mathbf{C}^*, f(c) \rangle \in exps^*$ and $A \in \mathbf{Params}(\mathbf{C})$, then
$$\varphi(A) \in \mathbf{Params}(\mathbf{C}^*).$$
8. For all nodes $a, b, c \in \mathcal{N}$, if $\langle c, a, b \rangle \in conns$, then
$$\langle f(c), f(a), f(b) \rangle \in conns^*.$$
9. For all nodes $e, d \in \mathcal{N}$, if $\langle e, d \rangle \in decs$, then
$$\langle f(e), f(d) \rangle \in decs^*.$$

Given a lifting map we can define a measure of how *close* the source expansion structure is to the target expansion structure.

DEFINITION 8.4.4. Let $\mathcal{Q}$ and $\mathcal{Q}^*$ be expansion structures and $\mathsf{f} = \langle \varphi, f \rangle$ be a lifting map from $\mathcal{Q}$ to $\mathcal{Q}^*$.

- Let $\mathbf{Im}(f; \mathcal{Q}) := \{f(a) \in \mathcal{N}^{\mathcal{Q}^*} \mid a \in \mathcal{N}^{\mathcal{Q}}\}$.

- Let $\|\mathsf{f} : \mathcal{Q} \to \mathcal{Q}^*\|$ be the ordinal

$$\omega^2 \cdot |\mathcal{N}^{\mathcal{Q}^*} \setminus \mathbf{Im}(f; \mathcal{Q})| + \omega \cdot |\mathcal{N}^{\mathcal{Q}} \setminus \mathbf{Dom}(succ^{\mathcal{Q}})| + |\varphi|_l^{\mathcal{Q}}.$$

As the definition of $\|\mathsf{f} : \mathcal{Q} \to \mathcal{Q}^*\|$ indicates, there are three situations in which the measure decreases. The first is when the image of the function contains a new node of $\mathcal{Q}^*$. The second is when a new node is extended. Finally, the measure decreases if the substitution φ is not as long. We handle the first two situations in the next lemma.

LEMMA 8.4.5. *Let $\mathcal{Q}$, $\mathcal{Q}'$ and $\mathcal{Q}^*$ be extensional expansion dags, $\mathsf{f} = \langle \varphi, f \rangle$ be a lifting map from $\mathcal{Q}$ to $\mathcal{Q}^*$ and $\mathsf{f}' = \langle \varphi', f' \rangle$ be a lifting map from $\mathcal{Q}'$ to $\mathcal{Q}^*$. Suppose $\mathcal{Q} \trianglelefteq \mathcal{Q}'$ and $f'|_{\mathcal{N}^{\mathcal{Q}}} = f$. If either $\mathcal{N}^{\mathcal{Q}'} \setminus \mathcal{N}^{\mathcal{Q}}$ is nonempty or $\mathbf{Dom}(succ^{\mathcal{Q}'}) \setminus \mathbf{Dom}(succ^{\mathcal{Q}})$ is nonempty, then*

$$\|\mathsf{f}' : \mathcal{Q}' \to \mathcal{Q}^*\| < \|\mathsf{f} : \mathcal{Q} \to \mathcal{Q}^*\|.$$

PROOF. See Appendix A.6. □

For any lifting map $\langle \varphi, f \rangle$ from $\mathcal{Q}$ to $\mathcal{Q}^*$, f is a graph homomorphism from $\langle \mathcal{N}^{\mathcal{Q}}, \overset{\mathcal{Q}}{\rightarrow} \rangle$ to $\langle \mathcal{N}^{\mathcal{Q}^*}, \overset{\mathcal{Q}^*}{\rightarrow} \rangle$.

LEMMA 8.4.6. *Let $\mathcal{Q}$ and $\mathcal{Q}^*$ be extensional expansion dags, $\langle \varphi, f \rangle$ be a lifting map from $\mathcal{Q}$ to $\mathcal{Q}^*$ and $a, b \in \mathcal{N}^{\mathcal{Q}}$ be nodes. If $a \overset{\mathcal{Q}}{\rightarrow} b$, then $f(a) \overset{\mathcal{Q}^*}{\rightarrow} f(b)$.*

PROOF. This follows from conditions (2), (3), (6), (8) and (9) of Definition 8.4.3. □

It will be useful to be able to extend a lifting map using the following lemma. While there are many conditions to check in order to apply Lemma 8.4.7, each of the conditions is relative to the *new* parts of the source extensional expansion dag. Since we will usually only add one new element to an extensional expansion dag at a time (e.g., an expansion arc, a connection, etc.) most of the conditions hold vacuously in each particular case.

LEMMA 8.4.7. *Let*

$$\mathcal{Q} = \langle \mathcal{N}, succ, label, sel, exps, conns, decs \rangle,$$

$$\mathcal{Q}' = \langle \mathcal{N}', succ', label', sel', exps', conns', decs' \rangle$$

and

$$\mathcal{Q}^* = \langle \mathcal{N}^*, succ^*, label^*, sel^*, exps^*, conns^*, decs^* \rangle$$

be extensional expansion dags, $\langle \varphi, f \rangle$ be a lifting map from $\mathcal{Q}$ to $\mathcal{Q}^*$, φ' be a substitution such that $\mathbf{Dom}(\varphi') \subset \mathcal{V} \cup \mathcal{P}$ and $f' : \mathcal{N}' \to \mathcal{N}^*$ be a function. Suppose $\mathcal{Q} \trianglelefteq \mathcal{Q}'$, $f'|_{\mathcal{N}} = f$ and $\varphi'(u) = \varphi(u)$ for every $u \in (\mathbf{Free}(\mathcal{Q}) \cup \mathbf{Params}(\mathcal{Q}))$. Assume the following conditions:

1. *For all $x \in \mathbf{Free}(\mathcal{Q}') \setminus \mathbf{Free}(\mathcal{Q})$, $\varphi'(x)$ is a closed long $\beta\eta$-form.*
2. *For all $A \in \mathbf{Params}(\mathcal{Q}') \setminus \mathbf{Params}(\mathcal{Q})$, $\varphi'(A)$ is a parameter.*
3. *For all $A \in \mathbf{Params}(\mathcal{Q}') \setminus \mathbf{Params}(\mathcal{Q})$ and $B \in \mathbf{Params}(\mathcal{Q}')$, if $\varphi'(A) = \varphi'(B)$, then $A = B$.*
4. *For every $a \in \mathcal{N}' \setminus \mathcal{N}$, $f'(a) \notin \mathbf{Im}(f; \mathcal{Q})$.*
5. *For all $a, b \in \mathcal{N}' \setminus \mathcal{N}$, if $f'(a) = f'(b)$, then $a = b$.*
6. *For every $a \in \mathcal{N}' \setminus \mathcal{N}$, if $label'(a) = \langle k, p, \mathbf{M} \rangle$, then*
$$label^*(f'(a)) = \langle k, p, (\varphi'(\mathbf{M}))^{\downarrow} \rangle.$$
7. *For every $a \in \mathbf{Dom}(succ') \setminus \mathbf{Dom}(succ)$, $f'(a) \in \mathbf{Dom}(succ^*)$.*
8. *For every $a \in \mathbf{Dom}(succ') \setminus \mathbf{Dom}(succ)$, if*
$$succ'(a) = \langle a^1, \ldots, a^n \rangle,$$
then $succ^*(f'(a)) = \langle f'(a^1), \ldots, f'(a^n) \rangle$.
9. *For every $a \in \mathbf{Dom}(sel') \setminus \mathbf{Dom}(sel)$, $f'(a) \in \mathbf{Dom}(sel^*)$ and*
$$sel^*(f'(a)) = \varphi'(sel'(a)).$$
10. *For all $A \in \mathbf{Params}(\mathcal{Q}') \setminus \mathbf{Params}(\mathcal{Q})$ and $a^* \in \mathbf{Dom}(sel^*)$, if $\varphi'(A) = sel^*(a^*)$, then there is some $a \in \mathbf{Dom}(sel')$ such that $f'(a) = a^*$ and $sel'(a) = A$.*
11. *For all $e, c \in \mathcal{N}'$ and every term $\mathbf{C}$, if $\langle e, \mathbf{C}, c \rangle \in exps' \setminus exps$, then*
$$\langle f'(e), \mathbf{C}^*, f'(c) \rangle \in exps^*$$
for some term $\mathbf{C}^*$ such that $\varphi'(\mathbf{C}) \stackrel{\beta\eta}{=} \mathbf{C}^*$.

12. *For all* $e, c \in \mathcal{N}'$, *every term* $\mathbf{C}$ *and every parameter* A, *if* $\langle e, \mathbf{C}, c\rangle \in exps' \setminus exps$, $\langle f(e), \mathbf{C}^*, f(c)\rangle \in exps^*$ *and*

$$A \in \mathbf{Params}(\mathbf{C}),$$

then

$$\varphi'(A) \in \mathbf{Params}(\mathbf{C}^*).$$

13. *For all nodes* $a, b, c \in \mathcal{N}'$, *if* $\langle c, a, b\rangle \in conns' \setminus conns$, *then*

$$\langle f'(c), f'(a), f'(b)\rangle \in conns^*.$$

14. *For all nodes* $e, d \in \mathcal{N}'$, *if* $\langle e, d\rangle \in decs' \setminus decs$, *then*

$$\langle f'(e), f'(d)\rangle \in decs^*.$$

Then $\langle \varphi', f'\rangle$ *is a lifting map from* $\mathcal{Q}'$ *to* $\mathcal{Q}^*$.

PROOF. The conditions for $\langle \varphi', f'\rangle$ to be a lifting map follow from the fact that $\langle \varphi, f\rangle$ is a lifting map and the assumptions of the lemma. □

LEMMA 8.4.8. *Let* $\mathcal{Q}$ *and* $\mathcal{Q}^*$ *be extensional expansion dags and* $\langle \varphi, f\rangle$ *be a lifting map from* $\mathcal{Q}$ *to* $\mathcal{Q}^*$. *For any* $a \in \mathcal{N}^{\mathcal{Q}}$ *if* $f(a)$ *is an atomic node in* $\mathcal{Q}^*$, *then* a *is either a flexible node or an atomic node in* $\mathcal{Q}$.

PROOF. Suppose $a \in \mathcal{N}^{\mathcal{Q}}$ and $f(a)$ is an atomic node in $\mathcal{Q}^*$. Then $f(a)$ is a primary node with shallow formula $[P^* \, \overline{\mathbf{B}^m}]$ for some parameter P^*. By Definition 8.4.3:1, a is a primary node with shallow formula $\mathbf{M}$ such that

$$(\varphi(\mathbf{M}))^{\downarrow} = [P^* \, \overline{\mathbf{B}^m}].$$

Hence we must have $\mathbf{M} \in wff_o$. Since $\mathbf{M}$ is $\beta\eta$-normal (as are all shallow formulas of nodes), $\mathbf{M}$ must be $[p \, \overline{\mathbf{A}^n}]$ for some $p \in \mathcal{V} \cup \mathcal{P} \cup \mathcal{S}$. Note that $\mathbf{Dom}(\varphi) \subset \mathcal{V} \cup \mathcal{P}$ since φ is a lifting substitution (cf. Definition 8.4.3 and 8.4.1). Assume p is a logical constant. Then $\varphi(p) = p$ and

$$[p \, \varphi(\mathbf{A}^1) \cdots \varphi(\mathbf{A}^n)] = \varphi([p \, \overline{\mathbf{A}^n}]) \stackrel{\beta\eta}{=} [P^* \, \overline{\mathbf{B}^m}].$$

Hence $p = P^*$, contradicting the fact that P^* is a parameter. Thus either p is a variable or a parameter and so a is a flexible node or an atomic node. □

LEMMA 8.4.9. *Let $\mathcal{Q}$ and $\mathcal{Q}^*$ be extensional expansion dags, $\mathsf{f} = \langle \varphi, f \rangle$ be a lifting map from $\mathcal{Q}$ to $\mathcal{Q}^*$ and $A, B \in \mathbf{SelPars}(\mathcal{Q})$ be selected parameters of $\mathcal{Q}$. If $A \prec_0^{\mathcal{Q}} B$, then $\varphi(A) \prec_0^{\mathcal{Q}^*} \varphi(B)$.*

PROOF. Suppose $A \prec_0^{\mathcal{Q}} B$. There is some $\langle e, \mathbf{C}, c \rangle \in exps^{\mathcal{Q}}$ such that $A \in \mathbf{Params}(\mathbf{C})$ and c dominates $\mathbf{SelNode}^{\mathcal{Q}}(B)$ in $\mathcal{Q}$. Let b be $\mathbf{SelNode}^{\mathcal{Q}}(B)$. Since f is a lifting map, $\langle f(e), \mathbf{C}^*, f(c) \rangle \in exps^{\mathcal{Q}^*}$, $\varphi(B) \in \mathbf{SelPars}(\mathcal{Q}^*)$ and $f(b)$ must be the node $\mathbf{SelNode}^{\mathcal{Q}^*}(\varphi(B))$ by conditions (6) and (4) of Definition 8.4.3. Also, we must have $\varphi(A) \in \mathbf{Params}(\mathbf{C}^*)$ by condition (7) of Definition 8.4.3. By Lemma 8.4.6, $f(c)$ dominates $f(b)$ in $\mathcal{Q}^*$. Thus $\langle f(e), \mathbf{C}^*, f(c) \rangle$ witnesses $\varphi(A) \prec_0^{\mathcal{Q}^*} \varphi(B)$. □

DEFINITION 8.4.10. Let $\mathcal{Q}$ and $\mathcal{Q}^*$ be extensional expansion dags and $\mathsf{f} = \langle \varphi, f \rangle$ be a lifting map from $\mathcal{Q}$ to $\mathcal{Q}^*$. For each $a \in \mathcal{N}^{\mathcal{Q}}$ and $b^* \in \mathcal{N}^{\mathcal{Q}^*}$, we say the pair $\langle a, b^* \rangle$ is a *schema for an edge with respect to* f if $b^* \notin \mathbf{Im}(f; \mathcal{Q})$ and $f(a) \overset{\mathcal{Q}^*}{\rightarrow} b^*$. Let $\mathbf{Edges}^-(\mathsf{f})$ be the set of all schemata for edges with respect to f.

DEFINITION 8.4.11. Let $\mathcal{Q}$ and $\mathcal{Q}^*$ be extensional expansion dags, f be a lifting map $\langle \varphi, f \rangle$ from $\mathcal{Q}$ to $\mathcal{Q}^*$ and

$$k \in \{\mathbf{mate}, \mathbf{eunif}, \mathbf{eunif}^{sym}\}$$

be a kind. For each $a, b \in \mathcal{N}^{\mathcal{Q}}$ and $c^* \in \mathcal{N}^{\mathcal{Q}^*}$, we say the triple $\langle c^*, a, b \rangle$ is a *schema for a k-connection with respect to* f if $\langle a, b \rangle$ is not k-connected in $\mathcal{Q}$, $kind^{\mathcal{Q}^*}(c^*) = k$ and $\langle c^*, f(a), f(b) \rangle \in conns^{\mathcal{Q}^*}$. Let $\mathbf{Conns}^-(\mathsf{f}; k)$ be the set of all schemata for k-connections with respect to f.

LEMMA 8.4.12. *Let $\mathcal{Q}$ be a developed extensional expansion dag, $\mathcal{Q}^*$ be an extensional expansion dag, $\langle \varphi, f \rangle$ be a lifting map from $\mathcal{Q}$ to $\mathcal{Q}^*$ and $a \in \mathcal{N}^{\mathcal{Q}}$ be a node.*

1. *If $f(a)$ is disjunctive in $\mathcal{Q}^*$, then a is disjunctive in $\mathcal{Q}$.*
2. *If $f(a)$ is neutral in $\mathcal{Q}^*$ and $f(a)$ is not a deepening node of $\mathcal{Q}^*$, then a is neutral in $\mathcal{Q}$.*

PROOF. This follows by a case analysis on the node a using condition (1) of Definition 8.4.3 and the assumption that $\mathcal{Q}$ is developed. □

The image of a flexible node under a lifting map may be a flexible node, an atomic node or a deepening node. In particular, we do not know the preimage of a deepening node is a deepening node. This corresponds to the fact that we may need to apply a primitive substitution during search. Under certain conditions we will be able to guarantee that the preimage of a deepening node is a deepening node. During search we will use these conditions to partition the search so that primitive substitutions are performed on variables at the head of a flexible node in a u-part before we consider adding connections or performing unification steps corresponding to the u-part.

First it is helpful to introduce the notion of an h-node.

DEFINITION 8.4.13. Let $\mathcal{Q}$ be an expansion structure, $a \in \mathcal{N}^{\mathcal{Q}}$ be a node and $h \in \mathcal{V} \cup \mathcal{P} \cup \mathcal{S}$ be a variable, parameter or logical constant. We say a is an *h-node of $\mathcal{Q}$* if $sh^{\mathcal{Q}}(a)$ is $[h\,\overline{\mathbf{A}^n}]$ for some $\mathbf{A}^1, \ldots, \mathbf{A}^n$.

Note that an h-node is a flexible node if h is a variable, an atomic node if h is a parameter and a deepening node if h is a logical constant. We can use this terminology to easily define the set of set variables of a collection of nodes.

DEFINITION 8.4.14. Let $\mathcal{Q}$ be an expansion structure and $\mathcal{R} \subseteq \mathcal{Q}$ be a set of nodes. We say p is a *set variable from $\mathcal{R}$ of $\mathcal{Q}$* if there is some flexible p-node $a \in \mathcal{R}$. We define $\mathbf{SetVars}^{\mathcal{Q}}(\mathcal{R})$ to be the set of all set variables from $\mathcal{R}$.

We say p is a *set variable of $\mathcal{Q}$* if p is a set variable from $\mathcal{N}^{\mathcal{Q}}$ of $\mathcal{Q}$.

We can use a lifting map to determine that a variable does not need to be instantiated with a primitive substitution. We will refer to this as being *almost atomic.*

DEFINITION 8.4.15. Let $\mathcal{Q}$ and $\mathcal{Q}^*$ be extensional expansion dags and f be a lifting map $\langle \varphi, f \rangle$ from $\mathcal{Q}$ to $\mathcal{Q}^*$. We say a free variable $p \in \mathbf{Free}(\mathcal{Q})$ is *almost atomic with respect to* f if p has type $(o\alpha^n \cdots \alpha^1)$ for some $n \geq 0$ and types $\alpha^1, \ldots, \alpha^n$ and the long $\beta\eta$-form $\varphi(p)$ is of the form $[\lambda \overline{x^n}\, [h\, \overline{\mathbf{C}^l}]]$ where $h \in \mathcal{P}$.

We say a set of variables $\mathcal{U} \subseteq \mathbf{Free}(\mathcal{Q})$ is *almost atomic with respect to* f if every variable in $\mathcal{U}$ is almost atomic with respect to f.

We can use the assumption that the set variables from a collection of nodes is almost atomic to determine that some preimages of deepening nodes are guaranteed to be deepening nodes.

LEMMA 8.4.16. *Let $\mathcal{Q}$ and $\mathcal{Q}^*$ be extensional expansion dags, $\mathsf{f} = \langle \varphi, f \rangle$ be a lifting map from $\mathcal{Q}$ to $\mathcal{Q}^*$ and $\mathcal{R} \subseteq \mathcal{N}^{\mathcal{Q}}$ be a set of nodes in $\mathcal{Q}$. Suppose* $\mathbf{SetVars}^{\mathcal{Q}}(\mathcal{R})$ *is almost atomic with respect to* f. *For every $a \in \mathcal{R}$ if $f(a)$ is a deepening node in $\mathcal{Q}^*$, then a is a deepening node in $\mathcal{Q}$.*

PROOF. See Appendix A.6. □

We now combine Lemmas 8.4.16 and 8.4.12 to obtain a result about preimages of neutral nodes.

LEMMA 8.4.17. *Let $\mathcal{Q}$ be a developed extensional expansion dag, $\mathcal{Q}^*$ be an extensional expansion dag, $\mathsf{f} = \langle \varphi, f \rangle$ be a lifting map from $\mathcal{Q}$ to $\mathcal{Q}^*$ and $\mathcal{R} \subseteq \mathcal{N}^{\mathcal{Q}}$ be a set of nodes in $\mathcal{Q}$. Suppose* $\mathbf{SetVars}^{\mathcal{Q}}(\mathcal{R})$ *is almost atomic with respect to* f. *For every $a \in \mathcal{R}$ if $f(a)$ is a neutral node in $\mathcal{Q}^*$, then a is a neutral node in $\mathcal{Q}$.*

PROOF. If $f(a)$ is not a deepening node, then we know a is a neutral node by Lemma 8.4.12:2. If $f(a)$ is a deepening node, then we know a is a deepening node by Lemma 8.4.16. Since $\mathcal{Q}$ is developed, a is extended and hence neutral. □

In order to handle information about duplications, we define the set of expansion nodes in a collection of nodes.

DEFINITION 8.4.18. Let $\mathcal{Q}$ be an expansion structure and $\mathcal{R} \subseteq \mathcal{Q}$ be a set of nodes. We define $\mathbf{ExpNodes}^{\mathcal{Q}}(\mathcal{R})$ to be the set of expansion nodes of $\mathcal{Q}$ in $\mathcal{R}$.

DEFINITION 8.4.19. Let $\mathcal{Q}$ and $\mathcal{Q}^*$ be extensional expansion dags and $\mathsf{f} = \langle \varphi, f \rangle$ be a lifting map from $\mathcal{Q}$ to $\mathcal{Q}^*$. We say an expansion node $e \in \mathcal{N}^{\mathcal{Q}}$ is *fully duplicated with respect to* f if the following condition holds:

- For every node $c^* \in \mathcal{N}^{\mathcal{Q}^*}$, if $f(e) \overset{\mathcal{Q}^*}{\rightharpoondown} c^*$, then there exists a node $c \in \mathcal{N}^{\mathcal{Q}}$ such that $e \overset{\mathcal{Q}}{\rightharpoondown} c$ and $f(c) = c^*$.

We say a set of nodes $\mathcal{R} \subseteq \mathcal{N}^{\mathcal{Q}}$ is *fully duplicated with respect to* f if every expansion node in $\mathcal{R}$ is fully duplicated with respect to f.

LEMMA 8.4.20. *Let $\mathcal{Q}$, $\mathcal{Q}'$ and $\mathcal{Q}^*$ be extensional expansion dags, f be a lifting map $\langle\varphi, f\rangle$ from $\mathcal{Q}$ to $\mathcal{Q}^*$, and f' be a lifting map $\langle\varphi', f'\rangle$ from $\mathcal{Q}'$ to $\mathcal{Q}^*$. Suppose $\mathcal{Q} \trianglelefteq \mathcal{Q}'$, $f'|_{\mathcal{N}^{\mathcal{Q}}} = f$ and $\varphi'(u) = \varphi(u)$ for every $u \in (\mathbf{Free}(\mathcal{Q}) \cup \mathbf{Params}(\mathcal{Q}))$. Then we have the following:*

1. *If $\mathscr{D} \subseteq \mathcal{N}^{\mathcal{Q}}$ is fully duplicated with respect to f, then $\mathscr{D} \subseteq \mathcal{N}^{\mathcal{Q}'}$ is fully duplicated with respect to f'.*
2. *If $\mathscr{S} \subseteq \mathbf{Free}(\mathcal{Q})$ is almost atomic with respect to f, then $\mathscr{S} \subseteq \mathbf{Free}(\mathcal{Q}')$ is almost atomic with respect to f'.*

PROOF. Suppose $\mathscr{D}$ is fully duplicated with respect to f and $e \in \mathscr{D}$ is an expansion node of $\mathcal{Q}'$. Since $\mathscr{D} \subseteq \mathcal{N}^{\mathcal{Q}}$ and $\mathcal{Q} \trianglelefteq \mathcal{Q}'$, e is also an expansion node of $\mathcal{Q}$. Suppose $c^* \in \mathcal{N}^{\mathcal{Q}^*}$ and $f'(e) \stackrel{\mathcal{Q}^*}{\rightharpoondown} c^*$. Since $e \in \mathcal{N}^{\mathcal{Q}}$, $f'(e) = f(e)$. Since $f(e) \stackrel{\mathcal{Q}^*}{\rightharpoondown} c^*$ and $e \in \mathscr{D}$ is fully duplicated with respect to f, there is some $c \in \mathcal{N}^{\mathcal{Q}}$ such that $e \stackrel{\mathcal{Q}}{\rightharpoondown} c$ and $f(c) = c^*$. Hence $c \in \mathcal{N}^{\mathcal{Q}'}$ and $f'(c) = f(c) = c^*$. By Lemma 7.6.2, $e \stackrel{\mathcal{Q}'}{\rightharpoondown} c$. Thus e is fully duplicated with respect to f'. Therefore, $\mathscr{D}$ is fully duplicated with respect to f'.

Suppose $\mathscr{S} \subseteq \mathbf{Free}(\mathcal{Q})$ is almost atomic with respect to f and $p \in \mathscr{S}$. Hence p has a type of the form $(o\alpha^n \cdots \alpha^1)$ and the long $\beta\eta$-form $\varphi(p)$ is of the form $[\lambda \overline{x^n}\, [h\, \overline{\mathbf{C}^l}]]$ where $h \in \mathcal{P}$. Since $p \in \mathbf{Free}(\mathcal{Q})$, $\varphi'(p) = \varphi(p)$. Thus $\varphi'(p)$ also has this form and we conclude p is almost atomic with respect to f'. Therefore, $\mathscr{S}$ is almost atomic with respect to f'. □

LEMMA 8.4.21. *Let $\mathcal{Q}$ and $\mathcal{Q}^*$ be extensional expansion dags, θ be a substitution, $\mathsf{f} = \langle\varphi, f\rangle$ be a lifting map from $\mathcal{Q}$ to $\mathcal{Q}^*$, and $\mathsf{f}' = \langle\varphi', f'\rangle$ be a lifting map from $\theta(\mathcal{Q})$ to $\mathcal{Q}^*$. Suppose $f' = f$ and $\varphi'(u) = \varphi(u)$ for every $u \in (\mathbf{Free}(\mathcal{Q}) \cup \mathbf{Params}(\mathcal{Q})) \setminus \mathbf{Dom}(\theta)$. We have the following:*

1. *If $\mathscr{D} \subseteq \mathcal{N}^{\mathcal{Q}}$ is fully duplicated with respect to f, then $\mathscr{D} \subseteq \mathcal{N}^{\theta(\mathcal{Q})}$ is fully duplicated with respect to f'.*
2. *If $\mathscr{S} \subseteq \mathbf{Free}(\mathcal{Q})$ is almost atomic with respect to f, then*

$$(\mathscr{S} \setminus \mathbf{Dom}(\theta)) \subseteq \mathbf{Free}(\theta(\mathcal{Q}))$$

is almost atomic with respect to f'.

PROOF. See Appendix A.6. □

8.5. Strict Expansions

Expansion terms will be incrementally instantiated during the search. At each stage, we will ensure every variable in the expansion term has a strict occurrence (cf. [54]). This restriction will be vital when showing a lifting map always indicates a legal step for making progress towards an extensional expansion proof (cf. Lemma 8.8.4). In particular, there are two situations (cf. Definitions 8.7.16 and 8.7.20) when the lifting map will indicate that an expansion variable x should be instantiated by imitating a parameter A (cf. Definition 8.7.2). We will only allow such imitation steps when the parameter A is not banned for x (cf. Definitions 8.2.10, 8.7.15, 8.7.16, and 8.7.20). Consequently, when a lifting map indicates an imitation step is appropriate, we must be able to verify that the corresponding parameter is not banned for the corresponding expansion variable. We will verify this fact using Lemma 8.5.10. Lemma 8.5.10, in turn, is proved using the assumption that expansions are strict and Lemma 8.4.9. Furthermore, even the proof of Lemma 8.4.9 above made use of condition (7) of the definition of lifting maps (cf. Definition 8.4.3). To ensure condition (7) is preserved by the steps performed during search, we use the assumption that expansion terms are strict by applying Lemma 8.5.6.

DEFINITION 8.5.1. We define a set of variables $\mathbf{Free_{str}}(\mathbf{A}) \subset \mathcal{V}$ by induction on the length $|\mathbf{A}^{\updownarrow}|_l$. The long $\beta\eta$-form $\mathbf{A}^{\updownarrow}$ is always of the form $[\lambda \overline{z^n}\, [h\, \overline{\mathbf{A}^m}]_\beta]$ for some $n \geq 0$, $m \geq 0$, $\beta \in \{o, \iota\}$ and $h \in \mathcal{V} \cup \mathcal{P} \cup \mathcal{S}$. By Lemma 2.1.32, we know

$$|[\lambda \overline{z^n}\, \mathbf{A}^j]^{\updownarrow}|_l < |[\lambda\, \overline{z^n}\, [h\, \overline{\mathbf{A}^m}]]^{\updownarrow}|_l.$$

Assuming $\mathbf{A}^{\updownarrow}$ is $[\lambda \overline{z^n}\, [h\, \overline{\mathbf{A}^m}]]$ and $\mathbf{Free_{str}}([\lambda \overline{z^n}\, \mathbf{A}^j])$ is defined for each j with $1 \leq j \leq m$, we define the set $\mathbf{Free_{str}}(\mathbf{A})$ by cases:

1. If $h \in (\mathcal{P} \cup \mathcal{S} \cup \{z^1, \ldots, z^n\})$, let $\mathbf{Free_{str}}(\mathbf{A})$ be
$$\mathbf{Free_{str}}([\lambda \overline{z^n} \mathbf{A}^1]) \cup \cdots \cup \mathbf{Free_{str}}([\lambda \overline{z^n} \mathbf{A}^m])$$
2. If $h \in \mathcal{V} \setminus \{z^1, \ldots, z^n\}$, $(\mathbf{A}^j)^{\downarrow} \in \{z^1, \ldots, z^n\}$ for each j with $1 \leq j \leq m$, and $(\mathbf{A}^1)^{\downarrow}, \ldots, (\mathbf{A}^m)^{\downarrow}$ are distinct, let $\mathbf{Free_{str}}(\mathbf{A})$ be $\{h\}$.
3. Otherwise, let $\mathbf{Free_{str}}(\mathbf{A})$ be $\emptyset$.

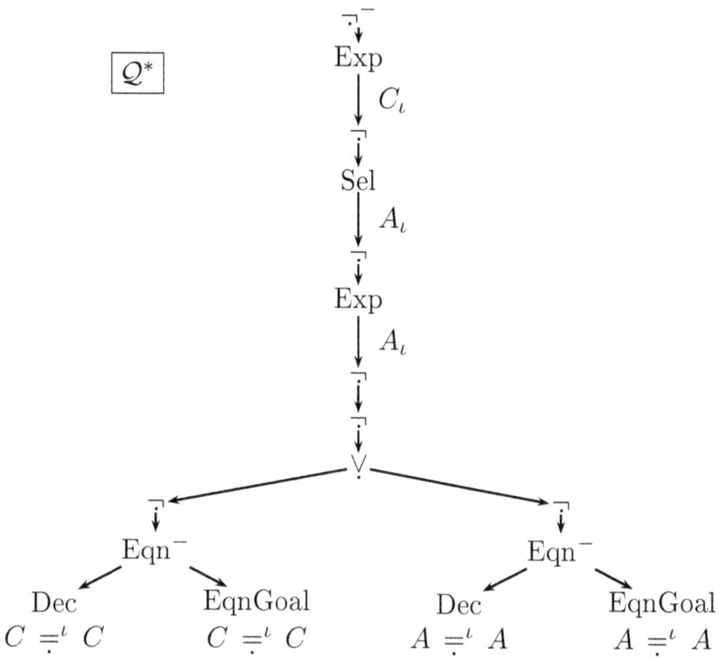

FIGURE 8.1. Extensional Expansion Proof

We say

x has a strict occurrence in a term $\mathbf{A}$

if $x \in \mathbf{Free_{str}}(\mathbf{A})$. We say a term $\mathbf{A}$ is *strict* if every $x \in \mathbf{Free}(\mathbf{A})$ has a strict occurrence in $\mathbf{A}$.

Note that

$$\mathbf{Free_{str}}(\mathbf{A}) = \mathbf{Free_{str}}(\mathbf{B})$$

whenever $\mathbf{A} \overset{\beta\eta}{=} \mathbf{B}$ since these sets are defined in terms of long $\beta\eta$-forms.

Before proving results regarding strictness, we consider a small example to demonstrate why we need the assumption that expansion terms are strict. The extensional expansion dag $\mathcal{Q}^*$ shown in Figure 8.1 provides an extensional expansion proof for the theorem

$$\dot{\exists} y_\iota \, \dot{\forall} a_\iota \, \dot{\exists} x_\iota \,\blacksquare\, y \mathrel{\dot{=}^\iota} C \mathrel{\dot{\wedge}} x \mathrel{\dot{=}^\iota} a.$$

The two expansion terms in this extensional expansion dag are the parameter C_ι and the selected parameter A_ι.

Assume while searching for a proof of

$$\dot{\exists} y_\iota \dot{\forall} a_\iota \dot{\exists} x_\iota \centerdot y \dot{=}^\iota C \dot{\wedge} x \dot{=}^\iota a$$

we obtain the extensional expansion dag $\mathcal{Q}$ shown in Figure 8.2. The expansion terms in this case are $[u_{\iota\iota}\, x_\iota]$ and x_ι where $u_{\iota\iota}$ and x_ι are expansion variables. Note that $[u\, x]$ is not a strict term since neither u nor x have a strict occurrence in $[u\, x]$. In this example the fact that x has no strict occurrence causes a problem in an attempt to lift the proof.

Let φ be the substitution given by $\varphi(u) := [\lambda z_\iota\, C_\iota]$ and $\varphi(x) := A_\iota$. Note that φ is a lifting substitution for $\mathcal{Q}$. After development, the instantiated extensional expansion dag $\varphi(\mathcal{Q})$ would be the extensional expansion proof $\mathcal{Q}^*$. In such a situation the lifting substitution φ indicates that we should instantiate the expansion variable x_ι by imitating the selected parameter A_ι. However, the selected parameter A is banned for x in $\mathcal{Q}$ since x occurs in the expansion term $[u\, x]$. We can avoid such situations by insisting all expansion terms are strict.

We could maintain the stronger property expansion terms are higher-order patterns (cf. [45]). However, requiring expansion terms to be strict is enough to prove completeness of search (cf. Theorem 8.8.6).

We now prove several results regarding strictness.

LEMMA 8.5.2. *Let* $\mathbf{A}$ *and* $\mathbf{C}_\gamma$ *be terms and* y_γ *and* x *be variables. If* $x \in \mathbf{Free_{str}}(\mathbf{A}) \setminus \{y\}$, *then* $x \in \mathbf{Free_{str}}([\mathbf{C}/y]\mathbf{A})$.

PROOF. The proof is by induction on the length of $\mathbf{A}^{\updownarrow}$ where x has a strict occurrence in $\mathbf{A}$. See Appendix A.6. □

DEFINITION 8.5.3. A term is called an *$\mathcal{S}$-general binding* if it has the form

$$[\lambda x^1_{\alpha^1} \cdots \lambda x^n_{\alpha^n} [R\, [w^1\, \overline{x^n}] \cdots [w^m\, \overline{x^n}]]]$$

where

- $R_{\beta\gamma^m\cdots\gamma^1} \in \mathcal{P} \cup \mathcal{S} \cup \{x^1, \ldots, x^n\}$,
- $\beta \in \{o, \iota\}$,
- $w^j_{\gamma^j\alpha^n\cdots\alpha^1} \in \mathcal{V} \setminus \{x^1, \ldots, x^n\}$ is a variable for each j $(1 \leq j \leq m)$ and

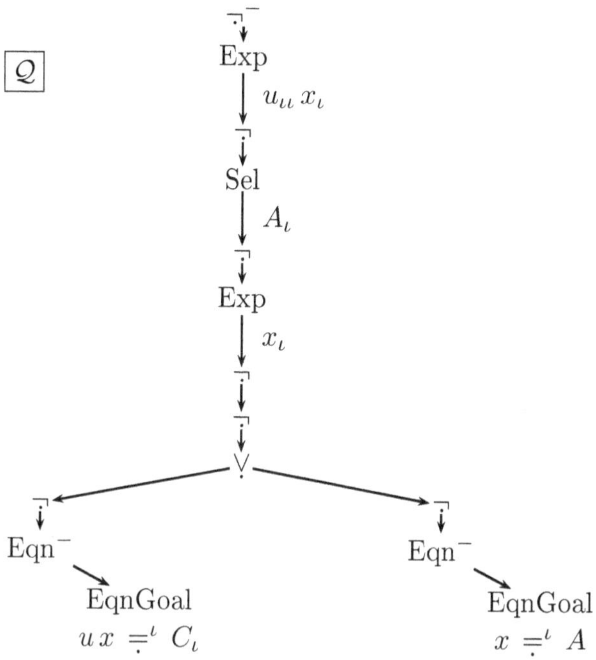

FIGURE 8.2. Example Without Strict Expansion Terms

- the variables $w^1, \ldots, w^m$ are distinct.

We often refer to $\mathcal{S}$-general bindings simply as *general bindings*, leaving the signature implicit.

LEMMA 8.5.4. *Let* $\mathbf{A}$ *be a term,* y_γ *be a variable and* $\mathbf{G}_\gamma$ *be a general binding. If* $y \in \mathbf{Free_{str}}(\mathbf{A})$ *and* $w \in \mathbf{Free}(\mathbf{G})$*, then*

$$w \in \mathbf{Free_{str}}([\mathbf{G}/y]\mathbf{A}).$$

PROOF. The proof is by induction on the length of $\mathbf{A}^\updownarrow$ where y has a strict occurrence in $\mathbf{A}$. See Appendix A.6. □

LEMMA 8.5.5. *Let* $\mathbf{A}$ *be a strict term,* y_γ *be a variable and* $\mathbf{G}_\gamma$ *be a general binding. If* $\mathbf{A}$ *is strict, then* $[\mathbf{G}/y]\mathbf{A}$ *is strict.*

PROOF. Suppose $x \in \mathbf{Free}([\mathbf{G}/y]\mathbf{A})$. Hence $x \in (\mathbf{Free}(\mathbf{A}) \setminus \{y\})$ or $x \in \mathbf{Free}(\mathbf{G})$. If $x \in (\mathbf{Free}(\mathbf{A}) \setminus \{y\})$, then x has a strict occurrence in $[\mathbf{G}/y]\mathbf{A}$ by Lemma 8.5.2. If $x \in \mathbf{Free}(\mathbf{G})$, then x has a strict occurrence in $[\mathbf{G}/y]\mathbf{A}$ by Lemma 8.5.4. □

LEMMA 8.5.6. *Let* $\mathbf{A}$ *be a term,* φ *be a substitution,* y_γ *be a variable and* $\mathbf{G}_\gamma$ *be the general binding*

$$[\lambda x^1_{\alpha^1} \cdots \lambda x^n_{\alpha^n} [R\, [w^1\, \overline{x^n}] \cdots [w^m\, \overline{x^n}]]]$$

where $R_{\beta\gamma^m\ldots\gamma^1} \in \mathcal{P}$ *and* $\beta \in \{o, \iota\}$. *If* y *has a strict occurrence in* $\mathbf{A}$, φ *respects parameters and* $\mathbf{Dom}(\varphi) \subset (\mathcal{V} \cup \mathcal{P})$, *then*

$$\varphi(R) \in \mathbf{Params}((\varphi([\mathbf{G}/y]\mathbf{A}))^{\downarrow}).$$

PROOF. The proof is by induction on the length of $\mathbf{A}^{\updownarrow}$ where y has a strict occurrence in $\mathbf{A}$. See Appendix A.6. □

LEMMA 8.5.7. *Let* $\mathbf{A}$ *be a term,* φ *be a substitution and* x *be a variable which has a strict occurrence in* $\mathbf{A}$. *Suppose* φ *respects parameters and* $\mathbf{Dom}(\varphi) \subset \mathcal{V} \cup \mathcal{P}$. *For any parameter* $B \in \mathbf{Params}((\varphi(x))^{\updownarrow})$, *we have* $B \in \mathbf{Params}((\varphi(\mathbf{A}))^{\updownarrow})$.

PROOF. The proof is by induction on the length of $\mathbf{A}^{\updownarrow}$ where x has a strict occurrence in $\mathbf{A}$. See Appendix A.6. □

DEFINITION 8.5.8. We say an expansion structure $\mathcal{Q}$ *has strict expansions* if every expansion term $\mathbf{C} \in \mathbf{ExpTerms}(\mathcal{Q})$ is strict.

LEMMA 8.5.9. *Let* $\mathcal{Q}$ *be an expansion structure,* y_γ *be a variable and* $\mathbf{G}_\gamma$ *be a general binding. If* $\mathcal{Q}$ *has strict expansions, then* $([\mathbf{G}/y]\mathcal{Q})$ *has strict expansions.*

PROOF. Suppose $\mathbf{D} \in \mathbf{ExpTerms}(([\mathbf{G}/y]\mathcal{Q}))$ and $\mathcal{Q}$ has strict expansions. There is some strict expansion term $\mathbf{C} \in \mathbf{ExpTerms}(\mathcal{Q})$ such that $\mathbf{D}$ is $[\mathbf{G}/y]\mathbf{C}$ (cf. Definition 8.2.2). By Lemma 8.5.5, $\mathbf{D}$ is strict. Thus $([\mathbf{G}/y]\mathcal{Q})$ has strict expansions. □

LEMMA 8.5.10. *Let* $\mathcal{Q}$ *and* $\mathcal{Q}^*$ *be extensional expansion dags,* $\mathsf{f} = \langle \varphi, f \rangle$ *be a lifting map from* $\mathcal{Q}$ *to* $\mathcal{Q}^*$, $x \in \mathbf{Free}(\mathcal{Q})$ *and* B *be a parameter. If*

- $\mathcal{Q}$ *has strict expansions,*
- $\prec_0^{\mathcal{Q}^*}$ *is acyclic and*
- $\varphi(B) \in \mathbf{Params}(\varphi(x))$,

then B *is not banned for* x *in* $\mathcal{Q}$.

PROOF. Assume B is banned for x in $\mathcal{Q}$. By Definition 8.2.10, there are parameters $A^1, \ldots, A^n \in \mathbf{SelPars}(\mathcal{Q})$ (where $n \geq 1$) and an expansion arc $\langle e, \mathbf{C}, c\rangle \in exps^{\mathcal{Q}}$ such that $x \in \mathbf{Free}(\mathbf{C})$, $A^n = B$,

$$A^1 \prec_0^{\mathcal{Q}} \cdots \prec_0^{\mathcal{Q}} A^n$$

and c dominates $\mathbf{SelNode}^{\mathcal{Q}}(A^1)$. Let a be $\mathbf{SelNode}^{\mathcal{Q}}(A^1)$. Since f is a lifting map, there is some term $\mathbf{C}^*$ such that $\mathbf{C}^* \stackrel{\beta\eta}{=} \varphi(\mathbf{C})$ and

$$\langle f(e), \mathbf{C}^*, f(c)\rangle \in exps^{\mathcal{Q}^*}$$

by condition (6) of Definition 8.4.3. We also know

$$\varphi(A^1), \ldots, \varphi(A^n) \in \mathbf{SelPars}(\mathcal{Q}^*)$$

and $f(a)$ is $\mathbf{SelNode}^{\mathcal{Q}^*}(\varphi(A^1))$ by condition (4) of Definition 8.4.3. By Lemma 8.4.6, $f(c)$ dominates $f(a)$ in $\mathcal{Q}^*$. By Lemma 8.4.9,

$$\varphi(A^1) \prec_0^{\mathcal{Q}^*} \cdots \prec_0^{\mathcal{Q}^*} \varphi(A^n).$$

Since $\mathcal{Q}$ has strict expansions, x has a strict occurrence in $\mathbf{C}$. Since $\varphi(B) \in \mathbf{Params}(\varphi(x))$ and $\varphi(x)$ is a long $\beta\eta$-form, we have

$$\varphi(B) \in \mathbf{Params}((\varphi(\mathbf{C}))^{\updownarrow})$$

by Lemma 8.5.7. Since η-expansions do not introduce parameters,

$$\varphi(B) \in \mathbf{Params}((\varphi(\mathbf{C}))^{\downarrow}).$$

Since $\mathbf{C}^* \stackrel{\beta\eta}{=} \varphi(\mathbf{C})$ and $\beta\eta$-reductions never introduce parameters, we must have $\varphi(A^n) = \varphi(B) \in \mathbf{Params}(\mathbf{C}^*)$. Hence the expansion arc $\langle f(e), \mathbf{C}^*, f(c)\rangle \in exps^{\mathcal{Q}^*}$ witnesses $\varphi(A^n) \prec_0^{\mathcal{Q}^*} \varphi(A^1)$. Thus we have the $\prec_0^{\mathcal{Q}^*}$-cycle

$$\varphi(A^1) \prec_0^{\mathcal{Q}^*} \cdots \prec_0^{\mathcal{Q}^*} \varphi(A^n) \prec_0^{\mathcal{Q}^*} \varphi(A^1),$$

contradicting acyclicity of $\prec_0^{\mathcal{Q}^*}$. □

8.6. Search States

For simplicity, we only consider extensional expansion dags with a unique specified root during search. We can impose a certain amount of order to the search space by keeping two sets of information with the extensional expansion dags. The first will be a set $\mathscr{D}$ of expansion nodes which we have stopped duplicating. The second will be a set $\mathscr{S}$ of set variables for which we have stopped making primitive substitutions.

DEFINITION 8.6.1 (Search State). Let $\mathcal{S}$ be a signature, $\mathcal{Q}$ be an $\mathcal{S}$-expansion structure, $r \in \mathcal{N}^{\mathcal{Q}}$ be a node, $\mathscr{D} \subseteq \mathbf{ExpNodes}^{\mathcal{Q}}(\mathcal{N}^{\mathcal{Q}})$ be a set of expansion nodes and $\mathscr{S} \subseteq \mathbf{SetVars}^{\mathcal{Q}}(\mathcal{N}^{\mathcal{Q}})$ be a set of variables. We say

$$\Xi = \langle \mathcal{Q}, r, \mathscr{D}, \mathscr{S} \rangle$$

is an $\mathcal{S}$-*search state* if $\mathcal{Q}$ is an extensional expansion dag, $\mathbf{Roots}(\mathcal{Q})$ is $\{r\}$, $sh^{\mathcal{Q}}(r)$ is closed, $\mathbf{RootSels}(\mathcal{Q})$ is empty, $\mathcal{Q}$ has strict expansions and $\prec_0^{\mathcal{Q}}$ is acyclic. We often refer to an $\mathcal{S}$-search state as a *search state* and leave the signature $\mathcal{S}$ implicit. We use the notation $\mathcal{Q}^{\Xi}$ to denote $\mathcal{Q}$.

DEFINITION 8.6.2. Let Ξ be a search state $\langle \mathcal{Q}, r, \mathscr{D}, \mathscr{S} \rangle$. We say Ξ is *developed* if $\mathcal{Q}$ is developed. We say Ξ is *pre-solved* if

$$\mathcal{U}parts^{-ps}(\mathcal{Q}; \{r\}) = \emptyset.$$

If a search state is pre-solved, then we can easily obtain an extensional expansion proof using Theorem 8.3.3.

THEOREM 8.6.3. *Let $\mathcal{S}$ be a signature and $\Xi = \langle \mathcal{Q}, r, \mathscr{D}, \mathscr{S} \rangle$ be a pre-solved $\mathcal{S}$-search state. There is an $(\mathcal{S} \cup \{\top\})$-extensional expansion dag $\mathcal{Q}'$ such that $\mathbf{Roots}(\mathcal{Q}') = \{r\}$, $\langle \mathcal{Q}', \mathbf{Roots}(\mathcal{Q}) \rangle$ is an $(\mathcal{S} \cup \{\top\})$-extensional expansion proof and*

$$label^{\mathcal{Q}'}(r) = label^{\mathcal{Q}}(r).$$

PROOF. Since Ξ is a search state we know $\mathcal{Q}$ is an $\mathcal{S}$-extensional expansion dag, $\mathbf{Roots}(\mathcal{Q}) = \{r\}$, $sh^{\mathcal{Q}}(r)$ is closed, $\mathbf{RootSels}(\mathcal{Q})$ is empty and $\prec_0^{\mathcal{Q}}$ is acyclic. Hence $\mathbf{Free}^{\mathcal{Q}}(\mathbf{Roots}(\mathcal{Q}))$ is empty. Since Ξ is pre-solved and $\mathbf{Roots}(\mathcal{Q}) = \{r\}$, we know $\mathcal{U}parts^{-ps}(\mathcal{Q}; \mathbf{Roots}(\mathcal{Q}))$ is empty. That is, every u-part in $\mathcal{U}parts(\mathcal{Q}; \mathbf{Roots}(\mathcal{Q}))$ is pre-solved. By Theorem 8.3.3, there is an $(\mathcal{S} \cup \{\top\})$-extensional expansion dag $\mathcal{Q}'$ such that $\mathbf{Roots}(\mathcal{Q}') = \{r\}$, $\langle \mathcal{Q}', \mathbf{Roots}(\mathcal{Q}) \rangle$ is an $(\mathcal{S} \cup \{\top\})$-extensional expansion proof and

$$label^{\mathcal{Q}'}(r) = label^{\mathcal{Q}}(r).$$

□

LEMMA 8.6.4. *Let $\mathcal{Q}$ and $\mathcal{Q}'$ be extensional expansion dags such that $\mathcal{Q} \trianglelefteq \mathcal{Q}'$. Then*

$$\mathbf{ExpNodes}^{\mathcal{Q}}(\mathcal{N}^{\mathcal{Q}}) \subseteq \mathbf{ExpNodes}^{\mathcal{Q}'}(\mathcal{N}^{\mathcal{Q}'})$$

and

$$\mathbf{SetVars}^{\mathcal{Q}}(\mathcal{N}^{\mathcal{Q}}) \subseteq \mathbf{SetVars}^{\mathcal{Q}'}(\mathcal{N}^{\mathcal{Q}'}).$$

PROOF. Suppose $e \in \mathbf{ExpNodes}^{\mathcal{Q}}(\mathcal{N}^{\mathcal{Q}})$. Then e is an expansion node in $\mathcal{Q}$. Since $e \in \mathcal{N}^{\mathcal{Q}} \subseteq \mathcal{N}^{\mathcal{Q}'}$ and $label^{\mathcal{Q}'}(e) = label^{\mathcal{Q}}(e)$, we conclude $e \in \mathbf{ExpNodes}^{\mathcal{Q}'}(\mathcal{N}^{\mathcal{Q}'})$.

Suppose $p \in \mathbf{SetVars}^{\mathcal{Q}}(\mathcal{N}^{\mathcal{Q}})$. Then p is a variable and there is a p-node $a \in \mathcal{N}^{\mathcal{Q}} \subseteq \mathcal{N}^{\mathcal{Q}'}$. Since $label^{\mathcal{Q}'}(a) = label^{\mathcal{Q}}(a)$, a is a p-node in $\mathcal{Q}'$ and so $p \in \mathbf{SetVars}^{\mathcal{Q}'}(\mathcal{N}^{\mathcal{Q}'})$. □

LEMMA 8.6.5. *Let* $\Xi = \langle \mathcal{Q}, r, \mathscr{D}, \mathscr{S} \rangle$ *be a search state and* $\mathcal{Q}'$ *be an extensional expansion dag. If* $\mathcal{Q} \trianglelefteq_r \mathcal{Q}'$, $\mathbf{RootSels}(\mathcal{Q}')$ *is empty,* $\prec_0^{\mathcal{Q}'}$ *is acyclic and every* $\mathbf{C} \in (\mathbf{ExpTerms}(\mathcal{Q}') \setminus \mathbf{ExpTerms}(\mathcal{Q}))$ *is strict, then* $\langle \mathcal{Q}', r, \mathscr{D}, \mathscr{S} \rangle$ *is a search state.*

PROOF. By Definition 8.6.1, we know $\mathbf{Roots}(\mathcal{Q}) = \{r\}$, $sh^{\mathcal{Q}}(r)$ is closed, $\mathcal{Q}$ has strict expansions,

$$\mathscr{D} \subseteq \mathbf{ExpNodes}^{\mathcal{Q}}(\mathcal{N}^{\mathcal{Q}}) \text{ and } \mathscr{S} \subseteq \mathbf{SetVars}^{\mathcal{Q}}(\mathcal{N}^{\mathcal{Q}}).$$

Every $\mathbf{C} \in \mathbf{ExpTerms}(\mathcal{Q})$ is strict since $\mathcal{Q}$ has strict expansions and every $\mathbf{C} \in (\mathbf{ExpTerms}(\mathcal{Q}') \setminus \mathbf{ExpTerms}(\mathcal{Q}))$ is strict by assumption. Hence $\mathcal{Q}'$ has strict expansions. Since $\mathcal{Q} \trianglelefteq_r \mathcal{Q}'$ we know $\mathbf{Roots}(\mathcal{Q}') = \mathbf{Roots}(\mathcal{Q}) = \{r\}$ and $sh^{\mathcal{Q}'}(r) = sh^{\mathcal{Q}}(r)$ is closed. By Lemma 8.6.4,

$$\mathscr{D} \subseteq \mathbf{ExpNodes}^{\mathcal{Q}}(\mathcal{N}^{\mathcal{Q}}) \subseteq \mathbf{ExpNodes}^{\mathcal{Q}'}(\mathcal{N}^{\mathcal{Q}'})$$

and

$$\mathscr{S} \subseteq \mathbf{SetVars}^{\mathcal{Q}}(\mathcal{N}^{\mathcal{Q}}) \subseteq \mathbf{SetVars}^{\mathcal{Q}'}(\mathcal{N}^{\mathcal{Q}'}).$$

We have assumed $\mathbf{RootSels}(\mathcal{Q}')$ is empty and $\prec_0^{\mathcal{Q}'}$ is acyclic. Therefore,

$$\langle \mathcal{Q}', r, \mathscr{D}, \mathscr{S} \rangle$$

is a search state. □

DEFINITION 8.6.6 (Lifting Target). Let $\mathcal{S}$ be a signature of logical constants, $\mathcal{Q}^*$ be an $\mathcal{S}$-extensional expansion dag and $r^* \in \mathbf{Roots}(\mathcal{Q}^*)$ be a root node in $\mathcal{Q}^*$. We say $\langle \mathcal{Q}^*, r^* \rangle$ is an $\mathcal{S}$-*lifting target* if $\langle \mathcal{Q}^*, \{r^*\} \rangle$ is an $\mathcal{S}$-extensional expansion proof, $\mathbf{Roots}(\mathcal{Q}^*) = \{r^*\}$, $\mathcal{Q}^*$ is developed and separated and $\mathcal{Q}^*$ has merged connections and decompositions. We often refer to an $\mathcal{S}$-lifting target as a *lifting target* and leave the signature $\mathcal{S}$ implicit.

DEFINITION 8.6.7 (Search Lifting Map). Let $\Xi = \langle \mathcal{Q}, r, \mathscr{D}, \mathscr{S} \rangle$ be a search state and $\langle \mathcal{Q}^*, r^* \rangle$ be a lifting target. We say $\mathsf{f} = \langle \varphi, f \rangle$ is a *search lifting map from* Ξ *to* $\langle \mathcal{Q}^*, r^* \rangle$ if f is a lifting map from $\mathcal{Q}$ to $\mathcal{Q}^*$, $f(r) = r^*$ and the following conditions hold:

1. The set $\mathscr{D}$ is fully duplicated with respect to f.
2. The set $\mathscr{S}$ is almost atomic with respect to f.

We can always start searching for a proof using an extensional expansion dag with a single (root) node corresponding to the sentence we wish to prove.

DEFINITION 8.6.8. For every $\beta\eta$-normal sentence $\mathbf{M}$ let

$$\mathbf{SearchInit}(\mathbf{M})$$

be the tuple

$$\langle \mathbf{EInit}(\langle\langle \mathbf{primary}, -1, \mathbf{M} \rangle\rangle), r, \emptyset, \emptyset \rangle$$

where $\mathsf{n}_1^{\emptyset} = \langle r \rangle$ (cf. Definition 7.7.2 and Remark 7.7.1).

The tuple $\mathbf{SearchInit}(\mathbf{M})$ is a search state which can always be embedded into a lifting target (appropriate for $\mathbf{M}$) via a search lifting map.

LEMMA 8.6.9. *Let* $\mathbf{M}$ *be a* $\beta\eta$*-normal sentence. Then*

$$\mathbf{SearchInit}(\mathbf{M})$$

is a search state. Furthermore, for any lifting target $\langle \mathcal{Q}^*, r^* \rangle$*, if* r^* *is a negative primary node of* $\mathcal{Q}^*$ *with shallow formula* $\mathbf{M}$*, then there exists a search lifting map* f *from* $\mathbf{SearchInit}(\mathbf{M})$ *to* $\langle \mathcal{Q}^*, r^* \rangle$.

PROOF. The proof is simply a matter of checking conditions. The search lifting map from $\mathbf{SearchInit}(\mathbf{M})$ to an appropriate lifting target $\langle \mathcal{Q}^*, r^* \rangle$ simply maps the unique node of

$$\mathbf{EInit}(\langle\langle \mathbf{primary}, -1, \mathbf{M} \rangle\rangle)$$

to r^*. See Appendix A.6. □

We can prove the converse of Lemma 8.4.6 for search lifting maps.

LEMMA 8.6.10. *Let* $\Xi = \langle \mathcal{Q}, r, \mathscr{D}, \mathscr{S} \rangle$ *be a search state,* $\langle \mathcal{Q}^*, r^* \rangle$ *be a lifting target and* $\langle \varphi, f \rangle$ *be a search lifting map from* Ξ *to* $\langle \mathcal{Q}^*, r^* \rangle$. *For all nodes* $a, b \in \mathcal{N}^{\mathcal{Q}}$*, if* $f(a) \overset{\mathcal{Q}^*}{\rightarrow} f(b)$*, then* $a \overset{\mathcal{Q}}{\rightarrow} b$.

PROOF. Suppose $f(a) \overset{\mathcal{Q}^*}{\rightarrow} f(b)$. Since $\langle \mathcal{Q}^*, r^* \rangle$ is a lifting target, $\mathcal{Q}^*$ is a separated extensional expansion dag. Using the facts that $\mathcal{Q}^*$ is separated, $f(r) = r^*$ and f is injective, we can verify $a \overset{\mathcal{Q}}{\rightarrow} b$. The full proof is in Appendix A.6. □

LEMMA 8.6.11. *Let $\mathcal{Q}$ and $\mathcal{Q}^*$ be extensional expansion dags, $\mathsf{f} = \langle \varphi, f \rangle$ be a lifting map from $\mathcal{Q}$ to $\mathcal{Q}^*$ and $a \in \mathcal{N}^{\mathcal{Q}}$ be a node which is not extended, not a flexible node and not a selection node. Let $\mathcal{Q}'$ be* **Extend**$(\mathcal{Q}, a)$. *Suppose $\mathcal{Q}^*$ is separated and $f(a) \in$* **Dom**$(succ^{\mathcal{Q}^*})$. *Then there exists an $n \geq 0$, distinct nodes $a^1, \ldots, a^n \in \mathcal{N}^{\mathcal{Q}'}$ and distinct nodes $b^1, \ldots, b^n \in \mathcal{N}^{\mathcal{Q}^*}$ such that*

1. $\mathcal{N}^{\mathcal{Q}'} = (\mathcal{N}^{\mathcal{Q}} \cup \{a^1, \ldots, a^n\})$,
2. $(\mathcal{N}^{\mathcal{Q}'} \setminus \mathcal{N}^{\mathcal{Q}}) = \{a^1, \ldots, a^n\}$,
3. $succ^{\mathcal{Q}'}(a) = \langle a^1, \ldots, a^n \rangle$,
4. $succ^{\mathcal{Q}^*}(f(a)) = \langle b^1, \ldots, b^n \rangle$ *and*
5. $label^{\mathcal{Q}^*}(b^i) = \langle kind^{\mathcal{Q}'}(a^i), pol^{\mathcal{Q}'}(a^i), (\varphi(sh^{\mathcal{Q}'}(a^i)))^{\downarrow} \rangle$ *for each* $i \in \{1, \ldots, n\}$.

PROOF. We must consider each case of Definition 7.7.4 except the case where a is a selection node. We consider a few sample cases here. To simplify the notation, assume $\mathcal{Q}$ is the extensional expansion dag

$$\langle \mathcal{N}, succ, label, sel, exps, conns, decs \rangle$$

and the extensional expansion dag $\mathcal{Q}^*$ is

$$\langle \mathcal{N}^*, succ^*, label^*, sel^*, exps^*, conns^*, decs^* \rangle.$$

In every case of Definition 7.7.4, $\mathcal{Q}'$ is

$$\langle \mathcal{N}', succ', label', sel', exps, conns, decs \rangle$$

for some $\mathcal{N}'$, $succ'$, $label'$ and sel'. Since we have assumed a is not a selection node, $sel' = sel$.

Suppose $label(a) = \langle \mathbf{primary}, p, [\mathbf{M} \vee \mathbf{N}] \rangle$. By Definition 7.7.4:3, we have $\mathcal{N}' = \mathcal{N} \cup \{a^1, a^2\}$, $succ'(a) = \langle a^1, a^2 \rangle$,

$$label'(a^1) = \langle \mathbf{primary}, p, \mathbf{M} \rangle$$

and

$$label'(a^2) = \langle \mathbf{primary}, p, \mathbf{N} \rangle$$

where $\mathsf{n}_2^{\mathcal{N}} = \langle a^1, a^2 \rangle$. We know a^1 and a^2 are distinct and $\{a^1, a^2\} \cap \mathcal{N}$ is empty by Remark 7.7.1. Hence $(\mathcal{N}' \setminus \mathcal{N}) = \{a^1, a^2\}$. Since f is a lifting map,

$$label^*(f(a)) = \langle \mathbf{primary}, p, (\varphi([\mathbf{M} \mathbin{\dot\vee} \mathbf{N}]))^{\downarrow} \rangle$$

by Definition 8.4.3:1. Since $(\varphi([\mathbf{M} \mathbin{\dot\vee} \mathbf{N}]))^{\downarrow}$ is $[(\varphi(\mathbf{M}))^{\downarrow} \mathbin{\dot\vee} (\varphi(\mathbf{N}))^{\downarrow}]$, $f(a)$ is a disjunction node or a conjunction node. Since $f(a)$ is extended in $\mathcal{Q}^*$, we know there exist $b^1, b^2 \in \mathcal{N}^*$ such that

$$succ^*(f(a)) = \langle b^1, b^2 \rangle,$$

$$label^*(b^1) = \langle \mathbf{primary}, p, (\varphi(\mathbf{M}))^{\downarrow} \rangle$$

and

$$label^*(b^2) = \langle \mathbf{primary}, p, (\varphi(\mathbf{N}))^{\downarrow} \rangle$$

by $\clubsuit_{\vee}$. Since $\mathcal{Q}^*$ is separated, b^1 and b^2 are distinct (cf. Definition 7.5.2:1), as desired.

Suppose $label(a) = \langle \mathbf{primary}, 1, [\dot\forall x_\alpha \, \mathbf{M}] \rangle$. By Definition 7.7.4:5, we have $\mathcal{N}' = \mathcal{N}$ and $succ'(a) = \langle \rangle$. Since f is a lifting map,

$$label^*(f(a)) = \langle \mathbf{primary}, 1, (\varphi([\dot\forall x_\alpha \, \mathbf{M}]))^{\downarrow} \rangle$$

by Definition 8.4.3:1. Hence $f(a)$ is an expansion node. Since $f(a)$ is extended in $\mathcal{Q}^*$, we know $succ^*(f(a)) = \langle \rangle$ by $\clubsuit_{ns}$. Thus we have the result with $n = 0$.

Suppose $label(a) = \langle \mathbf{primary}, p, [\mathbf{F} \mathbin{\dot=}^{\alpha\beta} \mathbf{G}] \rangle$. Using case 11 of Definition 7.7.4, we have $\mathcal{N}' = \mathcal{N} \cup \{a^1\}$, $succ'(a) = \langle a^1 \rangle$ and

$$label'(a^1) = \langle \mathbf{primary}, p, [\dot\forall y_\beta \centerdot [\mathbf{F}\, y] \mathbin{\dot=}^{\alpha\beta} [\mathbf{G}\, y]] \rangle$$

where $\mathsf{n}_1^{\mathcal{N}} = \langle a^1 \rangle$. We know $a^1 \notin \mathcal{N}$ by Remark 7.7.1. Consequently, we have $(\mathcal{N}' \setminus \mathcal{N}) = \{a^1\}$. Since f is a lifting map,

$$label^*(f(a)) = \langle \mathbf{primary}, p, (\varphi([\mathbf{F} \mathbin{\dot=}^{\alpha\beta} \mathbf{G}]))^{\downarrow} \rangle$$

by Definition 8.4.3:1. Hence $f(a)$ is a functional node. Since $f(a)$ is extended in $\mathcal{Q}^*$, we know $succ^*(f(a)) = \langle b^1 \rangle$ for some primary node b^1 with polarity p and shallow formula

$$[\dot\forall y_\beta \centerdot [(\varphi(\mathbf{F}))^{\downarrow} \, y] \mathbin{\dot=}^{\alpha} [(\varphi(\mathbf{G}))^{\downarrow} \, y]]^{\downarrow}$$

by $\clubsuit_{\dot=}^{\rightarrow}$. (We assume the bound variables are the same y_β with

$$y \notin (\mathbf{Free}(\mathbf{F}) \cup \mathbf{Free}(\mathbf{G}) \cup \mathbf{Free}(\varphi(\mathbf{F})) \cup \mathbf{Free}(\varphi(\mathbf{G})) \cup \mathbf{Dom}(\varphi))$$

by α-conversion.) Note that

$$[\forall y_\beta \centerdot [(\varphi(\mathbf{F}))^{\downarrow} y] \doteq^{\alpha} [(\varphi(\mathbf{G}))^{\downarrow} y]]^{\downarrow}$$
$$= (\varphi([\forall y_\beta \centerdot \mathbf{F}\, y \doteq^{\alpha} \mathbf{G}\, y]))^{\downarrow}.$$

Thus

$$label^*(b^1) = \langle \mathbf{primary}, p, (\varphi(sh^{\mathcal{Q}'}(a^1)))^{\downarrow} \rangle$$

as desired.

Suppose

$$label(a) = \langle \mathbf{dec}, -1, [[H\, \mathbf{A}^1 \cdots \mathbf{A}^n] \doteq^{\iota} [H\, \mathbf{B}^1 \cdots \mathbf{B}^n]] \rangle$$

where H is a parameter of type $(\iota\alpha^n \cdots \alpha^1)$. In this case we have $\mathcal{N}' = \mathcal{N} \cup \{a^1, \ldots a^n\}$, $succ'(a) = \langle a^1, \ldots, a^n \rangle$ and

$$label'(a^i) = \langle \mathbf{primary}, -1, [\mathbf{A}^i \doteq \mathbf{B}^i] \rangle$$

for each i $(1 \leq i \leq n)$ where $\mathsf{n}_n^{\mathcal{N}} = \langle a^1, \ldots, a^n \rangle$ by Definition 7.7.4:14. We know $a^1, \ldots, a^n$ are distinct and $\{a^1, \ldots, a^n\} \cap \mathcal{N}$ is empty by Remark 7.7.1. Hence we know $(\mathcal{N}' \setminus \mathcal{N}) = \{a^1, \ldots, a^n\}$. Since f is a lifting map,

$$label^*(f(a)) = \langle \mathbf{dec}, -1, (\varphi([[H\, \mathbf{A}^1 \cdots \mathbf{A}^n] \doteq^{\iota} [H\, \mathbf{B}^1 \cdots \mathbf{B}^n]]))^{\downarrow} \rangle$$

by Definition 8.4.3:1. Since φ is a lifting substitution, φ respects parameters and so $\varphi(H)$ is a parameter of type $(\iota\alpha^n \cdots \alpha^1)$. Hence $f(a)$ is a decomposition node with shallow formula

$$[[\varphi(H)\, (\varphi(\mathbf{A}^1))^{\downarrow} \cdots (\varphi(\mathbf{A}^n))^{\downarrow}] \doteq^{\iota} [\varphi(H)\, (\varphi(\mathbf{B}^1))^{\downarrow} \cdots (\varphi(\mathbf{B}^n))^{\downarrow}]].$$

Since $f(a)$ is extended in $\mathcal{Q}^*$, there exist nodes $b^1, \ldots, b^n \in \mathcal{N}^*$ such that $succ^*(f(a)) = \langle b^1, \ldots, b^n \rangle$ and

$$label^*(b^i) = \langle \mathbf{primary}, -1, [(\varphi(\mathbf{A}^i))^{\downarrow} \doteq (\varphi(\mathbf{B}^i))^{\downarrow}] \rangle$$

for each i $(1 \leq i \leq n)$ by $\clubsuit_{dec}$. Since $\mathcal{Q}^*$ is separated, $b^1, \ldots, b^n$ are distinct (cf. Definition 7.5.2:1), as desired. □

LEMMA 8.6.12. *Let $\mathcal{Q}$ and $\mathcal{Q}^*$ be extensional expansion dags, $\mathsf{f} = \langle \varphi, f \rangle$ be a lifting map from $\mathcal{Q}$ to $\mathcal{Q}^*$ and $a \in \mathcal{N}^{\mathcal{Q}}$ be a node which is not extended, not a flexible node and not a selection node. Let $\mathcal{Q}'$ be* $\mathbf{Extend}(\mathcal{Q}, a)$. *Suppose $\mathcal{Q}^*$ is separated and $f(a) \in \mathbf{Dom}(succ^{\mathcal{Q}^*})$. Then there is a lifting map $\mathsf{f}' = \langle \varphi', f' \rangle$ from $\mathcal{Q}'$ to $\mathcal{Q}^*$ such that $f'|_{\mathcal{N}^{\mathcal{Q}}} = f$ and $\varphi' = \varphi$.*

PROOF. We can apply Lemma 8.6.11 to obtain $n \geq 0$, distinct nodes $a^1, \ldots, a^n \in \mathcal{N}'$, and distinct nodes $b^1, \ldots, b^n \in \mathcal{N}^*$. The nodes $a^1, \ldots, a^n$ are the nodes in $\mathcal{Q}'$ which are not nodes in $\mathcal{Q}$. We extend the function $f : \mathcal{N}^{\mathcal{Q}} \to \mathcal{N}^{\mathcal{Q}^*}$ to be a function $f' : \mathcal{N}^{\mathcal{Q}'} \to \mathcal{N}^{\mathcal{Q}^*}$ by defining $f'(a^i) := b^i$ for each i $(1 \leq i \leq n)$. We can use Lemma 8.4.7 to verify $\langle \varphi, f' \rangle$ is a lifting map from $\mathcal{Q}'$ to $\mathcal{Q}^*$. The full proof is in Appendix A.6. □

LEMMA 8.6.13. *Let $\mathcal{Q}$ and $\mathcal{Q}^*$ be extensional expansion dags, $\mathsf{f} = \langle \varphi, f \rangle$ be a lifting map from $\mathcal{Q}$ to $\mathcal{Q}^*$ and $a \in \mathcal{N}^{\mathcal{Q}}$ be a selection node which is not extended in $\mathcal{Q}$. Let $\mathcal{Q}'$ be* **Extend**$(\mathcal{Q}, a)$. *Suppose $\mathcal{Q}^*$ is separated and $f(a) \in$* **Dom**$(succ^{\mathcal{Q}^*})$. *Then there is a lifting map $\mathsf{f}' = \langle \varphi', f' \rangle$ from $\mathcal{Q}'$ to $\mathcal{Q}^*$ such that $f'|_{\mathcal{N}^{\mathcal{Q}}} = f$ and $\varphi'(u) = \varphi(u)$ for all $u \in$ (**Free**$(\mathcal{Q}) \cup$ **Params**$(\mathcal{Q})$).*

PROOF. We extend f to include the new child of a in $\mathcal{Q}'$ and extend φ to map the new selected parameter to the corresponding selected parameter in $\mathcal{Q}^*$. See Appendix A.6. □

LEMMA 8.6.14. *Let $\Xi = \langle \mathcal{Q}, r, \mathscr{D}, \mathscr{S} \rangle$ be a search state and $a \in \mathcal{N}^{\mathcal{Q}}$ be a node in $\mathcal{Q}$ which is neither extended nor flexible. Then*

$$\Xi' = \langle \mathbf{Extend}(\mathcal{Q}, a), r, \mathscr{D}, \mathscr{S} \rangle$$

is a search state. Furthermore, for any lifting target $\langle \mathcal{Q}^*, r^* \rangle$ and search lifting map $\mathsf{f} = \langle \varphi, f \rangle$ from Ξ to $\langle \mathcal{Q}^*, r^* \rangle$, there is a search lifting map $\mathsf{f}' = \langle \varphi', f' \rangle$ from Ξ' to $\langle \mathcal{Q}^*, r^* \rangle$ such that $f'|_{\mathcal{N}^{\mathcal{Q}}} = f$ and

$$\|\mathsf{f}' : \mathbf{Extend}(\mathcal{Q}, a) \to \mathcal{Q}^*\| < \|\mathsf{f} : \mathcal{Q} \to \mathcal{Q}^*\|.$$

PROOF. Let $\mathcal{Q}'$ be **Extend**$(\mathcal{Q}, a)$. Since $\Xi = \langle \mathcal{Q}, r, \mathscr{D}, \mathscr{S} \rangle$ is a search state, $\mathcal{Q}$ is an extensional expansion dag, **RootSels**$(\mathcal{Q})$ is empty and $\prec_0^{\mathcal{Q}}$ is acyclic. By Lemma 7.7.7, $\mathcal{Q}'$ is an extensional expansion dag, $\mathcal{Q} \trianglelefteq_r \mathcal{Q}'$,

$$\mathbf{ExpTerms}(\mathcal{Q}) = \mathbf{ExpTerms}(\mathcal{Q}'),$$

$$\mathbf{RootSels}(\mathcal{Q}') = \mathbf{RootSels}(\mathcal{Q}) = \emptyset$$

and $\prec_0^{\mathcal{Q}'}$ is acyclic. Since **ExpTerms**$(\mathcal{Q}') \setminus$ **ExpTerms**$(\mathcal{Q})$ is empty,

$$\Xi' = \langle \mathcal{Q}', r, \mathscr{D}, \mathscr{S} \rangle$$

is a search state by Lemma 8.6.5.

Suppose $\langle \mathcal{Q}^*, r^* \rangle$ is a lifting target and $\mathsf{f} = \langle \varphi, f \rangle$ is a search lifting map from Ξ to $\langle \mathcal{Q}^*, r^* \rangle$. Since $\langle \mathcal{Q}^*, r^* \rangle$ is a lifting target, $\mathcal{Q}^*$ is separated and developed (cf. Definitions 8.6.6, 7.5.2 and 8.1.1). Hence $f(a)$ is extended in $\mathcal{Q}^*$

If a is a selection node, then we can apply Lemma 8.6.13 to obtain a lifting map $\mathsf{f}' = \langle \varphi', f' \rangle$ from $\mathcal{Q}'$ to $\mathcal{Q}^*$ such that $f'|_{\mathcal{N}^{\mathcal{Q}}}$ and $\varphi'(u) = \varphi(u)$ for all $u \in (\mathbf{Free}(\mathcal{Q}) \cup \mathbf{Params}(\mathcal{Q}))$. If a is not a selection node, then we can apply Lemma 8.6.12 to obtain a lifting map $\mathsf{f}' = \langle \varphi', f' \rangle$ from $\mathcal{Q}'$ to $\mathcal{Q}^*$ such that $f'|_{\mathcal{N}^{\mathcal{Q}}}$ and $\varphi' = \varphi$.

We next check f' is a search lifting map from Ξ' to $\langle \mathcal{Q}^*, r^* \rangle$. Since f is a search lifting map, we know $f(r) = r^*$, $\mathscr{D}$ is fully duplicated with respect to f and $\mathscr{S}$ is almost atomic with respect to f. Hence $f'(r) = f(r) = r^*$. Since $\mathcal{Q} \trianglelefteq \mathcal{Q}'$, $f'|_{\mathcal{N}^{\mathcal{Q}}} = f$ and $\varphi'(u) = \varphi(u)$ for every $u \in (\mathbf{Free}(\mathcal{Q}) \cup \mathbf{Params}(\mathcal{Q}))$, we can apply Lemma 8.4.20 to conclude $\mathscr{D}$ is fully duplicated with respect to f' and $\mathscr{S}$ is almost atomic with respect to f'. Thus f' is a search lifting map from Ξ' to $\langle \mathcal{Q}^*, r^* \rangle$.

Finally, we know

$$\|\mathsf{f}' : \mathcal{Q}' \to \mathcal{Q}^*\| < \|\mathsf{f} : \mathcal{Q} \to \mathcal{Q}^*\|$$

by Lemma 8.4.5 since $a \in \mathbf{Dom}(succ') \setminus \mathbf{Dom}(succ)$. □

LEMMA 8.6.15. *Let* $\Xi = \langle \mathcal{Q}, r, \mathscr{D}, \mathscr{S} \rangle$ *be a search state and* $a \in \mathcal{N}^{\mathcal{Q}}$ *be an extended decomposable node which is not decomposed in* $\mathcal{Q}$. *Then* $\Xi' = \langle \mathbf{Dec}(\mathcal{Q}, a), r, \mathscr{D}, \mathscr{S} \rangle$ *is a search state. Furthermore, for any lifting target* $\langle \mathcal{Q}^*, r^* \rangle$ *and search lifting map* f *from* Ξ *to* $\langle \mathcal{Q}^*, r^* \rangle$, *there is a search lifting map* f' *from* Ξ' *to* $\langle \mathcal{Q}^*, r^* \rangle$ *such that*

$$\|\mathsf{f}' : \mathbf{Dec}(\mathcal{Q}, a) \to \mathcal{Q}^*\| < \|\mathsf{f} : \mathcal{Q} \to \mathcal{Q}^*\|.$$

PROOF. See Appendix A.6. □

LEMMA 8.6.16. *Let* $\Xi = \langle \mathcal{Q}, r, \mathscr{D}, \mathscr{S} \rangle$ *be a search state,* $e \in \mathcal{N}^{\mathcal{Q}}$ *be an extended expansion node in* $\mathcal{Q}$ *with shallow formula* $[\forall x_\alpha \mathbf{M}]$ *and* y_α *be a variable. Suppose* $y_\alpha \notin \mathbf{Free}(\mathcal{Q})$. *Then*

$$\Xi' = \langle \mathbf{EExpand}(\mathcal{Q}, e, y), r, \mathscr{D}, \mathscr{S} \rangle$$

is a search state. Furthermore, for any lifting target $\langle \mathcal{Q}^*, r^* \rangle$, *search lifting map* f *from* Ξ *to* $\langle \mathcal{Q}^*, r^* \rangle$ *and node* $c^* \in \mathcal{N}^{\mathcal{Q}^*}$, *if*

$$\langle e, c^* \rangle \in \mathbf{Edges}^-(\mathsf{f}),$$

then there is a search lifting map f' *from* Ξ' *to* $\langle \mathcal{Q}^*, r^* \rangle$ *such that*

$$\|\mathsf{f}' : \mathbf{EExpand}(\mathcal{Q}, e, y) \to \mathcal{Q}^*\| < \|\mathsf{f} : \mathcal{Q} \to \mathcal{Q}^*\|.$$

PROOF. See Appendix A.6. □

LEMMA 8.6.17. *Let* $\Xi = \langle \mathcal{Q}, r, \mathscr{D}, \mathscr{S} \rangle$ *be a search state and* $a \in \mathcal{N}^{\mathcal{Q}}$ *be an undeveloped node in* $\mathcal{Q}$. *Then*

$$\Xi' = \langle \mathbf{OpDevelop}(\mathcal{Q}, a), r, \mathscr{D}, \mathscr{S} \rangle$$

is a search state. Furthermore, for any lifting target $\langle \mathcal{Q}^*, r^* \rangle$ *and search lifting map* f *from* Ξ *to* $\langle \mathcal{Q}^*, r^* \rangle$, *there is a search lifting map* f' *from* Ξ' *to* $\langle \mathcal{Q}^*, r^* \rangle$ *such that*

$$\|\mathsf{f}' : \mathbf{OpDevelop}(\mathcal{Q}, a) \to \mathcal{Q}^*\| < \|\mathsf{f} : \mathcal{Q} \to \mathcal{Q}^*\|.$$

PROOF. We consider each case for the undeveloped node a (cf. Definition 8.1.1). Let $\mathcal{Q}'$ be $\mathbf{OpDevelop}(\mathcal{Q}, a)$.

If a is neither extended nor flexible in $\mathcal{Q}$, then $\mathcal{Q}'$ is $\mathbf{Extend}(\mathcal{Q}, a)$ (cf. Definition 8.1.3) and we have the result by Lemma 8.6.14. If a is an extended decomposable node which is not decomposed in $\mathcal{Q}$, then $\mathcal{Q}'$ is $\mathbf{Dec}(\mathcal{Q}, a)$ (cf. Definition 8.1.3) and we have the result by Lemma 8.6.15.

Otherwise, a is an extended expansion leaf node in $\mathcal{Q}$ with shallow formula $[\forall x_\alpha \, \mathbf{M}]$. By Definition 8.1.3, $\mathcal{Q}'$ is $\mathbf{EExpand}(\mathcal{Q}, a, y_\alpha)$ where y_α is the first variable not in $\mathbf{Free}(\mathcal{Q})$ (cf. Remark 8.1.2). By Lemma 8.6.16, we know

$$\Xi' = \langle \mathcal{Q}', r, \mathscr{D}, \mathscr{S} \rangle$$

is a search state.

Suppose $\langle \mathcal{Q}^*, r^* \rangle$ is a lifting target and $\mathsf{f} = \langle \varphi, f \rangle$ is a search lifting map from Ξ to $\langle \mathcal{Q}^*, r^* \rangle$. Since f is a lifting map from $\mathcal{Q}$ to $\mathcal{Q}^*$, the node $f(a)$ is an extended expansion node of $\mathcal{Q}^*$ with shallow formula $[\forall x_\alpha \, (\varphi(\mathbf{M}))^{\downarrow}]$ (cf. Definition 8.4.3). Since $\langle \mathcal{Q}^*, r^* \rangle$ is a lifting target, $\mathcal{Q}^*$ is developed and hence $f(a)$ is developed in $\mathcal{Q}^*$. In particular, there is some $c^* \in \mathcal{N}^{\mathcal{Q}^*}$ such that $f(a) \overset{\mathcal{Q}^*}{\to} c^*$ (cf. Definition 8.1.1).

Assume $c^* \in \mathbf{Im}(f; \mathcal{Q})$ (cf. Definition 8.4.4). Then there is some $c \in \mathcal{N}^{\mathcal{Q}}$ such that $f(c) = c^*$. By Lemma 8.6.10, we know $a \overset{\mathcal{Q}}{\to} c$ since $f(a) \overset{\mathcal{Q}^*}{\to} f(c)$, contradicting the fact that a is a leaf node in $\mathcal{Q}$. Thus $c^* \notin \mathbf{Im}(f; \mathcal{Q})$ and so $\langle a, c^* \rangle \in \mathbf{Edges}^{-}(\mathsf{f})$.

We apply Lemma 8.6.16 with $\langle \mathcal{Q}^*, r^* \rangle$, f and $\langle a, c^* \rangle \in \mathbf{Edges}^-(\mathsf{f})$ to conclude existence of a search lifting map f' from Ξ' to $\langle \mathcal{Q}^*, r^* \rangle$ such that

$$\|\mathsf{f}' : \mathcal{Q}' \to \mathcal{Q}^*\| < \|\mathsf{f} : \mathcal{Q} \to \mathcal{Q}^*\|$$

as desired. □

LEMMA 8.6.18. *Let* $\Xi = \langle \mathcal{Q}, r, \mathscr{D}, \mathscr{S} \rangle$ *be a search state,* $a, b \in \mathcal{N}^{\mathcal{Q}}$ *be nodes,* $\mathcal{Q}'$ *be an extensional expansion dag,* k *be one of the kinds in* $\{\mathbf{mate}, \mathbf{eunif}, \mathbf{eunif}^{sym}\}$ *and* $c \in \mathcal{N}^{\mathcal{Q}'}$ *be a node of kind* k. *Suppose* $\mathcal{Q} \trianglelefteq_r \mathcal{Q}'$, $\mathbf{RootSels}(\mathcal{Q}')$ *is empty,* $\prec_0^{\mathcal{Q}'}$ *is acyclic,* $c \notin \mathcal{N}^{\mathcal{Q}}$,

$$\mathcal{N}^{\mathcal{Q}'} = \mathcal{N}^{\mathcal{Q}} \cup \{c\},$$

$$conns^{\mathcal{Q}'} = conns^{\mathcal{Q}} \cup \{\langle c, a, b \rangle\},$$

$$succ^{\mathcal{Q}'} = succ^{\mathcal{Q}},$$

$$sel^{\mathcal{Q}'} = sel^{\mathcal{Q}},$$

$$exps^{\mathcal{Q}'} = exps^{\mathcal{Q}}$$

and

$$decs^{\mathcal{Q}'} = decs^{\mathcal{Q}}.$$

Then $\Xi' = \langle \mathcal{Q}', r, \mathscr{D}, \mathscr{S} \rangle$ *is a search state. Furthermore, for any lifting target* $\langle \mathcal{Q}^*, r^* \rangle$, *search lifting map* f *from* Ξ *to* $\langle \mathcal{Q}^*, r^* \rangle$ *and node* $c^* \in \mathcal{N}^{\mathcal{Q}^*}$, *if* $\langle c^*, a, b \rangle \in \mathbf{Conns}^-(\mathsf{f}; k)$, *then there is a search lifting map* f' *from* Ξ' *to* $\langle \mathcal{Q}^*, r^* \rangle$ *such that*

$$\|\mathsf{f}' : \mathcal{Q}' \to \mathcal{Q}^*\| < \|\mathsf{f} : \mathcal{Q} \to \mathcal{Q}^*\|.$$

PROOF. See Appendix A.6. □

LEMMA 8.6.19. *Let* $\Xi = \langle \mathcal{Q}, r, \mathscr{D}, \mathscr{S} \rangle$ *be a search state,* x_α *be a variable in* $\mathbf{Free}(\mathcal{Q})$ *and* $\mathbf{G}_\alpha$ *be a general binding*

$$[\lambda \overline{x^n} \, [R \, [w^1 \, \overline{x^n}] \cdots [w^m \, \overline{x^n}]]]$$

where $R \in \mathcal{P} \cup \mathcal{S} \cup \{x^1, \ldots, x^n\}$. *Assume* $\{w^1, \ldots, w^m\} \cap \mathbf{Free}(\mathcal{Q})$ *is empty and if* R *is a parameter,* $R \in \mathbf{Params}(\mathcal{Q})$ *and* R *is not banned for* x *in* $\mathcal{Q}$. *Then*

$$\Xi' = \langle ([\mathbf{G}/x]\mathcal{Q}), r, \mathscr{D}, (\mathscr{S} \setminus \{x\}) \rangle$$

is a search state. Furthermore, for any lifting target $\langle \mathcal{Q}^*, r^* \rangle$, *search lifting map* $\mathsf{f} = \langle \varphi, f \rangle$ *from* Ξ *to* $\langle \mathcal{Q}^*, r^* \rangle$ *and ground substitution* θ *such that* $\mathbf{Dom}(\theta) = \{w^1, \ldots, w^m\}$, *if* $\theta(\varphi|_{\mathcal{P}}(\mathbf{G})) \stackrel{\beta\eta}{=} \varphi(x_\alpha)$, *then there* *is a search lifting map* $\mathsf{f}' = \langle \varphi', f' \rangle$ *from* Ξ' *to* $\langle \mathcal{Q}^*, r^* \rangle$ *such that* $f' = f$ *and*

$$\|\mathsf{f}' : \mathcal{Q}^{\Xi'} \to \mathcal{Q}^*\| < \|\mathsf{f} : \mathcal{Q} \to \mathcal{Q}^*\|.$$

PROOF. See Appendix A.6. □

8.7. Options

There will be three basic operations performed on search states during automated search for a proof. One operation duplicates expansion nodes in order to ensure there are enough expansion arcs from the given expansion node. Another operation involves instantiating an expansion variable. The remaining operation involves ensuring a pair of nodes is k-connected. These options are performed nondeterministically in a *don't know* fashion.

DEFINITION 8.7.1 (Option Genus). Let **dup**, **primsub**, **unif**, **mateop**, **frmateop** and **eunifop** be distinct values. Each of these values is called an *option genus*.

There are two cases where we may want to ensure a pair of nodes is k-connected. In each case we assume we have a positive node a and a negative node b which share some u-part (which is not pre-solved). The first case is when a and b are atoms with the same parameter at the head of their respective shallow formulas. In this case, we are concerned with ensuring $\langle a, b\rangle$ is **mate**-connected. The second case is when a is a positive equation node and b is an equation goal node. In this case, we can either ensure $\langle a, b\rangle$ is **eunif**-connected, $\mathbf{eunif}^{sym}$-connected or both.

There are two cases where we may want to instantiate an expansion variable x_α. The first case arises from unification and is necessary when there is a negative equation node (which is not a flex-flex node) with shallow formula $[\mathbf{C} \doteq^\iota \mathbf{D}]$ on a u-part (which is not pre-solved) where x is at the head of $\mathbf{C}$ or $\mathbf{D}$.[1] In the second case we instantiate x with a primitive substitution.

DEFINITION 8.7.2 (Imitations). Let $\mathcal{A}$ be a finite set of variables and $h_{\beta\alpha^n\cdots\alpha^1} \in \mathcal{S} \cup \mathcal{P}$ be a logical constant or parameter where β is a base type and $n \geq 0$. Let γ be a type of the form $(\beta\delta^m \cdots \delta^1)$ (where

[1]This is more properly referred to as pre-unification since we are not solving flex-flex pairs.

$m \geq 0$). The *imitation term for h at type γ avoiding $\mathcal{A}$* is the general binding

$$[\lambda u^1_{\delta^1} \cdots \lambda u^m_{\delta^m} \centerdot h\,[w^1\,\overline{u^m}] \cdots [w^n\,\overline{u^m}]]$$

where $w^j \in \mathcal{V}_{\alpha^i\delta^m\ldots\delta^1}$ is the first variable of type $(\alpha^i\delta^m\cdots\delta^1)$ such that $w^j \notin (\mathcal{A} \cup \{w^1,\ldots,w^{m-1}\})$ for each $j \in \{1,\ldots,n\}$ (cf. Remark 8.1.2).

DEFINITION 8.7.3. Let γ be a type $(\beta\delta^m\cdots\delta^1)$ where $\beta \in \{o,\iota\}$. We say *γ has i^{th} projections* if $1 \leq i \leq m$ and δ^i is $(\beta\alpha^n\cdots\alpha^1)$ for some types $\alpha^n,\ldots,\alpha^1$.

DEFINITION 8.7.4 (Projections). Let $\mathcal{A}$ be a finite set of variables, γ be a type $(\beta\delta^m\cdots\delta^1)$ and $1 \leq i \leq m$ where β is a base type δ^i is $(\beta\alpha^n\cdots\alpha^1)$. The *$i^{th}$-projection term at type γ avoiding $\mathcal{A}$* is the general binding

$$[\lambda u^1_{\delta^1} \cdots \lambda u^m_{\delta^m} \centerdot u^i\,[w^1\,\overline{u^m}] \cdots [w^n\,\overline{u^m}]]$$

where $w^j \in \mathcal{V}_{\alpha^i\delta^m\ldots\delta^1}$ is the first variable of type $(\alpha^i\delta^m\cdots\delta^1)$ such that $w^j \notin (\mathcal{A} \cup \{w^1,\ldots,w^{m-1}\})$ for each $j \in \{1,\ldots,n\}$ (cf. Remark 8.1.2).

We now define the set of options $\boldsymbol{Options}_{\mathcal{S}}(\Xi;\mathfrak{p})$ for a search state Ξ a u-part $\mathfrak{p}$ of $\mathcal{Q}^{\Xi}$. The set will be partitioned into duplication options, primitive substitution options, unification options, mating options and E-unification options.

DEFINITION 8.7.5 (Duplication Options). Let $\Xi = \langle \mathcal{Q}, r, \mathscr{D}, \mathscr{S}\rangle$ be a search state. A *duplication option in Ξ* is a pair $\langle \mathbf{dup}, e\rangle$ where e is an extended expansion node in $\mathcal{Q}$ and $e \notin \mathscr{D}$.

We define $\boldsymbol{Options}^{\mathbf{d}}(\Xi)$ to be the set of all duplication options in Ξ.

DEFINITION 8.7.6. Let $\Xi = \langle \mathcal{Q}, r, \mathscr{D}, \mathscr{S}\rangle$ be a search state, $\mathfrak{p}$ be a u-part of $\mathcal{Q}$ and $\mathscr{O} = \langle \mathbf{dup}, e\rangle \in \boldsymbol{Options}^{\mathbf{d}}(\Xi)$ be a duplication option in Ξ. Let $[\dot\forall x_\alpha\,\mathbf{M}]$ be the shallow formula of the expansion node e. We define $\mathbf{Search}(\Xi,\mathfrak{p},\mathscr{O})$ to be the tuple

$$\langle \mathbf{EExpand}(\mathcal{Q}, e, y_\alpha), r, \mathscr{D}, \mathscr{S}\rangle$$

where y_α is the first variable of type α such that $y_\alpha \notin \mathbf{Free}(\mathcal{Q})$ (cf. Remark 8.1.2).[2]

[2]In this case $\mathbf{Search}(\Xi,\mathfrak{p},\mathscr{O})$ does not depend on the u-part $\mathfrak{p}$. We will define $\mathbf{Search}(\Xi,\mathfrak{p},\mathscr{O})$ for other options below in a way which does depend on $\mathfrak{p}$.

LEMMA 8.7.7 (Duplication). *Let $\Xi = \langle \mathcal{Q}, r, \mathscr{D}, \mathscr{S} \rangle$ be a search state, $\mathfrak{p}$ be a u-part of $\mathcal{Q}$ and $\mathscr{O} = \langle \mathbf{dup}, e \rangle \in \boldsymbol{Options}^{\mathbf{d}}(\Xi)$ be a duplication option in Ξ. Then $\Xi' = \mathbf{Search}(\Xi, \mathfrak{p}, \mathscr{O})$ is a search state. Furthermore, for any lifting target $\langle \mathcal{Q}^*, r^* \rangle$, search lifting map f from Ξ to $\langle \mathcal{Q}^*, r^* \rangle$ and schema $\langle e, c^* \rangle \in \mathbf{Edges}^-(\mathsf{f})$ for an edge, there is a search lifting map f' from Ξ' to $\langle \mathcal{Q}^*, r^* \rangle$ such that*

$$\|\mathsf{f}' : \mathcal{Q}^{\Xi'} \to \mathcal{Q}^*\| < \|\mathsf{f} : \mathcal{Q} \to \mathcal{Q}^*\|.$$

PROOF. By Definition 8.7.6, Ξ' is

$$\langle \mathbf{EExpand}(\mathcal{Q}, e, y_\alpha), r, \mathscr{D}, \mathscr{S} \rangle$$

where $y \notin \mathbf{Free}(\mathcal{Q})$. By Lemma 8.6.16, Ξ' is a search state.

Suppose $\langle \mathcal{Q}^*, r^* \rangle$ is a lifting target, f is a search lifting map from Ξ to $\langle \mathcal{Q}^*, r^* \rangle$ and $\langle e, c^* \rangle \in \mathbf{Edges}^-(\mathsf{f})$. By Lemma 8.6.16, there is a search lifting map f' from Ξ' to $\langle \mathcal{Q}^*, r^* \rangle$ such that

$$\|\mathsf{f}' : \mathbf{EExpand}(\mathcal{Q}, e, y) \to \mathcal{Q}^*\| < \|\mathsf{f} : \mathcal{Q} \to \mathcal{Q}^*\|$$

as desired. □

DEFINITION 8.7.8 (Primitive Substitutions). Let $\mathcal{S}$ be a signature, $\mathcal{A}$ be a finite set of variables and $(o\alpha^n \cdots \alpha^1)$ be a type. We say a term $\mathbf{P} \in \mathit{wff}_{o\alpha^n \cdots \alpha^1}$ is a *term for a $\mathcal{S}$-primitive substitution of type $(o\alpha^n \cdots \alpha^1)$ avoiding $\mathcal{A}$* if either $\mathbf{P}$ is the imitation term for some logical constant $c \in \mathcal{S}$ at type $(o\alpha^n \cdots \alpha^1)$ avoiding $\mathcal{A}$ or $(o\alpha^n \cdots \alpha^1)$ has i^{th} projections and $\mathbf{P}$ is the ith projection term at type $(o\alpha^n \cdots \alpha^1)$ avoiding $\mathcal{A}$ for some i $(1 \leq i \leq n)$.

DEFINITION 8.7.9 (Primitive Substitution Options). Let $\mathcal{S}_1$ and $\mathcal{S}_2$ be signatures with $\mathcal{S}_1 \subseteq \mathcal{S}_2$ and $\Xi = \langle \mathcal{Q}, r, \mathscr{D}, \mathscr{S} \rangle$ be an $\mathcal{S}_2$-search state. An *$\mathcal{S}_1$-primitive substitution option in Ξ* is a triple of the form $\langle \mathbf{primsub}, p_{o\alpha^n \cdots \alpha^1}, \mathbf{P}_{o\alpha^n \cdots \alpha^1} \rangle$ where p is an expansion variable of $\mathcal{Q}$, $p \notin \mathscr{S}$ and $\mathbf{P}$ is a term for a $\mathcal{S}_1$-primitive substitution of type $(o\alpha^n \cdots \alpha^1)$ avoiding $\mathbf{Free}(\mathcal{Q})$.

We define $\boldsymbol{Options}^{\mathbf{p}}_{\mathcal{S}_1}(\Xi)$ to be the set of all $\mathcal{S}_1$-primitive substitution options in Ξ.

DEFINITION 8.7.10. Let $\mathcal{S}_1$ and $\mathcal{S}_2$ be signatures with $\mathcal{S}_1 \subseteq \mathcal{S}_2$, $\Xi = \langle \mathcal{Q}, r, \mathscr{D}, \mathscr{S} \rangle$ be an $\mathcal{S}_2$-search state, $\mathfrak{p} \subseteq \mathcal{N}^{\mathcal{Q}}$ be a u-part of $\mathcal{Q}$ and $\mathscr{O} = \langle \mathbf{primsub}, p_\alpha, \mathbf{P} \rangle \in \boldsymbol{Options}^{\mathbf{p}}_{\mathcal{S}_1}(\Xi)$ be a $\mathcal{S}_1$-primitive substitution option in Ξ. We define $\mathbf{Search}(\Xi, \mathfrak{p}, \mathscr{O})$ to be the tuple

$$\langle([\mathbf{P}/p]\mathcal{Q}), r, \mathscr{D} \cup \mathbf{ExpNodes}^{\mathcal{Q}}(\mathfrak{p}), \mathscr{S}\rangle.$$

LEMMA 8.7.11 (Primitive Substitution). *Let $\mathcal{S}_1$ and $\mathcal{S}_2$ be signatures with $\mathcal{S}_1 \subseteq \mathcal{S}_2$, $\Xi = \langle\mathcal{Q}, r, \mathscr{D}, \mathscr{S}\rangle$ be an $\mathcal{S}_2$-search state, $\mathfrak{p} \subseteq \mathcal{N}^{\mathcal{Q}}$ be a u-part of $\mathcal{Q}$ and $\mathscr{O} = \langle\mathbf{primsub}, p_\alpha, \mathbf{P}\rangle \in \boldsymbol{Options}^{\mathbf{P}}_{\mathcal{S}_1}(\Xi)$ be a $\mathcal{S}_1$-primitive substitution option in Ξ. Then $\Xi' = \mathbf{Search}(\Xi, \mathfrak{p}, \mathscr{O})$ is an $\mathcal{S}_2$-search state. Furthermore, for any $\mathcal{S}_2$-lifting target $\langle\mathcal{Q}^*, r^*\rangle$, search lifting map $\mathsf{f} = \langle\varphi, f\rangle$ from Ξ to $\langle\mathcal{Q}^*, r^*\rangle$ and ground substitution θ, if $\mathbf{Dom}(\theta) = \mathbf{Free}(\mathbf{P})$, $\theta(\mathbf{P}) \overset{\beta\eta}{=} \varphi(p_\alpha)$ and $\mathfrak{p}$ is fully duplicated with respect to f, then there is a search lifting map f' from Ξ' to $\langle\mathcal{Q}^*, r^*\rangle$ such that*

$$\|\mathsf{f}' : \mathcal{Q}^{\Xi'} \to \mathcal{Q}^*\| < \|\mathsf{f} : \mathcal{Q} \to \mathcal{Q}^*\|.$$

PROOF. Let $\mathscr{D}'$ be $(\mathscr{D} \cup \mathbf{ExpNodes}^{\mathcal{Q}}(\mathfrak{p}))$. By Definition 8.7.10, Ξ' is

$$\langle([\mathbf{P}/p]\mathcal{Q}), r, \mathscr{D}', \mathscr{S}\rangle.$$

By Definitions 8.7.9 and 8.7.8, $\mathbf{P}$ is an $\mathcal{S}_1$-primitive substitution of the form

$$[\lambda\overline{z^n}\,[h\,[w^1\,\overline{z^n}]\,\cdots\,[w^m\,\overline{z^n}]]]$$

where $h \in \mathcal{S} \cup \{z^1, \ldots, z^n\}$ and $w^1, \ldots, w^m$ are distinct free variables such that $\{w^1, \ldots, w^m\} \cap \mathbf{Free}(\mathcal{Q})$ is empty. Note that $\mathbf{P}$ is a general binding (cf. Definition 8.5.3).

Let Ξ^0 be the tuple

$$\langle\mathcal{Q}, r, \mathscr{D}', \mathscr{S}\rangle.$$

The fact that Ξ^0 is a search state follows directly from the fact that Ξ is a search state and

$$\mathbf{ExpNodes}^{\mathcal{Q}}(\mathfrak{p}) \subseteq \mathbf{ExpNodes}^{\mathcal{Q}}(\mathcal{N}^{\mathcal{Q}}).$$

Applying Lemma 8.6.19 to Ξ^0, we conclude Ξ' is a search state.

Suppose $\langle\mathcal{Q}^*, r^*\rangle$ is a lifting target, f is a search lifting map from Ξ to $\langle\mathcal{Q}^*, r^*\rangle$ and θ is a ground substitution such that the three conditions hold: $\mathbf{Dom}(\theta) = \mathbf{Free}(\mathbf{P})$, $\theta(\mathbf{P}) \overset{\beta\eta}{=} \varphi(p_\alpha)$ and $\mathfrak{p}$ is fully duplicated with respect to f. $\mathbf{ExpNodes}^{\mathcal{Q}}(\mathfrak{p})$ is fully duplicated with respect to f and so f is also a search lifting map from Ξ^0 to $\langle\mathcal{Q}^*, r^*\rangle$. Since $\mathbf{Params}(\mathbf{P})$ is empty,

$$\theta(\varphi\big|_{\mathcal{P}}(\mathbf{P})) = \theta(\mathbf{P}) = \mathbf{Free}(\mathbf{P}).$$

By Lemma 8.6.19, there is a search lifting map f' from Ξ' to $\langle \mathcal{Q}^*, r^* \rangle$ such that

$$\|\mathsf{f}' : \mathcal{Q}^{\Xi'} \to \mathcal{Q}^*\| = \|\mathsf{f}' : ([\mathbf{P}/p]\mathcal{Q}) \to \mathcal{Q}^*\| < \|\mathsf{f} : \mathcal{Q} \to \mathcal{Q}^*\|.$$

□

DEFINITION 8.7.12 (Flex-Rigid Pair). A *flex-rigid pair* is a pair of terms $\langle \mathbf{A}_\iota, \mathbf{B}_\iota \rangle$ where $\mathbf{A}$ is flexible and $\mathbf{B}$ is rigid.

DEFINITION 8.7.13 (Flex-Rigid Node). Let $\mathcal{Q}$ be an expansion structure. We say a node $a \in \mathcal{N}^{\mathcal{Q}}$ is a *flex-rigid node of* $\mathcal{Q}$ if a is a negative equation node with shallow formula $[\mathbf{A} \stackrel{.}{=}^\iota \mathbf{B}]$ where either $\langle \mathbf{A}, \mathbf{B} \rangle$ or $\langle \mathbf{B}, \mathbf{A} \rangle$ is a flex-rigid pair.

DEFINITION 8.7.14. Let $\mathcal{Q}$ be an expansion structure and $a \in \mathcal{N}^{\mathcal{Q}}$ be a flex-rigid node of $\mathcal{Q}$ with shallow formula $[\mathbf{A} \stackrel{.}{=}^\iota \mathbf{B}]$. We say the *flex-rigid pair of* a is $\langle \mathbf{A}, \mathbf{B} \rangle$ if $\mathbf{A}$ is flexible and $\langle \mathbf{B}, \mathbf{A} \rangle$ otherwise.

DEFINITION 8.7.15 (Unification Substitutions). Let $\mathcal{Q}$ be an expansion structure, $a \in \mathcal{N}^{\mathcal{Q}}$ be a flex-rigid node of $\mathcal{Q}$ with flex-rigid pair

$$\langle [x_{\iota\alpha^n\cdots\alpha^1}\, \overline{\mathbf{A}^n}], [H_{\iota\beta^m\cdots\beta^1}\, \overline{\mathbf{B}^m}] \rangle$$

and γ be the type $(\iota\alpha^n \cdots \alpha^1)$. We say a term $\mathbf{U}_\gamma$ is a *unification term for* a *in* $\mathcal{Q}$ if one of the following cases holds:

1. The parameter H is not banned for x in $\mathcal{Q}$ and $\mathbf{U}$ is the imitation term for H at type γ avoiding $\mathbf{Free}(\mathcal{Q})$.
2. There is some i $(1 \leq i \leq n)$ such that γ has i^{th} projections and $\mathbf{U}$ is the ith projection term at type γ avoiding $\mathbf{Free}(\mathcal{Q})$

DEFINITION 8.7.16 (Unification Options). Let $\Xi = \langle \mathcal{Q}, r, \mathcal{D}, \mathcal{S} \rangle$ be a search state and $\mathfrak{p} \subseteq \mathcal{N}^{\mathcal{Q}}$ be a u-part of $\mathcal{Q}$. A triple

$$\langle \mathbf{unif}, x_{\iota\alpha^n\cdots\alpha^1}, \mathbf{U} \rangle$$

is a *unification option* for $\mathfrak{p}$ in Ξ if there exists some flex-rigid node $a \in \mathfrak{p}$ with flex-rigid pair $\langle \mathbf{A}, \mathbf{B} \rangle$ such that x is the head of $\mathbf{A}$ and $\mathbf{U}$ is a unification term for a in $\mathcal{Q}$.

We define $\boldsymbol{Options}^{\mathbf{u}}(\Xi; \mathfrak{p})$ to be the set of all unification options for $\mathfrak{p}$ in Ξ.

DEFINITION 8.7.17. Let $\Xi = \langle \mathcal{Q}, r, \mathcal{D}, \mathcal{S} \rangle$ be a search state, $\mathfrak{p} \subseteq \mathcal{N}^{\mathcal{Q}}$ be a u-part of $\mathcal{Q}$ and $\mathcal{O} = \langle \mathbf{unif}, x_\alpha, \mathbf{U} \rangle \in \boldsymbol{Options}^{\mathbf{u}}(\Xi; \mathfrak{p})$

be a unification option of $\mathcal{Q}$ for $\mathfrak{p}$. We define $\mathbf{Search}(\Xi, \mathfrak{p}, \mathscr{O})$ to be the tuple

$$\langle([\mathbf{U}/x]\mathcal{Q}), r, \mathscr{D} \cup \mathbf{ExpNodes}^{\mathcal{Q}}(\mathfrak{p}), \mathscr{S} \cup \mathbf{SetVars}^{\mathcal{Q}}(\mathfrak{p})\rangle.$$

LEMMA 8.7.18 (Unification). *Let $\Xi = \langle \mathcal{Q}, r, \mathscr{D}, \mathscr{S}\rangle$ be a search state, $\mathfrak{p} \subseteq \mathcal{N}^{\mathcal{Q}}$ be a u-part of $\mathcal{Q}$ and*

$$\mathscr{O} = \langle \mathbf{unif}, x_\alpha, \mathbf{U}\rangle \in \boldsymbol{Options}^{\mathbf{u}}(\Xi; \mathfrak{p})$$

be a unification option for $\mathfrak{p}$ in Ξ. Then

$$\Xi' = \mathbf{Search}(\Xi, \mathfrak{p}, \mathscr{O})$$

is a search state. Furthermore, for any lifting target $\langle \mathcal{Q}^*, r^*\rangle$, search lifting map $\mathsf{f} = \langle \varphi, f\rangle$ from Ξ to $\langle \mathcal{Q}^*, r^*\rangle$ and ground substitution θ such that $\mathbf{Dom}(\theta) = \mathbf{Free}(\mathbf{U})$, if $\theta(\varphi|_{\mathcal{P}}(\mathbf{U})) \stackrel{\beta\eta}{=} \varphi(x_\alpha)$, $\mathfrak{p}$ is fully duplicated with respect to f and $\mathbf{SetVars}^{\mathcal{Q}}(\mathfrak{p})$ is almost atomic with respect to f, then there is a search lifting map f' from Ξ' to $\langle \mathcal{Q}^*, r^*\rangle$ such that

$$\|\mathsf{f}' : \mathcal{Q}^{\Xi'} \to \mathcal{Q}^*\| < \|\mathsf{f} : \mathcal{Q} \to \mathcal{Q}^*\|.$$

PROOF. Let

$$\mathscr{D}' \text{ be } (\mathscr{D} \cup \mathbf{ExpNodes}^{\mathcal{Q}}(\mathfrak{p})) \text{ and } \mathscr{S}' \text{ be } (\mathscr{S} \cup \mathbf{SetVars}^{\mathcal{Q}}(\mathfrak{p})).$$

By Definition 8.7.17, Ξ' is

$$\langle([\mathbf{U}/x]\mathcal{Q}), r, \mathscr{D}', \mathscr{S}'\rangle.$$

By Definitions 8.7.16 and 8.7.15, $\mathbf{U}$ is a unification term of the form

$$[\lambda \overline{z^n}\, [h\, [w^1\, \overline{z^n}] \cdots [w^m\, \overline{z^n}]]]$$

where $h \in \mathcal{P} \cup \{z^1, \ldots, z^n\}$ and $w^1, \ldots, w^m$ are distinct free variables such that $\{w^1, \ldots, w^m\} \cap \mathbf{Free}(\mathcal{Q})$ is empty. If $h \in \mathcal{P}$, then $h \in \mathbf{Params}(\mathcal{Q})$ and h is not banned for x in $\mathcal{Q}$ (cf. Definition 8.7.15).

Let Ξ^0 be the tuple

$$\langle \mathcal{Q}, r, \mathscr{D}', \mathscr{S}'\rangle.$$

The fact that Ξ^0 is a search state follows directly from the fact that Ξ is a search state,

$$\mathbf{ExpNodes}^{\mathcal{Q}}(\mathfrak{p}) \subseteq \mathbf{ExpNodes}^{\mathcal{Q}}(\mathcal{N}^{\mathcal{Q}})$$

and

$$\mathbf{SetVars}^{\mathcal{Q}}(\mathfrak{p}) \subseteq \mathbf{SetVars}^{\mathcal{Q}}(\mathcal{N}^{\mathcal{Q}}).$$

Applying Lemma 8.6.19 to Ξ^0, we conclude Ξ' is a search state.

Suppose $\langle \mathcal{Q}^*, r^* \rangle$ is a lifting target, f is a search lifting map from Ξ to $\langle \mathcal{Q}^*, r^* \rangle$ and θ is a ground substitution satisfying the four conditions: $\mathbf{Dom}(\theta) = \mathbf{Free}(\mathbf{U})$, $\theta(\varphi|_{\mathcal{P}}(\mathbf{U})) \stackrel{\beta\eta}{=} \varphi(x_\alpha)$, $\mathfrak{p}$ is fully duplicated with respect to f and $\mathbf{SetVars}^{\mathcal{Q}}(\mathfrak{p})$ is almost atomic with respect to f. Hence $\mathbf{ExpNodes}^{\mathcal{Q}}(\mathfrak{p})$ is fully duplicated with respect to f and so f is also a search lifting map from Ξ^0 to $\langle \mathcal{Q}^*, r^* \rangle$. By Lemma 8.6.19, there is a search lifting map f' from Ξ' to $\langle \mathcal{Q}^*, r^* \rangle$ such that

$$\|\mathsf{f}' : \mathcal{Q}^{\Xi'} \to \mathcal{Q}^*\| = \|\mathsf{f}' : ([\mathbf{U}/x]\mathcal{Q}) \to \mathcal{Q}^*\| < \|\mathsf{f} : \mathcal{Q} \to \mathcal{Q}^*\|.$$

□

DEFINITION 8.7.19 (Rigid Mating Options). Let $\Xi = \langle \mathcal{Q}, r, \mathscr{D}, \mathscr{S} \rangle$ be a search state and $\mathfrak{p} \subseteq \mathcal{N}^{\mathcal{Q}}$ be a u-part of $\mathcal{Q}$. A *rigid mating option for* $\mathfrak{p}$ *in* Ξ is a tuple $\langle \mathbf{mateop}, a, b \rangle$ where $a, b \in \mathfrak{p}$, $\langle a, b \rangle$ is not **mate**-connected in $\mathcal{Q}$, a is an extended positive atomic node and b is an extended negative atomic node such that a and b are both P-nodes for some parameter $P_{o\alpha^n\cdots\alpha^1} \in \mathcal{P}$. We define $\boldsymbol{Options}^{\mathbf{m}}(\Xi; \mathfrak{p})$ to be the set of all rigid mating options for $\mathfrak{p}$ in Ξ.

DEFINITION 8.7.20 (Flex-Rigid Mating Options). Let Ξ be a search state $\langle \mathcal{Q}, r, \mathscr{D}, \mathscr{S} \rangle$ and $\mathfrak{p} \subseteq \mathcal{N}^{\mathcal{Q}}$ be a u-part of $\mathcal{Q}$. A *flex-rigid mating option for* $\mathfrak{p}$ *in* Ξ is a tuple $\langle \mathbf{frmateop}, a, b, p_{o\alpha^n\cdots\alpha^1}, \mathbf{P} \rangle$ where $a, b \in \mathfrak{p}$, a is a positive flexible p-node, b is an extended negative atomic P-node for some parameter P which is not banned for p in $\mathcal{Q}$ and $\mathbf{P}$ is the imitation term for P at type $(o\alpha^n \cdots \alpha^1)$ avoiding $\mathbf{Free}(\mathcal{Q})$. We define $\boldsymbol{Options}^{\mathbf{frm}}(\Xi; \mathfrak{p})$ to be the set of all flex-rigid mating options for $\mathfrak{p}$ in Ξ.

DEFINITION 8.7.21 (E-unification Options). Let $\Xi = \langle \mathcal{Q}, r, \mathscr{D}, \mathscr{S} \rangle$ be a search state and $\mathfrak{p} \subseteq \mathcal{N}^{\mathcal{Q}}$ be a u-part of $\mathcal{Q}$. An *E-unification option for* $\mathfrak{p}$ *in* Ξ is a tuple $\langle \mathbf{eunifop}, k, a, b \rangle$ where k is either **eunif** or $\mathbf{eunif}^{sym}$, $a, b \in \mathfrak{p}$, $\langle a, b \rangle$ is not k-connected in $\mathcal{Q}$, a is an extended positive equation node and b is an extended equation goal node. We define $\boldsymbol{Options}^{\mathbf{e}}(\Xi; \mathfrak{p})$ to be the set of all E-unification options for $\mathfrak{p}$ in Ξ.

DEFINITION 8.7.22. Let $\Xi = \langle \mathcal{Q}, r, \mathscr{D}, \mathscr{S} \rangle$ be a search state, $\mathfrak{p} \subseteq \mathcal{N}^{\mathcal{Q}}$ be a u-part of $\mathcal{Q}$ and

$$\mathscr{O} = \langle \mathbf{mateop}, a, b \rangle \in \boldsymbol{Options}^{\mathbf{m}}(\Xi; \mathfrak{p})$$

be a rigid mating option of $\mathcal{Q}$ for $\mathfrak{p}$. We define $\mathbf{Search}(\Xi, \mathfrak{p}, \mathscr{O})$ to be the tuple

$$\langle \mathbf{Mate}(\mathcal{Q}, a, b), r, \mathscr{D} \cup \mathbf{ExpNodes}^{\mathcal{Q}}(\mathfrak{p}), \mathscr{S} \cup \mathbf{SetVars}^{\mathcal{Q}}(\mathfrak{p})\rangle.$$

DEFINITION 8.7.23. Let $\Xi = \langle \mathcal{Q}, r, \mathscr{D}, \mathscr{S}\rangle$ be a search state, $\mathfrak{p} \subseteq \mathcal{N}^{\mathcal{Q}}$ be a u-part of $\mathcal{Q}$ and

$$\mathscr{O} = \langle \mathbf{frmateop}, a, b, x, \mathbf{P}\rangle \in \boldsymbol{Options}^{\mathbf{frm}}(\Xi; \mathfrak{p})$$

be a flex-rigid mating option of $\mathcal{Q}$ for $\mathfrak{p}$. We define $\mathbf{Search}(\Xi, \mathfrak{p}, \mathscr{O})$ to be the tuple

$$\langle \mathcal{Q}', r, \mathscr{D} \cup \mathbf{ExpNodes}^{\mathcal{Q}}(\mathfrak{p}), (\mathscr{S} \setminus \{x\}) \cup \mathbf{SetVars}^{\mathcal{Q}}(\mathfrak{p})\rangle$$

where $\mathcal{Q}'$ is

$$\mathbf{Mate}((\mathbf{Extend}(([\mathbf{P}/x]\mathcal{Q}), a)), a, b).$$

DEFINITION 8.7.24. Let $\Xi = \langle \mathcal{Q}, r, \mathscr{D}, \mathscr{S}\rangle$ be a search state, $\mathfrak{p} \subseteq \mathcal{N}^{\mathcal{Q}}$ be a u-part of $\mathcal{Q}$ and $\mathscr{O} = \langle \mathbf{eunifop}, k, a, b\rangle \in \boldsymbol{Options}^{\mathbf{e}}(\Xi; \mathfrak{p})$ be an E-unification option of $\mathcal{Q}$ for $\mathfrak{p}$. We define $\mathbf{Search}(\Xi, \mathfrak{p}, \mathscr{O})$ to be the tuple

$$\langle \mathbf{EUnif}(k, \mathcal{Q}, a, b), r, \mathscr{D} \cup \mathbf{ExpNodes}^{\mathcal{Q}}(\mathfrak{p}), \mathscr{S} \cup \mathbf{SetVars}^{\mathcal{Q}}(\mathfrak{p})\rangle.$$

LEMMA 8.7.25 (Rigid Mating). *Let $\Xi = \langle \mathcal{Q}, r, \mathscr{D}, \mathscr{S}\rangle$ be a search state, $\mathfrak{p} \subseteq \mathcal{N}^{\mathcal{Q}}$ be a u-part of $\mathcal{Q}$, and*

$$\mathscr{O} = \langle \mathbf{mateop}, a, b\rangle \in \boldsymbol{Options}^{\mathbf{m}}(\Xi; \mathfrak{p})$$

be a rigid mating option of $\mathcal{Q}$ for $\mathfrak{p}$. Then $\Xi' = \mathbf{Search}(\Xi, \mathfrak{p}, \mathscr{O})$ is a search state. Furthermore, for any lifting target $\langle \mathcal{Q}^*, r^*\rangle$, search lifting map f from Ξ to $\langle \mathcal{Q}^*, r^*\rangle$ and $\langle c^*, a, b\rangle \in \mathbf{Conns}^-(\mathsf{f}; \mathbf{mate})$, if $\mathfrak{p}$ is fully duplicated with respect to f and $\mathbf{SetVars}^{\mathcal{Q}}(\mathfrak{p})$ is almost atomic with respect to f, then there is a search lifting map f' from Ξ' to $\langle \mathcal{Q}^*, r^*\rangle$ such that

$$\|\mathsf{f}' : \mathcal{Q}^{\Xi'} \to \mathcal{Q}^*\| < \|\mathsf{f} : \mathcal{Q} \to \mathcal{Q}^*\|.$$

PROOF. Let

$$\mathscr{D}' \text{ be } (\mathscr{D} \cup \mathbf{ExpNodes}^{\mathcal{Q}}(\mathfrak{p})) \text{ and } \mathscr{S}' \text{ be } (\mathscr{S} \cup \mathbf{SetVars}^{\mathcal{Q}}(\mathfrak{p})).$$

By Definition 8.7.22, Ξ' is

$$\langle \mathbf{Mate}(\mathcal{Q}, a, b), r, \mathscr{D}', \mathscr{S}'\rangle.$$

Let $\mathcal{Q}'$ be $\mathbf{Mate}(\mathcal{Q}, a, b)$. By Definition 7.7.11, there is some $c \notin \mathcal{N}^{\mathcal{Q}}$ such that

$$\mathcal{N}^{\mathcal{Q}'} = \mathcal{N}^{\mathcal{Q}} \cup \{c\},$$

$$conns^{\mathcal{Q}'} = conns^{\mathcal{Q}} \cup \{\langle c, a, b\rangle\},$$

$$succ^{\mathcal{Q}'} = succ^{\mathcal{Q}},$$

$$sel^{\mathcal{Q}'} = sel^{\mathcal{Q}},$$

$$exps^{\mathcal{Q}'} = exps^{\mathcal{Q}}$$

and

$$decs^{\mathcal{Q}'} = decs^{\mathcal{Q}}.$$

By Lemma 7.7.14, $\mathcal{Q}'$ is an extensional expansion dag, $\mathcal{Q} \trianglelefteq_r \mathcal{Q}'$, $\prec_0^{\mathcal{Q}'}$ is acyclic and $\mathbf{RootSels}(\mathcal{Q}')$ is empty.

Let Ξ^0 be the tuple

$$\langle \mathcal{Q}, r, \mathscr{D}', \mathscr{S}'\rangle.$$

The fact that Ξ^0 is a search state follows directly from the fact that Ξ is a search state,

$$\mathbf{ExpNodes}^{\mathcal{Q}}(\mathfrak{p}) \subseteq \mathbf{ExpNodes}^{\mathcal{Q}}(\mathcal{N}^{\mathcal{Q}})$$

and

$$\mathbf{SetVars}^{\mathcal{Q}}(\mathfrak{p}) \subseteq \mathbf{SetVars}^{\mathcal{Q}}(\mathcal{N}^{\mathcal{Q}}).$$

Applying Lemma 8.6.18 to Ξ^0, we conclude Ξ' is a search state.

Suppose $\langle \mathcal{Q}^*, r^*\rangle$ is a lifting target, f is a search lifting map from Ξ to $\langle \mathcal{Q}^*, r^*\rangle$ $\langle c^*, a, b\rangle \in \mathbf{Conns}^-(\mathsf{f}; \mathbf{mate})$, $\mathfrak{p}$ is fully duplicated with respect to f and $\mathbf{SetVars}^{\mathcal{Q}}(\mathfrak{p})$ is almost atomic with respect to f. Hence $\mathbf{ExpNodes}^{\mathcal{Q}}(\mathfrak{p})$ is fully duplicated with respect to f and so f is also a search lifting map from Ξ^0 to $\langle \mathcal{Q}^*, r^*\rangle$. Since

$$\langle c^*, a, b\rangle \in \mathbf{Conns}^-(\mathsf{f}; \mathbf{mate}),$$

we know $\langle a, b\rangle$ is not **mate**-connected in $\mathcal{Q}$ (cf. Definition 8.4.11). Applying Lemma 8.6.18 to Ξ^0 and the search lifting map f from Ξ^0 to $\langle \mathcal{Q}^*, r^*\rangle$, we obtain a search lifting map f' from Ξ' to $\langle \mathcal{Q}^*, r^*\rangle$ such that

$$\|\mathsf{f}' : \mathcal{Q}' \to \mathcal{Q}^*\| < \|\mathsf{f} : \mathcal{Q} \to \mathcal{Q}^*\|$$

as desired. □

LEMMA 8.7.26 (Flex-Rigid Mating). *Let* $\Xi = \langle \mathcal{Q}, r, \mathscr{D}, \mathscr{S} \rangle$ *be a search state,* $\mathfrak{p} \subseteq \mathcal{N}^{\mathcal{Q}}$ *be a u-part of* $\mathcal{Q}$, *and*

$$\mathscr{O} = \langle \mathbf{frmateop}, a, b, p, \mathbf{P} \rangle \in \boldsymbol{Options}^{\mathbf{m}}(\Xi; \mathfrak{p})$$

be a flex-rigid mating option of $\mathcal{Q}$ *for* $\mathfrak{p}$. *Then* $\Xi' = \mathbf{Search}(\Xi, \mathfrak{p}, \mathscr{O})$ *is a search state. Furthermore, for any lifting target* $\langle \mathcal{Q}^*, r^* \rangle$, *search lifting map* $\mathsf{f} = \langle \varphi, f \rangle$ *from* Ξ *to* $\langle \mathcal{Q}^*, r^* \rangle$, $\langle c^*, a, b \rangle \in \mathbf{Conns}^-(\mathsf{f}; \mathbf{mate})$ *and ground substitution* θ *such that* $\mathbf{Dom}(\theta)$ *is equal to* $\mathbf{Free}(\mathbf{P})$, *if* $\theta(\varphi|_{\mathcal{P}}(\mathbf{P})) \stackrel{\beta\eta}{=} \varphi(x_\alpha)$, $\mathfrak{p}$ *is fully duplicated with respect to* f *and* $\mathbf{SetVars}^{\mathcal{Q}}(\mathfrak{p})$ *is almost atomic with respect to* f, *then there is a search lifting map* f' *from* Ξ' *to* $\langle \mathcal{Q}^*, r^* \rangle$ *such that*

$$\|\mathsf{f}' : \mathcal{Q}^{\Xi'} \to \mathcal{Q}^*\| < \|\mathsf{f} : \mathcal{Q} \to \mathcal{Q}^*\|.$$

PROOF. Let

$$\mathscr{D}' \text{ be } (\mathscr{D} \cup \mathbf{ExpNodes}^{\mathcal{Q}}(\mathfrak{p})),\ \mathscr{S}' \text{ be } (\mathscr{S} \cup \mathbf{SetVars}^{\mathcal{Q}}(\mathfrak{p})),$$

$\mathcal{Q}^1$ be $([\mathbf{P}/p]\mathcal{Q})$, $\mathcal{Q}^2$ be $\mathbf{Extend}(\mathcal{Q}^1, a)$ and $\mathcal{Q}'$ be $\mathbf{Mate}(\mathcal{Q}^2, a, b)$. By Definition 8.7.23, Ξ' is

$$\langle \mathcal{Q}', r, \mathscr{D}', \mathscr{S}' \rangle.$$

By Definition 8.7.20, a is a positive flexible node with shallow formula $[p\,\overline{\mathbf{A}^n}]$, b is an extended negative atomic node with shallow formula $[P\,\overline{\mathbf{B}^m}]$, P is not banned for p in $\mathcal{Q}$ and $\mathbf{P}$ is of the form

$$[\lambda \overline{z^n}\, [P\, [w^1\, \overline{z^n}] \cdots [w^m\, \overline{z^n}]]]$$

where $w^1, \ldots, w^m$ are distinct free variables such that

$$(\{w^1, \ldots, w^m\} \cap \mathbf{Free}(\mathcal{Q})) = \emptyset.$$

Hence $\mathbf{Free}(\mathbf{P}) \cap \mathbf{Free}(\mathcal{Q})$ is empty and $\mathbf{Params}(\mathbf{P}) = \{P\}$.

We must first check that $\mathcal{Q}^2$ and $\mathcal{Q}'$ are well-defined expansion structures. Let $\mathbf{C}^i$ be $([\mathbf{P}/p]\mathbf{A}^i)^{\downarrow}$ for each i $(1 \leq i \leq n)$ and let $\mathbf{D}^j$ be $([\mathbf{P}/p]\mathbf{B}^j)^{\downarrow}$ for each j $(1 \leq j \leq m)$. By Definition 8.2.2,

$$succ^{\mathcal{Q}^1} = succ^{\mathcal{Q}},$$

$$\begin{aligned} sh^{\mathcal{Q}^1}(a) &= ([\mathbf{P}/p]sh^{\mathcal{Q}}(a))^{\downarrow} \\ &= [\mathbf{P}\,([\mathbf{P}/p]\mathbf{A}^1)\cdots([\mathbf{P}/p]\mathbf{A}^n)]^{\downarrow} \\ &= [\mathbf{P}\,\overline{\mathbf{C}^n}]^{\downarrow} \\ &= [P\,[w^1\,\overline{\mathbf{C}^n}]\cdots[w^m\,\overline{\mathbf{C}^n}]] \end{aligned}$$

and

$$\begin{aligned} sh^{\mathcal{Q}^1}(b) &= ([\mathbf{P}/p]sh^{\mathcal{Q}}(b))^{\downarrow} \\ &= [P\,([\mathbf{P}/p]\mathbf{B}^1)\cdots([\mathbf{P}/p]\mathbf{B}^m)]^{\downarrow} \\ &= [P\,\overline{\mathbf{D}^m}]. \end{aligned}$$

Hence a is a positive atomic P-node in $\mathcal{Q}^1$ and b is an extended negative atomic P-node in $\mathcal{Q}^1$. Since a is a flexible node in $\mathcal{Q}$, we know $a \notin \mathbf{Dom}(succ^{\mathcal{Q}})$ by $\clubsuit_f$. By Definition 8.2.2, $a \notin \mathbf{Dom}(succ^{\mathcal{Q}^1})$. That is, a is not extended in $\mathcal{Q}^1$. Thus $\mathcal{Q}^2 := \mathbf{Extend}(\mathcal{Q}^1, a)$ is well-defined by Definition 7.7.4. Also, $a, b \in \mathbf{Dom}(succ^{\mathcal{Q}^2})$,

$$label^{\mathcal{Q}^2}(a) = label^{\mathcal{Q}^1}(a) \text{ and } label^{\mathcal{Q}^2}(b) = label^{\mathcal{Q}^1}(b)$$

by Definition 7.7.4. Hence a is an extended positive atomic P-node in $\mathcal{Q}^2$ and b is an extended negative atomic P-node in $\mathcal{Q}^2$. Thus $\mathcal{Q}' := \mathbf{Mate}(\mathcal{Q}^2, a, b)$ is well-defined by Definition 7.7.11.

Let Ξ^0 be the tuple

$$\langle \mathcal{Q}, r, \mathscr{D}', \mathscr{S}' \rangle,$$

Ξ^1 be the tuple

$$\langle \mathcal{Q}^1, r, \mathscr{D}', \mathscr{S}' \rangle$$

and Ξ^2 be the tuple

$$\langle \mathcal{Q}^2, r, \mathscr{D}', \mathscr{S}' \rangle.$$

The fact that Ξ^0 is a search state follows directly from the fact that Ξ is a search state,

$$\mathbf{ExpNodes}^{\mathcal{Q}}(\mathfrak{p}) \subseteq \mathbf{ExpNodes}^{\mathcal{Q}}(\mathcal{N}^{\mathcal{Q}})$$

and

$$\mathbf{SetVars}^{\mathcal{Q}}(\mathfrak{p}) \subseteq \mathbf{SetVars}^{\mathcal{Q}}(\mathcal{N}^{\mathcal{Q}}).$$

Applying Lemma 8.6.19 to Ξ^0, we conclude Ξ^1 is a search state. Applying Lemma 8.6.14 to Ξ^1, we know Ξ^2 is a search state. In particular, $\mathcal{Q}^2$ is an extensional expansion dag, $\mathbf{RootSels}(\mathcal{Q}^2)$ is empty and $\prec_0^{\mathcal{Q}^2}$ is acyclic. By Lemma 7.7.14, $\mathcal{Q}'$ is an extensional expansion dag, $\mathcal{Q}^2 \trianglelefteq_r \mathcal{Q}'$, $\mathbf{RootSels}(\mathcal{Q}') = \mathbf{RootSels}(\mathcal{Q}^2)$ is empty and $\prec_0^{\mathcal{Q}'}$ is acyclic.

By Definition 7.7.11, there is some $c \in \mathcal{N}^{\mathcal{Q}'} \setminus \mathcal{N}^{\mathcal{Q}^2}$ of kind **mate** in $\mathcal{Q}'$ such that

$$\mathcal{N}^{\mathcal{Q}'} = \mathcal{N}^{\mathcal{Q}^2} \cup \{c\},$$
$$conns^{\mathcal{Q}'} = conns^{\mathcal{Q}^2} \cup \{\langle c, a, b\rangle\},$$
$$succ^{\mathcal{Q}'} = succ^{\mathcal{Q}^2},$$
$$sel^{\mathcal{Q}'} = sel^{\mathcal{Q}^2},$$
$$exps^{\mathcal{Q}'} = exps^{\mathcal{Q}^2}$$

and

$$decs^{\mathcal{Q}'} = decs^{\mathcal{Q}^2}.$$

Hence we can apply Lemma 8.6.18 to Ξ^2 and conclude Ξ' is a search state.

Suppose $\langle \mathcal{Q}^*, r^*\rangle$ is a lifting target, $\mathsf{f} = \langle \varphi, f\rangle$ is a search lifting map from Ξ to $\langle \mathcal{Q}^*, r^*\rangle$, $\langle c^*, a, b\rangle \in \mathbf{Conns}^-(\mathsf{f}; \mathbf{mate})$ and θ is a ground substitution where $\mathbf{Dom}(\theta) = \mathbf{Free}(\mathbf{P})$, $\theta(\varphi|_{\mathcal{P}}(\mathbf{P})) \stackrel{\beta\eta}{=} \varphi(x_\alpha)$, $\mathfrak{p}$ is fully duplicated with respect to f and $\mathbf{SetVars}^{\mathcal{Q}}(\mathfrak{p})$ is almost atomic with respect to f. Hence $\mathbf{ExpNodes}^{\mathcal{Q}}(\mathfrak{p})$ is fully duplicated with respect to f and so f is also a search lifting map from Ξ^0 to $\langle \mathcal{Q}^*, r^*\rangle$.

Applying Lemma 8.6.19 with Ξ^0, p, $\mathbf{P}$, f and θ, we obtain a search lifting map $\mathsf{f}^1 = \langle \varphi^1, f^1\rangle$ from Ξ^1 to $\langle \mathcal{Q}^*, r^*\rangle$ such that $f^1 = f$ and

(2) $$\|\mathsf{f}^1 : \mathcal{Q}^1 \to \mathcal{Q}^*\| < \|\mathsf{f} : \mathcal{Q} \to \mathcal{Q}^*\|.$$

Since a is not extended in $\mathcal{Q}^1$, we can apply Lemma 8.6.14 with Ξ^1, a and f^1 to obtain a search lifting map $\mathsf{f}^2 = \langle \varphi^2, f^2\rangle$ such that $f^2|_{\mathcal{N}^{\mathcal{Q}^1}} = f^1$ and

(3) $$\|\mathsf{f}^2 : \mathcal{Q}^2 \to \mathcal{Q}^*\| < \|\mathsf{f}^1 : \mathcal{Q}^2 \to \mathcal{Q}^*\|.$$

Hence $f^2(a) = f^1(a) = f(a)$ and $f^2(b) = f^1(b) = f(b)$.

Since $\langle c^*, a, b\rangle \in \mathbf{Conns}^-(\mathsf{f}; \mathbf{mate})$, we know $kind^{\mathcal{Q}^*}(c^*) = \mathbf{mate}$,

$$\langle c^*, f^2(a), f^2(b)\rangle = \langle c^*, f(a), f(b)\rangle \in conns^{\mathcal{Q}^*},$$

and $\langle a, b\rangle$ is not **mate**-connected in $\mathcal{Q}$. By Definitions 8.2.2 and 7.7.4, $conns^{\mathcal{Q}^2} = conns^{\mathcal{Q}}$. Hence $\langle a, b\rangle$ is not **mate**-connected in $\mathcal{Q}^2$. Thus $\langle c^*, a, b\rangle \in \mathbf{Conns}^-(\mathsf{f}^2; \mathbf{mate})$. Applying Lemma 8.6.18 to Ξ^2, a, b, $\mathcal{Q}'$, **mate**, $c \in \mathcal{N}^{\mathcal{Q}'}$, f^2 and $\langle c^*, a, b\rangle$, we know there is a search lifting map f' from Ξ' to $\langle \mathcal{Q}^*, r^*\rangle$ such that

$$\|\mathsf{f}' : \mathcal{Q}' \to \mathcal{Q}^*\| < \|\mathsf{f}^2 : \mathcal{Q}^2 \to \mathcal{Q}^*\|. \tag{4}$$

Combining the inequalities (2), (3) and (4), we have

$$\|\mathsf{f}' : \mathcal{Q}' \to \mathcal{Q}^*\| < \|\mathsf{f} : \mathcal{Q} \to \mathcal{Q}^*\|$$

as desired. □

LEMMA 8.7.27 (E-Unification). *Let $\Xi = \langle \mathcal{Q}, r, \mathscr{D}, \mathscr{S}\rangle$ be a search state, $\mathfrak{p} \subseteq \mathcal{N}^{\mathcal{Q}}$ be a u-part of $\mathcal{Q}$, $k \in \{\mathbf{eunif}, \mathbf{eunif}^{sym}\}$ and*

$$\mathscr{O} = \langle \mathbf{eunifop}, k, a, b\rangle \in \boldsymbol{Options}^{\mathbf{e}}(\Xi; \mathfrak{p})$$

be an E-unification option for $\mathfrak{p}$ in Ξ. Then $\Xi' = \mathbf{Search}(\Xi, \mathfrak{p}, \mathscr{O})$ is a search state. Furthermore, for any lifting target $\langle \mathcal{Q}^*, r^*\rangle$, search lifting map f from Ξ to $\langle \mathcal{Q}^*, r^*\rangle$ and $\langle c^*, a, b\rangle \in \mathbf{Conns}^-(\mathsf{f}; k)$, if $\mathfrak{p}$ is fully duplicated with respect to f and $\mathbf{SetVars}^{\mathcal{Q}}(\mathfrak{p})$ is almost atomic with respect to f, then there is a search lifting map f' from Ξ' to $\langle \mathcal{Q}^*, r^*\rangle$ such that

$$\|\mathsf{f}' : \mathcal{Q}^{\Xi'} \to \mathcal{Q}^*\| < \|\mathsf{f} : \mathcal{Q} \to \mathcal{Q}^*\|$$

PROOF. Let

$$\mathscr{D}' \text{ be } (\mathscr{D} \cup \mathbf{ExpNodes}^{\mathcal{Q}}(\mathfrak{p})) \text{ and } \mathscr{S}' \text{ be } (\mathscr{S} \cup \mathbf{SetVars}^{\mathcal{Q}}(\mathfrak{p})).$$

By Definition 8.7.24, Ξ' is

$$\langle \mathbf{EUnif}(k, \mathcal{Q}, a, b), r, \mathscr{D}', \mathscr{S}'\rangle.$$

Let $\mathcal{Q}'$ be $\mathbf{EUnif}(k, \mathcal{Q}, a, b)$. By Definition 7.7.12, there is some $c \notin \mathcal{N}^{\mathcal{Q}}$ such that

$$\mathcal{N}^{\mathcal{Q}'} = \mathcal{N}^{\mathcal{Q}} \cup \{c\},$$
$$conns^{\mathcal{Q}'} = conns^{\mathcal{Q}} \cup \{\langle c, a, b\rangle\},$$
$$succ^{\mathcal{Q}'} = succ^{\mathcal{Q}},$$
$$sel^{\mathcal{Q}'} = sel^{\mathcal{Q}},$$
$$exps^{\mathcal{Q}'} = exps^{\mathcal{Q}}$$

and

$$decs^{\mathcal{Q}'} = decs^{\mathcal{Q}}.$$

By Lemma 7.7.15, $\mathcal{Q}'$ is an extensional expansion dag, $\mathcal{Q} \trianglelefteq_r \mathcal{Q}'$, $\prec_0^{\mathcal{Q}'}$ is acyclic and $\mathbf{RootSels}(\mathcal{Q}')$ is empty.

Let Ξ^0 be the tuple

$$\langle \mathcal{Q}, r, \mathscr{D}', \mathscr{S}' \rangle.$$

The fact that Ξ^0 is a search state follows directly from the fact that Ξ is a search state,

$$\mathbf{ExpNodes}^{\mathcal{Q}}(\mathfrak{p}) \subseteq \mathbf{ExpNodes}^{\mathcal{Q}}(\mathcal{N}^{\mathcal{Q}})$$

and

$$\mathbf{SetVars}^{\mathcal{Q}}(\mathfrak{p}) \subseteq \mathbf{SetVars}^{\mathcal{Q}}(\mathcal{N}^{\mathcal{Q}}).$$

Applying Lemma 8.6.18 to Ξ^0, we conclude Ξ' is a search state.

Suppose $\langle \mathcal{Q}^*, r^* \rangle$ is a lifting target, f is a search lifting map from Ξ to $\langle \mathcal{Q}^*, r^* \rangle$ $\langle c^*, a, b \rangle \in \mathbf{Conns}^-(\mathsf{f}; k)$, $\mathfrak{p}$ is fully duplicated with respect to f and $\mathbf{SetVars}^{\mathcal{Q}}(\mathfrak{p})$ is almost atomic with respect to f. Hence $\mathbf{ExpNodes}^{\mathcal{Q}}(\mathfrak{p})$ is fully duplicated with respect to f and so f is also a search lifting map from Ξ^0 to $\langle \mathcal{Q}^*, r^* \rangle$. Since $\langle c^*, a, b \rangle \in \mathbf{Conns}^-(\mathsf{f}; k)$, we know $\langle a, b \rangle$ is not k-connected in $\mathcal{Q}$ (cf. Definition 8.4.11). Applying Lemma 8.6.18 to Ξ^0 and the search lifting map f from Ξ^0 to $\langle \mathcal{Q}^*, r^* \rangle$, we obtain a search lifting map f' from Ξ' to $\langle \mathcal{Q}^*, r^* \rangle$ such that

$$\|\mathsf{f}' : \mathcal{Q}' \to \mathcal{Q}^*\| < \|\mathsf{f} : \mathcal{Q} \to \mathcal{Q}^*\|$$

as desired. □

DEFINITION 8.7.28 (Options). Let $\mathcal{S}_1$ and $\mathcal{S}_2$ be signatures with $\mathcal{S}_1 \subseteq \mathcal{S}_2$, $\Xi = \langle \mathcal{Q}, r, \mathscr{D}, \mathscr{S} \rangle$ be an $\mathcal{S}_2$-search state and $\mathfrak{p} \subseteq \mathcal{N}^{\mathcal{Q}}$ be a u-part of $\mathcal{Q}$. We define $\boldsymbol{Options}_{\mathcal{S}_1}(\Xi; \mathfrak{p})$ to be the set

$$\boldsymbol{Options}_{\mathcal{S}_1}(\Xi; \mathfrak{p}) := \boldsymbol{Options}^{\mathbf{d}}(\Xi) \cup \boldsymbol{Options}^{\mathbf{p}}_{\mathcal{S}_1}(\Xi) \cup \boldsymbol{Options}^{\mathbf{u}}(\Xi; \mathfrak{p})$$

$$\cup \boldsymbol{Options}^{\mathbf{m}}(\Xi; \mathfrak{p}) \cup \boldsymbol{Options}^{\mathbf{frm}}(\Xi; \mathfrak{p}) \cup \boldsymbol{Options}^{\mathbf{e}}(\Xi; \mathfrak{p}).$$

An $\mathcal{S}_1$-*option for* $\mathfrak{p}$ *in* Ξ is any $\mathscr{O} \in \boldsymbol{Options}_{\mathcal{S}_1}(\Xi; \mathfrak{p})$.

LEMMA 8.7.29. *Let $\mathcal{S}_1$ and $\mathcal{S}_2$ be signatures with $\mathcal{S}_1 \subseteq \mathcal{S}_2$,*

$$\Xi = \langle \mathcal{Q}, r, \mathscr{D}, \mathscr{S} \rangle$$

be an $\mathcal{S}_2$-search state, $\mathfrak{p}$ be a u-part of $\mathcal{Q}$ and $\mathscr{O} \in \boldsymbol{Options}_{\mathcal{S}_1}(\Xi; \mathfrak{p})$ be an $\mathcal{S}_1$-option for $\mathfrak{p}$ in Ξ. Then $\mathbf{Search}(\Xi, \mathfrak{p}, \mathscr{O})$ is an $\mathcal{S}_2$-search state.

PROOF. This follows directly from Lemmas 8.7.7, 8.7.11, 8.7.18, 8.7.25 and 8.7.27. □

8.8. Completeness of Search

DEFINITION 8.8.1. A *part selector* is a partial function $\mathfrak{P}$ which chooses a u-part $\mathfrak{P}(\Xi) \in \mathcal{U}parts^{-ps}(\mathcal{Q}; \{r\})$ for every developed search state

$$\Xi = \langle \mathcal{Q}, r, \mathscr{D}, \mathscr{S} \rangle$$

unless $\mathcal{U}parts^{-ps}(\mathcal{Q}; \{r\})$ is empty.

Likewise, a *development strategy* is a function $\mathfrak{d}$ which chooses an undeveloped node of any undeveloped extensional expansion dag.

DEFINITION 8.8.2. Let $\mathfrak{P}$ be a part selector and $\mathfrak{d}$ be a development strategy. For any search states $\Xi = \langle \mathcal{Q}, r, \mathscr{D}, \mathscr{S} \rangle$ and Ξ' we say Ξ' is a *successor search state of* Ξ *with respect to* $\mathfrak{P}$ *and* $\mathfrak{d}$ if one of the following holds:

1. The search state Ξ is not developed and Ξ' is
$$\langle \mathbf{OpDevelop}(\mathcal{Q}, \mathfrak{d}(\mathcal{Q})), r, \mathscr{D}, \mathscr{S} \rangle.$$
2. The search state Ξ is developed, $\mathcal{U}parts^{-ps}(\mathcal{Q}; \{r\})$ is nonempty and there is some $\mathscr{O} \in \mathcal{O}ptions_{\mathcal{S}}(\Xi; \mathfrak{P}(\Xi))$ such that Ξ' is
$$\mathbf{Search}(\Xi, \mathfrak{P}(\Xi), \mathscr{O}).$$

We write

$$\Xi \overset{\mathfrak{P},\mathfrak{d}}{\longmapsto} \Xi'$$

when Ξ' is a successor search state of Ξ with respect to $\mathfrak{P}$ and $\mathfrak{d}$. We use $\overset{\mathfrak{P},\mathfrak{d}}{\longmapsto}_{*}$ to denote the reflexive, transitive closure of the $\overset{\mathfrak{P},\mathfrak{d}}{\longmapsto}$ relation on search states.

LEMMA 8.8.3. *Let* $\Xi = \langle \mathcal{Q}, r, \mathscr{D}, \mathscr{S} \rangle$ *be a developed search state,* $\langle \mathcal{Q}^*, r^* \rangle$ *be a lifting target,* $\mathsf{f} = \langle \varphi, f \rangle$ *be a search lifting map from* Ξ *to* $\langle \mathcal{Q}^*, r^* \rangle$*,* $\mathfrak{p} \in \mathcal{U}parts^{-ps}(\mathcal{Q}; \{r\})$ *be a u-part of* $\mathcal{Q}$ *including* $\{r\}$ *which is not pre-solved and*

$$\mathfrak{p}^* := \{f(a) \mid a \in \mathfrak{p}\}.$$

Suppose $\mathfrak{p}$ *is fully duplicated with respect to* f *and* $\mathbf{SetVars}^{\mathcal{Q}}(\mathfrak{p})$ *is almost atomic with respect to* f*. One of the following must hold:*

1. *There exist* $a, b \in \mathfrak{p}$ *and* $c^* \in \mathcal{N}^{\mathcal{Q}^*}$ *such that* a *is an extended positive atomic node,* b *is an extended negative atomic node and*
$$\langle c^*, a, b \rangle \in \mathbf{Conns}^-(\mathsf{f}; \mathbf{mate})$$

(i.e., $\langle c^*, a, b\rangle$ is a schema for a **mate***-connection with respect to* f*).*

2. *There exist $a, b \in \mathfrak{p}$ and $c^* \in \mathcal{N}^{\mathcal{Q}^*}$ such that a is a positive flexible node, b is an extended negative atomic node and*

$$\langle c^*, a, b\rangle \in \mathbf{Conns}^-(\mathsf{f}; \mathbf{mate})$$

(i.e., $\langle c^*, a, b\rangle$ is a schema for a **mate***-connection with respect to* f*).*

3. *There exist $a, b \in \mathfrak{p}$, $c^* \in \mathcal{N}^{\mathcal{Q}^*}$ and $k \in \{\mathbf{eunif}, \mathbf{eunif}^{sym}\}$ such that a is an extended positive equation node, b is an extended equation goal node and*

$$\langle c^*, a, b\rangle \in \mathbf{Conns}^-(\mathsf{f}; k)$$

(i.e., $\langle c^*, a, b\rangle$ is a schema for a k-connection with respect to f*).*

4. *There exists a node $e \in \mathfrak{p}$ such that e is an extended negative equation node, e is not decomposable in $\mathcal{Q}$ and $f(e)$ is decomposed in $\mathcal{Q}^*$.*

PROOF. Since $\langle \mathcal{Q}^*, r^*\rangle$ is a lifting target $\mathbf{Roots}(\mathcal{Q}^*) = \{r^*\}$ and $\{r^*\}$ is a sufficient set for $\mathcal{Q}^*$ (cf. Definitions 8.6.6 and 7.4.18). Since Ξ is a search state, we know $\mathbf{Roots}(\mathcal{Q}) = \{r\}$. Since f is a search lifting map, we must have $f(r) = r^*$ (cf. Definition 8.6.7). Hence $r^* \in \mathfrak{p}^*$. Since $\{r^*\}$ is sufficient, $\mathfrak{p}^*$ is not a u-part of $\mathcal{Q}^*$. Thus one of the conditions in Definition 7.4.4 must fail for $\mathfrak{p}^*$.

First we verify conditions (2) and (3) of Definition 7.4.4 do hold for $\mathfrak{p}^*$. Suppose $a^* \in \mathfrak{p}^*$ is disjunctive. There is some $a \in \mathfrak{p}$ such that $f(a) = a^*$. By Lemma 8.4.12:1, a is a disjunctive node of $\mathcal{Q}$. Hence there must be some $b \in \mathfrak{p}$ such that $a \xrightarrow{\mathcal{Q}} b$. Hence $f(b) \in \mathfrak{p}^*$ and $a^* \xrightarrow{\mathcal{Q}^*} f(b)$ by Lemma 8.4.6. Thus condition (2) of Definition 7.4.4 holds.

Suppose $a^* \in \mathfrak{p}^*$ is neutral and $succ^*(a^*) = \langle b^*\rangle$. There is some $a \in \mathfrak{p}$ such that $f(a) = a^*$. Since $\mathcal{Q}$ is developed, $\mathbf{SetVars}^{\mathcal{Q}}(\mathfrak{p})$ is almost atomic with respect to f and $a \in \mathfrak{p}$, we know a is neutral by Lemma 8.4.17. By Lemma 7.4.2, $succ(a) = \langle a^1\rangle$ for some $a^1 \in \mathcal{N}$. Thus $a^1 \in \mathfrak{p}$ since $a \in \mathfrak{p}$ and $\mathfrak{p}$ is a u-part of $\mathcal{Q}$. Since f is a lifting map, $succ^*(f(a)) = \langle f(a^1)\rangle$ and so $b^* = f(a^1) \in \mathfrak{p}^*$, as desired. Thus condition (3) of Definition 7.4.4 holds for $\mathfrak{p}^*$.

Therefore, either condition (1) or (4) of Definition 7.4.4 must fail for $\mathfrak{p}^*$. We summarize these cases here and present a detailed analysis in Appendix A.6.

Suppose there is a conjunctive node $a^* \in \mathfrak{p}^*$, $a^* \overset{\mathcal{Q}^*}{\rightarrow} b^*$ and $b^* \notin \mathfrak{p}^*$. Since $\mathcal{Q}$ is developed, the only way this can happen is if a^* is an expansion node or a^* is a decomposable node. We can rule out the case where a^* is an expansion node using the assumption that $\mathfrak{p}$ is fully duplicated with respect to f. If a^* is a decomposable node, then case (4) in the conclusion of the lemma holds.

Suppose there is a connection $\langle c^*, a^*, b^*\rangle \in conns^{\mathcal{Q}^*}$ such that $a^*, b^* \in \mathfrak{p}^*$ and $c^* \notin \mathfrak{p}^*$. In this case there exist $a, b \in \mathfrak{p}$ such that $f(a) = a^*$, $f(b) = b^*$ and $\langle c^*, a, b\rangle \in \mathbf{Conns}^-(\mathsf{f}; k)$ where k is $kind^{\mathcal{Q}^*}(c^*)$. If k is **mate**, then case (1) or case (2) in the conclusion of the lemma holds. If k is **eunif** or $\mathbf{eunif}^{sym}$, then case (3) in the conclusion of the lemma holds. □

We now establish a key lemma which implies we can always make progress during the lifting process.

LEMMA 8.8.4 (Progress). *Let $\mathcal{S}$ be a signature, $\Xi = \langle \mathcal{Q}, r, \mathscr{D}, \mathscr{S}\rangle$ be a developed search state, $\langle \mathcal{Q}^*, r^*\rangle$ be a lifting target and $\mathsf{f} = \langle \varphi, f\rangle$ be a search lifting map from Ξ to $\langle \mathcal{Q}^*, r^*\rangle$. For any $\mathfrak{p} \in \mathcal{U}parts^{-ps}(\mathcal{Q}; \{r\})$ there exists an option $\mathscr{O} \in \mathbf{Options}_{\mathcal{S}}(\mathcal{Q}; \mathfrak{p})$ of $\mathcal{Q}$ for $\mathfrak{p}$ and search lifting map f' from $\mathbf{Search}(\Xi, \mathfrak{p}, \mathscr{O})$ to $\langle \mathcal{Q}^*, r^*\rangle$ such that*

$$\|\mathsf{f}' : \mathcal{Q}^{\Xi'} \rightarrow \mathcal{Q}^*\| < \|\mathsf{f} : \mathcal{Q} \rightarrow \mathcal{Q}^*\|.$$

PROOF. We sketch the proof here. The full proof is in Appendix A.6. If $(\mathfrak{p} \setminus \mathscr{D})$ is not fully duplicated with respect to f, then we can use Lemma 8.7.7 to duplicate an expansion. Otherwise, we assume $(\mathfrak{p} \setminus \mathscr{D})$ is fully duplicated with respect to f. Hence $\mathfrak{p}$ is fully duplicated with respect to f. If $(\mathbf{SetVars}^{\mathcal{Q}}(\mathfrak{p}) \setminus \mathscr{S})$ is not almost atomic with respect to f, then we can apply a primitive substitution using Lemma 8.7.11. Otherwise, we assume $(\mathbf{SetVars}^{\mathcal{Q}}(\mathfrak{p}) \setminus \mathscr{S})$ is almost atomic with respect to f. Hence $\mathbf{SetVars}^{\mathcal{Q}}(\mathfrak{p})$ is almost atomic with respect to f.

Since $\mathfrak{p}$ is fully duplicated with respect to f and $\mathbf{SetVars}^{\mathcal{Q}}(\mathfrak{p})$ is almost atomic with respect to f, we can apply Lemma 8.8.3. We can then either apply a rigid mating option (using Lemma 8.7.25), a flex-rigid mating option (using Lemma 8.7.26), an E-unification option (using Lemma 8.7.27) or a unification option (using Lemma 8.7.18). □

We can now show that there is a successful choice of options leading to a pre-solved search state from any search state for which there is a lifting map to a lifting target.

LEMMA 8.8.5 (Lifting). *Let* $\mathbf{M}$ *be a sentence,* $\mathfrak{P}$ *be a part selector,* $\mathfrak{d}$ *be a development strategy,* $\langle \mathcal{Q}^*, r^* \rangle$ *be a lifting target and*

$$label^{\mathcal{Q}^*}(r^*) = \langle \mathbf{primary}, -1, \mathbf{M} \rangle.$$

For any search state Ξ *and lifting search map* f *from* Ξ *to* $\langle \mathcal{Q}^*, r^* \rangle$ *there exists a pre-solved search state* Ξ^{ps} *such that*

$$\Xi \overset{\mathfrak{P},\mathfrak{d}}{\longmapsto}_* \Xi^{ps}.$$

PROOF. We prove this by induction on the ordinal

$$\|\mathsf{f} : \mathcal{Q}^{\Xi} \to \mathcal{Q}^*\| < \omega^3.$$

Assume Ξ is $\langle \mathcal{Q}, r, \mathscr{D}, \mathscr{S} \rangle$ and f is $\langle \varphi, f \rangle$.

Suppose $\mathcal{Q}$ is developed and $\mathit{Uparts}^{-ps}(\mathcal{Q}; \{r\})$ is empty. In this case Ξ is pre-solved and

$$\Xi \overset{\mathfrak{P},\mathfrak{d}}{\longmapsto}_* \Xi.$$

Suppose Ξ is not developed. Let a be the node $\mathfrak{d}(\mathcal{Q})$ in $\mathbf{UnDev}(\mathcal{Q})$. Let Ξ' be $\langle \mathbf{OpDevelop}(\mathcal{Q}, a), r, \mathscr{D}, \mathscr{S} \rangle$. By Lemma 8.6.17, Ξ' is a search state and there is a search lifting map f' from Ξ' to $\langle \mathcal{Q}^*, r^* \rangle$ such that

$$\|\mathsf{f}' : \mathcal{Q}^{\Xi'} \to \mathcal{Q}^*\| < \|\mathsf{f} : \mathcal{Q} \to \mathcal{Q}^*\|.$$

By the inductive hypothesis, there is a pre-solved search state Ξ^{ps} such that

$$\Xi' \overset{\mathfrak{P},\mathfrak{d}}{\longmapsto}_* \Xi^{ps}.$$

By Definition 8.8.2,

$$\Xi \overset{\mathfrak{P},\mathfrak{d}}{\longmapsto} \Xi'$$

and so

$$\Xi \overset{\mathfrak{P},\mathfrak{d}}{\longmapsto}_* \Xi^{ps}.$$

Suppose Ξ is developed and $\mathit{Uparts}^{-ps}(\mathcal{Q}; \{r\})$ is nonempty. Let $\mathfrak{p}$ be the u-part $\mathfrak{P}(\Xi)$ of $\mathcal{Q}$ including $\{r\}$. By Lemma 8.8.4, there is an option $\mathscr{O} \in \boldsymbol{Options_S}(\mathcal{Q}; \mathfrak{p})$ of $\mathcal{Q}$ for $\mathfrak{p}$ and search lifting map f' from $\mathbf{Search}(\Xi, \mathfrak{p}, \mathscr{O})$ to $\langle \mathcal{Q}^*, r^* \rangle$ such that

$$\|\mathsf{f}' : \mathcal{Q}^{\mathbf{Search}(\Xi,\mathfrak{p},\mathscr{O})} \to \mathcal{Q}^*\| < \|\mathsf{f} : \mathcal{Q} \to \mathcal{Q}^*\|.$$

Let Ξ' be the search state $\mathbf{Search}(\Xi, \mathfrak{p}, \mathscr{O})$ (cf. Lemma 8.7.29). By the inductive hypothesis, there is a pre-solved search state Ξ^{ps} such that

$$\Xi' \overset{\mathfrak{P},\mathfrak{d}}{\longmapsto}_* \Xi^{ps}.$$

By Definition 8.8.2,

$$\Xi \overset{\mathfrak{P},\mathfrak{d}}{\longmapsto} \Xi'$$

and so

$$\Xi \overset{\mathfrak{P},\mathfrak{d}}{\longmapsto}_* \Xi^{ps}.$$

□

We can now prove the main result. Given any part selector and development strategy there are appropriate search steps which lead to a pre-solved search state (giving an extensional expansion proof by Theorem 8.3.3). Thus the operations described in this chapter form a (non-deterministic) basis for extensional higher-order theorem proving.

THEOREM 8.8.6. *Let $\mathfrak{P}$ be a part selector, $\mathfrak{d}$ be a development strategy and $\mathbf{M}$ be a sentence derivable in $\mathcal{G}^S_{\beta\mathfrak{fb}}$. There exists a pre-solved search state Ξ^{ps} such that*

$$\mathbf{SearchInit}(\mathbf{M}) \overset{\mathfrak{P},\mathfrak{d}}{\longmapsto}_* \Xi^{ps}.$$

PROOF. We know there is a merged, separated extensional expansion proof $\langle \mathcal{Q}^{**}, \{r^*\} \rangle$ where $\mathbf{Roots}(\mathcal{Q}^{**}) = \{r^*\}$, r^* is a negative primary node with shallow formula $\mathbf{M}$ by Corollary 7.10.13. Note that $\mathbf{RootSels}(\mathcal{Q}^{**})$ is empty, $\mathcal{Q}^{**}$ is merged, $\mathcal{Q}^{**}$ is separated and $\prec_0^{\mathcal{Q}^{**}}$ is acyclic. By Lemma 8.1.13, there exists a closed, developed extensional expansion dag $\mathcal{Q}^*$ such that $\mathcal{Q}^{**} \trianglelefteq_r \mathcal{Q}^*$, $\mathbf{RootSels}(\mathcal{Q}^*)$ is empty, $\mathcal{Q}^*$ is separated, $\mathcal{Q}^*$ has merged connections, $\mathcal{Q}^*$ has merged decompositions and $\prec_0^{\mathcal{Q}^*}$ is acyclic. Since $\mathcal{Q}^{**} \trianglelefteq_r \mathcal{Q}^*$, $\mathbf{Roots}(\mathcal{Q}^{**}) = \{r^*\}$. By Lemma 7.6.7, $\{r^*\}$ is sufficient for $\mathcal{Q}^*$ since $\{r^*\}$ is sufficient for $\mathcal{Q}^{**}$. Thus $\langle \mathcal{Q}^*, \{r^*\} \rangle$ is an extensional expansion proof and so $\langle \mathcal{Q}^*, r^* \rangle$ is a lifting target.

By Lemma 8.6.9, **SearchInit**(**M**) is a search state and there is a search lifting map f from Ξ to $\langle \mathcal{Q}^*, r^* \rangle$. By Lemma 8.8.5, there is a pre-solved search state Ξ^{ps} such that

$$\mathbf{SearchInit}(\mathbf{M}) \overset{\mathfrak{P},\mathfrak{d}}{\longmapsto}_* \Xi^{ps}$$

as desired. □

REMARK 8.8.7 (Lifting Algorithm). The proof of Lemma 8.8.5 provides an algorithm for using a given extensional expansion proof $\langle \mathcal{Q}^*, \mathcal{R} \rangle$ to guide the **MS04-2** search procedure by pruning options which do not correspond to $\langle \mathcal{Q}^*, \mathcal{R} \rangle$.

CHAPTER 9

Automated Search

We turn now to a description of search procedures for (fragments of) extensional type theory. In order to search for proofs, we make use of expansion variables to delay finding the expansion terms needed for the proof. We describe two fundamentally different search procedures: **MS04-2** and **MS03-7**. The search procedure **MS04-2** uses a bounded best-first search strategy with backtracking. Iterative deepening is used to ensure completeness of **MS04-2**. On the other hand, the search procedure **MS03-7** never backtracks. Instead, **MS03-7** follows more of a saturation approach by simply growing the extensional expansion dag during search.

9.1. The MS04-2 Search Procedure

The **MS04-2** search procedure uses a bounded best-first search strategy with iterative deepening to ensure completeness. The basic search procedure could be used with any logical signature $\mathcal{S}$. The current implementation assumes the signature has a certain form.

DEFINITION 9.1.1. Let $\mathcal{S}_s$ be the signature

$$\{\neg, \vee, =^{\iota}\} \cup \{\Pi^{\alpha} \mid \alpha \in \mathcal{T}_{gen}(\{\iota, (o\iota)\})\}.$$

Any signature which includes $\mathcal{S}_s$ is extensionally complete by Theorem 6.6.7. We will describe the procedure for an arbitrary signature $\mathcal{S}$ so long as $\mathcal{S} \setminus \mathcal{S}_s$ is finite.

We first give an abstract description of the procedure.

ALGORITHM 9.1.2 (Abstract Version of **MS04-2**). *Let Ξ be a search state, $d \in \mathbb{N}$ be a natural number (representing the depth) and $B \in \mathbb{N}$ be a natural number (representing the depth bound).*

1. *If $d > B$, then return* **FAILURE**.

2. *If* $\Xi = \langle \mathcal{Q}, r, \mathcal{D}, \mathcal{S}\rangle$, *then let* $\Xi' := \langle \mathcal{Q}', r, \mathcal{D}, \mathcal{S}\rangle$ *where* $\mathcal{Q}'$ *is developed (cf. Lemma 8.1.13).*

3. *If* $\mathcal{U}parts^{-ps}(\mathcal{Q}'; \{r\})$ *is empty, then we are done (cf. Theorem 8.3.3). Return* **SUCCESS**.

4. *Choose a u-part* $\mathfrak{p} \in \mathcal{U}parts^{-ps}(\mathcal{Q}'; \{r\})$.

5. *Generate the set of options* $\boldsymbol{Options}_{\mathcal{S}}(\Xi'; \mathfrak{p})$ *(cf. Definition 8.7.28).*

6. *Non-deterministically choose an option* $\mathcal{O} \in \boldsymbol{Options}_{\mathcal{S}}(\Xi'; \mathfrak{p})$ *and let* $\Xi'' := \mathbf{Search}(\Xi', \mathfrak{p}, \mathcal{O})$ *(cf. Definitions 8.7.6, 8.7.10, 8.7.17, 8.7.22, 8.7.23 and 8.7.24). Return to step (1) with* Ξ'' *and depth* $d + 1$.

Given a sentence $\mathbf{M}$ we wish to prove we can call Algorithm 9.1.2 with the initial search state (cf. Definition 8.6.8), depth 0 and some depth bound B. If the search fails, we call the procedure again with a higher depth bound (e.g., $B + 1$).

A number of comments about Algorithm 9.1.2 are in order.

To avoid non-determinism in step (2) we must choose a particular development strategy (cf. Definition 8.8.1). Since any such strategy will terminate (Lemma 8.1.12) with essentially the same developed extensional expansion dag, the particular development strategy is not important. One can simply recursively traverse the underlying dag considering the children in some order searching for the first node which is not developed (cf. Definition 8.1.1) and perform the appropriate operation to develop the node (cf. Definition 8.1.3).

When search succeeds in step (3) we would like to return an actual extensional expansion proof. The $\mathcal{S}$-extensional expansion dag $\mathcal{Q}'$ with no pre-solved u-parts can be instantiated as in the proof of Theorem 8.3.3 to obtain an $(\mathcal{S} \cup \{\top\})$-extensional expansion proof. In particular, we choose some arbitrary parameter C_ι (which is not a selected parameter) and substitute

$$[\lambda \overline{z^n}\, C_\iota] \text{ for any variable } x_{\iota\alpha^n\cdots\alpha^1} \in \mathbf{Free}(\mathcal{Q}')$$

and

$$[\lambda \overline{z^n}\, \top_o] \text{ for any variable } x_{o\alpha^n\cdots\alpha^1} \in \mathbf{Free}(\mathcal{Q}').$$

To avoid non-determinism in step (4) where the u-part is chosen, we need some part selector (cf. Definition 8.8.1). In the implementation **MS04-2** uses the first u-part (given some ordering of the nodes) which is not pre-solved and contains no flexible nodes (a *rigid u-part*) if such a u-part exists. Otherwise, **MS04-2** uses the first u-part which is not pre-solved (given some ordering of the nodes).

The set of options $\boldsymbol{Options}_{\mathcal{S}}(\Xi'; \mathfrak{p})$ generated in step (5) may be infinite. In particular, the set of primsub options may be infinite due to the fact that the set

$$\{\Pi^{\alpha} \mid \alpha \in \mathcal{T}_{gen}(\{\iota, (o\iota)\})\}$$

is infinite. This is, of course, a serious problem if we are really to consider this an algorithm. To avoid this problem, the actual algorithm must go into a type generation phase before considering a primsub for quantifiers. One can easily enumerate the types in $\mathcal{T}_{gen}(\{\iota, (o\iota)\})$ by a function

$$\tau : \mathbb{N} \to \mathcal{T}_{gen}(\{\iota, (o\iota)\}).$$

Instead of generating primsub options for quantifiers of arbitrary type, we generate an option to enter the type generation phase for the quantifier starting with $\tau(0)$. At each stage in the type generation phase with type $\tau(n)$ we either remain in the type generation phase with type $\tau(n+1)$ or perform a primitive substitution with the quantifier at type $\tau(n)$. In this manner, we can assure that the actual algorithm only has finitely many options at any step.

Let **typegen** be a value such that

$$\mathbf{typegen} \notin \{\mathbf{dup}, \mathbf{primsub}, \mathbf{unif}, \mathbf{mateop}, \mathbf{frmateop}, \mathbf{eunifop}\}.$$

That is, **typegen** is a value which is not an option genus (cf. Definition 8.7.1). We will use this value to identify the option of entering the type generation phase for a quantifier. To include these options, we define the notion of **MS04-2** *search option.*

DEFINITION 9.1.3. Let $\Xi = \langle \mathcal{Q}, r, \mathscr{D}, \mathscr{S} \rangle$ be a developed search state and $\mathfrak{p} \subseteq \mathcal{N}^{\mathcal{Q}}$ be a u-part of $\mathcal{Q}$. We say $\mathscr{O}$ is an **MS04-2** *search option* for $\mathfrak{p}$ in Ξ if $\mathscr{O}$ is an option for $\mathfrak{p}$ in Ξ (cf. Definition 8.7.28) or $\mathscr{O}$ is $\langle \mathbf{typegen}, p, P \rangle$ where $p \in \mathbf{SetVars}^{\mathcal{Q}}(\mathfrak{p})$ and $P \in \{\Pi, \Sigma\}$.

Finally, we must handle the non-determinism in step (6). Assume we have some method for giving natural number weights to each option. We can use such weights to order the set of options. Step (6) could then choose the first untried option, define Ξ'' using the option and return to step (1). If the recursive call to step (1) returns **SUCCESS**, then we are done. Otherwise, we choose the next untried option, define Ξ'' and return to step (1) again. This continues until we have tried all options or succeeded. If we try all options without success, then step (6) returns **FAILURE**. Furthermore, since we are assuming each option has a corresponding weight $W \in \mathbb{N}$, we can make recursive calls to step (1) with depth $d + 1 + W$.

In order to more closely model the implementation of **MS04-2** we separate certain phases of search. Namely, search alternates between an *option generation phase*, an *option attempt phase* and a *type generation phase* each of which is described below. The phases are mutually recursive with the *main phase*.

ALGORITHM 9.1.4 (Option Generation Phase). *Let*

$$\Xi = \langle \mathcal{Q}, r, \mathcal{D}, \mathcal{S} \rangle$$

be a search state, $\mathfrak{p} \in \boldsymbol{Uparts}^{-ps}(\mathcal{Q}; \{r\})$ be a u-part, $d \in \mathbb{N}$ be a natural number (representing the depth) and $B \in \mathbb{N}$ be a natural number (representing the depth bound).

1. *If $d > B$, then return* **FAILURE**.

2. *Let $\boldsymbol{Options}^{\mathbf{d}}_{MS04-2}$ be the (finite) set*

$$\{\langle \mathbf{dup}, e \rangle \in \boldsymbol{Options}^{\mathbf{d}}(\Xi) \mid e \in \mathfrak{p}\}$$

   of duplication options (cf. Definition 8.7.5). For each

$$\mathcal{O} \in \boldsymbol{Options}^{\mathbf{d}}_{MS04-2}$$

   assign a weight $W^{\mathcal{O}} \in \mathbb{N}$.

3. *Let $\mathcal{S}_0$ be*

$$(\{\neg, \vee, =^{\iota}\} \cap \mathcal{S}) \cup (\mathcal{S} \setminus \mathcal{S}_s),$$

   $\mathcal{U}$ be $\mathbf{SetVars}^{\mathcal{Q}}(\mathfrak{p})$ and $\boldsymbol{Options}^{\mathbf{P}}_{MS04-2}$ be the set

$$\{\langle \mathbf{typegen}, p, \Pi \rangle \mid p \in \mathcal{U}\}$$

$$\cup \{\langle \mathbf{primsub}, p, \mathbf{P} \rangle \in \boldsymbol{Options}^{\mathbf{P}}_{\mathcal{S}_0}(\Xi) \mid p \in \mathcal{U}\}.$$

   Since we have assumed $(\mathcal{S} \setminus \mathcal{S}_s)$ is finite, we know

$$\boldsymbol{Options}^{\mathbf{P}}_{MS04-2}$$

is a finite set. For each $\mathcal{O} \in \boldsymbol{Options}^{\mathbf{p}}_{MS04-2}$ assign a weight $W^{\mathcal{O}} \in \mathbb{N}$.

4. Let $\boldsymbol{Options}^{\mathbf{u}}_{MS04-2}$ be the (finite) set

$$\boldsymbol{Options}^{\mathbf{u}}(\Xi; \mathfrak{p})$$

of all unification options (cf. Definition 8.7.16) for $\mathfrak{p}$ in Ξ. Assign a natural number weight $W^{\mathcal{O}} \in \mathbb{N}$ to each option $\mathcal{O}$ in $\boldsymbol{Options}^{\mathbf{u}}_{MS04-2}$.

5. Let $\boldsymbol{Options}^{\mathbf{c}}_{MS04-2}$ be the (finite) set

$$\boldsymbol{Options}^{\mathbf{m}}(\Xi; \mathfrak{p}) \cup \boldsymbol{Options}^{\mathbf{frm}}(\Xi; \mathfrak{p}) \cup \boldsymbol{Options}^{\mathbf{e}}(\Xi; \mathfrak{p})$$

of all rigid mating options (cf. Definition 8.7.19), flex-rigid mating options (cf. Definition 8.7.20) and E-unification options (cf. Definition 8.7.21) for $\mathfrak{p}$ in Ξ. Assign a natural number weight $W^{\mathcal{O}} \in \mathbb{N}$ to each option $\mathcal{O}$ in $\boldsymbol{Options}^{\mathbf{c}}_{MS04-2}$.

6. Let $\boldsymbol{Options}^{\mathbf{a}}_{MS04-2}$ be the (finite) set

$$\boldsymbol{Options}^{\mathbf{d}}_{MS04-2} \cup \boldsymbol{Options}^{\mathbf{p}}_{MS04-2}$$
$$\cup \boldsymbol{Options}^{\mathbf{u}}_{MS04-2} \cup \boldsymbol{Options}^{\mathbf{c}}_{MS04-2}.$$

7. Order $\boldsymbol{Options}^{\mathbf{a}}_{MS04-2}$ using the weights as $\mathcal{O}^1, \ldots, \mathcal{O}^n$ such that

$$W^{\mathcal{O}^1} \leq \cdots \leq W^{\mathcal{O}^n}.$$

Let i be 1.

8. If $i > n$, then return **FAILURE**. Otherwise, continue.

9. Call the option attempt phase (Algorithm 9.1.5) with Ξ, $\mathfrak{p}$, $\mathcal{O}^i$, $d + W^{\mathcal{O}^i}$ and B. If this returns **SUCCESS**, then return **SUCCESS**. Otherwise, let i be $i + 1$ and goto step (8).

Next we describe an option attempt phase which tries a given option.

ALGORITHM 9.1.5 (Option Attempt Phase). Let

$$\Xi = \langle \mathcal{Q}, r, \mathcal{D}, \mathcal{S} \rangle$$

be a search state, $\mathfrak{p} \in \textbf{U}\text{parts}^{-ps}(\mathcal{Q}; \{r\})$ *be a u-part,* $d \in \textbf{N}$ *be a natural number (representing the depth),* $\mathcal{O}$ *be an* **MS04-2** *search option and* $B \in \textbf{N}$ *be a natural number (representing the depth bound).*

1. *If* $d > B$*, then return* **FAILURE**.

2. *If* $\mathcal{O}$ *is* ⟨**typegen**, p, P⟩*, then goto step* (3)*. Otherwise,*

$$\mathcal{O} \in \textbf{Options}_S(\Xi; \mathfrak{p}).$$

Goto step (4)*.*

3. *Call the type generation phase (Algorithm 9.1.6) with* p, P, 0, $d + 1$ *and* B*. (That is, we generate a type for* P *and* p*.) If this returns* **SUCCESS**, *then return* **SUCCESS**. *If this returns* **FAILURE**, *then return* **FAILURE**.

4. *Call the main phase with search state* **Search**$(\Xi, \mathfrak{p}, \mathcal{O})$, d *and* B. *If this returns* **SUCCESS**, *then return* **SUCCESS**. *If this returns* **FAILURE**, *then return* **FAILURE**.

The type generation phase determines a type α and substitutes with an imitation term for P^α where $P \in \{\Pi, \Sigma\}$. The type generation phase makes use of an enumeration τ of the types in $\mathcal{T}_{gen}(\{\iota, (\iota\iota)\})$. The implementation uses a particular function $\tau^{\mathcal{T}}$ defined as follows:

$$\tau^{\mathcal{T}}(t) := \begin{cases} \iota & \text{if } t = 0 \\ (\iota\iota) & \text{if } t = 1 \\ ((\tau^{\mathcal{T}}(n))(\tau^{\mathcal{T}}(m))) & \text{if } t = 2^n(2m+1)+1 \\ & \text{where } n \geq 0, m \geq 0 \end{cases}$$

Note that every $t \geq 2$ is $2^n(2m+1) + 1$ for a unique pair ⟨n, m⟩. (A similar construction works if there are base types other than ι. The implementation does account for this case.)

ALGORITHM 9.1.6 (Type Generation Phase). *Let*

$$\Xi = \langle \mathcal{Q}, r, \mathcal{D}, \mathcal{S} \rangle$$

be a search state, p *be a set variable,* $\mathcal{U}$ *be a set of expansion variables,* $P \in \{\Pi, \Sigma\}$, $t \in \textbf{N}$ *be a natural number (representing the generated type),* $d \in \textbf{N}$ *be a natural number (representing the depth) and* $B \in \textbf{N}$ *be a natural number (representing the depth bound).*

1. *If* $d > B$*, then return* **FAILURE**.

2. *Let* α *be* $\tau^{\mathcal{T}}(t)$. *If* $P^{\alpha} \in \mathcal{S}$, *then goto step* (3). *Otherwise, goto step* (4).

3. *Let* $\mathbf{P}$ *be the imitation term for* P^{α} *and call the main phase (Algorithm 9.1.7) with*

$$\langle [\mathbf{P}/p]\mathcal{Q}, r, \mathscr{D}, \mathscr{S} \rangle,$$

$d+1$ *and* B. (*That is, we* perform the primsub $\mathbf{P}$ for p.) *If this returns* **SUCCESS**, *then return* **SUCCESS**. *Otherwise, continue.*

4. *Call the type generation phase with* Ξ, p, $\mathcal{U}$, P, $t+1$,

$$d + W^{next}$$

and B (*where* W^{next} *is some natural number*). (*That is, we* generate the next quantifier type for p.)

These phases are mutually recursive with the *main phase* which calls the duplication phase, primitive substitution phase and part phase to perform search operations. The main phase will always terminate since we call the main phase with a finite depth bound B and the depth d increases each time the main phase calls another phase.

ALGORITHM 9.1.7 (Main Phase). *Let* Ξ *be a search state,* $d \in \mathbb{N}$ *be a natural number (representing the depth) and* $B \in \mathbb{N}$ *be a natural number (representing the depth bound).*

1. *If* $d > B$, *then return* **FAILURE**.

2. *If* $\Xi = \langle \mathcal{Q}, r, \mathscr{D}, \mathscr{S} \rangle$, *then let* $\Xi' := \langle \mathcal{Q}', r, \mathscr{D}, \mathscr{S} \rangle$ *where* $\mathcal{Q}'$ *is developed (cf. Lemma 8.1.13).*

3. *If* $\mathbf{Uparts}^{-ps}(\mathcal{Q}'; \{r\})$ *is empty, then we are done (cf. Theorem 8.6.3). Return* **SUCCESS**.

4. *Choose a u-part* $\mathfrak{p} \in \mathbf{Uparts}^{-ps}(\mathcal{Q}'; \{r\})$ *which is not pre-solved (cf. Definition 8.3.2).*

5. *Call the option generation phase (Algorithm 9.1.4) with* Ξ, $\mathfrak{p}$, $d+1$ *and* B.

Finally, the **MS04-2** search procedure proceeds by iterative deepening. We call the main phase with some initial depth bound. If the main phase does not find a proof within this depth bound, then we increase the depth bound and call the main phase again. This loop continues until a proof is found. In the actual implementation, the loop will terminate with **FAILURE** if the depth bound was never reached on an iteration.

ALGORITHM 9.1.8 (**MS04-2**). *Let $\mathcal{S}$ be a signature for which the set $(\mathcal{S} \setminus \mathcal{S}_s)$ is finite and let* $\mathbf{M}$ *be a $\mathcal{S}$-sentence.*

1. *Let B be the initial depth bound and Ξ be the initial search state (cf. Definition 8.6.8).*
2. *Call the main phase (Algorithm 9.1.7) with Ξ, depth 0 and depth bound B. If this returns* **SUCCESS**, *then return* **SUCCESS**. *Otherwise, continue.*
3. *Increase B and return to step (2).*

We can further bound calls to Algorithm 9.1.7 by bounding the number of duplications, primitive substitutions and connections allowed. In order to maintain completeness, we increase these bounds by 1 in step (3) of Algorithm 9.1.8.

9.2. Prenex Primitive Substitutions

One can also restrict the search so that only set substitutions in prenex-conjunctive normal form are generated. That is, we only generate set substitutions in the form

$$Q_1^{\alpha^1} \lambda x^1_{\alpha^1} \cdots Q_n^{\alpha^n} \lambda x^n_{\alpha^n} \centerdot \mathbf{A}_o$$

where $\mathbf{A}$ is in conjunctive normal form and $Q_i \in \{\Pi, \Sigma\}$ for each i $(1 \leq i \leq n)$. For prenex-conjunctive set substitutions to make sense, we assume the signature satisfies certain properties.

DEFINITION 9.2.1 (Prenex Appropriate Signatures). We say a signature $\mathcal{S}$ of logical constants is *prenex appropriate* if

- $\{\neg, \wedge, \vee\} \subseteq \mathcal{S}$,
- $\mathcal{S} \cap \{\top, \bot, \supset, \equiv\} = \emptyset$,
- $\mathcal{S} \cap \mathcal{S}^= = \{=^\iota\}$, and
- for each type α, $\Pi^\alpha \in \mathcal{S}$ iff $\Sigma^\alpha \in \mathcal{S}$.

DEFINITION 9.2.2. Let $\mathcal{S}_p$ be the signature

$$\{\neg, \vee, \wedge, =^\iota\} \cup \{\Pi^\alpha, \Sigma^\alpha \mid \alpha \in \mathcal{T}_{gen}(\{\iota, (o\iota)\})\}.$$

When searching for prenex set substitutions, we assume the signature $\mathcal{S}$ is prenex appropriate and use the signature $\mathcal{S}_p$ in place of $\mathcal{S}_s$ in a modified version of step (3) in the option generation phase (cf. Algorithm 9.1.4). The modified version of this step also generates an option $\langle \mathbf{typegen}, p, \Sigma \rangle$ in addition to $\langle \mathbf{typegen}, p, \Pi \rangle$.

The set $\mathscr{S}$ of a search state

$$\langle \mathcal{Q}, r, \mathscr{D}, \mathscr{S} \rangle$$

indicates set variables on which we will no longer perform primitive substitutions. Prenex-conjunctive set substitutions can be generated by adding the following sets of set variables to the search state:

$\mathscr{S}^{\Pi,\Sigma}$: set variables on which we will no longer perform primitive substitutions imitating quantifiers.

$\mathscr{S}^{\Pi,\Sigma,\wedge}$: set variables on which we will no longer perform primitive substitutions imitating quantifiers or $\wedge$.

$\mathscr{S}^{\Pi,\Sigma,\wedge,\vee}$: set variables on which we will no longer perform primitive substitutions imitating quantifiers, $\wedge$ or $\vee$.

All negations in a prenex-conjunctive set substitution occur in certain special positions: before projections, before equalities and before imitations. Thus we never explicitly generate negation primitive substitutions when searching for prenex set substitutions. Instead, in the final primsub phase we generate primitive substitutions using equalities at type ι, negations of equalities at type ι, projections and negations of projections.

In order to account for negations of imitations, **MS04-2** allows connections between flexible nodes and atomic nodes of the same polarity. Before mating the nodes, we apply a negation and an imitation to the head of the shallow formula of the flexible node. Since we need to consider such connections, we consider u-parts which contain negative flexible nodes (even though these are pre-solved). We still do not consider flex-flex u-parts (cf. Definition 8.3.2).

9.3. Optional Connections

Under certain flag settings **MS04-2** will consider certain connections which are not necessary for completeness, but can lead to faster solutions. In order to allow negations to be used in the instantiations below, we must assume $\neg \in \mathcal{S}$.

MS04-2 can optionally use projections (and negations of projections) on the head of a flexible node for the purpose of mating the flexible node to an atomic node. This is not necessary since projections and negations of projections are already generated by primitive substitutions.

MS04-2 can also optionally use substitutions introducing equalities (and negations of equalities) for the head of a flexible node for the purpose of making an E-unification connection between the flexible node and either a positive equation node or an equation goal node. This is not necessary since equalities and negations of equalities are already generated by primitive substitutions.

9.4. Part Constraints

We can augment the procedure by carrying a set of constraints during search. We define a part constraint to be a collection of nodes of the extensional expansion dag.

DEFINITION 9.4.1. Let $\mathcal{Q}$ be an extensional expansion dag. A *part constraint for* $\mathcal{Q}$ is any set $\mathfrak{c} \subseteq \mathcal{N}^{\mathcal{Q}}$ of nodes.

In Definition 7.4.4 we defined $\mathcal{U}parts(\mathcal{Q}; \mathfrak{c})$ to be the set of u-parts $\mathfrak{p}$ of $\mathcal{Q}$ which include $\mathfrak{c}$ (i.e., u-parts such that $\mathfrak{c} \subseteq \mathfrak{p}$). We now extend the notion of u-parts which include a part constraint to a notion of u-parts which include a set of part constraints.

DEFINITION 9.4.2. Let $\mathcal{Q}$ be an extensional expansion dag and $\mathscr{C}$ be a set of part constraints for $\mathcal{Q}$. We say a u-part $\mathfrak{p}$ is *spanned by* $\mathscr{C}$ if there exists some $\mathfrak{c} \in \mathscr{C}$ such that $\mathfrak{c} \subseteq \mathfrak{p}$.

If $\mathscr{C}$ is a set of (unordered) pairs of nodes, then our notion of $\mathscr{C}$ spanning $\mathfrak{p}$ corresponds directly to the notion of spanning described in both [21] and [7]. That is, $\mathscr{C}$ spans $\mathfrak{p}$ if there is some set $\{a, b\} \in \mathscr{C}$ (a *connection*, cf. [21; 7]) such that $a, b \in \mathfrak{p}$.

In order to search with part constraints, we assume we will be able to solve the u-parts spanned by a given part constraint set and apply operations based on the remaining u-parts. In the basic **MS04-2** search procedure (without constraints) we ignore pre-solved u-parts (cf. Definition 8.3.2). Suppose $\mathcal{Q}$ is an extensional expansion dag. Let $\mathscr{C}$ be

$$\{\{e\} \mid e \text{ is a flex-flex node or a positive flexible node}\}.$$

The u-parts spanned by $\mathscr{C}$ are precisely the pre-solved u-parts. Hence

$$\mathit{Uparts}^{-ps}(\mathcal{Q})$$

is the set of all u-parts which are not spanned by $\mathscr{C}$.

We introduce a notion of a *solution* to a set of part constraints by tracing the u-parts of extensional expansion dags as search operations are performed.

DEFINITION 9.4.3. Let $\mathcal{Q}$ and $\mathcal{Q}'$ be extensional expansion dags. We write $\mathcal{Q} \rightsquigarrow \mathcal{Q}'$ if either $\mathcal{Q} \trianglelefteq_r \mathcal{Q}'$ or $\mathcal{Q}' = \theta(\mathcal{Q})$ for some substitution θ. We use $\rightsquigarrow_*$ to denote the transitive closure of $\rightsquigarrow$.

LEMMA 9.4.4. *Let $\mathcal{Q}$ and $\mathcal{Q}'$ be extensional expansion dags. If $\mathcal{Q} \rightsquigarrow \mathcal{Q}'$, then $\mathcal{N}^{\mathcal{Q}} \subseteq \mathcal{N}^{\mathcal{Q}'}$, $\mathbf{Roots}(\mathcal{Q}) = \mathbf{Roots}(\mathcal{Q}')$ and*

$$(\mathfrak{p}' \cap (\mathcal{N}^{\mathcal{Q}})) \in \mathit{Uparts}(\mathcal{Q}; \mathbf{Roots}(\mathcal{Q}))$$

for every $\mathfrak{p}' \in \mathit{Uparts}(\mathcal{Q}'; \mathbf{Roots}(\mathcal{Q}'))$.

PROOF. The result follows from Definition 7.6.1 and Lemma 7.6.6 if $\mathcal{Q} \trianglelefteq_r \mathcal{Q}'$. The result follows from Definition 8.2.2 and Lemma 8.2.8 if $\mathcal{Q}'$ is $\theta(\mathcal{Q})$. $\square$

LEMMA 9.4.5. *Let $\mathcal{Q}$ and $\mathcal{Q}'$ be extensional expansion dags. If $\mathcal{Q} \rightsquigarrow_* \mathcal{Q}'$, then $\mathcal{N}^{\mathcal{Q}} \subseteq \mathcal{N}^{\mathcal{Q}'}$, $\mathbf{Roots}(\mathcal{Q}) = \mathbf{Roots}(\mathcal{Q}')$ and*

$$(\mathfrak{p}' \cap (\mathcal{N}^{\mathcal{Q}})) \in \mathit{Uparts}(\mathcal{Q}; \mathbf{Roots}(\mathcal{Q}))$$

for every $\mathfrak{p}' \in \mathit{Uparts}(\mathcal{Q}'; \mathbf{Roots}(\mathcal{Q}'))$.

PROOF. Since $\rightsquigarrow_*$ is simply the transitive closure of $\rightsquigarrow$, the result follows by induction using Lemma 9.4.4. $\square$

Using Lemma 9.4.5, we can define a *solution* of a set of part constraints.

DEFINITION 9.4.6. Let $\mathcal{Q}$ be an extensional expansion dag and $\mathscr{C}$ be a set of part constraints for $\mathcal{Q}$. We say $\mathcal{Q}'$ is a *solution for* $\mathscr{C}$ if $\mathcal{Q} \rightsquigarrow_* \mathcal{Q}'$ and for every u-part $\mathfrak{p} \in \mathit{Uparts}(\mathcal{Q}; \mathbf{Roots}(\mathcal{Q}))$ spanned by $\mathscr{C}$ there is no $\mathfrak{p}' \in \mathit{Uparts}(\mathcal{Q}'; \mathbf{Roots}(\mathcal{Q}'))$ such that $\mathfrak{p} = (\mathfrak{p}' \cap \mathcal{N}^{\mathcal{Q}})$.

EXAMPLE 9.4.7. Consider the extensional expansion dag $\mathcal{Q}$ in Figure 9.1. The only u-part which includes the root node of $\mathcal{Q}$ is the set of all nodes of $\mathcal{Q}$. Let $\mathfrak{p}$ be $\mathcal{N}^{\mathcal{Q}}$, $\mathfrak{c}$ be the singleton $\{\mathbf{EQN}\}$ and $\mathscr{C}$ be the singleton $\{\mathfrak{c}\}$. The set of constraints $\mathscr{C}$ spans $\mathfrak{p}$ since $\mathfrak{c} \subseteq \mathcal{N}^{\mathcal{Q}}$.

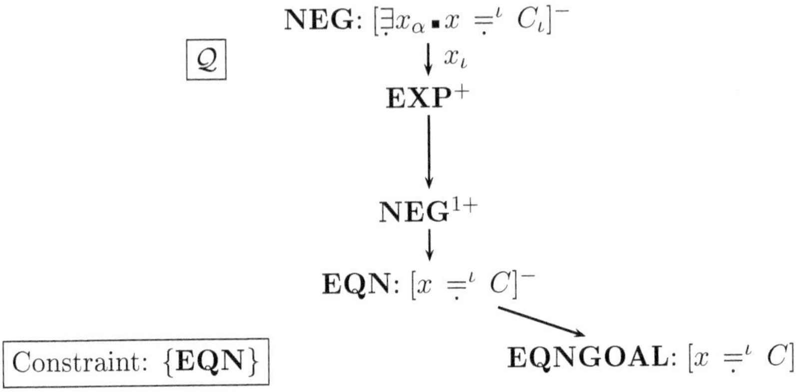

FIGURE 9.1. Extensional Expansion Dag with Constraint

The obvious method to solve the set $\mathscr{C}$ of constraints is simply to substitute C_ι for x_ι. To obtain a solution in the sense of Definition 9.4.6, let $\mathcal{Q}^1$ be $[C/x](\mathcal{Q})$ and $\mathcal{Q}^2$ be the result of developing $\mathcal{Q}^1$ by decomposing the node **EQN** in $\mathcal{Q}^1$ and extending the new decomposition node as shown in Figure 9.2. Note that $\mathcal{Q} \rightsquigarrow \mathcal{Q}^1 \rightsquigarrow \mathcal{Q}^2$ (cf. Definition 9.4.3). Also, there are no u-parts of $\mathcal{Q}^2$ which include the disjunctive leaf node **DEC**. Hence there are no u-parts of $\mathcal{Q}^2$ which include the node **EQN**. Consequently, there is no $\mathfrak{p}^2 \in \mathit{Uparts}(\mathcal{Q}^2; \mathbf{Roots}(\mathcal{Q}^2))$ such that $(\mathfrak{p}^2 \cap \mathcal{N}^{\mathcal{Q}})$ is $\mathfrak{p}$. Thus $\mathcal{Q}^2$ is a solution for $\mathscr{C}$.

Suppose $\mathscr{C}$ is a set of part constraints for an extensional expansion dag $\mathcal{Q}$ such that every u-part $\mathfrak{p} \in \mathit{Uparts}(\mathcal{Q}; \mathbf{Roots}(\mathcal{Q}))$ is spanned by $\mathscr{C}$. Suppose $\mathcal{Q}'$ is a solution for $\mathscr{C}$. In such a situation we can conclude $\mathit{Uparts}(\mathcal{Q}'; \mathbf{Roots}(\mathcal{Q}'))$ is empty since any u-part

$$\mathfrak{p}' \in \mathit{Uparts}(\mathcal{Q}'; \mathbf{Roots}(\mathcal{Q}'))$$

would induce a u-part $(\mathfrak{p}' \cap \mathcal{N}^{\mathcal{Q}}) \in \mathit{Uparts}(\mathcal{Q}; \mathbf{Roots}(\mathcal{Q}))$ (by Lemma 9.4.5), contradicting the assumption that $(\mathfrak{p}' \cap \mathcal{N}^{\mathcal{Q}})$ is spanned by $\mathscr{C}$ and $\mathcal{Q}'$ is a solution for $\mathscr{C}$.

We will use sets of constraints in two different ways. First, we will use *unification constraints* to delay unification. Second, we will use *set constraints* to solve for set variables.

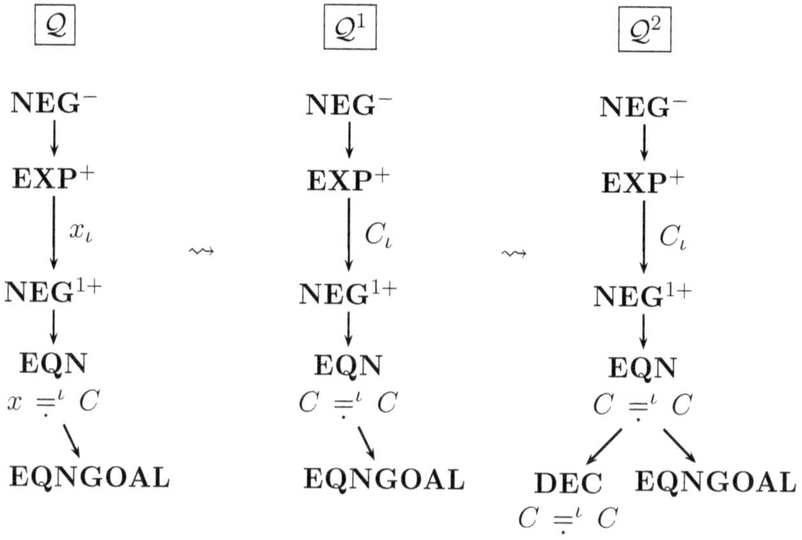

FIGURE 9.2. Solution for Constraints

9.5. Delaying Unification

Often near the end of a successful search, **MS04-2** must perform a final series of imitations and projections to solve a unification problem. If we try imitations and projections without viewing the disagreement pair as a member of a collection of disagreement pairs, then a high degree of branching occurs. To try to avoid this, **MS04-2** can optionally pass into a delayed unification phase. The delayed unification phase includes a set of unification constraints and a set of pairs of terms (disagreement pairs).

DEFINITION 9.5.1. Let $\mathcal{Q}$ be an extensional expansion dag. A set $\mathfrak{c}$ is a *unification constraint* if $\mathfrak{c}$ is a part constraint and one of the following holds:

1. $\mathfrak{c}$ is a singleton $\{e\}$ for some negative equation node e.
2. $\mathfrak{c}$ is $\{a, b\}$ for some positive atomic node a and negative atomic node b.
3. $\mathfrak{c}$ is $\{a, b\}$ for some positive equation node a and equation goal node b.

4. $\mathfrak{c}$ is $\{a, b\}$ for some flexible node a and atomic node b.

To search with unification constraints, we generate a new option in the option generation phase with a new genus **delayunif** for each flex-rigid node in a given u-part. We generate the new options by adding the following steps to the option generation phase (cf. Algorithm 9.1.4) between steps (3) and (4):

3.1 Let $\boldsymbol{Options}^{\mathbf{du}}_{MS04-2}$ be the set of all

$$\langle \mathbf{delayunif}, e\rangle$$

where $e \in \mathcal{N}^{\mathfrak{p}}$ is a flex-rigid node.

3.2 Assign a weight $W^{\mathcal{O}}$ to each

$$\mathcal{O} \in \boldsymbol{Options}^{\mathbf{du}}_{MS04-2}.$$

We also modify step (6) of Algorithm 9.1.4 to include the options in

$$\boldsymbol{Options}^{\mathbf{du}}_{MS04-2}$$

in the set $\boldsymbol{Options}^{\mathbf{a}}_{MS04-2}$. Furthermore, we change the option attempt phase (cf. Algorithm 9.1.5) by adding a case to handle an option $\langle \mathbf{delayunif}, e\rangle$. This is handled simply by adding the part constraint $\{e\}$ to the current set of unification constraints.

The nature of the search procedure changes once the current set of unification constraints is nonempty. In particular, (2), (3) and (4) of Algorithm 2 are skipped. That is, we no longer generate duplication options, primitive substitution options, type generation options and unification options. Also, in step (9) of the option attempt phase (cf. Algorithm 9.1.4) connection options in $\boldsymbol{Options}^{\mathbf{c}}_{MS04-2}$ are performed by adding the two nodes as a unification constraint.

Every time the set of unification constraints is extended, the search procedure performs pattern unification (cf. [45]) as well as some pre-unification up to a fixed depth. If the set of unification constraints is determined to be unsolvable at any stage, then **FAILURE** is returned, leading to backtracking to other options.

Once every pre-solved u-part is spanned by the set of unification constraints, the procedure continues by generating arbitrary u-parts and including more connections, flex-rigid nodes and flex-flex nodes until every u-part is spanned. When every u-part is spanned, the search procedure attempts to find a unifier using pre-unification up to a fixed depth. If a unifier is found, then search succeeds. (The solution $\mathcal{Q}'$ with $\mathcal{Q} \rightsquigarrow_* \mathcal{Q}'$ can be computed by instantiating using the unifier, making any delayed connections and developing the resulting extensional expansion dag.) Otherwise, search fails and backtracks.

In general, pre-unification for simply typed λ-calculus is undecidable. However, we can simply place bounds on the pre-unification procedure and thus guarantee termination. When pre-unification stops due to reaching a given bound, the value **MORE** is returned. When pre-unification stops because a pair cannot be unified, the value **FAILED** is returned. When pre-unification finds a pre-unifier, the pre-unifier is returned.

We still need to consider general imitation and projection steps to guarantee completeness. The delayed unification phase will only succeed if the remaining steps only involve making connections which can be unified without using equational or extensional reasoning (and the final pre-unification problem is not too difficult). This leads to a much faster completion of search, but may also lead to exploration of false paths before trying the necessary basic steps.

9.6. Set Constraints

As reported in [25], set constraints can often provide a method to solve for set variables. Upper and lower bound set constraints can be used to generate maximal and minimal solutions. Solutions to set constraints do not eliminate the need to enumerate set instantiations, but often lead to faster automatic proofs of theorems that involve set variables. Just as one can use unification to delay the instantiation of first-order (and some higher-order) variables, one can use set constraints to delay the instantiation of set variables.

Bledsoe [22; 23] used a method of finding maximal and minimal solutions for predicates to reduce higher-order problems to first-order problems when this is possible. Bledsoe's method applies to certain classes of set constraints. The problem of instantiating set variables also occurs in the Calculus of Constructions, and Felty [28; 29] extended Bledsoe's maximal method to this setting. The Calculus of Constructions includes (but is stronger than) intuitionistic higher-order logic via a Curry-Howard correspondence [30]. We extend Bledsoe's method in the following ways:

- Since we work in full higher-order logic, we need not reduce to a first-order problem.

- Using full higher-order logic, we can solve for sets with inductive (and coinductive) definitions.
- Instead of combining solutions, we combine constraints and solve them simultaneously. This is necessary in the recursive case where the desired solution cannot be found by solving the constraints independently.

Z-match [13] provides a method for instantiating set variables in the context of the first-order set theory theorem prover **&**. The **&** theorem prover searches in a sequent calculus with rules for (partially) instantiating set variables. Applying the Z-match rules is similar to solving certain kinds of set constraints. Incremental elaboration allows Z-match to establish a collection of sequents. This corresponds to solving a collection of constraints. Also, some of the extensions to Z-match proposed in [13] correspond to aspects of instantiating set variables using set constraints. In particular, in the present work, we find solutions by quantifying over "bad" variables (as defined in [13]), and we allow multiple "matching formulae" (as defined in [13]) by allowing set constraints to contain several literals.

A very different approach is used by the SCAN algorithm [49]. This algorithm reduces some second-order formulas to equivalent first-order formulas, avoiding the need to instantiate the set variables.

To illustrate proof search with a set variable, consider this simple theorem expressing the existence of the union of two sets $X_{o\iota}$ and $Y_{o\iota}$:

(5) $\underset{.}{\exists} v_{o\iota} \underset{.}{\forall} Z_\iota \,\blacksquare\, [v\, Z \underset{.}{\supset} [[X\, Z] \underset{.}{\vee} [Y\, Z]]] \underset{.}{\wedge} [X\, Z \underset{.}{\supset} v\, Z] \underset{.}{\wedge} [Y\, Z \underset{.}{\supset} v\, Z]$

MS04-2 starts by constructing a developed extensional expansion dag $\mathcal{Q}$ with a single negative primary root r with the sentence (5) as the shallow formula of r. After initializing the search, TPS begins considering u-parts.

REMARK 9.6.1. The set of u-parts can be viewed as a junctive-form, or jform (cf. [37; 38]) as shown in Figure 9.3. The u-parts correspond to the vertical paths in the jform. Jforms are not computed during **MS04-2** search since they are not actually used by the procedure. Instead, a function computes an unspanned u-part and either returns such a u-part or returns a value indicating that there are no such u-parts. Computing jforms for extensional expansion dags can be expensive since a technique similar to dissolution (cf. [48; 38]) must be used to handle connections. We sometimes refer to

$$\left[\left[\begin{array}{l}\boxed{1}\, v\, Z\\ \boxed{2}\, \neg X\, Z\\ \boxed{3}\, \neg Y\, Z\end{array}\right] \vee \left[\begin{array}{l}\boxed{4}\, X\, Z\\ \boxed{5}\, \neg v\, Z\end{array}\right] \vee \left[\begin{array}{l}\boxed{6}\, Y\, Z\\ \boxed{7}\, \neg v\, Z\end{array}\right]\right]$$

FIGURE 9.3. A Jform with a Set Variable

$$\left[\left[\begin{array}{c}\left[\boxed{1^1}\, v^1\, Z \vee \boxed{1^2}\, v^2\, Z\right]\\ \boxed{2}\, \neg X\, Z\\ \boxed{3}\, \neg Y\, Z\end{array}\right] \vee \left[\begin{array}{l}\boxed{4}\, X\, Z\\ \boxed{5^1}\, \neg v^1\, Z\\ \boxed{5^2}\, \neg v^2\, Z\end{array}\right] \vee \left[\begin{array}{l}\boxed{6}\, Y\, Z\\ \boxed{7^1}\, \neg v^1\, Z\\ \boxed{7^2}\, \neg v^2\, Z\end{array}\right]\right]$$

FIGURE 9.4. Jform after a Primitive Substitution

occurrences of formulas in a jform as *literals*. Each such literal corresponds to a node in the extensional expansion dag. Hence we use the term *literal* to refer to both an occurrence of a formula in a jform and the node corresponding to this occurrence. We use jforms in the following description of the search for the sake of clarity.

The jform for this example (after development) is shown in Figure 9.3. The nodes corresponding to $\boxed{1}$, $\boxed{5}$ and $\boxed{7}$ are flexible nodes with the set variable v at the head of each corresponding shallow formula.

In this simple example, there is no way to mate pairs of nodes to span every u-part. To see this, we consider the leftmost u-part which includes $\boxed{1}$, $\boxed{2}$, and $\boxed{3}$. We cannot connect $\boxed{2}$ and $\boxed{3}$ since they are both negative. If we connect $\boxed{1}$ with $\boxed{2}$, we must instantiate v with X. However, now there is no possible connection on the u-part $\boxed{6}$-$\boxed{7}$. Similarly, connecting $\boxed{1}$ with $\boxed{3}$ does not lead to a solution.

We can solve this example by instantiating v with the primitive substitution $\lambda z_\iota \centerdot v^1\, z \vee v^2\, z$ where $v^1_{o\iota}$ and $v^2_{o\iota}$ are new expansion variables. The new jform is shown in Figure 9.4. To solve the problem corresponding to this jform we can use flex-rigid mates for $\boxed{1^1\text{-}2}$ and $\boxed{1^2\text{-}3}$ (imitating v^1 using X and v^2 using Y) and rigid mates $\boxed{4\text{-}5^1}$, $\boxed{6\text{-}7^2}$. Two more projection steps lead to the proof.

An alternative to using a primitive substitution in this example is to view the u-parts of the jform in Figure 9.3 as constraints on the set variable v. Consider the u-part $\boxed{\text{1-2-3}}$ in the jform in Figure 9.3. We can represent this u-part as a two-sided sequent

$$v\,Z \rightarrow X\,Z, Y\,Z$$

with the positive literal $\boxed{1}$ on the left and the negative literals $\boxed{2}$ and $\boxed{3}$ on the right. Intuitively, this means that if Z is in v, then we want Z to be in X or Z to be in Y. Since the quantifier structure of the theorem does not allow the definition of v to depend on the variable Z, we read this sequent as saying that *any* Z in v must be in X or Y. We can make this independence of v on Z explicit in the notation by including it in a prefix to the sequent as

$$\forall Z\,\langle v\,Z \rightarrow X\,Z, Y\,Z\rangle. \tag{6}$$

We now define set constraints for an extensional expansion dag in general.

DEFINITION 9.6.2. Let $\mathcal{Q}$ be an extensional expansion dag and v be a set variable of $\mathcal{Q}$. A set $\mathfrak{c} \subseteq \mathcal{N}^{\mathcal{Q}}$ is an *upper bound set constraint for* v if there is a unique positive v-node in $\mathfrak{c}$. When $\mathfrak{c}$ is an upper bound set constraint for v we call this unique v-node $a \in \mathfrak{c}$ the *main positive* v*-node of* $\mathfrak{c}$ denoted by $\mathfrak{u}_v^{\mathcal{Q}}(\mathfrak{c})$ (or $\mathfrak{u}_v(\mathfrak{c})$ when $\mathcal{Q}$ is clear in context).

A set $\mathfrak{c} \subseteq \mathcal{N}^{\mathcal{Q}}$ is a *lower bound set constraint for* v if there is a unique negative flexible node $a \in \mathfrak{c}$ such that v is at the head of $sh(a)$. When $\mathfrak{c}$ is a lower bound set constraint for v we call this unique node $a \in \mathfrak{c}$ the *main negative* v*-node of* $\mathfrak{c}$ denoted by $\mathfrak{l}_v^{\mathcal{Q}}(\mathfrak{c})$ (or $\mathfrak{l}_v(\mathfrak{c})$ when $\mathcal{Q}$ is clear in context).

We say a set $\mathfrak{c} \subseteq \mathcal{N}^{\mathcal{Q}}$ is a *set constraint for* v if $\mathfrak{c}$ is either an upper bound set constraint for v or lower bound set constraint for v.

DEFINITION 9.6.3. Let $\mathcal{Q}$ be an extensional expansion dag, v be a set variable of $\mathcal{Q}$ and $\mathfrak{c}$ be a set constraint for v. We say a parameter Z is *banned for* v *in* $\mathfrak{c}$ if $Z \in \mathbf{Params}^{\mathcal{Q}}(\mathfrak{c})$ (cf. Definition 7.2.13) and Z is banned for v (cf. Definition 8.2.10). We define $\Psi_v^{\mathcal{Q}}(\mathfrak{c})$ to be the set of parameters banned for v in $\mathfrak{c}$. When the extensional expansion dag is clear in context we write $\Psi_v(\mathfrak{c})$ for $\Psi_v^{\mathcal{Q}}(\mathfrak{c})$.

When we are not concerned with the particular nodes in a set constraint, we can use the notation as in (6) above to describe the set constraint. In particular, suppose $\mathfrak{c}$ is an upper bound set constraint for v such that a is the main positive v-node,

$$(\mathfrak{c} \setminus \{a\}) = \{b^1, \ldots, b^n\}$$

and the set of parameters banned for v in $\mathfrak{c}$ is

$$\{Z^1, \ldots, Z^m\}.$$

We use the notation

$$\forall Z^1, \ldots, Z^m \langle sh(a) \rightarrow sh(b^1), \cdots, sh(b^n) \rangle.$$

Intuitively, this set constraint also corresponds to the proposition

$$\underset{\cdot}{\forall} x^1 \cdots \underset{\cdot}{\forall} x^m \centerdot \theta(sh(a)) \underset{\cdot}{\supset} \centerdot \theta(sh(b^1)) \underset{\cdot}{\vee} \cdots \underset{\cdot}{\vee} \theta(sh(b^n))$$

where $\theta(Z^j) = x^j$ for each j $(1 \leq j \leq m)$.

Likewise, we use the notation

$$\forall Z^1, \ldots, Z^m \langle sh(b^1), \cdots, sh(b^n) \rightarrow sh(a) \rangle$$

to describe a lower bound set constraint for v such that a is the main negative v-node,

$$(\mathfrak{c} \setminus \{a\}) = \{b^1, \ldots, b^n\}$$

and the set of parameters banned for v in $\mathfrak{c}$ is

$$\{Z^1, \ldots, Z^m\}.$$

The corresponding proposition in this case is

$$\underset{\cdot}{\forall} x^1 \cdots \underset{\cdot}{\forall} x^m \centerdot [\theta(sh(b^1)) \underset{\cdot}{\vee} \cdots \underset{\cdot}{\vee} \theta(sh(b^n))] \underset{\cdot}{\supset} \theta(sh(a))$$

where $\theta(Z^j) = x^j$ for each j $(1 \leq j \leq m)$.

In Bledsoe's case, the formulas $sh(b^1), \ldots, sh(b^n)$ are not allowed to depend on the set variable v, as this corresponds to making a recursive definition. In Z-match [13], n is required to be 1 (and $sh(b^1)$ is called the *matching formula*) and $sh(b^1)$ is not allowed to depend on v or any of the parameters $Z^1, \cdots, Z^m$ (corresponding to the *bad variables* in [13]).

Other set constraints for v can be extracted from the jform in Figure 9.3 by considering subsets of u-parts containing a flexible node. There is one positive literal $\boxed{1}$ with v at the head, and there are two negative literals $\boxed{5}$ and $\boxed{7}$ with v at the head. These literals and the subsets of the u-parts containing them give the following upper and lower bound constraints:

	Upper Bound Constraints
1	$\forall Z \langle v\, Z \rightarrow \cdot \rangle$
1-2	$\forall Z \langle v\, Z \rightarrow X\, Z \rangle$
1-3	$\forall Z \langle v\, Z \rightarrow Y\, Z \rangle$
1-2-3	$\forall Z \langle v\, Z \rightarrow X\, Z, Y\, Z \rangle$
	Lower Bound Constraints
5	$\forall Z \langle \cdot \rightarrow v\, Z \rangle$
4-5	$\forall Z \langle X\, Z \rightarrow v\, Z \rangle$
7	$\forall Z \langle \cdot \rightarrow v\, Z \rangle$
6-7	$\forall Z \langle Y\, Z \rightarrow v\, Z \rangle$

(where $\cdot$ indicates an empty list of formulas). Either the upper bound constraint 1-2-3 or the two lower bound constraints 4-5 and 6-7 will lead to a solution.

For any collection of lower bound constraints (or upper bound constraints) for an extensional expansion dag $\mathcal{Q}$ we can let θ be the substitution such that $\theta(v)$ is $[\lambda\overline{z}\top]$ (or $\lambda\overline{z}\bot$) and develop the extensional expansion dag to obtain an extensional expansion dag $\mathcal{Q}'$ which solves the constraints (cf. Definition 9.4.6). However, such trivial solutions are usually not the ones we want. Usually, we want solutions to constraints that satisfy additional properties as well. In particular, we are interested in minimal solutions (with respect to set inclusion) to lower bound constraints and maximal solutions to upper bound constraints.

First consider the case in which the lower bound constraint has description

$$\forall w^1 \cdots w^m \langle A^1, \ldots, A^l \rightarrow [v\, t^1 \cdots t^n] \rangle$$

where v does not occur in any A^i. As in Bledsoe and Feng [23], we can directly define the minimal solution as follows. We start by choosing new variables x^i of the same type as t^i for $i = 1, \ldots, n$. The minimal solution V can be defined by

$$(7)\ \lambda x^1 \cdots \lambda x^n \exists w^1 \cdots \exists w^m [x^1 = t^1 \wedge \cdots \wedge x^n = t^n \wedge A^1 \wedge \cdots \wedge A^l].$$

To verify this is a solution, we substitute V for v in the formula

$$\forall w^1 \cdots \forall w^m.[A^1 \wedge \cdots \wedge A^l] \supset v\, t^1 \cdots t^n$$

to obtain

$$\forall w^1 \cdots \forall w^m \mathbf{.} [A^1 \wedge \cdots \wedge A^l] \supset$$

$[\lambda x^1 \cdots \lambda x^n [\exists w^1 \cdots \exists w^m . x^1 = t^1 \wedge \cdots \wedge x^n = t^n \wedge A^1 \wedge \cdots \wedge A^l]] \, t^1 \cdots t^n$.
Although the w^j variables may occur in the terms t^i, this formula is easily derivable from

$$\forall w^1 \cdots \forall w^m \,\blacksquare\, [A^1 \wedge \cdots \wedge A^l] \supset [t^1 = t^1 \wedge \cdots \wedge t^n = t^n \wedge A^1 \wedge \cdots \wedge A^l].$$

When some members of Ψ occur as the t^j's, we can simplify this solution. Consider the lower bound constraint

$$\forall \, z_\iota \, w_\iota \, \langle \, R_{o\iota\iota} \, z \, w \;\rightarrow\; v \, z \, z \, [f_{\iota\iota\iota} \, z \, w] \, \rangle$$

for the set variable $v_{o\iota\iota\iota}$. The solution above is

$$\lambda x^1_\iota \, \lambda x^2_\iota \, \lambda x^3_\iota \,\blacksquare\, \exists z_\iota \, \exists w_\iota \, [[x^1 = z] \wedge [x^2 = z] \wedge [x^3 = [f \, z \, w]] \wedge R \, z \, w].$$

Since z occurs as an argument in the main literal $v \, z \, z \, [f_{\iota\iota\iota} \, z \, w]$, we have the simpler solution

$$\lambda z_\iota \, \lambda x^2_\iota \, \lambda x^3_\iota \,\blacksquare\, \exists w_\iota \, [[x^2 = z] \wedge [x^3 = [f \, z \, w]] \wedge R \, z \, w].$$

In general, we need to solve sets of constraints including the case where v occurs on both sides of some constraints. This may require making a recursive definition of v. This is best illustrated by the simple example of a set (N) of two constraints

$$(N) \left\{ \begin{array}{lcl} \cdot & \rightarrow & [v \, 0] \\ \forall w \, \langle \, [v \, w] & \rightarrow & [v \, [S \, w]] \rangle \end{array} \right.$$

Intuitively, the minimal solution to this is the least set containing 0 and closed under the function S. Using the expressive power of higher-order logic, a term giving the minimal solution can be defined by

$$\lambda x \forall p \,\blacksquare\, [[p \, 0] \wedge [\forall z \,\blacksquare\, [p \, z] \supset p \, [S \, z]]] \supset p \, x. \tag{8}$$

9.6.1. Generalized Solutions and Set Existence Lemmas. Making instantiations with large terms can be prohibitive in higher-order automated theorem proving since TPS may need to perform higher-order matching with the solution in some other part of the problem. This suggests that to prove the main theorem, it may be simpler to use lemmas asserting the existence of required sets without using explicit definitions of these sets. When using these lemmas, the sets are represented by selected parameters (*very* small terms). Another motivation for using such lemmas is that we gain more control over what properties of the sets are included in the lemmas. Using lemmas in this way corresponds to allowing special instances of cut in the proof.

To use these lemmas to prove the main theorem, we need a new notion of a solution to set constraints. The idea of a *generalized solution* to a set constraint for v is to find a formula $\varphi(v)$ where both

$$\forall v \mathbin{\blacksquare} \varphi(v) \supset \forall Z^1 \ldots \forall Z^n \mathbin{\blacksquare} A^1(v) \wedge \cdots \wedge A^m(v) \supset \mathbin{\blacksquare} B^1(v) \vee \cdots \vee B^k(v)$$

and $\exists v\, \varphi(v)$ are provable. The way we can make this precise is to add a positive root node to the extensional expansion dag corresponding to a *set existence lemma*

$$\forall \overline{y} \exists v\, \varphi(v)$$

where $\overline{y}$ are the free variables occurring in $\varphi(v)$ other than v.

DEFINITION 9.6.4. Let $\mathcal{Q}$ be an extensional expansion dag and $\mathscr{C}$ be a set of part constraints for $\mathcal{Q}$. We say $\langle \mathcal{Q}^1, a, \mathcal{Q}^2 \rangle$ is a *generalized solution for* $\mathscr{C}$ if a is a positive primary node of $\mathcal{Q}^1$, the shallow formula $sh^{\mathcal{Q}^1}(a)$ is a $\mathcal{G}^{\mathcal{S}}_{\beta\mathfrak{f}\mathfrak{b}}$-derivable sentence, $a \notin \mathcal{N}^{\mathcal{Q}}$, $\mathcal{Q} \trianglelefteq \mathcal{Q}^1$, $\mathbf{Roots}(\mathcal{Q}^1) = \mathbf{Roots}(\mathcal{Q}) \cup \{a\}$, $\mathcal{Q}^1 \rightsquigarrow_* \mathcal{Q}^2$ and for every u-part

$$\mathfrak{p} \in \mathit{Uparts}(\mathcal{Q}; \mathbf{Roots}(\mathcal{Q}))$$

spanned by $\mathscr{C}$ there is no $\mathfrak{p}' \in \mathit{Uparts}(\mathcal{Q}'; \mathbf{Roots}(\mathcal{Q}'))$ such that $\mathfrak{p} = (\mathfrak{p}' \cap \mathcal{N}^{\mathcal{Q}})$.

If $\mathscr{C}$ is a collection of set constraints for a variable v and $\langle \mathcal{Q}^1, a, \mathcal{Q}^2 \rangle$ is a generalized solution for $\mathscr{C}$, then we call $sh^{\mathcal{Q}^1}(a)$ the *set existence lemma for* $\mathscr{C}$.

Suppose $\mathscr{C}$ is a collection of set constraints for v in $\mathcal{Q}$ and there is a closed term V and a solution $\mathcal{Q}'$ such that

$$\mathcal{Q} \rightsquigarrow ([V/v]\mathcal{Q}) \rightsquigarrow_* \mathcal{Q}'.$$

We can modify this to be a generalized solution by taking $\mathcal{Q}^1$ to be an extensional expansion dag

$$\langle \mathcal{N}^{\mathcal{Q}} \cup \{a\}, \mathit{succ}^{\mathcal{Q}}, \mathit{label}', \mathit{sel}^{\mathcal{Q}}, \mathit{exps}^{\mathcal{Q}}, \mathit{conns}^{\mathcal{Q}}, \mathit{decs}^{\mathcal{Q}} \rangle$$

where $a \notin \mathcal{N}^{\mathcal{Q}}$, $\mathit{label}'|_{\mathcal{N}^{\mathcal{Q}}} = \mathit{label}^{\mathcal{Q}}$ and

$$\mathit{label}'(a) = \langle \mathbf{primary}, 1, [\dot{\exists}\, v \mathbin{\blacksquare} v \doteq V] \rangle.$$

Then we can modify the chain giving $([V/v]\mathcal{Q}) \rightsquigarrow_* \mathcal{Q}'$ to obtain an extensional expansion dag $\mathcal{Q}^2$ such that $\mathcal{Q}^1 \rightsquigarrow_* \mathcal{Q}^2$.

To construct generalized solutions to set constraints, we first need to consider what properties minimal and maximal solutions satisfy. Let $\mathbf{N}$ be the term shown above (8) defining the minimal solution to the constraint set (N). The most important property $\mathbf{N}$ has is that it satisfies the constraints. That is, $\mathbf{N}$ contains 0 and is closed under S. An inversion property also follows from the definition of $\mathbf{N}$. In particular, every element of $\mathbf{N}$ is either 0 or $S\,y$ for some $y \in \mathbf{N}$. Finally, $\mathbf{N}$ satisfies an induction principle which characterizes it as minimal. Intuitively, a minimal solution V to a collection of lower bound constraints satisfies three basic properties:

- *Constraint Satisfaction:* The set V satisfies all of the constraints.
- *Inversion Principle:* The only members of the set V are those forced to be a member by the constraints.
- *Induction Principle:* Any other set satisfying all of the constraints should contain V.

Identifying these properties as basic, we can create generalized solutions to lower bound constraints by conjoining the properties of a set v into a single formula $\varphi(v)$. Strictly speaking, only the property of Constraint Satisfaction is necessary to make $[\forall \overline{y}\, \exists v\, \varphi(v)]$ a set existence lemma which leads to a generalized solution. Other properties are explicitly included in the generalized solution to help prove the rest of the theorem. One may wonder why the Inversion Principle is included. It is, in fact, the case that the Inversion Principle follows from the Induction Principle. However, the *proof* of this fact involves a nontrivial instantiation of the set variable in the Induction Principle. So, when trying to use these properties to prove a theorem, it is useful to have the Inversion Principle explicitly given.

A maximal solution to a collection of upper bound constraints should satisfy dual properties, and generalized solutions to upper bound constraints are constructed by conjoining these dual properties. (A typical example would be the existence of a set of streams.)

For the constraint set (N), TPS finds the set existence lemma

$$\begin{gathered}
\underset{.}{\exists} v \centerdot v\, 0 \;\underset{.}{\wedge}\; \underset{.}{\forall} w\, [v\, w \;\underset{.}{\supset}\; v \centerdot S\, w] \\
\underset{.}{\wedge} \underset{.}{\forall} x\, [v\, x \;\underset{.}{\supset} \centerdot x \;\underset{.}{=}\; 0 \;\underset{.}{\vee}\; \underset{.}{\exists} w[x \;\underset{.}{=}\; S\, w \wedge v\, w]] \\
\underset{.}{\wedge} \underset{.}{\forall} p \centerdot [p\, 0 \;\underset{.}{\wedge}\; \underset{.}{\forall} W\, [p\, W \;\underset{.}{\supset}\; p \centerdot S\, W]] \;\underset{.}{\supset}\; \underset{.}{\forall} x \centerdot [v\, x] \;\underset{.}{\supset}\; [p\, x].
\end{gathered}$$

The set existence lemma in this case is $\exists v\, \varphi(v)$.

9.6.2. Proving Set Existence Lemmas. Along with any set existence lemma $[\forall \overline{y} \exists v\, \varphi(v)]$, TPS must also generate a proof of the set existence lemma. Fortunately, proving these lemmas does not require arbitrary theorem proving. Instead, the lemmas are generated along with their proofs (in the form of expansion proofs). The precise nature of the generated proofs depends on the nature of the constraints. If v only occurs once in each constraint (at the head of the main literal of each constraint), then the proof of the set existence lemma can be constructed in a fairly straightforward way. The definition of the set can be constructed directly as was the solution (7). In the general case, a proof of the set existence lemma may use the Knaster-Tarski Fixed Point Theorem. The Knaster-Tarski Theorem must be applied to a monotone set function. The monotone function of interest can be constructed from the Inversion Principle.

9.6.3. The Knaster-Tarski Fixed Point Theorem. The Knaster-Tarski Fixed Point Theorem states that monotone set functions have fixed points. There are more specific versions proving the least pre-fixed point is a fixed point and the greatest post-fixed point is a fixed point. (Other examples of using the Knaster-Tarski Theorem in higher-order logic can be found in [50; 51; 52; 15].)

DEFINITION 9.6.5 (Monotone Set Functions). Let A be a set and $\mathcal{P}(A)$ be the power set of A. A function $K : \mathcal{P}(A) \to \mathcal{P}(A)$ is *monotone* if $K(v) \subseteq K(w)$ whenever $v \subseteq w \subseteq A$.

DEFINITION 9.6.6 (Fixed Points). A function $K : \mathcal{P}(A) \to \mathcal{P}(A)$ where $\mathcal{P}(A)$ is the power set of A. A *pre-fixed point* of K is a set v such that $K(v) \subseteq v$. A *post-fixed point* of K is a set v such that $v \subseteq K(v)$. A *fixed point* of K is a set v satisfying $K(v) = v$.

We now state and prove the Knaster-Tarski Fixed Point Theorem.

THEOREM 9.6.7 (Knaster-Tarski Fixed Point Theorem). *Every monotone function $K : \mathcal{P}(A) \to \mathcal{P}(A)$ has a fixed point. Furthermore, there is a least pre-fixed point and a greatest post-fixed point and these are both fixed points.*

When the Induction Principle is not included in the set existence lemma, it is enough to use the existence of any fixed point. When the induction principle is included, more specific versions of the theorem are needed that determine the least pre-fixed point and the greatest post-fixed point of a monotone set function K are fixed points.

PROOF. Let $P_{pre} := \{X \subseteq A \mid K(X) \subseteq X\}$, the set of all pre-fixed points, and let $L := \bigcap P_{pre}$, the intersection of all pre-fixed points. To see L is a pre-fixed point, note that monotonicity of K implies $K(L) \subseteq K(X) \subseteq X$ for any pre-fixed point X. So, by the definition of L, $K(L) \subseteq L$. L is clearly the least pre-fixed point, as the intersection of all pre-fixed points. Since every fixed point is also a pre-fixed point, we already have $L \subseteq Y$ for every fixed point Y. To verify L is a fixed point, note that $K(L) \subseteq L$ implies $K(K(L)) \subseteq K(L)$. That is, $K(L)$ is a pre-fixed point. So, $K(L) \in P_{pre}$ and we have $L \subseteq K(L)$. Thus the least pre-fixed point L is also the least fixed point.

The proof that the greatest post-fixed point of K is also a fixed point is symmetric. We define the greatest fixed point of K by

$$\bigcup\{X \subseteq A \mid X \subseteq K(X)\}$$

and use monotonicity of K to verify this is a fixed point. □

These statements and proofs have a straightforward representation in type theory (e.g., **THM2** in the TPS library is a version of Knaster-Tarski). The same proof idea works regardless of the arity of the type of the variable v representing the set A. In the process of proving set existence lemmas where the set must be defined recursively, the appropriate version of the Knaster-Tarski Theorem is generated along with an expansion proof. The expansion proof is generated algorithmically, not by using automatic search. (No extensionality is required to represent the proof. Hence expansion proofs using expansion trees are used.)

When the Knaster-Tarski Theorem is used to prove a set existence lemma, the set function K must be chosen and proven monotone. Monotonicity can be guaranteed by restricting K to be of a certain syntactic form. In particular, we restrict the values of K to set functions represented by terms of the form $\lambda u \lambda \overline{z} P$ where P_o is u-positive as defined below (cf. Definition 9.6.8).

DEFINITION 9.6.8 (Positive and Negative Terms). Let u be a set variable. We define when a term is *u-positive* and *u-negative* inductively.

- P is u-positive and u-negative if u is not free in P.
- $[u\,\bar{t}]$ is u-positive, so long as u does not occur free in $\bar{t}$.
- $P_1 \vee P_2$ and $P_1 \wedge P_2$ are u-positive [u-negative] if both P_1 and P_2 are u-positive [u-negative].
- $P_1 \supset P_2$ is u-positive [u-negative] if P_1 is u-negative [u-positive] and P_2 is u-positive [u-negative].
- $\neg P$ is u-positive [u-negative] if P is u-negative [u-positive].
- $\forall x P$ and $\exists x P$ are u-positive [u-negative] if P is u-positive [u-negative].

When we apply the Knaster-Tarski Theorem to a term representing a set function (i.e., a function taking sets to sets), we can restrict ourselves to terms $\lambda u \lambda \bar{z} P$ where P_o is positive with respect to u. Such a term $\lambda u \lambda \bar{z} P$ can be proven monotone with respect to u. To see why positivity is important for monotonicity, consider the set function defined by taking complements,

$$\lambda u \lambda z \neg [u\, z].$$

Clearly, $\neg[u\, z]$ is not positive with respect to u. We can also see that this operator has no fixed point since no set is equal to its own complement (using the fact that all types are nonempty).

9.6.4. Combining MS04-2 With Set Constraints. In first-order theorem proving, one could in principle guess the instantiations for free variables and reduce the problem to propositional theorem proving. However, using unification to delay the instantiation of variables has been far more successful. In higher-order theorem proving, TPS does delay some parts of the instantiation of free variables for (higher-order) unification, but TPS has traditionally performed set instantiations (using general substitutions) in a preprocessing step before mating search begins. If mating search fails, TPS may start over with a different collection of general substitutions. In the **MS04-2**

search procedure, we have also adopted the idea of performing primitive substitutions before connections and unification steps (although primitive substitutions are not strictly limited to preprocessing since new set variables may be created during search). Using set constraints allows us to combine the **MS04-2** search procedure with a method for solving for sets.

To search with unification constraints, we first allow the part selector to select u-parts which contain both positive and negative flexible nodes. Given such a u-part $\mathfrak{p}$, we generate a new option

$$\langle \mathbf{setconstr}, a \rangle$$

in the option generation phase (cf. Algorithm 9.1.4) with a new genus **setconstr** for each flexible node $a \in \mathfrak{p}$.

Suppose an option $\langle \mathbf{setconstr}, a \rangle$ is passed to the option attempt phase (Algorithm 9.1.5). Since a is flexible, a is a v-node for some (set) variable v. TPS generates set constraints with main v-literal a. A set constraint is generated by selecting nodes from the set

$$\{b \in \mathfrak{p} \mid b \text{ is not a } v\text{-node with polarity } pol(a)\}.$$

The result is a set constraint $\mathfrak{c} \subseteq \mathfrak{p}$ such that $a \in \mathfrak{c}$ and every other v-node in $\mathfrak{c}$ has opposite polarity from a.

Once **MS04-2** has at least one upper [lower] bound set constraint for v, the option generation phase (cf. Algorithm 9.1.4) only generates two kinds of options: the option to generate another upper [lower] bound set constraint for v spanning an unspanned u-part or the option $\langle \mathbf{solvesets} \rangle$ to solve the current set of constraints (using a new genus **solvesets**).

The option attempt phase (Algorithm 9.1.5) solves a set $\mathscr{C}$ of set constraints by generating a set existence lemma. The set existence lemma **L** is included by adding a positive primary node to the set of roots of the extensional expansion dag, developing the new extensional expansion dag and adding connections between corresponding parts of the lemma and the nodes in the set constraints in $\mathscr{C}$. The result is an extensional expansion dag which no longer contains any of the u-parts which were spanned by the set constraints.

An example using delayed unification constraints and set constraints is discussed in Section 10.3.

9.7. The MS03-7 Search Procedure

The **MS03-7** search procedure predates **MS04-2**. Although the main procedure we are studying is **MS04-2**, we briefly describe **MS03-7** since the procedure has been implemented in TPS and does make use of extensional expansion dags. The **MS03-7** search procedure also searches with extensional expansion dags, but never backtracks (unlike **MS04-2**). The need to backtrack is removed by automatically creating a duplicate expansion arc before instantiating an expansion variable. We can consider **MS03-7** to be a kind of *saturation search* since TPS essentially grows the extensional expansion dag until it is an extensional expansion proof (though there will likely be many expansion arcs and connections which do not contribute to the proof). We have not proven **MS03-7** complete, but we conjecture its completeness.

MS03-7 duplicates an expansion variable x_α as follows: We create a new expansion variable y_α and for every expansion arc $\langle e, \mathbf{C}, c\rangle$ such that $x \in \mathbf{Free}(\mathbf{C})$ we add a new expansion arc $\langle e, [y/x]\mathbf{C}, c'\rangle$ to the extensional expansion dag.

The basic operations **MS03-7** performs are as follows:

Rigid Mate: Make a connection between a positive atomic node and negative atomic node which share the same parameter at the corresponding heads of their shallow formulas.

Flex-Rigid Mate: Make a connection between a flexible node and an atomic node after duplicating the head of the flexible node and imitating the head of the atomic node (including a negation if the polarities of the two nodes are the same).

E-unify: Make a connection between a positive equation node and an equation goal node using either an E-unification node or a symmetric E-unification node.

Imitate: If there is a flex-rigid node with flex-rigid pair

$$\langle [x\,\overline{\mathbf{A}^n}], [H\,\overline{\mathbf{B}^m}]\rangle$$

where H is not banned for x, then duplicate the variable x and substitute the imitation term for H into x.

Imitate Banned Var: If there is a flex-rigid node with flex-rigid pair $\langle [x\,\overline{\mathbf{A}^n}], [H\,\overline{\mathbf{B}^m}]\rangle$ where H is banned for x, then duplicate the variable x, let y be the expansion variable created by this duplication process and substitute the imitation term for H into y.

Project: If there is a flex-rigid node with flex-rigid pair
$$\langle [x\,\overline{\mathbf{A}^n}], [H\,\overline{\mathbf{B}^m}]\rangle,$$
x has type γ, $1 \le i \le n$ and γ has i^{th} projections, then duplicate the variable x and substitute the ith projection term into x.

Primsub: If x is a set variable of $\mathcal{Q}$, then generate a primitive substitution, duplicate the variable x and instantiate the primitive substitution for x.

In practice, extensional expansion dags generated by **MS03-7** grow large very quickly. Hence at each step we only consider a smaller extensional expansion dag obtained by removing many expansion arcs. During search, the included expansion arcs are occasionally changed. We always keep a master copy of the full extensional expansion dag in the background.

Set constraint reasoning has also been integrated with the **MS03-7** search procedure. When set existence lemmas are generated by solving constraints, these lemmas are included whenever an expansion arc containing the instantiated set variable is included.

CHAPTER 10

Examples

We now consider examples of theorems that TPS can prove using **MS04-2** and **MS03-7**. Many of these examples could not be proven by previous TPS search procedures. In particular, previous TPS search procedures only provided minimal support for extensional reasoning. Some of the examples we consider are also discussed in Chapter 8 of [18]. All the times reported here exclude the time spent performing garbage collection.

In TPS a *mode* is a collection of flag settings. For example, the mode BASIC-MS04-2-MODE uses the **MS04-2** search procedure with flag settings which are not optimized for any particular example.

10.1. Simple Examples

10.1.1. Simple Extensional Examples. We begin with simple equality-free extensional examples which demonstrate the search procedure **MS04-2**.

THM617: (cf. Definition 2.2.8)

$$[\neg A_o \mathbin{\dot\vee} \neg B_o \mathbin{\dot\vee} \neg[P_{oo}\, A] \mathbin{\dot\vee} [P\, B]]$$

E1EXT: ($\mathbf{E}_1^{ext}$ in [18])

$$[a_o \mathbin{\dot\equiv} b_o] \mathbin{\dot\supset} \dot\forall P_{oo} \centerdot P\, a \mathbin{\dot\supset} P\, b$$

E2EXT: ($\mathbf{E}_2^{ext}$ in [18])

$$p_{oo}\, [a_o \mathbin{\dot\wedge} b_o] \mathbin{\dot\supset} p \centerdot b \wedge a$$

E3EXT: ($\mathbf{E}_3^{ext}$ in [18])

$$p_{oo}\, a_o \mathbin{\dot\wedge} p\, b_o \mathbin{\dot\supset} p \centerdot b \mathbin{\dot\wedge} a$$

Note that each of these examples is a sentence in the empty signature. The search procedure **MS04-2** can prove all of these theorems quickly.

All of these theorems can be proven using the mode BASIC-MS04-2-MODE in less than 1 second. The mode BASIC-MS04-2-MODE searches to depth 100 on the first iteration of the search and increases the depth bound by 100 on each subsequent iteration. Proofs for examples **THM617**, **E1EXT** and **E2EXT** are found on the first iteration. A proof for the example **E3EXT** is found at depth 104 using the mode BASIC-MS04-2-MODE on the second iteration of the search. There is also a mode EASY-MS03-7-MODE using the search procedure **MS03-7** which proves each of **THM617**, **E1EXT**, **E2EXT** and **E3EXT** in a fraction of a second.

10.1.2. Simple Equality Examples. Direct support for reasoning with primitive equality is provided by extensional expansion dags since equation nodes either use extensionality or can be connected via E-unification and symmetric E-unification nodes. In order to give the flavor of this equational reasoning, we consider a few simple examples which use equality.

THM618: (cf. Definition 2.2.8)

$$[\dot{\neg}[C_\iota \doteq^\iota D_\iota] \dot{\vee} \dot{\neg}[Q_{o\iota}\, C] \dot{\vee} [Q\, D]]$$

THM619:

$$[\dot{\neg}[F_{\iota\iota} \doteq^{\iota\iota} G_{\iota\iota}] \dot{\vee} \dot{\neg}[Q_{o(\iota\iota)}\, F] \dot{\vee} [Q\, G]]$$

THM620:

$$[\dot{\neg}[C_{o\iota} \doteq^{o\iota} D_{o\iota}] \dot{\vee} \dot{\neg}[Q_{o(o\iota)}\, C] \dot{\vee} [Q\, D]]$$

THM621:

$$\dot{\neg}[A_o \doteq^o B_o] \dot{\vee} \dot{\neg}P_{oo}\, A \dot{\vee} P\, B$$

The examples **THM618**, **THM619**, **THM620** and **THM621** are all sentences in the empty signature. The mode BASIC-MS04-2-MODE can prove each of these in less than a second. In each case, the way equality nodes are treated in extensional expansion dags makes finding the proof very easy. The proofs are found on the first or second iteration of **MS04-2** search. The mode EASY-MS03-7-MODE can prove each of these in less than a second using the search procedure **MS03-7**.

The modes BASIC-MS04-2-MODE and EASY-MS03-7-MODE can also prove the following examples in less than a second:

E6EXT: ($\mathbf{E}_6^{ext}$ in [18])

$$\mathcal{P}_{o(o\alpha)(o\alpha)}\emptyset \doteq^{o(o\alpha)} \mathcal{U}\emptyset$$

THM615:

$$H_{\iota o}\,[H\top =^{\iota} H\bot] \doteq^{\iota} H\bot$$

The sentence **E6EXT** uses the following polymorphic abbreviations:

$\emptyset$: $\lambda x_\beta \bot$

$\mathcal{U}$: $\lambda x_\beta \lambda y_\beta \centerdot x =^\beta y$

$\mathcal{P}$: $\lambda P_{o\beta} \lambda R_{o\beta} \centerdot R \subseteq P$

$\subseteq$: $\lambda P_{o\beta} \lambda R_{o\beta}\, \forall x_\beta \centerdot P\, x \supset R\, x$ (Infix)

Note that the sentence **E6EXT** is a sentence over the signature

$$\{=^{o\alpha}, \Pi^\alpha, \supset, \bot\}$$

where we treat α as a base type (like ι). The sentence **THM615** is a sentence over the signature $\{\top, \bot, =^\iota\}$.

10.2. Leibniz Equality and Primitive Equality

Many theorems can be represented using either Leibniz equality or primitive equality. A disadvantage of Leibniz equality is that one may need to instantiate the set variable associated with Leibniz equality. On the other hand, the search procedures **MS04-2** and **MS03-7** provide direct support for equality reasoning. Here we consider examples represented in both ways and demonstrate that the formulations using primitive equality are significantly easier to prove automatically.

E4EXT: ($\mathbf{E}_4^{ext}$ in [18])

$$\dot\forall X_\iota \dot\forall P_{o\iota}\, [P\, [m_{\iota\iota}\, X] \dot\supset P \centerdot n_{\iota\iota}\, X]$$
$$\dot\supset \dot\forall Q_{o(\iota\iota)} \centerdot Q\, [\lambda X\, m\, X] \dot\supset Q \centerdot \lambda X\, n\, X$$

E4EXT1:

$$\dot\forall X_\iota\, [[m_{\iota\iota}\, X] \doteq^\iota \centerdot n_{\iota\iota}\, X] \dot\supset \dot\forall Q_{o(\iota\iota)} \centerdot Q\, [\lambda X\, m\, X] \dot\supset Q \centerdot \lambda X\, n\, X$$

E4EXT2:

$$\forall X_\iota\, [[m_{\iota\iota}\, X] \doteq^{\iota} \centerdot n_{\iota\iota}\, X] \supset \centerdot [\lambda X\, m\, X] \doteq^{\iota\iota} \centerdot \lambda X\, n\, X$$

E5EXT: ($\mathbf{E}_5^{ext}$ in [18])

$$\forall X_\iota \forall P_{o\iota}\, [P\, [m_{\iota\iota}\, X] \supset P \centerdot n_{\iota\iota}\, X] \supset \forall Q_{o(\iota\iota)} \centerdot Q\, m \supset Q\, n$$

E5EXT1:

$$\forall X_\iota\, [[m_{\iota\iota}\, X] \doteq^{\iota} \centerdot n_{\iota\iota}\, X] \supset \forall Q_{o(\iota\iota)} \centerdot Q\, m \supset Q\, n$$

E5EXT2:

$$\forall X_\iota\, [[m_{\iota\iota}\, X] \doteq^{\iota} \centerdot n_{\iota\iota}\, X] \supset \centerdot m \doteq^{\iota\iota} n$$

EDEC: ($\mathbf{E}^{Dec}$ in [18])

$$\forall X_{\alpha\alpha} \forall Y_\alpha\, [f_{\alpha\alpha(\alpha\alpha)}\, X\, Y \dot{\doteq}^{\alpha} g_{\alpha\alpha(\alpha\alpha)}\, X\, Y] \wedge \forall Z_\alpha\, [h_{\alpha\alpha}\, Z \dot{\doteq}^{\alpha} j_{\alpha\alpha}\, Z] \supset f\, h \dot{\doteq}^{\alpha\alpha} g\, j$$

EDEC1:

$$\forall X_{\alpha\alpha} \forall Y_\alpha\, [f_{\alpha\alpha(\alpha\alpha)}\, X Y \doteq^{\alpha} g_{\alpha\alpha(\alpha\alpha)}\, X\, Y] \wedge \forall Z_\alpha\, [h_{\alpha\alpha}\, Z \doteq^{\alpha} j_{\alpha\alpha}\, Z] \supset f\, h \dot{\doteq}^{\alpha\alpha} g\, j$$

EDEC2:

$$\forall X_{\alpha\alpha} \forall Y_\alpha\, [f_{\alpha\alpha(\alpha\alpha)}\, X Y \doteq^{\alpha} g_{\alpha\alpha(\alpha\alpha)}\, X\, Y] \wedge \forall Z_\alpha\, [h_{\alpha\alpha}\, Z \doteq^{\alpha} j_{\alpha\alpha}\, Z] \supset f\, h \doteq^{\alpha\alpha} g\, j$$

The theorems **E4EXT**, **E5EXT** and **EDEC** are formulated using Leibniz equality. The alternative formulations **E4EXT1**, **E5EXT1** and **EDEC1** replace the positive occurrences of Leibniz equality by primitive equality. The formulations **E4EXT2**, **E5EXT2** and **EDEC2** replace all occurrences of Leibniz equality by primitive equality. In Table 10.1, we document the time TPS takes proving each of these theorems automatically using different modes.

The mode BASIC-MS04-2-MODE is able to prove **E4EXT1**, **E4EXT2**, **E5EXT1** and **E5EXT2** in a fraction of a second. However, BASIC-MS04-2-MODE is not able to prove **E4EXT** and **E5EXT** quickly. The reason is that one must do a primitive substitution using $=^{\iota}$ and then several unification steps. A mode MODE-MS04-2-EQ-PRIMSUB allows TPS to prove the examples **E4EXT** and **E5EXT** in about 1 second. This mode tries a primitive substitution using $=^{\iota}$ early.

sentence	procedure	mode name	time
E4EXT	**MS04-2**	MODE-MS04-2-EQ-PRIMSUB	0.83 secs
E4EXT	**MS03-7**	EASY-MS03-7-EQ-MODE	2.61 secs
E4EXT1	**MS04-2**	BASIC-MS04-2-MODE	0.1 secs
E4EXT1	**MS03-7**	EASY-MS03-7-MODE	0.24 secs
E4EXT2	**MS04-2**	BASIC-MS04-2-MODE	0.05 secs
E4EXT2	**MS03-7**	EASY-MS03-7-MODE	0.11 secs
E5EXT	**MS04-2**	MODE-MS04-2-EQ-PRIMSUB	0.84 secs
E5EXT	**MS03-7**	EASY-MS03-7-EQ-MODE	2.25 secs
E5EXT1	**MS04-2**	BASIC-MS04-2-MODE	0.11 secs
E5EXT1	**MS03-7**	EASY-MS03-7-MODE	0.26 secs
E5EXT2	**MS04-2**	BASIC-MS04-2-MODE	0.05 secs
E5EXT2	**MS03-7**	EASY-MS03-7-MODE	0.09 secs
EDEC	**MS04-2**	MODE-MS04-2-EDEC	2.1 hours
EDEC1	**MS04-2**	MODE-MS04-2-EDEC2	0.18 secs
EDEC2	**MS04-2**	MODE-MS04-2-EDEC2	0.2 secs

TABLE 10.1. Search Times for Leibniz vs. Primitive Equality

The example **EDEC** makes the difference between Leibniz equality and primitive equality even clearer. The proof of **EDEC** using Leibniz equality requires two primitive substitutions using $=^{\iota}$ followed by unification steps including 8 imitations and 4 projections. The mode MODE-MS04-2-EDEC is optimized for this example (while still only allowing basic search operations), but still TPS takes over 2 hours to find the proof. On the other hand, the mode MODE-MS04-2-EDEC2 can prove the formulations **EDEC1** and **EDEC2** using primitive equality in less than a second. These formulations require no primitive substitutions and very little unification (3 imitation steps and 1 projection step).

The modes EASY-MS03-7-MODE and EASY-MS03-7-EQ-MODE use the search procedure **MS03-7**. The results shown in Table 10.1 for these modes similarly demonstrate that the formulations using primitive equality are easier to prove than the formulations using Leibniz equality.

10.3. Topology Theorems

We now turn to topology examples. In [25], we reported on the following theorem from topology originally discussed in [23]:

BLEDSOE-FENG-SV-10:

$$\dot\forall D_{o\iota}\dot\forall G_{o(o\iota)}\,[G \subseteq OPEN_{o(o\iota)} \mathrel{\dot\wedge} D \mathrel{\dot=}^{o(o\iota)} \bigcup G \mathrel{\dot\supset} OPEN\,D]$$
$$\mathrel{\dot\supset} \dot\forall B_{o\iota}.\dot\forall x_\iota\,[B\,x \mathrel{\dot\supset} \exists D\,\blacksquare OPEN\,D \mathrel{\dot\wedge} D\,x \mathrel{\dot\wedge} D \subseteq B] \mathrel{\dot\supset} OPEN\,B$$

Note that $OPEN_{o(o\iota)}$ is simply a predicate constant, not an abbreviation. The theorem state that if the collection $OPEN$ of "open" sets is closed under arbitrary unions, and B satisfies the neighborhood property (for each $x \in B$ there is an open $D \subseteq B$ with $x \in D$), then B is open.

REMARK 10.3.1. The sentence **BLEDSOE-FENG-SV-10** uses the (infix) abbreviation $\subseteq$ for

$$\lambda P_{o\beta}\lambda R_{o\beta}\forall x_\beta\,\blacksquare P\,x \supset R\,x$$

This abbreviation uses the logical constants $\supset$ and Π^β. However, this is not an essential use of $\supset$ and Π^β. The two occurrences of $\subseteq$ are $G \subseteq OPEN$ and $D \subseteq B$. The $\beta\eta$-normal forms of these terms are

$$\forall u_{o\alpha}\,\blacksquare G\,u \supset OPEN\,u$$

and

$$\forall x_\alpha\,\blacksquare D\,x \supset B\,x,$$

respectively. Since these terms are only used in the context of external propositions, we can consider $G \subseteq OPEN$ and $D \subseteq B$ as abbreviations for

$$\dot\forall u_{o\alpha}\,\blacksquare G\,u \mathrel{\dot\supset} OPEN\,u$$

and

$$\dot\forall x_\alpha\,\blacksquare D\,x \mathrel{\dot\supset} B\,x,$$

respectively. Note that these propositions do not use any logical constants. We will generally consider such uses of abbreviations as abbreviations for the corresponding external proposition whenever possible.

The hypothesis that $OPEN$ is closed under arbitrary unions is stated in awkward manner. Technically, the hypothesis says that if G is a collection of open sets and D is equal to the union of G, then D is open. The more natural formulation would simply say that the union of G is open whenever G is a collection of open sets. Consider the following alternative formulation of **BLEDSOE-FENG-SV-10**:

THM616: (**BLEDSOE-FENG-SV-10** - extensional version)

$$\dot{\forall} G_{o(o\iota)} [G \subseteq OPEN_{o(o\iota)} \dot{\supset} OPEN [\bigcup G]]$$
$$\dot{\supset} \dot{\forall} B_{o\iota} \centerdot \dot{\forall} x_\iota [B\, x \dot{\supset} \dot{\exists} D \centerdot OPEN\, D \dot{\wedge} D\, x \dot{\wedge} D \subseteq B] \dot{\supset} OPEN\, B$$

where $\bigcup$ abbreviates

$$\lambda G_{o(o\alpha)} \lambda x_\alpha \exists S_{o\alpha} \centerdot G\, S \wedge S\, x.$$

THM616 is only a theorem if extensionality is assumed. Note that **BLEDSOE-FENG-SV-10** is a sentence in the empty signature and **THM616** is a sentence in the signature $\{\wedge, \Sigma^{o\iota}\}$.

Using extensionality, we can define a topology as follows:

TOPOLOGY-X:

$$\lambda OPEN_{o(o\alpha)} \centerdot OPEN [\lambda y_\alpha \top]$$
$$\wedge \forall K_{o(o\alpha)} [K \subseteq OPEN \supset OPEN \centerdot \bigcup K]$$
$$\wedge \forall Y_{o\alpha} \forall Z_{o\alpha} \centerdot OPEN\, Y \wedge OPEN\, Z \supset OPEN \centerdot Y \cap Z$$

where $\cap$ abbreviates

$$\lambda Y_{o\iota} \lambda Z_{o\iota} \lambda u_\iota \centerdot Y\, u \wedge Z\, u$$

The definition of **TOPOLOGY-X** makes use of the logical constants $\top$, $\wedge$, $\Sigma^{o\alpha}$, $\Pi^{o(o\alpha)}$ and $\Pi^{o\alpha}$ (where α may be any type). As in Remark 10.3.1, any occurrence

$$[\textbf{TOPOLOGY-X}\, OPEN]$$

of the abbreviation at the level of external proposition can be viewed as an abbreviation for

$$OPEN [\lambda y_\alpha \top] \dot{\wedge} \dot{\forall} K_{o(o\alpha)} [K \subseteq OPEN \dot{\supset} OPEN \centerdot \bigcup OPEN]$$
$$\dot{\wedge} \dot{\forall} Y_{o\alpha} \dot{\forall} Z_{o\alpha} \centerdot X\, Y \dot{\wedge} X\, Z \dot{\supset} X \centerdot Y \cap Z$$

which uses only the logical constants $\top$, $\wedge$ and $\Sigma^{o\alpha}$.

We start by considering a very simple topology theorem:

THM625:

$$\textbf{TOPOLOGY-X}_{o(o(o\alpha))}\, OPEN_{o(o\alpha)} \dot{\supset} OPEN \centerdot \lambda y_\alpha \bot$$

$$\left[\begin{array}{c} \left[\left[\begin{array}{c}\boxed{1}\ g^0\, X^2 \\ \boxed{2}\ \neg\, OPEN\, X^2\end{array}\right] \vee \boxed{3}\ OPEN[\bigcup g^0]\right] \\ \left[\boxed{4}\ \neg\, B^0\, x^0 \vee \left[\begin{array}{c}\boxed{5}\ OPEN\, D^0 \\ \boxed{6}\ D^0\, x^0 \\ \left[\boxed{7}\ \neg D^0\, x^1 \vee \boxed{8}\ B^0\, x^1\right]\end{array}\right]\right] \\ \boxed{9}\ \neg\, OPEN\, B^0 \end{array}\right]$$

FIGURE 10.1. The Initial Jform for **THM616**

As in Remark 10.3.1, we can consider **THM625** to be a sentence in the signature $\{\bot, \top, \wedge, \Sigma^{o\alpha}\}$ where we treat α as a base type. Using the mode BASIC-MS04-2-MODE, TPS can prove **THM625** in about a second.

BLEDSOE-FENG-SV-10 could already be proven by the SET-VAR theorem prover [23]. TPS could also already prove **BLEDSOE-FENG-SV-10** using search procedures which do not include extensional reasoning. Using general substitutions and no extensional reasoning, TPS proves the theorem in 58 seconds. Using set constraints and no extensional reasoning, TPS proves the theorem in 5 seconds. The solution to **BLEDSOE-FENG-SV-10** using set constraints is described in [25].

BLEDSOE-FENG-SV-10 is Example 10 in [23]. The SET-VAR theorem prover instantiates the collection g to the collection of all open subsets of B. Then SET-VAR sends the rest of the problem to a first-order theorem prover (STR+VE). In TPS we do not need to instantiate all set variables until the formula is first-order. Also, it should be emphasized that we are naming the existing set by a selected parameter arising from a set existence lemma. That is, we do not instantiate g with a complex formula. The complex formula is used in the proof of the existence lemma.

Since **MS04-2** is an extensional search procedure, we analyze the extensional version **THM616** here. TPS can prove **THM616** in about 10 seconds using the mode THM616-MS04-MODE-1. This mode uses the search procedure **MS04-2** augmented with set constraints (cf. Section 9.6) and delayed unification (cf. Section 9.5).

The initial jform for **THM616** is shown in Figure 10.1. First TPS considers the leftmost rigid path. Note that $\boxed{1}$ is flexible. Hence

$$\left[\begin{array}{c}\left[\begin{array}{c}\boxed{10}\ B^0\,X^3\\ \left[\boxed{11}\ \neg\, g^0\, s^0 \vee \boxed{12}\ \neg\, s^0\, X^3\right]\end{array}\right] \vee \left[\begin{array}{c}\boxed{13}\ g^0\, S^1\\ \boxed{14}\ S^1\, X^3\\ \boxed{15}\ \neg\, B^0\, X^3\end{array}\right]\end{array}\right]$$

FIGURE 10.2. Jform Corresponding to Mate Node

the leftmost rigid path is the one including $\boxed{3}$, $\boxed{4}$ and $\boxed{9}$. TPS could either duplicate g^0, duplicate x^0 or mate the literals $\boxed{3}$ and $\boxed{9}$. TPS adds this connection and commits to not duplicating g^0 or x^0 in the future.

The connection between $\boxed{3}$ and $\boxed{9}$ introduces a mate node with shallow formula

$$OPEN_{o(o\iota)}\, B^0_{o\iota} \doteq^o OPEN \,\blacksquare\, \bigcup g^0_{o(o\iota)}.$$

After the new extensional expansion dag is developed, this mate node will have a single child which is a negative functional node with shallow formula

$$B^0_{o\iota} \doteq^{o\iota} \bigcup g^0_{o(o\iota)}.$$

Thus the u-parts containing this mate node lead to the jform shown in Figure 10.2. The new set of u-parts for the search state include the previous u-parts which do not pass include $\{\boxed{3}, \boxed{9}\}$. Also, the u-parts including $\{\boxed{3}, \boxed{9}\}$ must include the u-parts which include the mate node. The jform corresponding to the u-parts of the new search state can be obtained using a technique similar to dissolution (cf. [48]). This jform is shown in Figure 10.3.

Note that every vertical path of the jform in Figure 10.3 contains a flexible node since the literals $\boxed{1}$, $\boxed{11}$, $\boxed{12}$ and $\boxed{13}$ are all flexible. TPS chooses the u-part corresponding to the vertical path containing $\boxed{1}$, $\boxed{2}$, $\boxed{9}$ and $\boxed{4}$. The positive flexible node $\boxed{1}$ has the expansion variable $g^0_{o(o\iota)}$ as the head of its shallow formula.

TPS begins generating set constraints for g^0. TPS first generates the upper bound constraint

$$\forall X^2\, \langle g^0\, X^2 \rightarrow OPEN\, X^2 \rangle \tag{9}$$

using $\boxed{1}$ and $\boxed{2}$. TPS then finds the maximal solution of the constraint (9). Essentially, TPS takes g^0 to be $OPEN$, the collection of open sets. This path of search eventually fails and TPS backtracks.

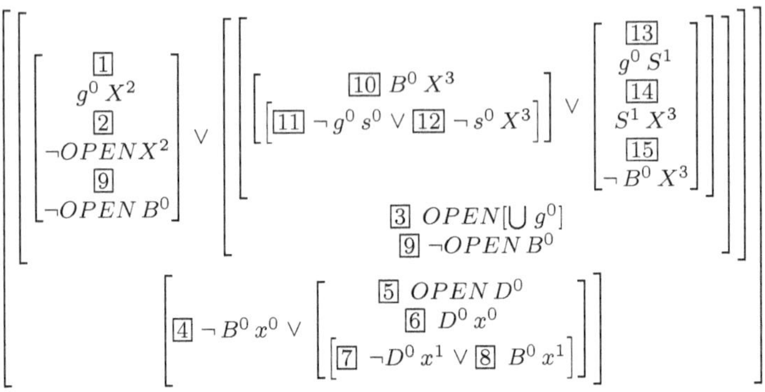

FIGURE 10.3. Jform after Connection

TPS uses 13 and 14 to generate the upper bound constraint

$$\forall S^1, X^3 \langle g^0 S^1 \rightarrow S^1 X^3 \rangle \tag{10}$$

which is combined with constraint (9). TPS solves the two constraints simultaneously, but this leads to a false path. Eventually TPS backtracks and generates the upper bound constraint

$$\forall S^1, X^3 \langle g^0 S^1 \rightarrow [S^1 X^3], [B^0 X^3] \rangle \tag{11}$$

using 13, 14 and 15.

TPS combines constraints (9) and (11). The maximal solution to these constraints is the desired solution where g^0 is taken to be the collection of open subsets of B^0. The set existence lemma corresponding to the solution to these constraints is

$$\begin{gathered}
\forall B_{o\iota} \exists G_{o(o\iota)} \centerdot \\
\forall V_{o\iota}[G V \supset OPEN V] \\
\land \forall x_\iota \forall S_{o\iota}[G S \supset B x \lor \neg S x] \\
\land \forall U_{o\iota} \centerdot OPEN U \land \forall y_\iota[B y \lor \neg U y] \supset G U
\end{gathered}$$

The quantifier for $B_{o\iota}$ corresponds to an expansion node. We choose the selected parameter B^0 from the original problem to be the expansion term. The quantifier for $G_{o(o\iota)}$ corresponds to a selection node

$$
\left[\begin{array}{c}
\left[\boxed{16}\ \neg G^0\, x^6 \vee \boxed{17}\ OPEN\, x^6\right] \\
\left[\boxed{18}\ \neg G^0\, X^2 \vee \boxed{19}\ OPEN\, X^2\right] \\
\left[\boxed{20}\ \neg G^0\, s^2 \vee \boxed{21}\ B^0\, x^5 \vee \boxed{22}\ \neg s^2\, x^5\right] \\
\left[\boxed{23}\ \neg G^0\, S^1 \vee \boxed{24}\ B^0\, X^2 \vee \boxed{25}\ \neg S^1\, X^2\right] \\
\left[\boxed{26}\ \neg OPEN\, u^1 \vee \left[\begin{array}{c}\boxed{27}\ \neg B^0\, X^4 \\ \boxed{28}\ u^1\, X^4\end{array}\right] \vee \boxed{29}\ G^0\, u^1\right] \\
\boxed{10}\ B^0\, X^3 \\
\left[\boxed{11}\ \neg\, G^0\, s^0 \vee \boxed{12}\ \neg\, s^0\, X^3\right] \\
\boxed{3}\ OPEN[\bigcup G^0] \\
\boxed{9}\ \neg OPEN\, B^0 \\
\left[\boxed{4}\ \neg\, B^0\, x^0 \vee \left[\begin{array}{c}\boxed{5}\ OPEN\, D^0 \\ \boxed{6}\ D^0\, x^0 \\ \left[\boxed{7}\ \neg D^0\, x^1 \vee \boxed{8}\ B^0\, x^1\right]\end{array}\right]\right]
\end{array}\right]
$$

FIGURE 10.4. Jform with Set Existence Lemma

with selected parameter $G^0_{o(o\iota)}$. The result is the proposition

$$
\begin{gathered}
\forall V_{o\iota}[G^0\, V \supset OPEN\, V] \\
\wedge \forall x_\iota \forall S_{o\iota}[G^0\, S \supset B^0\, x \vee \neg S\, x] \\
\wedge \forall U_{o\iota} \,\blacksquare\, OPEN\, U \wedge \forall y_\iota[B^0\, y \vee \neg U\, y] \supset G^0\, U
\end{gathered}
$$

The expansion variable g^0 is instantiated with the selected parameter G^0, which represents the solutions to the constraints. The set existence lemma is used to solve the u-parts which contain the literals from which the constraints were constructed. In particular, the u-parts containing $\boxed{1}$ and $\boxed{2}$ as well as the u-parts containing $\boxed{13}$, $\boxed{14}$ and $\boxed{15}$ disappear from the jform. The resulting jform is shown in Figure 10.4.

At this point in the search, TPS mates $\boxed{10}$ and $\boxed{4}$ leading to a flex-rigid node e correspond to proving x^0 equals X^2. Instead of unifying these two (by substituting X^2 for x^0), TPS goes delays unification using the set of unification constraints $\{\{e\}\}$ (cf. Section

9.5). Searching with unification constraints, TPS find the following unifiable complete mating:

$$\boxed{10} - \boxed{4} \quad \boxed{5} - \boxed{26} \quad \boxed{28} - \boxed{7}$$
$$\boxed{8} - \boxed{27} \quad \boxed{29} - \boxed{11} \quad \boxed{6} - \boxed{12}$$

TPS computes a unifier and translates the proof to natural deduction.

10.4. Variants of Knaster-Tarski

The following examples are variants of the Knaster-Tarski fixed point theorem:

THM1A:

$$\forall K_{o\iota(o\iota)} \,.\, \exists X_{o\iota} \exists Y_{o\iota} [X \subseteq Y \wedge \neg .\, K\, X \subseteq K\, Y] \vee \exists U_{o\iota} \,.\, K\, U =^{o\iota} U$$

THM2:

$$\forall K_{o\iota(o\iota)} \,.\, \forall x_{o\iota} \forall y_{o\iota} [x \subseteq y \supset K\, x \subseteq K\, y]$$
$$\supset \exists u_{o\iota} \,.\, K\, u \equiv^{s} u$$

where $K\, u \equiv^{s} u$ abbreviates

$$\forall x_{\iota} \,.\, K\, u\, x \equiv u\, x$$

THM598:

$$\forall K_{o\iota(o\iota)} \forall L_{o\iota(o\iota)} \,.\, \forall u_{o\iota} \forall v_{o\iota} [u \subseteq v \supset K\, u \subseteq K\, v]$$
$$\wedge \forall u \forall v [u \subseteq v \supset L\, u \subseteq L\, v]$$
$$\supset \exists w_{o\iota} \,.\, K\, w \cup L\, w =^{o\iota} w$$

where $\cup$ abbreviates

$$\lambda x_{o\iota} \lambda y_{o\iota} \lambda z_{\iota} \,.\, x\, z \vee y\, z$$

THM1A and **THM2** are simply versions of the Knaster-Tarski fixed point theorem. **THM598** states that if K and L are monotone, then $K \cup L$ has a fixed point, where $K \cup L$ applied to a set w is $[K\, w] \cup [L\, w]$.

Each of these examples are too difficult to prove using general substitutions. However, we can prove these examples using the search procedure **MS03-7** augmented with set constraints (cf. Table 10.2).

sentence	procedure	mode name	time
THM1A	MS03-7	QUIET-SV-MS03-MODE	6.34 secs
THM2	MS03-7	QUIET-SV-MS03-MODE	5.58 secs
THM598	MS03-7	QUIET-SV-MS03-MODE	26.56 secs

TABLE 10.2. Search Times for Variants of Knaster-Tarski

Part III

Appendix

APPENDIX A

Proofs

Here we present many proofs which were delayed in the body of the book.

A.1. Proofs from Chapter 3

PROOF OF THEOREM 3.2.18. We first check $\mathcal{E}^{W\mapsto \mathsf{a}}$ is an evaluation function.

1. $\mathcal{E}^{W\mapsto \mathsf{a}}_{\varphi}(y_\beta) = \mathcal{E}_{\varphi,[\mathsf{a}/x^y]}([x^y/W]y) = \mathcal{E}_{\varphi,[\mathsf{a}/x^y]}(y) = \varphi(y)$.
2. Let $\mathbf{F} \in \mathit{wff}_{\beta\gamma}(\mathcal{S})$ and $\mathbf{C} \in \mathit{wff}_{\gamma}(\mathcal{S})$ be given. Let x^1 denote $x^{[\mathbf{F}\,\mathbf{C}]}$. Thus

$$\begin{aligned}
\mathcal{E}^{W\mapsto \mathsf{a}}_{\varphi}([\mathbf{F}\,\mathbf{C}]) &= \mathcal{E}_{\varphi,[\mathsf{a}/x^1]}([x^1/W][\mathbf{F}\,\mathbf{C}]) \\
&= \mathcal{E}_{\varphi,[\mathsf{a}/x^1]}([x^1/W]\mathbf{F})@\mathcal{E}_{\varphi,[\mathsf{a}/x^1]}([x^1/W]\mathbf{C}) \\
\text{(using Lemma 3.2.17)} &= \mathcal{E}_{\varphi,[\mathsf{a}/x^{\mathbf{F}}]}([x^{\mathbf{F}}/W]\mathbf{F})@\mathcal{E}_{\varphi,[\mathsf{a}/x^{\mathbf{C}}]}([x^{\mathbf{C}}/W]\mathbf{C}) \\
&= \mathcal{E}^{W\mapsto \mathsf{a}}_{\varphi}(\mathbf{F})@\mathcal{E}^{W\mapsto \mathsf{a}}_{\varphi}(\mathbf{C}).
\end{aligned}$$

3. Suppose $\mathbf{B} \in \mathit{wff}_{\beta}(\mathcal{S})$ and φ and ψ are assignments which coincide on $\mathbf{Free}(\mathbf{B})$. Since $\varphi, [\mathsf{a}/x^{\mathbf{B}}]$ and $\psi, [\mathsf{a}/x^{\mathbf{B}}]$ coincide on $\mathbf{Free}([x^{\mathbf{B}}/W]\mathbf{B})$, we can compute

$$\mathcal{E}^{W\mapsto \mathsf{a}}_{\varphi}(\mathbf{B}) = \mathcal{E}_{\varphi,[\mathsf{a}/x^{\mathbf{B}}]}([x^{\mathbf{B}}/W]\mathbf{B}) = \mathcal{E}_{\psi,[\mathsf{a}/x^{\mathbf{B}}]}([x^{\mathbf{B}}/W]\mathbf{B}) = \mathcal{E}^{W\mapsto \mathsf{a}}_{\psi}(\mathbf{B}).$$

4. To verify $\mathcal{E}^{W\mapsto \mathsf{a}}$ respects β, let $\mathbf{B} \in \mathit{wff}_{\beta}(\mathcal{S})$ and x^2 be $x^{\mathbf{B}^{\downarrow\beta}}$.

$$\begin{aligned}
\mathcal{E}^{W\mapsto \mathsf{a}}_{\varphi}(\mathbf{B}) &= \mathcal{E}_{\varphi,[\mathsf{a}/x^{\mathbf{B}}]}([x^{\mathbf{B}}/W]\mathbf{B}) &&= \mathcal{E}_{\varphi,[\mathsf{a}/x^{\mathbf{B}}]}([x^{\mathbf{B}}/W]\mathbf{B}^{\downarrow\beta}) \\
&= \mathcal{E}_{\varphi,[\mathsf{a}/x^2]}([x^2/W]\mathbf{B}^{\downarrow\beta}) &&= \mathcal{E}^{W\mapsto \mathsf{a}}_{\varphi}(\mathbf{B}^{\downarrow\beta}).
\end{aligned}$$

Next suppose $\mathcal{J}$ is η-functional. We verify $(\mathcal{D}, @, \mathcal{E}^{W\mapsto \mathsf{a}})$ is η-functional by computing

$$\begin{aligned}\mathcal{E}_{\varphi}^{W\mapsto \mathsf{a}}(\mathbf{B}) &= \mathcal{E}_{\varphi,[\mathsf{a}/x^{\mathbf{B}}]}([x^{\mathbf{B}}/W]\mathbf{B}) &= \mathcal{E}_{\varphi,[\mathsf{a}/x^{\mathbf{B}}]}([x^{\mathbf{B}}/W]\mathbf{B}^{\downarrow}) \\ &= \mathcal{E}_{\varphi,[\mathsf{a}/x^{2}]}([x^{2}/W]\mathbf{B}^{\downarrow}) &= \mathcal{E}_{\varphi}^{W\mapsto \mathsf{a}}(\mathbf{B}^{\downarrow}).\end{aligned}$$

where x^2 is $x^{\mathbf{B}^{\downarrow}}$.

Suppose $\mathcal{J}$ is ξ-functional. Let $\mathbf{M}$ and $\mathbf{N}$ be terms of type γ and y be a variable of type β. Assume

$$\mathcal{E}_{\varphi,[\mathsf{b}/y]}^{W\mapsto \mathsf{a}}(\mathbf{M}) = \mathcal{E}_{\varphi,[\mathsf{b}/y]}^{W\mapsto \mathsf{a}}(\mathbf{N}) \text{ for every } \mathsf{b} \in \mathcal{D}_{\beta}.$$

Let x_{α} be a variable not in $\mathbf{FreeM} \cup \mathbf{FreeN} \cup \{y\}$. Using Lemma 3.2.17, we have

$$\mathcal{E}\varphi, [\mathsf{b}/y], [\mathsf{a}/x]([x/W]\mathbf{M}) = \mathcal{E}_{\varphi,[\mathsf{b}/y],[\mathsf{a}/x]}([x/W]\mathbf{N})$$

for every $\mathsf{b} \in \mathcal{D}_{\beta}$. By ξ-functionality of $\mathcal{J}$, we know

$$\mathcal{E}\varphi, [\mathsf{a}/x](\lambda y[x/W]\mathbf{M}) = \mathcal{E}_{\varphi,[\mathsf{a}/x]}(\lambda y[x/W]\mathbf{N}).$$

Using Lemma 3.2.17 again, we compute

$$\mathcal{E}_{\varphi}^{W\mapsto \mathsf{a}}(\lambda y\mathbf{M}) = \mathcal{E}_{\varphi,[\mathsf{a}/x]}([x/W](\lambda y\mathbf{M})) = \mathcal{E}_{\varphi,[\mathsf{a}/x]}(\lambda y[x/W]\mathbf{M})$$

and

$$\mathcal{E}_{\varphi}^{W\mapsto \mathsf{a}}(\lambda y\mathbf{N}) = \mathcal{E}_{\varphi,[\mathsf{a}/x]}([x/W](\lambda y\mathbf{N})) = \mathcal{E}_{\varphi,[\mathsf{a}/x]}(\lambda y[x/W]\mathbf{N}).$$

Hence

$$\mathcal{E}_{\varphi}^{W\mapsto \mathsf{a}}(\lambda y\mathbf{M}) = \mathcal{E}_{\varphi}^{W\mapsto \mathsf{a}}(\lambda y\mathbf{N})$$

as desired. Thus $(\mathcal{D}, @, \mathcal{E}^{W\mapsto \mathsf{a}})$ is ξ-functional.

Finally, note that $\mathcal{E}_{\varphi}^{W\mapsto \mathsf{a}}(W) = \mathcal{E}_{\varphi,[\mathsf{a}/x^{W}]}([x^{W}/W]W) = \mathsf{a}$ and

$$\mathcal{E}_{\varphi}^{W\mapsto \mathsf{a}}(\mathbf{B}) = \mathcal{E}_{\varphi,[\mathsf{a}/x^{\mathbf{B}}]}([x^{\mathbf{B}}/W]\mathbf{B}) = \mathcal{E}_{\varphi,[\mathsf{a}/x^{\mathbf{B}}]}(\mathbf{B}) = \mathcal{E}_{\varphi}(\mathbf{B})$$

if $W \notin \mathbf{Params}(\mathbf{B})$. □

PROOF OF LEMMA 3.2.20. Let $\mathcal{S}_2 := \mathcal{S} \cup \{c\}$. For each term $\mathbf{B} \in \mathit{wff}(\mathcal{S}_2)$, let $x_{\alpha}^{\mathbf{B}}$ be a variable with $x^{\mathbf{B}} \notin \mathbf{Free}(\mathbf{B})$. For each assignment φ into $(\mathcal{D}, @)$, define

$$\mathcal{E}'_{\varphi}(\mathbf{B}) := \mathcal{E}_{\varphi,[\mathsf{a}/x^{\mathbf{B}}]}([x^{\mathbf{B}}/c]\mathbf{B}).$$

First we check the conditions for $\mathcal{E}'$ to be an evaluation extension. Note that if $\mathbf{B} \in \mathit{wff}(\mathcal{S})$, then $([x^{\mathbf{B}}/c]\mathbf{B})$ is $\mathbf{B}$. Hence

$$\mathcal{E}'_{\varphi}(\mathbf{B}) = \mathcal{E}_{\varphi,[\mathsf{a}/x^{\mathbf{B}}]}([x^{\mathbf{B}}/c]\mathbf{B}) = \mathcal{E}_{\varphi,[\mathsf{a}/x^{\mathbf{B}}]}(\mathbf{B}) = \mathcal{E}_{\varphi}(\mathbf{B}).$$

Also,

$$\mathcal{E}'_\varphi(c) = \mathcal{E}_{\varphi,[\mathsf{a}/x^c]}([x^c/c]c) = \mathsf{a}.$$

Suppose $\mathcal{E}$ is η-functional. Then for each $\mathbf{B} \in \mathit{wff}(\mathcal{S}_2)$ we have

$$\mathcal{E}'_\varphi(\mathbf{B}) = \mathcal{E}_{\varphi,[\mathsf{a}/x^{\mathbf{B}}]}([x^{\mathbf{B}}/c]\mathbf{B}) = \mathcal{E}_{\varphi,[\mathsf{a}/x^{\mathbf{B}}]}(([x^{\mathbf{B}}/c]\mathbf{B})^{\downarrow}) = \mathcal{E}'_\varphi(\mathbf{B}^{\downarrow}).$$

Thus if $\mathcal{E}$ is η-functional, then $\mathcal{E}'$ is η-functional.

Now we check $\mathcal{E}'$ is an $\mathcal{S}_2$-evaluation. This follows the same pattern as the proof of Theorem 3.2.18.

1. $\mathcal{E}'_\varphi(y_\beta) = \mathcal{E}_{\varphi,[\mathsf{a}/x^y]}([x^y/c]y) = \mathcal{E}_{\varphi,[\mathsf{a}/x^y]}(y) = \varphi(y)$.
2. Let $\mathbf{F} \in \mathit{wff}_{\beta\gamma}(\mathcal{S})$ and $\mathbf{C} \in \mathit{wff}_\gamma(\mathcal{S})$ be given. Let x^1 denote $x^{[\mathbf{F}\,\mathbf{C}]}$. Thus

$$\begin{array}{rcl} \mathcal{E}'_\varphi([\mathbf{F}\,\mathbf{C}]) & = & \mathcal{E}_{\varphi,[\mathsf{a}/x^1]}([x^1/c][\mathbf{F}\,\mathbf{C}]) \\ & = & \mathcal{E}_{\varphi,[\mathsf{a}/x^1]}([x^1/c]\mathbf{F})@\mathcal{E}_{\varphi,[\mathsf{a}/x^1]}([x^1/c]\mathbf{C}) \\ \text{(using Lemma 3.2.17)} & = & \mathcal{E}_{\varphi,[\mathsf{a}/x^{\mathbf{F}}]}([x^{\mathbf{F}}/c]\mathbf{F})@\mathcal{E}_{\varphi,[\mathsf{a}/x^{\mathbf{C}}]}([x^{\mathbf{C}}/c]\mathbf{C}) \\ & = & \mathcal{E}'_\varphi(\mathbf{F})@\mathcal{E}'_\varphi(\mathbf{C}). \end{array}$$

3. Suppose $\mathbf{B} \in \mathit{wff}_\beta(\mathcal{S})$ and φ and ψ are assignments which coincide on $\mathbf{Free}(\mathbf{B})$. Since $\varphi, [\mathsf{a}/x^{\mathbf{B}}]$ and $\psi, [\mathsf{a}/x^{\mathbf{B}}]$ coincide on $\mathbf{Free}([x^{\mathbf{B}}/c]\mathbf{B})$, we can compute

$$\mathcal{E}'_\varphi(\mathbf{B}) = \mathcal{E}_{\varphi,[\mathsf{a}/x^{\mathbf{B}}]}([x^{\mathbf{B}}/c]\mathbf{B}) = \mathcal{E}_{\psi,[\mathsf{a}/x^{\mathbf{B}}]}([x^{\mathbf{B}}/c]\mathbf{B}) = \mathcal{E}'_\psi(\mathbf{B}).$$

4. To verify $\mathcal{E}'$ respects β, let $\mathbf{B} \in \mathit{wff}_\beta(\mathcal{S})$ and x^2 be $x^{\mathbf{B}^{\downarrow\beta}}$.

$$\begin{array}{rcl} \mathcal{E}'_\varphi(\mathbf{B}) & = & \mathcal{E}_{\varphi,[\mathsf{a}/x^{\mathbf{B}}]}([x^{\mathbf{B}}/c]\mathbf{B}) \;=\; \mathcal{E}_{\varphi,[\mathsf{a}/x^{\mathbf{B}}]}([x^{\mathbf{B}}/c]\mathbf{B}^{\downarrow\beta}) \\ & = & \mathcal{E}_{\varphi,[\mathsf{a}/x^2]}([x^2/c]\mathbf{B}^{\downarrow\beta}) \;=\; \mathcal{E}'_\varphi(\mathbf{B}^{\downarrow\beta}). \end{array}$$

□

Proof of Theorem 3.2.21. Define $\mathcal{S}_3 := \mathcal{S}_2 \setminus \mathcal{S}_1$. We know $\mathcal{S}_3$ is either finite or countably infinite. We first consider the finite case, then extend the result to the infinite case.

For the finite case, we prove the existence of $\mathcal{E}'$ by induction on the number of elements in $\mathcal{S}_3$. For the base case where $\mathcal{S}_3$ is empty, we can clearly choose $\mathcal{E}'$ to be $\mathcal{E}$, since $\mathcal{S}_1$ and $\mathcal{S}_2$ are the same. For the induction case, suppose $\mathcal{S}_3$ has $n + 1$ elements and choose some $c \in \mathcal{S}_3$. By the inductive hypothesis, we have an $(\mathcal{S}_2 \setminus \{c\})$-evaluation extension $\mathcal{E}^n$ of $\mathcal{E}$. By Lemma 3.2.20, we can extend $\mathcal{E}^n$ to $\mathcal{E}'$ by $\mathcal{E}'(c) := \mathcal{I}(c)$.

Assume $\mathcal{S}_3$ is countably infinite. Enumerate $\mathcal{S}_3$ as

$$\{c^1, c^2, \ldots, c^n, \ldots\}.$$

For each $n \geq 0$, let $\mathcal{S}^n := \mathcal{S}_1 \cup \{c^1, c^2, \ldots, c^n\}$. Hence

$$\mathcal{S}_1 = \mathcal{S}^0 \subset \mathcal{S}^1 \subset \cdots \subset \mathcal{S}^n \subset \mathcal{S}_2$$

and $\mathcal{S}_2 = \bigcup_{n \geq 0} \mathcal{S}^n$.

For each $n \geq 0$, we define an $\mathcal{S}^n$-evaluation function $\mathcal{E}^n$ such that

$$\mathcal{E}^n\big|_{\mathit{wff}(\mathcal{S}^i)} = \mathcal{E}^i$$

for each i $(1 \leq i \leq n)$ and

$$\mathcal{E}^n\big|_{\mathcal{S}^n \setminus \mathcal{S}_1} = \mathcal{I}\big|_{\mathcal{S}^n \setminus \mathcal{S}_1}.$$

Let $\mathcal{E}^0 := \mathcal{E}$. Assuming we have $\mathcal{E}^n$, Lemma 3.2.20 implies there is some $\mathcal{S}^{n+1}$-evaluation $\mathcal{E}^{n+1}$ extending $\mathcal{E}^n$ by $\mathcal{E}^{n+1}(c^{n+1}) := \mathcal{I}(c^{n+1})$.

Now, for each $\mathbf{B} \in \mathit{wff}(\mathcal{S}_2)$ we can define $\mathcal{E}'_\varphi(\mathbf{B}) := \mathcal{E}^n_\varphi(\mathbf{B})$ where n is the least natural number such that $\mathbf{Consts}(\mathbf{B}) \subseteq \mathcal{S}^n$. (Such an n must exist since $\mathbf{Consts}(\mathbf{B})$ is finite.) Note that for any $m \geq n$, we know $\mathcal{E}'_\varphi(\mathbf{B}) = \mathcal{E}^m_\varphi(\mathbf{B})$. Using this and the fact that each $\mathcal{E}^m$ is an $\mathcal{S}^m$-evaluation function, we can easily check that $\mathcal{E}'$ is an $\mathcal{S}_2$-evaluation function.

If $\mathcal{E}$ is η-functional, then we know every $\mathcal{E}^n$ is η-functional by the definition of evaluation extensions. From this we can check $\mathcal{E}'$ is η-functional as follows. For $\mathbf{B} \in \mathit{wff}(\mathcal{S}_2)$, let n be the least natural number such that $\mathbf{Consts}(\mathbf{B}) \subseteq \mathcal{S}^n$. Note that

$$\mathbf{Consts}(\mathbf{B}^\downarrow) \subseteq \mathbf{Consts}(\mathbf{B}) \subseteq \mathcal{S}^n.$$

Therefore,

$$\mathcal{E}'_\varphi(\mathbf{B}) = \mathcal{E}^n_\varphi(\mathbf{B}) = \mathcal{E}^n_\varphi(\mathbf{B}^\downarrow) = \mathcal{E}'_\varphi(\mathbf{B}^\downarrow).$$

□

PROOF OF THEOREM 3.3.14. To verify that $\mathcal{M}^{W \mapsto \mathsf{a}}$ is an $\mathcal{S}$-model, we must check $\mathfrak{L}_c(\mathcal{E}^{W \mapsto \mathsf{a}}(c))$ holds with respect to υ for each logical constant in $c \in \mathcal{S}$. This is immediate from the fact that $\mathcal{M}$ is an $\mathcal{S}$-model and $\mathcal{E}^{W \mapsto \mathsf{a}}(c) = \mathcal{E}(c)$ by Theorem 3.2.18.

The fact that $\mathcal{M}^{W\mapsto \mathsf{a}} \models_{\varphi} \mathbf{M}$ iff $\mathcal{M} \models_{\varphi,[\mathsf{a}/x]} [x/W]\mathbf{M}$ for any variable $x_\alpha \notin \mathbf{Free}(\mathbf{M})$ follows from a simple induction on $\mathbf{M}$. For the base cases, if $\mathbf{M} \in \mathit{wff}_o(\mathcal{S})$ and $x \notin \mathbf{Free}(\mathbf{M})$, then

$$\mathcal{E}_{\varphi}^{W\mapsto \mathsf{a}}(\mathbf{M}) = \mathcal{E}_{\varphi,[\mathsf{a}/x]}([x/W]\mathbf{M})$$

and so $v(\mathcal{E}_{\varphi}^{W\mapsto \mathsf{a}}(\mathbf{M})) = v(\mathcal{E}_{\varphi,[\mathsf{a}/x]}([x/W]\mathbf{M}))$. If $\mathbf{M}$ is $[\mathbf{C} \doteq^{\alpha} \mathbf{D}]$ and $x \notin \mathbf{Free}(\mathbf{M})$, then

$$\mathcal{E}_{\varphi}^{W\mapsto \mathsf{a}}(\mathbf{C}) = \mathcal{E}_{\varphi,[\mathsf{a}/x]}([x/W]\mathbf{C}) \text{ and } \mathcal{E}_{\varphi}^{W\mapsto \mathsf{a}}(\mathbf{D}) = \mathcal{E}_{\varphi,[\mathsf{a}/x]}([x/W]\mathbf{D}).$$

Thus $\mathcal{M}^{W\mapsto \mathsf{a}} \models_{\varphi} [\mathbf{C} \doteq^{\alpha} \mathbf{D}]$ iff $\mathcal{M} \models_{\varphi,[\mathsf{a}/x]} [[x/W]\mathbf{C} \doteq^{\alpha} [x/W]\mathbf{D}]$.

The induction cases for propositions constructed using $\dot{\top}$, $\dot{\neg}$, $\dot{\vee}$ and $\dot{\forall}$ are straightforward. We only check the $\dot{\forall}$ case. Assume $\mathbf{M}$ is $\dot{\forall} y_\beta\, \mathbf{N}$. Suppose $\mathcal{M} \models_{\varphi,[\mathsf{a}/x]} \dot{\forall} y_\beta\, [x/W]\mathbf{N}$ where $x \notin \mathbf{Free}(\mathbf{M})$. (By our implicit treatment of α-conversion, we can assume x and y are not the same variable.) Hence $\mathcal{M} \models_{\varphi,[\mathsf{b}/y],[\mathsf{a}/x]} [x/W]\mathbf{N}$ for every $\mathsf{b} \in \mathcal{D}_\beta$. By the inductive hypothesis, $\mathcal{M}^{W\mapsto \mathsf{a}} \models_{\varphi,[\mathsf{b}/y]} \mathbf{N}$ for every $\mathsf{b} \in \mathcal{D}_\beta$. Thus $\mathcal{M}^{W\mapsto \mathsf{a}} \models_{\varphi} \mathbf{M}$. Next suppose $\mathcal{M}^{W\mapsto \mathsf{a}} \models_{\varphi} \mathbf{M}$. Hence $\mathcal{M}^{W\mapsto \mathsf{a}} \models_{\varphi,[\mathsf{b}/y]} \mathbf{N}$ for every $\mathsf{b} \in \mathcal{D}_\beta$. By the inductive hypothesis, $\mathcal{M} \models_{\varphi,[\mathsf{b}/y],[\mathsf{a}/x]} [x/W]\mathbf{N}$ for every $\mathsf{b} \in \mathcal{D}_\beta$ and $x \notin \mathbf{Free}(\mathbf{N})$. Thus $\mathcal{M} \models_{\varphi,[\mathsf{a}/x]} [x/W]\mathbf{M}$ for every $\mathsf{b} \in \mathcal{D}_\beta$ and $x \notin \mathbf{Free}(\mathbf{M})$.

If $\mathcal{M}$ satisfies property η $[\xi]$, then $(\mathcal{D}, @, \mathcal{E})$ is η-functional $[\xi$-functional$]$. Thus $(\mathcal{D}, @, \mathcal{E}^{W\mapsto \mathsf{a}})$ is η-functional $[\xi$-functional$]$ by Theorem 3.2.18 and $\mathcal{M}^{W\mapsto \mathsf{a}}$ satisfies property η $[\xi]$. If $\mathcal{M}$ satisfies property $\mathfrak{f}$, then $(\mathcal{D}, @)$ is functional. Hence $\mathcal{M}^{W\mapsto \mathsf{a}}$ satisfies property $\mathfrak{f}$. Since the valuations of the two models are the same, if $\mathcal{M}$ satisfies property $\mathfrak{b}$, then v is injective and $\mathcal{M}^{W\mapsto \mathsf{a}}$ satisfies property $\mathfrak{b}$. □

PROOF OF LEMMA 3.5.3. Suppose $\mathbf{M} \in \mathit{wff}_o(\mathcal{S})$. Since j is a homomorphism from the evaluation $(\mathcal{D}^2, @^2, \mathcal{E}^2)$ to the evaluation $(\mathcal{D}^1, @^1, \mathcal{E}^1)$ (cf. Definition 3.2.25) we know $j(\mathcal{E}^2_{\varphi}(\mathbf{M})) = \mathcal{E}^1_{j\varphi}(\mathbf{M})$. Since j is a homomorphism from the model $\mathcal{M}^2$ to the model $\mathcal{M}^1$ (cf. Definition 3.5.1) we know $v^2(\mathcal{E}^2_{\varphi}(\mathbf{M})) = v^1(j(\mathcal{E}^2_{\varphi}(\mathbf{M})))$. Combining these two facts we obtain

$$v^2(\mathcal{E}^2_{\varphi}(\mathbf{M})) = v^1(j(\mathcal{E}^2_{\varphi}(\mathbf{M}))) = v^1(\mathcal{E}^1_{j\varphi}(\mathbf{M})).$$

Thus

$$\mathcal{M}^1 \models_{j\varphi} \mathbf{M} \quad \text{iff} \quad v^1(\mathcal{E}^1_{j\varphi}(\mathbf{M})) = \mathtt{T} \quad \text{iff} \quad v^2(\mathcal{E}^2_{\varphi}(\mathbf{M})) = \mathtt{T}$$
$$\text{iff} \quad \mathcal{M}^2 \models_{\varphi} \mathbf{M}.$$

Suppose $\mathbf{M}$ is of the form $[\mathbf{A}_\alpha \mathrel{\dot{=}^\alpha} \mathbf{B}_\alpha]$. We know

$$j(\mathcal{E}^2_\varphi(\mathbf{A})) = (\mathcal{E}^1_{j\varphi}(\mathbf{A})) \text{ and } j(\mathcal{E}^2_\varphi(\mathbf{B})) = \mathcal{E}^1_{j\varphi}(\mathbf{B})$$

since j is a homomorphism. Hence

$$\begin{aligned} \mathcal{M}^1 \models_{j\varphi} [\mathbf{A}_\alpha \mathrel{\dot{=}^\alpha} \mathbf{B}_\alpha] \quad & \text{iff} \quad \mathcal{E}^1_{j\varphi}(\mathbf{A}) = \mathcal{E}^1_{j\varphi}(\mathbf{B}) \\ & \text{iff} \quad j(\mathcal{E}^2_\varphi(\mathbf{A})) = j(\mathcal{E}^2_\varphi(\mathbf{B})) \\ \text{(since } j \text{ is injective)} \quad & \text{iff} \quad \mathcal{E}^2_\varphi(\mathbf{A})) = \mathcal{E}^2_\varphi(\mathbf{B}) \\ & \text{iff} \quad \mathcal{M}^2 \models_\varphi [\mathbf{A}_\alpha \mathrel{\dot{=}^\alpha} \mathbf{B}_\alpha]. \end{aligned}$$

The inductive cases where $\mathbf{M}$ is constructed using $\dot{\top}$, $\dot{\neg}$, $\dot{\vee}$ and $\dot{\forall}$ are easy. We only check the $\dot{\forall}$ case.

Suppose $\mathcal{M}^1 \models_{j\varphi} [\dot{\forall} x_\alpha \, \mathbf{N}]$. Then $\mathcal{M}^1 \models_{j\varphi,[j(\mathsf{a})/x]} \mathbf{N}$ for every $\mathsf{a} \in \mathcal{D}^2_\alpha$. By the inductive hypothesis, $\mathcal{M}^2 \models_{\varphi,[\mathsf{a}/x]} \mathbf{N}$ for every $\mathsf{a} \in \mathcal{D}^2_\alpha$. Thus $\mathcal{M}^2 \models_\varphi [\dot{\forall} x_\alpha \, \mathbf{N}]$.

Suppose $\mathcal{M}^2 \models_\varphi [\dot{\forall} x_\alpha \, \mathbf{N}]$. Let $\mathsf{a} \in \mathcal{D}^1_\alpha$ be given. Then

$$\mathcal{M}^2 \models_{\varphi,[i(\mathsf{a})/x]} \mathbf{N}$$

and so $\mathcal{M}^1 \models_{j\varphi,[ji(\mathsf{a})/x]} \mathbf{N}$ by the inductive hypothesis. Since $ji(\mathsf{a}) = \mathsf{a}$, we know $\mathcal{M}^1 \models_{j\varphi,[\mathsf{a}/x]} \mathbf{N}$. Generalizing over a, we conclude

$$\mathcal{M}^1 \models_{j\varphi} [\dot{\forall} x_\alpha \, \mathbf{N}].$$

□

Proof of Lemma 3.5.4. Let i be the homomorphism from $\mathcal{M}^1$ to $\mathcal{M}^2$ and j be its inverse.

1. This follows directly from Lemma 3.5.3.
2. Suppose $\mathcal{M}^1$ satisfies property η. To verify $\mathcal{M}^2$ satisfies η, let $\mathbf{A} \in \mathit{wff}_\alpha(\mathcal{S})$ and an assignment φ into $\mathcal{M}^2$ be given. We compute

$$\mathcal{E}^2_\varphi(\mathbf{A}) = ij(\mathcal{E}^2_\varphi(\mathbf{A})) = i(\mathcal{E}^1_{j\varphi}(\mathbf{A})) = i(\mathcal{E}^1_{j\varphi}(\mathbf{A}^\downarrow))$$

$$= ij(\mathcal{E}^2_\varphi(\mathbf{A}^\downarrow)) = \mathcal{E}^2_\varphi(\mathbf{A}^\downarrow).$$

   Thus $\mathcal{M}^2$ satisfies property η.
3. A similar argument verifies property ξ for $\mathcal{M}^2$ if it holds for $\mathcal{M}^1$.

4. Suppose $\mathcal{M}^1$ satisfies property $\mathfrak{f}$ and we are given $\mathsf{f}, \mathsf{g} \in \mathcal{D}^2_{\alpha\beta}$ for types α and β. Suppose further that $\mathsf{f}@^2\mathsf{b} = \mathsf{g}@^2\mathsf{b}$ for every $\mathsf{b} \in \mathcal{D}^2_\beta$. It is enough to check $\jmath(\mathsf{f}) = \jmath(\mathsf{g})$. This follows from property $\mathfrak{f}$ in $\mathcal{M}^1$ if we verify $\jmath(\mathsf{f})@^1\mathsf{c} = \jmath(\mathsf{g})@^1\mathsf{c}$ for every $\mathsf{c} \in \mathcal{D}^1_\beta$. Let $\mathsf{c} \in \mathcal{D}^1_\beta$ be given. We finish the proof by computing

$$\jmath(\mathsf{f})@^1\mathsf{c} = \jmath(\mathsf{f})@^1\jmath\imath(\mathsf{c}) = \jmath(\mathsf{f}@^2\imath(\mathsf{c})) = \jmath(\mathsf{g}@^2\imath(\mathsf{c}))$$

$$= \jmath(\mathsf{g})@^1\jmath\imath(\mathsf{c}) = \jmath(\mathsf{g})@^1\mathsf{c}.$$

5. Finally, suppose $\mathcal{M}^1$ satisfies property $\mathfrak{b}$. Then $\upsilon^2 = \upsilon^1 \circ \jmath_o$ is injective since both υ^1 and $\jmath_o$ are injective. Thus $\mathcal{M}^2$ satisfies property $\mathfrak{b}$.

□

Proof of Theorem 3.5.5. We define $\mathcal{M}^h := (\mathcal{D}^h, @^h, \mathcal{E}^h, \upsilon^h)$ by defining its components.

We first define the domains $\mathcal{D}^h$ for $\mathcal{M}^h$ by induction on types. We simultaneously define functions $\imath_\alpha : \mathcal{D}_\alpha \to \mathcal{D}^h_\alpha$ and $\jmath_\alpha : \mathcal{D}^h_\alpha \to \mathcal{D}_\alpha$ which will witness that the two models are isomorphic. At each step of the definition, we check that $\imath_\alpha$ and $\jmath_\alpha$ are mutual inverses.

At the base type ι, let $\mathcal{D}^h_\iota := \mathcal{D}_\iota$ and let $\imath_\iota$ and $\jmath_\iota$ be the identity functions.

By property $\mathfrak{b}$, we know υ is injective. There are three cases.

1. If $\mathcal{D}_o$ has one element $\overline{\mathtt{T}}$ and $\upsilon(\overline{\mathtt{T}}) = \mathtt{T}$, then we let

$$\mathcal{D}^h_o := \{\mathtt{T}\} \subseteq \{\mathtt{T}, \mathtt{F}\}$$

and $\jmath_o : \mathcal{D}^h_o \to \mathcal{D}_o$ be defined by $\jmath_o(\mathtt{T}) := \overline{\mathtt{T}}$.

2. If $\mathcal{D}_o$ has one element $\overline{\mathtt{F}}$ and $\upsilon(\overline{\mathtt{F}}) = \mathtt{F}$, then we let

$$\mathcal{D}^h_o := \{\mathtt{F}\} \subseteq \{\mathtt{T}, \mathtt{F}\}$$

and $\jmath_o : \mathcal{D}^h_o \to \mathcal{D}_o$ be defined by $\jmath_o(\mathtt{F}) := \overline{\mathtt{F}}$.

3. Otherwise, $\mathcal{D}_o = \{\overline{\mathtt{T}}, \overline{\mathtt{F}}\}$ where $\upsilon(\overline{\mathtt{T}}) = \mathtt{T}$ and $\upsilon(\overline{\mathtt{F}}) = \mathtt{F}$. In this case, we let $\mathcal{D}^h_o := \{\mathtt{T}, \mathtt{F}\}$ and $\jmath_o : \mathcal{D}^h_o \to \mathcal{D}_o$ be defined by $\jmath_o(\mathtt{T}) := \overline{\mathtt{T}}$ and $\jmath_o(\mathtt{F}) := \overline{\mathtt{F}}$

Note that in any case we know $\upsilon : \mathcal{D}_o \to \{\mathtt{T}, \mathtt{F}\}$ is also a map

$$\upsilon : \mathcal{D}_o \to \mathcal{D}^h_o.$$

We let $\imath_o : \mathcal{D}_o \to \mathcal{D}^h_o$ be υ and υ^h be the inclusion map from $\mathcal{D}^h_o$ into $\{\mathtt{T}, \mathtt{F}\}$. Clearly, $\imath_o$ and $\jmath_o$ are mutually inverse.

Given two types α and β, we assume we have $\mathcal{D}^h_\alpha$, mutual inverses $i_\alpha : \mathcal{D}_\alpha \to \mathcal{D}^h_\alpha$ and $j_\alpha : \mathcal{D}^h_\alpha \to \mathcal{D}_\alpha$, as well as $\mathcal{D}^h_\beta$ and mutual inverses $i_\beta : \mathcal{D}_\beta \to \mathcal{D}^h_\beta$ and $j_\beta : \mathcal{D}^h_\beta \to \mathcal{D}_\beta$. We define

$$\mathcal{D}^h_{\alpha\beta} := \{f : \mathcal{D}^h_\beta \to \mathcal{D}^h_\alpha \mid \exists \mathsf{f} \in \mathcal{D}_{\alpha\beta} \blacksquare \forall b \in \mathcal{D}^h_\beta \blacksquare f(b) = i_\alpha(\mathsf{f}@j_\beta(b))\}.$$

Note that $\mathcal{D}^h_{\alpha\beta} \subseteq (\mathcal{D}^h_\alpha)^{\mathcal{D}^h_\beta}$. For the map $i_{\alpha\beta} : \mathcal{D}_{\alpha\beta} \to \mathcal{D}^h_{\alpha\beta}$, for each $\mathsf{f} \in \mathcal{D}_{\alpha\beta}$, we let $i_{\alpha\beta}(\mathsf{f})$ be the function mapping each $b \in \mathcal{D}^h_\beta$ to the value $i_\alpha(\mathsf{f}@j_\beta(b))$. This choice for $i_{\alpha\beta}(\mathsf{f})$ is clearly in $\mathcal{D}^h_{\alpha\beta}$ by definition. To define the inverse map $j_{\alpha\beta}$ from $\mathcal{D}^h_{\alpha\beta}$ to $\mathcal{D}_{\alpha\beta}$, we must use the fact that $\mathcal{M}$ is functional. Given any $f \in \mathcal{D}^h_{\alpha\beta}$, by definition there is some $\mathsf{f} \in \mathcal{D}_{\alpha\beta}$ such that $f(b) = i_\alpha(\mathsf{f}@j_\beta(b))$ for every $b \in \mathcal{D}^h_\beta$. By functionality and the fact that the i and j at types α and β are already inverses, this f is unique, since if $i_\alpha(\mathsf{f}@j_\beta(b)) = i_\alpha(\mathsf{g}@j_\beta(b))$ for every $b \in \mathcal{D}^h_\beta$, then $\mathsf{f}@j_\beta(i_\beta(\mathsf{b})) = \mathsf{g}@j_\beta(i_\beta(\mathsf{b}))$ for every $\mathsf{b} \in \mathcal{D}^h_\beta$. That is, $\mathsf{f}@\mathsf{b} = \mathsf{g}@\mathsf{b}$ for every $\mathsf{b} \in \mathcal{D}^h_\beta$. For every $f \in \mathcal{D}^h_{\alpha\beta}$, we define $j_{\alpha\beta}(f)$ to be the *unique* f such that $f(b) = i_\alpha(\mathsf{f}@j_\beta(b))$. It is easy to check that $i_{\alpha\beta}$ and $j_{\alpha\beta}$ are mutually inverse.

For the applicative structure $(\mathcal{D}^h, @^h)$ to be a frame structure, we are forced to define the application operator $@^h$ to be function application. That is, for every $f \in \mathcal{D}^h_{\alpha\beta}$ and $b \in \mathcal{D}^h_\beta$, $f@^h b := f(b)$. We will continue to write $f@^h b$ except when we are using the fact that this is function application. We define the evaluation function $\mathcal{E}^h$ simply by $\mathcal{E}^h_\varphi(\mathbf{A}) := i(\mathcal{E}_{j\varphi}(\mathbf{A}))$ for every $\mathbf{A} \in \mathit{wff}_\alpha(\mathcal{S})$ and assignment φ into the applicative structure $(\mathcal{D}^h, @^h)$.

Now that we have defined i_α and j_α at every type α, we will often omit the type as a subscript. For i and j to be homomorphisms, they will have to respect application. We check this now. Given any $\mathsf{f} \in \mathcal{D}_{\alpha\beta}$ and $\mathsf{b} \in \mathcal{D}_\beta$, by the definition of i at function types, we have $i(\mathsf{f})(i(\mathsf{b})) = i(\mathsf{f}@ji(\mathsf{b}))$. Hence $i(\mathsf{f})@^h i(\mathsf{b}) = i(\mathsf{f}@ji(\mathsf{b})) = i(\mathsf{f}@\mathsf{b})$, as desired. Given $f \in \mathcal{D}^h_{\alpha\beta}$ and $a \in \mathcal{D}^h_\beta$, by the definition of j at function types, we have $f(b) = i(j(f)@j(b))$. Applying j to both sides, we have $j(f@^h b) = j(f(b)) = j(f)@j(b)$, as desired.

We next check that $\mathcal{E}^h$ is a valid evaluation function.

1. $\mathcal{E}^h_\varphi(x_\alpha) = ij(\varphi(x)) = \varphi(x)$.

2. Since we have already checked that i respects application, we have

$$\mathcal{E}^h_\varphi(\mathbf{F}\,\mathbf{A}) = i(\mathcal{E}_{j\varphi}(\mathbf{F}\,\mathbf{A})) = i(\mathcal{E}_{j\varphi}(\mathbf{F})@\mathcal{E}_{j\varphi}(\mathbf{A}))$$
$$= i(\mathcal{E}_{j\varphi}(\mathbf{F}))@^h i(\mathcal{E}_{j\varphi}(\mathbf{A}))) = \mathcal{E}^h_\varphi(\mathbf{F})@^h\mathcal{E}^h_\varphi(\mathbf{A})$$

for any $\mathbf{F} \in \mathit{wff}_{\alpha\beta}$, $\mathbf{A} \in \mathit{wff}_\beta$ and assignment φ into $\mathcal{D}^h$.

3. Suppose $\mathbf{A} \in \mathit{wff}_\alpha(\mathcal{S})$ and φ and ψ are assignments into the applicative structure $(\mathcal{D}^h, @^h)$ which agree on $\mathbf{Free}(\mathbf{A})$. This implies $j\varphi$ and $j\psi$ are assignments into $\mathcal{M}$ which agree on $\mathbf{Free}(\mathbf{A})$. Thus

$$\mathcal{E}^h_\varphi(\mathbf{A}) = i(\mathcal{E}_{j\varphi}(\mathbf{A})) = i(\mathcal{E}_{j\psi}(\mathbf{A})) = \mathcal{E}^h_\psi(\mathbf{A}).$$

4. Finally, $\mathcal{E}^h_\varphi(\mathbf{A}) = i(\mathcal{E}_{j\varphi}(\mathbf{A})) = i(\mathcal{E}_{j\varphi}(\mathbf{A}^{\downarrow\beta})) = \mathcal{E}^h_\varphi(\mathbf{A}^{\downarrow\beta})$.

To check $\mathcal{M}^h$ is a model, we must verify the conditions for the (inclusion) valuation υ^h relative to $\mathcal{M}^h$. We freely use the fact that i and j respect application and $\upsilon = i_o$. Note also that for any logical constant $c \in \mathcal{S}$ we know $\mathcal{E}^h(c) = i(\mathcal{E}(c))$.

$\top$: $\upsilon^h(\mathcal{E}^h(\top)) = \mathcal{E}^h(\top) = i(\mathcal{E}(\top)) = \upsilon(\mathcal{E}(\top)) = \mathtt{T}$.

$\bot$: $\upsilon^h(\mathcal{E}^h(\bot)) = \mathcal{E}^h(\bot) = i(\mathcal{E}(\bot)) = \upsilon(\mathcal{E}(\bot)) = \mathtt{F}$.

$\neg$: For $\mathsf{a} \in \mathcal{D}^h_o$, we have

$$\upsilon^h(\mathcal{E}^h(\neg)@^h\mathsf{a}) = \mathtt{T} \text{ iff } i(\mathcal{E}(\neg)@j(\mathsf{a})) = \mathtt{T} \text{ iff } \upsilon(\mathcal{E}(\neg)@j(\mathsf{a})) = \mathtt{T}$$
$$\text{iff } \upsilon(j(\mathsf{a})) = \mathtt{F} \text{ iff } i(j(\mathsf{a})) = \mathtt{F} \text{ iff } \mathsf{a} = \mathtt{F} \text{ iff } \upsilon^h(\mathsf{a}) = \mathtt{F}.$$

$\vee$: For $\mathsf{a}, \mathsf{b} \in \mathcal{D}^h_o$, we have

$$\upsilon^h(\mathcal{E}^h(\vee)@^h\mathsf{a}@^h\mathsf{b}) = \mathtt{T} \text{ iff } i(\mathcal{E}(\vee)@j(\mathsf{a})@j(\mathsf{b})) = \mathtt{T}$$
$$\text{iff } \upsilon(\mathcal{E}(\vee)@j(\mathsf{a})@j(\mathsf{b})) = \mathtt{T} \text{ iff } \mathsf{a} = \upsilon(j(\mathsf{a})) = \mathtt{T} \text{ or } \mathsf{b} = \upsilon(j(\mathsf{b})) = \mathtt{T}.$$

$\wedge,\supset,\equiv$: These cases are analogous to $\vee$.

Π^α: Let $f \in \mathcal{D}^h_{o\alpha}$ be given. Suppose $\upsilon^h(\mathcal{E}^h(\Pi^\alpha)@f) = \mathtt{T}$. We compute

$$\upsilon^h(\mathcal{E}^h(\Pi^\alpha))@f = i(\mathcal{E}(\Pi^\alpha)@j(f)) = \upsilon(\mathcal{E}(\Pi^\alpha)@j(f)).$$

Hence $\upsilon(\mathcal{E}(\Pi^\alpha)@j(f)) = \mathtt{T}$. Since $\mathcal{M}$ is an $\mathcal{S}$-model, this implies $\upsilon(j(f)@\mathsf{a}) = \mathtt{T}$ for every $\mathsf{a} \in \mathcal{D}_\alpha$. We therefore have

$$\upsilon^h(f@^h a) = f@^h a = ij(f@^h a)) = \upsilon(j(f)@j(a)) = \mathtt{T}$$

for any $a \in \mathcal{D}^h_\alpha$.

For the converse, suppose $\upsilon^h(f@^h a) = \mathtt{T}$ for every $a \in \mathcal{D}^h_\alpha$. For every $\mathsf{a} \in \mathcal{D}_\alpha$, since $\upsilon^h(f@^h i(\mathsf{a})) = i(j(f)@\mathsf{a}) = \upsilon(j(f)@\mathsf{a})$, this implies $\upsilon(j(f)@\mathsf{a}) = \mathtt{T}$. Hence $\upsilon(\mathcal{E}(\Pi^\alpha)@j(f)) = \mathtt{T}$. Since

$$\mathcal{E}^h(\Pi^\alpha)@^h f = ij(i(\mathcal{E}(\Pi^\alpha))@^h f) = \upsilon(\mathcal{E}(\Pi^\alpha)@j(f)),$$

we have $\upsilon^h(\mathcal{E}^h(\Pi^\alpha)@^h f) = \mathtt{T}$ and we are done.

Σ^α: This case is analogous to Π^α.

$=^\alpha$: Let $a, b \in \mathcal{D}^h_\alpha$ be given. We have

$$\upsilon^h(\mathcal{E}^h(=^\alpha)@^h a@^h b) = \mathtt{T} \text{ iff } i(\mathcal{E}(=^\alpha)@j(a)@j(b)) = \mathtt{T}$$
$$\text{iff } \upsilon(\mathcal{E}(=^\alpha)@j(a)@j(b)) = \mathtt{T} \text{ iff } j(a) = j(b) \text{ iff } a = b.$$

Finally, we check that i and j are homomorphisms. We have already checked that application is preserved. For i to be a homomorphism, we need $i(\mathcal{E}_\varphi(\mathbf{A})) = \mathcal{E}^h_{i\varphi}(\mathbf{A})$. This follows from

$$\mathcal{E}^h_{i\varphi}(\mathbf{A}) = i(\mathcal{E}_{ji\varphi}(\mathbf{A})) = i(\mathcal{E}_\varphi(\mathbf{A})).$$

For j to be a homomorphism, we need $j(\mathcal{E}^h_\varphi(\mathbf{A})) = \mathcal{E}_{j\varphi}(\mathbf{A})$. This follows from

$$j(\mathcal{E}^h_\varphi(\mathbf{A})) = ji(\mathcal{E}_{j\varphi}(\mathbf{A})) = \mathcal{E}_{j\varphi}(\mathbf{A}).$$

The facts that $\upsilon(\mathsf{a}) = \upsilon^h(i(\mathsf{a}))$ for $\mathsf{a} \in \mathcal{D}_o$ and $\upsilon^h(a) = \upsilon(j(a))$ for $a \in \mathcal{D}^h_o$ follow directly from $\upsilon = i_o$, the fact that υ^h is the inclusion function, and the fact that i_o and j_o are mutually inverse. □

PROOF OF LEMMA 3.6.11. We prove this by induction on $\mathbf{A}$, using the definition of $\lambda^* x$ (cf. Definition 3.6.2). If $x \notin \mathbf{Free}(\mathbf{A})$, then

$$\mathcal{E}^\natural_\varphi([\lambda^* x_\beta(\mathbf{A})])@\mathsf{b} = \mathcal{E}^\natural_\varphi([\mathbf{K}\,\mathbf{A}])@\mathsf{b} = \mathcal{E}^\natural(\mathbf{K})@\mathcal{E}^\natural_\varphi(\mathbf{A})@\mathsf{b}$$
$$= \mathcal{E}^\natural_\varphi(\mathbf{A}) = \mathcal{E}^\natural_{\varphi,[\mathsf{b}/x]}(\mathbf{A}).$$

If $\mathbf{A}$ is x_β, then

$$\mathcal{E}^\natural_\varphi(\lambda^* x_\beta(x))@\mathsf{b} = \mathcal{E}^\natural_\varphi([\mathbf{S}\,\mathbf{K}\,\mathbf{K}])@\mathsf{b} = \mathcal{E}^\natural(\mathbf{S})@\mathcal{E}^\natural(\mathbf{K})@\mathcal{E}^\natural(\mathbf{K})@\mathsf{b}$$
$$= \mathcal{E}^\natural(\mathbf{K})@\mathsf{b}@(\mathcal{E}^\natural(\mathbf{K})@\mathsf{b}) = \mathsf{b} = \mathcal{E}^\natural_{\varphi,[\mathsf{b}/x]}(x).$$

If **A** is $[\mathbf{F}\,\mathbf{D}]$, then

$$\begin{aligned}
\mathcal{E}^{\natural}_{\varphi}(\lambda^* x_\beta([\mathbf{F}\,\mathbf{D}]))@\mathsf{b} &= \mathcal{E}^{\natural}_{\varphi}([\mathbf{S}\,\lambda^* x(\mathbf{F})\,\lambda^* x(\mathbf{D})])@\mathsf{b} \\
&= \mathcal{E}^{\natural}(\mathbf{S})\,@\,\mathcal{E}^{\natural}_{\varphi}(\lambda^* x(\mathbf{F}))\,@\,\mathcal{E}^{\natural}_{\varphi}(\lambda^* x(\mathbf{D}))\,@\,\mathsf{b} \\
&= (\mathcal{E}^{\natural}_{\varphi}(\lambda^* x(\mathbf{F}))@\mathsf{b})@\,(\mathcal{E}^{\natural}_{\varphi}(\lambda^* x(\mathbf{D}))@\mathsf{b}) \\
\text{(inductive hypothesis)} \quad &= \mathcal{E}^{\natural}_{\varphi,[\mathsf{b}/x]}(\mathbf{F})@\mathcal{E}^{\natural}_{\varphi,[\mathsf{b}/x]}(\mathbf{D}) \\
&= \mathcal{E}^{\natural}_{\varphi,[\mathsf{b}/x]}([\mathbf{F}\,\mathbf{D}]).
\end{aligned}$$

□

A.2. Proofs from Chapter 4

PROOF OF THEOREM 4.1.8. We define $\mathcal{E}$ by induction on terms. At each stage, we ensure that the first component of $\mathcal{E}_\varphi(\mathbf{A}_\alpha)$ is $\varphi_1^*(\mathbf{A})$, i.e., $\mathcal{E}_\varphi(\mathbf{A}) \in \mathcal{D}_\alpha^{\varphi_1(\mathbf{A})}$.

- For each constant c_α, we must let $\mathcal{E}_\varphi(c) := \mathcal{I}(c) \in \mathcal{D}_\alpha^c$.
- For each constant W_α, we must let $\mathcal{E}_\varphi(W) := \mathcal{I}(W) \in \mathcal{D}_\alpha^W$.
- For variables x_α, let $\mathcal{E}_\varphi(x) := \varphi(x) \in \mathcal{D}_\alpha^{\varphi_1(x)}$. This ensures that $\mathcal{E}$ satisfies the first condition to be an evaluation function.
- For application, we must define

  $$\mathcal{E}_\varphi([\mathbf{G}_{\alpha\beta}\,\mathbf{B}_\beta]) := \mathcal{E}_\varphi(\mathbf{G})@\mathcal{E}_\varphi(\mathbf{B}).$$

  This definition ensures the second condition for $\mathcal{E}$ to be an evaluation function. We also must verify

  $$\mathcal{E}_\varphi([\mathbf{G}\,\mathbf{B}]) \in \mathcal{D}_\alpha^{\varphi_1([\mathbf{G}\,\mathbf{B}])}.$$

  By the inductive hypothesis,

  $$\mathcal{E}_\varphi(\mathbf{G}) \in \mathcal{D}_{\alpha\beta}^{\varphi_1(\mathbf{G})} \text{ and } \mathcal{E}_\varphi(\mathbf{B}) \in \mathcal{D}_\beta^{\varphi_1(\mathbf{B})}.$$

  Hence $\mathcal{E}_\varphi(\mathbf{G}) = \langle \varphi_1^*(\mathbf{G}), g\rangle$ for some g which restricts to a function

  $$g : \mathcal{D}_\beta^{\varphi_1(\mathbf{B})} \to \mathcal{D}_\alpha^{[\varphi_1(\mathbf{G})\,\varphi_1(\mathbf{B})]}.$$

  We conclude

  $$\mathcal{E}_\varphi([\mathbf{G}\,\mathbf{B}]) = g(\langle \mathbf{B}, b\rangle) \in \mathcal{D}_\beta^{\varphi_1([\mathbf{G}\,\mathbf{B}])}$$

  since $\varphi_1([\mathbf{G}\,\mathbf{B}]) = [\varphi_1(\mathbf{G})\,\varphi_1(\mathbf{B})]$.

- For abstraction, suppose we have

$$\mathcal{E}_{\varphi,[\langle \mathbf{B},b\rangle/x]}(\mathbf{A}) \in \mathcal{D}_\alpha^{\varphi_1([\mathbf{B}/x]\mathbf{A})}$$

for each $\langle \mathbf{B}, b\rangle \in \mathcal{D}_\beta$. Using this assumption, we can define a function g from $\mathcal{D}_\beta$ to $\mathcal{D}_\alpha$ by $g(\langle \mathbf{B}, b\rangle) := \mathcal{E}_{\varphi,[\langle \mathbf{B},b\rangle/x]}(\mathbf{A})$. Note that g restricts to be a function

$$g : \mathcal{D}_\beta^{\varphi_1(\mathbf{B})} \to \mathcal{D}_\alpha^{\varphi_1([\mathbf{B}/x]\mathbf{A})}$$

for each $\mathbf{B}$. Let $\mathcal{E}_\varphi([\lambda x_\beta\, \mathbf{A}]) := \langle (\varphi_1([\lambda x_\beta\, \mathbf{A}]))^{\downarrow *}, g\rangle$. We have ensured

$$\mathcal{E}_\varphi([\lambda x\, \mathbf{A}]) \in \mathcal{D}_{\alpha\beta}^{\varphi_1([\lambda x\, \mathbf{A}])}$$

by choosing the first component to be $(\varphi_1([\lambda x_\beta\, \mathbf{A}]))^{\downarrow *}$.

To complete the verification that $\mathcal{E}$ is an evaluation function, we must check two more conditions.

- Suppose φ and ψ coincide on $\mathbf{Free}(\mathbf{A})$. An easy induction using the definition of $\mathcal{E}$ proves $\mathcal{E}_\varphi(\mathbf{A}) = \mathcal{E}_\psi(\mathbf{A})$.
- To verify $\mathcal{E}$ respects β-reduction, we check $\mathcal{E}$ respects a single reduction, then use induction on the number of reductions.

  We first verify $\mathcal{E}_{\varphi,[\mathcal{E}_\varphi(\mathbf{B})/x]}(\mathbf{A}) = \mathcal{E}_\varphi([\mathbf{B}/x]\mathbf{A})$ by induction on $\mathbf{A}$. If $\mathbf{A}$ is x, then $\mathcal{E}_{\varphi,[\mathcal{E}_\varphi(\mathbf{B})/x]}(x) = \mathcal{E}_\varphi(\mathbf{B}) = \mathcal{E}_\varphi([\mathbf{B}/x]x)$. If $\mathbf{A}$ is a constant, parameter or any variable other than x, then

$$\mathcal{E}_{\varphi,[\mathcal{E}_\varphi(\mathbf{B})/x]}(\mathbf{A}) = \mathcal{E}_\varphi(\mathbf{A}) = \mathcal{E}_\varphi([\mathbf{B}/x]\mathbf{A}).$$

  If $\mathbf{A}$ is an application $[\mathbf{F}\, \mathbf{C}]$, then the inductive hypothesis implies

$$\mathcal{E}_{\varphi,[\mathcal{E}_\varphi(\mathbf{B})/x]}(\mathbf{F}\, \mathbf{C}) = \mathcal{E}_{\varphi,[\mathcal{E}_\varphi(\mathbf{B})/x]}(\mathbf{F})@\mathcal{E}_{\varphi,[\mathcal{E}_\varphi(\mathbf{B})/x]}(\mathbf{C})$$

$$= \mathcal{E}_\varphi([\mathbf{B}/x]\mathbf{F})@\mathcal{E}_\varphi([\mathbf{B}/x]\mathbf{C}) = \mathcal{E}_\varphi([\mathbf{B}/x][\mathbf{F}\, \mathbf{C}]).$$

  If $\mathbf{A}$ is an abstraction $[\lambda y_\beta\, \mathbf{C}_\gamma]$, then we check that the first and second components are equal. Let θ be the substitution with domain $(\mathbf{Free}(\mathbf{A}) \cup \mathbf{Free}(\mathbf{B}))$ such that $\theta(z)$ is the first component of $\varphi(z)$ for each $z \in (\mathbf{Free}(\mathbf{A}) \cup \mathbf{Free}(\mathbf{B}))$. Note that θ is a ground substitution. The first component of $\mathcal{E}_\varphi([\mathbf{B}/x]\mathbf{A})$ is $(\theta([[\mathbf{B}/x]\mathbf{A}]))^{\downarrow *}$. The first component of $\mathcal{E}_{\varphi,[\mathcal{E}_\varphi(\mathbf{B})/x]}(\mathbf{A})$ is $((\theta, [\theta(\mathbf{B})/x])(\mathbf{A}))^{\downarrow *}$. Since θ is a ground substitution, $\theta(\mathbf{B})$ is closed and so we can conclude $\theta([[\mathbf{B}/x]\mathbf{A}]) = \theta([\theta(\mathbf{B})/x]\mathbf{A})$.

Hence the first components are equal. The second component of $\mathcal{E}_{\varphi,[\mathcal{E}_\varphi(\mathbf{B})/x]}(\mathbf{A})$ is the function $g : \mathcal{D}_\beta \to \mathcal{D}_\gamma$ such that $g(\mathsf{b}) = \mathcal{E}_{\varphi,[\mathcal{E}_\varphi(\mathbf{B})/x],[\mathsf{b}/y]}(\mathbf{C})$ for every $\mathsf{b} \in \mathcal{D}_\beta$. The second component of $\mathcal{E}_\varphi([\mathbf{B}/x]\mathbf{A})$ is the function $h : \mathcal{D}_\beta \to \mathcal{D}_\gamma$ such that $h(\mathsf{b}) = \mathcal{E}_{\varphi,[\mathsf{b}/y]}([\mathbf{B}/x]\mathbf{C})$ for every $\mathsf{b} \in \mathcal{D}_\beta$. The inductive hypothesis implies $g(\mathsf{b}) = h(\mathsf{b})$ for every $\mathsf{b} \in \mathcal{D}_\beta$. That is, $g = h$. Thus the second components are also equal and we are done.

Now, using the definition of $\mathcal{E}$ on applications and abstractions, we have

$$\begin{aligned}&\mathcal{E}_\varphi([\lambda x\, \mathbf{A}]\,\mathbf{B}) = \mathcal{E}_\varphi(\lambda x \mathbf{A})@\mathcal{E}_\varphi(\mathbf{B})\\ &= \mathcal{E}_{\varphi,[\mathcal{E}_\varphi(\mathbf{B})/x]}(\mathbf{A}) = \mathcal{E}_\varphi([\mathbf{B}/x]\mathbf{A}).\end{aligned}$$

Next, if $\mathbf{C}$ β-reduces to $\mathbf{D}$ in a single step, then induction on the position of the redex in $\mathbf{C}$ proves $\mathcal{E}_\varphi(\mathbf{C}) = \mathcal{E}_\varphi(\mathbf{D})$. Finally, induction on the number of β-reduction steps proves $\mathcal{E}_\varphi(\mathbf{C}) = \mathcal{E}_\varphi(\mathbf{C}^{\downarrow\beta})$.

Assume $\downarrow^*$ is $\beta\eta$-normalization. We prove η-functionality of the evaluation $(\mathcal{D}, @, \mathcal{E})$. We first verify $\mathcal{E}_\varphi([\lambda x_\beta \,\centerdot\, \mathbf{A}_{\alpha\beta}\, x]) = \mathcal{E}_\varphi(\mathbf{A})$ under the assumption that $x \notin \mathbf{Free}(\mathbf{A})$. Let θ be the substitution with domain $\mathbf{Free}(\mathbf{A})$ where $\theta(z)$ is the first component of $\varphi(z)$ for each $z \in \mathbf{Free}(\mathbf{A})$. The first components of $\mathcal{E}_\varphi([\lambda x_\beta \,\centerdot\, \mathbf{A}_{\alpha\beta}\, x])$ and $\mathcal{E}_\varphi(\mathbf{A})$ are equal since

$$\theta([\lambda x \,\centerdot\, \mathbf{A}\, x])^{\downarrow} = \theta([\lambda x \,\centerdot\, \mathbf{A}\, x]^{\downarrow})^{\downarrow} = \theta(\mathbf{A}^{\downarrow})^{\downarrow} = \theta(\mathbf{A})^{\downarrow}.$$

Let $g : \mathcal{D}_\beta \to \mathcal{D}_\alpha$ be the second component of $\mathcal{E}_\varphi([\lambda x_\beta \,\centerdot\, \mathbf{A}\, x])$ and let $h : \mathcal{D}_\beta \to \mathcal{D}_\alpha$ be the second component of $\mathcal{E}_\varphi(\mathbf{A})$. By the definition of $\mathcal{E}$, we know

$$g(\mathsf{b}) = \mathcal{E}_{\varphi,[\mathsf{b}/x]}([\mathbf{A}\, x]) = \mathcal{E}_{\varphi,[\mathsf{b}/x]}(\mathbf{A})@\mathcal{E}_{\varphi,[\mathsf{a}/b]}(x) = \mathcal{E}_\varphi(\mathbf{A})@\mathsf{b} = h(\mathsf{b})$$

for each $\mathsf{b} \in \mathcal{D}_\beta$. Thus $g = h$.

Now, just as in the β-reduction case, we can prove $\mathcal{E}_\varphi(\mathbf{C}) = \mathcal{E}_\varphi(\mathbf{D})$ whenever $\mathbf{C}$ η-reduces to $\mathbf{D}$ in one step by induction on the position of the η-redex in $\mathbf{C}$. Then we have $\mathcal{E}_\varphi(\mathbf{A}) = \mathcal{E}_\varphi(\mathbf{A}^{\downarrow})$ by induction on the number of $\beta\eta$-reductions. □

PROOF OF LEMMA 4.2.7. We must verify $\sim$ is symmetric and transitive on each $\mathcal{D}_\alpha$. We can prove this by induction on α.

$\alpha \in \{o, \iota\}$: This is immediate since $\sim$ agrees with $\sim_\alpha$ which was assumed to be a per.

$\alpha = (\gamma\beta)$: Suppose $\mathsf{g} \sim \mathsf{h}$ in $\mathcal{D}_{\gamma\beta}$. Let $\mathsf{a} \sim \mathsf{b}$ in $\mathcal{D}_\beta$ be given. By the inductive hypothesis, $\sim$ is symmetric on $\mathcal{D}_\beta$, so $\mathsf{b} \sim \mathsf{a}$. By the definition of $\sim$ on type $\gamma\beta$, $\mathsf{g@b} \sim \mathsf{h@a}$. We know $\mathsf{h@a} \sim \mathsf{g@b}$ since $\sim$ is symmetric on $\mathcal{D}_\gamma$ by the inductive hypothesis. Generalizing on a and b, we have $\mathsf{h} \sim \mathsf{g}$, verifying symmetry.

For transitivity, suppose $\mathsf{g} \sim \mathsf{h}$ and $\mathsf{h} \sim \mathsf{k}$ in $\mathcal{D}_{\gamma\beta}$. Let $\mathsf{a} \sim \mathsf{b}$ in $\mathcal{D}_\beta$ be given. We know $\mathsf{a} \sim \mathsf{b} \sim \mathsf{a}$ since $\sim$ is symmetric and transitive on $\mathcal{D}_\beta$ by the inductive hypothesis. Thus $\mathsf{a} \sim \mathsf{a}$ and so $\mathsf{g@a} \sim \mathsf{h@a} \sim \mathsf{k@b}$. We conclude $\mathsf{g@a} \sim \mathsf{k@b}$ since $\sim$ is transitive on $\mathcal{D}_\gamma$ by the inductive hypothesis. Generalizing on a and b, we have $\mathsf{g} \sim \mathsf{k}$, verifying transitivity.

It is immediate by the definition at function types that $\partial^@$ and $\partial^\mathfrak{f}$ will hold for $\sim$ over $(\mathcal{D}, @)$ if $\sim$ is the functional per extension of any pers on $\mathcal{D}_o$ and $\mathcal{D}_\iota$. □

Proof of Theorem 4.2.11. First, we know $\partial^{\neq\emptyset}$ holds since $\mathcal{E}(W_\alpha) \sim \mathcal{E}(W_\alpha)$ holds for any parameter W_α. By $\partial^{\neq\emptyset}$ and $\partial^@$, we can let $(\mathcal{D}^\sim, @^\sim)$ be the $\sim$ per quotient structure of $(\mathcal{D}, @)$ (cf. Definition 4.2.3 and Lemma 4.2.4). It is helpful to choose representatives of the equivalence classes $\mathsf{A} \in \mathcal{D}^\sim_\alpha$. For each $\mathsf{A} \in \mathcal{D}^\sim_\alpha$, choose some representative $\mathsf{A}^* \in \mathsf{A}$. Note that $\mathsf{A}^* \in \overline{\mathcal{D}^\sim_\alpha}$ and $\mathsf{A} = [\![\mathsf{A}^*]\!]_\sim$ for any $\mathsf{A} \in \mathcal{D}^\sim_\alpha$. It follows that if $\mathsf{A}, \mathsf{B} \in \mathcal{D}^\sim_\alpha$ and $\mathsf{A}^* \sim \mathsf{B}^*$, then

$$\mathsf{A} = [\![\mathsf{A}^*]\!]_\sim = [\![\mathsf{B}^*]\!]_\sim = \mathsf{B}.$$

Also, for any $\mathsf{a} \in \overline{\mathcal{D}^\sim_\alpha}$, $\mathsf{a} \in [\![\mathsf{a}]\!]_\sim$ and $[\![\mathsf{a}]\!]^*_\sim \in [\![\mathsf{a}]\!]_\sim$, so we have $\mathsf{a} \sim [\![\mathsf{a}]\!]^*_\sim$. For each assignment ψ into $\mathcal{D}^\sim$, we can define ψ^* to be an assignment taking variables x_α to $\overline{\mathcal{D}^\sim_\alpha}$ by $\psi^*(x) := \psi(x)^*$. Note that for any $\sim$-assignment φ into $\mathcal{D}$, $\varphi^{\sim *} \sim \varphi$ since $\varphi^{\sim *}(x) = [\![\varphi(x)]\!]^*_\sim \sim \varphi(x)$ for each variable x.

We know $@^\sim$ satisfies

$$[\![\mathsf{g}]\!]_\sim @^\sim [\![\mathsf{b}]\!]_\sim = [\![\mathsf{g@b}]\!]_\sim$$

for each $\mathsf{g} \in \overline{\mathcal{D}^\sim_{\alpha\beta}}$ and $\mathsf{b} \in \overline{\mathcal{D}^\sim_\beta}$ by Definition 4.2.3. Hence

$$\mathsf{G} @^\sim \mathsf{B} = [\![\mathsf{G}^*]\!]_\sim @^\sim [\![\mathsf{B}^*]\!]_\sim = [\![\mathsf{G}^* @ \mathsf{B}^*]\!]_\sim$$

for $\mathsf{G} \in \mathcal{D}^\sim_{\alpha\beta}$ and $\mathsf{B} \in \mathcal{D}^\sim_\beta$.

By Lemma 4.2.8, ∂^{sub} holds. Hence $\mathcal{E}_{\psi^*}(\mathbf{A}_\alpha) \in \overline{\mathcal{D}_\alpha^\sim}$ for each $\mathbf{A} \in \mathit{wff}_\alpha(\mathcal{S})$ and assignment ψ into $\mathcal{A}^\sim$ since $\psi^* \sim \psi^*$. Thus we can define

$$\mathcal{E}_\psi^\sim(\mathbf{A}) := [\![\mathcal{E}_{\psi^*}(\mathbf{A})]\!]_\sim.$$

For any $\sim$-assignment φ into $\mathcal{D}$ we have

$$\mathcal{E}_{\varphi^\sim}^\sim(\mathbf{A}) = [\![\mathcal{E}_{\varphi^{\sim *}}(\mathbf{A})]\!]_\sim = [\![\mathcal{E}_\varphi(\mathbf{A})]\!]_\sim$$

by ∂^{sub} since $\varphi^{\sim *} \sim \varphi$. Thus $\mathcal{E}_\varphi(\mathbf{A}) \in \mathcal{E}_{\varphi^\sim}^\sim(\mathbf{A})$.

We check $\mathcal{E}^\sim$ is an evaluation function.

1. For each variable x_α, $\mathcal{E}_\psi^\sim(x) = [\![\psi^*(x)]\!]_\sim = \psi(x)$.
2. $\mathcal{E}^\sim$ preserves application since

$$\begin{aligned}\mathcal{E}_\psi^\sim([\mathbf{G}\,\mathbf{C}]) &= [\![\mathcal{E}_{\psi^*}([\mathbf{G}\,\mathbf{C}])]\!]_\sim = [\![\mathcal{E}_{\psi^*}(\mathbf{G})@\mathcal{E}_{\psi^*}(\mathbf{C})]\!]_\sim \\ &= [\![\mathcal{E}_{\psi^*}(\mathbf{G})]\!]_\sim @^\sim [\![\mathcal{E}_{\psi^*}(\mathbf{C})]\!]_\sim = \mathcal{E}_\psi^\sim(\mathbf{G}) @^\sim \mathcal{E}_\psi^\sim(\mathbf{C}).\end{aligned}$$

3. If φ and ψ coincide on the free variables of $\mathbf{A}$, then φ^* and ψ^* also coincide on these variables. Hence $\mathcal{E}_{\varphi^*}(\mathbf{A}) = \mathcal{E}_{\psi^*}(\mathbf{A})$ since $\mathcal{E}$ is an evaluation function. Thus $\mathcal{E}_\varphi^\sim(\mathbf{A}) = \mathcal{E}_\psi^\sim(\mathbf{A})$.
4. Let ψ be an assignment and $\mathbf{A} \in \mathit{wff}_\alpha(\mathcal{S})$. Since $\mathcal{E}$ is an evaluation function, we have $\mathcal{E}_{\psi^*}(\mathbf{A}) = \mathcal{E}_{\psi^*}(\mathbf{A}^{\downarrow\beta})$. By taking equivalence classes we obtain $\mathcal{E}_\psi^\sim(\mathbf{A}) = \mathcal{E}_\psi^\sim(\mathbf{A}^{\downarrow\beta})$, as desired.

Thus $\mathcal{E}^\sim$ is an evaluation function.

We define $\upsilon^\sim : \mathcal{D}_o^\sim \to \{\mathsf{T}, \mathsf{F}\}$ by

$$\upsilon^\sim(\mathsf{A}) := \upsilon(\mathsf{A}^*)$$

for $\mathsf{A} \in \mathcal{D}_o^\sim$. For each $\mathsf{a} \in \overline{\mathcal{D}_o^\sim}$, by ∂^υ we have

$$\upsilon^\sim([\![\mathsf{a}]\!]_\sim) = \upsilon([\![\mathsf{a}]\!]_\sim^*) = \upsilon(\mathsf{a}).$$

For any φ be a $\sim$-assignment into $\mathcal{D}$, we know $\mathcal{E}_\varphi(\mathbf{A}) \in \mathcal{E}_{\varphi^\sim}^\sim(\mathbf{A})$ and so

$$\upsilon^\sim(\mathcal{E}_{\varphi^\sim}^\sim(\mathbf{A})) = \upsilon(\mathcal{E}_\varphi(\mathbf{A})).$$

Let $\mathcal{M}$ be $(\mathcal{D}^\sim, @^\sim, \mathcal{E}^\sim, \upsilon^\sim)$. To check that $\mathcal{M}$ is a model, we verify $\mathfrak{L}_c(\mathcal{E}^\sim(c))$ (cf. Table 3.1) with respect to $\upsilon^\sim$ for each $c \in \mathcal{S}$. In each case we will use the fact that

$$\upsilon^\sim(\mathcal{E}^\sim(c)@^\sim\mathsf{A}_1@^\sim \cdots @^\sim\mathsf{A}_n) = \upsilon(\mathcal{E}(c)@\mathsf{A}_1^*@\cdots@\mathsf{A}_n^*)$$

for any $c_{o\alpha^n\cdots\alpha^1} \in \mathcal{S}$ and $\mathsf{A}_i \in \mathcal{D}^{\sim}_{\alpha^i}$ for each i $(1 \leq i \leq n)$. This equation follows from

$$\begin{aligned} \mathcal{E}^{\sim}(c)@^{\sim}\mathsf{A}_1@^{\sim}\cdots@^{\sim}\mathsf{A}_n &= [\![\mathcal{E}(c)]\!]_{\sim}@^{\sim}[\![\mathsf{A}_1^*]\!]_{\sim}@^{\sim}\cdots@^{\sim}[\![\mathsf{A}_n^*]\!]_{\sim} \\ &= [\![\mathcal{E}(c)@\mathsf{A}_1^*@\cdots@\mathsf{A}_n^*]\!]_{\sim}. \end{aligned}$$

$\top$: By $\mathfrak{L}^{\partial}_{\top}(\mathcal{E}(\top))$, $\upsilon^{\sim}(\mathcal{E}^{\sim}(\top)) = \upsilon(\mathcal{E}(\top)) = \mathtt{T}$.
$\bot$: By $\mathfrak{L}^{\partial}_{\bot}(\mathcal{E}(\bot))$, $\upsilon^{\sim}(\mathcal{E}^{\sim}(\bot)) = \upsilon(\mathcal{E}(\bot)) = \mathtt{F}$.
$\neg$: Let $\mathsf{A} \in \mathcal{D}^{\sim}_o$ be given. Using $\mathfrak{L}^{\partial}_{\neg}(\mathcal{E}(\neg))$ we know

$$\begin{aligned} \upsilon^{\sim}(\mathcal{E}^{\sim}(\neg)@^{\sim}\mathsf{A}) = \mathtt{T} \quad &\text{iff} \quad \upsilon(\mathcal{E}(\neg)@\mathsf{A}^*) = \mathtt{T} \\ &\text{iff} \quad \upsilon(\mathsf{A}^*) = \mathtt{F} \\ &\text{iff} \quad \upsilon^{\sim}(\mathsf{A}) = \mathtt{F}. \end{aligned}$$

$\vee$: Let $\mathsf{A}, \mathsf{B} \in \mathcal{D}^{\sim}_o$ be given. Using $\mathfrak{L}^{\partial}_{\vee}(\mathcal{E}(\vee))$ we have

$$\begin{aligned} \upsilon^{\sim}(\mathcal{E}^{\sim}(\vee)@^{\sim}\mathsf{A}@^{\sim}\mathsf{B}) = \mathtt{T} \quad &\text{iff} \quad \upsilon(\mathcal{E}(\vee)@\mathsf{A}^*@\mathsf{B}^*) = \mathtt{T} \\ &\text{iff} \quad \upsilon(\mathsf{A}^*) = \mathtt{T} \text{ or } \upsilon(\mathsf{B}^*) = \mathtt{T} \\ &\text{iff} \quad \upsilon^{\sim}(\mathsf{A}) = \mathtt{T} \text{ or } \upsilon^{\sim}(\mathsf{B}) = \mathtt{T}. \end{aligned}$$

$\wedge,\supset,\equiv$: These cases are analogous to $\vee$.
Π^α: Let $\mathsf{P} \in \mathcal{D}^{\sim}_{o\alpha}$ be given. Using $\mathfrak{L}^{\partial}_{\Pi^\alpha}(\mathcal{E}(\Pi^\alpha))$ we have

$$\begin{aligned} \upsilon^{\sim}(\mathcal{E}^{\sim}(\Pi^\alpha)@^{\sim}\mathsf{P}) = \mathtt{T} \quad &\text{iff} \quad \upsilon(\mathcal{E}(\Pi^\alpha)@\mathsf{P}^*) = \mathtt{T} \\ &\text{iff} \quad \upsilon(\mathsf{P}^*@\mathsf{a}) = \mathtt{T} \text{ for each } \mathsf{a} \in \overline{\mathcal{D}^{\sim}_\alpha} \\ &\text{iff} \quad \upsilon^{\sim}(\mathsf{P}@^{\sim}\mathsf{A}) = \mathtt{T} \text{ for each } \mathsf{A} \in \mathcal{D}^{\sim}_\alpha. \end{aligned}$$

This last equivalence is true since $\upsilon(\mathsf{P}^*@\mathsf{a}) = \upsilon^{\sim}(\mathsf{P}@^{\sim}[\![\mathsf{a}]\!]_{\sim})$ for every $\mathsf{a} \in \overline{\mathcal{D}^{\sim}_\alpha}$ and $\upsilon(\mathsf{P}^*@\mathsf{A}^*) = \upsilon^{\sim}(\mathsf{P}@^{\sim}\mathsf{A})$ for every $\mathsf{A} \in \mathcal{D}^{\sim}_\alpha$.
Σ^α: This case is analogous to Π^α.
$=^\alpha$: Let $\mathsf{A}, \mathsf{B} \in \mathcal{D}^{\sim}_\alpha$ be given. Using $\mathfrak{L}^{\partial}_{=^\alpha}(\mathcal{E}(=^\alpha))$ we have

$$\begin{aligned} \upsilon^{\sim}(\mathcal{E}^{\sim}(=^\alpha)@^{\sim}\mathsf{A}@^{\sim}\mathsf{B}) = \mathtt{T} \quad &\text{iff} \quad \upsilon(\mathcal{E}(=^\alpha)@\mathsf{A}^*@\mathsf{B}^*) = \mathtt{T} \\ &\text{iff} \quad \mathsf{A}^* \sim \mathsf{B}^* \\ &\text{iff} \quad \mathsf{A} = \mathsf{B}. \end{aligned}$$

Thus we have the desired model $\mathcal{M}$.

We next check that $\mathcal{M}$ satisfies property $\mathfrak{f}$. Let $\mathsf{G}, \mathsf{H} \in \mathcal{D}^{\sim}_{\beta\alpha}$ be such that for every $\mathsf{A} \in \mathcal{D}^{\sim}_\alpha$ we have $\mathsf{G}@^{\sim}\mathsf{A} = \mathsf{H}@^{\sim}\mathsf{A}$. This implies $\mathsf{G}^*@\mathsf{a} \sim \mathsf{H}^*@\mathsf{a}$ for every $\mathsf{a} \in \overline{\mathcal{D}^{\sim}_\alpha}$. If we take any $\mathsf{a}, \mathsf{b} \in \mathcal{D}_\alpha$ with $\mathsf{a} \sim \mathsf{b}$, then we have $\mathsf{G}^*@\mathsf{a} \sim \mathsf{H}^*@\mathsf{a} \sim \mathsf{H}^*@\mathsf{b}$ by $\partial^@$ since $\mathsf{H}^* \in \overline{\mathcal{D}^{\sim}_{\beta\alpha}}$. By $\partial^{\mathfrak{f}}$, we have $\mathsf{G}^* \sim \mathsf{H}^*$. Hence $\mathsf{G} = \mathsf{H}$, as desired.

To check $\mathcal{M}$ satisfies property $\mathfrak{b}$, suppose $\mathsf{A}, \mathsf{B} \in \mathcal{E}^{\sim}_{o}$ and $\upsilon^{\sim}(\mathsf{A}) = \upsilon^{\sim}(\mathsf{B})$. That is, $\upsilon(\mathsf{A}^*) = \upsilon(\mathsf{B}^*)$. By $\partial^{\mathfrak{b}}$, we know $\mathsf{A}^* \sim \mathsf{B}^*$ and so $\mathsf{A} = \mathsf{B}$. Hence $\upsilon^{\sim}$ is injective and $\mathcal{M}$ satisfies property $\mathfrak{b}$.

Finally, we prove by induction on $\mathbf{M} \in prop(\mathcal{S})$ that $\mathcal{M} \models_{\varphi^{\sim}} \mathbf{M}$ iff $\mathscr{P} \models\!\!\approx_{\varphi} \mathbf{M}$ for all $\sim$-assignments φ.

- Suppose $\mathbf{M} \in \mathit{wff}_o(\mathcal{S})$. For any $\sim$-assignment φ,

$$\begin{aligned} \mathcal{M} \models_{\varphi^{\sim}} \mathbf{M} \quad &\text{iff} \quad \upsilon^{\sim}(\mathcal{E}^{\sim}_{\varphi^{\sim}}(\mathbf{M})) = \mathtt{T} \\ &\text{iff} \quad \upsilon(\mathcal{E}_{\varphi}(\mathbf{M})) = \mathtt{T} \\ &\text{iff} \quad \mathscr{P} \models\!\!\approx_{\varphi} \mathbf{M}. \end{aligned}$$

- Suppose $\mathbf{M}$ is of the form $[\mathbf{A}_\alpha \doteq^{\alpha} \mathbf{B}_\alpha]$. For any $\sim$-assignment φ,

$$\begin{aligned} \mathcal{M} \models_{\varphi^{\sim}} [\mathbf{A}_\alpha \doteq^{\alpha} \mathbf{B}_\alpha] \quad &\text{iff} \quad \mathcal{E}^{\sim}_{\varphi^{\sim}}(\mathbf{A}) = \mathcal{E}^{\sim}_{\varphi^{\sim}}(\mathbf{B}) \\ &\text{iff} \quad \mathcal{E}_{\varphi}(\mathbf{A}) \sim \mathcal{E}_{\varphi}(\mathbf{B}) \\ &\text{iff} \quad \mathscr{P} \models\!\!\approx_{\varphi} [\mathbf{A}_\alpha \doteq^{\alpha} \mathbf{B}_\alpha]. \end{aligned}$$

- Suppose $\mathbf{M}$ is $\dot{\top}$. The equivalence is trivial since $\mathcal{M} \models_{\varphi^{\sim}} \dot{\top}$ and $\mathscr{P} \models\!\!\approx_{\varphi} \dot{\top}$ for any $\sim$-assignment φ,
- Suppose $\mathbf{M}$ is of the form $[\dot{\neg}\, \mathbf{N}]$. For any $\sim$-assignment φ,

$$\begin{aligned} \mathcal{M} \models_{\varphi^{\sim}} [\dot{\neg}\, \mathbf{N}] \quad &\text{iff} \quad \mathcal{M} \not\models_{\varphi^{\sim}} \mathbf{N} \\ &\text{iff} \quad \mathscr{P} \not\models\!\!\approx_{\varphi} \mathbf{N} \\ &\text{iff} \quad \mathscr{P} \models\!\!\approx_{\varphi} [\dot{\neg}\, \mathbf{N}]. \end{aligned}$$

- Suppose $\mathbf{M}$ is of the form $[\mathbf{N} \dot{\vee} \mathbf{P}]$. For any $\sim$-assignment φ,

$$\begin{aligned} \mathcal{M} \models_{\varphi^{\sim}} [\mathbf{N} \dot{\vee} \mathbf{P}] \quad &\text{iff} \quad \mathcal{M} \models_{\varphi^{\sim}} \mathbf{N} \text{ or } \mathcal{M} \models_{\varphi^{\sim}} \mathbf{P} \\ &\text{iff} \quad \mathscr{P} \models\!\!\approx_{\varphi} \mathbf{N} \text{ or } \mathscr{P} \models\!\!\approx_{\varphi} \mathbf{P} \\ &\text{iff} \quad \mathscr{P} \models\!\!\approx_{\varphi} [\mathbf{N} \dot{\vee} \mathbf{P}]. \end{aligned}$$

- Suppose $\mathbf{M}$ is of the form $[\dot{\forall} x_\alpha \mathbf{N}]$. Let φ be a $\sim$-assignment. Suppose $\mathcal{M} \models_{\varphi^{\sim}} [\dot{\forall} x_\alpha \mathbf{N}]$. Then $\mathcal{M} \models_{\varphi^{\sim}, [\mathsf{A}/x]} \mathbf{N}$ for every $\mathsf{A} \in \mathcal{D}^{\sim}_{\alpha}$. Let $\mathsf{a} \in \overline{\mathcal{D}^{\sim}_{\alpha}}$ be given. Note that

$$(\varphi, [\mathsf{a}/x])^{\sim} \text{ is } \varphi^{\sim}, [\mathsf{A}/x]$$

where A is $[\![\mathsf{a}]\!]_\sim$. By the inductive hypothesis, $\mathscr{P} \models\!\!\!\approx_{\varphi,[\mathsf{a}/x]}\mathbf{N}$. Generalizing over a, we know $\mathscr{P} \models\!\!\!\approx_{\varphi}[\dot{\forall} x_\alpha\, \mathbf{N}]$.

Suppose $\mathscr{P} \models\!\!\!\approx_{\varphi}[\dot{\forall} x_\alpha\, \mathbf{N}]$. Then for every $\mathsf{a} \in \overline{\mathcal{D}_\alpha^\sim}$,

$$\mathscr{P} \models\!\!\!\approx_{\varphi,[\mathsf{a}/x]}\mathbf{N}.$$

Hence $\mathscr{P} \models\!\!\!\approx_{\varphi,[\mathsf{A}^*/x]}\mathbf{N}$ for any $\mathsf{A} \in \overline{\mathcal{D}_\alpha^\sim}$. By the inductive hypothesis, we know $\mathcal{M} \models_{\varphi^\sim,[\mathsf{A}/x]} \mathbf{N}$. Thus $\mathcal{M} \models_{\varphi^\sim} [\dot{\forall} x_\alpha\, \mathbf{N}]$.

In particular, if $\mathcal{M} \models \mathbf{M}$, then $\mathcal{M} \models_{\varphi^\sim} \mathbf{M}$ and so $\mathscr{P} \models\!\!\!\approx_{\varphi}\mathbf{M}$ for every $\sim$-assignment φ. On the other hand, if $\mathscr{P} \models\!\!\!\approx\mathbf{M}$, then for any assignment ψ into $\mathcal{M}$, $\mathscr{P} \models\!\!\!\approx_{\psi^*}\mathbf{M}$ and so $\mathcal{M} \models_\psi \mathbf{M}$ (since $(\psi^*)^\sim = \psi$). Therefore, $\mathcal{M} \models \mathbf{M}$ iff $\mathscr{P} \models\!\!\!\approx\mathbf{M}$ for every $\mathcal{S}$-proposition $\mathbf{M}$. □

Proof of Lemma 4.3.6. We prove this in three steps.

1. We first prove by induction on $\mathbf{C}$ that

$$\mathcal{E}_{\varphi,[\mathcal{E}_\varphi(\mathbf{D})/x]}(\mathbf{C}) = \mathcal{E}_\varphi([\mathbf{D}/x]\mathbf{C}).$$

If $\mathbf{C}$ is x, then $\mathcal{E}_{\varphi,[\mathcal{E}_\varphi(\mathbf{D})/x]}(x) = \mathcal{E}_\varphi(\mathbf{D}) = \mathcal{E}_\varphi([\mathbf{D}/x]x)$. If $x \notin \mathbf{C}$, then $\mathcal{E}_{\varphi,[\mathcal{E}_\varphi(\mathbf{D})/x]}(\mathbf{C}) = \mathcal{E}_\varphi(\mathbf{C}) = \mathcal{E}_\varphi([\mathbf{D}/x]\mathbf{C})$. If $\mathbf{C}$ is an application of the form $[\mathbf{F}\,\mathbf{U}]$, then

$$\begin{aligned}\mathcal{E}_{\varphi,[\mathcal{E}_\varphi(\mathbf{D})/x]}([\mathbf{F}\,\mathbf{U}]) &= \mathcal{E}_{\varphi,[\mathcal{E}_\varphi(\mathbf{D})/x]}(\mathbf{F})@\mathcal{E}_{\varphi,[\mathcal{E}_\varphi(\mathbf{D})/x]}(\mathbf{U})\\ &= \mathcal{E}_\varphi([\mathbf{D}/x]\mathbf{F})@\mathcal{E}_\varphi([\mathbf{D}/x]\mathbf{U})\\ &= \mathcal{E}_\varphi([[\mathbf{D}/x]\mathbf{F}\,[\mathbf{D}/x]\mathbf{U}]) = \mathcal{E}_\varphi([\mathbf{D}/x][\mathbf{F}\,\mathbf{U}]).\end{aligned}$$

by the inductive hypothesis and the definition of $\mathcal{E}$.

Suppose $\mathbf{C}$ is a λ-abstraction $[\lambda y_\gamma\, \mathbf{U}_\delta]$ (where x and y are distinct). By the definition of $\mathcal{E}$, we know

$$\mathcal{E}_{\varphi,[\mathcal{E}_\varphi(\mathbf{D})/x]}([\lambda y\, \mathbf{U}]) = i(f) \text{ and } \mathcal{E}_\varphi([\lambda y\, [\mathbf{D}/x]\mathbf{U}]) = i(g)$$

where $f, g : \mathcal{D}_\gamma \to \mathcal{D}_\delta$ are defined by

$$f(\mathsf{c}) := \mathcal{E}_{\varphi,[\mathcal{D}_\varphi(\mathbf{D})/x],[\mathsf{c}/y]}(\mathbf{U})$$

and

$$g(\mathsf{c}) := \mathcal{E}_{\varphi,[\mathsf{c}/y]}([\mathbf{D}/x]\mathbf{U})$$

for $\mathsf{c} \in \mathcal{D}_\gamma$. We must check $i(f) = i(g)$. It suffices to determine $f = g$ (as functions). By the inductive hypothesis, we know

$$f(\mathsf{c}) = \mathcal{E}_{\varphi,[\mathcal{D}_\varphi(\mathbf{D})/x],[\mathsf{c}/y]}(\mathbf{U}) = \mathcal{E}_{\varphi,[\mathsf{c}/y]}([\mathbf{D}/x]\mathbf{U}) = g(\mathsf{c})$$

and we are done.

2. We next prove by induction on the position of the β-redex in $\mathbf{A}$ that if $\mathbf{A}$ β-reduces to $\mathbf{B}$ in one step, then $\mathcal{E}_\varphi(\mathbf{A}) = \mathcal{E}_\varphi(\mathbf{B})$. If $\mathbf{A}$ is the β-redex $[[\lambda x\, \mathbf{C}]\, \mathbf{D}]$, then by the result above and the definition of $\mathcal{E}$ we know

$$\mathcal{E}_\varphi([[\lambda x\, \mathbf{C}]\, \mathbf{D}]) = \mathcal{E}_{\varphi,[\mathcal{E}_\varphi(\mathbf{D})/x]}(\mathbf{C}) = \mathcal{E}_\varphi([\mathbf{D}/x]\mathbf{C}).$$

If $\mathbf{A}$ is an application (and not the β-redex), then we can simply apply the inductive hypothesis to the smaller position where the β-redex occurs.

Suppose $\mathbf{A}$ is a λ-abstraction $[\lambda y_\gamma\, \mathbf{U}_\delta]$. Then $\mathbf{B}$ is of the form $[\lambda y_\gamma\, \mathbf{V}_\delta]$ where $\mathbf{U}$ β-reduces in one step to $\mathbf{V}$. By the inductive hypothesis,

$$\mathcal{E}_{\varphi,[\mathsf{c}/y]}(\mathbf{U}) = \mathcal{E}_{\varphi,[\mathsf{c}/y]}(\mathbf{V})$$

for every c. Let $f : \mathcal{D}_\gamma \to \mathcal{D}_\delta$ be

$$f(\mathsf{c}) := \mathcal{E}_{\varphi,[\mathsf{c}/y]}(\mathbf{U}) = \mathcal{E}_{\varphi,[\mathsf{c}/y]}(\mathbf{V})$$

for every c. Hence $\mathcal{E}_\varphi([\lambda y_\gamma\, \mathbf{U}_\delta]) = i(f) = \mathcal{E}_\varphi([\lambda y_\gamma\, \mathbf{V}_\delta])$.

3. The full result now follows by induction on the number of β-reductions.

□

A.3. Proofs from Chapter 5

PROOF OF LEMMA 5.2.9. The proof is by induction on α.

ι: By $\nabla^u_{\doteq}$, either $\{\neg[\mathbf{A}^\downarrow \doteq^\iota \mathbf{C}^\downarrow]\} \in \mathscr{C}$ or $\{\neg[\mathbf{B}^\downarrow \doteq^\iota \mathbf{D}^\downarrow]\} \in \mathscr{C}$. We know $\{\neg[\mathbf{A}^\downarrow \doteq^\iota \mathbf{C}^\downarrow]\} \notin \mathscr{C}$ since $\mathbf{A} \parallel \mathbf{C}$. Hence we must have $\{\neg[\mathbf{B}^\downarrow \doteq^\iota \mathbf{D}^\downarrow]\} \in \mathscr{C}$. Thus $\mathbf{B}$ and $\mathbf{D}$ are incompatible.

o: By $\nabla_{\mathfrak{b}}$ and $\nabla^o_{\doteq}$, one of the four cases must hold:

- $\{\mathbf{A}^\downarrow, \mathbf{B}^\downarrow, \neg\mathbf{C}^\downarrow, \mathbf{D}^\downarrow\} \in \mathscr{C}$: This case contradicts $\mathbf{A} \parallel \mathbf{C}$.
- $\{\mathbf{A}^\downarrow, \mathbf{B}^\downarrow, \mathbf{C}^\downarrow, \neg\mathbf{D}^\downarrow\} \in \mathscr{C}$: In this case $\mathbf{B}$ and $\mathbf{D}$ are incompatible, as desired.
- $\{\neg\mathbf{A}^\downarrow, \neg\mathbf{B}^\downarrow, \neg\mathbf{C}^\downarrow, \mathbf{D}^\downarrow\} \in \mathscr{C}$: In this case $\mathbf{B}$ and $\mathbf{D}$ are incompatible, as desired.
- $\{\neg\mathbf{A}^\downarrow, \neg\mathbf{B}^\downarrow, \mathbf{C}^\downarrow, \neg\mathbf{D}^\downarrow\} \in \mathscr{C}$: This case contradicts $\mathbf{A} \parallel \mathbf{C}$.

$(\gamma\beta)$: By $\nabla^w_{\mathfrak{f}}$, $\nabla^{\to}_{\doteq}$ and $\nabla_{\beta\eta}$, we know there is a parameter W_β such that

$$\{[[\mathbf{A}\, W]^\downarrow \doteq [\mathbf{B}\, W]^\downarrow], \neg[[\mathbf{C}\, W]^\downarrow \doteq [\mathbf{D}\, W]^\downarrow]\} \in \mathscr{C}.$$

By Lemma 5.2.8:2, we have $W \parallel W$. Hence $[\mathbf{A}\, W] \parallel [\mathbf{C}\, W]$. By the inductive hypothesis at type γ, $[\mathbf{B}\, W]$ and $[\mathbf{D}\, W]$ must be incompatible. Thus $\mathbf{B}$ and $\mathbf{D}$ are incompatible.

□

PROOF OF LEMMA 5.2.10.

$\top$: Assume $\top$ is $\mathscr{C}$-incompatible with itself, i.e., $\{\top, \dot{\neg}\top\} \in \mathscr{C}$. By $\nabla^\sharp$ and subset closure, $\{\dot{\neg}\dot{\top}\} \in \mathscr{C}$, contradicting $\nabla_\bot$.

$\bot$: Assume $\bot$ is $\mathscr{C}$-incompatible with itself, i.e., $\{\bot, \dot{\neg}\bot\} \in \mathscr{C}$. By $\nabla^\sharp$ and subset closure, $\{\dot{\bot}\} \in \mathscr{C}$. Since $\dot{\bot}$ is $\dot{\neg}\dot{\top}$, this contradicts $\nabla_\bot$.

$\neg$: Assume $\neg$ is $\mathscr{C}$-incompatible with itself. There must exist $\mathbf{A}_o \parallel \mathbf{B}_o$ such that $\neg\mathbf{A}$ and $\neg\mathbf{B}$ are $\mathscr{C}$-incompatible. Without loss of generality, $\{\neg\mathbf{A}, \dot{\neg}\neg\mathbf{B}\} \in \mathscr{C}$. Hence $\{\dot{\neg}\mathbf{A}, \mathbf{B}\} \in \mathscr{C}$ by $\nabla^\sharp$ and $\nabla_\neg$. This contradicts $\mathbf{A}_o \parallel \mathbf{B}_o$.

$\vee$: Assume $\vee$ is $\mathscr{C}$-incompatible with itself. There must exist $\mathbf{A}_o \parallel \mathbf{C}_o$ and $\mathbf{B}_o \parallel \mathbf{D}_o$ such that $\{[\mathbf{A} \vee \mathbf{B}], \dot{\neg}[\mathbf{C} \vee \mathbf{D}]\} \in \mathscr{C}$. By $\nabla^\sharp$, $\nabla_\vee$ and $\nabla_\wedge$, we have either $\{\mathbf{A}, \dot{\neg}\mathbf{C}, \dot{\neg}\mathbf{D}\} \in \mathscr{C}$ (contradicting $\mathbf{A} \parallel \mathbf{C}$) or $\{\mathbf{B}, \dot{\neg}\mathbf{C}, \dot{\neg}\mathbf{D}\} \in \mathscr{C}$ (contradicting $\mathbf{B} \parallel \mathbf{D}$).

$\wedge,\supset,\equiv$: These cases are analogous to $\vee$.

Π^α: Assume Π^α is $\mathscr{C}$-incompatible with itself. There exist some $\mathbf{F}_{o\alpha}$ and $\mathbf{G}_{o\alpha}$ such that $\mathbf{F} \parallel \mathbf{G}$ and $\{[\Pi^\alpha\, \mathbf{F}], \dot{\neg}[\Pi^\alpha\, \mathbf{G}]\} \in \mathscr{C}$. By $\nabla^\sharp$, $\nabla_\exists^w$, $\nabla_\forall$ and $\nabla_{\beta\eta}$, there is a parameter W_α such that

$$\{[\mathbf{F}\, W]^\downarrow, \dot{\neg}[\mathbf{G}\, W]^\downarrow\} \in \mathscr{C}.$$

On the other hand, we know $W \parallel W$ by Lemma 5.2.8:2. Thus $\mathbf{F} \parallel \mathbf{G}$ implies $[\mathbf{F}\, W] \parallel [\mathbf{G}\, W]$, a contradiction.

Σ^α: This case is analogous to Π^α.

$=^\alpha$: Suppose $\mathbf{A}_\alpha \parallel \mathbf{C}_\alpha$, $\mathbf{B}_\alpha \parallel \mathbf{D}_\alpha$, but $[\mathbf{A} =^\alpha \mathbf{B}]$ and $[\mathbf{C} =^\alpha \mathbf{D}]$ are $\mathscr{C}$-incompatible. Without loss of generality (using symmetry of compatibility), we may assume

$$\{[\mathbf{A}^\downarrow = \mathbf{B}^\downarrow], \dot{\neg}[\mathbf{C}^\downarrow = \mathbf{D}^\downarrow]\} \in \mathscr{C}.$$

Hence $\{[\mathbf{A}^\downarrow \doteq \mathbf{B}^\downarrow], \dot{\neg}[\mathbf{C}^\downarrow \doteq \mathbf{D}^\downarrow]\} \in \mathscr{C}$ by $\nabla^\sharp$. By Lemma 5.2.9, we conclude $\mathbf{B}$ and $\mathbf{D}$ are $\mathscr{C}$-incompatible, a contradiction.

□

PROOF OF LEMMA 5.3.1. If $\mathbf{A}$ is an atom, this follows immediately from ∇_c. Otherwise, the head of $\mathbf{A}$ must be a logical constant.

Since $\mathcal{S}$ is assumed to be equality-free, the head of $\mathbf{A}$ cannot be $=^\alpha$ for any type α.

If $\mathbf{A}$ is $\top$, then $\{\top, \neg\top\} \notin \mathscr{C}$ by $\nabla^\sharp$ and $\nabla_\bot$. If $\mathbf{A}$ is $\bot$, then $\{\bot, \neg\bot\} \notin \mathscr{C}$ by $\nabla^\sharp$ and $\nabla_\bot$. Suppose $\mathbf{A}$ is $\neg\mathbf{B}$ for some $\mathbf{B} \in cwff_o(\mathcal{S})$ and $\{\neg\mathbf{B}, \neg\neg\mathbf{B}\} \in \mathscr{C}$. By $\nabla^\sharp$, $\nabla_\neg$ and closure under subsets, we have $\{\neg\mathbf{B}, \mathbf{B}\} \in \mathscr{C}$, contradicting the inductive hypothesis for $\mathbf{B}$. Suppose $\mathbf{A}$ is $[\mathbf{B} \vee \mathbf{C}]$ for some $\mathbf{B}, \mathbf{C} \in cwff_o(\mathcal{S})$ and $\{[\mathbf{B} \vee \mathbf{C}], \neg[\mathbf{B} \vee \mathbf{C}]\}$ is in $\mathscr{C}$. By $\nabla^\sharp$, $\nabla_\vee$, $\nabla_\wedge$ and closure under subsets, we have either $\{\mathbf{B}, \neg\mathbf{B}\} \in \mathscr{C}$ or $\{\mathbf{C}, \neg\mathbf{C}\} \in \mathscr{C}$, contradicting the inductive hypotheses for $\mathbf{B}$ and $\mathbf{C}$. The arguments in the $\wedge$, $\supset$ and $\equiv$ cases are analogous.

Suppose $\mathbf{A}$ is $[\Pi^\alpha \mathbf{F}]$ for some $\mathbf{F} \in cwff_{o\alpha}(\mathcal{S})$ and

$$\{[\Pi^\alpha \mathbf{F}], \neg[\Pi^\alpha \mathbf{F}]\} \in \mathscr{C}.$$

Let $W_\alpha \in \mathcal{P}_\alpha$ be a parameter which does not occur in $\mathbf{A}$. We do know there are fewer occurrences of logical constants in $[\mathbf{F}\, W]$ than in $[\Pi^\alpha \mathbf{F}]$. However, we cannot directly apply the inductive hypothesis as $[\mathbf{F}\, W]$ may not be β-normal. Since $\mathbf{F}$ is β-normal the only way $[\mathbf{F}\, W]$ can fail to be β-normal is if $\mathbf{F}$ has the form $[\lambda x_\alpha\, \mathbf{C}]$ for some $\mathbf{C} \in wff_o(\mathcal{S})$ where $\mathbf{Free}(\mathbf{C}) \subseteq \{x_\alpha\}$. In this case it is easy to verify the reduct $[W/x]\mathbf{C}$ is β-normal and contains the same number of occurrences of logical constants as $[\mathbf{F}\, W]$. In either case we can let $\mathbf{N}$ be the β-normal form or $\beta\eta$-normal form (depending on whether ∇_β or $\nabla_{\beta\eta}$ holds) of $[\mathbf{F}\, W]$ and apply the inductive hypothesis to obtain $\{\mathbf{N}, \neg\mathbf{N}\} \notin \mathscr{C}$. On the other hand, $\nabla_\exists$, $\nabla_\forall$, ∇_β (or $\nabla_{\beta\eta}$) and closure under subsets implies $\{\mathbf{N}, \neg\mathbf{N}\} \in \mathscr{C}$, a contradiction. The argument when $\mathbf{A}$ is $[\Sigma^\alpha \mathbf{F}]$ is analogous. □

Proof of Theorem 5.3.4. We prove this by induction on the construction of $\mathbf{M}$.

If $\mathbf{M}$ is $\mathbf{A} \in cwff_o(\mathcal{S})$, $\{\mathbf{A}, \neg\mathbf{A}\} \not\subseteq \Phi$ follows from subset closure and Lemma 5.3.1 (using ∇_β or $\nabla_{\beta\eta}$) or Lemma 5.3.3.

Suppose $\mathbf{M}$ is $[\mathbf{C}_\alpha \doteq^\alpha \mathbf{D}_\alpha]$. If $* \in \{\beta, \beta\eta\}$, then this follows from the definition of $\mathfrak{Acc}_*(\mathcal{S})$ since $\Phi \in \mathscr{C}$ must be equality-free (and so $\mathbf{M} \notin \Phi$ and $\neg\mathbf{M} \notin \Phi$). Assume $*$ is $\beta\mathfrak{fb}$. Let

$$\mathscr{C}^f \text{ be } \{\Phi \in \mathscr{C} \mid \Phi \text{ is finite}\}$$

and $\|$ be the compatibility relation induced by $\mathscr{C}^f$. $\mathscr{C}^f$ is an $\mathcal{S}$-abstract compatibility class by Theorem 5.2.7. Suppose

$$[\mathbf{C}_\alpha \doteq^\alpha \mathbf{D}_\alpha] \in \Phi \text{ and } \dot\neg[\mathbf{C}_\alpha \doteq^\alpha \mathbf{D}_\alpha] \in \Phi.$$

By subset closure, the finite set $\{[\mathbf{C}_\alpha \doteq^\alpha \mathbf{D}_\alpha], \dot\neg[\mathbf{C}_\alpha \doteq^\alpha \mathbf{D}_\alpha]\}$ is in $\mathscr{C}^f$. By Theorem 5.2.11, we know $\mathbf{C} \| \mathbf{C}$ and $\mathbf{D} \| \mathbf{D}$. This contradicts Lemma 5.2.9.

If $\mathbf{M}$ is $\dot\top$, then $\dot\neg\dot\top \notin \Phi$ by $\nabla_\bot$.

Suppose $\mathbf{M}$ is $\dot\neg\mathbf{N}$ and $\{\dot\neg\mathbf{N}, \dot\neg\dot\neg\mathbf{N}\} \subseteq \Phi$. By $\nabla_\neg$, $\Phi * \mathbf{N} \in \mathscr{C}$, contradicting the inductive hypothesis for $\mathbf{N}$.

Suppose $\mathbf{M}$ is $[\mathbf{N} \dot\vee \mathbf{P}]$ and $\{[\mathbf{N} \dot\vee \mathbf{P}], \dot\neg[\mathbf{N} \dot\vee \mathbf{P}]\} \subseteq \Phi$. By $\nabla_\wedge$ and $\nabla_\vee$, either $\Phi * \dot\neg\mathbf{N} * \dot\neg\mathbf{P} * \mathbf{N} \in \mathscr{C}$ or $\Phi * \dot\neg\mathbf{N} * \dot\neg\mathbf{P} * \mathbf{P} \in \mathscr{C}$. Either case contradicts the inductive hypothesis.

Suppose $\mathbf{M}$ is $[\dot\forall x_\alpha \mathbf{N}]$ and $\{\mathbf{M}, \dot\neg\mathbf{M}\} \subseteq \Phi$. By subset closure, we know $\{\mathbf{M}, \dot\neg\mathbf{M}\} \in \mathscr{C}$. Let W_α be a parameter with $W \notin \mathbf{Params}(\mathbf{M})$. By $\nabla_\exists$, $\nabla_\forall$ and subset closure, $\{[W/x]\mathbf{N}, \dot\neg[W/x]\mathbf{N}\} \in \mathscr{C}$, contradicting the inductive hypothesis for $[W/x]\mathbf{N}$. □

Proof of Theorem 5.4.7. To complete the proof of Theorem 5.4.7 we first check $\nabla_\bot$, $\nabla_\neg$, $\nabla_\forall$ and $\nabla^\sharp$ hold in each $\mathscr{C}_*^{\mathcal{S}}$.

$\nabla_\bot$: Assume $\Phi \in \mathscr{C}_*^{\mathcal{S}}$ and $\dot\neg\dot\top \in \Phi$. This is contradicted by the Φ-sequent $\dot\neg\dot\neg\dot\top$ with derivation

$$\cfrac{\cfrac{}{\dot\top}\,\mathcal{G}(\dot\top)}{\dot\neg\dot\neg\dot\top}\,\mathcal{G}(\dot\neg\dot\neg)$$

$\nabla_\neg$: Assume $\Phi \in \mathscr{C}_*^{\mathcal{S}}$, $\dot\neg\dot\neg\mathbf{M} \in \Phi$ and $\Phi * \mathbf{M} \notin \mathscr{C}_*^{\mathcal{S}}$. By Lemma 5.4.4, there is a Φ-sequent Γ and a $\mathcal{G}_*^{\mathcal{S}}$-derivation $\mathcal{D}$ of $\Gamma, \dot\neg\mathbf{M}$. This is contradicted by the derivation

$$\cfrac{\begin{array}{c}\mathcal{D}\\ \Gamma, \dot\neg\mathbf{M}\end{array}}{\Gamma, \dot\neg\dot\neg\dot\neg\mathbf{M}}\,\mathcal{G}(\dot\neg\dot\neg)$$

of the Φ-sequent $\Gamma, \dot\neg\dot\neg\dot\neg\mathbf{M}$.

$\nabla_\forall$: Assume $\Phi \in \mathscr{C}_*^{\mathcal{S}}$, $[\dot\forall x_\alpha \mathbf{M}] \in \Phi$ and $\Phi * [\mathbf{C}/x]\mathbf{M} \notin \mathscr{C}_*^{\mathcal{S}}$ where $\mathbf{C}$ is a closed term of type α. By Lemma 5.4.4, there is a Φ-sequent Γ and a derivation $\mathcal{D}$ of $\Gamma, \dot\neg[\mathbf{C}/x]\mathbf{M}$. Applying $\mathcal{G}(\dot\neg\dot\forall, \mathcal{S})$ we have a derivation of the Φ-sequent $\Gamma, \dot\neg[\dot\forall x_\alpha \mathbf{M}]$, a contradiction.

$\nabla^\sharp$: Suppose $\Phi \in \mathcal{C}_*^{\mathcal{S}}$ and $\mathbf{A} \in cwff_o(\mathcal{S})$. Assume we have $\mathbf{A} \in \Phi$ and $\Phi * \mathbf{A}^\sharp \notin \mathcal{C}_*^{\mathcal{S}}$. By Lemma 5.4.4, there is a Φ-sequent Γ and a derivation of $\Gamma, \neg\mathbf{A}^\sharp$. Applying $\mathcal{G}(\neg\sharp)$ we have a derivation of the Φ-sequent $\Gamma, \neg\mathbf{A}$, a contradiction.[1]

Assume $\neg\mathbf{A} \in \Phi$ and $\Phi * \neg\mathbf{A}^\sharp \notin \mathcal{C}_*^{\mathcal{S}}$. By Lemma 5.4.6, there is a Φ-sequent Γ and a derivation of $\Gamma, \mathbf{A}^\sharp$. Applying $\mathcal{G}(\sharp)$ and $\mathcal{G}(\neg\neg)$ we have a derivation of the Φ-sequent $\Gamma, \neg\neg\mathbf{A}$, a contradiction.

We next verify $\nabla_{\beta\eta}$ holds in $\mathcal{C}_{\beta\eta}^{\mathcal{S}}$ and $\mathcal{C}_{\beta\mathfrak{fb}}^{\mathcal{S}}$.

$\nabla_{\beta\eta}$: Assume $\mathbf{M} \in \Phi$, $\Phi \in \mathcal{C}_*^{\mathcal{S}}$ and $\Phi * \mathbf{M}^\downarrow \notin \mathcal{C}_*^{\mathcal{S}}$ where $*$ is $\beta\eta$ or $\beta\mathfrak{fb}$. By Lemma 5.4.4, there is a Φ-sequent Γ and a $\mathcal{G}_*^{\mathcal{S}}$-derivation $\mathcal{D}$ of $\Gamma, \neg\mathbf{M}^\downarrow$. The fact that $\Gamma, \neg\mathbf{M}$ is a Φ-sequent with derivation

$$\dfrac{\begin{matrix}\mathcal{D}\\ \Gamma, \neg\mathbf{M}^\downarrow\end{matrix}}{\Gamma, \neg\mathbf{M}}\,\mathcal{G}(\beta\eta)$$

contradicts $\Phi \in \mathcal{C}_*^{\mathcal{S}}$.

Finally, we check that ∇_{dec}, $\nabla_{\mathfrak{f}}$, $\nabla_{=}^{o}$, $\nabla_{=}^{\rightarrow}$ and $\nabla_{=}^{u}$ hold in $\mathcal{C}_{\beta\mathfrak{fb}}^{\mathcal{S}}$.

∇_{dec}: Suppose H is a parameter and $\neg[[H\,\overline{\mathbf{A}^n}] \doteq^{\iota} [H\,\overline{\mathbf{B}^n}]] \in \Phi$. Assume $\Phi \in \mathcal{C}_{\beta\mathfrak{fb}}^{\mathcal{S}}$ and $\Phi * \neg[\mathbf{A}^i \doteq \mathbf{B}^i] \notin \mathcal{C}_{\beta\mathfrak{fb}}^{\mathcal{S}}$ for each i with $1 \leq i \leq n$. By Lemma 5.4.6, and weakening there is some Φ-sequent Γ such that there are derivations $\mathcal{D}_i$ for $\Gamma, [\mathbf{A}^i \doteq \mathbf{B}^i]$. The derivation

$$\dfrac{\dfrac{\begin{matrix}\mathcal{D}_1\\ \Gamma, [\mathbf{A}^1 \doteq \mathbf{B}^1]\end{matrix} \quad \cdots \quad \begin{matrix}\mathcal{D}_n\\ \Gamma, [\mathbf{A}^n \doteq \mathbf{B}^n]\end{matrix}}{\Gamma, [[H\,\overline{\mathbf{A}^n}] \doteq [H\,\overline{\mathbf{B}^n}]]}\,\mathcal{G}(Dec)}{\Gamma, \neg\neg[[H\,\overline{\mathbf{A}^n}] \doteq [H\,\overline{\mathbf{B}^n}]]}\,\mathcal{G}(\neg\neg)$$

of the Φ-sequent $\Gamma, \neg\neg[[H\,\overline{\mathbf{A}^n}] \doteq^{\iota} [H\,\overline{\mathbf{B}^n}]]$ contradicts $\Phi \in \mathcal{C}_{\beta\mathfrak{fb}}^{\mathcal{S}}$.

$\nabla_{\mathfrak{f}}$: Suppose $\Phi \in \mathcal{C}_{\beta\mathfrak{fb}}^{\mathcal{S}}$, W_β is a parameter with $W \notin \mathbf{Params}(\Phi)$ and $\neg[\mathbf{G} \doteq^{\alpha\beta} \mathbf{H}] \in \Phi$. Assume $\Phi * \neg[[\mathbf{G}\,W] \doteq^{\alpha} [\mathbf{H}\,W]] \notin \mathcal{C}_{\beta\mathfrak{fb}}^{\mathcal{S}}$.

[1] Note that we can legally apply the rules $\mathcal{G}(\sharp)$ and $\mathcal{G}(\neg\sharp)$ even if the head of $\mathbf{A}$ is not a logical constant. However, if the head of $\mathbf{A}$ is not a logical constant, then the rules $\mathcal{G}(\sharp)$ and $\mathcal{G}(\neg\sharp)$ are trivial since $\mathbf{A}^\sharp$ is the same as $\mathbf{A}$.

By Lemma 5.4.6, there is a Φ-sequent Γ such that

$$\Gamma, [[\mathbf{G}\,W] \doteq^{\alpha} [\mathbf{H}\,W]]$$

is derivable. Using $\mathcal{G}(\mathfrak{f}^W)$ and $\mathcal{G}(\neg\neg)$ we can derive the Φ-sequent $\Gamma, \neg\neg[\mathbf{G} \doteq^{\alpha\beta} \mathbf{H}]$, a contradiction.

$\nabla_{=}^{o}$: Suppose $\Phi \in \mathscr{C}_{\beta\mathfrak{fb}}^{\mathcal{S}}$ and $[\mathbf{A} \doteq^{o} \mathbf{B}] \in \Phi$. Assume

$$\Phi * \mathbf{A} * \mathbf{B} \notin \mathscr{C}_{\beta\mathfrak{fb}}^{\mathcal{S}} \text{ and} \Phi * \neg\mathbf{A} * \neg\mathbf{B} \notin \mathscr{C}_{\beta\mathfrak{fb}}^{\mathcal{S}}.$$

By Lemma 5.4.5 and weakening, there is a Φ-sequent Γ such that $\Gamma, \neg\mathbf{A}, \neg\mathbf{B}$ and $\Gamma, \neg\neg\mathbf{A}, \neg\neg\mathbf{B}$ are derivable. Using Lemma 2.2.12 there is a derivation of of $\Gamma, \mathbf{A}, \mathbf{B}$. Using $\mathcal{G}(\neg \doteq^{o})$ we can derive the Φ-sequent $\Gamma, \neg[\mathbf{A} \doteq^{o} \mathbf{B}]$, a contradiction.

$\nabla_{=}^{\rightarrow}$: Suppose $\Phi \in \mathscr{C}_{\beta\mathfrak{fb}}^{\mathcal{S}}$, $\mathbf{B} \in cwff_{\beta}(\mathcal{S})$ and $[\mathbf{G}_{\alpha\beta} \doteq^{\alpha\beta} \mathbf{H}_{\alpha\beta}] \in \Phi$. Assume $\Phi * [[\mathbf{G}\,\mathbf{B}] \doteq^{\alpha} [\mathbf{H}\,\mathbf{B}]] \notin \mathscr{C}_{\beta\mathfrak{fb}}^{\mathcal{S}}$. By Lemma 5.4.4, there is a Φ-sequent Γ such that $\Gamma, \neg[[\mathbf{G}\,\mathbf{B}] \doteq^{\alpha} [\mathbf{H}\,\mathbf{B}]]$ is derivable. We can derive the Φ-sequent $\Gamma, \neg[\mathbf{G}_{\alpha\beta} \doteq^{\alpha\beta} \mathbf{H}_{\alpha\beta}]$ using the rule $\mathcal{G}(\neg \doteq^{\rightarrow}, \mathcal{S})$, a contradiction.

$\nabla_{=}^{u}$: Suppose $\Phi \in \mathscr{C}_{\beta\mathfrak{fb}}^{\mathcal{S}}$, $\mathbf{A}, \mathbf{B}, \mathbf{C}, \mathbf{D} \in cwff_{\iota}(\mathcal{S})$,

$$[\mathbf{A}_{\iota} \doteq^{\iota} \mathbf{B}_{\iota}] \in \Phi \text{ and } \neg[\mathbf{C}_{\iota} \doteq^{\iota} \mathbf{D}_{\iota}] \in \Phi.$$

Assume $\Phi * \neg[\mathbf{A} \doteq^{\iota} \mathbf{C}] \notin \mathscr{C}_{\beta\mathfrak{fb}}^{\mathcal{S}}$ and $\Phi * \neg[\mathbf{B} \doteq^{\iota} \mathbf{D}] \notin \mathscr{C}_{\beta\mathfrak{fb}}^{\mathcal{S}}$. By Lemma 5.4.6 and weakening, there is a Φ-sequent Γ such that there are derivations $\mathcal{D}_1$ and $\mathcal{D}_2$ of

$$\Gamma, [\mathbf{A} \doteq^{\iota} \mathbf{C}] \text{ and } \Gamma, [\mathbf{B} \doteq^{\iota} \mathbf{D}],$$

respectively. The derivation

$$\cfrac{\cfrac{\begin{matrix}\mathcal{D}_1 \\ \Gamma, [\mathbf{A} \doteq^{\iota} \mathbf{C}]\end{matrix} \qquad \begin{matrix}\mathcal{D}_2 \\ \Gamma, [\mathbf{B} \doteq^{\iota} \mathbf{D}]\end{matrix}}{\Gamma, \neg[\mathbf{A}_{\iota} \doteq^{\iota} \mathbf{B}_{\iota}], [\mathbf{C}_{\iota} \doteq^{\iota} \mathbf{D}_{\iota}]}\ \mathcal{G}(EUnif_1)}{\Gamma, \neg[\mathbf{A}_{\iota} \doteq^{\iota} \mathbf{B}_{\iota}], \neg\neg[\mathbf{C}_{\iota} \doteq^{\iota} \mathbf{D}_{\iota}]}\ \mathcal{G}(\neg\neg)$$

of the Φ-sequent $\Gamma, \neg[\mathbf{A}_{\iota} \doteq^{\iota} \mathbf{B}_{\iota}], \neg\neg[\mathbf{C}_{\iota} \doteq^{\iota} \mathbf{D}_{\iota}]$, a contradiction.

Next assume

$$\Phi * \neg[\mathbf{A} \doteq^{\iota} \mathbf{D}] \notin \mathscr{C}_{\beta\mathfrak{fb}}^{\mathcal{S}} \text{ and } \Phi * \neg[\mathbf{B} \doteq^{\iota} \mathbf{C}] \notin \mathscr{C}_{\beta\mathfrak{fb}}^{\mathcal{S}}.$$

By Lemma 5.4.6 and weakening, there is a Φ-sequent Γ such that there are derivations $\mathcal{D}_1$ and $\mathcal{D}_2$ of

$$\Gamma, [\mathbf{A} \doteq^\iota \mathbf{D}] \text{ and } \Gamma, [\mathbf{B} \doteq^\iota \mathbf{C}],$$

respectively. The derivation

$$\dfrac{\dfrac{\begin{array}{c}\mathcal{D}_1\\ \Gamma, [\mathbf{A} \doteq^\iota \mathbf{D}]\end{array} \qquad \begin{array}{c}\mathcal{D}_2\\ \Gamma, [\mathbf{B} \doteq^\iota \mathbf{C}]\end{array}}{\Gamma, \neg[\mathbf{A}_\iota \doteq^\iota \mathbf{B}_\iota], [\mathbf{C}_\iota \doteq^\iota \mathbf{D}_\iota]}\,\mathcal{G}(EUnif_2)}{\Gamma, \neg[\mathbf{A}_\iota \doteq^\iota \mathbf{B}_\iota], \neg\neg[\mathbf{C}_\iota \doteq^\iota \mathbf{D}_\iota]}\,\mathcal{G}(\neg\neg)$$

of the Φ-sequent $\Gamma, \neg[\mathbf{A}_\iota \doteq^\iota \mathbf{B}_\iota], \neg\neg[\mathbf{C}_\iota \doteq^\iota \mathbf{D}_\iota]$ again yielding a contradiction.

□

Proof of Lemma 5.5.6. In the following argument, note that α, β and γ are types as usual while δ, ϵ, σ and τ are ordinals.

We have assumed there is an infinite cardinal $\aleph_s$ which is the cardinality of $\mathcal{P}_\alpha$ for each type α (cf. Remark 2.1.2). This easily implies $cwff_\alpha(\mathcal{S})$ is of cardinality $\aleph_s$ for each type α. Let ϵ be the first ordinal of this cardinality. (In the countable case, ϵ is ω.) This also implies the cardinality of $sent_*(\mathcal{S})$ is $\aleph_s$. We use the well-ordering principle to enumerate $sent_*(\mathcal{S})$ as $(\mathbf{M}^\delta)_{\delta<\epsilon}$.

Let α be a type. For each $\delta < \epsilon$ let U_α^δ be the set of parameters which occur in a sentence in the set $\{\mathbf{M}^\sigma \mid \sigma \leq \delta\}$. The set $\{\mathbf{M}^\sigma \mid \sigma \leq \delta\}$ has cardinality less than $\aleph_s$ since $\delta < \epsilon$. Hence U_α^δ has cardinality less than $\aleph_s$. By sufficient purity, we know there is a set of parameters $P_\alpha \subseteq \mathcal{S}_\alpha$ of cardinality $\aleph_s$ such that the parameters in P_α do not occur in Φ. Since, considering cardinality, we cannot have $P_\alpha \subseteq U_\alpha^\delta$ for any $\delta < \epsilon$, we know $P_\alpha \setminus U_\alpha^\delta$ is non-empty for each $\delta < \epsilon$. Using the axiom of choice, we can find a sequence $(W_\alpha^\delta)_{\delta<\epsilon}$ where $W_\alpha^\delta \in P_\alpha \setminus (U_\alpha^\delta \cup \{W_\alpha^\sigma \mid \sigma < \delta\})$ for each $\delta < \epsilon$. Hence for each type α we know W_α^δ is a parameter of type α which does not occur in $\Phi \cup \{\mathbf{M}^\sigma \mid \sigma \leq \delta\}$. As a consequence, if W_α^δ occurs in $\mathbf{M}^\sigma$, then $\delta < \sigma$. Also, we have ensured that if $W_\alpha^\delta = W_\alpha^\sigma$, then $\delta = \sigma$ for any $\delta, \sigma < \epsilon$.

The parameters W_α^δ are intended to serve as witnesses. To ease the argument we define two sequences of witnessing sentences related to the sequence $(\mathbf{M}^\delta)_{\delta<\epsilon}$. For each $\delta < \epsilon$, define

$$\mathbf{E}^\delta := \begin{cases} \dot{\neg}[W_\alpha^\delta/x]\mathbf{N} & \text{if } \mathbf{M}^\delta \text{ is of the form } \dot{\neg}[\dot{\forall} x_\alpha \mathbf{N}] \\ \mathbf{M}^\delta & \text{otherwise.} \end{cases}$$

If $*$ is $\beta\mathfrak{fb}$, then we define

$$\mathbf{X}^\delta := \begin{cases} \dot{\neg}[[\mathbf{G}\, W_\beta^\delta] \doteq^\alpha [\mathbf{H}\, W_\beta^\delta]] & \text{if } \mathbf{M}^\delta \text{ is of the form } \dot{\neg}[\mathbf{G} \doteq^{\alpha\beta} \mathbf{H}] \\ \mathbf{M}^\delta & \text{otherwise.} \end{cases}$$

If $* \in \{\beta, \beta\eta\}$, then let $\mathbf{X}^\delta := \mathbf{M}^\delta$. Note that for every $\delta < \epsilon$, either $\mathbf{E}^\delta = \mathbf{M}^\delta$ or $\mathbf{X}^\delta = \mathbf{M}^\delta$ since $\mathbf{M}^\delta$ cannot simultaneously be of the forms $\dot{\neg}[\dot{\forall} x_\alpha \mathbf{N}]$ and $\dot{\neg}[\mathbf{G} \doteq^{\alpha\beta} \mathbf{H}]$. (This is not true if one considers equality to mean Leibniz equality.)

We construct $\mathcal{H}$ by inductively giving a transfinite sequence $(\mathcal{H}^\delta)_{\delta<\epsilon}$ such that $\mathcal{H}^\delta \in \mathscr{C}$ for each $\delta < \epsilon$. Then the $\mathcal{S}$-Hintikka set will be the union of this sequence. We define $\mathcal{H}^0 := \Phi$. For limit ordinals δ, we define $\mathcal{H}^\delta := \bigcup_{\sigma<\delta} \mathcal{H}^\sigma$. We finally define $\mathcal{H} := \bigcup_{\delta<\epsilon} \mathcal{H}^\delta$.

In the successor case, if $\mathcal{H}^\delta * \mathbf{M}^\delta \in \mathscr{C}$, then we let

$$\mathcal{H}^{\delta+1} := \mathcal{H}^\delta * \mathbf{M}^\delta * \mathbf{E}^\delta * \mathbf{X}^\delta.$$

If $\mathcal{H}^\delta * \mathbf{M}^\delta \notin \mathscr{C}$, we let $\mathcal{H}^{\delta+1} := \mathcal{H}^\delta$.

We prove by induction that for every $\delta < \epsilon$, type α and parameter W_α^τ which occurs in some sentence in $\mathcal{H}^\delta$, we have $\tau < \delta$. The base case holds since no W_α^τ occurs in any sentence in $\mathcal{H}^0 = \Phi$. For any limit ordinal δ, if W_α^τ occurs in some sentence in $\mathcal{H}^\delta$, then W_α^τ already occurs in some sentence in $\mathcal{H}^\sigma$ for some $\sigma < \delta$ by definition of $\mathcal{H}^\delta$. Hence $\tau < \sigma < \delta$.

For any successor ordinal $\delta + 1$, suppose W_α^τ occurs in some sentence in $\mathcal{H}^{\delta+1}$. If it already occurred in a sentence in $\mathcal{H}^\delta$, then we have $\tau < \delta < \delta + 1$ by the inductive hypothesis. Consequently, we need only consider the case where W_α^τ occurs in a sentence in $(\mathcal{H}^{\delta+1} \setminus \mathcal{H}^\delta) \subseteq \{\mathbf{M}^\delta, \mathbf{E}^\delta, \mathbf{X}^\delta\}$. If τ is δ, then $\tau = \delta < \delta + 1$. If W_α^τ is any parameter with $\tau \neq \delta$ occurring in $\mathbf{E}^\delta$ or $\mathbf{X}^\delta$, then it must also occur in $\mathbf{M}^\delta$ (by noting that $W_\alpha^\tau \neq W_\alpha^\delta$ and inspecting the possible definitions of $\mathbf{E}^\delta$ and $\mathbf{X}^\delta$) and so $\tau < \delta < \delta + 1$ (by the choice of W_α^τ).

In particular, we now know W_α^δ does not occur in any sentence in $\mathcal{H}^\delta$ for any $\delta < \epsilon$ and type α.

Next we prove by induction that $\mathcal{H}^\delta \in \mathscr{C}$ for all $\delta < \epsilon$. The base case holds by the assumption that $\mathcal{H}^0 = \Phi \in \mathscr{C}$. For any limit ordinal

δ assume $\mathcal{H}^\sigma \in \mathscr{C}$ for every $\sigma < \delta$. We have $\mathcal{H}^\delta = \bigcup_{\sigma<\delta} \mathcal{H}^\sigma \in \mathscr{C}$ since any finite subset of $\mathcal{H}^\delta$ is a subset of $\mathcal{H}^\sigma$ for some $\sigma < \delta$.

For any successor ordinal $\delta + 1$, we assume $\mathcal{H}^\delta \in \mathscr{C}$ and verify $\mathcal{H}^{\delta+1} \in \mathscr{C}$. This is trivial in case $\mathcal{H}^\delta * \mathbf{M}^\delta \notin \mathscr{C}$ (for all abstract consistency classes) since $\mathcal{H}^{\delta+1} = \mathcal{H}^\delta$. Suppose $\mathcal{H}^\delta * \mathbf{M}^\delta \in \mathscr{C}$. We consider three cases:

1. If $\mathbf{E}^\delta = \mathbf{M}^\delta$ and $\mathbf{X}^\delta = \mathbf{M}^\delta$, then $\mathcal{H}^\delta * \mathbf{M}^\delta * \mathbf{E}^\delta * \mathbf{X}^\delta \in \mathscr{C}$ since $\mathcal{H}^\delta * \mathbf{M}^\delta \in \mathscr{C}$.
2. If $\mathbf{E}^\delta \neq \mathbf{M}^\delta$, then $\mathbf{M}^\delta$ is of the form $\dot{\neg}[\dot{\forall} x_\alpha \mathbf{N}]$ and $\mathbf{E}^\delta$ is equal to $\dot{\neg}[W_\alpha^\delta/x]\mathbf{N}$. We conclude that $\mathcal{H}^\delta * \mathbf{M}^\delta * \mathbf{E}^\delta \in \mathscr{C}$ by $\nabla_\exists$ since W_α^δ does not occur in $\mathbf{M}^\delta$ or any sentence in $\mathcal{H}^\delta$. Since $\mathbf{X}^\delta = \mathbf{M}^\delta$, we have $\mathcal{H}^\delta * \mathbf{M}^\delta * \mathbf{E}^\delta * \mathbf{X}^\delta \in \mathscr{C}$.
3. If $\mathbf{X}^\delta \neq \mathbf{M}^\delta$, then $*$ is $\beta\mathfrak{fb}$ (by the definition of $\mathbf{X}^\delta$). W_α^δ does not occur in any sentence in $\mathcal{H}^\delta * \mathbf{M}^\delta$. Thus we have $\mathcal{H}^\delta * \mathbf{M}^\delta * \mathbf{X}^\delta \in \mathcal{H}$ by $\nabla_\mathfrak{f}$. Since $\mathbf{E}^\delta = \mathbf{M}^\delta$ in this case, we have $\mathcal{H}^\delta * \mathbf{M}^\delta * \mathbf{E}^\delta * \mathbf{X}^\delta \in \mathcal{H}$.

Since $\mathscr{C}$ has finite character, we also have $\mathcal{H} \in \mathscr{C}$.

Now we know that our inductively defined set $\mathcal{H}$ is indeed in $\mathscr{C}$ and that $\Phi \subseteq \mathcal{H}$. In order to apply Lemma 5.5.5, we must check $\mathcal{H}$ is maximal and satisfies $\vec{\nabla}_\exists$ and $\vec{\nabla}_\mathfrak{f}$ (if $*$ is $\beta\mathfrak{fb}$). It is immediate from the construction that $\vec{\nabla}_\exists$ holds since if $\dot{\neg}[\dot{\forall} x_\alpha \mathbf{N}]$, then $\dot{\neg}[W_\alpha^\delta/x]\mathbf{N} \in \mathcal{H}$ where δ is the ordinal such that $\mathbf{M}^\delta = \dot{\neg}[\dot{\forall} x_\alpha \mathbf{N}]$. We have ensured $\vec{\nabla}_\mathfrak{f}$ holds when $*$ is $\beta\mathfrak{fb}$ since $\dot{\neg}[[\mathbf{G}\, W_\beta^\delta] \doteq^\alpha [\mathbf{H}\, W_\beta^\delta]] \in \mathcal{H}$ whenever $\dot{\neg}[\mathbf{G} \doteq^{\alpha\beta} \mathbf{H}] \in \mathcal{H}$ where δ is the ordinal such that $\mathbf{M}^\delta$ is the sentence $\dot{\neg}[\mathbf{G} \doteq^{\alpha\beta} \mathbf{H}]$.

It only remains to prove $\mathcal{H}$ is maximal in $\mathscr{C}$. Suppose $\mathcal{H} * \mathbf{M} \in \mathscr{C}$ where $\mathbf{M} \in \mathit{sent}(\mathcal{S})$. If $* \in \{\beta, \beta\eta\}$, then $\mathbf{M}$ is equality-free since $\mathcal{H} * \mathbf{M} \in \mathscr{C}$. Thus $\mathbf{M} = \mathbf{M}^\delta$ for some $\delta < \epsilon$. Since $\mathcal{H}$ is closed under subsets we know that $\mathcal{H}^\delta * \mathbf{M}^\delta \in \mathscr{C}$. By definition of $\mathcal{H}^{\delta+1}$, we conclude that $\mathbf{M}^\delta \in \mathcal{H}^{\delta+1}$ and hence $\mathbf{M} \in \mathcal{H}$.

Therefore, $\mathcal{H} \in \mathfrak{Hint}_*(\mathcal{S})$ by Lemma 5.5.5. □

Proof of Lemma 5.7.5. First, $\mathbf{A} \in S^{\mathbf{A}}$ by Theorem 5.2.11. Suppose $\mathbf{B} \notin S^{\mathbf{A}}$. Then either

$$\dot{\neg}[\mathbf{A}^\downarrow \doteq^\iota \mathbf{B}^\downarrow] \in \mathcal{H} \text{ or } \dot{\neg}[\mathbf{B}^\downarrow \doteq^\iota \mathbf{A}^\downarrow] \in \mathcal{H}.$$

By $\vec{\nabla}^u_=$ and $\vec{\nabla}_{\beta\eta}$, in either case we can infer that $\dot\neg[\mathbf{A}^\downarrow \mathbin{\dot=^\iota} \mathbf{A}^\downarrow] \in \mathcal{H}$ (contradicting $\vec{\nabla}^r_=$) or $\dot\neg[\mathbf{B}^\downarrow \mathbin{\dot=^\iota} \mathbf{B}^\downarrow] \in \mathcal{H}$ (contradicting $\vec{\nabla}^r_=$). Thus $\mathbf{A} \parallel \mathbf{B}$ and $\mathbf{B} \in S^{\mathbf{A}}$.

Next suppose $S^{\mathbf{A}}$ is not $\mathcal{H}$-compatible. Then there must exist some $\mathbf{C}, \mathbf{D} \in S^{\mathbf{A}}$ with $\dot\neg[\mathbf{C}^\downarrow \mathbin{\dot=^\iota} \mathbf{D}^\downarrow] \in \mathcal{H}$. By $\vec{\nabla}^u_=$ and $\vec{\nabla}_{\beta\eta}$, we can conclude that either $\dot\neg[\mathbf{A}^\downarrow \mathbin{\dot=^\iota} \mathbf{C}^\downarrow] \in \mathcal{H}$ or $\dot\neg[\mathbf{B}^\downarrow \mathbin{\dot=^\iota} \mathbf{D}^\downarrow] \in \mathcal{H}$. However, $\mathbf{A} \parallel \mathbf{C}$ implies $\dot\neg[\mathbf{A}^\downarrow \mathbin{\dot=^\iota} \mathbf{C}^\downarrow] \notin \mathcal{H}$, so we must have $\dot\neg[\mathbf{B}^\downarrow \mathbin{\dot=^\iota} \mathbf{D}^\downarrow] \in \mathcal{H}$. By $\vec{\nabla}^u_=$ and $\vec{\nabla}_{\beta\eta}$, we have $\dot\neg[\mathbf{B}^\downarrow \mathbin{\dot=} \mathbf{B}^\downarrow] \in \mathcal{H}$ (contradicting $\vec{\nabla}^r_=$) or $\dot\neg[\mathbf{A}^\downarrow \mathbin{\dot=} \mathbf{D}^\downarrow] \in \mathcal{H}$ (contradicting $\mathbf{D} \in S^{\mathbf{A}}$). In either case we have a contradiction. Hence $S^{\mathbf{A}}$ is an $\mathcal{H}$-compatible set with $\mathbf{A} \in S$.

To prove $S^{\mathbf{A}}$ is the unique maximally $\mathcal{H}$-compatible set containing $\mathbf{A}$, suppose $T \subseteq \mathit{cwff}_\iota(\mathcal{S})$ is an $\mathcal{H}$-compatible set with $\mathbf{A} \in T$. For every $\mathbf{C} \in T$, $\mathbf{A} \parallel \mathbf{C}$ implies $\mathbf{C} \in S^{\mathbf{A}}$. Thus $T \subseteq S^{\mathbf{A}}$. □

Proof of Lemma 5.7.12. The proof is by induction on the type α.

ι: This follows from Lemma 5.7.6 and the definition of $\mathcal{D}_\iota$.

o: By $\vec{\nabla}^o_=$, we have either $\mathbf{A}_o \in \mathcal{H}$ or $\dot\neg\mathbf{A}_o \in \mathcal{H}$. If $\mathbf{A}_o \in \mathcal{H}$, then $a = \mathtt{T} = b$. If $\dot\neg\mathbf{A}_o \in \mathcal{H}$, then $a = \mathtt{F} = b$. In either case $a = b$.

$\gamma\beta$: Let $\langle \mathbf{C}, c\rangle \in \mathcal{D}_\beta$. We must prove $a(\langle \mathbf{C}, c\rangle) = b(\langle \mathbf{C}, c\rangle)$. We know $a(\langle \mathbf{C}, c\rangle) = \langle [\mathbf{A}\,\mathbf{C}]^\downarrow, d\rangle$ and $b(\langle \mathbf{C}, c\rangle) = \langle [\mathbf{A}\,\mathbf{C}]^\downarrow, e\rangle$ for some d and e. By $\vec{\nabla}^\rightarrow_=$, we know $[[\mathbf{A}\,\mathbf{C}]^\downarrow \mathbin{\dot=^\gamma} [\mathbf{B}\,\mathbf{C}]^\downarrow] \in \mathcal{H}$. By the inductive hypothesis at γ, we know $d = e$. Hence

$$a(\langle \mathbf{C}, c\rangle) = b(\langle \mathbf{C}, c\rangle) \text{ for each } \langle \mathbf{C}, c\rangle \in \mathcal{D}_\beta$$

and so $a = b$ as functions.

□

Proof of Lemma 5.7.14. The proof is by induction on the type α.

ι: By definition of $\mathcal{D}_\iota$, we know a is a maximally $\mathcal{H}$-compatible set with $\mathbf{A} \in a$ and b is a maximally $\mathcal{H}$-compatible set with $\mathbf{B} \in b$. Hence $a = b$ by Lemma 5.7.6. Thus $\langle \mathbf{A}, a\rangle \sim \langle \mathbf{B}, b\rangle$.

o: By $\vec{\nabla}^o_=$, either $\mathbf{A}, \mathbf{B} \in \mathcal{H}$ or $\dot\neg\mathbf{A}, \dot\neg\mathbf{B} \in \mathcal{H}$. Hence either

$$a = \mathtt{T} = b \text{ or } a = \mathtt{F} = b.$$

Thus $\langle \mathbf{A}, a\rangle \sim \langle \mathbf{B}, b\rangle$.

$\gamma\beta$: In order to verify $\langle \mathbf{A}, a\rangle \sim \langle \mathbf{B}, b\rangle$, we must check

$$a(\langle \mathbf{C}, c\rangle) \sim b(\langle \mathbf{D}, d\rangle) \text{ whenever } \langle \mathbf{C}, c\rangle \sim \langle \mathbf{D}, d\rangle.$$

Let $\langle \mathbf{C}, c\rangle \sim \langle \mathbf{D}, d\rangle$ be given. By $\vec{\nabla}_{\dot{=}}^{\rightarrow}$, $[[\mathbf{A}\,\mathbf{D}]^{\downarrow} \dot{=}^{\gamma} [\mathbf{B}\,\mathbf{D}]^{\downarrow}] \in \mathcal{H}$. By the inductive hypothesis, $a(\langle \mathbf{D}, d\rangle) \sim b(\langle \mathbf{D}, d\rangle)$. By Lemma 5.7.13, we have $\langle \mathbf{A}, a\rangle \sim \langle \mathbf{A}, a\rangle$ and so $a(\langle \mathbf{C}, c\rangle) \sim a(\langle \mathbf{D}, d\rangle)$. Thus

$$a(\langle \mathbf{C}, c\rangle) \sim a(\langle \mathbf{D}, d\rangle) \sim b(\langle \mathbf{D}, d\rangle)$$

as desired.

□

PROOF OF LEMMA 5.7.15. If $\mathbf{M}$ is $\dot{\top}$, then we certainly have $\mathscr{P} \approx_{\varphi} \dot{\top}$. If $\mathbf{M}$ is $\dot{\neg}\dot{\top}$, then we have $\dot{\neg}\dot{\top} \notin \mathcal{H}$ by $\vec{\nabla}_{\bot}$.

Suppose $\mathbf{M} \in \mathit{wff}_o(\mathcal{S})$ and $\varphi_1^{\beta\eta}(\mathbf{M}) \in \mathcal{H}$. Since $(\mathcal{D}, @, \mathcal{E})$ is a possible values evaluation, we know the first component of $\mathcal{E}_{\varphi}(\mathbf{M})$ is $\varphi_1^{\beta\eta}(\mathbf{M})$. Hence $\mathcal{B}_{\mathcal{H}}^{\varphi_1^{\beta\eta}(\mathbf{M})} = \{\mathtt{T}\}$ yielding $\mathcal{E}_{\varphi}(\mathbf{M}) = \langle \varphi_1^{\beta\eta}(\mathbf{M}), \mathtt{T}\rangle$ and $\upsilon(\mathcal{E}_{\varphi}(\mathbf{M})) = \mathtt{T}$. Thus $\mathscr{P} \approx_{\varphi} \mathbf{M}$.

Suppose $\mathbf{M}$ is $\dot{\neg}\mathbf{N}$, $\mathbf{N} \in \mathit{wff}_o(\mathcal{S})$ and $\varphi_1(\dot{\neg}\mathbf{N}) \in \mathcal{H}$. Since $(\mathcal{D}, @, \mathcal{E})$ is a possible values evaluation, we know the first component of $\mathcal{E}_{\varphi}(\mathbf{N})$ is $\varphi_1^{\beta\eta}(\mathbf{N})$. Since $\dot{\neg}\varphi_1^{\beta\eta}(\mathbf{N}) \in \mathcal{H}$ we know $\mathcal{B}_{\mathcal{H}}^{\varphi_1^{\beta\eta}(\mathbf{N})} = \{\mathtt{F}\}$. Thus we conclude $\upsilon(\mathcal{E}_{\varphi}(\mathbf{N})) = \mathtt{F}$ and $\mathscr{P} \approx_{\varphi} \dot{\neg}\mathbf{N}$.

We next consider the inductive cases in which $\mathbf{M}$ is of the form $\dot{\neg}\mathbf{N}$.

Suppose $\mathbf{N}$ is $\dot{\neg}\mathbf{P}$ and $\dot{\neg}\dot{\neg}\varphi_1^{\beta\eta}(\mathbf{P}) \in \mathcal{H}$. By $\vec{\nabla}_{\neg}$, $\varphi_1^{\beta\eta}(\mathbf{P}) \in \mathcal{H}$. Since $|\mathbf{P}|_{\mathbf{e}} < |\mathbf{M}|_{\mathbf{e}}$ we have $\mathscr{P} \approx_{\varphi} \mathbf{P}$ and so $\mathscr{P} \approx_{\varphi} \dot{\neg}\dot{\neg}\mathbf{P}$.

Suppose $\mathbf{N}$ is $[\mathbf{P} \dot{\vee} \mathbf{Q}]$ and $\dot{\neg}\varphi_1^{\beta\eta}([\mathbf{P} \dot{\vee} \mathbf{Q}]) \in \mathcal{H}$. By $\vec{\nabla}_{\wedge}$, we know $\dot{\neg}\varphi_1^{\beta\eta}(\mathbf{P}) \in \mathcal{H}$ and $\dot{\neg}\varphi_1^{\beta\eta}(\mathbf{Q}) \in \mathcal{H}$. By Lemma 2.1.28, $|\dot{\neg}\mathbf{P}|_{\mathbf{e}} < |\mathbf{M}|_{\mathbf{e}}$ and $|\dot{\neg}\mathbf{Q}|_{\mathbf{e}} < |\mathbf{M}|_{\mathbf{e}}$. By the inductive hypothesis, we have $\mathscr{P} \approx_{\varphi} \dot{\neg}\mathbf{P}$ and $\mathscr{P} \approx_{\varphi} \dot{\neg}\mathbf{Q}$. Hence $\mathscr{P} \not\approx_{\varphi} \mathbf{P}$ and $\mathscr{P} \not\approx_{\varphi} \mathbf{Q}$. Thus $\mathscr{P} \not\approx_{\varphi} [\mathbf{P} \dot{\vee} \mathbf{Q}]$ and $\mathscr{P} \approx_{\varphi} \mathbf{M}$.

Suppose $\mathbf{N}$ is $[\dot{\forall} x_{\alpha}\, \mathbf{P}]$ and $\dot{\neg}\varphi_1^{\beta\eta}(\mathbf{N}) \in \mathcal{H}$. By $\vec{\nabla}_{\exists}$ and $\vec{\nabla}_{\beta\eta}$, there is some parameter W_{α} such that $\dot{\neg}\varphi_1^{\beta\eta}([W/x]\mathbf{P}) \in \mathcal{H}$. By Lemma 2.1.28, we know $|\dot{\neg}[W/x]\mathbf{P}|_{\mathbf{e}} < |\mathbf{M}|_{\mathbf{e}}$. By the inductive hypothesis, we have $\mathscr{P} \models\!\!\approx_{\varphi} \dot{\neg}[W/x]\mathbf{P}$. Let $\mathsf{w} := \mathcal{E}(W)$. Hence $\mathscr{P} \not\models\!\!\approx_{\varphi,[\mathsf{w}/x]}\mathbf{P}$ and $\mathscr{P} \not\models\!\!\approx_{\varphi} \dot{\forall} x_\alpha \mathbf{P}$. Thus $\mathscr{P} \models\!\!\approx_{\varphi}\mathbf{M}$.

Finally, we consider the cases in which $\mathbf{M}$ is not of the form $\dot{\neg}\mathbf{N}$.

Suppose $\mathbf{M}$ is $[\mathbf{P} \dot{\vee} \mathbf{Q}]$ and $\varphi_1^{\beta\eta}([\mathbf{P} \dot{\vee} \mathbf{Q}]) \in \mathcal{H}$. By $\vec{\nabla}_{\vee}$, either $\varphi_1^{\beta\eta}(\mathbf{P}) \in \mathcal{H}$ or $\varphi_1^{\beta\eta}(\mathbf{Q}) \in \mathcal{H}$. Since $|\mathbf{P}|_{\mathbf{e}} < |\mathbf{M}|_{\mathbf{e}}$ and $|\mathbf{Q}|_{\mathbf{e}} < |\mathbf{M}|_{\mathbf{e}}$, we know $\mathscr{P} \models\!\!\approx_{\varphi}\mathbf{P}$ or $\mathscr{P} \models\!\!\approx_{\varphi}\mathbf{Q}$. Thus $\mathscr{P} \models\!\!\approx_{\varphi}\mathbf{M}$.

Suppose $\mathbf{M}$ is $[\dot{\forall} x_\alpha \mathbf{P}]$ and $\varphi_1^{\beta\eta}(\mathbf{M}) \in \mathcal{H}$. Let $\langle \mathbf{A}, a\rangle \in \mathcal{D}_\alpha$ be given. Note that $(\varphi, [\langle \mathbf{A}, a\rangle/x])_1^{\beta\eta}(\mathbf{P})$ is $\varphi_1^{\beta\eta}([\mathbf{A}/x]\mathbf{P})$ (since $\mathbf{A}$ is closed). By $\vec{\nabla}_{\forall}$ and $\vec{\nabla}_{\beta\eta}$, we have $\varphi_1^{\beta\eta}([\mathbf{A}/x]\mathbf{P}) \in \mathcal{H}$. Since $|\mathbf{P}|_{\mathbf{e}} < |\mathbf{M}|_{\mathbf{e}}$ we know $\mathscr{P} \models\!\!\approx_{\varphi,[\langle \mathbf{A},a\rangle/x]}\mathbf{P}$. Thus $\mathscr{P} \models\!\!\approx_{\varphi}\mathbf{M}$. □

PROOF OF THEOREM 5.7.16. We define $\mathcal{I}(c)$ for the remaining logical constants $c \in \mathcal{S}$.

$\top$: Suppose $\top \in \mathcal{S}$. We know $\mathtt{T}$ is a possible value for $\top$ since otherwise $\dot{\neg}\top \in \mathcal{H}$, contradicting $\vec{\nabla}^{\sharp}$ and $\vec{\nabla}_{\bot}$. Let $\mathcal{I}(\top) := \langle \top, \mathtt{T}\rangle$. Clearly $\mathcal{I}(\top) \sim \mathcal{I}(\top)$ since $\sim$ is total on $\mathcal{D}_o$. We know $\mathfrak{L}^{\partial}_{\top}(\mathcal{I}(\top))$ holds since $\upsilon(\mathcal{I}(\top)) = \upsilon(\langle \top, \mathtt{T}\rangle) = \mathtt{T}$.

$\bot$: Suppose $\bot \in \mathcal{S}$. We know $\mathtt{F}$ is a possible value for $\bot$ since otherwise $\bot \in \mathcal{H}$, contradicting $\vec{\nabla}^{\sharp}$ and $\vec{\nabla}_{\bot}$. Let $\mathcal{I}(\bot) := \langle \bot, \mathtt{F}\rangle$. Again we have $\mathcal{I}(\bot) \sim \mathcal{I}(\bot)$ since $\sim$ is total on $\mathcal{D}_o$. We know $\mathfrak{L}^{\partial}_{\bot}(\mathcal{I}(\bot))$ holds since $\upsilon(\mathcal{I}(\bot)) = \upsilon(\langle \bot, \mathtt{F}\rangle) = \mathtt{F}$.

$\neg$: Let $n : \mathcal{D}_o \to \mathcal{D}_o$ be the function taking $\langle \mathbf{A}, a\rangle$ to $\langle \neg\mathbf{A}, b\rangle$ where $b = \mathtt{F}$ if $a = \mathtt{T}$ and $b = \mathtt{T}$ if $a = \mathtt{F}$. It is easy to check using $\vec{\nabla}^{\sharp}$, $\vec{\nabla}_{\neg}$ and $\vec{\nabla}_{c}$ that $\langle \neg, n\rangle \in \mathcal{D}_{oo}$. It is also easy to check that $\langle \neg, n\rangle \in \overline{\mathcal{D}_{oo}^{\sim}}$ using the definition of $\sim$. Let $\mathcal{I}(\neg) := \langle \neg, n\rangle$. $\mathfrak{L}^{\partial}_{\neg}(\mathcal{I}(\neg))$ holds by the definition of n.

Π^α: We let $\langle \Pi^\alpha, \pi^\alpha\rangle$ where $\pi^\alpha : \mathcal{D}_{o\alpha} \to \mathcal{D}_o$ is defined by

$$\pi^\alpha(\langle \mathbf{F}, f\rangle) := \begin{cases} \langle \Pi^\alpha\mathbf{F}, \mathtt{T}\rangle & \upsilon(f(\mathsf{a})) = \mathtt{T} \text{ for every } \mathsf{a} \in \overline{\mathcal{D}_\alpha^{\sim}}. \\ \langle \Pi^\alpha\mathbf{F}, \mathtt{F}\rangle & \text{otherwise.} \end{cases}$$

Note that we have relativized the Π^α quantifier to $\overline{\mathcal{D}_\alpha^{\sim}}$.

In order to verify π^α is well-defined, let $\langle \mathbf{F}, f\rangle \in \mathcal{D}_{o\alpha}$ be given. We must check $\pi^\alpha(\langle \mathbf{F}, f\rangle) \in \mathcal{D}_o$.

First suppose $\upsilon(f(\mathsf{a})) = \mathtt{T}$ for every $\mathsf{a} \in \overline{\mathcal{D}_\alpha^\sim}$. In this case $\pi^\alpha(\langle \mathbf{F}, f\rangle)$ is $\langle [\Pi^\alpha\, \mathbf{F}], \mathtt{T}\rangle$. Assume $\langle [\Pi^\alpha\, \mathbf{F}], \mathtt{T}\rangle \notin \mathcal{D}_o$. Hence we have $\dot\neg[\Pi^\alpha\, \mathbf{F}] \in \mathcal{H}$. By $\vec{\nabla}^\sharp$, $\vec{\nabla}_\exists$ and $\vec{\nabla}_{\beta\eta}$, there is some parameter W_α such that $\dot\neg[\mathbf{F}\, W]^\downarrow \in \mathcal{H}$. Since $\mathcal{I}(W) \in \overline{\mathcal{D}_\alpha^\sim}$ we must have $\upsilon(f(\mathcal{I}(W))) = \mathtt{T}$. This is impossible, since $\langle [\mathbf{F}\, W]^\downarrow, \mathtt{T}\rangle \notin \mathcal{D}_o$.

On the other hand, suppose there is some $\mathsf{a} = \langle \mathbf{A}, a\rangle \in \overline{\mathcal{D}_\alpha^\sim}$ with $\upsilon(f(\mathsf{a})) = \mathtt{F}$. Assume $\langle [\Pi^\alpha\, \mathbf{F}], \mathtt{F}\rangle \notin \mathcal{D}_o$. Hence $[\Pi^\alpha\, \mathbf{F}] \in \mathcal{H}$. By $\vec{\nabla}^\sharp$, $\vec{\nabla}_\forall$ and $\vec{\nabla}_{\beta\eta}$, we have $[\mathbf{F}\, \mathbf{A}]^\downarrow \in \mathcal{H}$. However, we must have $\langle [\mathbf{F}\, \mathbf{A}]^\downarrow, \mathtt{F}\rangle \in \mathcal{D}_o$ since $\upsilon(f(\langle \mathbf{A}, a\rangle)) = \mathtt{F}$. This implies that $[\mathbf{F}\, \mathbf{A}]^\downarrow \notin \mathcal{H}$, a contradiction.

Thus π^α is well-defined and $\langle \Pi^\alpha, \pi^\alpha\rangle \in \mathcal{D}_{o(o\alpha)}$. We must check $\langle \Pi^\alpha, \pi^\alpha\rangle \sim \langle \Pi^\alpha, \pi^\alpha\rangle$. Let $\langle \mathbf{F}, f\rangle \sim \langle \mathbf{G}, g\rangle$ be given. By the definition of $\sim$, $f(\mathsf{a}) \sim g(\mathsf{a})$ and so $\upsilon(f(\mathsf{a})) = \upsilon(g(\mathsf{a}))$ for every $\mathsf{a} \in \overline{\mathcal{D}_\alpha^\sim}$. Hence $\upsilon(\pi^\alpha(\langle \mathbf{F}, f\rangle)) = \upsilon(\pi^\alpha(\langle \mathbf{G}, g\rangle))$ which means $\pi^\alpha(\langle \mathbf{F}, f\rangle) \sim \pi^\alpha(\langle \mathbf{G}, g\rangle)$. Let $\mathcal{I}(\Pi^\alpha) := \langle \Pi^\alpha, \pi^\alpha\rangle$. It is easy to check $\mathfrak{L}^\partial_{\Pi^\alpha}(\mathcal{I}(\Pi^\alpha))$ holds using the definition of π^α.

Σ^α: We define $\mathcal{I}(\Sigma^\alpha)$ in a manner analogous to $\mathcal{I}(\Pi^\alpha)$.

□

A.4. Proofs from Chapter 6

PROOF OF THEOREM 6.1.3.

1. We have the following $\mathcal{G}^\emptyset_{\beta\eta}$-derivation of $\mathbf{M}$:

$$\dfrac{\dfrac{\dfrac{\dfrac{\dfrac{}{\dot\neg[P\,F], [P\,F]}\,\mathcal{G}(Init)}{\dot\neg[P\,F], [P\,\lambda x \,\text{.}\, F\,x]}\,\mathcal{G}(\beta\eta)}{[P\,F] \mathrel{\dot\supset} [P_{o(\iota\iota)}\,\lambda x \,\text{.}\, F\,x]}\,\mathcal{G}(\dot\supset)}{F \doteq^{\iota\iota} \lambda x \,\text{.}\, F\,x}\,\mathcal{G}(\dot\forall^P)}{\mathbf{M}}\,\mathcal{G}(\dot\forall^F)$$

2. Let $\mathcal{M} = (\mathcal{D}, @, \mathcal{E}, \upsilon) \in \mathfrak{M}_\beta(\mathcal{S}_{all})$ be the retraction model from Example 4.3.14. Let $g : \mathcal{D}_\iota \to \mathcal{D}_\iota$ be the trivial function (since $\mathcal{D}_\iota$ is a singleton). Let $p : \mathcal{D}_{\iota\iota} \to \mathcal{D}_o$ be defined by $p(\langle 1, g\rangle) := \mathtt{F}$ and $p(\langle 2, g\rangle) := \mathtt{T}$.

Note that

$$\mathcal{E}_{\varphi,[\langle 2,g\rangle/f],[\langle 1,p\rangle/q]}(f) = \langle 2, g\rangle$$

and

$$\mathcal{E}_{\varphi,[\langle 2,g\rangle/f],[\langle 1,p\rangle/q]}([\lambda x_\iota \,\blacksquare\, f\, x]) = i(g) = \langle 1, g\rangle.$$

Hence

$$\mathcal{E}_{\varphi,[\langle 2,g\rangle/f],[\langle 1,p\rangle/q]}([q\, f]) = p(\langle 2, g\rangle) = \mathtt{T}$$

and

$$\mathcal{E}_{\varphi,[\langle 2,g\rangle/f],[\langle 1,p\rangle/q]}([q\, [\lambda x_\iota \,\blacksquare\, f\, x]]) = p(\langle 1, g\rangle) = \mathtt{F}.$$

Thus

$$\mathcal{M} \not\models_{\varphi,[\langle 2,g\rangle/f],[\langle 1,p\rangle/q]} \underset{\cdot}{\neg}[q\, f] \,\underset{\cdot}{\vee}\, [q\, [\lambda x_\iota \,\blacksquare\, f\, x]]$$

and so

$$\mathcal{M} \not\models_{\varphi,[\langle 2,g\rangle/f]} f \underset{\cdot}{=}^{\iota\iota} \lambda x_\iota \,\blacksquare\, f\, x.$$

Hence $\mathcal{M} \not\models \mathbf{M}$. By soundness (Theorem 3.4.6), we know $\mathbf{M}$ is not $\mathcal{G}_\beta^{\mathcal{S}_{all}}$-derivable.

3. We have the following $\mathcal{G}_{\beta\mathfrak{f}\mathfrak{b}}^{\emptyset}$-derivation of $\mathbf{N}$:

$$\cfrac{\cfrac{\cfrac{\cfrac{\cfrac{\cfrac{\cfrac{\begin{matrix}\mathcal{D} \\ \underset{\cdot}{\neg}[A \underset{\cdot}{\equiv} B], \underset{\cdot}{\neg}B, A\end{matrix} \qquad \begin{matrix}\mathcal{E} \\ \underset{\cdot}{\neg}[A \underset{\cdot}{\equiv} B], \underset{\cdot}{\neg}A, B\end{matrix}}{\underset{\cdot}{\neg}[A \underset{\cdot}{\equiv} B], [B \underset{\cdot}{=}^{o} A]}\,\mathcal{G}(\mathfrak{b})}{\underset{\cdot}{\neg}[A \underset{\cdot}{\equiv} B], \underset{\cdot}{\neg}[P\, A], [P\, B]}\,\mathcal{G}(Init^{\underset{\cdot}{=}})}{\underset{\cdot}{\neg}[A \underset{\cdot}{\equiv} B], [[P\, A] \underset{\cdot}{\supset} [P\, B]]}\,\mathcal{G}(\underset{\cdot}{\supset})}{\underset{\cdot}{\neg}[A \underset{\cdot}{\equiv} B], [A \underset{\cdot}{=}^{o} B]}\,\mathcal{G}(\underset{\cdot}{\forall}^{P})}{[A \underset{\cdot}{\equiv} B] \underset{\cdot}{\supset} \blacksquare\, A \underset{\cdot}{=}^{o} B}\,\mathcal{G}(\underset{\cdot}{\supset})}{\underset{\cdot}{\forall} y \,\blacksquare\, [A \underset{\cdot}{\equiv} y] \underset{\cdot}{\supset} A \underset{\cdot}{=}^{o} y}\,\mathcal{G}(\underset{\cdot}{\forall}^{B})}{\mathbf{N}}\,\mathcal{G}(\underset{\cdot}{\forall}^{A})$$

where $\mathcal{D}$ is the derivation

$$\cfrac{\cfrac{}{\underset{\cdot}{\neg}A, \underset{\cdot}{\neg}B, \underset{\cdot}{\neg}B, A}\,\mathcal{G}(Init^{\underset{\cdot}{=}}) \qquad \cfrac{}{A, B, \underset{\cdot}{\neg}B, A}\,\mathcal{G}(Init^{\underset{\cdot}{=}})}{\underset{\cdot}{\neg}[A \underset{\cdot}{\equiv} B], \underset{\cdot}{\neg}B, A}\,\mathcal{G}(\underset{\cdot}{\neg} \underset{\cdot}{\equiv})$$

and $\mathcal{E}$ is the derivation

$$\dfrac{\dfrac{}{\neg A, \neg B, \neg A, B}\,\mathcal{G}(Init^{\dot{=}}) \qquad \dfrac{}{A, B, \neg A, B}\,\mathcal{G}(Init^{\dot{=}})}{\neg[A \stackrel{.}{\equiv} B], \neg A, B}\,\mathcal{G}(\neg \stackrel{.}{\equiv})$$

using the admissible rule $\mathcal{G}(\neg \stackrel{.}{\equiv})$ (cf. Corollary 5.4.8).

4. Let $\mathcal{M} = (\mathcal{D}, @, \mathcal{E}, \upsilon) \in \mathfrak{M}_{\beta\eta}(\mathcal{S}_{all})$ be the retraction model from Example 4.3.13. Let $p : \mathcal{D}_o \to \mathcal{D}_o$ be defined by

$$p(\langle 1, \mathsf{b}\rangle) := \langle 1, \mathtt{F}\rangle \text{ and } p(\langle 2, \mathsf{b}\rangle) := \langle 1, \mathtt{T}\rangle$$

for each $\mathsf{b} \in \{\mathtt{T}, \mathtt{F}\}$.

Note that

$$\mathcal{E}_{\varphi,[\langle 2,\mathtt{T}\rangle/x],[\langle 1,\mathtt{T}\rangle/y],[\langle 1,p\rangle/q]}([q\,x]) = p(\langle 2, \mathtt{T}\rangle) = \langle 1, \mathtt{T}\rangle$$

and

$$\mathcal{E}_{\varphi,[\langle 2,\mathtt{T}\rangle/x],[\langle 1,\mathtt{T}\rangle/y],[\langle 1,p\rangle/q]}([q\,y]) = p(\langle 1, \mathtt{T}\rangle) = \langle 1, \mathtt{F}\rangle.$$

Hence

$$\mathcal{M} \not\models_{\varphi,[\langle 2,\mathtt{T}\rangle/x],[\langle 1,\mathtt{T}\rangle/y]} [x \doteq^o y].$$

On the other hand, since $\upsilon(\langle 1, \mathtt{T}\rangle) = \mathtt{T} = \upsilon(\langle 2, \mathtt{T}\rangle)$ we know $\mathcal{M} \models_\psi x$ and $\mathcal{M} \models_\psi y$ where ψ is $\varphi, [\langle 2, \mathtt{T}\rangle/x], [\langle 1, \mathtt{T}\rangle/y]$. Using Lemma 3.3.4 we conclude

$$\mathcal{M} \models_{\varphi,[\langle 2,\mathtt{T}\rangle/x],[\langle 1,\mathtt{T}\rangle/y]} [x \stackrel{.}{\equiv} y].$$

Thus $\mathcal{M} \not\models \mathbf{N}$. Therefore, we know $\mathbf{M}$ is not $\mathcal{G}^{\mathcal{S}_{all}}_{\beta\eta}$-derivable by soundness (Theorem 3.4.9).

□

Proof of Lemma 6.3.10. Let (g, f) be an internal retraction from $\mathcal{D}_\alpha$ onto $\mathcal{D}_\beta$. Let $f_{\alpha\beta}$, $g_{\beta\alpha}$, $h_{\alpha\gamma}$, $k_{\beta\gamma}$, $p_{\gamma\alpha}$, $q_{\gamma\beta}$, x_α, y_β and z_γ be distinct variables and φ be an assignment with $\varphi(f) = \mathsf{f}$ and $\varphi(g) = \mathsf{g}$.

We prove $(\mathcal{E}_\varphi([\lambda h_{\alpha\gamma} \lambda z_\gamma \,\blacksquare\, g \,\blacksquare\, h\, z]), \mathcal{E}_\varphi([\lambda k_{\beta\gamma} \lambda z_\gamma \,\blacksquare\, f \,\blacksquare\, k\, z]))$ is an internal retraction from $\mathcal{D}_{\alpha\gamma}$ onto $\mathcal{D}_{\beta\gamma}$. For any $\mathsf{k} \in \mathcal{D}_{\beta\gamma}$ and $\mathsf{c} \in \mathcal{D}_\gamma$, we compute

$$\begin{aligned}
& \mathcal{E}_{\varphi}([\lambda h_{\alpha\gamma}\lambda z_{\gamma} \centerdot g \centerdot h\, z])@(\mathcal{E}_{\varphi}([\lambda k_{\beta\gamma}\lambda z_{\gamma} \centerdot f \centerdot k\, z])@\mathsf{k})@\mathsf{c} \\
= \; & \mathcal{E}_{\varphi,[\mathsf{k}/k]}([\lambda h_{\alpha\gamma}\lambda z_{\gamma} \centerdot g \centerdot h\, z])@\mathcal{E}_{\varphi,[\mathsf{k}/k]}([\lambda z_{\gamma} \centerdot f \centerdot k\, z])@\mathsf{c} \\
= \; & \mathcal{E}_{\varphi,[\mathsf{k}/k]}([[\lambda h_{\alpha\gamma}\lambda z_{\gamma} \centerdot g \centerdot h\, z]\,[\lambda z_{\gamma} \centerdot f \centerdot k\, z]])@\mathsf{c} \\
= \; & \mathcal{E}_{\varphi,[\mathsf{k}/k]}([\lambda z_{\gamma} \centerdot g \centerdot [\lambda z_{\gamma} \centerdot f \centerdot k\, z]\, z])@\mathsf{c} \\
= \; & \mathcal{E}_{\varphi,[\mathsf{k}/k],[\mathsf{c}/z]}([g \centerdot f \centerdot k\, z]) \\
= \; & \mathsf{g}@(\mathsf{f}@(\mathsf{k}@\mathsf{c})) = \mathsf{k}@\mathsf{c}.
\end{aligned}$$

By functionality, we conclude

$$\mathcal{E}_{\varphi}([\lambda h_{\alpha\gamma}\lambda z_{\gamma} \centerdot g \centerdot h\, z])@(\mathcal{E}_{\varphi}([\lambda k_{\beta\gamma}\lambda z_{\gamma} \centerdot f \centerdot k\, z])@\mathsf{k}) = \mathsf{k}.$$

Thus we have an internal retraction from $\mathcal{D}_{\alpha\gamma}$ onto $\mathcal{D}_{\beta\gamma}$.

We prove $(\mathcal{E}_{\varphi}([\lambda p_{\gamma\alpha}\lambda y_{\beta} \centerdot p \centerdot f\, y]), \mathcal{E}_{\varphi}([\lambda q_{\gamma\beta}\lambda x_{\alpha} \centerdot q \centerdot g\, x]))$ is an internal retraction from $\mathcal{D}_{\gamma\alpha}$ onto $\mathcal{D}_{\gamma\beta}$. For any $\mathsf{q} \in \mathcal{D}_{\gamma\beta}$ and $\mathsf{b} \in \mathcal{D}_{\beta}$, we compute

$$\begin{aligned}
& \mathcal{E}_{\varphi}([\lambda p_{\gamma\alpha}\lambda y_{\beta} \centerdot p \centerdot f\, y])@(\mathcal{E}_{\varphi}([\lambda q_{\gamma\beta}\lambda x_{\alpha} \centerdot q \centerdot g\, x])@\mathsf{q})@\mathsf{b} \\
= \; & \mathcal{E}_{\varphi,[\mathsf{q}/q]}([\lambda p_{\gamma\alpha}\lambda y_{\beta} \centerdot p \centerdot f\, y])@\mathcal{E}_{\varphi,[\mathsf{q}/q]}([\lambda x_{\alpha} \centerdot q \centerdot g\, x])@\mathsf{b} \\
= \; & \mathcal{E}_{\varphi,[\mathsf{q}/q]}([[\lambda p_{\gamma\alpha}\lambda y_{\beta} \centerdot p \centerdot f\, y]\,[\lambda x_{\alpha} \centerdot q \centerdot g\, x]])@\mathsf{b} \\
= \; & \mathcal{E}_{\varphi,[\mathsf{q}/q]}([\lambda y_{\beta} \centerdot [\lambda x_{\alpha} \centerdot q \centerdot g\, x] \centerdot f\, y])@\mathsf{b} \\
= \; & \mathcal{E}_{\varphi,[\mathsf{q}/q]}([\lambda y_{\beta} \centerdot q \centerdot g \centerdot f\, y])@\mathsf{b} \\
= \; & \mathcal{E}_{\varphi,[\mathsf{q}/q],[\mathsf{b}/y]}([q \centerdot g \centerdot f\, y]) \\
= \; & \mathsf{q}@(\mathsf{g}@(\mathsf{f}@\mathsf{b})) = \mathsf{q}@\mathsf{b}.
\end{aligned}$$

By functionality, we conclude

$$\mathcal{E}_{\varphi}([\lambda p_{\gamma\alpha}\lambda y_{\beta} \centerdot p \centerdot f\, y])@(\mathcal{E}_{\varphi}([\lambda q_{\gamma\beta}\lambda x_{\alpha} \centerdot q \centerdot g\, x])@\mathsf{q}) = \mathsf{q}.$$

Thus we have an internal retraction from $\mathcal{D}_{\gamma\alpha}$ onto $\mathcal{D}_{\gamma\beta}$. □

PROOF OF LEMMA 6.6.5. To verify $\mathcal{M}$ realizes $\top$, let P_o be any parameter of type o. By $\mathfrak{L}_{=^o}(=^o)$, we know $\mathfrak{L}_{\top}(\mathcal{E}(P =^o P))$ holds. Let $\mathsf{t} := \mathcal{E}(P =^o P) \in \mathcal{D}_o$.

We briefly consider two cases.

1. Suppose $\mathcal{M}$ realizes $\neg$. In this case there is some $\mathsf{n} \in \mathcal{D}_{oo}$ such that $\mathfrak{L}_{\neg}(\mathsf{n})$ holds. Let $\mathsf{f} := \mathsf{n}@\mathsf{t}$. By $\mathfrak{L}_{\neg}(\mathsf{n})$ and $\mathfrak{L}_{\top}(\mathsf{t})$, we know $\mathfrak{L}_{\bot}(\mathsf{f})$ holds.

2. Suppose $\mathcal{M}$ realizes $\bot$. In this case we have some $\mathsf{f} \in \mathcal{D}_o$ such that $\mathfrak{L}_\bot(\mathsf{f})$ holds. Let f_o be a variable and φ be any assignment with $\varphi(f) = \mathsf{f}$. Here, we can let $\mathsf{n} := \mathcal{E}_\varphi(\lambda x_o\, [f_o =^o x])$. Using $\mathfrak{L}_\bot(\mathsf{f})$ and $\mathfrak{L}_{=^o}(\mathcal{E}(=^o))$ (and property $\mathfrak{b}$), we know $\mathfrak{L}_\neg(\mathsf{n})$.

In either case we have $\mathsf{f} \in \mathcal{D}_o$ and $\mathsf{n} \in \mathcal{D}_{oo}$ such that $\mathfrak{L}_\bot(\mathsf{f})$ and $\mathfrak{L}_\neg(\mathsf{n})$ hold. Hence $\mathcal{M}$ realizes $\top$, $\bot$ and $\neg$.

Next, we prove $\mathcal{M}$ realizes $\wedge$ using the definition in [9]. Let t_o be a variable and φ be an assignment with $\varphi(t) = \mathsf{t}$. We define

$$\mathsf{c} := \mathcal{E}_\varphi([\lambda x_o \lambda y_o \centerdot [\lambda g_{ooo}\, g\, x\, y] = [\lambda g_{ooo}\, g\, t\, t]])$$

(where x and y are distinct and different from t).

To check $\mathfrak{L}_\wedge(\mathsf{c})$, let $\mathsf{a}, \mathsf{b} \in \mathcal{D}_o$ be given. Define

$$\psi := (\varphi, [\mathsf{a}/x], [\mathsf{b}/y]).$$

Suppose $\upsilon(\mathsf{c}@\mathsf{a}@\mathsf{b}) = \mathtt{T}$. By $\mathfrak{L}_{=^{o(ooo)}}(\mathcal{E}(=^{o(ooo)}))$, we have

$$\mathcal{E}_\psi(\lambda g_{ooo}\, g\, x\, y) = \mathcal{E}_\psi(\lambda g_{ooo}\, g\, t\, t).$$

We compute

$$\begin{aligned} \mathsf{a} = \psi(x) &= \mathcal{E}_\psi(\lambda g_{ooo}\, g\, x\, y)@\mathcal{E}_\psi(\lambda x_o \lambda y_o x) \\ &= \mathcal{E}_\psi(\lambda g_{ooo}\, g\, t\, t)@\mathcal{E}_\psi(\lambda x_o \lambda y_o x) = \psi(t) = \mathsf{t} \end{aligned}$$

and

$$\begin{aligned} \mathsf{b} = \psi(y) &= \mathcal{E}_\psi(\lambda g_{ooo}\, g\, x\, y)@\mathcal{E}_\psi(\lambda x_o \lambda y_o y) \\ &= \mathcal{E}_\psi(\lambda g_{ooo}\, g\, t\, t)@\mathcal{E}_\psi(\lambda x_o \lambda y_o y) = \psi(t) = \mathsf{t}. \end{aligned}$$

That is, $\mathsf{a} = \mathsf{b} = \mathsf{t}$. In particular, $\upsilon(\mathsf{a}) = \upsilon(\mathsf{b}) = \upsilon(\mathsf{t}) = \mathtt{T}$.

To prove the converse, we must use extensionality. Suppose

$$\upsilon(\mathsf{a}) = \mathtt{T} \text{ and } \upsilon(\mathsf{b}) = \mathtt{T}.$$

By property $\mathfrak{b}$, we know $\mathsf{a} = \mathsf{b} = \mathsf{t}$. Hence

$$\mathcal{E}_{\psi,[\mathsf{g}/g]}(g\, x\, y) = \mathsf{g}@\mathsf{a}@\mathsf{b} = \mathsf{g}@\mathsf{t}@\mathsf{t} = \mathcal{E}_{\psi,[\mathsf{g}/g]}(g\, t\, t)$$

for any $\mathsf{g} \in \mathcal{D}_{ooo}$. By functionality, we have

$$\mathcal{E}_\psi(\lambda g_{ooo}\, g\, x\, y) = \mathcal{E}_\psi(\lambda g_{ooo}\, g\, t\, t).$$

We use $\mathfrak{L}_{=^{o(ooo)}}(\mathcal{E}(=^{o(ooo)}))$ to conclude $\upsilon(\mathsf{c}@\mathsf{a}@\mathsf{b}) = \mathtt{T}$, as desired.

Since $\mathcal{M}$ realizes $\neg$ and $\wedge$, we know $\mathcal{M}$ realizes $\mathcal{S}_{Bool}$ by Lemma 6.4.1. All that remains is to prove $\mathcal{M}$ realizes all quantifiers. For any type α, since $\mathcal{M}$ is extensional and realizes $\top$ and $=^{o\alpha}$, we can use Lemma 6.2.2 to conclude $\mathcal{M}$ realizes Π^α. Finally, Lemma 6.2.3 implies $\mathcal{M}$ realizes Σ^α for every type α.

Thus $\mathcal{M}$ realizes $\mathcal{S}_{all}$. □

PROOF OF LEMMA 6.7.4. Suppose $f \in \mathcal{R}_\iota$. Assume there exists x and y with $-1 < x < y < 1$ and $f(y) < f(x)$. Hence

$$x \in f^{-1}((f(y), 1)) \in \mathcal{B}.$$

Since every interval in $\mathcal{B}$ is upward closed, we must have

$$y \in f^{-1}((f(y), 1)),$$

a contradiction. Thus f is nondecreasing. Next, if f is not left continuous, then there is some $a \in (-1, 1)$ and increasing sequence

$$-1 < x_1 < x_2 < \cdots < a$$

such that $x_n \to a$ but $f(x_n) \not\to f(a)$ as $n \to \infty$. Since f is nondecreasing, $\{f(x_n)\}_n$ is a bounded, monotone sequence and hence convergent. Let L be its limit. Since $L \neq f(a)$ and each $-1 < f(x_n) \leq f(a)$, we must have $-1 < L < f(a)$. Hence we know $a \in f^{-1}((L, 1)) \in \mathcal{B}$. There is some $b \in [-1, 1]$ such that $f^{-1}((L, 1)) = (b, 1)$. We know $a > b$ since $a \in (b, 1)$. Choose some $c \in (b, a) \subseteq f^{-1}((L, 1))$. Since $f(x_n) \leq L < f(c)$ and f is nondecreasing, $x_n < c$ for every n. Since $x_n \to a$ as $n \to \infty$, we conclude $a \leq c$, contradicting our choice of c. Thus f is left continuous.

Conversely, suppose f is nondecreasing and left continuous. For any $b \in [-1, 1]$, we prove $f^{-1}((b, 1)) \in \mathcal{B}$. If $f^{-1}((b, 1))$ is empty, then we are done. Hence we can assume $f^{-1}((b, 1))$ is nonempty. Let

$$a := \inf(\{y \mid f(y) > b\}) \in [-1, 1).$$

Then for every $x \in (a, 1)$, there is some $y \in (a, x)$ such that $f(y) > b$. Since f is nondecreasing, $f(x) \geq f(y) > b$ and so $x \in f^{-1}((b, 1))$. Thus $(a, 1) \subseteq f^{-1}((b, 1)) \subseteq (-1, 1)$. If $a = -1$, then we are done. Assume $a > -1$. To verify $f^{-1}((b, 1)) \subseteq (a, 1)$, assume there is some $x \in (-1, a]$ with $f(x) > b$. Since f is left continuous, there is some $c \in (-1, x)$ such that $f(c) > b$. This $c < a$ contradicts the definition of a as the infimum of such values. □

PROOF OF LEMMA 6.7.5. Since $g \in \mathcal{R}_{o\iota}$, $g(x) \in \mathcal{D}_{o\iota}$. By Lemma 4.5.4, there is some $a \in [-1,1]$ such that $g(x) = \chi_{(a,1)}$. Since

$$\chi_{(a,1)}(y) = g(x)(y) = \mathtt{T},$$

we know $y > a$. Hence for any $y' \in (y,1)$ we have

$$g(x)(y') = \chi_{(a,1)}(y') = \mathtt{T}.$$

By Lemma 4.4.4, we know $K(y) \in \mathcal{R}_\iota$. Since $g \in \mathcal{R}_{o\iota}$ we must have $S(g, K(y)) \in \mathcal{R}_o$. That is, there is some $b \in [-1,1]$ such that

$$\{z \in (-1,1) \mid g(z)(y) = \mathtt{T}\}$$
$$= \{z \in (-1,1) \mid S(g, K(y))(z) = \mathtt{T}\} = (b,1).$$

Since $g(x)(y) = \mathtt{T}$, we have $x > b$. Hence

$$g(x')(y) = \mathtt{T} \text{ for any } x' \in (x,1).$$

□

PROOF OF LEMMA 6.7.6.

(1)⇒(2): This follows from $\mathcal{D}_{o\iota\iota} \subseteq \mathcal{R}_{o\iota}$ (cf. Lemma 4.5.5).

(2)⇒(3): Suppose $g \in \mathcal{R}_{o\iota}$. Then $g(x) \in \mathcal{D}_{o\iota}$ for each $x \in (-1,1)$. By Lemma 4.5.4, for every x there is some $l(x) \in [-1,1]$ such that

$$g(x) = \chi_{(l(x),1)}.$$

Clearly, $g(x)(y) = \mathtt{T}$ iff $l(x) < y$ for every $x, y \in (-1,1)$.

First we check l is nonincreasing. Suppose $-1 < a < b < 1$ and $-1 \leq l(a) < l(b) \leq 1$. Choose some $c \in (l(a), l(b))$. Since $l(a) < c$ and $l(b) \not< c$, we know $g(a)(c) = \mathtt{T}$ and $g(b)(c) = \mathtt{F}$, contradicting Lemma 6.7.5.

To prove l is left continuous, suppose there is an $a \in (-1,1)$ and increasing sequence $-1 < x_1 < x_2 < \cdots < a$ where $x_n \to a$ and $l(x_n) \not\to l(a)$ as $n \to \infty$. Since l is nonincreasing,

$$l(x_1) \geq l(x_2) \geq \cdots$$

is a monotone sequence and has a limit $L \in [-1,1]$. Also, $l(x_n) \geq l(a)$ for every n and $L \neq l(a)$, so we must have $L > l(a)$. Choose some $c \in (l(a), L)$. Since $g \in \mathcal{R}_{o\iota}$ and $K(c) \in \mathcal{R}_\iota$, we know $S(g, K(c)) \in \mathcal{R}_o$. Hence there is some $d \in [-1,1]$ such that $S(g, K(c)) = \chi_{(d,1)}$. Note that for any $x \in (-1, 1)$, we have

$$\begin{aligned} x > d \quad &\text{iff} \quad S(g, K(c))(x) = \mathtt{T} \\ &\text{iff} \quad g(x)(c) = \mathtt{T} \\ &\text{iff} \quad l(x) < c. \end{aligned}$$

By the choice of c, we have $l(a) < c$ and so (by the equivalence) $a > d$. Since $x_n \to a$, there must be some N with $x_N > d$. Consequently, $l(x_N) < c$ by the equivalence. This contradicts $l(x_N) \geq L > c$. Thus l is left continuous.

(3)⇒(1): Suppose $l : (-1, 1) \to \Re$ is a nonincreasing left continuous function such that $g(x)(y) = \mathtt{T}$ iff $l(x) > y$. We first must check $g : \mathcal{D}_\iota \to \mathcal{D}_{o\iota}$. This follows from Lemma 4.5.4 since

$$g(x) = \chi_{(l(x),1)} \in \mathcal{D}_{o\iota}$$

for each $x \in \mathcal{D}_\iota$. To check $g \in \mathcal{D}_{o\iota\iota}$, let $f_1 \in \mathcal{R}_\iota$ be given. To check $g \circ f_1 \in \mathcal{R}_{o\iota}$, let $f_2 \in \mathcal{R}_\iota$ be given. By Lemma 6.7.4, we know f_1 and f_2 are nondecreasing and left continuous. We must check $S(g \circ f_1, f_2) \in \mathcal{R}_o$. That is,

$$\{x \in (-1, 1) \mid g(f_1(x))(f_2(x)) = \mathtt{T}\} \in \mathcal{B}.$$

By our assumption about l, we know

$$\begin{aligned} &\{x \in (-1, 1) \mid g(f_1(x))(f_2(x)) = \mathtt{T}\} \\ &= \{x \in (-1, 1) \mid l(f_1(x)) < f_2(x)\}. \end{aligned}$$

If $\{x \in (-1, 1) \mid l(f_1(x)) < f_2(x)\}$ is empty, then we are done. Assume $\{x \in (-1, 1) \mid l(f_1(x)) < f_2(x)\}$ is nonempty and let

$$a := \inf\{y \in (-1, 1) \mid l(f_1(y)) < f_2(y)\}.$$

We will determine

$$\{x \in (-1, 1) \mid l(f_1(x)) < f_2(x)\} = (a, 1) \in \mathcal{B}.$$

Suppose $x \in (a, 1)$. By our choice of a, there is a $y \in (a, x)$ with $l(f_1(y)) < f_2(y)$. Since $y < x$, f_1 and f_2 are nondecreasing and l is nonincreasing, $f_1(y) \leq f_1(x)$, $l(f_1(x)) \leq l(f_1(y))$ and $f_2(y) \leq f_2(x)$. Hence

$$l(f_1(x)) \leq l(f_1(y)) < f_2(y) \leq f_2(x).$$

Thus

$$(a,1) \subseteq \{x \in (-1,1) \mid l(f_1(x)) < f_2(x)\} \subseteq (-1,1).$$

To verify $\{x \in (-1,1) \mid l(f_1(x)) < f_2(x)\} \subseteq (a,1)$ assume there is some $x \in (-1,a]$ with $l(f_1(x)) < f_2(x)$. That is,

$$l(f_1(x)) - f_2(x) < 0.$$

Since l, f_1 and f_2 are left continuous, there is some $y \in (-1,x)$ such that $l(f_1(y)) - f_2(y) < 0$. This $y < x \leq a$ contradicts the choice of a as the infimum of such values.

□

PROOF OF LEMMA 6.7.13. Suppose $f \in \mathcal{R}_\iota$ and f is not eventually constant. We will prove f is uniformly unbounded. Given $j \in \mathbb{N}$, since $\{0,\ldots,j\} \in \mathcal{B}$ we must have $f^{-1}(\{0,\ldots,j\}) \in \mathcal{B}$. Assume $f^{-1}(\{0,\ldots,j\})$ is cofinite. Then there is an $L \in \{0,\ldots,j\}$ such that $f^{-1}(L)$ is cofinite. That is, $(\mathbb{N} \setminus f^{-1}(L))$ is finite. There must be some $N \in \mathbb{N}$ such that

$$(\mathbb{N} \setminus f^{-1}(L)) \subseteq \{0,\ldots,N\}.$$

However, this implies $f(n) = L$ for every $n > N$, contradicting our assumption that f is not eventually constant. Hence

$$f^{-1}(\{0,\ldots,j\}) \in \mathcal{B}$$

must be finite and there must be some $I \in \mathbb{N}$ such that

$$f^{-1}(\{0,\ldots,j\}) \subseteq \{0,\ldots,I\}.$$

This means for any $i > I$ we have $f(i) > j$. Thus f is uniformly unbounded.

Next suppose f is eventually constant. There exist $N, L \in \mathbb{N}$ such that $f(n) = L$ for all $n > N$. In this case, for any $X \subseteq \mathbb{N}$, $f^{-1}(X)$ is finite if $L \notin X$ and $f^{-1}(X)$ is cofinite if $L \in X$. In either case $f^{-1}(X) \in \mathcal{B}$. In particular, this holds for $X \in \mathcal{B}$ and so $f \in \mathcal{R}_\iota$.

Finally, suppose f is uniformly unbounded. For every $j \in \mathbb{N}$ there is some $I \in \mathbb{N}$ such that $f(i) > j$ for every $i > I$. To verify $f \in \mathcal{R}_\iota$, let $X \in \mathcal{B}$ be given. Suppose X is finite and let $j_1 \in \mathbb{N}$ be such that $X \subseteq \{0,\ldots,j_1\}$. Let $I_1 \in \mathbb{N}$ be such that $f(i) > j_1$ for every $i > I_1$. This implies $f^{-1}(X) \subseteq \{0,\cdots,I_1\}$. Hence $f^{-1}(X)$ is finite and $f^{-1}(X) \in \mathcal{B}$. Suppose X is cofinite and let $j_2 \in \mathbb{N}$ be such that

$(\mathbb{N} \setminus X) \subseteq \{0, \ldots, j_2\}$. Let $I_2 \in \mathbb{N}$ be such that $f(i) > j_2$ for every $i > I_2$. This implies

$$(\mathbb{N} \setminus f^{-1}(X)) = f^{-1}(\mathbb{N} \setminus X) \subseteq \{0, \cdots, I_2\}.$$

Hence $f^{-1}(X)$ is cofinite and $f^{-1}(X) \in \mathcal{B}$. Thus $f \in \mathcal{R}_\iota$. □

PROOF OF LEMMA 6.7.20. Let $p \in \mathcal{R}_{o\iota}$ be given. Assume for every $m, n \in \mathbb{N}$ there exist $m' > m$ and $n' > n$ such that

$$p(m')(n') \neq p(m)(n).$$

Using this assumption, we can construct increasing sequences

$$m_0 < m_1 < \cdots \text{ and } n_0 < n_1 < \cdots$$

in $\mathbb{N}$ such that $m_0 = n_0 = 0$ and

$$p(m_{i+1})(n_{i+1}) \neq p(m_i)(n_i).$$

Let $f : \mathbb{N} \to \mathbb{N}$ be defined by $f(m) := n_I$ where I is the least i such that $m_i \leq m < m_{i+1}$. By Lemma 6.7.13, we can prove $f \in \mathcal{R}_\iota$ by verifying it is uniformly unbounded. Given any $j \in \mathbb{N}$, since $(n_i)_{i \in \mathbb{N}}$ is an increasing sequence there is some $k \in \mathbb{N}$ such that $j < n_k$. Since $f(m) > n_k > j$ for each $m > m_{k+1}$, we know f is uniformly unbounded and $f \in \mathcal{R}_\iota$. Hence $S(p, f) \in \mathcal{R}_o$ and is eventually constant by Lemma 6.7.11. On the other hand, by construction we have

$$\begin{aligned} S(p, f)(m_{i+1}) = p(m_{i+1})(f(m_{i+1})) &= p(m_{i+1})(n_{i+1}) \neq p(m_i)(n_i) \\ &= p(m_i)(f(m_i)) = S(p, f)(m_i), \end{aligned}$$

a contradiction. □

PROOF OF LEMMA 6.7.21. Let $p \in \mathcal{R}_{o\iota}$ be given. By Lemma 6.7.20, there exist $M, N \in \mathbb{N}$ such that $p(m)(n) = p(M)(N)$ for every $m > M$ and $n > N$. Consider each $n \in \{0, \ldots, N\}$. By Lemma 4.4.4, we know the constant function $K(n)$ is in $\mathcal{R}_\iota$. Since $p \in \mathcal{R}_{o\iota}$ we know $S(p, K(n)) \in \mathcal{R}_o$ is eventually constant by Lemma 6.7.11. Hence there exist M_n and $\mathsf{a}_n \in \mathcal{D}_o$ such that $p(m)(n) = \mathsf{a}_n$ for every $m > M_n$.

Define $M' := \max(M, M_0, \ldots, M_N)$ and $g := p(M' + 1)$. We will prove $p(m) = g$ for any $m > M'$. Let $n \in \mathbb{N}$ be given. If $n \leq N$, then

$$p(m)(n) = \mathsf{a}_n = p(M' + 1)(n) = g(n)$$

since $m > M' \geq M_n$ and $M' + 1 > M' \geq M_n$. If $n > N$, then

$$p(m)(n) = p(M)(N) = p(M' + 1)(n) = g(n)$$

since $m > M' \geq M$ and $M' + 1 > M' \geq M$. Thus p is eventually constant. □

PROOF OF LEMMA 6.7.26.

1. Suppose $f \in \mathcal{R}_{\alpha\beta}$. Let $\mathsf{b} \in \mathcal{D}_\beta$ be given. By Lemma 4.4.4, we know $K(\mathsf{b}) \in \mathcal{R}_\beta$. Hence $S(f, K(\mathsf{b})) \in \mathcal{R}_\alpha \subseteq \mathcal{K}_\alpha$. This means there is some $u_\mathsf{b} \in \mathcal{D}_\alpha$ and $N \in \mathbb{N}$ such that $f(n)(\mathsf{b}) = u_\mathsf{b}$ for every $n > N$. Define $u : \mathcal{D}_\beta \to \mathcal{D}_\alpha$ by $u(\mathsf{b}) := u_\mathsf{b}$. Clearly f converges pointwise to u. Thus $f \in \mathcal{K}^p_{\alpha\beta}$.

2. This follows from part 1 since we know $\mathcal{R}_o \subseteq \mathcal{K}_o$ by Lemma 6.7.11.

3. Suppose $\mathcal{R}_\beta \subseteq \mathcal{K}_\beta$ and $f \in \mathcal{K}^p_{\alpha\beta}$. Hence f converges pointwise to some $u : \mathcal{D}_\beta \to \mathcal{D}_\alpha$. Let $h \in \mathcal{R}_\beta \subseteq \mathcal{K}_\beta$ be given. We must check $S(f, h)$ is in $\mathcal{R}_\alpha$. By Lemma 6.7.17, we know $\mathcal{K}_\alpha \subseteq \mathcal{R}_\alpha$. It is enough to check $S(f, h) \in \mathcal{K}_\alpha$. Since $h \in \mathcal{K}_\beta$, we know there is some $N_1 \in \mathbb{N}$ and $\mathsf{b} \in \mathcal{D}_\beta$ such that $h(n) = \mathsf{b}$ for every $n > N_1$. Hence $S(f, h)(n) = f(n)(h(n)) = f(n)(\mathsf{b})$ for every $n > N_1$. Since f converges pointwise to u, there is some N_2 such that $f(n)(\mathsf{b}) = u(\mathsf{b})$ for every $n > N_2$. Thus we have

$$S(f, h)(n) = f(n)(\mathsf{b}) = u(\mathsf{b})$$

for every $n > \max(N_1, N_2)$, proving $S(f, h) \in \mathcal{K}_\alpha \subseteq \mathcal{R}_\alpha$, as desired.

□

PROOF OF LEMMA 6.7.27. Define functions $f_1, f_2 : \mathbb{N} \to \mathcal{D}_\alpha^{\mathcal{D}_\beta}$ by

$$f_1(m) := \begin{cases} u^n & \text{if } m \text{ is even and } m = m^n \\ u & \text{otherwise} \end{cases}$$

and

$$f_2(m) := \begin{cases} u^n & \text{if } m \text{ is odd and } m = m^n \\ u & \text{otherwise.} \end{cases}$$

We first prove $f_1(n) \to^p_n u$ and $f_2(n) \to^p_n u$. Suppose $\mathsf{b} \in \mathcal{D}_\beta$. There is some $N \in \mathbb{N}$ such that $u^n(\mathsf{b}) = u(\mathsf{b})$ for every $n > N$ since $u^n \to^p_n u$. Since $(m^n)_{n\in\mathbb{N}}$ is an increasing sequence, there is some $M \in \mathbb{N}$ such that $n > N$ whenever $m^n > M$. Consequently, for any $m > M$ and $i \in \{1, 2\}$, we know

$$f_i(m)(\mathsf{b}) = u^n(\mathsf{b}) = u(\mathsf{b})$$

for some $n > N$ or

$$f_i(m)(\mathsf{b}) = u(\mathsf{b}).$$

For any $m > M$, $f_1(m)(\mathsf{b}) = f_2(m)(\mathsf{b}) = u(\mathsf{b})$ by a simple case analysis. Hence $f_1(n)(\mathsf{b}) \rightarrow_n u(\mathsf{b})$ and $f_2(n)(\mathsf{b}) \rightarrow_n u(\mathsf{b})$. Thus $f_1(n) \rightarrow_n^p u$ and $f_2(n) \rightarrow_n^p u$.

Since f_1 and f_2 converge pointwise to u, we know $f_1 \in \mathcal{K}^p_{\alpha\beta}$ and $f_2 \in \mathcal{K}^p_{\alpha\beta}$. Since $\mathcal{R}_\beta \subseteq \mathcal{K}_\beta$ we have $\mathcal{K}^p_{\alpha\beta} \subseteq \mathcal{R}_{\alpha\beta}$ by Lemma 6.7.26:3. Thus $f_1 \in \mathcal{R}_{\alpha\beta}$ and $f_2 \in \mathcal{R}_{\alpha\beta}$.

Let $p \in \mathcal{R}_{\gamma(\alpha\beta)}$ be given. By Lemma 6.7.26:1, $\mathcal{R}_{\gamma(\alpha\beta)} \subseteq \mathcal{K}^p_{\gamma(\alpha\beta)}$ since $\mathcal{R}_\gamma \subseteq \mathcal{K}_\gamma$. Hence p converges pointwise to some $h : \mathcal{D}_{\alpha\beta} \rightarrow \mathcal{D}_\gamma$. Thus there exists some $M_0 \in \mathbb{N}$ such that $p(m)(u) = h(u)$ for every $m > M_0$. Let $\mathsf{c} := h(u)$.

Since $(m^n)_{n \in \mathbb{N}}$ is a (strictly) increasing sequence, we know $m^n \geq n$ for every $n \in \mathbb{N}$. Hence $p(m^n)(u) = \mathsf{c}$ for every $n > M_0$ and so

$$p(m^n)(u) \rightarrow_n \mathsf{c}.$$

All that remains is to verify $p(m^n)(u^n) \rightarrow_n \mathsf{c}$.

Since $p \in \mathcal{R}_{\gamma(\alpha\beta)}$ and $f_1, f_2 \in \mathcal{R}_{\alpha\beta}$, we know $S(p, f_1) \in \mathcal{R}_\gamma \subseteq \mathcal{K}_\gamma$ and $S(p, f_2) \in \mathcal{R}_\gamma \subseteq \mathcal{K}_\gamma$. Thus there exist $\mathsf{c}_1 \in \mathcal{D}_\gamma$ and $\mathsf{c}_2 \in \mathcal{D}_\gamma$ and $M_1, M_2 \in \mathbb{N}$ such that

$$p(m)(f_1(m)) = \mathsf{c}_1 \text{ for all } m > M_1$$

and

$$p(m)(f_2(m)) = \mathsf{c}_2 \text{ for all } m > M_2.$$

Let $M := \max(M_0, M_1, M_2)$. We can prove $\mathsf{c}_1 = \mathsf{c}_2 = \mathsf{c}$ using the definition of f_1 on odd numbers and f_2 on even numbers to compute

$$\mathsf{c}_1 = p(2M + 1)(f_1(2M + 1)) = p(2M + 1)(u) = h(u) = \mathsf{c}$$

and

$$\mathsf{c}_2 = p(2M + 2)(f_2(2M + 2)) = p(2M + 2)(u) = h(u) = \mathsf{c}.$$

Let $n > M$ be given. Hence

$$m^n \geq n > M \geq M_1 \text{ and } m^n \geq n > M \geq M_2.$$

If m^n is even, then

$$p(m^n)(u^n) = p(m^n)(f_1(m^n)) = \mathsf{c}_1 = \mathsf{c}$$

since $m^n > M_1$. If m^n is odd, then

$$p(m^n)(u^n) = p(m^n)(f_2(m^n)) = \mathsf{c}_2 = \mathsf{c}$$

since $m^n > M_2$. Therefore, $p(m^n)(u^n) \to_n \mathsf{c}$, as desired. □

PROOF OF LEMMA 6.7.29.

(1)⇒(2): Suppose $B \in \mathscr{B}_\beta$ and let $w \in B$ be given. Then there exist $V^1, \ldots, V^n \in \mathscr{S}_\beta$ such that $B = V^1 \cap \cdots \cap V^n$. For each i $(1 \leq i \leq n)$ there is some $\mathsf{b}_i \in \mathcal{D}_\beta$ and $\mathsf{c}_i \in \mathcal{D}_o$ such that

$$V^i = \{u \in \mathcal{D}_o^{\mathcal{D}_\beta} \mid u(\mathsf{b}_i) = \mathsf{c}_i\}.$$

Since $w \in B$ we must have $w(\mathsf{b}_i) = \mathsf{c}_i$ for each i $(1 \leq i \leq n)$. Let X be the finite set $\{\mathsf{b}_i \mid 1 \leq i \leq n\}$. Suppose $u \in \mathcal{D}_o^{\mathcal{D}_\beta}$. If $u \in B$, then $u(\mathsf{b}_i) = \mathsf{c}_i = w(\mathsf{b}_i)$ for each $\mathsf{b}_i \in X$ and so

$$u\big|_X = w\big|_X.$$

If $u\big|_X = w\big|_X$, then $u(\mathsf{b}_i) = w(\mathsf{b}_i) = \mathsf{c}_i$ for each $\mathsf{b}_i \in X$ and so $u \in (V^1 \cap \cdots \cap V^n) = B$.

(2)⇒(1): If B is empty, then B is a basic set since

$$proj_{\mathsf{b}}^{-1}(\mathtt{F}) \cap proj_{\mathsf{b}}^{-1}(\mathtt{T})$$

is empty (where $\mathsf{b} \in \mathcal{D}_\beta$ is arbitrary). Assume B is nonempty and choose some $w \in B$ arbitrarily. By assumption, there is a finite set $X \subseteq \mathcal{D}_\beta$ such that forall $u \in \mathcal{D}_o^{\mathcal{D}_\beta}$ we have $u \in B$ iff $u\big|_X = w\big|_X$. Enumerate the elements of X as $\{\mathsf{b}_1, \ldots, \mathsf{b}_n\}$. For each i $(1 \leq i \leq n)$ let

$$V^i := \{u \in \mathcal{D}_o^{\mathcal{D}_\beta} \mid u(\mathsf{b}_i) = w(\mathsf{b}_i)\} \in \mathscr{S}_\beta.$$

Clearly $u \in V^1 \cap \cdots \cap V^n$ iff $u\big|_X = w\big|_X$ for every $u \in \mathcal{D}_o^{\mathcal{D}_\beta}$. Hence $u \in V^1 \cap \cdots \cap V^n$ iff $u \in B$ for every $u \in \mathcal{D}_o^{\mathcal{D}_\beta}$. Thus $B = V^1 \cap \cdots \cap V^n$ and so $B \in \mathscr{B}_\beta$.

□

PROOF OF LEMMA 6.7.30. Since $\mathcal{D}_\beta$ is countably infinite, we can enumerate it as

$$\mathcal{D}_\beta = \{\mathsf{b}_0, \mathsf{b}_1, \mathsf{b}_2, \ldots\}.$$

Assume there is no such finite set and natural number. In particular, applying this assumption to the empty set and 0, there is some $u^0 \in \mathcal{D}_{o\beta}$ and $m^0 > 0$ such that $p(m^0)(u^0) \neq p(m^0)(w)$. Suppose we have defined $u^0, \ldots, u^n \in \mathcal{D}_{o\beta}$ and $m^0 < \cdots < m^n$ such that $p(m^i)(u^i) \neq p(m^i)(w)$ and $u^i(\mathsf{b}_j) = w(\mathsf{b}_j)$ for each i $(0 \leq i \leq n)$ and j $(0 \leq j < i)$. Applying our assumption to the finite set $\{\mathsf{b}_0, \ldots, \mathsf{b}_n\}$ and m^n we know there is some $u^{n+1} \in \mathcal{D}_{o\beta}$ and $m^{n+1} > m^n$ such that $p(m^{n+1})(u^{n+1}) \neq p(m^{n+1})(w)$ and $u^{n+1}(\mathsf{b}_j) = w(\mathsf{b}_j)$ for $0 \leq j \leq n$.

The sequence u^n converges pointwise to w since for any fixed $m \in \mathbb{N}$ we know $u^n(\mathsf{b}_m) = w(\mathsf{b}_m)$ for every $n > m$. By Lemma 6.7.27, there is some $\mathsf{c} \in \mathcal{R}_o$ such that $p(m^n)(w) \to_n \mathsf{c}$ and $p(m^n)(u^n) \to_n \mathsf{c}$. This contradicts $p(m^n)(u^n) \neq p(m^n)(w)$ for every $n \in \mathbb{N}$. □

PROOF OF LEMMA 6.7.31. Let $p \in \mathcal{R}_{o(o\beta)}$ and $w \in \mathcal{R}_{o\beta}$ be given. By Lemma 6.7.30, there is a finite set $X \subseteq \mathcal{D}_\beta$ and an $N \in \mathbb{N}$ such that for all $u \in \mathcal{D}_{o\beta}$ and $n > N$ if $u\big|_X = w\big|_X$, then $p(n)(u) = p(n)(w)$. Define

$$B := \{u \in \mathcal{D}_{o\beta} \mid u\big|_X = w\big|_X\}.$$

Hence $p(n)(u) = p(n)(w)$ for every $u \in B$. Note that $w \in B$. We use Lemma 6.7.29 to conclude B is a basic open set. (Since $\mathcal{R}_\beta \subseteq \mathcal{K}_\beta$ we know $\mathcal{D}_{o\beta} = \mathcal{D}_o^{\mathcal{D}_\beta}$ by Lemma 6.7.18.)

Let $w' \in B$ be given. By definition, this means $w'\big|_X = w\big|_X$. Hence for any $u \in \mathcal{D}_{o\beta}$ we know $u \in B$ iff $u\big|_X = w'\big|_X$. Thus B is a basic open set. □

PROOF OF LEMMA 6.7.32. We apply Lemma 6.7.31 to obtain a basic neighborhood B^w and an $N^w \in \mathbb{N}$ for each $w \in \mathcal{D}_{o\beta}$ such that for every $n > N^w$ and every $u \in B^w$ we have $p(n)(u) = p(n)(w)$. This family $(B^w)_w$ is an open cover of the space. By compactness, there is a finite subcover. That is, there exist $w^1, \ldots, w^m \in \mathcal{D}_{o\beta}$ such that for every $u \in \mathcal{D}_{o\beta}$ we have $u \in B^{w^i}$ for some i $(1 \leq i \leq m)$. By Lemma 6.7.29, there are finite sets $X^1, \ldots, X^m$ of $\mathcal{D}_\beta$ such that for each i $(1 \leq i \leq m)$ and $u \in \mathcal{D}_o^{\mathcal{D}_\beta}$ we have $u \in B^{w^i}$ iff $u\big|_{X^i} = w^i\big|_{X^i}$. Let X be the finite set $X^1 \cup \cdots \cup X^m$ and N be the maximum of $N^{w^1}, \ldots, N^{w^m}$.

Suppose $n > N$, $u, v \in \mathcal{D}_{o\beta}$ and $u|_X = v|_X$. There exists some $i \in \{1, \ldots, m\}$ such that $u \in B^{w^i}$. Hence $u|_{X^i} = w^i|_{X^i}$. We also know $v|_{X^i} = u|_{X^i} = w^i|_{X^i}$ since $X \subseteq X^i$. Hence $v \in B^{w^i}$. Since $n > N \geq N^{w^i}$ and $u, v \in B^{w^i}$ we can conclude

$$p(n)(u) = p(n)(w^i) = p(n)(v)$$

as desired. □

PROOF OF LEMMA 6.7.33. Let $p \in \mathcal{R}_{o(o\beta)}$ be given. We must prove $p : \mathbb{N} \to \mathcal{D}_{o(o\beta)}$ is eventually constant. By Lemma 6.7.32, there exists a finite set $X \subseteq \mathcal{D}_\beta$ and an $N_0 \in \mathbb{N}$ such that for every $u, v \in \mathcal{D}_\beta$ and $n > N_0$ if $u|_X = v|_X$, then $p(n)(u) = p(n)(v)$.

For each $Y \subseteq X$, let $\chi_Y : \mathcal{D}_\beta \to \mathcal{D}_o$ be defined by

$$\chi_Y(\mathsf{b}) := \begin{cases} \mathsf{T} & \text{if } \mathsf{b} \in Y \\ \mathsf{F} & \text{otherwise.} \end{cases}$$

By Lemma 6.7.26:2, we know $\mathcal{R}_{o(o\beta)} \subseteq \mathcal{K}^p_{o(o\beta)}$. Hence $p \in \mathcal{K}^p_{o(o\beta)}$ and p converges pointwise. Thus for each $Y \subseteq X$ there is some $N^Y \in \mathbb{N}$ and $\mathsf{c}^Y \in \mathcal{D}_o$ such that $p(n)(\chi_Y) = \mathsf{c}^Y$ for all $n > N^Y$.

There are only finitely many $Y \subseteq X$ since X is finite. Define

$$N := \max(\{N_0\} \cup \{N^Y \mid Y \subseteq X\}) + 1.$$

We will prove $p(n) = p(N)$ for all $n > N$. Let $n > N$ and $u \in \mathcal{D}_{o\beta}$ be given. Let $Y := u^{-1}(\mathsf{T}) \cap X \subseteq X$. Clearly, $u|_X = \chi_Y|_X$. Since $n > N > N_0$, we know $p(n)(u) = p(n)(\chi_Y)$ and $p(N)(u) = p(N)(\chi_Y)$. Since $n > N > N^Y$, we know $p(n)(\chi_Y) = \mathsf{c}^Y = p(N)(\chi_Y)$. Combining these equations, we have

$$p(n)(u) = p(n)(\chi_Y) = \mathsf{c}^Y = p(N)(\chi_Y) = p(N)(u).$$

Thus $p(n) = p(N)$ for all $n > N$ and hence $p \in \mathcal{K}_{o(o\beta)}$. □

PROOF OF LEMMA 6.7.35. By Lemma 3.6.28, $\mathcal{P}\mathfrak{roj}_\beta \subseteq \mathcal{D}_{o(o\beta)}$. Let $\overline{\mathcal{P}\mathfrak{roj}_\beta}$ be the propositional closure of $\mathcal{P}\mathfrak{roj}_\beta$. By Theorem 3.6.33, we know $\overline{\mathcal{P}\mathfrak{roj}_\beta} \subseteq \mathcal{D}_{o(o\beta)}$.

To prove $\mathcal{D}_{o(o\beta)} \subseteq \overline{\mathcal{P}\mathfrak{roj}_\beta}$, let $g \in \mathcal{D}_{o(o\beta)}$ be given. By Lemma 6.7.34, there is a finite set $X \subseteq \mathcal{D}_\beta$ such that $g(u) = g(v)$ whenever $u|_X = v|_X$ for $u, v \in \mathcal{D}_{o\beta}$. Without loss of generality, we can assume

X is nonempty (otherwise g is constant and we are done). Enumerate X as $X = \{\mathsf{b}_1, \ldots, \mathsf{b}_n\}$ where $n \geq 1$. Enumerate the power set of X by $\{Y_1, Y_2, \ldots, Y_m\}$ where $m = 2^n$. For each j $(1 \leq j \leq m)$, let $\mathsf{u}_j : \mathcal{D}_\beta \to \mathcal{D}_o$ the characteristic function of Y_j defined by $\mathsf{u}_j(\mathsf{b}) = \mathtt{T}$ iff $\mathsf{b} \in Y_j$. By Lemma 6.7.18, we know $\mathsf{u}_j \in \mathcal{D}_{o\beta}$.

Let $k_\mathtt{F} : \mathcal{D}_{o\beta} \to \mathcal{D}_o$ be the constant $\mathtt{F}$ function $k_\mathtt{F}(w) := \mathtt{F}$. For each j $(1 \leq j \leq m)$ and i $(1 \leq i \leq n)$ we define $\mathsf{f}_j^i \in \overline{\mathfrak{Proj}_\beta}$ by cases.

- If $\mathsf{b}_i \in Y_j$ and $g(\mathsf{u}_j) = \mathtt{T}$, let $\mathsf{f}_j^i := proj_{\mathsf{b}_i} \in \overline{\mathfrak{Proj}_\beta}$.
- If $\mathsf{b}_i \notin Y_j$ and $g(\mathsf{u}_j) = \mathtt{T}$, let $\mathsf{f}_j^i := \overline{proj_{\mathsf{b}_i}} \in \overline{\mathfrak{Proj}_\beta}$.
- If $g(\mathsf{u}_j) = \mathtt{F}$, let $\mathsf{f}_j^i := k_\mathtt{F} \in \overline{\mathfrak{Proj}_\beta}$.

For each j $(1 \leq j \leq m)$, let

$$\mathsf{h}_j := \mathsf{f}_j^1 \cap \cdots \cap \mathsf{f}_j^n \in \overline{\mathfrak{Proj}_\beta}.$$

Finally, let

$$\mathsf{g}' := \mathsf{h}_1 \cup \cdots \cup \mathsf{h}_m \in \overline{\mathfrak{Proj}_\beta}.$$

We complete the proof by verifying $g = \mathsf{g}' \in \overline{\mathfrak{Proj}_\beta}$.

Let $v \in \mathcal{D}_{o\beta}$ be given. Let $V := \{\mathsf{b}_i \in X \mid v(\mathsf{b}_i) = \mathtt{T}\}$. Since V is a subset of X, there is some j_0 $(1 \leq j_0 \leq m)$ such that $V = Y_{j_0}$. By the definition of u_{j_0}, we have $v(\mathsf{b}_i) = \mathsf{u}_{j_0}(\mathsf{b}_i)$ for each $\mathsf{b}_i \in X$. Since $v\big|_X = \mathsf{u}_{j_0}\big|_X$, we know $g(v) = g(\mathsf{u}_{j_0})$.

Suppose $g(v) = g(\mathsf{u}_{j_0}) = \mathtt{T}$. If $v(\mathsf{b}_i) = \mathtt{T}$, then $\mathsf{b}_i \in V = Y_{j_0}$ and

$$\mathsf{f}_{j_0}^i(v) = proj_{\mathsf{b}_i}(v) = v(\mathsf{b}_i) = \mathtt{T}.$$

If $v(\mathsf{b}_i) = \mathtt{F}$, then $\mathsf{b}_i \notin Y_{j_0}$ and

$$\mathsf{f}_{j_0}^i(v) = \overline{proj_{\mathsf{b}_i}}(v) = \mathtt{T}.$$

Hence $\mathsf{h}_{j_0}(v) = \mathtt{T}$ and so $\mathsf{g}'(v) = \mathtt{T}$. Thus $g(v) = g(\mathsf{u}_{j_0}) = \mathtt{T} = \mathsf{g}'(v)$ in this case.

Suppose $g(v) = g(\mathsf{u}_{j_0}) = \mathtt{F}$. We must prove $\mathsf{g}'(v) = \mathtt{F}$. This follows if we verify $\mathsf{h}_j(v) = \mathtt{F}$ for each j $(1 \leq j \leq m)$. For any j with $1 \leq j \leq m$ and $g(\mathsf{u}_j) = \mathtt{F}$, we know $\mathsf{f}_j^i = k_\mathtt{F}$ for each i $(1 \leq i \leq n)$. Thus $\mathsf{h}_j(v) = \mathtt{F}$, as desired.

Assume $1 \leq j \leq m$ and $g(\mathsf{u}_j) = \mathtt{T}$. Since $g(\mathsf{u}_{j_0}) = \mathtt{F}$, we know $j \neq j_0$. Hence $Y_j \neq Y_{j_0}$. Let $\mathsf{b}_i \in X$ be such that $\mathsf{b}_i \in Y_j \setminus Y_{j_0}$ or $\mathsf{b}_i \in Y_{j_0} \setminus Y_j$. If $\mathsf{b}_i \in Y_j \setminus Y_{j_0}$, then $\mathsf{b}_i \notin V$ and

$$\mathsf{f}_j^i(v) = proj_{\mathsf{b}_i}(v) = v(\mathsf{b}_i) = \mathtt{F}.$$

If $\mathsf{b}_i \in Y_{j_0} \setminus Y_j$, then $\mathsf{b}_i \in V$ giving $proj_{\mathsf{b}_i}(v) = v(\mathsf{b}_i) = \mathtt{T}$ and

$$\mathsf{f}_j^i(v) = \overline{proj_{\mathsf{b}_i}}(v) = \mathtt{F}$$

In either case $\mathsf{h}_j(v) = \mathtt{F}$, as desired. □

A.5. Proofs from Chapter 7

PROOF OF LEMMA 7.6.4. We know $\langle \mathcal{N}^{\mathcal{Q}}, \overset{\mathcal{Q}'}{\rightarrow} \rangle$ is a dag by Lemmas 7.6.2 and 7.1.11. If $a \in \mathbf{Dom}(sel^{\mathcal{Q}}) \subseteq \mathbf{Dom}(\mathbf{SelPars}(\mathcal{Q}'))$, then a is a selection node of $\mathcal{Q}'$ and hence a is a selection node of $\mathcal{Q}$. Also, $sel^{\mathcal{Q}}$ is injective since $sel^{\mathcal{Q}'}$ is injective, $\mathbf{Dom}(sel^{\mathcal{Q}}) \subseteq \mathbf{Dom}(sel^{\mathcal{Q}'})$ and $sel^{\mathcal{Q}} = sel^{\mathcal{Q}'}|_{\mathbf{Dom}(sel^{\mathcal{Q}})}$. The relevant $\clubsuit_*$ properties relative to $\mathcal{Q}$ follow directly from the assumed $\clubsuit_*$ properties of $\mathcal{Q}'$. We only check a few properties here.

Suppose $\langle e, d\rangle \in decs^{\mathcal{Q}} \subseteq decs^{\mathcal{Q}'}$. By $\clubsuit^p_{dec}$ in $\mathcal{Q}'$, e is an extended negative equation node in $\mathcal{Q}'$, d is a decomposition node in $\mathcal{Q}'$ and the shallow formulas of e and d are the same (with respect to $\mathcal{Q}'$). Since e is an equation node, $sh^{\mathcal{Q}'}(e)$ is of the form $[\mathbf{A} \overset{\cdot}{=}^\iota \mathbf{B}]$. Thus

$$label^{\mathcal{Q}}(e) = label^{\mathcal{Q}'}(e) = \langle \mathbf{primary}, -1, [\mathbf{A} \overset{\cdot}{=}^\iota \mathbf{B}]\rangle$$

and

$$label^{\mathcal{Q}}(d) = label^{\mathcal{Q}'}(d) = \langle \mathbf{dec}, -1, [\mathbf{A} \overset{\cdot}{=}^\iota \mathbf{B}]\rangle.$$

Hence e is a negative equation node in $\mathcal{Q}$, d is a decomposition node in $\mathcal{Q}$ and $sh^{\mathcal{Q}}(e) = [\mathbf{A} \overset{\cdot}{=}^\iota \mathbf{B}] = sh^{\mathcal{Q}}(d)$. Since e is extended in $\mathcal{Q}'$, we know

$$e \in (\mathbf{Dom}(succ^{\mathcal{Q}'}) \cap \mathcal{N}^{\mathcal{Q}}) = \mathbf{Dom}(succ^{\mathcal{Q}}).$$

Hence e is extended in $\mathcal{Q}$. Thus $\langle e, d\rangle$ satisfies $\clubsuit^p_{dec}$ with respect to $\mathcal{Q}$.

Suppose $a \in \mathbf{Dom}(succ^{\mathcal{Q}})$ is a selection node with shallow formula $[\overset{\cdot}{\forall} x_\alpha \mathbf{M}]$ in $\mathcal{Q}$. Then a is an extended selection node with shallow formula $[\overset{\cdot}{\forall} x_\alpha \mathbf{M}]$ in $\mathcal{Q}'$. By $\clubsuit_{sel}$ in $\mathcal{Q}'$, $a \in \mathbf{Dom}(sel^{\mathcal{Q}'})$. Let A be $sel^{\mathcal{Q}'}(a)$. By $\clubsuit_{sel}$ in $\mathcal{Q}'$, there is some negative primary node $a^1 \in \mathcal{N}^{\mathcal{Q}'}$ such that $succ^{\mathcal{Q}'}(a) = \langle a^1\rangle$ and $sh^{\mathcal{Q}'}(a^1) = [A/x]\mathbf{M}$. Since we have assumed

$$\mathbf{Dom}(sel^{\mathcal{Q}}) = (\mathbf{Dom}(sel^{\mathcal{Q}'}) \cap \mathcal{N}^{\mathcal{Q}}),$$

we know $a \in \mathbf{Dom}(sel^{\mathcal{Q}})$. We know A is $sel^{\mathcal{Q}}(a)$, $succ^{\mathcal{Q}}(a) = \langle a^1 \rangle$ and $sh^{\mathcal{Q}}(a^1) = [A/x]\mathbf{M}$ by $\mathcal{Q} \trianglelefteq \mathcal{Q}'$. Thus a satisfies $\clubsuit_{sel}$ with respect to $\mathcal{Q}$.

Suppose $a \in \mathbf{Dom}(succ^{\mathcal{Q}})$ is a flexible node in $\mathcal{Q}$. Then

$$a \in \mathbf{Dom}(succ^{\mathcal{Q}}) \subseteq \mathbf{Dom}(succ^{\mathcal{Q}'})$$

and a is a flexible node in $\mathcal{Q}'$, contradicting $\clubsuit_f$ in $\mathcal{Q}'$. Thus a satisfies $\clubsuit_f$ with respect to $\mathcal{Q}$.

The remaining $\clubsuit_*$ properties are checked in the same straightforward way. □

PROOF OF LEMMA 7.6.5. Suppose $\langle a, \mathbf{C}, c \rangle \in exps^{\mathcal{Q}}$. By $\clubsuit_{exp}$ in $\mathcal{Q}$, there exists some variable x_α and proposition $\mathbf{M}$ such that a is an extended expansion node of $\mathcal{Q}$ with

$$sh^{\mathcal{Q}}(a) = [\dot{\forall} x_\alpha \, \mathbf{M}]$$

and c is a positive primary node of $\mathcal{Q}$ with

$$sh^{\mathcal{Q}}(c) = ([\mathbf{C}/x]\mathbf{M})^{\downarrow}.$$

Hence a is extended in $\mathcal{Q}'$,

$$label^{\mathcal{Q}'}(a) = label^{\mathcal{Q}}(a) = \langle \mathbf{primary}, 1, [\dot{\forall} x_\alpha \, \mathbf{M}] \rangle$$

and

$$label^{\mathcal{Q}'}(c) = label^{\mathcal{Q}}(c) = \langle \mathbf{primary}, 1, ([\mathbf{C}/x]\mathbf{M})^{\downarrow} \rangle.$$

Thus a is an extended expansion node of $\mathcal{Q}'$ with $sh^{\mathcal{Q}'}(a) = [\dot{\forall} x_\alpha \, \mathbf{M}]$ and c is a positive primary node of $\mathcal{Q}'$ with $sh^{\mathcal{Q}'}(c) = ([\mathbf{C}/x]\mathbf{M})^{\downarrow}$. Therefore, $\langle a, \mathbf{C}, c \rangle$ satisfies $\clubsuit_{exp}$ with respect to $\mathcal{Q}'$.

Suppose $a \in \mathbf{Dom}(succ^{\mathcal{Q}})$ is a selection node with shallow formula $[\dot{\forall} x_\alpha \, \mathbf{M}]$ in $\mathcal{Q}'$. Then a is a selection node with shallow formula $[\dot{\forall} x_\alpha \, \mathbf{M}]$ in $\mathcal{Q}$. By $\clubsuit_{sel}$ in $\mathcal{Q}$, $a \in \mathbf{Dom}(sel^{\mathcal{Q}}) \subseteq \mathbf{Dom}(sel^{\mathcal{Q}'})$. Let A be $sel^{\mathcal{Q}}(a)$. Note that A is also $sel^{\mathcal{Q}'}(a)$. By $\clubsuit_{sel}$ in $\mathcal{Q}$, there is some negative primary node $a^1 \in \mathcal{N}^{\mathcal{Q}}$ such that $succ^{\mathcal{Q}}(a) = \langle a^1 \rangle$ and $sh^{\mathcal{Q}}(a^1) = [A/x]\mathbf{M}$. Hence a^1 is a negative primary node in $\mathcal{Q}'$ such that $succ^{\mathcal{Q}'}(a) = \langle a^1 \rangle$ and $sh^{\mathcal{Q}'}(a^1) = [A/x]\mathbf{M}$. Thus a satisfies $\clubsuit_{sel}$ with respect to $\mathcal{Q}'$.

The remaining properties are checked in the same straightforward way using the fact that $\mathcal{Q} \trianglelefteq \mathcal{Q}'$. □

PROOF OF LEMMA 7.6.6. Suppose $\mathfrak{p}'$ is a u-part of $\mathcal{Q}'$. Let $\mathcal{N}$ be $\mathcal{N}^{\mathcal{Q}}$ and $\mathfrak{p}$ be $\mathfrak{p}' \cap \mathcal{N}$.

Suppose $c \in \mathfrak{p}$ is conjunctive in $\mathcal{Q}$ and $c \overset{\mathcal{Q}}{\rightharpoondown} b$ where $b \in \mathcal{N}$. Then $c \in \mathfrak{p}'$ is conjunctive in $\mathcal{Q}'$ with $c \overset{\mathcal{Q}'}{\rightharpoondown} b$ by Lemma 7.6.2. Hence $b \in \mathfrak{p}'$ and so $b \in \mathfrak{p}$.

Suppose $d \in \mathfrak{p}$ is disjunctive in $\mathcal{Q}$. Let $d^1, \ldots, d^n$ be such that

$$succ^{\mathcal{Q}}(d) = \langle d^1, \ldots, d^n \rangle.$$

Then $d \in \mathfrak{p}'$ is disjunctive and $succ^{\mathcal{Q}'}(d) = \langle d^1, \ldots, d^n \rangle$. Hence there is some $c' \in \mathfrak{p}'$ such that $d \overset{\mathcal{Q}'}{\rightharpoondown} c'$. By Lemma 7.4.3, there is some i $(1 \leq i \leq n)$ such that $c' = d^i \in \mathcal{N}$. Thus $d^i \in \mathfrak{p}$ where $d \overset{\mathcal{Q}}{\rightharpoondown} d^i$.

Suppose $d \in \mathfrak{p}$ is neutral in $\mathcal{Q}$. By Lemma 7.4.2, there is some $d^1 \in \mathcal{N}$ such that $succ^{\mathcal{Q}}(d) = \langle d^1 \rangle$. Hence $d \in \mathfrak{p}'$ and $succ^{\mathcal{Q}'}(d) = \langle d^1 \rangle$. Since $\mathfrak{p}'$ is a u-part of $\mathcal{Q}'$, $d^1 \in \mathfrak{p}'$. Thus $d^1 \in \mathfrak{p}$.

Suppose $\langle c, a, b \rangle \in conns^{\mathcal{Q}}$ where $a, b \in \mathfrak{p}$. Hence $a, b \in \mathfrak{p}'$ and

$$\langle c, a, b \rangle \in conns^{\mathcal{Q}} \subseteq conns^{\mathcal{Q}'}.$$

Thus $c \in \mathfrak{p}'$ and so $c \in \mathfrak{p}$. □

PROOF OF LEMMA 7.6.8. Suppose $d \in \mathfrak{p}$ is disjunctive in $\mathcal{Q}'$ (and hence extended in $\mathcal{Q}'$). By our assumption, d is extended in $\mathcal{Q}$. Thus d is disjunctive in $\mathcal{Q}$ since $label^{\mathcal{Q}}(d) = label^{\mathcal{Q}'}(d)$. Let $d^1, \ldots, d^n$ be such that $succ^{\mathcal{Q}}(d) = \langle d^1, \ldots, d^n \rangle$. Hence $succ^{\mathcal{Q}'}(d) = \langle d^1, \ldots, d^n \rangle$ and so $d \overset{\mathcal{Q}}{\rightharpoondown} d^i$ for each i $(1 \leq i \leq n)$. By Lemma 7.4.3, there is some i $(1 \leq i \leq n)$ such that $d^i \in \mathfrak{p}$.

Suppose $d \in \mathfrak{p}$ is neutral in $\mathcal{Q}'$ (hence extended in $\mathcal{Q}'$). By our assumption, d is extended in $\mathcal{Q}$. Hence d is neutral in $\mathcal{Q}$ since $label^{\mathcal{Q}}(d) = label^{\mathcal{Q}'}(d)$. By Lemma 7.4.2, $succ^{\mathcal{Q}}(d) = \langle d^1 \rangle$ for some $d^1 \in \mathcal{N}$. Thus $succ^{\mathcal{Q}'}(d) = \langle d^1 \rangle$ and $d^1 \in \mathfrak{p}$.

Suppose $d \in \mathfrak{p}$ is conjunctive in $\mathcal{Q}'$ and $d \overset{\mathcal{Q}'}{\rightharpoondown} c'$. Since d is extended in $\mathcal{Q}'$, we know d is extended in $\mathcal{Q}$ by assumption. Hence d is conjunctive in $\mathcal{Q}$ since $label^{\mathcal{Q}}(d) = label^{\mathcal{Q}'}(d)$. Let $d^1, \ldots, d^n$ be such that $succ^{\mathcal{Q}}(d) = \langle d^1, \ldots, d^n \rangle$. Hence $succ^{\mathcal{Q}'}(d) = \langle d^1, \ldots, d^n \rangle$. We consider three cases separately.

First assume d is an expansion node. By Lemma 7.3.11, there is some $\mathbf{D}$ with $\langle d, \mathbf{D}, c' \rangle \in exps^{\mathcal{Q}'}$. In this situation, we have assumed $c' \in \mathcal{N}$ and $d \overset{\mathcal{Q}}{\rightharpoondown} c'$. Thus $c' \in \mathfrak{p}$.

Next assume d is a negative equation node in $\mathcal{Q}'$. By Lemma 7.3.12, either $\langle d^1 \rangle = succ^{\mathcal{Q}'}(d) = \langle c' \rangle$ or $\langle d, c' \rangle \in decs^{\mathcal{Q}'}$. In the first case $c' = d^1 \in \mathfrak{p}$. In the second case we have assumed $c' \in \mathcal{N}$ and $d \overset{\mathcal{Q}}{\rightharpoondown} c'$. Hence $c' \in \mathfrak{p}$ in this case as well.

For any other conjunctive node d, we can apply Lemma 7.3.13 to conclude $c' = d^i$ for some i $(1 \leq i \leq n)$. Thus $c' \in \mathfrak{p}$.

Finally assume $\langle c', a, b \rangle \in conns^{\mathcal{Q}'}$ and $a, b \in \mathfrak{p}$. We have assumed in such a case that $c' \in \mathcal{N}$ and $\langle c', a, b \rangle \in conns^{\mathcal{Q}}$. Since $\mathfrak{p}$ is a u-part of $\mathcal{Q}$, $c' \in \mathfrak{p}$. □

Proof of Lemma 7.6.13. We know $\mathcal{Q} \trianglelefteq \mathcal{Q}'$ by Definition 7.6.11. In order to prove $\mathcal{Q} \trianglelefteq_r \mathcal{Q}'$ we must verify

$$\mathbf{Roots}(\mathcal{Q}) = \mathbf{Roots}(\mathcal{Q}').$$

Assume there is some $a \in (\mathbf{Roots}(\mathcal{Q}) \setminus \mathbf{Roots}(\mathcal{Q}'))$. Then there is some $a' \in \mathcal{N}^{\mathcal{Q}'}$ such that $a' \overset{\mathcal{Q}'}{\rightharpoondown} a$. Since $\mathcal{Q}'$ augments $\mathcal{Q}$, either

$$a' \overset{\mathcal{Q}}{\rightharpoondown} a \text{ (contradicting } a \in \mathbf{Roots}(\mathcal{Q}))$$

or

$$a \notin \mathcal{N}^{\mathcal{Q}} \text{ (contradicting } a \in \mathbf{Roots}(\mathcal{Q})).$$

Hence $\mathbf{Roots}(\mathcal{Q}) \subseteq \mathbf{Roots}(\mathcal{Q}')$.

Assume there is some $a' \in (\mathbf{Roots}(\mathcal{Q}') \setminus \mathbf{Roots}(\mathcal{Q}))$. Since $\mathcal{Q}'$ augments $\mathcal{Q}$, we have $a' \in \mathbf{Roots}(\mathcal{Q}') \subseteq \mathcal{N}^{\mathcal{Q}}$. Since $a' \notin \mathbf{Roots}(\mathcal{Q})$, there is some $a \in \mathcal{N}^{\mathcal{Q}}$ such that $a \overset{\mathcal{Q}}{\rightharpoondown} a'$. Since $\mathcal{Q} \trianglelefteq \mathcal{Q}'$ we have $a \overset{\mathcal{Q}'}{\rightharpoondown} a'$ by Lemma 7.6.2 (contradicting $a' \in \mathbf{Roots}(\mathcal{Q}')$). Hence we have $\mathbf{Roots}(\mathcal{Q}) = \mathbf{Roots}(\mathcal{Q}')$ and so $\mathcal{Q} \trianglelefteq_r \mathcal{Q}'$, as desired. □

Proof of Lemma 7.6.15. Suppose there is a cycle

$$A^1 \prec_0^{\mathcal{Q}'} \cdots \prec_0^{\mathcal{Q}'} A^n \prec_0^{\mathcal{Q}'} A^1$$

for $A^1, \ldots, A^n \in \mathbf{SelPars}(\mathcal{Q}')$. Let A^{n+1} be A^1.

Assume $1 \leq i \leq n$ and $A^i \notin \mathbf{SelPars}(\mathcal{Q})$. By our assumption there is no expansion arc $\langle e, \mathbf{C}, c \rangle$ in $\mathcal{Q}'$ such that $A^i \in \mathbf{Params}(\mathbf{C})$, contradicting $A^i \prec_0^{\mathcal{Q}'} A^{i+1}$. Thus $A^i \in \mathbf{SelPars}(\mathcal{Q})$ for every i where $1 \leq i \leq n$.

Since $\prec_0^{\mathcal{Q}}$ is acyclic there is some $i \in \{1, \ldots, n\}$ such that

$$A^i \not\prec_0^{\mathcal{Q}} A^{i+1}.$$

Let $a \in \mathcal{N}^{\mathcal{Q}}$ be $\mathbf{SelNode}^{\mathcal{Q}}(A^{i+1})$. Since $\mathcal{Q} \trianglelefteq \mathcal{Q}'$, a also happens to be $\mathbf{SelNode}^{\mathcal{Q}'}(A^{i+1})$. Since $A^i \prec_0^{\mathcal{Q}'} A^{i+1}$ does hold, there is some $\langle e, \mathbf{C}, c\rangle \in exps^{\mathcal{Q}'}$ such that $A^i \in \mathbf{Params}(\mathbf{C})$ and c dominates a in $\mathcal{Q}'$. Let $\langle c^1, \ldots, c^m\rangle$ be a walk in $\mathcal{Q}'$ from c to a. Note that $c^m = a$ and so $c^m \in \mathcal{N}^{\mathcal{Q}}$. By Lemma 7.6.14, $\langle c^1, \ldots, c^m\rangle$ is a walk in $\mathcal{Q}$ from c to a. Thus c dominates a in $\mathcal{Q}$ and so $A^i \prec_0^{\mathcal{Q}} A^{i+1}$, a contradiction. Therefore, $\prec_0^{\mathcal{Q}'}$ is acyclic. □

PROOF OF LEMMA 7.6.16. By Lemma 7.6.12, every $a \in \mathcal{N}' \setminus \mathcal{N}$ is a leaf node of $\mathcal{Q}'$. Thus $\langle \mathcal{N}^{\mathcal{Q}'}, \overset{\mathcal{Q}'}{\rightarrow}\rangle$ is a dag by Lemma 7.1.12. Every $a \in \mathbf{Dom}(sel)$ is a selection node of $\mathcal{Q}'$ since $\mathcal{Q}$ is an extensional expansion dag and $\mathcal{Q} \trianglelefteq \mathcal{Q}'$. Every $a \in \mathbf{Dom}(sel') \setminus \mathbf{Dom}(sel)$ is a selection node of $\mathcal{Q}'$ by assumption (1).

Suppose $a, b \in \mathbf{Dom}(sel')$ and $sel'(a) = sel'(b)$. If a and b are in $\mathbf{Dom}(sel)$, then $a = b$ since sel is injective and

$$sel(a) = sel'(a) = sel'(b) = sel(b).$$

Otherwise, without loss of generality $a \in \mathbf{Dom}(sel') \setminus \mathbf{Dom}(sel)$. By assumption (2), we have $a = b$. Thus sel' is injective.

By Lemma 7.6.5, every expansion arc in $exps$ satisfies $\clubsuit_{exp}$ with respect to $\mathcal{Q}'$. Every expansion arc in $exps' \setminus exps$ satisfies $\clubsuit_{exp}$ with respect to $\mathcal{Q}'$ by assumption (3). Every connection in $conns$ satisfies $\clubsuit_c$, $\clubsuit_m^p$, $\clubsuit_{eu}^p$ and $\clubsuit_{eu}^{sp}$ with respect to $\mathcal{Q}'$ by Lemma 7.6.5. Every connection in $conns' \setminus conns$ satisfies $\clubsuit_c$, $\clubsuit_m^p$, $\clubsuit_{eu}^p$ and $\clubsuit_{eu}^{sp}$ with respect to $\mathcal{Q}'$ by assumption (4). Every decomposition arc in $decs$ satisfies $\clubsuit_{dec}^p$ with respect to $\mathcal{Q}'$ by Lemma 7.6.5. Every decomposition arc in $decs' \setminus decs$ satisfies $\clubsuit_{dec}^p$ with respect to $\mathcal{Q}'$ by assumption (5). Every node $a \in \mathbf{Dom}(succ)$ satisfies $\clubsuit_{ns}$, $\clubsuit_f$, $\clubsuit_\neg$, $\clubsuit_\vee$, $\clubsuit_{sel}$, $\clubsuit_\sharp$, $\clubsuit_{dec}$, $\clubsuit_=^-$, $\clubsuit_=^\rightarrow$, $\clubsuit_=^{-o}$, $\clubsuit_=^{+o}$, $\clubsuit_m$ and $\clubsuit_{eu}$ with respect to $\mathcal{Q}'$ by Lemma 7.6.5. Every node $a \in \mathbf{Dom}(succ') \setminus \mathbf{Dom}(succ)$ satisfies $\clubsuit_{ns}$, $\clubsuit_f$, $\clubsuit_\neg$, $\clubsuit_\vee$, $\clubsuit_{sel}$, $\clubsuit_\sharp$, $\clubsuit_{dec}$, $\clubsuit_=^-$, $\clubsuit_=^\rightarrow$, $\clubsuit_=^{-o}$, $\clubsuit_=^{+o}$, $\clubsuit_m$ and $\clubsuit_{eu}$ with respect to $\mathcal{Q}'$ by assumption (6). Therefore, $\mathcal{Q}'$ is an extensional expansion dag. □

PROOF OF LEMMA 7.6.17. We check each condition in Definition 7.5.2 by either using an assumption of the lemma or the fact that $\mathcal{Q}$ is separated.

Suppose $a \in \mathcal{N}^{\mathcal{Q}'}$ is extended in $\mathcal{Q}'$ and $succ^{\mathcal{Q}'}(a) = \langle a^1, \ldots, a^n\rangle$. If $a \in \mathbf{Dom}(succ^{\mathcal{Q}})$, then $succ^{\mathcal{Q}}(a) = \langle a^1, \ldots, a^n\rangle$ since $\mathcal{Q} \trianglelefteq \mathcal{Q}'$ and

$a^1, \ldots, a^n$ are distinct since $\mathcal{Q}$ is separated. If

$$a \in (\mathbf{Dom}(succ^{\mathcal{Q}}) \setminus \mathbf{Dom}(succ^{\mathcal{Q}})),$$

then $a^1, \ldots, a^n$ are distinct by assumption (1).

Suppose $b \in \mathcal{N}^{\mathcal{Q}'}$ is not a connection node of $\mathcal{Q}'$ and two nodes a and a' are parents of b in $\mathcal{Q}'$. We know $a, a' \in \mathcal{N}^{\mathcal{Q}}$ since $\mathcal{Q}'$ augments $\mathcal{Q}$. If $b \in \mathcal{N}^{\mathcal{Q}}$, then $a \stackrel{\mathcal{Q}}{\to} b$ and $a' \stackrel{\mathcal{Q}}{\to} b$ since $\mathcal{Q}$ augments $\mathcal{Q}'$ and hence $a = a'$ since $\mathcal{Q}$ separated. If $b \notin \mathcal{N}^{\mathcal{Q}}$, then we know $a = a'$ by assumption (2).

Suppose $c \in \mathcal{N}^{\mathcal{Q}'}$ is a connection node of $\mathcal{Q}'$ and

$$\langle a, b\rangle, \langle a', b'\rangle \in \mathcal{N}^{\mathcal{Q}'} \times \mathcal{N}^{\mathcal{Q}'}$$

are two pairs of nodes such that $\langle c, a, b\rangle, \langle c, a', b'\rangle \in conns^{\mathcal{Q}'}$. We know $a, b, a', b' \in \mathcal{N}^{\mathcal{Q}}$ since $\mathcal{Q}'$ augments $\mathcal{Q}$. If $c \notin \mathcal{N}^{\mathcal{Q}}$, then $\langle a, b\rangle = \langle a', b'\rangle$ by assumption (3).

Assume $c \in \mathcal{N}^{\mathcal{Q}}$. Since $\mathcal{Q}$ augments $\mathcal{Q}'$, we know $a \stackrel{\mathcal{Q}}{\to} c$, $b \stackrel{\mathcal{Q}}{\to} c$, $a' \stackrel{\mathcal{Q}}{\to} c$ and $b' \stackrel{\mathcal{Q}}{\to} c$. By Lemma 7.3.5, a and a' are positive potentially connecting nodes of $\mathcal{Q}'$. Since $\mathcal{Q} \trianglelefteq \mathcal{Q}'$, a and a' are positive potentially connecting nodes of $\mathcal{Q}$. By Lemma 7.3.8, there exist negative potentially connecting nodes $b'', b''' \in \mathcal{N}^{\mathcal{Q}}$ such that

$$\langle c, a, b''\rangle, \langle c, a', b'''\rangle \in conns^{\mathcal{Q}}.$$

Since $\mathcal{Q}$ is separated, $a = a'$. Similarly, b and b' are negative potentially connecting nodes of $\mathcal{Q}$ by Lemma 7.3.5 and $\mathcal{Q} \trianglelefteq \mathcal{Q}'$. By Lemma 7.3.8, there exist positive potentially connecting nodes $a'', a''' \in \mathcal{N}^{\mathcal{Q}}$ such that $\langle c, a'', b\rangle, \langle c, a''', b'\rangle \in conns^{\mathcal{Q}}$. Since $\mathcal{Q}$ is separated, $b = b'$. Hence $\langle a, b\rangle = \langle a', b'\rangle$.

Thus $\mathcal{Q}'$ is separated. □

PROOF OF LEMMA 7.7.3. By Definition 7.7.2, $\mathcal{Q}$ is

$$\langle \mathcal{N}, \emptyset, label, \emptyset, \emptyset, \emptyset, \emptyset\rangle$$

where $\mathsf{n}_n^{\emptyset} = \langle a^1, \ldots, a^n\rangle$, $\mathcal{N} := \{a^1, \ldots, a^n\}$ and $label(a^i) := lab^i$. Also, since $\mathbf{Dom}(succ^{\mathcal{Q}}) = \emptyset$, $exps^{\mathcal{Q}} = \emptyset$, $conns^{\mathcal{Q}} = \emptyset$ and $decs^{\mathcal{Q}} = \emptyset$, there are no edges in the graph $\langle \mathcal{N}, \stackrel{\mathcal{Q}}{\to}\rangle$. Hence the graph $\langle \mathcal{N}, \stackrel{\mathcal{Q}}{\to}\rangle$ is trivially a dag and every node is both a leaf node and a root node. In particular,

$$\mathbf{Roots}(\mathcal{Q}) = \mathcal{N}^{\mathcal{Q}} = \{a^1, \ldots, a^n\}.$$

The conditions (2), (3), (4), (5), (6) and (7) in Definition 7.3.2 are vacuous since $\mathbf{Dom}(sel^{\mathcal{Q}}) = \emptyset$, $exps^{\mathcal{Q}} = \emptyset$, $conns^{\mathcal{Q}} = \emptyset$, $decs^{\mathcal{Q}} = \emptyset$ and $\mathbf{Dom}(succ^{\mathcal{Q}}) = \emptyset$. Therefore, $\mathcal{Q}$ is an extensional expansion dag.

The conditions of Definition 7.5.3 are vacuous since $\mathcal{Q}$ has no expansion arcs, no connections and no decomposition arcs. Thus $\mathcal{Q}$ is merged.

Similarly, condition (1) of Definition 7.5.2 is vacuous since $\mathbf{Dom}(succ^{\mathcal{Q}})$ is empty. Conditions (2) and (3) of Definition 7.5.2 hold since every node of $\mathcal{Q}$ is a root node. Thus $\mathcal{Q}$ is separated. □

PROOF OF LEMMA 7.7.5. In each case of Definition 7.7.4 there is an $n \geq 0$ such that $succ^{\mathcal{Q}'}(a) = \langle a^1, \ldots, a^n \rangle$ where $\mathsf{n}_n^{\mathcal{N}^{\mathcal{Q}}} = \langle a^1, \ldots, a^n \rangle$. Furthermore,

$$(\mathcal{N}^{\mathcal{Q}'} \setminus \mathcal{N}^{\mathcal{Q}}) = \{a^1, \ldots, a^n\}.$$

Suppose $c \in \mathbf{Dom}(succ^{\mathcal{Q}'})$, $succ^{\mathcal{Q}'}(c) = \langle c^1, \ldots, c^m \rangle$ and

$$d \in \{c^1, \ldots, c^m\}.$$

Then either $c \in \mathbf{Dom}(succ^{\mathcal{Q}})$ and $succ^{\mathcal{Q}}(c) = \langle c^1, \ldots, c^m \rangle$ or $c = a$, $m = n$ and $c^i = a^i$ for each i $(1 \leq i \leq n)$. In the first case we have $c \in \mathcal{N}^{\mathcal{Q}}$,

$$d \in \{c^1, \ldots, c^m\} \subseteq \mathcal{N}^{\mathcal{Q}}$$

and $c \xrightarrow{\mathcal{Q}} d$. In the second case we have $c = a$ and

$$d \in \{a^1, \ldots, a^n\} \subseteq (\mathcal{N}^{\mathcal{Q}'} \setminus \mathcal{N}^{\mathcal{Q}}).$$

In every other case (cf. Definition 7.2.11) we know $c, d \in \mathcal{N}^{\mathcal{Q}}$ and $c \xrightarrow{\mathcal{Q}} d$ since $\mathcal{Q}$ and $\mathcal{Q}'$ have the same expansion arcs, connections and decomposition arcs (cf. Definition 7.7.4). □

PROOF OF LEMMA 7.7.7. Let $\mathcal{N}$ be $\mathcal{N}^{\mathcal{Q}}$. In each case of Definition 7.7.4 there is some $n \geq 0$ such that $succ^{\mathcal{Q}'}(a) = \langle a^1, \ldots, a^n \rangle$ and

$$(\mathcal{N}^{\mathcal{Q}'} \setminus \mathcal{N}) = \{a^1, \ldots, a^n\} \text{ where } \mathsf{n}_n^{\mathcal{N}} = \langle a^1, \ldots, a^n \rangle.$$

Let D be $\mathbf{Dom}(succ^{\mathcal{Q}})$ and S be $\mathbf{Dom}(sel^{\mathcal{Q}})$. We verify $\mathcal{Q} \trianglelefteq \mathcal{Q}'$ by noting the following:

$$\mathcal{N}^{\mathcal{Q}} \subseteq \mathcal{N} \cup \{a^1, \ldots, a^n\} = \mathcal{N}^{\mathcal{Q}'},$$

$$\mathbf{Dom}(succ^{\mathcal{Q}}) = D \subseteq D \cup \{a\} = \mathbf{Dom}(succ^{\mathcal{Q}'}),$$

$$succ^{\mathcal{Q}'}\big|_D = succ^{\mathcal{Q}},$$

$$label^{\mathcal{Q}'}\big|_{\mathcal{N}} = label^{\mathcal{Q}},$$
$$\mathbf{Dom}(sel^{\mathcal{Q}}) \subseteq \mathbf{Dom}(sel^{\mathcal{Q}'}),$$
$$sel^{\mathcal{Q}'}\big|_{S} = sel^{\mathcal{Q}},$$
$$exps^{\mathcal{Q}} = exps^{\mathcal{Q}'},$$
$$conns^{\mathcal{Q}} = conns^{\mathcal{Q}'}$$

and

$$decs^{\mathcal{Q}} = decs^{\mathcal{Q}'}.$$

Since a is a parent of a^i, a^i is not a root node of $\mathcal{Q}'$ for each i $(1 \leq i \leq n)$, Hence $\mathbf{Roots}(\mathcal{Q}') \subseteq \mathcal{N}$. For any $c, d \in \mathcal{N}^{\mathcal{Q}'}$ if $c \overset{\mathcal{Q}'}{\rightharpoondown} d$, then either $c, d \in \mathcal{N}^{\mathcal{Q}}$ and $c \overset{\mathcal{Q}}{\rightharpoondown} d$ or $c \in \mathcal{N}^{\mathcal{Q}}$ and $d \notin \mathcal{N}^{\mathcal{Q}}$ by Lemma 7.7.5. Thus $\mathcal{Q}'$ augments $\mathcal{Q}$. We know $\mathcal{Q} \trianglelefteq_r \mathcal{Q}'$ by Lemma 7.6.13. We can conclude $\mathcal{Q}'$ is an extensional expansion dag using Lemma 7.6.16 once we verify the assumptions of the lemma hold. Assumptions (3), (4) and (5) of Lemma 7.6.16 are vacuous since $exps^{\mathcal{Q}} = exps^{\mathcal{Q}'}$, $conns^{\mathcal{Q}} = conns^{\mathcal{Q}'}$ and $decs^{\mathcal{Q}} = decs^{\mathcal{Q}'}$. If a is not a selection node, then $\mathbf{Dom}(sel^{\mathcal{Q}}) = \mathbf{Dom}(sel^{\mathcal{Q}'})$ and so assumptions (1) and (2) of Lemma 7.6.16 hold vacuously. If a is a selection node, then

$$(\mathbf{Dom}(sel^{\mathcal{Q}'}) \setminus \mathbf{Dom}(sel^{\mathcal{Q}})) = \{a\}$$

(cf. Definition 7.7.4:4) and so assumptions (1) and (2) hold since a is a selection node and $sel^{\mathcal{Q}'}(a) = \mathsf{A}_{\alpha}^{\mathbf{Params}(\mathcal{Q})} \notin \mathbf{SelPars}(\mathcal{Q})$. In each case in Definition 7.7.4 we have

$$(\mathbf{Dom}(succ^{\mathcal{Q}'}) \setminus \mathbf{Dom}(succ^{\mathcal{Q}})) = \{a\}.$$

By Lemma 7.7.6, a satisfies $\clubsuit_{ns}$, $\clubsuit_{f}$, $\clubsuit_{\neg}$, $\clubsuit_{\vee}$, $\clubsuit_{sel}$, $\clubsuit_{\sharp}$, $\clubsuit_{dec}$, $\clubsuit_{=}^{-}$, $\clubsuit_{=}^{\rightarrow}$, $\clubsuit_{=}^{-o}$, $\clubsuit_{=}^{+o}$, $\clubsuit_{m}$ and $\clubsuit_{eu}$ with respect to $\mathcal{Q}'$. That is, assumption (6) of Lemma 7.6.16 holds. Therefore, Lemma 7.6.16 allows us to conclude $\mathcal{Q}'$ is an extensional expansion dag.

Since $a \in \mathbf{Dom}(succ^{\mathcal{Q}'})$, a is extended in $\mathcal{Q}'$. Since $exps^{\mathcal{Q}'}$ equals $exps^{\mathcal{Q}}$, we know $\mathbf{ExpTerms}(\mathcal{Q}') = \mathbf{ExpTerms}(\mathcal{Q})$. We know

$$\mathbf{Roots}(\mathcal{Q}') = \mathbf{Roots}(\mathcal{Q})$$

since we have already established $\mathcal{Q} \trianglelefteq_r \mathcal{Q}'$. If a is not a selection node of $\mathcal{Q}$, then $\mathbf{SelPars}(\mathcal{Q}') = \mathbf{SelPars}(\mathcal{Q})$ and so

$$\mathbf{RootSels}(\mathcal{Q}') = \mathbf{RootSels}(\mathcal{Q}).$$

Assume a is a selection node of $\mathcal{Q}$ with shallow formula $[\forall x_\alpha \mathbf{M}]$. By Definition 7.7.4:4, $\mathbf{SelPars}(\mathcal{Q}') = \mathbf{SelPars}(\mathcal{Q}) \cup \{A_\alpha\}$ where A_α is $\mathsf{A}_\alpha^{\mathbf{Params}(\mathcal{Q})}$. That is, A_α is a parameter such that $A \notin \mathbf{Params}(\mathcal{Q})$. Hence $A_\alpha \notin \mathbf{RootParams}(\mathcal{Q})$ and so $A_\alpha \notin \mathbf{RootParams}(\mathcal{Q}')$. Thus we know $A \notin \mathbf{RootSels}(\mathcal{Q}')$. Hence

$$\mathbf{RootSels}(\mathcal{Q}') = \mathbf{RootSels}(\mathcal{Q}).$$

Suppose $\mathcal{Q}$ is closed. By Lemma 7.7.6,

$$\mathbf{Free}(sh^{\mathcal{Q}'}(a^i)) \subseteq \mathbf{Free}(sh^{\mathcal{Q}}(a)) \subseteq \emptyset$$

for each i $(1 \leq i \leq n)$. Hence every node in $\mathcal{Q}'$ has a closed shallow formula. Furthermore, $\mathbf{ExpVars}(\mathcal{Q}') = \mathbf{ExpVars}(\mathcal{Q}) = \emptyset$. Thus $\mathcal{Q}'$ is closed.

Suppose $\mathcal{Q}$ is separated. We prove $\mathcal{Q}'$ is separated using Lemma 7.6.17. Assumption (1) of Lemma 7.6.17 holds since

$$(succ^{\mathcal{Q}'} \setminus succ^{\mathcal{Q}}) = \{a\} \text{ and } succ^{\mathcal{Q}'}(a) = \langle a^1, \ldots, a^n \rangle$$

where $a^1, \ldots, a^n$ were chosen to be distinct. To check assumptions (2) and (3) of Lemma 7.6.17 note that

$$(\mathcal{N}^{\mathcal{Q}'} \setminus \mathcal{N}) = \{a^1, \ldots, a^n\}.$$

By examining each case in Definition 7.7.4, we know a^i is either a primary node or an equation goal node of $\mathcal{Q}'$. In particular, no node of $(\mathcal{N}^{\mathcal{Q}'} \setminus \mathcal{N})$ is a connection node and so assumption (3) holds vacuously. To check assumption (2) consider a particular $a^i \in \mathcal{N}^{\mathcal{Q}'} \setminus \mathcal{N}$. We clearly have $a \overset{\mathcal{Q}'}{\rightharpoondown} a^i$. By Lemma 7.7.5, we know $b = a$ for any $b \in \mathcal{N}^{\mathcal{Q}}$ such that $b \overset{\mathcal{Q}'}{\rightharpoondown} a^i$ since $a^i \notin \mathcal{N}$. Thus Lemma 7.6.17 applies and we conclude $\mathcal{Q}'$ is separated.

If $\mathcal{Q}$ has merged expansions [connections,decompositions], then $\mathcal{Q}'$ has merged expansions [connections,decompositions] since $\mathcal{Q}$ and $\mathcal{Q}'$ have the same expansion arcs, connections and decomposition arcs. In particular, if $\mathcal{Q}$ is merged, then $\mathcal{Q}'$ is merged.

Suppose $\prec_0^{\mathcal{Q}}$ is acyclic. If a is not a selection node, then we know $\prec_0^{\mathcal{Q}'}$ is acyclic by Lemma 7.6.15 since $\mathcal{Q}'$ augments $\mathcal{Q}$ and $sel^{\mathcal{Q}'} = sel^{\mathcal{Q}}$. Assume a is a selection node. By Definition 7.7.4:4,

$\mathbf{SelPars}(\mathcal{Q}') = \mathbf{SelPars}(\mathcal{Q}) \cup \{A_\alpha\}$ where $A_\alpha \notin \mathbf{Params}(\mathcal{Q})$. In particular, we know $A \notin \mathbf{Params}(\mathbf{C})$ for every expansion arc

$$\langle e, \mathbf{C}, c\rangle \in \mathit{exps}^{\mathcal{Q}'} = \mathit{exps}^{\mathcal{Q}}.$$

Hence Lemma 7.6.15 also applies in this case and we can conclude $\prec_0^{\mathcal{Q}'}$ is acyclic. □

PROOF OF LEMMA 7.7.10. Let $\mathcal{N}$ be $\mathcal{N}^{\mathcal{Q}}$ and c be such that $\mathsf{n}_1^{\mathcal{N}} = \langle c\rangle$. We omit the easy verification that $\mathcal{Q} \trianglelefteq \mathcal{Q}'$. Since c is not a root node of $\mathcal{Q}'$, we know $\mathbf{Roots}(\mathcal{Q}') \subseteq \mathcal{N}$. For any $a, b \in \mathcal{N}^{\mathcal{Q}'}$ if $a \overset{\mathcal{Q}'}{\rightarrow} b$, then either $a, b \in \mathcal{N}^{\mathcal{Q}}$ and $a \overset{\mathcal{Q}}{\rightarrow} b$ or $a \in \mathcal{N}^{\mathcal{Q}}$ and $b = c \notin \mathcal{N}^{\mathcal{Q}}$ by Lemma 7.7.9. Thus $\mathcal{Q}'$ augments $\mathcal{Q}$. We know $\mathcal{Q} \trianglelefteq_r \mathcal{Q}'$ by Lemma 7.6.13. We can conclude $\mathcal{Q}'$ is an extensional expansion dag using Lemma 7.6.16 once we verify the assumptions of the lemma hold. Assumptions (1), (2), (4), (5) and (6) of Lemma 7.6.16 are vacuous since $\mathit{sel}^{\mathcal{Q}} = \mathit{sel}^{\mathcal{Q}'}$, $\mathit{conns}^{\mathcal{Q}} = \mathit{conns}^{\mathcal{Q}'}$, $\mathit{decs}^{\mathcal{Q}} = \mathit{decs}^{\mathcal{Q}'}$ and $\mathit{succ}^{\mathcal{Q}} = \mathit{succ}^{\mathcal{Q}'}$. Note that e is an extended expansion node with shallow formula $[\forall x_\alpha \mathbf{M}]$ and $\mathcal{Q}'$ is defined so that c is a positive primary node with shallow formula $([\mathbf{C}/x]\mathbf{M})^{\downarrow}$. Hence $\langle e, \mathbf{C}, c\rangle$ satisfies $\clubsuit_{exp}$ in $\mathcal{Q}'$ and so assumption (3) of Lemma 7.6.16 holds. Thus $\mathcal{Q}'$ is an extensional expansion dag.

Since $\langle e, \mathbf{C}, c\rangle \in \mathit{exps}^{\mathcal{Q}'}$ and the shallow formula of c is $([\mathbf{C}/x]\mathbf{M})^{\downarrow}$, we know $([\mathbf{C}/x]\mathbf{M})^{\downarrow} \in \mathbf{ExpInsts}^{\mathcal{Q}'}(e)$. Since

$$\mathit{exps}^{\mathcal{Q}'} = (\mathit{exps}^{\mathcal{Q}} \cup \{\langle e, \mathbf{C}, c\rangle\}),$$

we know

$$\mathbf{ExpTerms}(\mathcal{Q}') = \mathbf{ExpTerms}(\mathcal{Q}) \cup \{\mathbf{C}\}.$$

Since $\mathbf{Roots}(\mathcal{Q}') = \mathbf{Roots}(\mathcal{Q})$ and $\mathbf{SelPars}(\mathcal{Q}') = \mathbf{SelPars}(\mathcal{Q})$, we know $\mathbf{RootSels}(\mathcal{Q}') = \mathbf{RootSels}(\mathcal{Q})$.

Suppose $\mathcal{Q}$ is closed and $\mathbf{C}$ is closed. Then the shallow formula of c in $\mathcal{Q}'$ is the closed proposition $([\mathbf{C}/x]\mathbf{M})^{\downarrow}$. Hence every node in $\mathcal{Q}'$ has a closed shallow formula. Furthermore,

$$\mathbf{ExpVars}(\mathcal{Q}') = \mathbf{ExpVars}(\mathcal{Q}) \cup \mathbf{Free}(\mathbf{C}) = \emptyset.$$

Thus $\mathcal{Q}'$ is closed.

Suppose $\mathcal{Q}$ is separated. We prove $\mathcal{Q}'$ is separated using Lemma 7.6.17. Assumption (1) holds vacuously since $\mathit{succ}^{\mathcal{Q}'} = \mathit{succ}^{\mathcal{Q}}$. The only node in $\mathcal{N}^{\mathcal{Q}'} \setminus \mathcal{N}$ is c. Since c is a primary node (not a connection node), assumption (3) holds vacuously. Assumption (2) holds since we know e is the unique node in $\mathcal{N}^{\mathcal{Q}}$ such that $e \overset{\mathcal{Q}'}{\rightharpoondown} c$ by Lemma 7.7.9. Thus Lemma 7.6.17 applies and we conclude $\mathcal{Q}'$ is separated.

Note that $\mathcal{Q}$ and $\mathcal{Q}'$ have the same connections and decomposition arcs. Thus if $\mathcal{Q}$ has merged connections [decompositions], then $\mathcal{Q}'$ has merged connections [decompositions].

Suppose $\mathcal{Q}$ has merged expansions and

$$([\mathbf{C}/x]\mathbf{M})^{\downarrow} \notin \mathbf{ExpInsts}^{\mathcal{Q}}(e).$$

Assume $\langle e', \mathbf{A}^1_\alpha, c^1\rangle, \langle e', \mathbf{A}^2_\alpha, c^2\rangle \in exps^{\mathcal{Q}'}$, $sh^{\mathcal{Q}'}(c^1) = sh^{\mathcal{Q}'}(c^2)$ and either $\mathbf{A}^1 \neq \mathbf{A}^2$ or $c^1 \neq c^2$. Since $\mathcal{Q}$ has merged expansions and

$$exps^{\mathcal{Q}'} = exps^{\mathcal{Q}} \cup \{\langle e, \mathbf{C}, c\rangle\},$$

one of the two expansion arcs must be $\langle e, \mathbf{C}, c\rangle$ and the other must be in $exps^{\mathcal{Q}}$. Without loss of generality, assume $\langle e', \mathbf{A}^1, c^1\rangle \in exps^{\mathcal{Q}}$ and $\langle e', \mathbf{A}^2, c^2\rangle$ is $\langle e, \mathbf{C}, c\rangle$. Note that $e' = e$ and

$$sh^{\mathcal{Q}}(c^1) = sh^{\mathcal{Q}'}(c^1) = sh^{\mathcal{Q}'}(c^2) = sh^{\mathcal{Q}'}(c) = ([\mathbf{C}/x]\mathbf{M})^{\downarrow}$$

and so $([\mathbf{C}/x]\mathbf{M})^{\downarrow} \notin \mathbf{ExpInsts}^{\mathcal{Q}}(e)$, contradicting our assumption. Thus $\mathcal{Q}'$ has merged expansions.

In particular, if $\mathcal{Q}$ is merged and $([\mathbf{C}/x]\mathbf{M})^{\downarrow} \notin \mathbf{ExpInsts}^{\mathcal{Q}}(e)$, then $\mathcal{Q}'$ is merged.

Suppose $\prec^{\mathcal{Q}}_0$ is acyclic. Since $\mathcal{Q}'$ augments $\mathcal{Q}$ and $sel^{\mathcal{Q}'} = sel^{\mathcal{Q}}$, we know $\prec^{\mathcal{Q}'}_0$ is acyclic by Lemma 7.6.15. □

PROOF OF LEMMA 7.7.14. Let $\mathcal{N}$ be $\mathcal{N}^{\mathcal{Q}}$ and c be such that $\mathsf{n}^{\mathcal{N}}_1 = \langle c\rangle$. We clearly have $\mathcal{Q} \trianglelefteq \mathcal{Q}'$. Since c is not a root node of $\mathcal{Q}'$, we know $\mathbf{Roots}(\mathcal{Q}') \subseteq \mathcal{N}$. For any $d, e \in \mathcal{N}^{\mathcal{Q}'}$ if $d \overset{\mathcal{Q}'}{\rightharpoondown} e$, then either $d, e \in \mathcal{N}^{\mathcal{Q}}$ and $d \overset{\mathcal{Q}}{\rightharpoondown} e$ or $d \in \{a, b\} \subseteq \mathcal{N}^{\mathcal{Q}}$ and $e = c \notin \mathcal{N}^{\mathcal{Q}}$ by Lemma 7.7.13. Thus $\mathcal{Q}'$ augments $\mathcal{Q}$. We know $\mathcal{Q} \trianglelefteq_r \mathcal{Q}'$ by Lemma 7.6.13. We can conclude $\mathcal{Q}'$ is an extensional expansion dag using Lemma 7.6.16 once we verify the assumptions of the lemma hold. Assumptions (1), (2), (3), (5) and (6) of Lemma 7.6.16 are vacuous since $sel^{\mathcal{Q}} = sel^{\mathcal{Q}'}$, $exps^{\mathcal{Q}} = exps^{\mathcal{Q}'}$, $decs^{\mathcal{Q}} = decs^{\mathcal{Q}'}$ and $succ^{\mathcal{Q}} = succ^{\mathcal{Q}'}$. The properties $\clubsuit_c$ and $\clubsuit^p_m$ hold for $\langle c, a, b\rangle$ since c is a mate node with shallow formula $[[P\,\overline{\mathbf{B}^n}] \doteq^o [P\,\overline{\mathbf{A}^n}]]$ (by Definition 7.7.11), a is an extended positive atomic node with shallow formula $[P\,\overline{\mathbf{A}^n}]$ (by assumption) and b is an extended negative atomic node with shallow formula $[P\,\overline{\mathbf{B}^n}]$ (by assumption). The properties $\clubsuit^p_{eu}$ and $\clubsuit^{sp}_{eu}$ trivially hold for $\langle c, a, b\rangle$ since c is neither an E-unification node nor a symmetric E-unification node. Hence assumption (4) of Lemma 7.6.16 holds. Applying Lemma 7.6.16, we conclude $\mathcal{Q}'$ is an extensional expansion dag.

Since c is a mate node and $\langle c, a, b\rangle \in conns^{\mathcal{Q}'}$, we know $\langle a, b\rangle$ is **mate**-connected in $\mathcal{Q}'$. Since $exps^{\mathcal{Q}'} = exps^{\mathcal{Q}}$, we know

$$\mathbf{ExpTerms}(\mathcal{Q}') = \mathbf{ExpTerms}(\mathcal{Q}).$$

Since $\mathbf{Roots}(\mathcal{Q}') = \mathbf{Roots}(\mathcal{Q})$ and $\mathbf{SelPars}(\mathcal{Q}') = \mathbf{SelPars}(\mathcal{Q})$, we know $\mathbf{RootSels}(\mathcal{Q}') = \mathbf{RootSels}(\mathcal{Q})$.

Suppose $\mathcal{Q}$ is closed. Note that

$$\begin{aligned}\mathbf{Free}(sh^{\mathcal{Q}'}(c)) &= \bigcup_{1\leq i\leq n}(\mathbf{Free}(\mathbf{A}^i) \cup \mathbf{Free}(\mathbf{B}^i)) \\ &= \mathbf{Free}(sh^{\mathcal{Q}}(a)) \cup \mathbf{Free}(sh^{\mathcal{Q}}(b)) = \emptyset.\end{aligned}$$

Hence every node in $\mathcal{Q}'$ has a closed shallow formula. Furthermore,

$$\mathbf{ExpVars}(\mathcal{Q}') = \mathbf{ExpVars}(\mathcal{Q}) = \emptyset.$$

Thus $\mathcal{Q}'$ is closed.

Suppose $\mathcal{Q}$ is separated. We prove $\mathcal{Q}'$ is separated using Lemma 7.6.17. Assumption (1) holds vacuously since $succ^{\mathcal{Q}'} = succ^{\mathcal{Q}}$. The only node in $\mathcal{N}^{\mathcal{Q}'} \setminus \mathcal{N}$ is the connection node c. Hence assumption (2) holds vacuously. To verify assumption (3) holds suppose

$$\langle a', b'\rangle, \langle a'', b''\rangle \in \mathcal{N}^{\mathcal{Q}} \times \mathcal{N}^{\mathcal{Q}}$$

and

$$\langle c, a', b'\rangle, \langle c, a'', b''\rangle \in conns^{\mathcal{Q}'}.$$

Hence $a' \overset{\mathcal{Q}'}{\rightharpoondown} c$, $b' \overset{\mathcal{Q}'}{\rightharpoondown} c$, $a'' \overset{\mathcal{Q}'}{\rightharpoondown} c$ and $b'' \overset{\mathcal{Q}'}{\rightharpoondown} c$. By Lemma 7.7.13,

$$\{a', b', a'', b''\} \subseteq \{a, b\}.$$

By Lemma 7.3.5, a' and a'' are positive nodes of $\mathcal{Q}'$ while b' and b'' are negative nodes of $\mathcal{Q}'$. Hence $a' = a = a''$ and $b' = b = b''$. Thus Lemma 7.6.17 applies and we conclude $\mathcal{Q}'$ is separated.

Since $\mathcal{Q}$ and $\mathcal{Q}'$ have the same expansion arcs and decomposition arcs, if $\mathcal{Q}$ has merged expansions [decompositions], then $\mathcal{Q}'$ has merged expansions [decompositions].

Suppose $\mathcal{Q}$ has merged connections and $\langle a,b\rangle$ is not **mate**-connected in $\mathcal{Q}$. Assume $\langle c^1,a',b'\rangle,\langle c^2,a',b'\rangle \in conns^{\mathcal{Q}'}$, $c^1 \neq c^2$ and $kind^{\mathcal{Q}'}(c^1) = kind^{\mathcal{Q}'}(c^2)$. Since $\mathcal{Q}$ has merged connections and

$$conns^{\mathcal{Q}'} = conns^{\mathcal{Q}} \cup \{\langle c,a,b\rangle\},$$

one of the two connections must be $\langle c,a,b\rangle$ and the other must be in $conns^{\mathcal{Q}}$. Without loss of generality, assume $\langle c^1,a',b'\rangle \in conns^{\mathcal{Q}}$ and $\langle c^2,a',b'\rangle$ is $\langle c,a,b\rangle$. Thus $a'=a$, $b'=b$ and

$$kind^{\mathcal{Q}}(c^1) = kind^{\mathcal{Q}'}(c) = \mathbf{mate}.$$

However, $\langle c^1,a,b\rangle \in conns^{\mathcal{Q}}$ contradicts our assumption that $\langle a,b\rangle$ is not **mate**-connected in $\mathcal{Q}$. Therefore, $\mathcal{Q}'$ has merged connections.

In particular, if $\mathcal{Q}$ is merged and $\langle a,b\rangle$ is not **mate**-connected in $\mathcal{Q}$, then $\mathcal{Q}'$ is merged.

Suppose $\prec_0^{\mathcal{Q}}$ is acyclic. Since $\mathcal{Q}'$ augments $\mathcal{Q}$ and $sel^{\mathcal{Q}'} = sel^{\mathcal{Q}}$, we know $\prec_0^{\mathcal{Q}'}$ is acyclic by Lemma 7.6.15. □

PROOF OF LEMMA 7.7.18. Let $\mathcal{N}$ be $\mathcal{N}^{\mathcal{Q}}$ and d be such that $\mathsf{n}_1^{\mathcal{N}} = \langle d\rangle$. Clearly, $\mathcal{Q} \trianglelefteq \mathcal{Q}'$. Since d is not a root node of $\mathcal{Q}'$, we know $\mathbf{Roots}(\mathcal{Q}') \subseteq \mathcal{N}$. For any $a,b \in \mathcal{N}^{\mathcal{Q}'}$ if $a \overset{\mathcal{Q}'}{\rightharpoondown} b$, then either $a,b \in \mathcal{N}^{\mathcal{Q}}$ and $a \overset{\mathcal{Q}}{\rightharpoondown} b$ or $a \in \mathcal{N}^{\mathcal{Q}}$ and $b = d \notin \mathcal{N}^{\mathcal{Q}}$ by Lemma 7.7.17. Thus $\mathcal{Q}'$ augments $\mathcal{Q}$. We know $\mathcal{Q} \trianglelefteq_r \mathcal{Q}'$ by Lemma 7.6.13. We can conclude $\mathcal{Q}'$ is an extensional expansion dag using Lemma 7.6.16 once we verify the assumptions of the lemma hold. Assumptions (1), (2), (3), (4) and (6) of Lemma 7.6.16 are vacuous since $sel^{\mathcal{Q}} = sel^{\mathcal{Q}'}$, $exps^{\mathcal{Q}} = exps^{\mathcal{Q}'}$, $conns^{\mathcal{Q}} = conns^{\mathcal{Q}'}$ and $succ^{\mathcal{Q}} = succ^{\mathcal{Q}'}$. The property $\clubsuit^p_{dec}$ holds for $\langle e,d\rangle$ since we have assumed e is an extended decomposable (negative equation) node and $\mathcal{Q}'$ is defined so that d is a decomposition node with the same shallow formula as e. Hence assumption (5) of Lemma 7.6.16 holds. Applying Lemma 7.6.16, we conclude $\mathcal{Q}'$ is an extensional expansion dag.

Since $\langle e,d\rangle \in decs^{\mathcal{Q}'}$, we know e is decomposed in $\mathcal{Q}'$. Since $\mathcal{Q}$ and $\mathcal{Q}'$ have the same expansion arcs, $\mathbf{ExpTerms}(\mathcal{Q}') = \mathbf{ExpTerms}(\mathcal{Q})$. Since $\mathbf{Roots}(\mathcal{Q}') = \mathbf{Roots}(\mathcal{Q})$ and $\mathbf{SelPars}(\mathcal{Q}') = \mathbf{SelPars}(\mathcal{Q})$, we know $\mathbf{RootSels}(\mathcal{Q}') = \mathbf{RootSels}(\mathcal{Q})$.

Suppose $\mathcal{Q}$ is closed. Then the shallow formula of d in $\mathcal{Q}'$ is closed since e and d have the same shallow formula in $\mathcal{Q}'$. Hence every node in $\mathcal{Q}'$ has a closed shallow formula. Furthermore,

$$\mathbf{ExpVars}(\mathcal{Q}') = \mathbf{ExpVars}(\mathcal{Q}) = \emptyset.$$

Thus $\mathcal{Q}'$ is closed.

Suppose $\mathcal{Q}$ is separated. We prove $\mathcal{Q}'$ is separated using Lemma 7.6.17. Assumption (1) holds vacuously since $succ^{\mathcal{Q}'} = succ^{\mathcal{Q}}$. The only node in $\mathcal{N}^{\mathcal{Q}'} \setminus \mathcal{N}$ is the decomposition node d. Hence assumption (3) holds vacuously. Assumption (2) holds since e is the only member of $\mathcal{N}$ such that $e \overset{\mathcal{Q}'}{\rightharpoondown} d$ by Lemma 7.7.17. Thus Lemma 7.6.17 applies and we conclude $\mathcal{Q}'$ is separated.

If $\mathcal{Q}$ has merged expansions [connections], then $\mathcal{Q}'$ has merged expansions [connections] since $\mathcal{Q}$ and $\mathcal{Q}'$ have the same expansion arcs and connections.

Suppose $\mathcal{Q}$ has merged decompositions and e is not decomposed in $\mathcal{Q}$. Assume $\langle e', d^1\rangle, \langle e', d^2\rangle \in decs^{\mathcal{Q}'}$ and $d^1 \neq d^2$. Since $\mathcal{Q}$ has merged decompositions and $decs^{\mathcal{Q}'} = decs^{\mathcal{Q}} \cup \{\langle e, d\rangle\}$, one of the two decomposition arcs must be $\langle e, d\rangle$ and the other must be in $decs^{\mathcal{Q}}$. Without loss of generality, assume $\langle e', d^1\rangle \in decs^{\mathcal{Q}}$ and $\langle e', d^2\rangle$ is $\langle e, d\rangle$. However, $\langle e, d^1\rangle \in decs^{\mathcal{Q}}$ contradicts our assumption that e is not decomposed in $\mathcal{Q}$. Therefore, $\mathcal{Q}'$ has merged decompositions.

In particular, if $\mathcal{Q}$ is merged and e is not decomposed in $\mathcal{Q}$, then $\mathcal{Q}'$ is merged.

Suppose $\prec_0^{\mathcal{Q}}$ is acyclic. Since $\mathcal{Q}'$ augments $\mathcal{Q}$ and $sel^{\mathcal{Q}'} = sel^{\mathcal{Q}}$, we know $\prec_0^{\mathcal{Q}'}$ is acyclic by Lemma 7.6.15. □

Proof of Lemma 7.8.10. Let $\mathcal{Q}^-$ be $\mathcal{Q}_a^{-\mathbf{r}}$ and $\mathcal{N}$ be $\mathcal{N}^{\mathcal{Q}}$. We know $\mathcal{Q}^-$ is an extensional expansion dag and the embedding relation $\prec_0^{\mathcal{Q}^-}$ is acyclic by Lemma 7.8.9.

Assume $\mathfrak{p}$ is a u-part of $\mathcal{Q}^-$ which includes $\mathcal{R}$. Note that $\mathcal{R}$ is a subset of $\mathcal{N} \setminus \{a\}$ since we have assumed $a \notin \mathcal{R}$. By Lemma 7.8.9:3, $\mathfrak{p}$ is a u-part of $\mathcal{Q}$ including $\mathcal{R}$, contradicting sufficiency of $\mathcal{R}$ with respect to $\mathcal{Q}$. Hence $\mathcal{R}$ is sufficient for $\mathcal{Q}^-$.

We know $\mathbf{SelPars}^{\mathcal{Q}}(\mathcal{R})$ is empty since $\langle \mathcal{Q}, \mathcal{R}\rangle$ is an extensional expansion proof. Since every node in $\mathcal{R}$ has the same shallow formula with respect to both $\mathcal{Q}$ and $\mathcal{Q}^-$, $\mathbf{Params}^{\mathcal{Q}^-}(\mathcal{R}) = \mathbf{Params}^{\mathcal{Q}}(\mathcal{R})$. We also know $\mathbf{SelPars}^{\mathcal{Q}^-}(\mathcal{R})$ is empty since $\mathbf{SelPars}(\mathcal{Q}^-) \subseteq \mathbf{SelPars}(\mathcal{Q})$ and $\mathbf{Params}^{\mathcal{Q}^-}(\mathcal{R}) = \mathbf{Params}^{\mathcal{Q}}(\mathcal{R})$.

Thus $\langle \mathcal{Q}^-, \mathcal{R}\rangle$ is an extensional expansion proof. □

PROOF OF LEMMA 7.8.11. Let $\mathcal{Q}^-$ be $\mathcal{Q}_a^{-\mathbf{r}}$ and $\mathcal{R}^-$ be $(\mathcal{R} \setminus \{a\}) \cup \{b\}$. By Lemma 7.8.9, we know $\mathcal{Q}^-$ is an extensional expansion dag and the embedding relation $\prec_0^{\mathcal{Q}^-}$ is acyclic.

Assume $\mathfrak{p}$ is a u-part of $\mathcal{Q}^-$ which includes $\mathcal{R}^-$. By Lemma 7.8.9:3, $\mathfrak{p}$ is a u-part of $\mathcal{Q}$ including $\mathcal{R}^-$. Since $b \in \mathfrak{p}$, $\mathfrak{p} \cup \{a\}$ is a u-part of $\mathcal{Q}$ including $\mathcal{R} \subseteq \mathcal{R}^- \cup \{a\}$ by Lemma 7.4.8. This contradicts sufficiency of $\mathcal{R}$ for $\mathcal{Q}$. Hence $\mathcal{R}^-$ is sufficient for $\mathcal{Q}^-$.

By Lemma 7.4.2, $succ^{\mathcal{Q}}(a) = \langle b \rangle$. We know $\mathbf{SelPars}^{\mathcal{Q}}(\mathcal{R}) = \emptyset$ and must check $\mathbf{SelPars}^{\mathcal{Q}^-}(\mathcal{R}^-)$ is also empty.

Assume a is not a selection node. By Lemma 7.3.16,

$$\mathbf{Params}(sh^{\mathcal{Q}}(b)) \subseteq \mathbf{Params}(sh^{\mathcal{Q}}(a)).$$

Hence

$$\mathbf{Params}^{\mathcal{Q}^-}(\mathcal{R}^-) = \mathbf{Params}^{\mathcal{Q}}(\mathcal{R}^-) \subseteq \mathbf{Params}^{\mathcal{Q}}(\mathcal{R}).$$

Since $\mathbf{SelPars}(\mathcal{Q}^-) = \mathbf{SelPars}(\mathcal{Q})$, we have

$$\mathbf{SelPars}^{\mathcal{Q}^-}(\mathcal{R}^-) \subseteq \mathbf{SelPars}^{\mathcal{Q}}(\mathcal{R}) \subseteq \emptyset.$$

Next assume a is a selection node. By $\clubsuit_{sel}$, there is some A_α, variable x_α and proposition $\mathbf{M}$ such that $sel^{\mathcal{Q}}(a) = A$, the shallow formula of a is $[\forall x_\alpha \, \mathbf{M}]$ and the shallow formula of b is $[A/x]\mathbf{M}$. In this case $\mathbf{SelPars}(\mathcal{Q}^-) = \mathbf{SelPars}(\mathcal{Q}) \setminus \{A\}$ since $sel^{\mathcal{Q}}$ is injective and

$$\begin{gathered}\mathbf{Params}^{\mathcal{Q}^-}(\{b\}) = \mathbf{Params}([A/x]\mathbf{M}) \\ \subseteq \mathbf{Params}(\mathbf{M}) \cup \{A\} = \mathbf{Params}^{\mathcal{Q}}(\{a\}) \cup \{A\}.\end{gathered}$$

Hence

$$\begin{aligned}\mathbf{Params}^{\mathcal{Q}^-}(\mathcal{R}^-) &= \mathbf{Params}^{\mathcal{Q}^-}(\mathcal{R} \setminus \{a\}) \cup \mathbf{Params}^{\mathcal{Q}^-}(\{b\}) \\ &\subseteq \mathbf{Params}^{\mathcal{Q}}(\mathcal{R} \setminus \{a\}) \cup \mathbf{Params}^{\mathcal{Q}}(\{a\}) \cup \{A\} \\ &= \mathbf{Params}^{\mathcal{Q}}(\mathcal{R}) \cup \{A\}.\end{aligned}$$

We conclude

$$\begin{aligned}\mathbf{SelPars}^{\mathcal{Q}^-}(\mathcal{R}^-) &\subseteq \mathbf{SelPars}(\mathcal{Q}^-) \cap \mathbf{Params}^{\mathcal{Q}^-}(\mathcal{R}^-) \\ &\subseteq (\mathbf{SelPars}(\mathcal{Q}) \setminus \{A\}) \cap (\mathbf{Params}^{\mathcal{Q}}(\mathcal{R}) \cup \{A\}) \\ &\subseteq \mathbf{SelPars}^{\mathcal{Q}}(\mathcal{R}) \subseteq \emptyset\end{aligned}$$

as desired.

Thus $\langle \mathcal{Q}^-, \mathcal{R}^- \rangle$ is an extensional expansion proof. □

PROOF OF LEMMA 7.8.12. Let

- $\mathcal{Q}^-$ be $\mathcal{Q}_a^{-\mathbf{r}}$,
- $\mathcal{K}$ be $\{b \in \mathcal{N}^{\mathcal{Q}} \mid a \overset{\mathcal{Q}}{\rightarrow} b\}$ and
- $\mathcal{R}^-$ be $(\mathcal{R} \setminus \{a\}) \cup \mathcal{K}$.

By Lemma 7.8.9, $\mathcal{Q}^-$ is an extensional expansion dag and the embedding relation $\prec_0^{\mathcal{Q}^-}$ is acyclic.

Suppose $\mathfrak{p}$ is a u-part of $\mathcal{Q}^-$ which includes $\mathcal{R}^-$. By Lemma 7.8.9:3, $\mathfrak{p}$ is a u-part of $\mathcal{Q}$ including $\mathcal{R}^-$. Since every child of a is in $\mathfrak{p}$ (i.e., $\mathcal{K} \subseteq \mathfrak{p}$), $\mathfrak{p} \cup \{a\}$ is a u-part of $\mathcal{Q}$ including $\mathcal{R} \subseteq \mathcal{R}^- \cup \{a\}$ by Lemma 7.4.8. This contradicts sufficiency of $\mathcal{R}$ for $\mathcal{Q}$. Thus $\mathcal{R}^-$ is sufficient for $\mathcal{Q}^-$.

Since a is an accessible node, it is not an expansion node and not a potentially connecting node. Since a is conjunctive, it is not a selection node. Consequently, Lemma 7.3.17 applies and we conclude

$$\mathbf{Params}(sh^{\mathcal{Q}}(b)) \subseteq \mathbf{Params}(sh^{\mathcal{Q}}(a))$$

for each $b \in \mathcal{K}$. Also, the fact that a is not a selection node means

$$\mathbf{SelPars}(\mathcal{Q}^-) = \mathbf{SelPars}(\mathcal{Q}).$$

Hence

$$\mathbf{SelPars}^{\mathcal{Q}^-}(\mathcal{R}^-) = \mathbf{SelPars}^{\mathcal{Q}}(\mathcal{R}^-) \subseteq \mathbf{SelPars}^{\mathcal{Q}}(\mathcal{R}) \subseteq \emptyset$$

since the nodes in $\mathcal{Q}^-$ have the same shallow formula in $\mathcal{Q}^-$ as in $\mathcal{Q}$.

Therefore, $\langle \mathcal{Q}^-, \mathcal{R}^- \rangle$ is an extensional expansion proof. □

PROOF OF LEMMA 7.8.14. Let $\mathcal{Q}^-$ be $\mathcal{Q}_a^{-\mathbf{r}}$. By Lemma 7.8.9, $\mathcal{Q}^-$ is an extensional expansion dag and the embedding relation $\prec_0^{\mathcal{Q}^-}$ is acyclic.

Let $i \in \{1, \ldots, n\}$ be given and let $\mathcal{R}^i$ be $(\mathcal{R} \setminus \{a\}) \cup \{a^i\}$.

Suppose $\mathfrak{p}$ is a u-part of $\mathcal{Q}^-$ which includes $\mathcal{R}^i$. By Lemma 7.8.9:3, $\mathfrak{p}$ is a u-part of $\mathcal{Q}$ including $\mathcal{R}^i$. Since a is a disjunctive node and the child a^i is in $\mathfrak{p}$, $\mathfrak{p} \cup \{a\}$ is a u-part of $\mathcal{Q}$ including $\mathcal{R}$ by Lemma 7.4.8. This contradicts sufficiency of $\mathcal{R}$ for $\mathcal{Q}$. Thus $\mathcal{R}^i$ is a sufficient set for $\mathcal{Q}^-$.

Since a is a disjunctive node, Lemma 7.3.17 applies and we conclude $\mathbf{Params}(sh^{\mathcal{Q}}(a^i)) \subseteq \mathbf{Params}(sh^{\mathcal{Q}}(a))$. Also, the fact that a is not a selection node means $\mathbf{SelPars}(\mathcal{Q}^-) = \mathbf{SelPars}(\mathcal{Q})$. We conclude

$$\mathbf{SelPars}^{\mathcal{Q}^-}(\mathcal{R}^i) = \mathbf{SelPars}^{\mathcal{Q}}(\mathcal{R}^i) \subseteq \mathbf{SelPars}^{\mathcal{Q}}(\mathcal{R}) \subseteq \emptyset.$$

Hence $\langle \mathcal{Q}^-, \mathcal{R}^i \rangle$ is an extensional expansion proof for each i such that $1 \leq i \leq n$. □

PROOF OF LEMMA 7.8.17. Let $\mathcal{Q}^-$ be $\mathcal{Q}^{-\mathbf{e}}_{\langle e,\mathbf{C},c\rangle}$. By Lemma 7.8.16, we know $\mathcal{Q}^-$ is an extensional expansion dag and the embedding relation $\prec_0^{\mathcal{Q}^-}$ is acyclic.

Suppose $\mathfrak{p}$ is a u-part of $\mathcal{Q}^-$ which includes $\mathcal{R} \cup \{c\}$. Hence $\mathcal{R} \subseteq \mathfrak{p}$ and $b \in \mathfrak{p}$. By Lemma 7.4.7, $\mathfrak{p} \setminus \{e\}$ is a u-part of $\mathcal{Q}^-$. By Lemma 7.8.16:3, $\mathfrak{p} \setminus \{e\}$ is a u-part of $\mathcal{Q}$.

Since $e \in \mathcal{R}$, $e \in \mathfrak{p}$ and so every child of e in $\mathcal{Q}^-$ is in $\mathfrak{p}$. Every child of e in $\mathcal{Q}$ is either c or a child of e in $\mathcal{Q}^-$. Hence every child of e in $\mathcal{Q}$ is in $\mathfrak{p}$. By Lemma 7.4.8, $\mathfrak{p}$ is a u-part of $\mathcal{Q}$ including $\mathcal{R}$, contradicting sufficiency of $\mathcal{R}$ for $\mathcal{Q}$. Thus $\mathcal{R} \cup \{c\}$ is sufficient for $\mathcal{Q}^-$.

Note that both expansion structures $\mathcal{Q}$ and $\mathcal{Q}^-$ have the same selected parameters and every node has the same shallow formula with respect to both $\mathcal{Q}$ and $\mathcal{Q}^-$. By $\clubsuit_{exp}$, the shallow formula of e is $[\dot{\forall} x_\alpha \mathbf{M}]$ and the shallow formula of c is $[[\mathbf{C}/x]\mathbf{M}]^{\downarrow}$ for some x_α and $\mathbf{M}$. Hence

$$\begin{aligned}\mathbf{Params}(\{c\}) &= \mathbf{Params}([[\mathbf{C}/x]\mathbf{M}]^{\downarrow}) \\ &\subseteq (\mathbf{Params}([\dot{\forall} x_\alpha \mathbf{M}]) \cup \mathbf{Params}(\mathbf{C})) \\ &= (\mathbf{Params}(\{e\}) \cup \mathbf{Params}(\mathbf{C})).\end{aligned}$$

Since $e \in \mathcal{R}$, we have

$$\begin{aligned}\mathbf{Params}(\mathcal{R} \cup \{c\}) &= (\mathbf{Params}(\mathcal{R}) \cup \mathbf{Params}(\{c\})) \\ &\subseteq (\mathbf{Params}(\mathcal{R}) \cup \mathbf{Params}(\{e\}) \cup \mathbf{Params}(\mathbf{C})) \\ &= (\mathbf{Params}(\mathcal{R}) \cup \mathbf{Params}(\mathbf{C})).\end{aligned}$$

Since $\langle e, \mathbf{C}, c\rangle$ is an accessible expansion,

$$\mathbf{SelPars}(\mathcal{Q}) \cap \mathbf{Params}(\mathbf{C}) = \emptyset.$$

Hence

$$\begin{aligned}
&\mathbf{SelPars}^{\mathcal{Q}^-}(\mathcal{R} \cup \{c\}) \\
&\subseteq (\mathbf{SelPars}(\mathcal{Q}) \cap (\mathbf{Params}(\mathcal{R}) \cup \mathbf{Params}(\mathbf{C})) \\
&\subseteq \mathbf{SelPars}^{\mathcal{Q}}(\mathcal{R}) \cup (\mathbf{SelPars}(\mathcal{Q}) \cap \mathbf{Params}(\mathbf{C})) \\
&\subseteq \emptyset \cup \emptyset = \emptyset.
\end{aligned}$$

Therefore, $\langle \mathcal{Q}^-, \mathcal{R} \cup \{c\} \rangle$ is an extensional expansion proof. □

PROOF OF LEMMA 7.8.20. Let $\mathcal{Q}^-$ be $\mathcal{Q}^{-\mathbf{c}}_{\langle c,a,b \rangle}$ and $\mathcal{R}^-$ be $\mathcal{R} \cup \{c\}$. By Lemma 7.8.19, we know $\mathcal{Q}^-$ is an extensional expansion dag and the embedding relation $\prec_0^{\mathcal{Q}^-}$ is acyclic.

Assume $a, b \in \mathcal{R}$ and $\mathfrak{p}$ is a u-part of $\mathcal{Q}^-$ including $\mathcal{R}^-$. Hence $\mathcal{R} \subseteq \mathfrak{p}$ and $c \in \mathfrak{p}$. By Lemma 7.4.7, $\mathfrak{p} \setminus \{a\}$ is a u-part of $\mathcal{Q}^-$. By Lemma 7.8.19:3, $\mathfrak{p} \setminus \{a\}$ is a u-part of $\mathcal{Q}$. Since $c \in \mathfrak{p} \setminus \{a\}$, $\mathfrak{p}$ is a u-part of $\mathcal{Q}$ by Lemma 7.4.8. This contradicts sufficiency of $\mathcal{R}$ for $\mathcal{Q}$ since $\mathcal{R} \subseteq \mathfrak{p}$. Thus $\mathcal{R}^-$ is sufficient for $\mathcal{Q}^-$.

By Lemma 7.3.18,

$$\mathbf{Params}(sh(c)) = \mathbf{Params}(sh(a)) \cup \mathbf{Params}(sh(b)).$$

Also, both expansion structures have the same selected parameters and the same shallow formulas associated with nodes. Hence

$$\mathbf{SelPars}^{\mathcal{Q}^-}(\mathcal{R}^-) = \mathbf{SelPars}^{\mathcal{Q}}(\mathcal{R}) = \emptyset.$$

Therefore, $\langle \mathcal{Q}^-, \mathcal{R}^- \rangle$ is an extensional expansion proof. □

PROOF OF THEOREM 7.9.1. We first check the remaining cases where $a \in \mathbf{Roots}(\mathcal{Q})$ is an accessible node with $a \in \mathcal{R}$. Recall that $\mathbf{M}$ is the dual shallow formula of a, $\Gamma(\mathcal{R})$ is $\Gamma', \mathbf{M}$ and Γ' is $\Gamma(\mathcal{R} \setminus \{a\})$

- Suppose a is a positive negation node. Consequently, $\mathbf{M}$ is of the form $\neg\neg\mathbf{N}$ (since a is positive). By $\clubsuit_\neg$, there is some negative node b with dual shallow formula $\mathbf{N}$ with $succ^{\mathcal{Q}}(a) = \langle b \rangle$. By Lemma 7.8.11, $\langle \mathcal{Q}_a^{-\mathbf{r}}, (\mathcal{R} \setminus \{a\}) \cup \{b\} \rangle$ is an extensional expansion proof. By the inductive hypothesis, there is a derivation $\mathcal{D}_1$ of $\Gamma', \mathbf{N}$. We complete the derivation $\mathcal{D}$ by

$$\cfrac{\begin{array}{c}\mathcal{D}_1 \\ \Gamma', \mathbf{N}\end{array}}{\Gamma', \neg\neg\mathbf{N}}\ \mathcal{G}(\neg\neg)$$

- Suppose a is a negative negation node. In this case, $\mathbf{M}$ is of the form $\neg\mathbf{N}$. By $\clubsuit_{\neg}$, there is a positive node b with dual shallow formula $\neg\mathbf{N}$ such that $succ^{\mathcal{Q}}(a) = \langle b \rangle$. By Lemma 7.8.11, $\langle \mathcal{Q}_a^{-\mathbf{r}}, (\mathcal{R} \setminus \{a\}) \cup \{b\} \rangle$ is an extensional expansion proof. By the inductive hypothesis, there is a derivation $\mathcal{D}$ of $\Gamma', \neg\mathbf{N}$, as desired.
- Suppose a is a negative [positive] deepening node. Consequently, the shallow formula of a is $[c\,\overline{\mathbf{A}^n}]$ where $c \in \mathcal{S}$. By $\clubsuit_{\sharp}$, there is a negative [positive] node b with shallow formula $([c\,\overline{\mathbf{A}^n}]^{\sharp})^{\downarrow}$ where $succ^{\mathcal{Q}}(a) = \langle b \rangle$. Let $\mathbf{N}$ be the dual shallow formula of b. Lemma 7.8.11 implies $\langle \mathcal{Q}_a^{-\mathbf{r}}, (\mathcal{R} \setminus \{a\}) \cup \{b\} \rangle$ is an extensional expansion proof. By the inductive hypothesis, there is a derivation $\mathcal{D}_1$ of $\Gamma', \mathbf{N}$. Applying $\mathcal{G}(\sharp)$ $[\mathcal{G}(\neg\sharp)]$, we have a derivation $\mathcal{D}$ of $\Gamma', \mathbf{M}$.
- Suppose a is a negative equation node. By $\clubsuit_{=}^{-}$, there is some negative node b^0 such that $succ^{\mathcal{Q}}(a) = \langle b^0 \rangle$ and b^0 has the same shallow formula $\mathbf{M}$ as a. Let $b^1, \ldots, b^m$ be such that

$$\{b^1, \ldots, b^m\} = \{b \in \mathcal{N} \mid \langle a, b \rangle \in decs\}$$

where $m \geq 0$. By $\clubsuit_{dec}^{p}$, each b^i is a negative node with shallow formula $\mathbf{M}$. By Lemma 7.3.12, $\{b^0, \ldots, b^m\}$ are all the children of a. By Lemma 7.8.12,

$$\langle \mathcal{Q}_a^{-\mathbf{r}}, ((\mathcal{R} \setminus \{a\}) \cup \{b^0, \ldots, b^m\}) \rangle$$

is an extensional expansion proof. By the inductive hypothesis, there is a derivation $\mathcal{D}_1$ of

$$\Gamma', \overbrace{\mathbf{M}, \ldots, \mathbf{M}}^{m+1}$$

We complete the derivation $\mathcal{D}$ of $\Gamma', \mathbf{M}$ using m applications of the contraction rule $\mathcal{G}(Contr)$.
- Suppose a is a positive Boolean node. Consequently, $\mathbf{M}$ is of the form $\neg[\mathbf{A} \doteq^{o} \mathbf{B}]$. By $\clubsuit_{=}^{+o}$, there are negative primary nodes b and c where $succ^{\mathcal{Q}}(a) = \langle b, c \rangle$, $sh(b) = [\mathbf{A} \dot{\vee} \mathbf{B}]$ and $sh(c) = [\neg\mathbf{A} \dot{\vee} \neg\mathbf{B}]$. Since a is disjunctive, Lemma 7.8.14 implies $\langle \mathcal{Q}_a^{-\mathbf{r}}, (\mathcal{R} \setminus \{a\}) \cup \{b\} \rangle$ and $\langle \mathcal{Q}_a^{-\mathbf{r}}, (\mathcal{R} \setminus \{a\}) \cup \{c\} \rangle$ are extensional expansion proofs.

Assume either b or c is not extended in $\mathcal{Q}$ and call this node $d \in \{b, c\}$. Note that d is also not extended in $\mathcal{Q}_a^{-\mathbf{r}}$. In this case $\mathcal{R} \setminus \{a\}$ is a sufficient set for $\mathcal{Q}_a^{-\mathbf{r}}$ by Lemma 7.4.12. Also,

$$\mathbf{SelPars}^{\mathcal{Q}_a^{-\mathbf{r}}}(\mathcal{R} \setminus \{a\}) \subseteq \mathbf{SelPars}^{\mathcal{Q}_a^{-\mathbf{r}}}((\mathcal{R} \setminus \{a\}) \cup \{d\}) \subseteq \emptyset$$

and so $\langle \mathcal{Q}_a^{-\mathbf{r}}, (\mathcal{R} \setminus \{a\} \rangle$ is an extensional expansion proof. Applying the inductive hypothesis, we have a derivation $\mathcal{D}_0$ of Γ'. By Lemma 2.2.10 (Weakening), there is a derivation $\mathcal{D}$ of $\Gamma(\mathcal{R})$ as desired.

Assume both b and c are extended. By Definition 7.2.7, both b and c are conjunction nodes. By $\clubsuit_\vee$, there exist negative nodes b^1 and b^2 with shallow formulas $\mathbf{A}$ and $\mathbf{B}$, respectively, such that $succ^{\mathcal{Q}}(b) = \langle b^1, b^2 \rangle$. Also, $succ^{\mathcal{Q}}(c) = \langle c^1, c^2 \rangle$ for some negative nodes c^1 and c^2 with shallow formulas $\dot{\neg}\mathbf{A}$ and $\dot{\neg}\mathbf{B}$, respectively.

Applying Lemma 7.8.13 with b and $(\mathcal{R} \setminus \{a\}) \cup \{b\}$, we know

$$\langle (\mathcal{Q}_a^{-\mathbf{r}})_b^{-\mathbf{r}}, (\mathcal{R} \setminus \{a\}) \cup \{b^1, b^2\} \rangle$$

is an extensional expansion proof. Applying Lemma 7.8.13 with c and $(\mathcal{R} \setminus \{a\}) \cup \{c\}$, we know

$$\langle (\mathcal{Q}_a^{-\mathbf{r}})_c^{-\mathbf{r}}, (\mathcal{R} \setminus \{a\}) \cup \{c^1, c^2\} \rangle$$

is an extensional expansion proof.

There is a derivation $\mathcal{D}_1$ of $\Gamma', \mathbf{A}, \mathbf{B}$ by the inductive hypothesis applied to $\langle (\mathcal{Q}_a^{-\mathbf{r}})_b^{-\mathbf{r}}, (\mathcal{R} \setminus \{a\}) \cup \{b^1, b^2\} \rangle$. Applying the inductive hypothesis to $\langle (\mathcal{Q}_a^{-\mathbf{r}})_c^{-\mathbf{r}}, (\mathcal{R} \setminus \{a\}) \cup \{c^1, c^2\} \rangle$, we have a derivation $\mathcal{D}_2$ of $\Gamma', \dot{\neg}\mathbf{A}, \dot{\neg}\mathbf{B}$. We complete the derivation $\mathcal{D}$ by

$$\frac{\begin{array}{cc} \mathcal{D}_1 & \mathcal{D}_2 \\ \Gamma', \mathbf{A}, \mathbf{B} & \Gamma', \dot{\neg}\mathbf{A}, \dot{\neg}\mathbf{B} \end{array}}{\Gamma', \dot{\neg}[\mathbf{A} \doteq^o \mathbf{B}]} \mathcal{G}(\dot{\neg} \doteq^o)$$

Finally, we check the case where $\langle c, a, b \rangle$ is an accessible connection, $a, b \in \mathcal{R}$ and c is an extended symmetric E-unification node. By $\clubsuit_{eu}$ and $\clubsuit_{eu}^{sp}$, $succ^{\mathcal{Q}}(c) = \langle c^1, c^2 \rangle$, $\mathbf{M}$ is of the form

$$[[\mathbf{A} \doteq^\iota \mathbf{D}] \dot{\wedge} [\mathbf{B} \doteq^\iota \mathbf{C}]],$$

$$\overline{sh}(c^1) = [\mathbf{A} \doteq^\iota \mathbf{D}], \overline{sh}(c^2) = [\mathbf{B} \doteq^\iota \mathbf{C}],$$

$$\overline{sh}(a) = \dot{\neg}[\mathbf{A} \doteq^{\iota} \mathbf{B}] \text{ and } \overline{sh}(b) = [\mathbf{C} \doteq^{\iota} \mathbf{D}].$$

Hence $\mathcal{D}_1$ is a derivation of $\Gamma'', [\mathbf{A} \doteq^{\iota} \mathbf{D}]$ and $\mathcal{D}_2$ is a derivation of $\Gamma'', [\mathbf{B} \doteq^{\iota} \mathbf{C}]$. We complete the derivation $\mathcal{D}$ as follows:

$$\dfrac{\dfrac{\dfrac{\begin{matrix}\mathcal{D}_1 \\ \Gamma'', [\mathbf{A} \doteq^{\iota} \mathbf{D}]\end{matrix} \qquad \begin{matrix}\mathcal{D}_2 \\ \Gamma'', [\mathbf{B} \doteq^{\iota} \mathbf{C}]\end{matrix}}{\Gamma', \dot{\neg}[\mathbf{A} \doteq^{\iota} \mathbf{B}], [\mathbf{C} \doteq^{\iota} \mathbf{D}], \dot{\neg}[\mathbf{A} \doteq^{\iota} \mathbf{B}], [\mathbf{C} \doteq^{\iota} \mathbf{D}]}\,\mathcal{G}(EUnif_2)}{\Gamma', [\mathbf{C} \doteq^{\iota} \mathbf{D}], \dot{\neg}[\mathbf{A} \doteq^{\iota} \mathbf{B}], [\mathbf{C} \doteq^{\iota} \mathbf{D}]}\,\mathcal{G}(Contr)}{\Gamma', \dot{\neg}[\mathbf{A} \doteq^{\iota} \mathbf{B}], [\mathbf{C} \doteq^{\iota} \mathbf{D}]}\,\mathcal{G}(Contr)$$

□

PROOF OF LEMMA 7.10.11. We check the cases for the remaining sequent calculus rules.

Suppose $\mathcal{D}$ is of the form

$$\dfrac{\begin{matrix}\mathcal{D}_1 \\ \Gamma, [[\mathbf{G}\, W] \doteq^{\alpha} [\mathbf{H}\, W]]\end{matrix}}{\Gamma, [\mathbf{G} \doteq^{\alpha\beta} \mathbf{H}]}\,\mathcal{G}(\mathfrak{f}^{W})$$

where $\mathbf{M}$ is $[\mathbf{G} \doteq^{\alpha\beta} \mathbf{H}]$ and W_β is a parameter such that W does not occur in any proposition in Γ and does not occur in $\mathbf{M}$. Consequently, we know W does not occur in $\mathbf{N}^i$ for each i $(1 \leq i \leq n-1)$. Let $\mathbf{N}^n$ be $[[\mathbf{G}\, W] \doteq^{\alpha} [\mathbf{H}\, W]]$. Since the dual shallow formula of a is $\theta([\mathbf{G} \doteq \mathbf{H}]^{\downarrow})$, a must be negative. Also, a must be a functional node by Lemma 7.2.9:5. By Lemma 7.10.2, there is some $\mathcal{Q}^0 \in \mathfrak{Nice}_{\mathcal{W}}$ such that $\mathcal{Q} \trianglelefteq_r \mathcal{Q}^0$ and a is extended in $\mathcal{Q}^0$. By $\clubsuit_{\doteq}^{\rightarrow}$ (in $\mathcal{Q}^0$), there is a negative primary node $c \in \mathcal{N}^{\mathcal{Q}^0}$ and variable y_β such that $succ^{\mathcal{Q}^0}(a) = \langle c \rangle$ and c has shallow formula

$$[\dot{\forall} y_\beta \centerdot [\theta(\mathbf{G})\, y]^{\downarrow} \doteq^{\alpha} [\theta(\mathbf{H})\, y]^{\downarrow}].$$

By Definition 7.2.7, we know c must be a selection node. By Lemma 7.10.2, there is some $\mathcal{Q}^1 \in \mathfrak{Nice}_{\mathcal{W}}$ such that $\mathcal{Q}^0 \trianglelefteq_r \mathcal{Q}^1$ and c is extended in $\mathcal{Q}^1$. By $\clubsuit_{sel}$ (in $\mathcal{Q}^1$), $c \in \mathbf{Dom}(sel^{\mathcal{Q}^1})$. Let B_β be the selected parameter $sel^{\mathcal{Q}^1}(c)$ for c. By $\clubsuit_{sel}$ (in $\mathcal{Q}^1$), there exists a negative primary node b^n with shallow formula

$$[B/y][[\theta(\mathbf{G})\, y]^{\downarrow} \doteq^{\alpha} [\theta(\mathbf{H})\, y]^{\downarrow}]$$

such that $succ^{\mathcal{Q}^1}(c) = \langle b^n \rangle$. Let ψ be the parameter substitution $\theta, [B/W]$. Since $W \notin (\mathbf{Params}(\mathbf{G}) \cup \mathbf{Params}(\mathbf{H}))$, the dual shallow formula of b^n is $\psi([\mathbf{G}\,W \doteq \mathbf{H}\,W]^{\downarrow})$, i.e., $\psi((\mathbf{N}^n)^{\downarrow})$. For each i $(1 \leq i \leq n-1)$, we know $W \notin \mathbf{Params}((\mathbf{N}^i)^{\downarrow})$ and hence the dual shallow formula of b^i is $\psi((\mathbf{N}^i)^{\downarrow})$. By the inductive hypothesis for $\mathcal{D}_1$, ψ, $\mathcal{Q}^1$, $\mathbf{N}^1, \ldots, \mathbf{N}^n$ and $b^1, \ldots, b^n$, we obtain a $\mathcal{Q}' \in \mathfrak{Nice}_{\mathcal{W}}$ such that $\mathcal{Q}^1 \trianglelefteq_r \mathcal{Q}'$ and $\{b^i \mid 1 \leq i \leq n\}$ is a sufficient set for $\mathcal{Q}'$. Hence we have

$$\mathcal{Q} \trianglelefteq_r \mathcal{Q}^0 \trianglelefteq_r \mathcal{Q}^1 \trianglelefteq_r \mathcal{Q}'.$$

By Lemma 7.4.15, $\{b^i \mid 1 \leq i \leq n-1\} \cup \{c\}$ is a sufficient set for $\mathcal{Q}'$. By Lemma 7.4.15 again, $\{b^i \mid 1 \leq i \leq n-1\} \cup \{a\}$ is a sufficient set for $\mathcal{Q}'$. Since $\{b^i \mid 1 \leq i \leq n-1\} \cup \{a\}$ is the same as $\{a^i \mid 1 \leq i \leq n\}$ we are done.

Suppose $\mathcal{D}$ is of the form

$$\dfrac{\begin{array}{c}\mathcal{D}_1 \\ \Gamma, \mathbf{P}\end{array}}{\Gamma, \dot{\neg}\dot{\neg}\mathbf{P}}\ \mathcal{G}(\dot{\neg}\dot{\neg})$$

where $\mathbf{M}$ is $\dot{\neg}\dot{\neg}\mathbf{P}$. Hence $\overline{sh}^{\mathcal{Q}}(a)$ is $[\dot{\neg}\dot{\neg}\theta(\mathbf{P}^{\downarrow})]$. By Lemma 7.10.4, there is some $\mathcal{Q}^0 \in \mathfrak{Nice}_{\mathcal{W}}$ and a positive primary node $c \in \mathcal{N}^{\mathcal{Q}^0}$ such that $\mathcal{Q} \trianglelefteq_r \mathcal{Q}^0$, c is a positive counterpart of a with respect to $\mathcal{Q}^0$ and c has shallow formula $[\dot{\neg}\theta(\mathbf{P}^{\downarrow})]$. Hence the primary node c is a negation node (cf. Definition 7.2.7). By Lemma 7.10.2, there is some $\mathcal{Q}^1 \in \mathfrak{Nice}_{\mathcal{W}}$ such that $\mathcal{Q}^0 \trianglelefteq_r \mathcal{Q}^1$ and c is extended in $\mathcal{Q}^1$. By $\clubsuit_{\neg}$ (in $\mathcal{Q}^1$), there is some negative primary node b^n with shallow formula $\theta(\mathbf{P}^{\downarrow})$ such that $succ^{\mathcal{Q}^1}(c) = \langle b^n \rangle$. We apply the inductive hypothesis to $\mathcal{D}_1$ with θ, $\mathcal{Q}^1$, $\mathbf{N}^1, \ldots, \mathbf{N}^{n-1}, \mathbf{P}$ and $b^1, \ldots, b^n$ to obtain some $\mathcal{Q}' \in \mathfrak{Nice}_{\mathcal{W}}$ such that $\mathcal{Q}^1 \trianglelefteq_r \mathcal{Q}'$, and $\{b^1, \ldots, b^{n-1}, b^n\}$ is sufficient for $\mathcal{Q}'$. Hence

$$\mathcal{Q} \trianglelefteq_r \mathcal{Q}^0 \trianglelefteq_r \mathcal{Q}^1 \trianglelefteq_r \mathcal{Q}'.$$

By Lemma 7.4.15, $\{b^1, \ldots, b^{n-1}, c\}$ is sufficient for $\mathcal{Q}'$. By Lemma 7.10.5, $\{b^1, \ldots, b^{n-1}, a\}$ is sufficient for $\mathcal{Q}'$. That is, $\{a^1, \ldots, a^n\}$ is sufficient for $\mathcal{Q}'$.

Suppose $\mathcal{D}$ is of the form

$$\dfrac{\begin{array}{c}\mathcal{D}_1\\ \Gamma, \neg \mathbf{P}^{\sharp}\end{array}}{\Gamma, \neg \mathbf{P}}\ \mathcal{G}(\neg\sharp)$$

where $\mathbf{M}$ is $\neg\mathbf{P}$ and $\mathbf{P} \in \mathit{cwff}_o$. If $\mathbf{P}$ is not of the form $[c\,\overline{\mathbf{A}^m}]$ for some $c \in \mathcal{S}$, then $\mathbf{P}^{\sharp}$ is the same as $\mathbf{P}$ and we can apply the inductive hypothesis to $\mathcal{D}_1$ with θ, $\mathcal{Q}$, $\mathbf{M}^1, \ldots, \mathbf{M}^n$ and $a^1, \ldots, a^n$ to obtain the desired $\mathcal{Q}'$. Consequently, we may assume $\mathbf{P}$ is of the form $[c\,\overline{\mathbf{A}^m}]$ for some $c \in \mathcal{S}$. By Lemma 7.10.4, there is some $\mathcal{Q}^0 \in \mathfrak{Nice}_{\mathcal{W}}$ and a positive primary node $d \in \mathcal{N}^{\mathcal{Q}^0}$ such that $\mathcal{Q} \trianglelefteq_r \mathcal{Q}^0$, d is a positive counterpart of a with respect to $\mathcal{Q}^0$ and d has shallow formula $\theta([c\,\overline{\mathbf{A}^m}]^{\downarrow})$. Since $c \notin \mathbf{Dom}(\theta)$ and θ is a parameter substitution, we know $\theta([c\,\overline{\mathbf{A}^m}]^{\downarrow})$ is the same as $[c\,\theta(\mathbf{A}^1)\cdots\theta(\mathbf{A}^m)]^{\downarrow}$. By Definition 7.2.7, we know d is a deepening node. By Lemma 7.10.2, there is some $\mathcal{Q}^1 \in \mathfrak{Nice}_{\mathcal{W}}$ such that $\mathcal{Q}^0 \trianglelefteq_r \mathcal{Q}^1$ and d is extended in $\mathcal{Q}^1$. By $\clubsuit_{\sharp}$, there is a positive primary node $b^n \in \mathcal{N}^{\mathcal{Q}^1}$ with shallow formula $(\theta(\mathbf{P}^{\downarrow}))^{\sharp\downarrow}$ such that $\mathit{succ}^{\mathcal{Q}^1}(d) = \langle b^n \rangle$. Using Lemma 2.1.40 and the fact that θ is a parameter substitution, we can conclude the dual shallow formula of b^n is $\theta([\neg\,\mathbf{P}^{\sharp}]^{\downarrow})$. We apply the inductive hypothesis to $\mathcal{D}_1$ with θ, $\mathcal{Q}^1$, $\mathbf{N}^1, \ldots, \mathbf{N}^{n-1}, [\neg\,\mathbf{P}^{\sharp}]$ and $b^1, \ldots, b^n$ to obtain some $\mathcal{Q}' \in \mathfrak{Nice}_{\mathcal{W}}$ such that $\mathcal{Q}^1 \trianglelefteq_r \mathcal{Q}'$ and $\{b^i \mid 1 \leq i \leq n\}$ is a sufficient set for $\mathcal{Q}'$. Hence

$$\mathcal{Q} \trianglelefteq_r \mathcal{Q}^0 \trianglelefteq_r \mathcal{Q}^1 \trianglelefteq_r \mathcal{Q}'.$$

By Lemma 7.4.15, $\{b^1, \ldots, b^n, d\}$ is a sufficient set for $\mathcal{Q}'$. By Lemma 7.10.5, $\{a^i \mid 1 \leq i \leq n\}$ is a sufficient set for $\mathcal{Q}'$.

Suppose $\mathcal{D}$ is of the form

$$\dfrac{\Gamma, \neg[[\mathbf{G}\,\mathbf{B}] \doteq^{\alpha} [\mathbf{H}\,\mathbf{B}]] \quad \mathbf{B} \in \mathcal{W}_{\beta}}{\Gamma, \neg[\mathbf{G} \doteq^{\alpha\beta} \mathbf{H}]}\ \mathcal{G}(\neg \doteq^{\rightarrow}_{re}, \mathcal{W})$$

where $\mathbf{M}$ is $\neg\theta([\mathbf{G} \doteq^{\alpha\beta} \mathbf{H}]^{\downarrow})$. By Lemma 7.10.4, we know there exists some $\mathcal{Q}^0 \in \mathfrak{Nice}_{\mathcal{W}}$ and a positive primary node $e \in \mathcal{N}^{\mathcal{Q}^0}$ such that $\mathcal{Q} \trianglelefteq_r \mathcal{Q}^0$, e is a positive counterpart of a with respect to $\mathcal{Q}^0$ and e has shallow formula $\theta([\mathbf{G} \doteq^{\alpha\beta} \mathbf{H}]^{\downarrow})$. By Definition 7.2.7, we know e is a functional node. By Lemma 7.10.2, there is some $\mathcal{Q}^1 \in \mathfrak{Nice}_{\mathcal{W}}$ such that $\mathcal{Q}^0 \trianglelefteq_r \mathcal{Q}^1$ and a is extended in $\mathcal{Q}'$. By $\clubsuit_{\doteq}^{\rightarrow}$, there is a positive primary node e^1 and variable y_β such that $succ^{\mathcal{Q}^1}(e) = \langle e^1 \rangle$ and e^1 has shallow formula $[\forall y_\beta \centerdot \theta(\mathbf{G}^{\downarrow})\, y \stackrel{.}{=}^{\alpha} \theta(\mathbf{H}^{\downarrow})\, y]^{\downarrow}$. Since $\theta(\mathbf{B}) \in \mathcal{W}$, we can apply Lemma 7.10.7 to obtain some $\mathcal{Q}^2 \in \mathfrak{Nice}_{\mathcal{W}}$, term $\mathbf{D} \in \mathcal{W}$ and $d \in \mathcal{N}^{\mathcal{Q}^2}$ such that $\mathcal{Q}^1 \trianglelefteq_r \mathcal{Q}^2$, $\langle e^1, \mathbf{D}, d \rangle \in exps^{\mathcal{Q}^2}$ and $sh^{\mathcal{Q}^2}(d)$ is $[[\theta(\mathbf{G})\, \theta(\mathbf{B})] \stackrel{.}{=}^{\alpha} [\theta(\mathbf{H})\, \theta(\mathbf{B})]]^{\downarrow}$. Let $\mathbf{N}^n$ be $\neg[[\mathbf{G}\,\mathbf{B}] \stackrel{.}{=}^{\alpha} [\mathbf{H}\,\mathbf{B}]]$. By $\clubsuit_{exp}$, d is a positive primary node. Hence d has dual shallow formula $\theta((\mathbf{N}^n)^{\downarrow})$. Applying the inductive hypothesis to $\mathcal{D}_1$ with θ, $\mathcal{Q}^2$, $\mathbf{N}^1, \ldots, \mathbf{N}^{n-1}, \mathbf{N}^n$ and $b^1, \ldots, b^{n-1}, d$, we obtain some $\mathcal{Q}' \in \mathfrak{Nice}_{\mathcal{W}}$ such that $\mathcal{Q}^2 \trianglelefteq_r \mathcal{Q}'$ and $\{b^1, \ldots, b^{n-1}, d\}$ is sufficient for $\mathcal{Q}'$. Hence

$$\mathcal{Q} \trianglelefteq_r \mathcal{Q}^0 \trianglelefteq_r \mathcal{Q}^1 \trianglelefteq_r \mathcal{Q}^2 \trianglelefteq_r \mathcal{Q}'.$$

Let

$$\mathcal{K} := \{z \in \mathcal{N}^{\mathcal{Q}'} \mid e^1 \stackrel{\mathcal{Q}'}{\rightharpoondown} z\}.$$

Since $e^1 \stackrel{\mathcal{Q}^2}{\rightharpoondown} d$, we know $e^1 \stackrel{\mathcal{Q}'}{\rightharpoondown} d$ and hence $d \in \mathcal{K}$ by Lemma 7.6.2. By Lemma 7.4.10, $\{b^1, \ldots, b^{n-1}\} \cup \mathcal{K}$ is sufficient for $\mathcal{Q}'$. Since e^1 is conjunctive, $\{b^1, \ldots, b^{n-1}, e^1\}$ is sufficient for $\mathcal{Q}'$ by Lemma 7.4.13. Since $\mathcal{Q}^1 \trianglelefteq_r \mathcal{Q}'$ and $e \stackrel{\mathcal{Q}^1}{\rightharpoondown} e^1$, we know e is a neutral (functional) node of $\mathcal{Q}'$ and $e \stackrel{\mathcal{Q}'}{\rightharpoondown} e^1$ by Lemma 7.6.2. Thus $\{b^1, \ldots, b^{n-1}, e\}$ is sufficient for $\mathcal{Q}'$ by Lemma 7.4.15. By Lemma 7.10.5, $\{a^i \mid 1 \leq i \leq n\}$ is a sufficient set for $\mathcal{Q}'$.

Suppose $\mathcal{D}$ is of the form

$$\frac{\begin{array}{cc}\mathcal{D}_1 & \mathcal{D}_2 \\ \Gamma, \mathbf{A}, \mathbf{B} & \Gamma, \neg\mathbf{A}, \neg\mathbf{B}\end{array}}{\Gamma, \neg[\mathbf{A} \stackrel{.}{=}^{o} \mathbf{B}]}\ \mathcal{G}(\neg \stackrel{.}{=}^{o})$$

where $\mathbf{M}$ is $\neg[\mathbf{A} \stackrel{.}{=}^{o} \mathbf{B}]$ and $\mathbf{A}, \mathbf{B} \in cwff_o$. By Lemma 7.10.4, there is some $\mathcal{Q}^0 \in \mathfrak{Nice}_{\mathcal{W}}$ and a positive primary node $e \in \mathcal{N}^{\mathcal{Q}^0}$ such that $\mathcal{Q} \trianglelefteq_r \mathcal{Q}^0$, e is a positive counterpart of a with respect to $\mathcal{Q}^0$ and e has shallow formula $\theta([\mathbf{A} \stackrel{.}{=}^{o} \mathbf{B}]^{\downarrow})$. By Definition 7.2.7, we know e is a Boolean node. By Lemma 7.10.2, there is some $\mathcal{Q}^1 \in \mathfrak{Nice}_{\mathcal{W}}$ such that $\mathcal{Q}^0 \trianglelefteq_r \mathcal{Q}^1$ and e is extended in $\mathcal{Q}^1$. By $\clubsuit_{\stackrel{.}{=}}^{+o}$, there are negative primary nodes c and d such that $succ^{\mathcal{Q}^1}(e) = \langle c, d \rangle$, $sh^{\mathcal{Q}^1}(c)$ is $[\theta(\mathbf{A}^{\downarrow}) \stackrel{.}{\vee} \theta(\mathbf{B}^{\downarrow})]$ and $sh^{\mathcal{Q}^1}(d)$ is $[[\neg \theta(\mathbf{A}^{\downarrow})] \stackrel{.}{\vee} [\neg \theta(\mathbf{B}^{\downarrow})]]$. By Definition 7.2.7, we know c and d are conjunction nodes. Applying Lemma 7.10.2 twice to the nodes c and d, there is some $\mathcal{Q}^2 \in \mathfrak{Nice}_{\mathcal{W}}$ such that

$\mathcal{Q}^1 \trianglelefteq_r \mathcal{Q}^2$ and the nodes c and d are extended in $\mathcal{Q}^2$. By $\clubsuit_\vee$, there exist negative primary nodes c^1, c^2, d^1 and d^2 such that

$$succ^{\mathcal{Q}^2}(c) = \langle c^1, c^2\rangle,\ succ^{\mathcal{Q}^2}(d) = \langle d^1, d^2\rangle,$$
$$sh^{\mathcal{Q}^2}(c^1) = \theta(\mathbf{A}^\downarrow),\ sh^{\mathcal{Q}^2}(c^2) = \theta(\mathbf{B}^\downarrow),$$
$$sh^{\mathcal{Q}^2}(d^1) = [\neg\,\theta(\mathbf{A}^\downarrow)] \text{ and } sh^{\mathcal{Q}^2}(d^2) = [\neg\,\theta(\mathbf{B}^\downarrow)].$$

We first apply the inductive hypothesis to $\mathcal{D}_1$ with θ, $\mathcal{Q}^2$,

$$\mathbf{N}^1, \ldots, \mathbf{N}^{n-1}, \mathbf{A}, \mathbf{B} \text{ and } b^1, \ldots, b^{n-1}, c^1, c^2$$

to obtain some $\mathcal{Q}^3 \in \mathfrak{Nice}_{\mathcal{W}}$ such that $\mathcal{Q}^2 \trianglelefteq_r \mathcal{Q}^3$, and the set of nodes $\{b^1, \ldots, b^{n-1}, c^1, c^2\}$ is sufficient for $\mathcal{Q}^3$. Since $\mathcal{Q}^2 \trianglelefteq_r \mathcal{Q}^3$, we know c and d are conjunction nodes of $\mathcal{Q}^3$,

$$succ^{\mathcal{Q}^3}(c) = \langle c^1, c^2\rangle,\ succ^{\mathcal{Q}^3}(d) = \langle d^1, d^2\rangle,$$
$$\overline{sh}^{\mathcal{Q}^3}(d^1) = [\neg\,\theta(\mathbf{A}^\downarrow)] \text{ and } \overline{sh}^{\mathcal{Q}^3}(d^2) = [\neg\,\theta(\mathbf{B}^\downarrow)].$$

Note that

$$\mathcal{Q} \trianglelefteq_r \mathcal{Q}^0 \trianglelefteq_r \mathcal{Q}^1 \trianglelefteq_r \mathcal{Q}^2 \trianglelefteq_r \mathcal{Q}^3.$$

Since $\mathcal{Q} \trianglelefteq_r \mathcal{Q}^3$, we know $b^i \in \mathcal{N}^{\mathcal{Q}^3}$ and $\overline{sh}^{\mathcal{Q}^3}(b^i)$ is $\theta((\mathbf{N}^i)^\downarrow)$ for each i $(1 \le i \le n-1)$. By Lemma 7.3.13, c^1 and c^2 are the only children of c in $\mathcal{Q}^3$. By Lemma 7.4.13, we can conclude $\{b^1, \ldots, b^{n-1}, c\}$ is sufficient for $\mathcal{Q}^3$. We apply the inductive hypothesis to $\mathcal{D}_2$ with θ, $\mathcal{Q}^3$, $\mathbf{N}^1, \ldots, \mathbf{N}^{n-1}, \neg\mathbf{A}, \neg\mathbf{B}$ and $b^1, \ldots, b^{n-1}, d^1, d^2$ to obtain some $\mathcal{Q}' \in \mathfrak{Nice}_{\mathcal{W}}$ such that $\mathcal{Q}^3 \trianglelefteq_r \mathcal{Q}'$ and $\{b^1, \ldots, b^{n-1}, d^1, d^2\}$ is sufficient for $\mathcal{Q}'$. Hence

$$\mathcal{Q} \trianglelefteq_r \mathcal{Q}^0 \trianglelefteq_r \mathcal{Q}^1 \trianglelefteq_r \mathcal{Q}^2 \trianglelefteq_r \mathcal{Q}^3 \trianglelefteq_r \mathcal{Q}'.$$

Since $\mathcal{Q}^3 \trianglelefteq_r \mathcal{Q}'$, d is a conjunction node of $\mathcal{Q}'$ such that $succ^{\mathcal{Q}'}(d)$ is $\langle d^1, d^2\rangle$. By Lemma 7.3.13, d^1 and d^2 are the only children of d in $\mathcal{Q}'$. Thus $\{b^1, \ldots, b^n, d\}$ is sufficient for $\mathcal{Q}'$. Since $\mathcal{Q}^3 \trianglelefteq_r \mathcal{Q}'$, we know $\{b^1, \ldots, b^{n-1}, c\}$ is sufficient for $\mathcal{Q}'$ by Lemma 7.6.7. Since $\mathcal{Q}^1 \trianglelefteq_r \mathcal{Q}'$, $succ^{\mathcal{Q}'}(e) = \langle c, d\rangle$. By Lemma 7.3.13, c and d are the only children of e in $\mathcal{Q}'$. By Lemma 7.4.14, $\{b^1, \ldots, b^{n-1}, e\}$ is sufficient for $\mathcal{Q}'$. By Lemma 7.10.5, $\{a^i \mid 1 \le i \le n\}$ is a sufficient set for $\mathcal{Q}'$.

Suppose $\mathcal{D}$ is of the form

$$\frac{\begin{array}{cc}\mathcal{D}_1 & \mathcal{D}_2 \\ \Gamma, [\mathbf{A} \doteq^\iota \mathbf{D}] & \Gamma, [\mathbf{B} \doteq^\iota \mathbf{C}]\end{array}}{\Gamma, \neg[\mathbf{A} \doteq^\iota \mathbf{B}], [\mathbf{C} \doteq^\iota \mathbf{D}]}\ \mathcal{G}(EUnif_2)$$

where $\mathbf{A}, \mathbf{B}, \mathbf{C}, \mathbf{D} \in cwff_\iota$, $\mathbf{M}$ is $\neg[\mathbf{A} \doteq^\iota \mathbf{B}]$ and $\mathbf{Q}$ is $[\mathbf{C} \doteq^\iota \mathbf{D}]$. Consequently, the dual shallow formulas of a' and b are $\neg\theta([\mathbf{A} \doteq^\iota \mathbf{B}]^\downarrow)$ and $\theta([\mathbf{C} \doteq^\iota \mathbf{D}]^\downarrow)$, respectively. By Lemma 7.10.4, there is some $\mathcal{Q}^0 \in \mathfrak{Nice}_\mathcal{W}$ and a positive primary node $a \in \mathcal{N}^{\mathcal{Q}^0}$ such that $\mathcal{Q} \trianglelefteq_r \mathcal{Q}^0$, a is a positive counterpart of a' with respect to $\mathcal{Q}^0$ and a has shallow formula $\theta([\mathbf{A} \doteq^\iota \mathbf{B}]^\downarrow)$. By Definition 7.2.7, the primary node a is a positive equation node. By Lemma 7.2.9:4, the primary node b is a negative equation node. There is some $\mathcal{Q}^1 \in \mathfrak{Nice}_\mathcal{W}$ such that $\mathcal{Q}^0 \trianglelefteq_r \mathcal{Q}^1$ and b is extended in $\mathcal{Q}^1$ by Lemma 7.10.2. By $\clubsuit_{\doteq}^{-}$, there is an equation goal node e such that

$$succ^{\mathcal{Q}^1}(b) = \langle e \rangle \text{ and } sh^{\mathcal{Q}^1}(e) = sh^{\mathcal{Q}^1}(b).$$

By Lemma 7.10.9, we know there is some $\mathcal{Q}^2 \in \mathfrak{Nice}_\mathcal{W}$ and symmetric E-unification node $c \in \mathcal{N}^{\mathcal{Q}^2}$ such that $\mathcal{Q}^1 \trianglelefteq_r \mathcal{Q}^2$ and $\langle c, a, e \rangle$ is in $conns^{\mathcal{Q}^2}$. By $\clubsuit_{eu}^{sp}$, the shallow formula of c must be

$$\theta([[\mathbf{A} \doteq^\iota \mathbf{D}] \mathbin{\dot\wedge} [\mathbf{B} \doteq^\iota \mathbf{C}]]^\downarrow).$$

By Lemma 7.10.2, there is some $\mathcal{Q}^3 \in \mathfrak{Nice}_\mathcal{W}$ such that $\mathcal{Q}^2 \trianglelefteq_r \mathcal{Q}^3$ and c is extended in $\mathcal{Q}^3$. By $\clubsuit_{eu}$, there are negative equation nodes c^1 and c^2 in $\mathcal{Q}^3$ such that $succ^{\mathcal{Q}^3}(c) = \langle c^1, c^2 \rangle$, $sh^{\mathcal{Q}^3}(c^1)$ is $\theta([\mathbf{A} \doteq^\iota \mathbf{D}]^\downarrow)$ and $sh^{\mathcal{Q}^3}(c^2)$ is $\theta([\mathbf{B} \doteq^\iota \mathbf{C}]^\downarrow)$. Applying the inductive hypothesis to $\mathcal{D}_1$ with θ, $\mathcal{Q}^3$, $\mathbf{N}^1, \ldots, \mathbf{N}^{n-2}, [\mathbf{A} \doteq^\iota \mathbf{D}]$ and $b^1, \ldots, b^{n-2}, c^1$, we obtain some $\mathcal{Q}^4 \in \mathfrak{Nice}_\mathcal{W}$ such that $\mathcal{Q}^3 \trianglelefteq_r \mathcal{Q}^4$ and $\{b^1, \ldots, b^{n-2}, c^1\}$ is sufficient for $\mathcal{Q}^4$. Applying the inductive hypothesis to $\mathcal{D}_2$ with θ, $\mathcal{Q}^4$, $\mathbf{N}^1, \ldots, \mathbf{N}^{n-2}, [\mathbf{B} \doteq^\iota \mathbf{C}]$ and $b^1, \ldots, b^{n-2}, c^2$, we obtain some $\mathcal{Q}' \in \mathfrak{Nice}_\mathcal{W}$ such that $\mathcal{Q}^4 \trianglelefteq_r \mathcal{Q}'$ and $\{b^1, \ldots, b^{n-2}, c^2\}$ is sufficient for $\mathcal{Q}'$. Note that

$$\mathcal{Q} \trianglelefteq_r \mathcal{Q}^0 \trianglelefteq_r \mathcal{Q}^1 \trianglelefteq_r \mathcal{Q}^2 \trianglelefteq_r \mathcal{Q}^3 \trianglelefteq_r \mathcal{Q}^4 \trianglelefteq_r \mathcal{Q}'.$$

By Lemma 7.6.7, $\{b^1, \ldots, b^{n-2}, c^1\}$ is sufficient for $\mathcal{Q}'$. Since c is a disjunctive node, $\{b^1, \ldots, b^{n-2}, c\}$ is sufficient for $\mathcal{Q}'$ by Lemmas 7.4.14 and 7.4.3. By Lemma 7.4.16, $\{b^1, \ldots, b^{n-2}, a, e\}$ is sufficient for $\mathcal{Q}'$. Since b is a conjunctive node and $b \xrightarrow{\mathcal{Q}'} e$, $\{b^1, \ldots, b^{n-2}, a, b\}$ is sufficient for $\mathcal{Q}'$ by Lemmas 7.4.13 and 7.4.10. Finally, $\{b^1, \ldots, b^{n-2}, a', b\}$ is sufficient by Lemma 7.10.5. □

A.6. Proofs from Chapter 8

PROOF OF LEMMA 8.2.3. Let $\mathcal{Q}'$ be $[\mathbf{A}/x]\mathcal{Q}$. Suppose y is in $\mathbf{Free}(\mathcal{Q}')$. By Definition 7.2.16, we know either $y \in \mathbf{Free}^{\mathcal{Q}'}(\mathcal{N})$ or $y \in \mathbf{ExpVars}(\mathcal{Q}')$. If $y \in \mathbf{Free}^{\mathcal{Q}'}(\mathcal{N})$, then there is a node $a \in \mathcal{N}$ such that $y \in \mathbf{Free}(sh^{\mathcal{Q}'}(a))$ and so either $y \in \mathbf{Free}(\mathbf{A})$ or

$$y \in (\mathbf{Free}(sh^{\mathcal{Q}}(a)) \setminus \{x\}) \subseteq (\mathbf{Free}(\mathcal{Q}) \setminus \{x\})$$

since $sh^{\mathcal{Q}'}(a)$ is $([\mathbf{A}/x]sh^{\mathcal{Q}}(a))^{\downarrow}$. If $y \in \mathbf{ExpVars}(\mathcal{Q}')$, then there is an expansion arc $\langle e, \mathbf{C}, c\rangle \in exps^{\mathcal{Q}}$ such that $y \in \mathbf{Free}([\mathbf{A}/x]\mathbf{C})$ and so either $y \in \mathbf{Free}(\mathbf{A})$ or

$$y \in (\mathbf{Free}(\mathbf{C}) \setminus \{x\}) \subseteq (\mathbf{Free}(\mathcal{Q}) \setminus \{x\}).$$

Suppose $P \in \mathbf{Params}(\mathcal{Q}')$. By Definition 7.2.16, we know either $P \in \mathbf{Params}^{\mathcal{Q}'}(\mathcal{N})$, $P \in \mathbf{ExpParams}(\mathcal{Q}')$ or $P \in \mathbf{SelPars}(\mathcal{Q}')$. If $P \in \mathbf{Params}^{\mathcal{Q}'}(\mathcal{N})$, then there exists some node $a \in \mathcal{N}$ such that $P \in \mathbf{Params}(sh^{\mathcal{Q}'}(a))$ and so either $P \in \mathbf{Params}(\mathbf{A})$ or

$$P \in \mathbf{Params}(sh^{\mathcal{Q}}(a)) \subseteq \mathbf{Params}(\mathcal{Q})$$

since $sh^{\mathcal{Q}'}(a)$ is $([\mathbf{A}/x]sh^{\mathcal{Q}}(a))^{\downarrow}$. If $P \in \mathbf{ExpParams}(\mathcal{Q}')$, then there is an expansion arc $\langle e, \mathbf{C}, c\rangle \in exps^{\mathcal{Q}}$ such that $P \in \mathbf{Params}([\mathbf{A}/x]\mathbf{C})$ and so $P \in \mathbf{Params}(\mathbf{A})$ or

$$P \in \mathbf{Params}(\mathbf{C}) \subseteq \mathbf{Params}(\mathcal{Q}).$$

If $P \in \mathbf{SelPars}(\mathcal{Q}')$, then

$$P \in \mathbf{SelPars}(\mathcal{Q}') = \mathbf{SelPars}(\mathcal{Q}) \subseteq \mathbf{Params}(\mathcal{Q})$$

since $\mathcal{Q}$ and $\mathcal{Q}'$ share the same selected parameter function. □

PROOF OF LEMMA 8.2.6. Let

$$\mathcal{Q} = \langle \mathcal{N}, succ, label, sel, exps, conns, decs\rangle$$

be the given extensional expansion dag. Let $\mathcal{Q}'$ be $\theta(\mathcal{Q})$. By Definition 8.2.2, we know $\mathcal{Q}'$ is

$$\langle \mathcal{N}, succ, label', sel, exps', conns, decs\rangle$$

where

$$label'(a) := \langle kind^{\mathcal{Q}}(a), pol^{\mathcal{Q}}(a), (\theta(sh^{\mathcal{Q}}(a)))^{\downarrow}\rangle$$

for every $a \in \mathcal{N}$ and

$$exps' := \{\langle e, \theta(\mathbf{C}), c\rangle \mid \langle e, \mathbf{C}, c\rangle \in exps.$$

The underlying graphs $\langle \mathcal{N}, \xrightarrow{\mathcal{Q}} \rangle$ and $\langle \mathcal{N}, \xrightarrow{\mathcal{Q}'} \rangle$ are the same by Lemma 8.2.5. In particular, $\langle \mathcal{N}, \xrightarrow{\mathcal{Q}'} \rangle$ is a dag. Also, note that a node $a \in \mathcal{N}$ is extended in $\mathcal{Q}$ iff it is extended in $\mathcal{Q}'$ since the two structures share the same successor function.

Let $a \in \mathbf{Dom}(sel)$ be given. Since $\mathcal{Q}$ is an extensional expansion dag, a is a selection node of $\mathcal{Q}$. Hence

$$label(a) = \langle \mathbf{primary}, -1, [\dot{\forall} x_\alpha\, \mathbf{M}] \rangle$$

for some variable x_α and proposition $\mathbf{M} \in prop(\mathcal{S})^{\downarrow}$. Thus

$$label'(a) = \langle \mathbf{primary}, -1, [\dot{\forall} x_\alpha\, (\theta(\mathbf{M}))^{\downarrow}] \rangle$$

and so a is a selection node of $\mathcal{Q}'$. We also know sel is also an injective partial function since $\mathcal{Q}$ is an extensional expansion dag.

It remains to check the $\clubsuit_*$ properties. We check a few sample cases only. The remaining cases are analogous.

Consider an expansion arc $\langle e, (\theta(\mathbf{C})), c \rangle \in exps'$ of $\mathcal{Q}'$ induced by an expansion arc $\langle e, \mathbf{C}, c \rangle \in exps$ of $\mathcal{Q}$. By $\clubsuit_{exp}$ in $\mathcal{Q}$, we know e is an extended expansion node with shallow formula $[\dot{\forall} x_\alpha\, \mathbf{M}]$ (for some x_α and $\mathbf{M}$) and c is a positive primary node with shallow formula $([\mathbf{C}/x]\mathbf{M})^{\downarrow}$. Hence e is an extended expansion node of $\mathcal{Q}'$ with shallow formula $[\dot{\forall} x_\alpha\, (\theta(\mathbf{M}))^{\downarrow}]$ and c is a positive primary node of $\mathcal{Q}'$ with shallow formula $(\theta([\mathbf{C}/x]\mathbf{M}))^{\downarrow}$. Since $(\theta([\mathbf{C}/x]\mathbf{M}))^{\downarrow}$ is

$$([\theta(\mathbf{C})/x]((\theta(\mathbf{M}))^{\downarrow}))^{\downarrow}$$

we know $\langle e, \theta(\mathbf{C}), c \rangle$ satisfies $\clubsuit_{exp}$ with respect to $\mathcal{Q}'$.

Let $\langle c, a, b \rangle \in conns$ be a connection in $\mathcal{Q}'$ (hence in $\mathcal{Q}$). Suppose c is a mate node in $\mathcal{Q}'$ with shallow formula

$$[[P\, \mathbf{B}^1_{\alpha^i} \cdots \mathbf{B}^n_{\alpha^n}] \doteq^o [P\, \mathbf{A}^1_{\alpha^1} \cdots \mathbf{A}^n_{\alpha^n}]].$$

Then c is a mate node in $\mathcal{Q}$ with shallow formula

$$[[P\, \mathbf{D}^1_{\alpha^i} \cdots \mathbf{D}^n_{\alpha^n}] \doteq^o [P\, \mathbf{C}^1_{\alpha^1} \cdots \mathbf{C}^n_{\alpha^n}]]$$

where $\mathbf{A}^i$ is $(\theta(\mathbf{C}^i))^{\downarrow}$ and $\mathbf{B}^i$ is $(\theta(\mathbf{D}^i))^{\downarrow}$ for each i $(1 \leq i \leq n)$. By $\clubsuit^p_m$ in $\mathcal{Q}$, we know a is an extended positive atomic node in $\mathcal{Q}$ with shallow formula $[P\, \mathbf{C}^1 \cdots \mathbf{C}^n]$ and b is an extended negative atomic node in $\mathcal{Q}$ with shallow formula $[P\, \mathbf{D}^1 \cdots \mathbf{D}^n]$. Thus a is an extended positive atomic node in $\mathcal{Q}'$ with shallow formula $[P\, \mathbf{A}^1 \cdots \mathbf{A}^n]$ and

b is an extended negative atomic node in $\mathcal{Q}'$ with shallow formula $[P\,\mathbf{B}^1 \cdots \mathbf{B}^n]$. Therefore, $\langle c, a, b\rangle$ satisfies $\clubsuit^p_m$ with respect to $\mathcal{Q}'$.

Let $a \in \mathbf{Dom}(succ)$ be an extended node in $\mathcal{Q}'$ (hence in $\mathcal{Q}$). We must verify a satisfies the $\clubsuit_*$ properties for extended nodes in $\mathcal{Q}'$.

Suppose a is a true node, false node, expansion node or potentially connecting node in $\mathcal{Q}'$. By Lemma 8.2.4, a is a true node, false node, expansion node or potentially connecting node in $\mathcal{Q}'$. By $\clubsuit_{ns}$ in $\mathcal{Q}$, we know $succ(a) = \langle\rangle$ and so $\clubsuit_{ns}$ holds for a with respect to $\mathcal{Q}'$.

Note that a is not flexible in $\mathcal{Q}$ by $\clubsuit_f$ since a is extended in $\mathcal{Q}$. Hence $(\theta(sh^{\mathcal{Q}}(a)))^{\downarrow}$ cannot be of the form $[p\,\overline{\mathbf{A}^n}]$ where p is a variable. Thus a is not flexible in $\mathcal{Q}'$ and so $\clubsuit_f$ holds for a with respect to $\mathcal{Q}'$.

Suppose a is a negation node in $\mathcal{Q}'$ with polarity p and shallow formula $\neg\mathbf{M}'$. Then a is a negation node in $\mathcal{Q}$ with polarity p and shallow formula $\neg\mathbf{M}$ where $\mathbf{M}'$ is $(\theta(\mathbf{M}))^{\downarrow}$. By $\clubsuit_\neg$ in $\mathcal{Q}$, there is a node $a^1 \in \mathcal{N}$ such that $succ(a) = \langle a^1\rangle$ and $label(a^1) = \langle\mathbf{primary}, -p, \mathbf{M}\rangle$. Hence

$$label'(a^1) = \langle\mathbf{primary}, -p, \mathbf{M}'\rangle$$

and so a satisfies $\clubsuit_\neg$ with respect to $\mathcal{Q}'$.

Suppose a is a selection node of $\mathcal{Q}'$ with shallow formula $[\dot\forall x_\alpha\,\mathbf{M}']$. Then a is a selection node of $\mathcal{Q}$ with shallow formula $[\dot\forall x_\alpha\,\mathbf{M}]$ where $\mathbf{M}'$ is $(\theta(\mathbf{M}))^{\downarrow}$.[2] By $\clubsuit_{sel}$ in $\mathcal{Q}$, $a \in \mathbf{Dom}(sel)$ and there is some negative primary node a^1 such that $succ(a) = \langle a^1\rangle$ and a^1 has shallow formula $[A/x]\mathbf{M}$ where A is $sel(a)$. Thus a^1 is a negative primary node of $\mathcal{Q}'$ with shallow formula $(\theta([A/x]\mathbf{M}))^{\downarrow}$. Note that

$$(\theta([A/x]\mathbf{M}))^{\downarrow} = ([A/x](\theta(\mathbf{M})))^{\downarrow} = [A/x]\mathbf{M}'$$

since $\mathbf{Dom}(\theta) \subset \mathcal{V}$. Hence a satisfies $\clubsuit_{sel}$ with respect to $\mathcal{Q}'$.

Suppose a is a deepening node of $\mathcal{Q}'$ with polarity p and shallow formula $[c\,\overline{\mathbf{A}^n}]$. By Lemma 8.2.4:5, a is a deepening node of $\mathcal{Q}$. Thus a must have polarity p and shallow formula $[c\,\overline{\mathbf{B}^n}]$ where $\mathbf{A}^i$ is $(\theta(\mathbf{B}^i))^{\downarrow}$ for each i such that $1 \leq i \leq n$. By $\clubsuit_\sharp$ in $\mathcal{Q}$, there is a node a^1 such that $succ(a) = \langle a^1\rangle$ and

$$label(a^1) = \langle\mathbf{primary}, p, ([c\,\overline{\mathbf{B}^n}]^\sharp)^{\downarrow}\rangle.$$

Thus

$$label'(a^1) = \langle\mathbf{primary}, p, (\theta(([c\,\overline{\mathbf{B}^n}]^\sharp)^{\downarrow}))^{\downarrow}\rangle.$$

[2]As usual, we assume x_α is chosen to avoid any conflict.

We compute

$$
\begin{aligned}
(\theta(([c\,\overline{\mathbf{B}^n}]^\sharp)^\downarrow))^\downarrow &= (\theta([c\,\overline{\mathbf{B}^n}]^\sharp))^\downarrow \\
&= ((\theta([c\,\overline{\mathbf{B}^n}]))^\sharp)^\downarrow \quad \text{(by Lemma 2.1.39)} \\
&= (((\theta[c\,\mathbf{B}^n])^\downarrow)^\sharp)^\downarrow \quad \text{(by Lemma 2.1.40)} \\
&= ([c\,\mathbf{A}^n]^\sharp)^\downarrow.
\end{aligned}
$$

Hence a satisfies $\clubsuit_\sharp$ with respect to $\mathcal{Q}'$.

Suppose a is a decomposition node of $\mathcal{Q}'$ with shallow formula

$$[[H\,\mathbf{A}^1_{\alpha^1}\cdots\mathbf{A}^n_{\alpha^n}] \mathrel{\dot{=}^\iota} [H\,\mathbf{B}^1_{\alpha^1}\cdots\mathbf{B}^n_{\alpha^n}]].$$

Then a is a decomposition node of $\mathcal{Q}$ with shallow formula

$$[[K\,\mathbf{C}^1_{\alpha^1}\cdots\mathbf{C}^m_{\alpha^m}] \mathrel{\dot{=}^\iota} [K\,\mathbf{D}^1_{\alpha^1}\cdots\mathbf{D}^m_{\alpha^m}]]$$

where

$$(\theta([K\,\overline{\mathbf{C}^m}]))^\downarrow = [H\,\overline{\mathbf{A}^n}]$$

and

$$(\theta([K\,\overline{\mathbf{D}^m}]))^\downarrow = [H\,\overline{\mathbf{B}^n}].$$

Since $\theta(K) = K$ is a parameter, we must have $K = H$ and $m = n$. Hence $\mathbf{A}^i$ is $(\theta(\mathbf{C}^i))^\downarrow$ and $\mathbf{B}^i$ is $(\theta(\mathbf{D}^i))^\downarrow$ for each i $(1 \leq i \leq n)$. By $\clubsuit_{dec}$ in $\mathcal{Q}$, there exist negative primary nodes $a^1, \ldots, a^n$ in $\mathcal{Q}$ such that

$$succ(a) = \langle a^1, \ldots, a^n\rangle \text{ and } sh^{\mathcal{Q}}(a^i) = [\mathbf{C}^i \mathrel{\dot{=}^{\alpha^i}} \mathbf{D}^i]$$

for each i $(1 \leq i \leq n)$. Thus $a^1, \ldots, a^n$ are negative primary nodes in $\mathcal{Q}'$ and $sh^{\mathcal{Q}}(a^i) = [\mathbf{A}^i \mathrel{\dot{=}^{\alpha^i}} \mathbf{B}^i]$ for each i $(1 \leq i \leq n)$. Hence a satisfies $\clubsuit_{dec}$ with respect to $\mathcal{Q}'$.

The remaining $\clubsuit_*$ properties follow analogously. Therefore, $\mathcal{Q}'$ is an extensional expansion dag. Finally, we check the remaining conditions.

1. For every node $r \in \mathcal{N}$, r is a root node of $\mathcal{Q}$ iff r is a root node of $\mathcal{Q}'$ by Lemma 8.2.5. Thus $\mathbf{Roots}(\mathcal{Q}) = \mathbf{Roots}(\mathcal{Q}')$.
2. We know $\mathbf{SelPars}(\mathcal{Q}) = \mathbf{SelPars}(\mathcal{Q}')$ since $\mathcal{Q}$ and $\mathcal{Q}'$ share the same selected parameter function.
3. Suppose θ is a ground substitution and $\mathbf{Free}(\mathcal{Q}) \subseteq \mathbf{Dom}(\theta)$. For every $a \in \mathcal{N}$,

$$sh^{\theta(\mathcal{Q})}(a) = (\theta(sh^{\mathcal{Q}}(a)))^\downarrow$$

is closed. Hence $\mathbf{Free}^{\theta(\mathcal{Q})}(\mathcal{N})$ is empty. Every expansion arc of $\mathcal{Q}$ is $\langle e, \theta(\mathbf{C}), c\rangle$ where $\langle e, \mathbf{C}, c\rangle \in exps$. Hence $\mathbf{ExpVars}(\theta(\mathcal{Q}))$ is empty. Thus $\theta(\mathcal{Q})$ is closed.

□

PROOF OF LEMMA 8.2.12. Let $\mathcal{Q}'$ be $([\mathbf{A}/x]\mathcal{Q})$. Assume $\prec_0^{\mathcal{Q}}$ is acyclic and there is a $\prec_0^{\mathcal{Q}'}$-cycle

$$B^1 \prec_0^{\mathcal{Q}'} \cdots \prec_0^{\mathcal{Q}'} B^n \prec_0^{\mathcal{Q}'} B^1.$$

Let B^{n+1} be B^1. Since $\prec_0^{\mathcal{Q}}$ is ayclic, there must be some i $(1 \leq i \leq n)$ such that $B^i \not\prec_0^{\mathcal{Q}} B^{i+1}$. Reordering the sequence if necessary, assume i is n. By Lemma 8.2.11, $B^n \in \mathbf{Params}(\mathbf{A})$ and there is an expansion arc $\langle e, \mathbf{C}, c\rangle$ in $\mathcal{Q}$ such that $x \in \mathbf{Free}(\mathbf{C})$ and c dominates $\mathbf{SelNode}^{\mathcal{Q}}(B^1)$ in $\mathcal{Q}$. Since

$$\mathbf{Params}(\mathbf{A}) \subseteq \{B\} \text{ and } B^n \in \mathbf{Params}(\mathbf{A}),$$

B^n must be B and $B \in \mathbf{Params}(\mathbf{A})$. Let k be the least member of $\{1, \ldots, n\}$ such that $B^k = B$. Since

$$\{B^1, \ldots, B^{k-1}\} \cap \mathbf{Params}(\mathbf{A}) = \emptyset,$$

we must have $B^j \prec_0^{\mathcal{Q}} B^{j+1}$ for each j $(1 \leq j \leq k-1)$ by Lemma 8.2.11. Thus $B \in \mathbf{Params}(\mathbf{A})$, c dominates $\mathbf{SelNode}^{\mathcal{Q}}(B^1)$ in $\mathcal{Q}$ and

$$B^1 \prec_0^{\mathcal{Q}} \cdots \prec_0^{\mathcal{Q}} B^{k-1} \prec_0^{\mathcal{Q}} B^k = B.$$

That is, B is banned for x, contradicting our assumption. □

PROOF OF THEOREM 8.3.3. Choose some parameter $C_\iota \in \mathcal{P}$ such that $C_\iota \notin \mathbf{Params}(\mathcal{Q})$. Define the substitution θ as

$$\theta(x_{\beta\alpha^n\cdots\alpha^1}) := \begin{cases} [\lambda\overline{z^n}\, C_\iota] & \text{if } \beta = \iota \\ [\lambda\overline{z^n}\, \top] & \text{if } \beta = o. \end{cases}$$

for any $x_{\beta\alpha^n\cdots\alpha^1} \in \mathbf{Free}(\mathcal{Q})$ where $\beta \in \{o, \iota\}$. Note that

$$\mathbf{Dom}(\theta) = \mathbf{Free}(\mathcal{Q}) \subset \mathcal{V}$$

and θ is a ground substitution.

Since $\mathcal{Q}$ is an $\mathcal{S}$-extensional expansion dag, $\mathcal{Q}$ is also an $(\mathcal{S} \cup \{\top\})$-extensional expansion dag. Let $\mathcal{Q}^0$ be $\theta(\mathcal{Q})$. By Lemma 8.2.6, $\mathcal{Q}^0$ is a closed $(\mathcal{S} \cup \{\top\})$-extensional expansion dag,

$$\mathbf{SelPars}(\mathcal{Q}) = \mathbf{SelPars}(\mathcal{Q}^0) \text{ and } \mathbf{Roots}(\mathcal{Q}) = \mathbf{Roots}(\mathcal{Q}^0).$$

Since $\mathbf{Free}^{\mathcal{Q}}(\mathbf{Roots}(\mathcal{Q}))$ is empty, $sh^{\mathcal{Q}}(r)$ is closed for every $r \in \mathbf{Roots}(\mathcal{Q})$. Hence

$$label^{\mathcal{Q}^0}(r) = label^{\mathcal{Q}}(r)$$

and $sh^{\mathcal{Q}^0}(r)$ is closed for each $r \in \mathbf{Roots}(\mathcal{Q})$ by Definition 8.2.2. Thus $\mathbf{Free}^{\mathcal{Q}^0}(\mathbf{Roots}(\mathcal{Q}^0))$ is empty. Since $\mathcal{Q}$ and $\mathcal{Q}^0$ share the same selected parameter function and $\mathbf{RootSels}(\mathcal{Q})$ is empty, we know

$$\begin{aligned} &\mathbf{Params}(sh^{\mathcal{Q}^0}(r)) \cap \mathbf{SelPars}(\mathcal{Q}^0) \\ &= \mathbf{Params}(sh^{\mathcal{Q}}(r)) \cap \mathbf{SelPars}(\mathcal{Q}) = \emptyset \end{aligned}$$

for every $r \in \mathbf{Roots}(\mathcal{Q})$. Hence $\mathbf{RootSels}(\mathcal{Q}^0) = \emptyset$. Since C_ι is not a selected parameter of $\mathcal{Q}$ we can apply Lemma 8.2.9 to conclude $\prec_0^{\mathcal{Q}^0}$ is acyclic.

By Lemma 8.1.13, there is an $(\mathcal{S} \cup \{\top\})$-extensional expansion dag $\mathcal{Q}'$ which is closed and developed where $\mathcal{Q}^0 \trianglelefteq_r \mathcal{Q}'$, $\mathbf{RootSels}(\mathcal{Q}')$ is empty and $\prec_0^{\mathcal{Q}'}$ is acyclic. Since $\mathcal{Q}^0 \trianglelefteq_r \mathcal{Q}'$,

$$\mathbf{Roots}(\mathcal{Q}') = \mathbf{Roots}(\mathcal{Q}^0) = \mathbf{Roots}(\mathcal{Q}) \tag{12}$$

and for every $r \in \mathbf{Roots}(\mathcal{Q})$ we have

$$label^{\mathcal{Q}'}(r) = label^{\mathcal{Q}^0}(r) = label^{\mathcal{Q}}(r).$$

In order to prove $\langle \mathcal{Q}', \mathbf{Roots}(\mathcal{Q}) \rangle$ is an extensional expansion proof, we need only verify $\mathbf{Roots}(\mathcal{Q})$ is a sufficient set for $\mathcal{Q}'$.

Assume there is a u-part $\mathfrak{p}'$ of $\mathcal{Q}'$ which includes $\mathbf{Roots}(\mathcal{Q})$. Let $\mathfrak{p}$ be $\mathfrak{p}' \cap \mathcal{N}^{\mathcal{Q}^0}$. By Lemma 7.6.6, $\mathfrak{p}$ is a u-part of $\mathcal{Q}^0$. By Lemma 8.2.8, $\mathfrak{p}$ is a u-part of $\mathcal{Q}$. Since $\mathbf{Roots}(\mathcal{Q}) \subseteq \mathfrak{p}$, we know $\mathfrak{p}$ is pre-solved in $\mathcal{Q}$ by assumption. That is, there is either a flex-flex node (of $\mathcal{Q}$) in $\mathfrak{p}$ or a negative flexible node (of $\mathcal{Q}$) in $\mathfrak{p}$.

Suppose $a \in \mathfrak{p} \subseteq \mathfrak{p}'$ is a flex-flex node of $\mathcal{Q}$. That is, a is a negative equation node with shallow formula $[[f\,\overline{\mathbf{A}^n}] \doteq^\iota [g\,\overline{\mathbf{B}^m}]]$ for some $f, g \in \mathcal{V}$. Hence a is a negative equation node of $\mathcal{Q}^0$ with shallow formula

$$(\theta([[f\,\overline{\mathbf{A}^n}] \doteq^\iota [g\,\overline{\mathbf{B}^m}]]))^\downarrow = [C_\iota \doteq^\iota C].$$

Since $\mathcal{Q}^0 \trianglelefteq_r \mathcal{Q}'$, a is a negative equation node of $\mathcal{Q}'$ with shallow formula $[C \doteq^\iota C]$. Thus a is decomposable in $\mathcal{Q}'$. Since $\mathcal{Q}'$ is developed, a is extended and decomposed in $\mathcal{Q}'$. Hence a is conjunctive in $\mathcal{Q}'$ and there is a node $d \in \mathcal{N}^{\mathcal{Q}'}$ such that $\langle a, d \rangle \in decs^{\mathcal{Q}'}$.

By $\clubsuit^p_{dec}$, d is a decomposition node with shallow formula $[C \doteq^\iota C]$. Since a is a conjunctive node of $\mathcal{Q}'$ and $\mathfrak{p}'$ is a u-part of $\mathcal{Q}'$, we have $d \in \mathfrak{p}'$. Since $\mathcal{Q}'$ is developed, d is extended in $\mathcal{Q}'$ and hence disjunctive. Thus there must be some $d^1 \in \mathfrak{p}'$ such that $d \overset{\mathcal{Q}'}{\rightarrow} d^1$. By Lemma 7.3.13, $d^1 \in \{a^1, \ldots, a^n\}$ where $succ^{\mathcal{Q}'}(d) = \langle a^1, \ldots, a^n \rangle$. However, $succ^{\mathcal{Q}'}(d) = \langle\rangle$ by $\clubsuit_{dec}$. This contradiction proves there is no flex-flex node of $\mathcal{Q}$ in $\mathfrak{p}$.

Suppose $a \in \mathfrak{p} \subseteq \mathfrak{p}'$ is a negative flexible node of $\mathcal{Q}$. Hence a is a negative primary node in $\mathcal{Q}$ with shallow formula $[p\,\overline{\mathbf{D}^m}]$ for some $p \in \mathcal{V}$. Thus a is a negative primary node in $\mathcal{Q}^0$ with shallow formula

$$(\theta([p\,\overline{\mathbf{D}^m}]))^{\downarrow} = \top$$

and so a is a negative deepening node in $\mathcal{Q}^0$. Since $\mathcal{Q}^0 \trianglelefteq_r \mathcal{Q}'$, a is a negative deepening node in $\mathcal{Q}'$ with shallow formula $\top$. Since $\mathcal{Q}'$ is developed, a is extended in $\mathcal{Q}'$ and by $\clubsuit_\sharp$ there is a negative primary node $a^1 \in \mathcal{N}^{\mathcal{Q}'}$ with shallow formula $\dot{\top}$ such that $succ^{\mathcal{Q}'}(a) = \langle a^1 \rangle$. Hence a^1 is a false node of $\mathcal{Q}'$. Since $a \in \mathfrak{p}'$ is a neutral node of $\mathcal{Q}'$ and $\mathfrak{p}'$ is a u-part of $\mathcal{Q}'$, we have $a^1 \in \mathfrak{p}'$. Since $\mathcal{Q}'$ is developed, a^1 is extended and hence disjunctive. Since $a^1 \in \mathfrak{p}'$, there must be some $b \in \mathfrak{p}'$ such that $a^1 \overset{\mathcal{Q}'}{\rightarrow} b$. By Lemma 7.3.13, $b \in \{b^1, \ldots, b^n\}$ where $succ^{\mathcal{Q}'}(a^1) = \langle b^1, \ldots, b^n \rangle$. However, $succ^{\mathcal{Q}'}(a') = \langle\rangle$ by $\clubsuit_{ns}$, a contradiction.

Therefore, there is no u-part of $\mathcal{Q}'$ including $\mathbf{Roots}(\mathcal{Q})$ and we conclude $\langle \mathcal{Q}', \mathbf{Roots}(\mathcal{Q}) \rangle$ is an $(\mathcal{S} \cup \{\top\})$-extensional expansion proof. □

PROOF OF LEMMA 8.4.5. If $a^* \in \mathbf{Im}(f; \mathcal{Q})$, then there is a node $a \in \mathcal{N}^{\mathcal{Q}} \subseteq \mathcal{N}^{\mathcal{Q}'}$ such that $f'(a) = f(a) = a^*$ and so $a^* \in \mathbf{Im}(f'; \mathcal{Q}')$. Hence $\mathbf{Im}(f; \mathcal{Q}) \subseteq \mathbf{Im}(f'; \mathcal{Q}')$ and so

$$(\mathcal{N}^{\mathcal{Q}^*} \setminus \mathbf{Im}(f'; \mathcal{Q}')) \subseteq (\mathcal{N}^{\mathcal{Q}^*} \setminus \mathbf{Im}(f; \mathcal{Q})) \tag{13}$$

Suppose there is some $a' \in \mathcal{N}^{\mathcal{Q}'} \setminus \mathcal{N}^{\mathcal{Q}}$. Then $f'(a') \in \mathbf{Im}(f'; \mathcal{Q}')$. Since f' is injective, $f'(a') \neq f(a)$ for any $a \in \mathcal{N}^{\mathcal{Q}}$. Hence

$$f'(a') \notin \mathbf{Im}(f; \mathcal{Q}).$$

Thus the inclusion (13) is proper since

$$f'(a') \in (\mathcal{N}^{\mathcal{Q}^*} \setminus \mathbf{Im}(f; \mathcal{Q})) \text{ and } f'(a') \notin (\mathcal{N}^{\mathcal{Q}^*} \setminus \mathbf{Im}(f'; \mathcal{Q}')).$$

Therefore,

$$|\mathcal{N}^{\mathcal{Q}^*} \setminus \mathbf{Im}(f'; \mathcal{Q}')| < |\mathcal{N}^{\mathcal{Q}^*} \setminus \mathbf{Im}(f; \mathcal{Q})|$$

and so

$$\|\mathsf{f}' : \mathcal{Q}' \to \mathcal{Q}^*\| < \|\mathsf{f} : \mathcal{Q} \to \mathcal{Q}^*\|.$$

Assume $\mathcal{N}^{\mathcal{Q}'} \setminus \mathcal{N}^{\mathcal{Q}}$ is empty. That is, $\mathcal{N}^{\mathcal{Q}} = \mathcal{N}^{\mathcal{Q}'}$.

Suppose there is some $a' \in \mathbf{Dom}(succ^{\mathcal{Q}'}) \setminus \mathbf{Dom}(succ^{\mathcal{Q}})$. Since $\mathcal{Q} \trianglelefteq \mathcal{Q}'$, we know $\mathbf{Dom}(succ^{\mathcal{Q}}) \subseteq \mathbf{Dom}(succ^{\mathcal{Q}'})$. The node a' witnesses that the inclusion is proper. Since $\mathcal{N}^{\mathcal{Q}} = \mathcal{N}^{\mathcal{Q}'}$, we know $a' \in \mathcal{N}^{\mathcal{Q}} \setminus \mathbf{Dom}(succ^{\mathcal{Q}})$ and the inclusion

$$(\mathcal{N}^{\mathcal{Q}'} \setminus \mathbf{Dom}(succ^{\mathcal{Q}'})) \subset (\mathcal{N}^{\mathcal{Q}} \setminus \mathbf{Dom}(succ^{\mathcal{Q}}))$$

is proper. Thus

$$|\mathcal{N}^{\mathcal{Q}'} \setminus \mathbf{Dom}(succ^{\mathcal{Q}'})| < |\mathcal{N}^{\mathcal{Q}} \setminus \mathbf{Dom}(succ^{\mathcal{Q}})|$$

We also know

$$|\mathcal{N}^{\mathcal{Q}^*} \setminus \mathbf{Im}(f'; \mathcal{Q}')| \leq |\mathcal{N}^{\mathcal{Q}^*} \setminus \mathbf{Im}(f; \mathcal{Q})|.$$

by the inclusion (13). (Actually, we have equality since we are assuming $\mathcal{N}^{\mathcal{Q}} = \mathcal{N}^{\mathcal{Q}'}$ in this case of the proof.) Therefore,

$$\|\mathsf{f}' : \mathcal{Q}' \to \mathcal{Q}^*\| < \|\mathsf{f} : \mathcal{Q} \to \mathcal{Q}^*\|.$$

□

PROOF OF LEMMA 8.4.16. Suppose $a \in \mathcal{R}$ and $f(a)$ is a deepening node of $\mathcal{Q}^*$ with shallow formula $[c\,\overline{\mathbf{B}^m}]$ where c is a logical constant. Since f is a lifting map, we know a is a primary node with shallow formula $\mathbf{M}$ where

$$\varphi(\mathbf{M})^{\downarrow} = [c\,\overline{\mathbf{B}^m}].$$

Thus $\mathbf{M}$ must be a $\beta\eta$-normal term of type o. Hence $\mathbf{M}$ has the form $[p\,\overline{\mathbf{A}^n}]$ for some $p \in \mathcal{V} \cup \mathcal{P} \cup \mathcal{S}$. If p is a parameter, then $\varphi(p)$ is a parameter since φ respects parameters (cf. Definitions 8.4.3, 8.4.1 and 2.1.9). If p is a logical constant, then $\varphi(p) = p$ since $\mathbf{Dom}(\varphi) \subset \mathcal{V} \cup \mathcal{P}$ (cf. Definition 8.4.1). Consequently, if $p \in \mathcal{P} \cup \mathcal{S}$, then

$$[\varphi(p)\,(\varphi(\mathbf{A}^1))^{\downarrow} \cdots (\varphi(\mathbf{A}^n))^{\downarrow}] = \varphi([p\,\overline{\mathbf{A}^n}])^{\downarrow} = [c\,\overline{\mathbf{B}^m}]$$

and so $\varphi(p) = c$. Thus $p \in \mathcal{V} \cup \{c\}$. Assume $p \in \mathcal{V}$. Since $a \in \mathcal{R}$ is a p-node, $p \in \mathbf{SetVars}^{\mathcal{Q}}(\mathcal{R})$ (cf. Definition 8.4.14). Since $\mathbf{SetVars}^{\mathcal{Q}}(\mathcal{R})$ is assumed to be almost atomic with respect to f (cf. Definition

8.4.15), p is almost atomic with respect to f. Hence the long $\beta\eta$-form $\varphi(p)$ (cf. Definition 8.4.1) is $[\lambda\overline{x^n}\,[h\,\overline{\mathbf{C}^k}]]$ for some $h \in \mathcal{P}$. Let θ be the substitution defined by $\theta(x^i) := \varphi(\mathbf{A}^i)$ for each i $(1 \leq i \leq n)$. For each j $(1 \leq j \leq k)$, let $\mathbf{D}^j$ be $(\theta(\mathbf{C}^j))^{\downarrow}$. Since $h \notin \{x^1, \ldots, x^n\}$, we know $(\theta([h\,\overline{\mathbf{C}^k}]))^{\downarrow} = [h\,\overline{\mathbf{D}^k}]$. Thus

$$\begin{aligned} [h\,\overline{\mathbf{D}^k}] &\overset{\beta\eta}{=} [[\lambda\overline{x^n}\,[h\,\overline{\mathbf{C}^k}]]\,\varphi(\mathbf{A}^1)\,\cdots\,\varphi(\mathbf{A}^n)] \\ &= [\varphi(p)\,\varphi(\mathbf{A}^1)\,\cdots\,\varphi(\mathbf{A}^n)] \\ &= \varphi([p\,\overline{\mathbf{A}^n}]) \overset{\beta\eta}{=} [c\,\overline{\mathbf{B}^m}]. \end{aligned}$$

Hence $h = c$, contradicting $h \in \mathcal{P}$ (since $c \in \mathcal{S}$). Thus we must have $p \notin \mathcal{V}$ and so $p = c$. Therefore, a is a deepening node in $\mathcal{Q}$. □

PROOF OF LEMMA 8.4.21. Suppose $\mathscr{D}$ is fully duplicated with respect to f and $e \in \mathscr{D}$ is an expansion node of $\theta(\mathcal{Q})$. By Lemma 8.2.4:3, e is an expansion node of $\mathcal{Q}$. Hence e is fully duplicated with respect to f. Suppose $c^* \in \mathcal{N}^{\mathcal{Q}^*}$ and $f'(e) \overset{\mathcal{Q}^*}{\rightarrow} c^*$. Since $f' = f$, $f(e) \overset{\mathcal{Q}^*}{\rightarrow} c^*$ and so there is some $c \in \mathcal{N}^{\mathcal{Q}} = \mathcal{N}^{\mathcal{Q}'}$ such that $e \overset{\mathcal{Q}}{\rightarrow} c$ and $f'(c) = f(c) = c^*$. By Lemma 8.2.5, $e \overset{\theta(\mathcal{Q})}{\rightarrow} c$. Hence e is fully duplicated with respect to f'. Thus $\mathscr{D}$ is fully duplicated with respect to f'.

Suppose $\mathscr{S} \subseteq \mathbf{Free}(\mathcal{Q})$ is almost atomic with respect to f and

$$p \in (\mathscr{S} \setminus \mathbf{Dom}(\theta)).$$

Hence p has a type of the form $(o\alpha^n \cdots \alpha^1)$ and the long $\beta\eta$-form $\varphi(p)$ is of the form $[\lambda\overline{x^n}\,[h\,\overline{\mathbf{C}^l}]]$ where $h \in \mathcal{P}$. Since $p \in \mathbf{Free}(\mathcal{Q}) \setminus \mathbf{Dom}(\theta)$, we know $\varphi'(p) = \varphi(p)$. Thus p is almost atomic with respect to f'. Therefore, $(\mathscr{S} \setminus \mathbf{Dom}(\theta))$ is almost atomic with respect to f'. □

PROOF OF LEMMA 8.5.2. Suppose $x \in \mathbf{Free_{str}}(\mathbf{A})$ and $x \neq y$. We prove $x \in \mathbf{Free_{str}}([\mathbf{C}/y]\mathbf{A})$ by induction on $\mathbf{A}^{\updownarrow}$. The long $\beta\eta$-form $\mathbf{A}^{\updownarrow}$ is of the form $[\lambda\overline{z^n}\,[h\,\overline{\mathbf{A}^m}]_\beta]$ where $\beta \in \{o, \iota\}$.

Suppose $h \in \mathcal{V}$. Since $x \in \mathbf{Free_{str}}(\mathbf{A})$, we must have $h = x$ and $(\mathbf{A}^j)^{\downarrow} \in \{z^1, \ldots, z^n\}$ for each j $(1 \leq j \leq m)$ by Definition 8.5.1. In particular, x is the only free variable in $\mathbf{A}^{\downarrow}$. Since $x \neq y$, we know

$$([\mathbf{C}/y]\mathbf{A})^{\updownarrow} = ([\mathbf{C}/y](\mathbf{A}^{\downarrow}))^{\updownarrow} = \mathbf{A}^{\updownarrow}.$$

Thus $x \in \mathbf{Free_{str}}([\mathbf{C}/y]\mathbf{A})$.

Otherwise, $h \in (\mathcal{P} \cup \mathcal{S} \cup \{z^1, \ldots, z^n\})$. Then $\mathbf{Free_{str}}(\mathbf{A})$ is

$$\mathbf{Free_{str}}([\lambda \overline{z^n} \mathbf{A}^1]) \cup \cdots \cup \mathbf{Free_{str}}([\lambda \overline{z^n} \mathbf{A}^m])$$

and there is some j $(1 \leq j \leq m)$ such that $x \in \mathbf{Free_{str}}([\lambda \overline{z^n}\, \mathbf{A}^j])$. We know

$$|[\lambda \overline{z^n}\, \mathbf{A}^j]^{\updownarrow}|_l < |\mathbf{A}^{\updownarrow}|_l$$

by Lemma 2.1.32. By the inductive hypothesis,

$$x \in \mathbf{Free_{str}}([\mathbf{C}/y][\lambda \overline{z^n}\, \mathbf{A}^j]).$$

Let $\mathbf{B}^i$ be $([\mathbf{C}/y]\mathbf{A}^i)^{\updownarrow}$ for each i $(1 \leq i \leq m)$. Note that

$$x \in \mathbf{Free_{str}}([\lambda \overline{z^n}\, \mathbf{B}^j]) \text{ since } [\lambda \overline{z^n}\, \mathbf{B}^j]^{\updownarrow} = ([\mathbf{C}/y][\lambda \overline{z^n}\, \mathbf{A}^j])^{\updownarrow}.$$

We compute

$$\begin{aligned} [\mathbf{C}/y]\mathbf{A} &\stackrel{\beta\eta}{=} [\lambda \overline{z^n}\, [h\, ([\mathbf{C}/y]\mathbf{A}^1) \cdots ([\mathbf{C}/y]\mathbf{A}^m)]] \\ &\stackrel{\beta\eta}{=} [\lambda \overline{z^n}\, [h\, \overline{\mathbf{B}^m}]]. \end{aligned}$$

Hence

$$([\mathbf{C}/y]\mathbf{A})^{\updownarrow} = [\lambda \overline{z^n}\, [h\, \overline{\mathbf{B}^m}]]$$

and $\mathbf{Free_{str}}([\mathbf{C}/y]\mathbf{A})$ is

$$\mathbf{Free_{str}}([\lambda \overline{z^n} \mathbf{B}^1]) \cup \cdots \cup \mathbf{Free_{str}}([\lambda \overline{z^n} \mathbf{B}^m]).$$

Thus

$$x \in \mathbf{Free_{str}}([\lambda \overline{z^n}\, \mathbf{B}^j]) \subseteq \mathbf{Free_{str}}([\mathbf{C}/y]\mathbf{A})$$

as desired. □

Proof of Lemma 8.5.4. Suppose

$$y \in \mathbf{Free_{str}}(\mathbf{A}) \text{ and } w \in \mathbf{Free}(\mathbf{G}).$$

We prove $w \in \mathbf{Free_{str}}([\mathbf{G}/y]\mathbf{A})$ by induction on $\mathbf{A}^{\updownarrow}$. The long $\beta\eta$-form $\mathbf{A}^{\updownarrow}$ is of the form $[\lambda \overline{z^n}\, [h\, \overline{\mathbf{A}^m}]]$.

Suppose $h \in \mathcal{V}$. By Definition 8.5.1, $h = y$, $(\mathbf{A}^j)^{\downarrow} \in \{z^1, \ldots, z^n\}$ for each j such that $1 \leq j \leq m$, and $(\mathbf{A}^1)^{\downarrow}, \ldots, (\mathbf{A}^m)^{\downarrow}$ must be distinct since $y \in \mathbf{Free_{str}}(\mathbf{A})$. Let u^j be $(\mathbf{A}^j)^{\downarrow}$ and $\mathbf{U}^j$ be $(\mathbf{A}^j)^{\updownarrow}$ for each j such that $1 \leq j \leq m$. We compute

$$\begin{aligned}[\mathbf{G}/y]\mathbf{A} &\stackrel{\beta\eta}{=} [\mathbf{G}/y][\lambda\overline{z^n}\,[y\,\overline{u^m}]] \\ &\stackrel{\beta\eta}{=} [\lambda\overline{z^n}\,[\mathbf{G}\,\overline{u^m}]].\end{aligned}$$

As a general binding, $\mathbf{G}$ must be of the form

$$[\lambda x^1_{\alpha^1} \cdots \lambda x^m_{\alpha^m}\,[R\,[w^1\,\overline{x^m}] \cdots [w^k\,\overline{x^m}]]]$$

where $R_{\beta\gamma^k\ldots\gamma^1} \in \mathcal{P} \cup \mathcal{S} \cup \{x^1, \ldots, x^m\}$ and $w^1, \ldots, w^k$ are variables. If $R \in \mathcal{P} \cup \mathcal{S}$, then let R^* be R. For each j $(1 \leq j \leq m)$, if $R = x^j$, then let R^* be u^j. In either case $([\mathbf{G}/y]\mathbf{A})^{\updownarrow}$ is

$$[\lambda\overline{z^n}\,[R^*\,[w^1\,\overline{\mathbf{U}^m}] \cdots [w^k\,\overline{\mathbf{U}^m}]]]$$

where

$$R^* \in \mathcal{P} \cup \mathcal{S} \cup \{u^1, \ldots, u^m\} \subseteq \mathcal{P} \cup \mathcal{S} \cup \{z^1, \ldots, z^n\}.$$

Hence $\mathbf{Free_{str}}([\mathbf{G}/y]\mathbf{A})$ is

$$\mathbf{Free_{str}}([\lambda\overline{z^n}\,[w^1\,\overline{\mathbf{U}^m}]]) \cup \cdots \cup \mathbf{Free_{str}}([\lambda\overline{z^n}\,[w^k\,\overline{\mathbf{U}^m}]]).$$

Since $(\mathbf{U}^j)^{\downarrow} = u^j$ for each j $(1 \leq j \leq m)$, $\{u^1, \ldots, u^m\} \subseteq \{z^1, \ldots, z^n\}$ and $u^1, \ldots, u^m$ are distinct,

$$\mathbf{Free_{str}}([\lambda\overline{z^n}\,[w^i\,\overline{u^m}]]) = \{w^i\}$$

for each i $(1 \leq i \leq k)$. Thus

$$\mathbf{Free_{str}}([\mathbf{G}/y]\mathbf{A}) = \{w^1, \ldots, w^k\}.$$

Since $w \in \mathbf{Free}(\mathbf{G})$, we conclude $w = w^i \in \mathbf{Free_{str}}([\mathbf{G}/y]\mathbf{A})$ for some i $(1 \leq i \leq k)$, as desired.

Otherwise, $h \in \mathcal{P} \cup \mathcal{S} \cup \{z^1, \ldots, z^n\}$. Then $\mathbf{Free_{str}}(\mathbf{A})$ is

$$\mathbf{Free_{str}}([\lambda\overline{z^n}\mathbf{A}^1]) \cup \cdots \cup \mathbf{Free_{str}}([\lambda\overline{z^n}\mathbf{A}^m])$$

and there is some j $(1 \leq j \leq m)$ such that $y \in \mathbf{Free_{str}}([\lambda\overline{z^n}\mathbf{A}^j])$. We know

$$|[\lambda\overline{z^n}\,\mathbf{A}^j]^{\updownarrow}|_l < |\mathbf{A}^{\updownarrow}|_l$$

by Lemma 2.1.32. By the inductive hypothesis,

$$w \in \mathbf{Free_{str}}([\mathbf{G}/y][\lambda\overline{z^n}\,\mathbf{A}^j]).$$

Let $\mathbf{B}^i$ be $([\mathbf{G}/y]\mathbf{A}^i)^{\updownarrow}$ for each i $(1 \leq i \leq m)$. Note that

$$w \in \mathbf{Free_{str}}([\lambda \overline{z^n} \mathbf{B}^j])$$

and

$$[\mathbf{G}/y]\mathbf{A} \stackrel{\beta\eta}{=} [\lambda \overline{z^n}\,[h\,([\mathbf{G}/y]\mathbf{A}^1) \cdots ([\mathbf{G}/y]\mathbf{A}^m)]] \stackrel{\beta\eta}{=} [\lambda \overline{z^n}\,[h\,\overline{\mathbf{B}^m}]]$$

Hence $\mathbf{Free_{str}}([\mathbf{G}/y]\mathbf{A})$ is

$$\mathbf{Free_{str}}([\lambda \overline{z^n} \mathbf{B}^1]) \cup \cdots \cup \mathbf{Free_{str}}([\lambda \overline{z^n} \mathbf{B}^m])$$

and so $w \in \mathbf{Free_{str}}([\lambda \overline{z^n}\, \mathbf{B}^j]) \subseteq \mathbf{Free_{str}}([\mathbf{G}/y]\mathbf{A})$, as desired. □

PROOF OF LEMMA 8.5.6. Suppose y has a strict occurrence in $\mathbf{A}$, φ respects parameters and $\mathbf{Dom}(\varphi) \subset \mathcal{V} \cup \mathcal{P}$. We prove

$$\varphi(R) \in \mathbf{Params}((\varphi([\mathbf{G}/y]\mathbf{A}))^{\downarrow})$$

by induction on $\mathbf{A}^{\updownarrow}$. Note that $\varphi(R)$ is a parameter since φ respects parameters. The long $\beta\eta$-form $\mathbf{A}^{\updownarrow}$ is of the form $[\lambda \overline{z^s}\,[h\,\overline{\mathbf{A}^k}]]$.

Suppose $h \in \mathcal{P} \cup \mathcal{S} \cup \{z^1, \ldots, z^s\}$. Then $\mathbf{Free_{str}}(\mathbf{A})$ is

$$\mathbf{Free_{str}}([\lambda \overline{z^s} \mathbf{A}^1]) \cup \cdots \cup \mathbf{Free_{str}}([\lambda \overline{z^s} \mathbf{A}^k])$$

and there is some j $(1 \leq j \leq k)$ such that $y \in \mathbf{Free_{str}}([\lambda \overline{z^s} \mathbf{A}^j])$. We know

$$|[\lambda \overline{z^s}\, \mathbf{A}^j]^{\updownarrow}|_l < |\mathbf{A}^{\updownarrow}|_l.$$

by Lemma 2.1.32. By the inductive hypothesis,

$$\varphi(R) \in \mathbf{Params}((\varphi([\mathbf{G}/y][\lambda \overline{z^s}\, \mathbf{A}^j]))^{\downarrow}).$$

Let $\mathbf{B}^j$ be $(\varphi([\mathbf{G}/y]\mathbf{A}^j))^{\downarrow}$ and note $\varphi(R) \in \mathbf{Params}(\mathbf{B}^j)$. If $h \in \mathcal{P}$, let H be $\varphi(h)$. Otherwise, let H be h. Since $\mathbf{Dom}(\varphi) \subset \mathcal{V} \cup \mathcal{P}$ and φ respects parameters, we can compute

$$\begin{aligned} \varphi([\mathbf{G}/y]\mathbf{A}) &\stackrel{\beta\eta}{=} \varphi([\mathbf{G}/y][\lambda \overline{z^s}\,[h\,\overline{\mathbf{A}^k}]]) \\ &= [\lambda \overline{z^s}\,[H\ \ \varphi([\mathbf{G}/y]\mathbf{A}^1) \cdots \varphi([\mathbf{G}/y]\mathbf{A}^k)]] \\ &\stackrel{\beta\eta}{=} [\lambda \overline{z^s}\,[H\,\overline{\mathbf{B}^k}]] \end{aligned}$$

Hence

$$(\varphi([\mathbf{G}/y]\mathbf{A}))^{\downarrow} = [\lambda \overline{z^{s-r}}\,[H\,\mathbf{B}^1 \cdots \mathbf{B}^{k-r}]]$$

for some r $(0 \leq r \leq \min(s,k))$ such that $\mathbf{B}^{k-t} = z^{s-t}$ for each t $(0 \leq t < r)$. Since $\varphi(R) \in \mathbf{Params}(\mathbf{B}^j)$, we must have $j \leq k - r$. Thus $\varphi(R) \in (\varphi([\mathbf{G}/y]\mathbf{A}))^{\downarrow}$, as desired.

Next suppose $h \in \mathcal{V}$. Since $y \in \mathbf{Free_{str}}(\mathbf{A})$, we must have $h = y$ and $(\mathbf{A}^j)^{\downarrow} \in \{z^1, \ldots, z^s\}$ for each j $(1 \leq j \leq k)$ by Definition 8.5.1. The type of y is the same as the type of $\mathbf{G}$, namely $(\beta\alpha^n \cdots \alpha^1)$ where $\beta \in \{o, \iota\}$. Hence $k = n$. Let u^j be $(\mathbf{A}^j)^{\downarrow}$ and $\mathbf{U}^j$ be $(u^j)^{\updownarrow}$ for each j $(1 \leq j \leq n)$. Then $\mathbf{A}^{\updownarrow}$ is $[\lambda\overline{z^s}\,[y\,\overline{\mathbf{U}^n}]]$. We compute

$$\begin{aligned}
\varphi([\mathbf{G}/y]\mathbf{A}) &\stackrel{\beta\eta}{=} [\lambda\overline{z^s}\,[\varphi(\mathbf{G})\,\overline{u^n}]] \\
&= [\lambda\overline{z^s}\,[\lambda\overline{x^n}\,[\varphi(R)\,[\varphi(w^1)\,\overline{x^n}] \cdots [\varphi(w^m)\,\overline{x^n}]]]\,\overline{u^n}] \\
&\stackrel{\beta\eta}{=} [\lambda\overline{z^s}\,[\varphi(R)\,[(\varphi(w^1)\,\overline{u^n}] \cdots [\varphi(w^m)\,\overline{u^n}]]].
\end{aligned}$$

Let $\mathbf{B}^j$ be $[\varphi(w^j)\,\overline{u^n}]^{\downarrow}$ for each j $(1 \leq j \leq m)$. Thus

$$(\varphi([\mathbf{G}/y]\mathbf{A}))^{\downarrow} = [\lambda\overline{z^{s-r}}\,[\varphi(R)\,\overline{\mathbf{B}^{m-r}}]]$$

for some r $(0 \leq r \leq \min(s,m))$ such that $\mathbf{B}^{k-t} = z^{s-t}$ for each t $(0 \leq t < r)$. Therefore, $\varphi(R) \in \mathbf{Params}((\varphi([\mathbf{G}/y]\mathbf{A}))^{\downarrow})$. □

PROOF OF LEMMA 8.5.7. Suppose $B \in \mathbf{Params}((\varphi(x))^{\updownarrow})$. We prove $B \in \mathbf{Params}((\varphi(\mathbf{A}))^{\updownarrow})$ by induction on $\mathbf{A}^{\updownarrow}$. Let $(\beta\alpha^k \cdots \alpha^1)$ be the type of x where $\beta \in \{o, \iota\}$. The long $\beta\eta$-form $\mathbf{A}^{\updownarrow}$ is of the form $[\lambda\overline{z^n}\,[h\,\overline{\mathbf{A}^m}]]$.

Suppose $h \in \mathcal{V}$. Since $x \in \mathbf{Free_{str}}(\mathbf{A})$, we have $h = x$, $k = m$, $(\mathbf{A}^j)^{\downarrow} \in \{z^1, \ldots, z^n\}$ for each j $(1 \leq j \leq m)$, and $(\mathbf{A}^1)^{\downarrow}, \ldots, (\mathbf{A}^m)^{\downarrow}$ must be distinct by Definition 8.5.1. Let u^j be $(\mathbf{A}^j)^{\downarrow}$ and $\mathbf{U}^j$ be $(\mathbf{A}^j)^{\updownarrow}$ for each j $(1 \leq j \leq m)$. Hence

$$\varphi(\mathbf{A}) \stackrel{\beta\eta}{=} [\lambda\overline{z^n}\,[\varphi(x)\,\overline{u^m}]]$$

The long $\beta\eta$-form $(\varphi(x))^{\updownarrow}$ must be of the form

$$[\lambda\overline{y^m}\,\mathbf{D}]$$

for some β-normal $\mathbf{D}_\beta$ such that $B \in \mathbf{Params}(\mathbf{D})$. Let θ be the substitution $\theta(y^j) = u^j$ for each j $(1 \leq j \leq m)$. Since θ maps variables to variables and $\mathbf{Dom}(\theta) \subset \mathcal{V}$, a simple induction on the term $\mathbf{D}$ verifies $\theta(\mathbf{D})$ is β-normal and $B \in \mathbf{Params}(\theta(\mathbf{D}))$. Hence

$$B \in \mathbf{Params}([\lambda \overline{z^n}\, \theta(\mathbf{D})]).$$

Since $[\lambda \overline{z^n}\, \theta(\mathbf{D})]$ is β-normal and η-reductions never introduce new parameters, $B \in \mathbf{Params}([\lambda \overline{z^n}\, \theta(\mathbf{D})]^{\updownarrow})$. Since

$$\varphi(\mathbf{A}) \stackrel{\beta\eta}{=} [\lambda \overline{z^n}\, \theta(\mathbf{D})],$$

we have $B \in \mathbf{Params}((\varphi(\mathbf{A}))^{\updownarrow})$, as desired.

Otherwise, $h \in \mathcal{P} \cup \mathcal{S} \cup \{z^1, \ldots, z^n\}$. Then $\mathbf{Free_{str}}(\mathbf{A})$ is

$$\mathbf{Free_{str}}([\lambda \overline{z^n} \mathbf{A}^1]) \cup \cdots \cup \mathbf{Free_{str}}([\lambda \overline{z^n} \mathbf{A}^m])$$

and there is some j $(1 \leq j \leq m)$ such that $x \in \mathbf{Free_{str}}([\lambda \overline{z^n} \mathbf{A}^j])$. We know

$$|[\lambda \overline{z^n}\, \mathbf{A}^j]^{\updownarrow}|_l < |\mathbf{A}^{\updownarrow}|_l.$$

by Lemma 2.1.32. By the inductive hypothesis,

$$B \in \mathbf{Params}((\varphi([\lambda \overline{z^n}\, \mathbf{A}^j]))^{\updownarrow}).$$

Note that

$$(\varphi([\lambda \overline{z^n} \mathbf{A}^j]))^{\updownarrow} = [\lambda \overline{z^n}\, (\varphi(\mathbf{A}^j))^{\updownarrow}]$$

and so

$$B \in \mathbf{Params}((\varphi(\mathbf{A}^j))^{\updownarrow}).$$

Since $\mathbf{Dom}(\varphi) \subset \mathcal{V} \cup \mathcal{P}$ and φ respects parameters, $(\varphi(\mathbf{A}))^{\updownarrow}$ is

$$[\lambda \overline{z^n}\, [H^*\, (\varphi(\mathbf{A}^1))^{\updownarrow} \cdots (\varphi(\mathbf{A}^m))^{\updownarrow}]]$$

for some $H^* \in \mathcal{P} \cup \mathcal{S} \cup \{z^1, \ldots, z^n\}$ of the same type as h. Thus

$$B \in \mathbf{Params}((\varphi(\mathbf{A}^j))^{\updownarrow}) \subseteq \mathbf{Params}((\varphi(\mathbf{A}))^{\updownarrow})$$

as desired. □

PROOF OF LEMMA 8.6.9. Let $\mathcal{Q}$ be $\mathbf{EInit}(\langle\langle \mathbf{primary}, -1, \mathbf{M}\rangle\rangle)$. By Definition 7.7.2, $\mathcal{Q}$ is

$$\langle \mathcal{N}, \emptyset, label, \emptyset, \emptyset, \emptyset, \emptyset\rangle$$

where $\mathcal{N} = \{r\}$, $label(r) = \langle \mathbf{primary}, -1, \mathbf{M}\rangle$ and $\mathsf{n}_1^{\emptyset} = \langle r\rangle$. By Definition 8.6.8, $\mathbf{SearchInit}(\mathbf{M})$ is the tuple $\langle \mathcal{Q}, r, \emptyset, \emptyset\rangle$. By Lemma 7.7.3, $\mathcal{Q}$ is an extensional expansion dag and $\mathbf{Roots}(\mathcal{Q}) = \{r\}$. Since $\mathbf{M}$ is closed (as a sentence), $sh^{\mathcal{Q}}(r)$ is closed. Since $sel^{\mathcal{Q}}$ is the empty partial function, $\mathbf{SelPars}(\mathcal{Q}) = \emptyset$ and so $\mathbf{RootSels}(\mathcal{Q})$ is empty and

$\prec_0^{\mathcal{Q}}$ is trivially acyclic. Since $exps^{\mathcal{Q}}$ is empty, $\mathcal{Q}$ trivially has strict expansions. Therefore, **SearchInit**(**M**) is a search state.

Suppose $\langle \mathcal{Q}^*, r^* \rangle$ is a lifting target and r^* is a negative primary node of $\mathcal{Q}^*$ with shallow formula **M**. That is,

$$label^{\mathcal{Q}^*}(r^*) = \langle \mathbf{primary}, -1, \mathbf{M} \rangle.$$

Note that $\mathbf{Params}(\mathcal{Q}) = \mathbf{Params}(\mathbf{M})$ and $\mathbf{Free}(\mathcal{Q}) = \emptyset$ since **M** is closed. Let φ be the identity substitution on $\mathbf{Params}(\mathcal{Q})$ and let $f : \{r\} \to \mathcal{N}^{\mathcal{Q}^*}$ be the function given by $f(r) := r^*$. We prove $\langle \varphi, f \rangle$ is a lifting map from **SearchInit**(**M**) to $\langle \mathcal{Q}^*, r^* \rangle$.

First $\mathbf{Dom}(\varphi) = \mathbf{Params}(\mathcal{Q}) \subset \mathcal{V} \cup \mathcal{P}$, φ respects parameters and φ is injective on (all) parameters. Also, $\varphi(x)$ is a closed long $\beta\eta$-form for every $x \in \mathbf{Free}(\mathcal{Q})$ vacuously since $\mathbf{Free}(\mathcal{Q})$ is empty. Hence φ is a lifting substitution for $\mathcal{Q}$. The function f is clearly injective. Since φ is the identity substitution and **M** is $\beta\eta$-normal, we have

$$label^*(f(r^*)) = \langle \mathbf{primary}, -1, (\varphi(\mathbf{M}))^{\downarrow} \rangle$$

as desired. Conditions (2), (3), (4), (6), (7), (8) and (9) of Definition 8.4.3 are all vacuous since $succ^{\mathcal{Q}}$ and $sel^{\mathcal{Q}}$ are both the empty function and the sets $exps^{\mathcal{Q}}$, $conns^{\mathcal{Q}}$ and $decs^{\mathcal{Q}}$ are all empty.

It only remains to check condition (5). Suppose $A \in \mathbf{Params}(\mathcal{Q})$, $a^* \in \mathbf{Dom}(sel^{\mathcal{Q}^*})$ and $\varphi(A) = sel^{\mathcal{Q}^*}(a^*)$. Then $A \in \mathbf{SelPars}(\mathcal{Q}^*)$ since $\varphi(A) = A$. However,

$$A \in \mathbf{Params}(\mathcal{Q}) = \mathbf{Params}(\mathbf{M}).$$

Since $\langle \mathcal{Q}^*, r^* \rangle$ is a lifting target, $\langle \mathcal{Q}^*, \{r^*\} \rangle$ is an extensional expansion proof and so

$$\mathbf{SelPars}^{\mathcal{Q}^*}(\{r^*\}) = (\mathbf{Params}(\mathbf{M}) \cap \mathbf{SelPars}(\mathcal{Q}^*)) = \emptyset.$$

This contradiction verifies condition (5) also holds.

Therefore, $\langle \varphi, f \rangle$ is a lifting map from $\mathcal{Q}$ to $\mathcal{Q}^*$. We know $f(r) = r^*$ by the definition of f. The conditions (1) and (2) of Definition 8.6.7 hold vacuously relative to

$$\langle \mathcal{Q}, r, \emptyset, \emptyset \rangle.$$

Thus $\langle \varphi, f \rangle$ is a search lifting map from **SearchInit**(**M**) to $\langle \mathcal{Q}^*, r^* \rangle$. □

PROOF OF LEMMA 8.6.10. Suppose $f(a) \xrightarrow{\mathcal{Q}^*} f(b)$. We start by proving b is not a root node of $\mathcal{Q}$. Assume $b \in \mathbf{Roots}(\mathcal{Q})$. Since

Roots$(\mathcal{Q}) = \{r\}$ and $f(r) = r^*$, $f(b)$ must be the root node r^*, contradicting $f(a) \overset{\mathcal{Q}^*}{\rightharpoondown} f(b)$. Thus b is not a root node of $\mathcal{Q}$ and so there is some $a' \in \mathcal{N}^{\mathcal{Q}}$ such that $a' \overset{\mathcal{Q}}{\rightharpoondown} b$. By Lemma 8.4.6, $f(a') \overset{\mathcal{Q}^*}{\rightharpoondown} f(b)$. Since $\langle \mathcal{Q}^*, r^* \rangle$ is a lifting target, $\mathcal{Q}^*$ is separated.

Suppose b is not a connection node. Hence $f(b)$ is not a connection node (cf. Definition 8.4.3:1). Since $f(a) \overset{\mathcal{Q}^*}{\rightharpoondown} f(b)$, $f(a') \overset{\mathcal{Q}^*}{\rightharpoondown} f(b)$ and $\mathcal{Q}^*$ is separated, we know $f(a) = f(a')$ (cf. Definition 7.5.2:2). Thus $a = a'$ and $a \overset{\mathcal{Q}}{\rightharpoondown} b$ since f is injective.

Suppose b is a connection node. Hence $f(b)$ is a connection node (cf. Definition 8.4.3:1). Applying Lemma 7.3.15 to $a' \overset{\mathcal{Q}}{\rightharpoondown} b$ and $f(a) \overset{\mathcal{Q}^*}{\rightharpoondown} f(b)$, there are connections

$$\langle b, d^1, d^2 \rangle \in conns^{\mathcal{Q}} \text{ and } \langle f(b), e^1, e^2 \rangle \in conns^{\mathcal{Q}^*}$$

such that $a' \in \{d^1, d^2\}$ and $f(a) \in \{e^1, e^2\}$. We have

$$\langle f(b), f(d^1), f(d^2) \rangle \in conns^{\mathcal{Q}^*}$$

by Definition 8.4.3:8. Since $\mathcal{Q}^*$ is separated, $f(d^1) = e^1$ and $f(d^2) = e^2$ (cf. Definition 7.5.2:3). Hence $f(a) = e^i = f(d^i)$ for some $i \in \{1, 2\}$. Thus $a = d^i$. Since $d^i \overset{\mathcal{Q}}{\rightharpoondown} b$, we conclude $a \overset{\mathcal{Q}}{\rightharpoondown} b$. □

PROOF OF LEMMA 8.6.12. To simplify the notation, assume $\mathcal{Q}$ is the extensional expansion dag

$$\langle \mathcal{N}, succ, label, sel, exps, conns, decs \rangle$$

and the extensional expansion dag $\mathcal{Q}^*$ is

$$\langle \mathcal{N}^*, succ^*, label^*, sel^*, exps^*, conns^*, decs^* \rangle.$$

In every case of Definition 7.7.4, $\mathcal{Q}'$ is

$$\langle \mathcal{N}', succ', label', sel', exps, conns, decs \rangle$$

for some $\mathcal{N}'$, $succ'$, $label'$ and sel'. Since we have assumed a is not a selection node, $sel' = sel$.

By Lemma 8.6.11, there exists an $n \geq 0$, distinct nodes

$$a^1, \ldots, a^n \in \mathcal{N}'$$

and distinct nodes

$$b^1, \ldots, b^n \in \mathcal{N}^*$$

such that $\mathcal{N}' = (\mathcal{N} \cup \{a^1, \ldots, a^n\})$, $(\mathcal{N}' \setminus \mathcal{N}) = \{a^1, \ldots, a^n\}$,

$$succ'(a) = \langle a^1, \ldots, a^n \rangle,$$

$$succ^*(f(a)) = \langle b^1, \ldots, b^n \rangle, \tag{14}$$

and

$$label^*(b^i) = \langle kind^{\mathcal{Q}'}(a^i), pol^{\mathcal{Q}'}(a^i), (\varphi(sh^{\mathcal{Q}'}(a^i)))^{\downarrow} \rangle \tag{15}$$

for each i $(1 \leq i \leq n)$.

Let $f' : \mathcal{N}' \to \mathcal{N}^*$ be the function

$$f'(c) := \begin{cases} f(c) & \text{if } c \in \mathcal{N} \\ b^i & \text{if } c = a^i \text{ for } 1 \leq i \leq n. \end{cases}$$

Note that $f'|_{\mathcal{N}} = f$.

Let f' be $\langle \varphi, f' \rangle$. By Lemma 7.7.7, $\mathcal{Q}'$ is an extensional expansion dag and $\mathcal{Q} \trianglelefteq \mathcal{Q}'$ (since $\mathcal{Q} \trianglelefteq_r \mathcal{Q}'$). We can use Lemma 8.4.7 to conclude f' is a lifting map from $\mathcal{Q}'$ to $\mathcal{Q}^*$ once we check the conditions of the lemma. Conditions (9), (11), (12), (13) and (14) of Lemma 8.4.7 are vacuous since $\mathcal{Q}$ and $\mathcal{Q}'$ share the same selected parameter function, expansion arcs, connections and decomposition arcs. We know $\mathbf{ExpVars}(\mathcal{Q}) = \mathbf{ExpVars}(\mathcal{Q}')$ since $\mathcal{Q}$ and $\mathcal{Q}'$ have the same expansion arcs. By Lemma 7.7.6,

$$\mathbf{Free}(sh^{\mathcal{Q}'}(b)) \subseteq \mathbf{Free}(sh^{\mathcal{Q}}(a))$$

for every $b \in \mathcal{N}' \setminus \mathcal{N}$. Hence $\mathbf{Free}^{\mathcal{Q}'}(\mathcal{N}') = \mathbf{Free}^{\mathcal{Q}}(\mathcal{N})$ and so

$$\mathbf{Free}(\mathcal{Q}') = \mathbf{Free}(\mathcal{Q}).$$

Thus condition (1) is holds vacuously. By Lemma 7.3.16,

$$(\mathbf{Params}(sh^{\mathcal{Q}'}(a^1)) \cup \cdots \cup \mathbf{Params}(sh^{\mathcal{Q}'}(a^n)))$$
$$\subseteq \mathbf{Params}(sh^{\mathcal{Q}}(a)).$$

Hence

$$\mathbf{Params}(\mathcal{Q}') = \mathbf{Params}(\mathcal{Q})$$

and so conditions (2), (3) and (10) of Lemma 8.4.7 also hold vacuously.

To check condition (4) of Lemma 8.4.7 we must verify that $f'(a^i) \notin \mathbf{Im}(f; \mathcal{Q})$ for each i $(1 \leq i \leq n)$. Suppose $f(c) = f'(a^i) = b^i$

for some $c \in \mathcal{N}$. Since $f(a) \overset{\mathcal{Q}^*}{\rightarrow} b^i$, we know $a \overset{\mathcal{Q}}{\rightarrow} c$ by Lemma 8.6.10. This contradicts Lemma 7.3.6 since a is not extended in $\mathcal{Q}$. Thus

$$(\{f'(a^1), \ldots, f'(a^n)\} \cap \mathbf{Im}(f; \mathcal{Q})) = \emptyset$$

and condition (4) holds. Since $b^1, \ldots, b^n$ are distinct, $f(a^i) = f(a^j)$ implies $i = j$ for all $a^i, a^j \in \mathcal{N}' \setminus \mathcal{N}$. Hence condition (5) holds. Condition (6) holds since

$$label'(a^i) = \langle kind^{\mathcal{Q}'}(a^i), pol^{\mathcal{Q}'}(a^i), sh^{\mathcal{Q}'}(a^i)\rangle$$

and

$$label^*(f'(a^i)) = label^*(b^i) = \langle kind^{\mathcal{Q}'}(a^i), pol^{\mathcal{Q}'}(a^i), (\varphi(sh^{\mathcal{Q}'}(a^i)))^{\downarrow}\rangle$$

for each $a^i \in \mathcal{N}' \setminus \mathcal{N}$ by equation (15) above. Conditions (7) and (8) hold since $(\mathbf{Dom}(succ') \setminus \mathbf{Dom}(succ)) = \{a\}$, $f(a) \in \mathbf{Dom}(succ^*)$ (i.e., $f(a)$ is extended in $\mathcal{Q}^*$) and

$$succ^*(f(a)) = \langle b^1, \ldots, b^n\rangle = \langle f'(a^1), \ldots, f'(a^n)\rangle$$

by equation (14) above.

Thus Lemma 8.4.7 applies and we conclude f' is a lifting map from $\mathcal{Q}'$ to $\mathcal{Q}^*$. □

PROOF OF LEMMA 8.6.13. To simplify the notation, assume $\mathcal{Q}$ is the extensional expansion dag

$$\langle \mathcal{N}, succ, label, sel, exps, conns, decs\rangle$$

and the extensional expansion dag $\mathcal{Q}^*$ is

$$\langle \mathcal{N}^*, succ^*, label^*, sel^*, exps^*, conns^*, decs^*\rangle.$$

Suppose a is a selection node with shallow formula $[\forall x_\alpha \mathbf{M}]$. By Definition 7.7.4, $\mathcal{Q}'$ is

$$\langle \mathcal{N}', succ', label', sel', exps, conns, decs\rangle$$

where $\mathcal{N}' = \mathcal{N} \cup \{a^1\}$,

$$succ'(b) = \begin{cases} succ(b) & \text{if } b \in \mathbf{Dom}(succ) \\ \langle a^1\rangle & \text{if } b = a \end{cases}$$

$$label'(b) = \begin{cases} label(b) & \text{if } b \in \mathcal{N} \\ \langle \mathbf{primary}, -1, ([A_\alpha/x_\alpha]\mathbf{M})^{\downarrow}\rangle & \text{if } b = a^1 \end{cases}$$

and

$$sel'(b) = \begin{cases} sel(b) & \text{if } b \in \mathbf{Dom}(sel) \\ A_\alpha & \text{if } b = a \end{cases}$$

for some $a^1 \notin \mathcal{N}$ and parameter $A_\alpha \notin \mathbf{Params}(\mathcal{Q})$.

Since f is a lifting map, $f(a)$ is a negative primary node and

$$sh^{\mathcal{Q}^*}(f(a)) = [\dot{\forall} x_\alpha\, (\varphi(\mathbf{M}))^{\downarrow}].$$

Hence $f(a)$ is a selection node in $\mathcal{Q}^*$. By $\clubsuit_{sel}$, there is a negative primary node $b^1 \in \mathcal{N}^*$ such that $f(a) \in \mathbf{Dom}(sel^*)$, $succ^*(f(a)) = \langle b^1 \rangle$ and $sh^{\mathcal{Q}^*}(b^1) = [sel^*(f(a))/x](\varphi(\mathbf{M}))^{\downarrow}$. Let B be $sel^*(f(a))$. Hence we have

$$(16) \qquad label^*(b^1) = \langle \mathbf{primary}, -1, ((\varphi, [B/x])\mathbf{M})^{\downarrow} \rangle$$

since B is a parameter and we assume x not to be vulnerable.

Let φ' be $\varphi, [B/A]$ and $f' : \mathcal{N}' \to \mathcal{N}^*$ be the function

$$f'(c) := \begin{cases} f(c) & \text{if } c \in \mathcal{N} \\ b^1 & \text{if } c = a^1 \end{cases}$$

Note that $f'|_{\mathcal{N}} = f$ and $\varphi'(u) = \varphi(u)$ for all

$$u \in (\mathbf{Free}(\mathcal{Q}) \cup \mathbf{Params}(\mathcal{Q}))$$

(since we know $A \notin \mathbf{Params}(\mathcal{Q})$).

Let f' be $\langle \varphi', f' \rangle$. By Lemma 7.7.7, $\mathcal{Q}'$ is an extensional expansion dag and $\mathcal{Q} \trianglelefteq \mathcal{Q}'$ (since $\mathcal{Q} \trianglelefteq_r \mathcal{Q}'$). We can use Lemma 8.4.7 to conclude f' is a lifting map from $\mathcal{Q}'$ to $\mathcal{Q}^*$ once we check the conditions of the lemma. Conditions (11), (12), (13) and (14) of Lemma 8.4.7 are vacuous since $\mathcal{Q}$ and $\mathcal{Q}'$ share the same expansion arcs, connections and decomposition arcs. Since $\mathcal{Q}$ and $\mathcal{Q}'$ have the same expansion arcs, $\mathbf{ExpVars}(\mathcal{Q}) = \mathbf{ExpVars}(\mathcal{Q}')$. By Lemma 7.7.6, $\mathbf{Free}(sh^{\mathcal{Q}'}(b)) \subseteq \mathbf{Free}(sh^{\mathcal{Q}}(a))$ for every $b \in \mathcal{N}' \setminus \mathcal{N}$. Hence $\mathbf{Free}^{\mathcal{Q}'}(\mathcal{N}') = \mathbf{Free}^{\mathcal{Q}}(\mathcal{N})$ and so $\mathbf{Free}(\mathcal{Q}') = \mathbf{Free}(\mathcal{Q})$. Thus condition (1) also holds vacuously.

Note that $\mathbf{SelPars}(\mathcal{Q}') = \mathbf{SelPars}(\mathcal{Q}) \cup \{A\}$. Also,

$$\mathbf{Params}(sh^{\mathcal{Q}'}(a^1)) \subseteq \mathbf{Params}(\mathbf{M}) \cup \{A\}$$

and so

$$\mathbf{Params}^{\mathcal{Q}}(\mathcal{N}) \subseteq \mathbf{Params}^{\mathcal{Q}'}(\mathcal{N}') \subseteq \mathbf{Params}^{\mathcal{Q}}(\mathcal{N}) \cup \{A\}.$$

Since $\mathcal{Q}$ and $\mathcal{Q}'$ share the same expansion arcs,

$$\mathbf{ExpParams}(\mathcal{Q}') = \mathbf{ExpParams}(\mathcal{Q}).$$

Hence $\mathbf{Params}(\mathcal{Q}') = \mathbf{Params}(\mathcal{Q}) \cup \{A\}$. Since $\varphi'(A) = B \in \mathcal{P}$, condition (2) of Lemma 8.4.7 holds. Since $(\mathbf{Dom}(sel') \setminus \mathbf{Dom}(sel))$ is $\{a\}$,

$$f'(a) = f(a) \in \mathbf{Dom}(sel^*)$$

and

$$sel^*(f'(a)) = B = \varphi'(A) = \varphi'(sel'(a)),$$

condition (9) holds. Condition (10) holds since $f'(a)$ is the unique node in $\mathcal{Q}^*$ such that $sel^*(f'(a)) = \varphi'(A)$.

Suppose $C \in \mathbf{Params}(\mathcal{Q}')$, $\varphi'(C) = B$ and $C \neq A$. Then

$$C \in \mathbf{Params}(\mathcal{Q}) \subseteq \mathbf{Params}(\mathcal{Q})$$

and $\varphi(C) = B$. Since f is a lifting map, there must exist some $c \in \mathbf{Dom}(sel)$ such that $f(c) = f(a)$ by condition (5) of Definition 8.4.3. Since f is injective, we must have $c = a$, contradicting the fact that a is not in $\mathbf{Dom}(sel)$. Hence if $C \in \mathbf{Params}(\mathcal{Q}')$ and $\varphi'(C) = B$, then $C = A$. Thus condition (3) of Lemma 8.4.7 holds for f'.

To check condition (4) of Lemma 8.4.7 we prove $f'(a^1)$ is not in $\mathbf{Im}(f;\mathcal{Q})$. Suppose $f(c) = f'(a^1) = b^1$ for some $c \in \mathcal{N}$. Since $f(a) \overset{\mathcal{Q}^*}{\rightharpoondown} b^1$, we know $a \overset{\mathcal{Q}}{\rightharpoondown} c$ by Lemma 8.6.10. This contradicts Lemma 7.3.6 since a is not extended in $\mathcal{Q}$. Thus

$$f'(a^1) \notin \mathbf{Im}(f;\mathcal{Q})$$

and condition (4) holds. Condition (5) holds trivially since $\mathcal{N}' \setminus \mathcal{N}$ has only the one element a^1. Conditions (7) and (8) hold since

$$(\mathbf{Dom}(succ') \setminus \mathbf{Dom}(succ)) = \{a\},$$

$f(a) \in \mathbf{Dom}(succ^*)$ (i.e., $f(a)$ is extended in $\mathcal{Q}^*$) and

$$succ^*(f(a)) = \langle b^1 \rangle = \langle f'(a^1) \rangle.$$

Recall

$$label'(a^1) = \langle \mathbf{primary}, -1, ([A/x]\mathbf{M}) \rangle$$

and

$$label^*(f'(a^1)) = label^*(b^1) = \langle \mathbf{primary}, -1, ([B/x]\varphi(\mathbf{M}))^{\downarrow} \rangle$$

(cf. equation (16) above). We compute

$$((\varphi, [B/x])\mathbf{M})^{\downarrow} = (\varphi'([A/x]\mathbf{M}))^{\downarrow}$$

and so condition (6) holds.

Thus Lemma 8.4.7 applies and we conclude f' is a lifting map from $\mathcal{Q}'$ to $\mathcal{Q}^*$. □

PROOF OF LEMMA 8.6.15. Let $\mathcal{Q}'$ be $\mathbf{Dec}(\mathcal{Q}, a)$. Since Ξ is a search state, $\mathcal{Q}$ is an extensional expansion dag, $\mathbf{RootSels}(\mathcal{Q})$ is empty and $\prec_0^{\mathcal{Q}}$ is acyclic. By Lemma 7.7.18, $\mathcal{Q}'$ is an extensional expansion dag, $\mathcal{Q} \trianglelefteq_r \mathcal{Q}'$,

$$\mathbf{ExpTerms}(\mathcal{Q}') = \mathbf{ExpTerms}(\mathcal{Q}),$$

$$\mathbf{RootSels}(\mathcal{Q}') = \mathbf{RootSels}(\mathcal{Q}) = \emptyset$$

and $\prec_0^{\mathcal{Q}'}$ is acyclic. Since $(\mathbf{ExpTerms}(\mathcal{Q}') \setminus \mathbf{ExpTerms}(\mathcal{Q})) = \emptyset$,

$$\Xi' = \langle \mathcal{Q}', r, \mathscr{D}, \mathscr{S} \rangle$$

is a search state by Lemma 8.6.5.

Suppose $\langle \mathcal{Q}^*, r^* \rangle$ is a lifting target and $\mathsf{f} = \langle \varphi, f \rangle$ is a search lifting map from Ξ to $\langle \mathcal{Q}^*, r^* \rangle$. To simplify the notation, assume $\mathcal{Q}$ is the extensional expansion dag

$$\langle \mathcal{N}, succ, label, sel, exps, conns, decs \rangle$$

and the extensional expansion dag $\mathcal{Q}^*$ is

$$\langle \mathcal{N}^*, succ^*, label^*, sel^*, exps^*, conns^*, decs^* \rangle.$$

By Definition 7.7.16, $\mathcal{Q}'$ is

$$\langle \mathcal{N}', succ, label', sel, exps, conns, decs' \rangle$$

where $\mathcal{N}' := \mathcal{N} \cup \{d\}$, $d \notin \mathcal{N}$, $decs' := decs \cup \{\langle e, d \rangle\}$ and

$$label'(b) := \begin{cases} label(b) & \text{if } b \in \mathcal{N} \\ \langle \mathbf{dec}, -1, sh^{\mathcal{Q}}(a) \rangle & \text{if } b = d. \end{cases}$$

Since a is decomposable in $\mathcal{Q}$, a is a negative primary node with shallow formula $[[H\,\overline{\mathbf{A}^n}] \doteq^{\iota} [H\,\overline{\mathbf{B}^n}]]$ for some parameter H. Since f is a lifting map from $\mathcal{Q}$ to $\mathcal{Q}^*$, $f(a)$ is an extended negative primary node in $\mathcal{Q}^*$ with shallow formula $(\varphi([[H\,\overline{\mathbf{A}^n}] \doteq^{\iota} [H\,\overline{\mathbf{B}^n}]]))^{\downarrow}$. Since φ

respects parameters, $\varphi(H)$ is a parameter and the shallow formula of $f(a)$ in $\mathcal{Q}^*$ is

$$[[\varphi(H)\,(\varphi(\mathbf{A}^1))^{\downarrow}\cdots(\varphi(\mathbf{A}^n))^{\downarrow}] \doteq^{\iota} [\varphi(H)\,(\varphi(\mathbf{B}^1))^{\downarrow}\cdots(\varphi(\mathbf{B}^n))^{\downarrow}]]$$

Thus $f(a)$ is a decomposable node in $\mathcal{Q}^*$. Since $\langle \mathcal{Q}^*, r^* \rangle$ is a lifting target, $\mathcal{Q}^*$ is developed and so $f(a)$ is decomposed. That is, there is some node $d^* \in \mathcal{N}^*$ such that $\langle f(a), d^* \rangle \in decs^*$.

Let $f' : \mathcal{N}' \to \mathcal{N}^*$ be the function

$$f'(b) := \begin{cases} f(b) & \text{if } b \in \mathcal{N} \\ d^* & \text{if } b = d. \end{cases}$$

Since $sh^{\mathcal{Q}'}(d) = sh^{\mathcal{Q}}(a)$,

$$\mathbf{Free}^{\mathcal{Q}'}(\mathcal{N}') = \mathbf{Free}^{\mathcal{Q}}(\mathcal{N})$$

and

$$\mathbf{Params}^{\mathcal{Q}'}(\mathcal{N}') = \mathbf{Params}^{\mathcal{Q}}(\mathcal{N}).$$

Since $sel^{\mathcal{Q}'} = sel = sel^{\mathcal{Q}}$ and $exps^{\mathcal{Q}'} = exps = sel^{\mathcal{Q}}$, we have

$$\mathbf{Free}(\mathcal{Q}') = \mathbf{Free}(\mathcal{Q})$$

and

$$\mathbf{Params}(\mathcal{Q}') = \mathbf{Params}(\mathcal{Q}).$$

Let f' be $\langle \varphi, f' \rangle$. We can use Lemma 8.4.7 to conclude f' is a lifting map from $\mathcal{Q}'$ to $\mathcal{Q}^*$. Note that $\mathcal{Q} \trianglelefteq \mathcal{Q}'$ and $f'|_{\mathcal{N}} = f$ by the definition of f'. Conditions (1), (2), (3) and (10) of Lemma 8.4.7 are vacuous since $\mathbf{Free}(\mathcal{Q}') \setminus \mathbf{Free}(\mathcal{Q})$ and $\mathbf{Params}(\mathcal{Q}') \setminus \mathbf{Params}(\mathcal{Q})$ are empty. Condition (5) is trivial since $\mathcal{N}' \setminus \mathcal{N}$ contains only one element, namely d. Conditions (7), (8), (9), (11), (12) and (13) are vacuous since $\mathcal{Q}$ and $\mathcal{Q}'$ share the same successor function, selected parameter function, expansion arcs and connections.

To check condition (4) of Lemma 8.4.7 we prove $d^* \notin \mathbf{Im}(f; \mathcal{Q})$. Suppose $f(b) = d^*$ for some $b \in \mathcal{N}$. Since $\langle f(a), d^* \rangle \in decs^*$, we know $f(a) \overset{\mathcal{Q}^*}{\rightarrow} f(b)$ and so $a \overset{\mathcal{Q}}{\rightarrow} b$ by Lemma 8.6.10. Also, b is a decomposition node as

$$kind^{\mathcal{Q}}(b) = kind^{\mathcal{Q}^*}(f(b)) = kind^{\mathcal{Q}^*}(d^*) = \mathbf{dec}$$

since d^* is a decomposition node. In particular, b is not an equation goal node and so $\langle a, b \rangle \in decs$ by Lemma 7.3.12. This contradicts our assumption that a is not decomposed in $\mathcal{Q}$. Thus $d^* \notin \mathbf{Im}(f; \mathcal{Q})$ and condition (4) of Lemma 8.4.7 holds.

Condition (14) of Lemma 8.4.7 holds since

$$\langle f'(a), f'(d) \rangle = \langle f(a), d^* \rangle \in decs^*.$$

Condition (6) holds since $label'(d) = \langle \mathbf{dec}, -1, sh^{\mathcal{Q}}(a) \rangle$,

$$label^*(f'(d)) = label^*(d^*) = \langle \mathbf{dec}, -1, sh^{\mathcal{Q}^*}(f(a)) \rangle$$

and $sh^{\mathcal{Q}^*}(f(a)) = (\varphi(sh^{\mathcal{Q}}(a)))^{\downarrow}$. Thus Lemma 8.4.7 applies and we conclude f' is a lifting map from $\mathcal{Q}'$ to $\mathcal{Q}^*$.

We next check $\mathsf{f}' = \langle \varphi, f' \rangle$ is a search lifting map from Ξ' to $\langle \mathcal{Q}^*, r^* \rangle$. Since f is a search lifting map, we know $f(r) = r^*$, $\mathscr{D}$ is fully duplicated with respect to f and $\mathscr{S}$ is almost atomic with respect to f. Hence

$$f'(r) = f(r) = r^*.$$

Since $\mathcal{Q} \trianglelefteq \mathcal{Q}'$ and $f'|_{\mathcal{N}} = f$, we can apply Lemma 8.4.20 to conclude $\mathscr{D}$ is fully duplicated with respect to f' and $\mathscr{S}$ is almost atomic with respect to f'. Thus f' is a search lifting map from Ξ' to $\langle \mathcal{Q}^*, r^* \rangle$.

Finally, we know

$$\|\mathsf{f}' : \mathcal{Q}' \to \mathcal{Q}^*\| < \|\mathsf{f} : \mathcal{Q} \to \mathcal{Q}^*\|$$

by Lemma 8.4.5 since $d \in \mathcal{N}' \setminus \mathcal{N}$. □

PROOF OF LEMMA 8.6.16. Let $\mathcal{Q}'$ be $\mathbf{EExpand}(\mathcal{Q}, e, y)$. Since Ξ is a search state, $\mathcal{Q}$ is an extensional expansion dag, $\mathbf{RootSels}(\mathcal{Q})$ is empty and $\prec_0^{\mathcal{Q}}$ is acyclic. By Lemma 7.7.10, $\mathcal{Q}'$ is an extensional expansion dag, $\mathcal{Q} \trianglelefteq_r \mathcal{Q}'$,

$$\mathbf{ExpTerms}(\mathcal{Q}') = \mathbf{ExpTerms}(\mathcal{Q}) \cup \{y\},$$

$$\mathbf{RootSels}(\mathcal{Q}') = \mathbf{RootSels}(\mathcal{Q}) = \emptyset$$

and $\prec_0^{\mathcal{Q}'}$ is acyclic. By Definition 8.5.1, the variable y is strict since

$$\mathbf{Free_{str}}(y) = \mathbf{Free_{str}}([\lambda \overline{z^n}\, [y\, \overline{z^n}]]) = \{y\}$$

where y has type $(\beta\alpha^n \cdots \alpha^1)$ with $\beta \in \{o, \iota\}$. By Lemma 8.6.5,

$$\Xi' = \langle \mathcal{Q}', r, \mathscr{D}, \mathscr{S} \rangle$$

is a search state.

Suppose $\langle \mathcal{Q}^*, r^* \rangle$ is a lifting target, $\mathsf{f} = \langle \varphi, f \rangle$ is a search lifting map from Ξ to $\langle \mathcal{Q}^*, r^* \rangle$, $c^* \in \mathcal{N}^{\mathcal{Q}^*}$ is a node and $\langle e, c^* \rangle \in \mathbf{Edges}^-(\mathsf{f})$. To simplify the notation, assume the extensional expansion dag $\mathcal{Q}$ is

$$\langle \mathcal{N}, succ, label, sel, exps, conns, decs \rangle$$

and the extensional expansion dag $\mathcal{Q}^*$ is

$$\langle \mathcal{N}^*, succ^*, label^*, sel^*, exps^*, conns^*, decs^* \rangle.$$

By Definition 7.7.8, there is some $c \notin \mathcal{N}$ and variable $y \notin \mathbf{Free}(\mathcal{Q})$ such that $\mathcal{Q}'$ is

$$\langle \mathcal{N}', succ, label', sel, exps', conns, decs \rangle$$

where $\mathcal{N}' := \mathcal{N} \cup \{c\}$, $exps' = exps \cup \{\langle e, y, c \rangle\}$ and

$$label'(a) := \begin{cases} label(a) & \text{if } a \in \mathcal{N} \\ \langle \mathbf{primary}, 1, ([y/x]\mathbf{M})^{\downarrow} \rangle & \text{if } a = c. \end{cases}$$

Since $\langle e, c^* \rangle \in \mathbf{Edges}^-(\mathsf{f})$, we know $c^* \notin \mathbf{Im}(f; \mathcal{Q})$ and $f(e) \overset{\mathcal{Q}^*}{\rightarrow} c^*$. Since e is an extended expansion node of $\mathcal{Q}$ with shallow formula $[\dot{\forall} x_\alpha \mathbf{M}]$, $f(e)$ is an extended expansion node of $\mathcal{Q}^*$ with shallow formula $[\dot{\forall} x_\alpha (\varphi(\mathbf{M}))^{\downarrow}]$ (cf. Definition 8.4.3:1). By Lemma 7.3.11, there is some term $\mathbf{C}$ (of type α by $\clubsuit_{exp}$) such that $\langle f(e), \mathbf{C}, c^* \rangle \in exps^*$.

Let φ' be the substitution $(\varphi, [\mathbf{C}^{\updownarrow}/y])$ and $f' : \mathcal{N}' \to \mathcal{N}^*$ be the function

$$f'(a) := \begin{cases} f(a) & \text{if } a \in \mathcal{N} \\ c^* & \text{if } a = c. \end{cases}$$

Let f' be $\langle \varphi', f' \rangle$.

We can use Lemma 8.4.7 to conclude f' is a lifting map from $\mathcal{Q}'$ to $\mathcal{Q}^*$. We already know $\mathcal{Q} \trianglelefteq \mathcal{Q}'$ and

$$\mathbf{Dom}(\varphi') \subseteq \mathbf{Dom}(\varphi) \cup \{y\} \subset \mathcal{V} \cup \mathcal{P}.$$

Also, $f'|_{\mathcal{N}} = f$ and

$$\varphi'(u) = \varphi(u) \text{ for every } u \in (\mathbf{Free}(\mathcal{Q}) \cup \mathbf{Params}(\mathcal{Q}))$$

(since $y \notin \mathbf{Free}(\mathcal{Q})$) by the definitions of f' and φ'.

Note that

$$\mathbf{Free}(([y/x]\mathbf{M})^{\downarrow}) \subseteq \mathbf{Free}([\dot{\forall} x_\alpha \mathbf{M}]) \cup \{y\}$$

and

$$\mathbf{Params}(([y/x]\mathbf{M})^{\downarrow}) \subseteq \mathbf{Params}([\dot{\forall} x_\alpha \mathbf{M}]).$$

Hence $\mathbf{Free}(\mathcal{Q}') \subseteq \mathbf{Free}(\mathcal{Q}) \cup \{y\}$ and $\mathbf{Params}(\mathcal{Q}') \subseteq \mathbf{Params}(\mathcal{Q})$. Since $y \notin \mathbf{Free}(\mathcal{Q})$, $(\mathbf{Free}(\mathcal{Q}') \setminus \mathbf{Free}(\mathcal{Q}))$ is $\{y\}$. Since $\langle \mathcal{Q}^*, r^* \rangle$ is a lifting target, $\mathcal{Q}^*$ is closed and so $\mathbf{C}$ is closed. Hence condition (1) of Lemma 8.4.7 holds since $\varphi'(y)$ is the closed long $\beta\eta$-form $\mathbf{C}^{\updownarrow}$. Conditions (2), (3) and (10) are vacuous since $(\mathbf{Params}(\mathcal{Q}') \setminus \mathbf{Params}(\mathcal{Q}))$ is empty. Recall $(\mathcal{N}' \setminus \mathcal{N}) = \{c\}$ and $f'(c) = c^*$. Since $\langle e, c^* \rangle$ is in $\mathbf{Edges}^-(\mathsf{f})$, we know $c^* \notin \mathbf{Im}(f; \mathcal{Q})$ (cf. Definition 8.4.10). Hence condition (4) holds. Condition (5) is trivial.

Conditions (7), (8), (9), (13), and (14) of Lemma 8.4.7 are vacuous since $\mathcal{Q}$ and $\mathcal{Q}'$ share the same successor function, selected parameter function, connections and decomposition arcs. Condition (11) holds since $\langle f'(e), \mathbf{C}, f'(c) \rangle = \langle f(e), \mathbf{C}, c^* \rangle \in \mathit{exps}^*$ and $\varphi'(y) = \mathbf{C}^{\updownarrow} \stackrel{\beta\eta}{=} \mathbf{C}$. Condition (12) holds vacuously since $\mathbf{Params}(y)$ is empty. Lastly, we check condition (6) for $(\mathcal{N}' \setminus \mathcal{N}) = \{c\}$. Note that

$$\mathit{label}'(c) = \langle \mathbf{primary}, 1, ([y/x]\mathbf{M})^{\downarrow} \rangle.$$

Since $[\dot{\forall} x_\alpha \, (\varphi(\mathbf{M}))^{\downarrow}]$ is the shallow formula of $f(e)$ and $\langle f(e), \mathbf{C}, c^* \rangle$ is in exps^*, c^* must be a positive primary node with shallow formula $([\mathbf{C}/x]((\varphi(\mathbf{M}))^{\downarrow}))^{\downarrow}$ by $\clubsuit_{exp}$. Since $y \notin \mathbf{Free}(\mathcal{Q})$, $y \notin \mathbf{Free}(\mathbf{M})$. Assuming the bound variable x is not vulnerable for φ (cf. Remark 2.1.7), we compute

$$\begin{array}{rcll} [\mathbf{C}/x]((\varphi(\mathbf{M}))^{\downarrow}) & \stackrel{\beta\eta}{=} & [\mathbf{C}/x](\varphi(\mathbf{M})) & \\ & = & (\varphi, [\mathbf{C}/x])\mathbf{M} & \\ & = & (\varphi, [\mathbf{C}/y])([y/x]\mathbf{M})) & \text{since } y \notin \mathbf{Free}(\mathbf{M}) \\ & \stackrel{\beta\eta}{=} & (\varphi, [\mathbf{C}^{\updownarrow}/y])([y/x]\mathbf{M})) & \\ & = & \varphi'([y/x]\mathbf{M}). & \end{array}$$

Hence

$$\mathit{label}^*(c^*) = \langle \mathbf{primary}, 1, (\varphi'([y/x]\mathbf{M}))^{\downarrow} \rangle$$

and so condition (6) of Lemma 8.4.7 holds. Thus Lemma 8.4.7 applies and we conclude f' is a lifting map from $\mathcal{Q}'$ to $\mathcal{Q}^*$.

We next check $\mathsf{f}' = \langle \varphi, f' \rangle$ is a search lifting map from Ξ' to $\langle \mathcal{Q}^*, r^* \rangle$. Since f is a search lifting map, we know $f(r) = r^*$, $\mathscr{D}$ is fully duplicated with respect to f and $\mathscr{S}$ is almost atomic with respect to f. Hence

$$f'(r) = f(r) = r^*.$$

Since $\mathcal{Q} \trianglelefteq \mathcal{Q}'$ and $f'|_{\mathcal{N}} = f$, we can apply Lemma 8.4.20 to conclude $\mathscr{D}$ is fully duplicated with respect to f' and $\mathscr{S}$ is almost atomic with respect to f'. Therefore, f' is a search lifting map from Ξ' to $\langle \mathcal{Q}^*, r^* \rangle$.

Finally, we know

$$\|\mathsf{f}' : \mathcal{Q}' \to \mathcal{Q}^*\| < \|\mathsf{f} : \mathcal{Q} \to \mathcal{Q}^*\|$$

by Lemma 8.4.5 since $c \in \mathcal{N}' \setminus \mathcal{N}$. □

PROOF OF LEMMA 8.6.18. To ease notation, assume $\mathcal{Q}$ is the extensional expansion dag

$$\langle \mathcal{N}, succ, label, sel, exps, conns, decs \rangle.$$

By our assumptions about $\mathcal{Q}'$, $\mathcal{Q}'$ is

$$\langle \mathcal{N}', succ, label', sel, exps, conns', decs \rangle$$

where $\mathcal{N}' = \mathcal{N} \cup \{c\}$ and $conns' = conns \cup \{\langle c, a, b \rangle\}$. We assumed $\mathcal{Q} \trianglelefteq_r \mathcal{Q}'$, $\mathbf{RootSels}(\mathcal{Q}')$ is empty and $\prec_0^{\mathcal{Q}'}$ is acyclic. Note that $(exps^{\mathcal{Q}'} \setminus exps^{\mathcal{Q}})$ is empty. Thus we can apply Lemma 8.6.5 to conclude

$$\Xi' = \langle \mathcal{Q}', r, \mathscr{D}, \mathscr{S} \rangle$$

is a search state.

Suppose $\langle \mathcal{Q}^*, r^* \rangle$ is a lifting target and $\mathsf{f} = \langle \varphi, f \rangle$ is a search lifting map from Ξ to $\langle \mathcal{Q}^*, r^* \rangle$. To ease notation, assume $\mathcal{Q}^*$ is the extensional expansion dag

$$\langle \mathcal{N}^*, succ^*, label^*, sel^*, exps^*, conns^*, decs^* \rangle.$$

Suppose $c^* \in \mathcal{N}^*$ and $\langle c^*, a, b \rangle \in \mathbf{Conns}^-(\mathsf{f}; k)$. We construct a search lifting map f' from Ξ' to $\langle \mathcal{Q}^*, r^* \rangle$.

Let $f' : \mathcal{N}' \to \mathcal{N}^*$ be the function

$$f'(d) := \begin{cases} f(d) & \text{if } d \in \mathcal{N} \\ c^* & \text{if } d = c. \end{cases}$$

Let f' be $\langle \varphi, f' \rangle$.

We can use Lemma 8.4.7 to conclude f' is a lifting map from $\mathcal{Q}'$ to $\mathcal{Q}^*$. Note that $\mathcal{Q} \trianglelefteq \mathcal{Q}'$ and $f'|_{\mathcal{N}} = f$ by the definition of f'.

Since $\mathcal{Q}$ and $\mathcal{Q}'$ share the same selected parameter function and expansion arcs,

$$\mathbf{ExpVars}(\mathcal{Q}') = \mathbf{ExpVars}(\mathcal{Q}),$$

$$\mathbf{ExpParams}(\mathcal{Q}') = \mathbf{ExpParams}(\mathcal{Q})$$

and

$$\mathbf{SelPars}(\mathcal{Q}') = \mathbf{SelPars}(\mathcal{Q}).$$

By Lemma 7.3.18,

$$\begin{aligned}\mathbf{Params}(sh^{\mathcal{Q}'}(c)) &= \mathbf{Params}(sh^{\mathcal{Q}}(a)) \cup \mathbf{Params}(sh^{\mathcal{Q}}(b)) \\ &\subseteq \mathbf{Params}(\mathcal{Q})\end{aligned}$$

and

$$\mathbf{Free}(sh^{\mathcal{Q}'}(c)) = \mathbf{Free}(sh^{\mathcal{Q}}(a)) \cup \mathbf{Free}(sh^{\mathcal{Q}}(b)) \subseteq \mathbf{Free}(\mathcal{Q}).$$

Thus $\mathbf{Free}(\mathcal{Q}') = \mathbf{Free}(\mathcal{Q})$ and $\mathbf{Params}(\mathcal{Q}') = \mathbf{Params}(\mathcal{Q})$. Conditions (1), (2), (3) and (10) of Lemma 8.4.7 are all vacuous since $\mathbf{Free}(\mathcal{Q}') \setminus \mathbf{Free}(\mathcal{Q})$ and $\mathbf{Params}(\mathcal{Q}') \setminus \mathbf{Params}(\mathcal{Q})$ are both empty.

Condition (5) of Lemma 8.4.7 is trivial since $\mathcal{N}' \setminus \mathcal{N}$ contains only one element, namely c. Conditions (7), (8), (9), (11), (12) and (14) are vacuous since $\mathcal{Q}$ and $\mathcal{Q}'$ share the same successor function, selected parameter function, expansion arcs and decomposition arcs.

To check condition (4) of Lemma 8.4.7 we prove $c^* \notin \mathbf{Im}(f; \mathcal{Q})$. Assume $c^* \in \mathbf{Im}(f; \mathcal{Q})$. There is some $c^0 \in \mathcal{N}$ such that $f(c^0) = c^*$. Since $\langle c^*, a, b\rangle \in \mathbf{Conns}^-(\mathsf{f}; k)$ and f is a lifting map, we know

$$kind^{\mathcal{Q}'}(c^0) = kind^{\mathcal{Q}^*}(c^*) = k,$$

$$\langle c^*, f(a), f(b)\rangle \in conns^*$$

and $\langle a, b\rangle$ is not k-connected in $\mathcal{Q}$. Note that $f(c^0)$ is not a root node since $f(a) \stackrel{\mathcal{Q}^*}{\rightarrow} f(c^0)$. Thus $c^0 \neq r$ and c^0 is not a root node of $\mathcal{Q}$. Using Lemma 7.3.15, we know there is a connection $\langle c^0, a^0, b^0\rangle \in conns$. Since f is a lifting map,

$$\langle c^*, f(a^0), f(b^0)\rangle = \langle f(c^0), f(a^0), f(b^0)\rangle \in conns^*$$

(cf. Definition 8.4.3:8). Since $\mathcal{Q}^*$ is separated (cf. Definition 7.5.2:3) and $\langle c^*, f(a), f(b)\rangle \in conns^*$, we conclude $\langle c^0, a, b\rangle \in conns$, contradicting the assumption that $\langle a, b\rangle$ is not k-connected in $\mathcal{Q}$. Thus $c^* \notin \mathbf{Im}(f; \mathcal{Q})$ and condition (4) of Lemma 8.4.7 holds.

Condition (13) of Lemma 8.4.7 holds since

$$\langle f'(c), f'(a), f'(b)\rangle = \langle c^*, f(a), f(b)\rangle \in conns^*.$$

The only remaining condition of Lemma 8.4.7 to check is (6). To check this condition we must distinguish cases for k.

Suppose k is **mate**. Then c is a mate node of $\mathcal{Q}'$ with shallow formula $[\mathbf{B} \dot{=}^o \mathbf{A}]$ for some $\mathbf{A}, \mathbf{B} \in \mathit{wff}_o$. By $\clubsuit^p_m$ in $\mathcal{Q}'$, a has shallow formula $\mathbf{A}$ and b has shallow formula $\mathbf{B}$. Since f is a lifting map, $f(a)$ has shallow formula $(\varphi(\mathbf{A}))^\downarrow$ and $f(b)$ has shallow formula $(\varphi(\mathbf{B}))^\downarrow$. By $\clubsuit^p_m$ in $\mathcal{Q}^*$, c^* has shallow formula $[(\varphi(\mathbf{B}))^\downarrow \dot{=}^o (\varphi(\mathbf{A}))^\downarrow]$. Thus

$$label'(c) = \langle \mathbf{mate}, -1, [\mathbf{B} \dot{=}^o \mathbf{A}]\rangle$$

and

$$label^*(f'(c)) = \langle \mathbf{mate}, -1, (\varphi([\mathbf{B} \dot{=}^o \mathbf{A}]))^\downarrow\rangle$$

as desired.

Suppose k is **eunif**. Then c is an E-unification node with shallow formula

$$[[\mathbf{A} \dot{=}^\iota \mathbf{C}] \dot{\wedge} [\mathbf{B} \dot{=}^\iota \mathbf{D}]]$$

for some $\mathbf{A}, \mathbf{B}, \mathbf{C}, \mathbf{D} \in \mathit{wff}_\iota$. By $\clubsuit^p_{eu}$ in $\mathcal{Q}'$, a has shallow formula $[\mathbf{A} \dot{=}^\iota \mathbf{B}]$ and b has shallow formula $[\mathbf{C} \dot{=}^\iota \mathbf{D}]$. Since f is a lifting map,

$$sh^{\mathcal{Q}^*}(f(a)) = (\varphi([\mathbf{A} \dot{=}^\iota \mathbf{B}]))^\downarrow = [(\varphi(\mathbf{A}))^\downarrow \dot{=}^\iota (\varphi(\mathbf{B}))^\downarrow]$$

and

$$sh^{\mathcal{Q}^*}(f(b)) = (\varphi([\mathbf{C} \dot{=}^\iota \mathbf{D}]))^\downarrow = [(\varphi(\mathbf{C}))^\downarrow \dot{=}^\iota (\varphi(\mathbf{D}))^\downarrow].$$

By $\clubsuit^p_{eu}$ in $\mathcal{Q}^*$,

$$sh^{\mathcal{Q}^*}(c^*) = [[(\varphi(\mathbf{A}))^\downarrow \dot{=}^\iota (\varphi(\mathbf{C}))^\downarrow] \dot{\wedge} [(\varphi(\mathbf{B}))^\downarrow \dot{=}^\iota (\varphi(\mathbf{D}))^\downarrow]].$$

Thus

$$label'(c) = \langle \mathbf{eunif}, -1, [[\mathbf{A} \dot{=}^\iota \mathbf{C}] \dot{\wedge} [\mathbf{B} \dot{=}^\iota \mathbf{D}]]\rangle$$

and

$$label^*(f'(c)) = \langle \mathbf{eunif}, -1, (\varphi([[\mathbf{A} \dot{=}^\iota \mathbf{C}] \dot{\wedge} [\mathbf{B} \dot{=}^\iota \mathbf{D}]]))^\downarrow\rangle$$

as desired.

Suppose k is $\mathbf{eunif}^{sym}$. Then c is a symmetric E-unification node with shallow formula

$$[[\mathbf{A} \dot{=}^\iota \mathbf{D}] \dot{\wedge} [\mathbf{B} \dot{=}^\iota \mathbf{C}]]$$

for some $\mathbf{A}, \mathbf{B}, \mathbf{C}, \mathbf{D} \in \mathit{wff}_\iota$. By $\clubsuit^{sp}_{eu}$ in $\mathcal{Q}'$, a has shallow formula $[\mathbf{A} \dot{=}^\iota \mathbf{B}]$ and b has shallow formula $[\mathbf{C} \dot{=}^\iota \mathbf{D}]$. Since f is a lifting map,

$$sh^{\mathcal{Q}^*}(f(a)) = (\varphi([\mathbf{A} \dot{=}^\iota \mathbf{B}]))^\downarrow = [(\varphi(\mathbf{A}))^\downarrow \dot{=}^\iota (\varphi(\mathbf{B}))^\downarrow]$$

and

$$sh^{\mathcal{Q}^*}(f(b)) = (\varphi([\mathbf{C} \doteq^{\iota} \mathbf{D}]))^{\downarrow} = [(\varphi(\mathbf{C}))^{\downarrow} \doteq^{\iota} (\varphi(\mathbf{D}))^{\downarrow}].$$

By $\clubsuit^{sp}_{eu}$ in $\mathcal{Q}^*$,

$$sh^{\mathcal{Q}^*}(c^*) = [[(\varphi(\mathbf{A}))^{\downarrow} \doteq^{\iota} (\varphi(\mathbf{D}))^{\downarrow}] \dot{\wedge} [(\varphi(\mathbf{B}))^{\downarrow} \doteq^{\iota} (\varphi(\mathbf{C}))^{\downarrow}]].$$

Thus

$$label'(c) = \langle \mathbf{eunif}^{sym}, -1, [[\mathbf{A} \doteq^{\iota} \mathbf{D}] \dot{\wedge} [\mathbf{B} \doteq^{\iota} \mathbf{C}]] \rangle$$

and

$$label^*(f'(c)) = \langle \mathbf{eunif}^{sym}, -1, (\varphi([[\mathbf{A} \doteq^{\iota} \mathbf{D}] \dot{\wedge} [\mathbf{B} \doteq^{\iota} \mathbf{C}]]))^{\downarrow} \rangle$$

as desired.

Therefore, in any of these three cases condition (6) of Lemma 8.4.7 holds. Hence we can apply Lemma 8.4.7 to conclude f' is a lifting map from $\mathcal{Q}'$ to $\mathcal{Q}^*$.

We next check $\mathsf{f}' = \langle \varphi, f' \rangle$ is a search lifting map from Ξ' to $\langle \mathcal{Q}^*, r^* \rangle$. Since f is a search lifting map, we know $f(r) = r^*$, $\mathscr{D}$ is fully duplicated with respect to f and $\mathscr{S}$ is almost atomic with respect to f. Hence

$$f'(r) = f(r) = r^*.$$

Since $\mathcal{Q} \trianglelefteq \mathcal{Q}'$ and $f'|_{\mathcal{N}} = f$, we can apply Lemma 8.4.20 to conclude $\mathscr{D}$ is fully duplicated with respect to f' and $\mathscr{S}$ is almost atomic with respect to f'. Therefore, f' is a search lifting map from Ξ' to $\langle \mathcal{Q}^*, r^* \rangle$.

Finally, we know

$$\|\mathsf{f}' : \mathcal{Q}' \to \mathcal{Q}^*\| < \|\mathsf{f} : \mathcal{Q} \to \mathcal{Q}^*\|$$

by Lemma 8.4.5 since $c \in \mathcal{N}' \setminus \mathcal{N}$. □

PROOF OF LEMMA 8.6.19. To ease notation, assume $\mathcal{Q}$ is the extensional expansion dag

$$\langle \mathcal{N}, succ, label, sel, exps, conns, decs \rangle.$$

Let $\mathcal{Q}'$ be $[\mathbf{G}/x]\mathcal{Q}$. By Definition 8.2.2, $\mathcal{Q}'$ is

$$\langle \mathcal{N}, succ, label', sel, exps', conns, decs \rangle$$

where

$$label'(a) := \langle kind^{\mathcal{Q}}(a), pol^{\mathcal{Q}}(a), ([\mathbf{G}/x](sh^{\mathcal{Q}}(a)))^{\downarrow} \rangle$$

for every $a \in \mathcal{N}$ and

$$exps' := \{\langle e, ([\mathbf{G}/x]\mathbf{C}), c\rangle \mid \langle e, \mathbf{C}, c\rangle \in exps.$$

For each $\langle e, \mathbf{C}, c\rangle \in exps$, $\mathbf{C}$ is strict (since $\mathcal{Q}$ has strict expansions) and so $[\mathbf{G}/x]\mathbf{C}$ is strict by Lemma 8.5.5. Hence $\mathcal{Q}'$ has strict expansions.

We must verify

$$\Xi' = \langle \mathcal{Q}', r, \mathscr{D}, (\mathscr{S} \setminus \{x\})\rangle$$

is a search state. First we establish $\mathscr{D} \subseteq \mathbf{ExpNodes}^{\mathcal{Q}'}(\mathcal{N})$ and

$$(\mathscr{S} \setminus \{x\}) \subseteq \mathbf{SetVars}^{\mathcal{Q}'}(\mathcal{N}).$$

Suppose $e \in \mathscr{D} \subseteq \mathbf{ExpNodes}^{\mathcal{Q}}(\mathcal{N})$. Then e is a positive primary node in $\mathcal{Q}$ with shallow formula of the form $[\dot{\forall} y_\beta \, \mathbf{M}]$. Hence e is a positive primary node in $\mathcal{Q}'$ with shallow formula $[\dot{\forall} y_\beta \, ([\mathbf{G}/x]\mathbf{M})]$ and so e is an expansion node in $\mathcal{Q}'$. Thus $\mathscr{D} \subseteq \mathbf{ExpNodes}^{\mathcal{Q}'}(\mathcal{N})$.

Suppose $p \in (\mathscr{S} \setminus \{x\}) \subseteq \mathbf{SetVars}^{\mathcal{Q}}(\mathcal{N})$. Then there is a flexible node a in $\mathcal{Q}$ with shallow formula of the form $[p\,\overline{\mathbf{B}^n}]$. Hence the shallow formula of a in $\mathcal{Q}'$ has shallow formula $[\mathbf{G}/x]([p\,\overline{\mathbf{B}^n}])$. Since $p \neq x$, a is a p-node in $\mathcal{Q}'$ and so $p \in \mathbf{SetVars}^{\mathcal{Q}'}(\mathcal{N})$. Thus $(\mathscr{S} \setminus \{x\}) \subseteq \mathbf{SetVars}^{\mathcal{Q}'}(\mathcal{N})$, as desired.

By Lemma 8.2.6, $\mathcal{Q}'$ is an extensional expansion dag and

$$\mathbf{Roots}(\mathcal{Q}') = \mathbf{Roots}(\mathcal{Q}) = \{r\}.$$

Since $sh^{\mathcal{Q}}(r)$ is closed, $sh^{\mathcal{Q}'}(r) = sh^{\mathcal{Q}}(r)$ is closed and

$$\mathbf{Params}(sh^{\mathcal{Q}'}(r)) = \mathbf{Params}(sh^{\mathcal{Q}}(r)).$$

We know $\mathbf{SelPars}(\mathcal{Q}') = \mathbf{SelPars}(\mathcal{Q})$ since $\mathcal{Q}$ and $\mathcal{Q}'$ share the same selected parameter function. Thus

$$\begin{aligned} \mathbf{RootSels}(\mathcal{Q}') &= \mathbf{SelPars}^{\mathcal{Q}'}(\{r\}) \\ &= \mathbf{Params}(sh^{\mathcal{Q}'}(r)) \cap \mathbf{SelPars}(\mathcal{Q}') \\ &= \mathbf{Params}(sh^{\mathcal{Q}}(r)) \cap \mathbf{SelPars}(\mathcal{Q}) \\ &= \mathbf{RootSels}(\mathcal{Q}) = \emptyset. \end{aligned}$$

If R is a parameter, let B be R. Otherwise, let $B \in \mathcal{P} \setminus \mathbf{Params}(\mathcal{Q})$ be arbitrary. In either case $\mathbf{Params}(\mathbf{G}) \subseteq \{B\}$ where B is not banned for x in $\mathcal{Q}$. Thus we know $\prec_0^{\mathcal{Q}'}$ is acyclic by Lemma 8.2.12 since $\prec_0^{\mathcal{Q}}$ is acyclic. Therefore, Ξ' is a search state.

Suppose $\langle \mathcal{Q}^*, r^* \rangle$ is a lifting target, $\mathsf{f} = \langle \varphi, f \rangle$ is a search lifting map from Ξ to $\langle \mathcal{Q}^*, r^* \rangle$, θ is a ground substitution, $\mathbf{Dom}(\theta)$ is $\{w^1, \ldots, w^m\}$, $\mathbf{Dom}(\varphi) \cap \{w^1, \ldots, w^m\}$ is $\emptyset$ and $\theta(\varphi(\mathbf{G})) \stackrel{\beta\eta}{=} \varphi(x_\alpha)$. To ease notation, assume $\mathcal{Q}^*$ is the extensional expansion dag

$$\langle \mathcal{N}^*, succ^*, label^*, sel^*, exps^*, conns^*, decs^* \rangle.$$

We construct a lifting substitution φ' for $\mathcal{Q}'$ such that $\langle \varphi', f \rangle$ is a search lifting map from Ξ' to $\langle \mathcal{Q}^*, r^* \rangle$. Define φ' as follows:

$$\varphi'(u) := \begin{cases} \varphi(u) & \text{if } u \in (\mathbf{Free}(\mathcal{Q}) \cup \mathbf{Params}(\mathcal{Q})) \\ (\theta(u))^{\downarrow} & \text{if } u \in \{w^1, \ldots, w^m\}. \end{cases}$$

This is well-defined since we have assumed $\{w^1, \ldots, w^m\} \cap \mathbf{Free}(\mathcal{Q})$ is empty and so $\{w^1, \ldots, w^m\} \cap (\mathbf{Free}(\mathcal{Q}) \cup \mathbf{Params}(\mathcal{Q}))$ is empty. By Lemma 8.2.3, we know

$$\mathbf{Free}(\mathcal{Q}') \subseteq (\mathbf{Free}(\mathcal{Q}) \setminus \{x\}) \cup \{w^1, \ldots, w^n\} \tag{17}$$

and

$$\mathbf{Params}(\mathcal{Q}') \subseteq \mathbf{Params}(\mathcal{Q}) \cup \mathbf{Params}(\mathbf{G}) \subseteq \mathbf{Params}(\mathcal{Q}) \tag{18}$$

since we have assumed $R \in \mathbf{Params}(\mathcal{Q})$ if R is a parameter.

Let f' be $\langle \varphi', f \rangle$. We must prove f' is a search lifting map from Ξ' to $\langle \mathcal{Q}^*, r^* \rangle$. We cannot apply Lemma 8.4.7 in this case since we do not have $\mathcal{Q} \trianglelefteq \mathcal{Q}'$. Instead we directly check φ' is a lifting substitution for $\mathcal{Q}^*$, f' is a lifting map from $\mathcal{Q}'$ to $\mathcal{Q}^*$ and finally that f' is a search lifting map from Ξ' to $\langle \mathcal{Q}^*, r^* \rangle$.

Note that

$$\mathbf{Dom}(\varphi') \subseteq (\mathbf{Free}(\mathcal{Q}) \cup \mathbf{Params}(\mathcal{Q}) \cup \mathbf{Free}(\mathbf{G})) \subset \mathcal{V} \cup \mathcal{P}$$

by definition. For any $D \in \mathbf{Params}(\mathcal{Q}')$, $\varphi'(D)$ is the parameter $\varphi(D)$ (since φ is a lifting substitution). Hence φ' respects parameters. For any

$$C, D \in \mathbf{Params}(\mathcal{Q}') \subseteq \mathbf{Params}(\mathcal{Q}),$$

if $\varphi'(C) = \varphi'(D)$, then $\varphi(C) = \varphi(D)$ and so $C = D$ (since φ is injective on parameters of $\mathcal{Q}$). For any $v \in \mathbf{Free}(\mathcal{Q}')$, either $v \in \mathbf{Free}(\mathcal{Q})$ or $v \in \mathbf{Free}(\mathbf{G})$. If $v \in \mathbf{Free}(\mathcal{Q})$, then $\varphi'(v)$ is the closed long $\beta\eta$-form

$\varphi(v)$. If $v \in \mathbf{Free}(\mathbf{G})$, then $\varphi'(v)$ is the closed long $\beta\eta$-form $(\theta(v))^{\updownarrow}$. Thus φ' is a lifting substitution for $\mathcal{Q}'$.

We know $f : \mathcal{N} \to \mathcal{N}^*$ is injective since f is a lifting map. We next check the enumerated conditions of Definition 8.4.3 to prove f' is a lifting map from $\mathcal{Q}'$ to $\mathcal{Q}^*$. Conditions (2), (3), (8) and (9) follow easily from the fact that f is a lifting map from $\mathcal{Q}$ to $\mathcal{Q}^*$ and the fact that $\mathcal{Q}$ and $\mathcal{Q}'$ share the same successor function, connections and decomposition arcs.

Suppose $a \in \mathbf{Dom}(sel)$. Since f is a lifting map from $\mathcal{Q}$ to $\mathcal{Q}^*$ and $sel(a) \in \mathbf{Params}(\mathcal{Q})$, we know

$$\varphi'(sel(a)) = \varphi(sel(a)) = sel^*(f(a)).$$

Thus condition (4) of Definition 8.4.3 holds for f'.

Suppose $B \in \mathbf{Params}(\mathcal{Q}')$, $b^* \in \mathbf{Dom}(sel^*)$ and $\varphi'(B) = sel^*(b^*)$. Then $B \in \mathbf{Params}(\mathcal{Q})$ and so $\varphi(B) = \varphi'(B) = sel^*(b^*)$. Since f is a lifting map from $\mathcal{Q}$ to $\mathcal{Q}^*$, there is a node $b \in \mathbf{Dom}(sel)$ such that $f(b) = b^*$ and $sel(b) = B$ (cf. Definition 8.4.3:5). Thus condition (5) of Definition 8.4.3 holds for f'.

Before checking conditions (6), (7) and (1) of Definition 8.4.3, we relate φ' to φ. If R is a parameter, then $R \in \mathbf{Params}(\mathcal{Q})$ (by assumption) and so $\varphi'(R) = \varphi(R)$. If R is a parameter, let R^* be $\varphi(R)$. Otherwise, let R^* be R. We compute

$$\begin{aligned}
\varphi'(\mathbf{G}) &= \varphi'([\lambda \overline{x^n}\,[R\,[w^1\,\overline{x^n}]\ \cdots\ [w^m\,\overline{x^n}]]]) \\
&= [\lambda \overline{x^n}\,[R^*\,[\theta(w^1)\,\overline{x^n}]\ \cdots\ [\theta(w^m)\,\overline{x^n}]]] && \text{by choice of } R^* \\
&= \theta([\lambda \overline{x^n}\,[R^*\,[w^1\,\overline{x^n}]\ \cdots\ [w^m\,\overline{x^n}]]]) \\
&= \theta(\varphi|_{\mathcal{P}}([\lambda \overline{x^n}\,[R\,[w^1\,\overline{x^n}]\ \cdots\ [w^m\,\overline{x^n}]]])) && \text{by choice of } R^* \text{ and Definition 8.4.1} \\
&= \theta(\varphi|_{\mathcal{P}}(\mathbf{G})) \stackrel{\beta\eta}{=} \varphi(x) && \text{by assumption.}
\end{aligned}$$

Thus we have

$$\varphi(x) \stackrel{\beta\eta}{=} [\lambda \overline{x^n}\,[R^*\,[\theta(w^1)\,\overline{x^n}]\ \cdots\ [\theta(w^m)\,\overline{x^n}]]] \tag{19}$$

and

$$\varphi'(\mathbf{G}) \stackrel{\beta\eta}{=} \varphi(x). \tag{20}$$

Using (20), we can prove

$$(21) \qquad \varphi'([\mathbf{G}/x]\mathbf{C}) \stackrel{\beta\eta}{=} \varphi(\mathbf{C})$$

for any term $\mathbf{C}$ such that

$$\mathbf{Params}(\mathbf{C}) \subseteq \mathbf{Params}(\mathcal{Q})$$

and

$$\mathbf{Free}(\mathbf{C}) \subseteq \mathbf{Free}(\mathcal{Q})$$

by a simple induction on the term $\mathbf{C}$. Another induction on propositions allows us to conclude

$$(22) \qquad \varphi'([\mathbf{G}/x]\mathbf{M}) \stackrel{\beta\eta}{=} \varphi(\mathbf{M})$$

whenever $\mathbf{Params}(\mathbf{M}) \subseteq \mathbf{Params}(\mathcal{Q})$ and $\mathbf{Free}(\mathbf{M}) \subseteq \mathbf{Free}(\mathcal{Q})$.

Suppose $\langle e, \mathbf{C}', c\rangle \in exps'$. Then there is some term $\mathbf{C}$ such that

$$\langle e, \mathbf{C}, c\rangle \in exps$$

and $\mathbf{C}'$ is $[\mathbf{G}/x]\mathbf{C}$. Since f is a lifting map, $\langle f(e), \mathbf{C}^*, f(c)\rangle$ is in $exps^*$ for some term $\mathbf{C}^*$ such that $\mathbf{C}^* \stackrel{\beta\eta}{=} \varphi(\mathbf{C})$. By (21),

$$\mathbf{C}^* \stackrel{\beta\eta}{=} \varphi'([\mathbf{G}/x]\mathbf{C}) = \varphi'(\mathbf{C}').$$

Thus condition (6) of Definition 8.4.3 holds for f'. Suppose $A \in \mathbf{Params}(\mathbf{C}')$. Either $A \in \mathbf{Params}(\mathbf{C})$ or both $x \in \mathbf{Free}(\mathbf{C})$ and $A \in \mathbf{Params}(\mathbf{G})$. If $A \in \mathbf{Params}(\mathbf{C})$, then $\varphi(A) \in \mathbf{Params}(\mathbf{C}^*)$ by condition (7) of Definition 8.4.3 for f. Otherwise, $x \in \mathbf{Free}(\mathbf{C})$ and $A \in \mathbf{Params}(\mathbf{G})$. This is only possible if R is a parameter and $A = R$. Since $\mathcal{Q}$ has strict expansions, $\mathbf{C}$ is strict and so x has a strict occurrence in $\mathbf{C}$. By Lemma 8.5.6, $\varphi(R) \in \mathbf{Params}(([\mathbf{G}/x]\mathbf{C})^{\downarrow})$. Since $\mathbf{C}^* \stackrel{\beta\eta}{=} [\mathbf{G}/x]\mathbf{C}$ and $\beta\eta$-reductions do not introduce parameters, we must have $\varphi(R) \in \mathbf{Params}(\mathbf{C}^*)$. Thus condition (7) of Definition 8.4.3 holds for f'.

Suppose $a \in \mathcal{N}$ and $label'(a) = \langle k, p, \mathbf{M}'\rangle$. Let $\mathbf{M}$ be the shallow formula of a in $\mathcal{Q}$. Then $\mathbf{M}'$ is $([\mathbf{G}/x]\mathbf{M})^{\downarrow}$. Since f is a lifting map from $\mathcal{Q}$ to $\mathcal{Q}^*$,

$$label^*(f(a)) = \langle k, p, (\varphi(\mathbf{M}))^{\downarrow}\rangle.$$

By (22),

$$(\varphi(\mathbf{M}))^{\downarrow} = (\varphi'([\mathbf{G}/x]\mathbf{M}))^{\downarrow} = (\varphi'(\mathbf{M}'))^{\downarrow}.$$

Hence

$$label^*(f(a)) = \langle k, p, (\varphi'(\mathbf{M}'))^{\downarrow}\rangle.$$

Thus condition (1) of Definition 8.4.3 holds for f'.

Therefore, f' is a lifting map from $\mathcal{Q}'$ to $\mathcal{Q}^*$. Note that $f(r) = r^*$ since f is a search lifting map from Ξ to $\langle \mathcal{Q}^*, r^* \rangle$. By Lemma 8.4.21, $\mathscr{D}$ is fully duplicated with respect to f' and $(\mathscr{S} \setminus \{x\})$ is almost atomic with respect to f' since f is a search lifting map from Ξ to $\langle \mathcal{Q}^*, r^* \rangle$. Thus f' is a search lifting map from Ξ' to $\langle \mathcal{Q}^*, r^* \rangle$.

Finally, we check the measure of f'. By (17) and (18) above,

$$\begin{aligned} &(\mathbf{Free}(\mathcal{Q}') \cup \mathbf{Params}(\mathcal{Q}')) \\ &\subseteq ((\mathbf{Free}(\mathcal{Q}) \cup \mathbf{Params}(\mathcal{Q})) \setminus \{x\}) \cup \{w^1, \ldots, w^m\}. \end{aligned}$$

By (19) above,

$$(\varphi(x))^{\updownarrow} = [\lambda \overline{x^n}\, [R^*\, \overline{\mathbf{B}^m}]]$$

where $R^* \in \mathcal{P} \cup \mathcal{S} \cup \{x^1, \ldots, x^n\}$ and $\mathbf{B}^j$ is $[\theta(w^j)\, \overline{x^n}]^{\updownarrow}$ for each j $(1 \leq j \leq m)$. By Lemma 2.1.32,

$$\left(\sum_{j=1}^{m} |[\lambda \overline{x^n}\, \mathbf{B}^j]|_l \right) < |(\varphi(x))|_l.$$

Applying Lemma 2.1.36 n times,

$$|(\theta(w^j))^{\updownarrow}|_l = |[(\theta(w^j))\, \overline{x^n}]^{\updownarrow}|_l.$$

for each j $(1 \leq j \leq m)$. Hence

$$\left(\sum_{j=1}^{m} |(\theta(w^j))^{\updownarrow}|_l \right) < |(\varphi(x))|_l.$$

Thus

$$|\varphi'|_l^{\mathcal{Q}'} = \sum_{v \in \mathbf{Free}(\mathcal{Q}')} |\varphi'(v)|_l$$

$$\leq \left(\sum_{v \in (\mathbf{Free}(\mathcal{Q}) \setminus \{x\})} |\varphi(v)|_l \right) + \left(\sum_{j=1}^{m} |(\theta(w^j))^{\updownarrow}|_l \right)$$

$$< \left(\sum_{v \in (\mathbf{Free}(\mathcal{Q}) \setminus \{x\})} |\varphi(v)|_l \right) + |\varphi(x)|_l$$

$$= \sum_{v \in \mathbf{Free}(\mathcal{Q})} |\varphi(v)|_l$$

$$= |\varphi|_l^{\mathcal{Q}}$$

Note also that

$$(\mathcal{N}^{\mathcal{Q}'} \setminus \mathbf{Dom}(succ^{\mathcal{Q}'})) = (\mathcal{N}^{\mathcal{Q}} \setminus \mathbf{Dom}(succ^{\mathcal{Q}}))$$

since $\mathcal{N}^{\mathcal{Q}'} = \mathcal{N}^{\mathcal{Q}}$ and $succ^{\mathcal{Q}'} = succ^{\mathcal{Q}}$. Therefore,

$$\|\mathsf{f}' : \mathcal{Q}' \to \mathcal{Q}^*\|$$

$$= \omega^2 \cdot |\mathcal{N}^{\mathcal{Q}^*} \setminus \mathbf{Im}(f; \mathcal{Q})| + \omega \cdot |\mathcal{N}^{\mathcal{Q}'} \setminus \mathbf{Dom}(succ^{\mathcal{Q}'})| + |\varphi'|_l^{\mathcal{Q}'}$$

$$< \omega^2 \cdot |\mathcal{N}^{\mathcal{Q}^*} \setminus \mathbf{Im}(f; \mathcal{Q})| + \omega \cdot |\mathcal{N}^{\mathcal{Q}} \setminus \mathbf{Dom}(succ^{\mathcal{Q}})| + |\varphi|_l^{\mathcal{Q}}$$

$$= \|\mathsf{f} : \mathcal{Q} \to \mathcal{Q}^*\|.$$

□

PROOF OF LEMMA 8.8.3. We know $\mathfrak{p}^*$ is not a u-part of $\mathcal{Q}^*$ and consider each case in Definition 7.4.4 which may fail for $\mathfrak{p}^*$. To simplify the notation, assume the extensional expansion dag $\mathcal{Q}$ is

$$\langle \mathcal{N}, succ, label, sel, exps, conns, decs \rangle$$

and the extensional expansion dag $\mathcal{Q}^*$ is

$$\langle \mathcal{N}^*, succ^*, label^*, sel^*, exps^*, conns^*, decs^* \rangle.$$

Suppose condition (1) of Definition 7.4.4 fails for $\mathfrak{p}^*$. That is, there is a conjunctive node $a^* \in \mathfrak{p}^*$ and a node $b^* \in \mathcal{N}^*$ such that $a^* \stackrel{\mathcal{Q}^*}{\rightarrow} b^*$ and $b^* \notin \mathfrak{p}^*$. Since a^* is not a leaf, we know a^* is not a true node by $\clubsuit_{ns}$ and Lemma 7.3.13. Thus a^* is a conjunction node, an expansion node or a negative equation node. Since $a^* \in \mathfrak{p}^*$ and $\langle \varphi, f \rangle$ is a lifting map, there is a node $a \in \mathfrak{p}$ such that $f(a) = a^*$, $kind^{\mathcal{Q}}(a) = kind^{\mathcal{Q}^*}(a)$, $pol^{\mathcal{Q}}(a) = pol^{\mathcal{Q}^*}(a)$, and

$$(\varphi(sh^{\mathcal{Q}}(a)))^{\downarrow} = sh^{\mathcal{Q}^*}(a^*).$$

Suppose a^* is a negative equation node. We use a as a witness to verify case (4) of the lemma holds. That is, we determine a is an extended negative equation node such that a is not decomposable in $\mathcal{Q}$ and a^* is decomposed in $\mathcal{Q}^*$. As a negative equation node, a^* is a primary node with shallow formula $[\mathbf{C}^* \stackrel{.}{=}^{\iota} \mathbf{D}^*]$. Hence a is a negative primary node with shallow formula $[\mathbf{C} \stackrel{.}{=}^{\iota} \mathbf{D}]$ such that $(\varphi(\mathbf{C}))^{\downarrow} = \mathbf{C}^*$ and $(\varphi(\mathbf{D}))^{\downarrow} = \mathbf{D}^*$. Thus a is a negative equation node in $\mathcal{Q}$, as desired. Since $\mathcal{Q}$ is developed, a is extended and so a is conjunctive. By $\clubsuit_{\stackrel{.}{=}}^{-}$, there is an equation goal node $a_1 \in \mathcal{N}$ such that $succ(a) = \langle a_1 \rangle$ and shallow formula of a_1 is $[\mathbf{C} \stackrel{.}{=}^{\iota} \mathbf{D}]$. Since $\mathfrak{p}$ is a u-part of $\mathcal{Q}$ and $a \in \mathfrak{p}$, we must have $a_1 \in \mathfrak{p}$. Since f is a lifting map, we know $succ^*(a^*) = \langle f(a_1) \rangle$ and $f(a_1)$ is an equation goal node. Thus $f(a_1) \in \mathfrak{p}^*$ and hence $b^* \neq f(a_1)$. By Lemma 7.3.12, b^* must be a decomposition node with shallow formula $[\mathbf{C}^* \stackrel{.}{=}^{\iota} \mathbf{D}^*]$ and $\langle a^*, b^* \rangle \in decs^*$. In particular, a^* is decomposed in $\mathcal{Q}^*$, as desired. Finally, we must verify a is not decomposable in $\mathcal{Q}$. Assume a is decomposable in $\mathcal{Q}$. Since $\mathcal{Q}$ is developed, a is decomposed in $\mathcal{Q}$. That is, there is a node $b \in \mathcal{N}$ such that $\langle a, b \rangle \in decs$. Since a is conjunctive and $\mathfrak{p}$ is a u-part of $\mathcal{Q}$, $b \in \mathfrak{p}$. Since f is a lifting map, $\langle f(a), f(b) \rangle \in decs^*$ (cf. Definition 8.4.3:9). Since $\mathcal{Q}^*$ has merged decompositions and $f(a) = a^*$, we know $b^* = f(b) \in \mathfrak{p}^*$, contradicting our assumption that $b^* \notin \mathfrak{p}^*$. Therefore, a is not decomposable in $\mathcal{Q}$ and (4) holds, as desired.

Suppose a^* is an expansion node. Then a^* is a positive primary node with shallow formula $[\forall x_\alpha \mathbf{M}^*]$ for some x_α and $\mathbf{M}^*$. Hence a is a positive primary node in $\mathcal{Q}$ with shallow formula $[\forall x_\alpha \mathbf{M}]$ where

$(\varphi(\mathbf{M}))^{\downarrow} = \mathbf{M}^*$. Thus a is an expansion node in $\mathcal{Q}$. Since $\mathcal{Q}$ is developed, a is extended and so a is conjunctive. Since $a^* \overset{\mathcal{Q}^*}{\rightharpoondown} b^*$, $\langle a^*, \mathbf{B}^*, b^* \rangle \in exps^*$ for some term $\mathbf{B}^*_\alpha$. by Lemma 7.3.11. Since $a \in \mathfrak{p}$ and $\mathfrak{p}$ is fully duplicated with respect to f, there is some $b \in \mathcal{N}$ such that $a \overset{\mathcal{Q}}{\rightharpoondown} b$ and $f(b) = b^*$ (cf. Definitions 8.4.19 and 8.4.4). Since $\mathfrak{p}$ is a u-part of $\mathcal{Q}$, $a \in \mathfrak{p}$ and a is conjunctive, we have $b \in \mathfrak{p}$. Hence $b^* = f(b) \in \mathfrak{p}^*$, contradicting our assumption that $b^* \notin \mathfrak{p}^*$.

Otherwise, a^* is a conjunction node. We prove this case is contradictory. By Lemma 7.3.13, there exist nodes $a_1^*, \ldots, a_n^* \in \mathcal{N}^*$ such that

$$succ^*(a^*) = \langle a_1^*, \ldots, a_n^* \rangle$$

and $b^* \in \{a_1^*, \ldots, a_n^*\}$. As a conjunction node, a^* is a negative primary node with shallow formula $[\mathbf{M}^* \dot{\vee} \mathbf{N}^*]$. Hence a is a negative primary node with shallow formula $[\mathbf{M} \dot{\vee} \mathbf{N}]$ for some $\mathbf{M}$ and $\mathbf{N}$ such that $(\varphi(\mathbf{M}))^{\downarrow}$ is $\mathbf{M}^*$ and $(\varphi(\mathbf{N}))^{\downarrow}$ is $\mathbf{N}^*$. Thus a is a conjunction node in $\mathcal{Q}$. Since $\mathcal{Q}$ is developed, a is extended in $\mathcal{Q}$. By $\clubsuit_{\vee}$, there exist nodes $a_1, a_2 \in \mathcal{N}$ such that $succ(a) = \langle a_1, a_2 \rangle$. Since $\langle \varphi, f \rangle$ is a lifting map, $succ^*(a^*) = \langle f(a_1), f(a_2) \rangle$ (cf. Definition 8.4.3:3). That is, $n = 2$ and $b^* \in \{f(a_1), f(a_2)\}$. Since $\mathfrak{p}$ is a u-part of $\mathcal{Q}$ and $a \in \mathfrak{p}$, we know $a_1, a_2 \in \mathfrak{p}$. Hence $b^* \in \{f(a_1), f(a_2)\} \subseteq \mathfrak{p}^*$, contradicting our assumption that $b^* \notin \mathfrak{p}^*$.

Suppose condition (4) of Definition 7.4.4 fails for $\mathfrak{p}^*$. That is, there is a connection $\langle c^*, a^*, b^* \rangle \in conns^*$ such that $a^*, b^* \in \mathfrak{p}^*$ and $c^* \notin \mathfrak{p}^*$. Since $a^*, b^* \in \mathfrak{p}^*$, there exist nodes $a, b \in \mathfrak{p}$ such that $f(a) = a^*$ and $f(b) = b^*$. By $\clubsuit_c$, c^* is a connection node in $\mathcal{Q}^*$. That is,

$$kind^{\mathcal{Q}^*}(c^*) \in \{\mathbf{mate}, \mathbf{eunif}, \mathbf{eunif}^{sym}\}.$$

Let k be $kind^{\mathcal{Q}^*}(c^*)$.

We begin by proving $\langle c^*, a, b \rangle \in \mathbf{Conns}^-(\mathsf{f}; k)$. Assume not. Since we already know $\langle c^*, f(a), f(b) \rangle \in conns^*$, the only possibility is that the pair $\langle a, b \rangle$ is k-connected in $\mathcal{Q}$ (cf. Definition 8.4.11). That is, there is some $c \in \mathcal{N}$ such that $\langle c, a, b \rangle \in conns$. Since f is a lifting map, $\langle f(c), f(a), f(b) \rangle \in conns$ (cf. Definition 8.4.3:8). Hence $f(c) = c^*$ since $\mathcal{Q}^*$ has merged connections (cf. Definitions 8.6.6 and 7.5.3:2). Since $a, b \in \mathfrak{p}$ and $\mathfrak{p}$ is a u-part, we must have $c \in \mathfrak{p}$.

Thus $c^* \in \mathfrak{p}^*$, contradicting our assumption about c^*. Therefore, $\langle c^*, a, b\rangle \in \mathbf{Conns}^-(\mathsf{f}; k)$.

Suppose $k \in \{\mathbf{eunif}, \mathbf{eunif}^{sym}\}$. We establish that case (3) of the lemma holds. That is, we determine a is a positive equation node and b is an equation goal node. By $\clubsuit^p_{eu}$ or $\clubsuit^{sp}_{eu}$, a^* is a positive equation node and b^* is an equation goal node. Since $kind^{\mathcal{Q}}(b) = kind^{\mathcal{Q}^*}(b^*)$, b is an equation goal node in $\mathcal{Q}$. Similarly, a is a positive primary node. The shallow formula of a^* (in $\mathcal{Q}^*$) is of the form $[\mathbf{A}^* \doteq^\iota \mathbf{B}^*]$ and so the shallow formula of a (in $\mathcal{Q}$) is of the form $[\mathbf{A} \doteq^\iota \mathbf{B}]$ where $(\varphi(\mathbf{A}))^\downarrow = \mathbf{A}^*$ and $(\varphi(\mathbf{B}))^\downarrow = \mathbf{B}^*$. Hence a is a positive equation node. Since $\mathcal{Q}$ is developed, a and b are extended in $\mathcal{Q}$ and so case (3) of the lemma holds.

Suppose k is **mate**. We establish that case (1) or case (2) of the lemma holds. By $\clubsuit^p_m$, a^* is a positive atomic node and b^* is a negative atomic node. Since f is a lifting map, a is a positive primary node and b is a negative primary node. By Lemma 8.4.8, b is either a flexible node or an atomic node in $\mathcal{Q}$. Since $b \in \mathfrak{p}$ and $\mathfrak{p}$ is a pre-solved u-part, b cannot be a negative flexible node. Thus b is a negative atomic node which is extended in $\mathcal{Q}$ since $\mathcal{Q}$ is developed. By Lemma 8.4.8, a is either a flexible node or an atomic node in $\mathcal{Q}$. If a is an atomic node in $\mathcal{Q}$, then a is an extended positive atomic node (since $\mathcal{Q}$ is developed), b is an extended negative atomic node and $\langle c^*, a, b\rangle \in \mathbf{Conns}^-(\mathsf{f}; \mathbf{mate})$ and so case (1) of the lemma holds. If a is a flexible node in $\mathcal{Q}$, then a is a positive flexible node, b is an extended negative atomic node and $\langle c^*, a, b\rangle \in \mathbf{Conns}^-(\mathsf{f}; \mathbf{mate})$ and so case (2) of the lemma holds. □

PROOF OF LEMMA 8.8.4. Let $\mathfrak{p} \in \mathcal{U}parts^{-ps}(\mathcal{Q}; \{r\})$ be given.

Suppose $(\mathfrak{p} \setminus \mathcal{D})$ is not fully duplicated with respect to f. Then there is some expansion node $e \in \mathfrak{p} \setminus \mathcal{D}$ and some $c^* \in \mathcal{N}^{\mathcal{Q}^*}$ such that $f(e) \stackrel{\mathcal{Q}^*}{\rightarrow} c^*$ and $c^* \notin \mathbf{Im}(f; \mathcal{Q})$. Hence $\langle e, c^*\rangle \in \mathbf{Edges}^-(\mathsf{f})$ is a schema for an edge with respect to f (cf. Definition 8.4.10). Note that e is extended since $\mathcal{Q}$ is developed. Let $\mathcal{O}$ be the duplication option $\langle \mathbf{dup}, e\rangle$ (cf. Definition 8.7.5). Let Ξ' be $\mathbf{Search}(\Xi, \mathfrak{p}, \mathcal{O})$ (cf. Definition 8.7.6). By Lemma 8.7.7, Ξ' is a search state and there is a search lifting map f' from Ξ' to $\langle \mathcal{Q}^*, r^*\rangle$ such that $\|\mathsf{f}' : \mathcal{Q}^{\Xi'} \rightarrow \mathcal{Q}^*\| < \|\mathsf{f} : \mathcal{Q} \rightarrow \mathcal{Q}^*\|$, as desired.

For the remainder of the proof we assume $(\mathfrak{p} \setminus \mathscr{D})$ is fully duplicated with respect to f. Since f is a search lifting map we know $\mathscr{D}$ is fully duplicated with respect to f. Thus $\mathfrak{p}$ is fully duplicated with respect to f.

Suppose $(\mathbf{SetVars}^{\mathcal{Q}}(\mathfrak{p}) \setminus \mathscr{S})$ is not almost atomic with respect to f. Then there is some $p_{o\alpha^n\cdots\alpha^1} \in (\mathbf{SetVars}^{\mathcal{Q}}(\mathfrak{p}) \setminus \mathscr{S})$ which is not almost atomic with respect to f and there is a flexible p-node $a \in \mathfrak{p}$ (cf. Definition 8.4.14). Since $\varphi(p)$ is a closed long $\beta\eta$-form, it must be of the form $[\lambda\overline{x^n}\,[h\,\overline{\mathbf{C}^m}]]$ for some $h \in \mathcal{P} \cup \mathcal{S} \cup \{x^1, \ldots, x^n\}$. We know $h \notin \mathcal{P}$ since p is not almost atomic. Hence $h \in \mathcal{S} \cup \{x^1, \ldots, x^n\}$.

First assume $h \in \mathcal{S}$. Since φ is a lifting substitution, $\varphi(p)$ is a closed. Hence $[\lambda\overline{x^n}\,\mathbf{C}^j]$ is closed for each j $(1 \leq j \leq m)$. Let $\mathbf{P}$ be the imitation term for h at type $(o\alpha^n \cdots \alpha^1)$ avoiding $\mathbf{Free}(\mathcal{Q})$ (cf. Definition 8.7.2). Hence $\mathbf{P}$ is of the form

$$[\lambda x^1_{\alpha^1} \cdots \lambda x^n_{\alpha^n} \centerdot h\,[w^1\,\overline{x^n}] \cdots [w^m\,\overline{x^n}]]$$

where $w^1_{\beta^1\alpha^n\cdots\alpha^1}, \ldots, w^m_{\beta^m\alpha^n\cdots\alpha^1}$ are distinct variables such that

$$\{w^1, \ldots, w^m\} \cap \mathbf{Free}(\mathcal{Q}) = \emptyset.$$

Define a ground substitution θ on $\{w^1, \ldots, w^m\}$ by $\theta(w^j) := [\lambda\overline{x^n}\,\mathbf{C}^j]$ for each j $(1 \leq j \leq m)$. Note that $\mathbf{Dom}(\theta) = \mathbf{Free}(\mathbf{P})$ and

$$\begin{aligned} \theta(\mathbf{P}) &= [\lambda x^1_{\alpha^1} \cdots \lambda x^n_{\alpha^n} \centerdot h\,[\theta(w^1)\,\overline{x^n}] \cdots [\theta(w^m)\,\overline{x^n}]] \\ &\stackrel{\beta\eta}{=} [\lambda x^1_{\alpha^1} \cdots \lambda x^n_{\alpha^n} \centerdot h\,\mathbf{C}^1 \cdots \mathbf{C}^m] \\ &= \varphi(p). \end{aligned}$$

Since $p \notin \mathscr{S}$ we can define $\mathscr{O}$ to be the primitive substitution option

$$\langle \mathbf{primsub}, p, \mathbf{P} \rangle \in \boldsymbol{Options}^{\mathfrak{p}}_{\mathcal{S}}(\Xi)$$

(cf. Definition 8.7.9). By Lemma 8.7.11, $\mathbf{Search}(\Xi, \mathfrak{p}, \mathscr{O})$ is a search state. Since $\mathfrak{p}$ is fully duplicated with respect to f, we also know there is a search lifting map f' from $\mathbf{Search}(\Xi, \mathfrak{p}, \mathscr{O})$ to $\langle \mathcal{Q}^*, r^* \rangle$ such that

$$\|\mathsf{f}' : \mathcal{Q}^{\mathbf{Search}(\Xi,\mathfrak{p},\mathscr{O})} \to \mathcal{Q}^*\| < \|\mathsf{f} : \mathcal{Q} \to \mathcal{Q}^*\|$$

by Lemma 8.7.11.

Next assume $h = x^i_{\alpha^i}$ for some i $(1 \leq i \leq n)$. In this case $\varphi(p)$ is of the form $[\lambda\overline{x^n}\,[x^i\,\overline{\mathbf{C}^m}]]$ where α^i is of the form $(o\gamma^m \cdots \gamma^1)$ and $[\lambda\overline{x^n}\,\mathbf{C}^j]$ is closed for each j $(1 \leq j \leq m)$. Let $\mathbf{P}$ be the ith projection term at type $(o\alpha^n \cdots \alpha^1)$ avoiding $\mathbf{Free}(\mathcal{Q})$ (cf. Definition 8.7.4). Hence $\mathbf{P}$ is of the form

$$[\lambda x^1_{\alpha^1} \cdots \lambda x^n_{\alpha^n} \blacksquare x^i \, [w^1 \, \overline{x^n}] \cdots [w^m \, \overline{x^n}]]$$

where $w^1_{\gamma^1\alpha^n\cdots\alpha^1}, \ldots, w^m_{\gamma^m\alpha^n\cdots\alpha^1}$ are distinct variables such that

$$\{w^1, \ldots, w^m\} \cap \mathbf{Free}(\mathcal{Q}) = \emptyset.$$

Define a ground substitution θ on $\{w^1, \ldots, w^m\}$ by $\theta(w^j) := [\lambda \overline{x^n} \, \mathbf{C}^j]$ for each j $(1 \leq j \leq m)$. Note that $\mathbf{Dom}(\theta) = \mathbf{Free}(\mathbf{P})$ and

$$\begin{aligned} \theta(\mathbf{P}) &= [\lambda x^1_{\alpha^1} \cdots \lambda x^n_{\alpha^n} \blacksquare x^i_{o\gamma^l\cdots\gamma^1} \, [\theta(w^1) \, \overline{x^n}] \cdots [\theta(w^m) \, \overline{x^n}]] \\ &\stackrel{\beta\eta}{=} [\lambda x^1_{\alpha^1} \cdots \lambda x^n_{\alpha^n} \blacksquare x^i \, \mathbf{C}^1 \cdots \mathbf{C}^m] \\ &= \varphi(p). \end{aligned}$$

Since $p \notin \mathscr{S}$ we can define $\mathscr{O}$ to be the primitive substitution option

$$\langle \mathbf{primsub}, p, \mathbf{P} \rangle \in \boldsymbol{Options}^{\mathfrak{p}}_{\mathcal{S}}(\Xi)$$

(cf. Definition 8.7.9). By Lemma 8.7.11, $\mathbf{Search}(\Xi, \mathfrak{p}, \mathscr{O})$ is a search state. Since $\mathfrak{p}$ is fully duplicated with respect to f, there is a search lifting map f' from $\mathbf{Search}(\Xi, \mathfrak{p}, \mathscr{O})$ to $\langle \mathcal{Q}^*, r^* \rangle$ such that

$$\|\mathsf{f}' : \mathcal{Q}^{\mathbf{Search}(\Xi,\mathfrak{p},\mathscr{O})} \to \mathcal{Q}^*\| < \|\mathsf{f} : \mathcal{Q} \to \mathcal{Q}^*\|$$

by Lemma 8.7.11.

For the remainder of the proof we assume $(\mathbf{SetVars}^{\mathcal{Q}}(\mathfrak{p}) \setminus \mathscr{S})$ is almost atomic with respect to f. Since f is a search lifting map we know $\mathscr{S}$ is almost atomic with respect to f. Thus $\mathbf{SetVars}^{\mathcal{Q}}(\mathfrak{p})$ is almost atomic with respect to f.

Let $\mathfrak{p}^*$ be the image

$$\{f(a) \mid a \in \mathfrak{p}\} \subseteq \mathcal{N}^{\mathcal{Q}^*}$$

of $\mathfrak{p}$ under f. Since $\mathfrak{p}$ is fully duplicated and $\mathbf{SetVars}^{\mathcal{Q}}(\mathfrak{p})$ is almost atomic with respect to f, we can apply Lemma 8.8.3. We consider the four possibilities in the conclusion of Lemma 8.8.3.

1. Suppose there exist $a, b \in \mathfrak{p}$ and $c^* \in \mathcal{N}^{\mathcal{Q}^*}$ such that a is an extended positive atomic node, b is an extended negative atomic node and

$$\langle c^*, a, b \rangle \in \mathbf{Conns}^-(\mathsf{f}; \mathbf{mate}).$$

Hence $\langle c^*, f(a), f(b)\rangle \in conns^{\mathcal{Q}^*}$. Also, the shallow formula of a in $\mathcal{Q}$ is of the form $[p\,\overline{\mathbf{A}^m}]$ and the shallow formula of b in $\mathcal{Q}$ is of the form $[q\,\overline{\mathbf{B}^l}]$ for parameters $p, q \in \mathcal{P}$. By $\clubsuit^p_m$, we know $f(a)$ is a positive atomic node with shallow formula $[R^*\,\mathbf{C}^1 \cdots \mathbf{C}^n]$ and $f(b)$ is a negative atomic node with shallow formula $[R^*\,\mathbf{D}^1 \cdots \mathbf{D}^n]$ where R^* is a parameter. Since $\langle \varphi, f\rangle$ is a lifting map,

$$(\varphi(sh^{\mathcal{Q}}(a)))^{\downarrow} = [R^*\,\mathbf{C}^1 \cdots \mathbf{C}^n]$$

and

$$(\varphi(sh^{\mathcal{Q}}(b)))^{\downarrow} = [R^*\,\mathbf{D}^1 \cdots \mathbf{D}^n].$$

Since φ respects parameters we must have $\varphi(p) \in \mathcal{P}$ and $\varphi(q) \in \mathcal{P}$ (cf. Definitions 8.4.3 and 8.4.1). Since

$$[\varphi(p)\,\varphi(\mathbf{A}^1)\ \cdots\ \varphi(\mathbf{A}^m)] = \varphi([p\,\overline{\mathbf{A}^m}]) \stackrel{\beta\eta}{=} [R^*\,\overline{\mathbf{C}^n}]$$

and

$$[\varphi(q)\,\varphi(\mathbf{B}^1)\ \cdots\ \varphi(\mathbf{B}^l)] = \varphi([q\,\overline{\mathbf{B}^l}]) \stackrel{\beta\eta}{=} [R^*\,\overline{\mathbf{D}^n}]$$

we conclude $\varphi(p) = R^* = \varphi(q)$ and $m = n = l$. Hence $p = q$ since $\varphi(p) = R^* = \varphi(q)$ and φ is a lifting substitution for $\mathcal{Q}$ (cf. Definitions 8.4.1, 8.4.3 and 8.6.7). Let $\mathscr{O}$ be the rigid mating option $\langle \mathbf{mateop}, a, b\rangle$ (cf. Definition 8.7.19). By Lemma 8.7.25, we know $\mathbf{Search}(\Xi, \mathfrak{p}, \mathscr{O})$ is a search state. Since $\langle c^*, a, b\rangle$ is in the set $\mathbf{Conns}^-(\mathsf{f}; \mathbf{mate})$, $\mathfrak{p}$ is fully duplicated with respect to f and $\mathbf{SetVars}^{\mathcal{Q}}(\mathfrak{p})$ is almost atomic with respect to f, there is a search lifting map f' from $\mathbf{Search}(\Xi, \mathfrak{p}, \mathscr{O})$ to $\langle \mathcal{Q}^*, r^*\rangle$ such that

$$\|\mathsf{f}' : \mathcal{Q}^{\mathbf{Search}(\Xi,\mathfrak{p},\mathscr{O})} \to \mathcal{Q}^*\| < \|\mathsf{f} : \mathcal{Q} \to \mathcal{Q}^*\|$$

by Lemma 8.7.25.

2. Suppose there exist $a, b \in \mathfrak{p}$ and $c^* \in \mathcal{N}^{\mathcal{Q}^*}$ such that a is a positive flexible node, b is an extended negative atomic node and

$$\langle c^*, a, b\rangle \in \mathbf{Conns}^-(\mathsf{f}; \mathbf{mate}).$$

Hence $\langle c^*, f(a), f(b)\rangle \in conns^{\mathcal{Q}^*}$. Also, the shallow formula of a in $\mathcal{Q}$ is of the form $[p\,\overline{\mathbf{A}^m}]$ and the shallow formula of b in $\mathcal{Q}$ is of the form $[q\,\overline{\mathbf{B}^l}]$ for a variable $p \in \mathcal{V}$ and a parameter $q \in \mathcal{P}$. As above, there is a parameter R^* such that $f(a)$ has shallow formula $[R^* \mathbf{C}^1 \cdots \mathbf{C}^n]$, $f(b)$ has shallow formula $[R^* \mathbf{D}^1 \cdots \mathbf{D}^n]$,

$$(\varphi([p\,\overline{\mathbf{A}^m}]))^{\downarrow} = (\varphi(sh^{\mathcal{Q}}(a)))^{\downarrow} = [R^* \mathbf{C}^1 \cdots \mathbf{C}^n]$$

and

$$(\varphi([q\,\overline{\mathbf{B}^l}]))^{\downarrow} = (\varphi(sh^{\mathcal{Q}}(b)))^{\downarrow} = [R^* \mathbf{D}^1 \cdots \mathbf{D}^n].$$

Since $\varphi(q)$ is a parameter, $\varphi(q)$ must be the parameter R^* and l must be n. Since $a \in \mathfrak{p}$ is a p-node, we know $p \in \mathbf{SetVars}^{\mathcal{Q}}(\mathfrak{p})$. Hence p is almost atomic with respect to f. Thus the closed long $\beta\eta$-normal form $\varphi(p)$ is of the form $[\lambda \overline{x^m}\,[R\,\overline{\mathbf{E}^r}]]$ where $R \in \mathcal{P}$ and $[\lambda \overline{x^m}\,\mathbf{E}^j]$ is closed for each j $(1 \leq j \leq r)$. Since $R \notin \{x^1, \ldots, x^m\}$, the head of $(\varphi([p\,\overline{\mathbf{A}^m}]))^{\downarrow}$ is R. Since

$$[\varphi(p)\,\varphi(\mathbf{A}^1)\,\cdots\,\varphi(\mathbf{A}^m)] = \varphi([p\,\overline{\mathbf{A}^m}]) \stackrel{\beta\eta}{=} [R^*\,\overline{\mathbf{C}^n}]$$

we must have $R = R^* \in \mathcal{P}$ and $r = n$. Hence $\varphi(p)$ is $[\lambda \overline{x^m}\,[R^*\,\overline{\mathbf{E}^n}]]$ and $R^* \in \mathbf{Params}(\varphi(p))$. The parameter q is not banned for p by Lemma 8.5.10 since $\varphi(q) = R^* \in \mathbf{Params}(\varphi(p))$. Let $\mathbf{P}$ be the imitation term for q at the type of p avoiding $\mathbf{Free}(\mathcal{Q})$ (cf. Definition 8.7.2). Hence $\mathbf{P}$ is of the form

$$[\lambda x^1_{\alpha^1} \cdots \lambda x^m_{\alpha^m} \,\blacksquare\, q\,[w^1\,\overline{x^m}] \cdots [w^n\,\overline{x^m}]]$$

where $w^1_{\beta^1\alpha^m\cdots\alpha^1}, \ldots, w^n_{\beta^n\alpha^m\cdots\alpha^1}$ are distinct variables such that

$$\{w^1, \ldots, w^n\} \cap \mathbf{Free}(\mathcal{Q}) = \emptyset.$$

Define a ground substitution θ on $\{w^1, \ldots, w^n\}$ by $\theta(w^j) := [\lambda \overline{x^m}\,\mathbf{E}^j]$ for each j $(1 \leq j \leq n)$. Note that $\mathbf{Dom}(\theta) = \mathbf{Free}(\mathbf{P})$ and

$$\theta(\varphi|_{\mathcal{P}}(\mathbf{P})) = [\lambda \overline{x^m} \,\blacksquare\, \varphi(q)\,[\theta(w^1)\,\overline{x^m}] \cdots [\theta(w^n)\,\overline{x^m}]]$$

$$\stackrel{\beta\eta}{=} [\lambda \overline{x^m}\,[R^*\,\overline{\mathbf{E}^n}]] = \varphi(p).$$

Let $\mathcal{O}$ be the flex-rigid mating option

$$\langle \mathbf{frmateop}, a, b, p, \mathbf{P} \rangle \in \boldsymbol{Options}^{\mathbf{frm}}(\Xi, \mathfrak{p}).$$

By Lemma 8.7.26, $\mathbf{Search}(\Xi, \mathfrak{p}, \mathcal{O})$ is a search state. Since

$$\langle c^*, a, b \rangle \in \mathbf{Conns}^-(\mathsf{f}; \mathbf{mate}),$$

$\mathfrak{p}$ is fully duplicated with respect to f and $\mathbf{SetVars}^{\mathcal{Q}}(\mathfrak{p})$ is almost atomic with respect to f, there is a search lifting map f' from $\mathbf{Search}(\Xi, \mathfrak{p}, \mathscr{O})$ to $\langle \mathcal{Q}^*, r^* \rangle$ such that

$$\|\mathsf{f}' : \mathcal{Q}^{\mathbf{Search}(\Xi,\mathfrak{p},\mathscr{O})} \to \mathcal{Q}^*\| < \|\mathsf{f} : \mathcal{Q} \to \mathcal{Q}^*\|$$

by Lemma 8.7.26.

3. Suppose there exist $a, b \in \mathfrak{p}$, $c^* \in \mathcal{N}^{\mathcal{Q}^*}$ and

$$k \in \{\mathbf{eunif}, \mathbf{eunif}^{sym}\}$$

such that a is an extended positive equation node, b is an extended equation goal node and $\langle c^*, a, b \rangle \in \mathbf{Conns}^-(\mathsf{f}; k)$. Hence $\langle a, b \rangle$ is not k-connected in $\mathcal{Q}$. Let $\mathscr{O}$ be $\langle \mathbf{eunifop}, k, a, b \rangle \in \boldsymbol{Options}^{\mathbf{e}}(\Xi; \mathfrak{p})$. By Lemma 8.7.27, $\mathbf{Search}(\Xi, \mathfrak{p}, \mathscr{O})$ is a search state. Since $\langle c^*, a, b \rangle$ is in the set $\mathbf{Conns}^-(\mathsf{f}; k)$, $\mathfrak{p}$ is fully duplicated with respect to f and $\mathbf{SetVars}^{\mathcal{Q}}(\mathfrak{p})$ is almost atomic with respect to f, there is a search lifting map f' from $\mathbf{Search}(\Xi, \mathfrak{p}, \mathscr{O})$ to $\langle \mathcal{Q}^*, r^* \rangle$ such that

$$\|\mathsf{f}' : \mathcal{Q}^{\mathbf{Search}(\Xi,\mathfrak{p},\mathscr{O})} \to \mathcal{Q}^*\| < \|\mathsf{f} : \mathcal{Q} \to \mathcal{Q}^*\|$$

by Lemma 8.7.27.

4. Suppose there is a node $e \in \mathfrak{p}$ such that e is an extended negative equation node which is not decomposable in $\mathcal{Q}$ such that $f(e)$ is decomposed in $\mathcal{Q}^*$. The shallow formula of the negative equation node e in $\mathcal{Q}$ is of the form

$$[[g_{\iota\alpha^n\cdots\alpha^1}\,\overline{\mathbf{A}^n}] \mathrel{\dot{=}^\iota} [h_{\iota\beta^m\cdots\beta^1}\,\overline{\mathbf{B}^m}]]$$

where $g, h \in \mathcal{V} \cup \mathcal{P}$. If g and h were both variables, then $e \in \mathfrak{p}$ would be a flex-flex node, contradicting the assumption that $\mathfrak{p}$ is not a pre-solved u-part (cf. Definition 8.3.2). Hence either g or h is a parameter.

Since $f(e)$ is decomposed in $\mathcal{Q}^*$, there is some $d^* \in \mathcal{N}^{\mathcal{Q}^*}$ such that $\langle f(e), d^* \rangle \in decs^{\mathcal{Q}^*}$. By $\clubsuit^p_{dec}$, $f(e)$ is a negative equation node in $\mathcal{Q}^*$ with shallow formula

$$[[K_{\iota\gamma^l\cdots\gamma^1}\,\overline{\mathbf{C}^l}] \mathrel{\dot{=}^\iota} [K_{\iota\gamma^l\cdots\gamma^1}\,\overline{\mathbf{D}^l}]]$$

for some parameter K. Since $\langle\varphi, f\rangle$ is a lifting map from $\mathcal{Q}$ to $\mathcal{Q}^*$, we know

$$[[K_{\iota\gamma^l\cdots\gamma^1}\,\overline{\mathbf{C}^l}] \mathrel{\dot{=}^\iota} [K_{\iota\gamma^l\cdots\gamma^1}\,\overline{\mathbf{D}^l}]] \stackrel{\beta\eta}{=} \varphi([[g\,\overline{\mathbf{A}^n}] \mathrel{\dot{=}^\iota} [h\,\overline{\mathbf{B}^m}]])$$
$$= [[\varphi(g)\,\varphi(\mathbf{A}^1)\,\cdots\,\varphi(\mathbf{A}^n)] \mathrel{\dot{=}^\iota} [\varphi(h)\,\varphi(\mathbf{B}^1)\,\cdots\,\varphi(\mathbf{B}^m)]].$$

Assume g and h are both parameters. In this case we must have $\varphi(g) = K = \varphi(h)$ and so $g = h$ since φ is a lifting substitution for $\mathcal{Q}$. Hence e is a decomposable node of $\mathcal{Q}$, contradicting our assumption. Thus either g or h is a variable and so e is a flex-rigid node of $\mathcal{Q}$.

Assume g is a variable and h is a parameter. In this case $\varphi(h) = K$ and $l = m$. Also, $\gamma^j = \beta^j$ and $(\varphi(\mathbf{B}^j))^{\downarrow} = \mathbf{D}^j$ for each j $(1 \leq j \leq m)$. Since $\varphi(g)$ is a long $\beta\eta$-form, it must be of the form

$$[\lambda\overline{x^n}\,[R_{\iota\delta^r\cdots 1}\,\overline{\mathbf{E}^r}]]$$

for some $R \in \mathcal{V} \cup \mathcal{P}$. Since $\varphi(g)$ is closed, $[\lambda\overline{x^n}\,\mathbf{E}^j]$ is closed for each j $(1 \leq j \leq r)$. Note that

$$[\varphi(g)\,\varphi(\mathbf{A}^1)\,\cdots\,\varphi(\mathbf{A}^n)]^{\downarrow} = [K\,\overline{\mathbf{C}^l}].$$

If $R \notin \{x^1, \ldots, x^n\}$, then the head of

$$[\varphi(g)\,\varphi(\mathbf{A}^1)\,\cdots\,\varphi(\mathbf{A}^n)]^{\downarrow}$$

must be R and so $R = K$. Thus we must have $R \in \{K, x^1, \ldots, x^n\}$.

Assume $R = K$. Then $r = m$, $\delta^j = \beta^j$ for all $j \in \{1, \ldots, m\}$, and the closed long $\beta\eta$-form $\varphi(g)$ is $[\lambda\overline{x^n}\,[K\,\overline{\mathbf{E}^m}]]$. Clearly, we have $K \in \mathbf{Params}(\varphi(g))$. Since $\varphi(h) = K \in \mathbf{Params}(\varphi(g))$, the parameter h is not banned for g by Lemma 8.5.10. Let $\mathbf{U}$ be the imitation term for h at type $(\iota\alpha^n\cdots\alpha^1)$ avoiding $\mathbf{Free}(\mathcal{Q})$ (cf. Definition 8.7.2). Hence $\mathbf{U}$ is of the form

$$[\lambda x^1_{\alpha^1}\,\cdots\,\lambda x^n_{\alpha^n}\,\blacksquare\, h\,[w^1\,\overline{x^n}]\,\cdots\,[w^m\,\overline{x^n}]]$$

where $w^1_{\beta^1\alpha^n\cdots\alpha^1}, \ldots, w^m_{\beta^m\alpha^n\cdots\alpha^1}$ are distinct variables such that

$$\{w^1, \ldots, w^m\} \cap \mathbf{Free}(\mathcal{Q}) = \emptyset.$$

Define a ground substitution θ on $\{w^1, \ldots, w^m\}$ by $\theta(w^j) := [\lambda\overline{x^n}\,\mathbf{E}^j]$ for each j $(1 \leq j \leq m)$. Note that $\mathbf{Dom}(\theta) = \mathbf{Free}(\mathbf{U})$ and

$$\theta(\varphi|_{\mathcal{P}}(\mathbf{U})) = [\lambda x^1_{\alpha^1}\,\cdots\,\lambda x^n_{\alpha^n}\,\blacksquare\,\varphi(h)\,[\theta(w^1)\,\overline{x^n}]\,\cdots\,[\theta(w^m)\,\overline{x^n}]]$$
$$\stackrel{\beta\eta}{=} [\lambda x^1_{\alpha^1}\,\cdots\,\lambda x^n_{\alpha^n}\,\blacksquare\, K\,\mathbf{E}^1\,\cdots\,\mathbf{E}^m] \stackrel{\beta\eta}{=} \varphi(g).$$

Let $\mathscr{O}$ be the unification option $\langle \mathbf{unif}, g, \mathbf{U}\rangle \in \boldsymbol{Options}^{\mathbf{u}}(\Xi; \mathfrak{p})$ (cf. Definition 8.7.16). By Lemma 8.7.18, $\mathbf{Search}(\Xi, \mathfrak{p}, \mathscr{O})$ is a search state. Since $\mathfrak{p}$ is fully duplicated with respect to f and $\mathbf{SetVars}^{\mathcal{Q}}(\mathfrak{p})$ is almost atomic with respect to f, there is a search lifting map f' from $\mathbf{Search}(\Xi, \mathfrak{p}, \mathscr{O})$ to $\langle \mathcal{Q}^*, r^*\rangle$. such that

$$\|\mathsf{f}' : \mathcal{Q}^{\mathbf{Search}(\Xi,\mathfrak{p},\mathscr{O})} \to \mathcal{Q}^*\| < \|\mathsf{f} : \mathcal{Q} \to \mathcal{Q}^*\|$$

by Lemma 8.7.18.

Assume $R = x^i$ for some i $(1 \leq i \leq n)$. In this case α^i is equal to $(\iota\delta^r \cdots \delta^1)$. Let $\mathbf{U}$ be the ith projection term at type $(\iota\alpha^n \cdots \alpha^1)$ avoiding $\mathbf{Free}(\mathcal{Q})$ (cf. Definition 8.7.4). Hence $\mathbf{U}$ is of the form

$$[\lambda x^1_{\alpha^1} \cdots \lambda x^n_{\alpha^n} \blacksquare x^i\, [w^1\, \overline{x^n}] \cdots [w^r\, \overline{x^n}]]$$

where $w^1_{\gamma^1\alpha^n\cdots\alpha^1}, \ldots, w^r_{\gamma^r\alpha^n\cdots\alpha^1}$ are distinct variables such that

$$\{w^1, \ldots, w^r\} \cap \mathbf{Free}(\mathcal{Q}) = \emptyset.$$

Define a ground substitution θ on $\{w^1, \ldots, w^r\}$ by $\theta(w^j) := [\lambda \overline{x^n}\, \mathbf{E}^j]$ for each j $(1 \leq j \leq r)$. Note that $\mathbf{Dom}(\theta) = \mathbf{Free}(\mathbf{U})$ and

$$\theta(\varphi|_{\mathcal{P}}(\mathbf{U})) = [\lambda x^1_{\alpha^1} \cdots \lambda x^n_{\alpha^n} \blacksquare x^i_{o\delta^r\cdots\delta^1}\, [\theta(w^1)\, \overline{x^n}] \cdots [\theta(w^r)\, \overline{x^n}]]$$
$$\stackrel{\beta\eta}{=} [\lambda x^1_{\alpha^1} \cdots \lambda x^n_{\alpha^n} \blacksquare x^i\, \mathbf{E}^1 \cdots \mathbf{E}^r] = \varphi(p).$$

Let $\mathscr{O}$ be the unification option $\langle \mathbf{unif}, g, \mathbf{U}\rangle \in \boldsymbol{Options}^{\mathbf{u}}(\Xi; \mathfrak{p})$ (cf. Definition 8.7.16). By Lemma 8.7.18, $\mathbf{Search}(\Xi, \mathfrak{p}, \mathscr{O})$ is a search state. Since $\mathfrak{p}$ is fully duplicated with respect to f and $\mathbf{SetVars}^{\mathcal{Q}}(\mathfrak{p})$ is almost atomic with respect to f, there is a search lifting map f' from $\mathbf{Search}(\Xi, \mathfrak{p}, \mathscr{O})$ to $\langle \mathcal{Q}^*, r^*\rangle$ such that

$$\|\mathsf{f}' : \mathcal{Q}^{\mathbf{Search}(\Xi,\mathfrak{p},\mathscr{O})} \to \mathcal{Q}^*\| < \|\mathsf{f} : \mathcal{Q} \to \mathcal{Q}^*\|$$

by Lemma 8.7.18.

The case where g is a parameter and h is a variable is analogous.

□

Bibliography

[1] P. B. Andrews. A reduction of the axioms for the theory of propositional types. *Fundamenta Mathematicae*, 52:345–350, 1963.

[2] P. B. Andrews. Resolution in type theory. *Journal of Symbolic Logic*, 36:414–432, 1971.

[3] P. B. Andrews. General models and extensionality. *Journal of Symbolic Logic*, 37:395–397, 1972.

[4] P. B. Andrews. General models, descriptions, and choice in type theory. *Journal of Symbolic Logic*, 37:385–394, 1972.

[5] P. B. Andrews. Resolution and the consistency of analysis. *Notre Dame Journal of Formal Logic*, 15(1):73–84, 1974.

[6] P. B. Andrews. Theorem proving via general matings. *Journal of the ACM*, 28:193–214, 1981.

[7] P. B. Andrews. On connections and higher-order logic. *Journal of Automated Reasoning*, 5:257–291, 1989.

[8] P. B. Andrews. Classical type theory. In A. Robinson and A. Voronkov, editors, *Handbook of Automated Reasoning*, volume 2, chapter 15, pages 965–1007. Elsevier Science, 2001.

[9] P. B. Andrews. *An Introduction to Mathematical Logic and Type Theory: To Truth Through Proof.* Kluwer Academic Publishers, second edition, 2002.

[10] P. B. Andrews and M. Bishop. On sets, types, fixed points, and checkerboards. In P. Miglioli, U. Moscato, D. Mundici, and M. Ornaghi, editors, *Theorem Proving with Analytic Tableaux and Related Methods. 5th International Workshop. (TABLEAUX '96)*, volume 1071 of *Lecture Notes in Artificial Intelligence*, pages 1–15, Terrasini, Italy, May 1996. Springer-Verlag.

[11] P. B. Andrews, M. Bishop, and C. E. Brown. System description: Tps: A theorem proving system for type theory. In D. McAllester, editor, *Proceedings of the 17th International Conference on Automated Deduction*, volume 1831 of *Lecture Notes in Artificial Intelligence*, pages 164–169, Pittsburgh, PA, USA, 2000. Springer-Verlag.

[12] P. B. Andrews, M. Bishop, S. Issar, D. Nesmith, F. Pfenning, and H. Xi. TPS: A theorem proving system for classical type theory. *Journal of Automated Reasoning*, 16:321–353, 1996.

[13] S. C. Bailin and D. Barker-Plummer. Z-match: An inference rule for incrementally elaborating set instantiations. *Journal of Automated Reasoning*, 11:391–428, 1993. Errata: JAR 12 (1994), 411–412.

[14] H. P. Barendregt. *The λ-Calculus.* Studies in logic and the foundations of mathematics, North-Holland, 1984.

[15] F. Bartels, A. Dold, F. W. v. Henke, H. Pfeifer, and H. Rueß. Formalizing Fixed-Point Theory in PVS. Ulmer Informatik-Berichte 96-10, Universität Ulm, Fakultät für Informatik, 1996.

[16] J. G. F. Belinfante. Computer Proofs in Gödel's class theory with equational definitions for composite and cross. *Journal of Automated Reasoning*, 22:311–339, 1999.

[17] J. L. Bell and A. B. Slomson. *Models and Ultraproducts: An Introduction.* North-Holland, Amsterdam, 1971.

[18] C. Benzmüller. *Equality and Extensionality in Automated Higher-Order Theorem Proving.* PhD thesis, Universität des Saarlandes, 1999.

[19] C. Benzmüller, C. E. Brown, and M. Kohlhase. Higher order semantics and extensionality. *Journal of Symbolic Logic*, 69:1027–1088, 2004.

[20] C. Benzmüller and M. Kohlhase. System description: LEO — a higher-order theorem prover. In C. Kirchner and H. Kirchner, editors, *Proceedings of the 15th International Conference on Automated Deduction*, volume 1421 of *Lecture Notes in Artificial Intelligence*, pages 139–143, Lindau, Germany, 1998. Springer-Verlag.

[21] W. Bibel. On matrices with connections. *Journal of the ACM*, 28(4):633–645, Oct. 1981.

[22] W. W. Bledsoe. A maximal method for set variables in automatic theorem proving. In J. E. Hayes, D. Michie, and L. I. Mikulich, editors, *Machine Intelligence 9*, pages 53–100. Ellis Harwood Ltd., Chichester, and John Wiley & Sons, 1979.

[23] W. W. Bledsoe and G. Feng. Set-Var. *Journal of Automated Reasoning*, 11:293–314, 1993.

[24] R. Boyer, E. Lusk, W. McCune, R. Overbeek, M. Stickel, and L. Wos. Set theory in first-order logic: Clauses for Gödel's axioms. *Journal of Automated Reasoning*, 2:287–327, 1986.

[25] C. E. Brown. Solving for set variables in higher-order theorem proving. In A. Voronkov, editor, *Proceedings of the 18th International Conference on Automated Deduction*, volume 2392 of *Lecture Notes in Artificial Intelligence*, pages 408–422, Copenhagen, Denmark, 2002. Springer-Verlag.

[26] A. Church. A formulation of the simple theory of types. *Journal of Symbolic Logic*, 5:56–68, 1940.

[27] H. B. Curry and R. Feys. *Combinatory Logic.* Studies in logic and the foundations of mathematics, North-Holland, 1958.

[28] A. Felty. Proof search with set variable instantiation in the calculus of constructions. In M. A. McRobbie and J. K. Slaney, editors, *Automated Deduction: CADE-13*, volume 1104 of *Lecture Notes in Artificial Intelligence*, pages 658–672. Springer, 1996.

[29] A. Felty. The calculus of constructions as a framework for proof search with set variable instantiation. *Theoretical Computer Science*, 232:187–229, 2000.

[30] J. H. Geuvers. The calculus of constructions and higher order logic. In P. de Groote, editor, *The Curry-Howard Isomorphism*, pages 139–191. Academia, Louvain-la-Neuve (Belgium), 1995.

[31] L. Henkin. Completeness in the theory of types. *Journal of Symbolic Logic*, 15:81–91, 1950.

[32] L. Henkin. An extension of the Craig-Lyndon interpolation theorem. *Journal of Symbolic Logic*, 28:201–216, 1963.

[33] J. R. Hindley. *Basic Simple Type Theory*. Cambridge University Press, 1997.

[34] J. R. Hindley and J. P. Seldin. *An Introduction to Combinators and the λ-Calculus*. Cambridge University Press, 1986.

[35] F. Honsell and M. Lenisa. Coinductive characterizations of applicative structures. *Mathematical Structures in Computer Science*, 9:403–435, 1999.

[36] G. P. Huet. A unification algorithm for typed λ-calculus. *Theoretical Computer Science*, 1:27–57, 1975.

[37] S. Issar. Path-focused duplication: A search procedure for general matings. In *AAAI–90. Proceedings of the Eighth National Conference on Artificial Intelligence*, volume 1, pages 221–226. AAAI Press/The MIT Press, 1990.

[38] S. Issar. *Operational Issues in Automated Theorem Proving Using Matings*. PhD thesis, Carnegie Mellon University, 1991. 147 pp.

[39] B. Jacobs. *Categorical Logic and Type Theory*. Elsevier, 1999.

[40] S. MacLane. *Categories for the Working Mathematician*. Springer-Verlag, second edition, 1998.

[41] S. MacLane and I. Moerdijk. *Sheaves in Geometry and Logic: A First Introduction to Topos Theory*. Springer-Verlag, 1992.

[42] C. McLarty. *Elementary Categories, Elementary Toposes*. Oxford University Press, 1995.

[43] D. A. Miller. *Proofs in Higher-Order Logic*. PhD thesis, Carnegie Mellon University, 1983. 81 pp.

[44] D. A. Miller. A compact representation of proofs. *Studia Logica*, 46(4):347–370, 1987.

[45] D. A. Miller. A logic programming language with lambda-abstraction, function variables, and simple unification. *Journal of Logic and Computation*, 1(4):497–536, 1991.

[46] J. C. Mitchell. *Foundations for Programming Languages*. Foundations of Computing. MIT Press, 1996.

[47] J. R. Munkres. *Topology: A First Course*. Prentice-Hall, 1975.

[48] N. V. Murray and E. Rosenthal. Dissolution: Making paths vanish. *Journal of the ACM*, 40(3):504–535, July 1993.

[49] H. J. Ohlbach. SCAN—elimination of predicate quantifiers. In M. A. McRobbie and J. K. Slaney, editors, *Automated Deduction: CADE-13*, volume 1104 of *Lecture Notes in Artificial Intelligence*, pages 161–165. Springer, 1996.

[50] L. C. Paulson. A fixedpoint approach to implementing (co)inductive definitions. In A. Bundy, editor, *Proceedings of the 12th International Conference on Automated Deduction*, pages 148–161, Nancy, France, June 1994. Springer-Verlag LNAI 814.

[51] L. C. Paulson. Set theory for verification: II. induction and recursion. *Journal of Automated Reasoning*, 15(2):167–215, 1995.

[52] L. C. Paulson. Mechanizing coinduction and corecursion in higher-order logic. *Journal of Logic and Computation*, 7(2):175–204, Mar. 1997.

[53] F. Pfenning. *Proof Transformations in Higher-Order Logic*. PhD thesis, Carnegie Mellon University, 1987. 156 pp.

[54] F. Pfenning and C. Schürmann. Algorithms for equality and unification in the presence of notational definitions. In T. Altenkirch, W. Naraschewski, and B. Reus, editors, *Types for Proofs and Programs*, pages 179–193, Kloster Irsee, Germany, Mar. 1998. Springer-Verlag LNCS 1657.

[55] G. D. Plotkin. Lambda-definability in the full type hierarchy. In J. P. Seldin and J. R. Hindley, editors, *To H. B. Curry: Essays on Combinatory Logic, Lambda Calculus and Formalism*, pages 363–373. Academic Press, 1980.

[56] D. Prawitz. Hauptsatz for higher order logic. *Journal of Symbolic Logic*, 33:452–457, 1968.

[57] C. Prehofer. *Solving Higher-Order Equations: From Logic to Programming*. Birkhäuser Boston, 1997.

[58] A. Quaife. Automated deduction in von Neumann-Bernays-Gödel set theory. *Journal of Automated Reasoning*, 8:91–147, 1992.

[59] K. Schütte. Syntactical and semantical properties of simple type theory. *Journal of Symbolic Logic*, 25(4):305–326, 1960.

[60] R. Statman. Logical relations and the typed λ-calculus. *Information and Control*, 65:85–97, 1985.

[61] M. Takahashi. A proof of cut-elimination theorem in simple type theory. *Journal of the Mathematical Society of Japan*, 19:399–410, 1967.

Index of Notation (In order of appearance)

Subject Index

www.ingramcontent.com/pod-product-compliance
Lightning Source LLC
LaVergne TN
LVHW012325060326
832902LV00011B/1732
* 9 7 8 1 9 0 4 9 8 7 5 7 4 *